Applied Calculus

THIRD EDITION

Titles of Related Interest

Albright, Winston, & Zappe   *Data Analysis and Decision Making*

Albright, Winston, & Zappe   *Data Analysis for Managers*

Berk & Carey   *Data Analysis with Microsoft Excel*

Clemen & Reilly   *Making Hard Decisions with Decision Tools*

Giordano et al.   *A First Course in Mathematical Modeling*

Goodman & Stampfli   *The Mathematics of Finance: Modeling and Hedging*

Gordon   *Succeeding in Applied Calculus: Algebra Essentials*

Graybill   *Matrices with Applications in Statistics*

Hoerl & Snee   *Statistical Thinking: Improving Business Performance*

Howell   *Fundamental Statistics for the Behavioral Sciences*

Kao   *Introduction to Stochastic Processes*

Keller & Warrack   *Statistics for Management and Economics*

Lapin & Whisler   *Quantitative Decision Making with Spreadsheet Applications*

Middleton   *Data Analysis Using Microsoft Excel*

Minh   *Applied Probability Models*

Neuwirth & Arganbright   *The Active Modeler: Mathematical Modeling with Excel*

Pagano & Gauvreau   *Principles of Biostatistics*

Pilant et al.   *Finite Math on the Web*

Rosner   *Fundamentals of Biostatistics*

Sandefur   *Elementary Mathematical Modeling—A Dynamic Approach*

Savage   *Decision Making with Insight*

Scheaffer   *Introduction to Probability and Its Applications*

Schrage   *Optimization Modeling Using LINDO*

Taylor   *Excel Essentials: Using Microsoft Excel for Data Analysis and Decision Making*

Waner & Costenoble   *Finite Mathematics*

Waner & Costenoble   *Finite Mathematics and Applied Calculus*

Weiers   *Introduction to Business Statistics*

Winston   *Introduction to Probability Models*

Winston & Albright   *Practical Management Science*

Winston & Venkataramanan   *Introduction to Mathematical Programming*

Applied Calculus

THIRD EDITION

STEFAN WANER

HOFSTRA UNIVERSITY

STEVEN R. COSTENOBLE

HOFSTRA UNIVERSITY

THOMSON

BROOKS/COLE

Australia • Canada • Mexico • Singapore • Spain
United Kingdom • United States

THOMSON

BROOKS/COLE

Publisher: Curt Hinrichs
Development Editor: Cheryll Linthicum
Assistant Editor: Ann Day
Editorial Assistant: Katherine Brayton
Technology Project Manager: Earl Perry
Marketing Manager: Karin Sandberg
Marketing Assistant: Jennifer Gee
Advertising Project Manager: Bryan Vann
Project Manager, Editorial Production: Sandra Craig
Print/Media Buyer: Jessica Reed
Permissions Editor: Kiely Sexton

Production: Cecile Joyner, The Cooper Company
Text Designer: Kathleen Cunningham
Photo Researcher: Terri Wright
Copy Editor: Betty Duncan
Indexer: Julie Shawvan
Proofreader: Amy Mayfield
Cover Designer: Rob Hugel
Cover Image: Getty Images
Cover Printer: The Lehigh Press, Inc.
Composition, Illustrations: Graphic World
Printer: R.R. Donnelley/Willard

Printed in the United States of America

1   2   3   4   5   6   7   07   06   05   04   03

For more information about our products, contact us at:
Thomson Learning Academic Resource Center
1-800-423-0563
For permission to use material from this text, contact us by:
Phone: 1-800-730-2214
Fax: 1-800-730-2215
Web: http://www.thomsonrights.com

Brooks/Cole–Thomson Learning
10 Davis Drive
Belmont, CA 94002
USA

Asia
Thomson Learning
5 Shenton Way #01-01
UIC Building
Singapore 068808

Australia/New Zealand
Thomson Learning
102 Dodds Street
Southbank, Victoria 3006
Australia

Canada
Nelson
1120 Birchmount Road
Toronto, Ontario M1K 5G4
Canada

Europe/Middle East/Africa
Thomson Learning
High Holborn House
50/51 Bedford Row
London WC1R 4LR
United Kingdom

Latin America
Thomson Learning
Seneca, 53
Colonia Polanco
11560 Mexico D.F.
Mexico

Library of Congress Control Number: 2002116967

Student Edition: ISBN 0-534-41958-5

Instructor's Edition: ISBN 0-534-41973-9

Contents

Chapter 8 FUNCTIONS OF SEVERAL VARIABLES   466

Chapter 9 TRIGONOMETRIC MODELS   538

Chapter S CALCULUS APPLIED TO PROBABILITY AND STATISTICS
              (OPTIONAL INTERNET TOPIC)

Appendix A ALGEBRA REVIEW   A1

Preface

Applied Calculus, Third Edition, is a careful revision of the Second Edition. This book is intended for a one- or two-term course for students majoring in business, the social sciences, or the liberal arts. Like the earlier editions, the Third Edition is designed to address the considerable challenge of generating enthusiasm and developing mathematical sophistication in an audience that is often under-prepared for and disaffected by traditional mathematics courses. Like the Second Edition, this book is supported by, and linked with, an extensive and highly developed Web site containing interactive tutorials, online technology, optional supplementary material, and much more.

Rationale for the Third Edition

While preparing the Third Edition, we have updated and increased the number of examples and exercises based on real data. We have expanded the collections of communication and reasoning exercises designed to help students articulate mathematical concepts and ideas. We have also responded to the growing use of technology in the classroom by expanding the discussions of Excel and graphing calculators as optional tools. These discussions are seamlessly integrated into the text for those instructors who desire them but carefully delineated so that instructors can easily skip all or parts of them without interrupting the flow of the material. Thus, this book supports a wide range of instructional paradigms: from settings incorporating little or no technology to courses taught in computerized classrooms, and from classes in which a single form of technology is used exclusively to those incorporating several technologies.

Our Approach and Hallmark Features

Real World Orientation We are particularly proud of the diversity, breadth and abundance of examples and exercises we have been able to include in this edition. A large number of these are based on real, referenced data from business, economics, the life sciences and the social sciences. Examples and exercises based on dated information have generally been replaced by more current versions; applications based on unique or historically interesting data have been kept.

Adapting real data for pedagogical use can be tricky; available data can be numerically complex, intimidating for students, or incomplete. We have modified and streamlined many of the real world applications, rendering them as tractable as any "made-up" application. At the same time, we have been careful to strike a pedagogically sound balance between applications based on real data and more traditional "generic" applications. Thus, the density and selection of real data-based applications has been tailored to the pedagogical goals and appropriate difficulty level for each section.

Readability We would like students to read this book. We would like students to *enjoy* reading this book. Thus, we have written the book in a conversational and student-oriented style, and have made frequent use of question-and-answer dialogues to encourage the development of the student's mathematical curiosity and intuition. We hope that this text will give the student insight into how a mathematician develops and thinks about mathematical ideas and their applications.

Five Elements of Mathematical Pedagogy The "Rule of Three" is a common theme in reform-oriented texts. Adapting this approach, we discuss many of the central concepts numerically, graphically and algebraically, and go to some lengths to clearly delineate these distinctions. As a fourth element, we incorporate verbal communication of mathematical concepts through our emphasis on verbalizing mathematical concepts, our discussions on translating English sentences into mathematical statements, and our communication and reasoning exercises at the end of each section. The fifth element, interactivity, is implemented within the printed text through expanded use of question-and-answer dialogs, but is seen most dramatically in the Web site. At the Web site, students can interact with the material in several ways: through true-false quizzes on every topic, interactive review exercises and questions and answers in the tutorials (including multiple choice items with detailed feedback and Javascript-based interactive elements), and online utilities that automate a variety of tasks, from graphing to regression and matrix algebra.

Exercise Sets The substantial collection of exercises provides a wealth of material that can be used to challenge students at almost every level of preparation, and includes everything from straightforward drill exercises to interesting and rather challenging applications. The exercise sets have been carefully graded to move from the straightforward to the challenging. We have also included, in virtually every section of every chapter, interesting applications based on real data, communication and reasoning exercises that help the student articulate mathematical concepts, exercises ideal for the use of technology, and amusing exercises.

Many of the scenarios used in application examples and exercises are revisited several times throughout the book. Thus, for instance, students will find themselves using a variety of techniques, from graphing through the use of derivatives and elasticity, to analyze the same application. Reusing scenarios and important functions provides unifying threads and shows students the complex texture of real-life problems.

New to This Edition

- Chapter 1 has been revised to include expanded coverage of linear regression.
- Trigonometric functions material has been placed in a separate optional chapter.
- A new (optional) section on logistic models and regression has been added to Chapter 2.
- New material has been added on regression for all the nonlinear functions in Chapter 2 and for the new Trigonometry chapter.
- An expanded treatment of limits includes an additional section on the concept of continuity. The discussion of the algebraic approach to limits is now in a section by itself.
- The significantly revised chapter on derivatives consolidates the graphical and numerical approaches and includes more on the algebraic approach.
- The integrated discussions of Excel and graphing calculators as optional tools have been expanded.
- The successful pedagogy from the second edition has been expanded with more "Quick Examples," more "Question and Answer Dialog" boxes, additional Excel and graphing calculator instructions, and additional Web site information. A new boxed feature called "Guidelines" appears at the ends of sections to reinforce concepts.
- The number of exercises has been increased by 15%–20% with additional easier computational exercises at the start of many exercise sets and significantly more Communication and Reasoning exercises. Exercises and examples based on dated information have generally been replaced by more current versions.

- A new arrangement of exercise sets reflects increasing level of difficulty.
- A new four-color design clearly delineates the pedagogical tools.

Additional Features

- **Case Studies** Each chapter begins with the statement of an interesting problem that is returned to at the end of that chapter in a section entitled "Case Study." This extended application uses and illustrates the central ideas of the chapter. The themes of these applications are varied and they are designed to be as nonintimidating as possible. We avoid pulling complicated formulas out of thin air, but focus instead on the development of mathematical models appropriate to the topics. These applications are ideal for assignment as projects, and to this end we have included groups of exercises at the end of each.

- **Question-and-Answer Dialogue** We frequently use informal question-and-answer dialogues that anticipate the kind of questions that may occur to the student and also guide the student through the development of new concepts.

- **Before We Go On** Most examples are followed by supplementary interpretive discussions under the heading "Before we go on." These discussions may include a check on the answer, a discussion of the feasibility and significance of a solution, or an in-depth look at what the solution means.

- **Quick Examples** Most definition boxes include one or more straightforward examples that a student can use to solidify each new concept as soon as it is encountered.

- **Communication and Reasoning Exercises for Writing and Discussion** These are exercises designed to broaden the student's grasp of the mathematical concepts. They include exercises in which the student is asked to provide his or her own examples to illustrate a point or design an application with a given solution. They also include "fill in the blank" type exercises and exercises that invite discussion and debate. These exercises often have no single correct answer.

- **Footnotes** We use footnotes throughout the text to provide interesting background, extended discussion, and various asides.

- **Thorough Integration of Spreadsheet and Graphing Technology** Guidance on the use of spreadsheets (we use Microsoft® Excel) and graphing calculators (we use the TI-83) is thoroughly integrated throughout the discussion, examples, and exercise sets. At the same time, this material is clearly delineated so that it can be skipped over when not used. In many examples we include a discussion of the use of spreadsheets or graphing technology to aid in the solution. Groups of exercises for which the use of technology is suggested or required appear throughout the exercise sets.

The Web Site: www.AppliedCalc.com

Our site at www.AppliedCalc.com has been evolving for several years with growing recognition. Students, raised in an environment in which computers suffuse both work and play, can use their familiar web browsers to engage the material in an active way. At the Web site students and faculty can find:

- **Interactive Tutorials** Highly interactive tutorials are included on many major topics, with guided exercises that parallel the text.

- **Interactive True/False Chapter Quizzes** Interactive true/false quizzes based on the material in each chapter help the student review all the pertinent concepts and avoid common pitfalls.

- **Chapter Review Exercises** The site includes a constantly growing collection of review questions, taken from tests and exams, that do not appear in the printed text.
- **Detailed Chapter Summaries** Comprehensive summaries with interactive elements review all the basic definitions and problem solving techniques discussed in each chapter. The summaries include additional examples for review and can be easily printed.
- **Downloadable Excel Tutorials** Detailed Excel tutorials are available for almost every section of the book. These interactive tutorials expand on the examples given in the text.
- **Online Utilities** Our collection of easy-to-use online utilities, written in Java™ and Javascript, allow students to solve many of the technology-based application exercises directly on the Web page. The utilities available include a function grapher, a function evaluator, regression tools, and a numerical integration tool. These utilities require nothing more than a standard, Java-capable Web browser such as the current versions of Netscape Navigator and Microsoft® Internet Explorer.
- **Downloadable Software** In addition to the Web-based utilities the site offers a suite of free and intuitive stand-alone Macintosh® programs, including one for function graphing.
- **Supplemental Topics** We include complete interactive text and exercise sets for a selection of topics not ordinarily included in printed texts, but often requested by instructors. The text refers to these topics at appropriate points, letting instructors decide whether to include this material in their courses.
- **Optional Internet Chapters** The Web site includes a complete additional chapter on Calculus Applied to Probability. These chapters incorporate the same pedagogy as the printed material, and can be readily printed and distributed.

Supplemental Material

For Students

Student Solutions Manual *by Waner and Costenoble*
ISBN: 0-534-41960-7
The student solutions manual provides worked-out solutions to the odd-numbered problems in the text as well as complete solutions to all the chapter review tests. Solutions have been completely rewritten and expanded by the authors.

Microsoft Excel Manual
ISBN: 0-534-41964-X
This distinctive, text-specific manual uses Excel instructions and formulas to reinforce vital concepts.

Graphing Calculator Manual
ISBN: 0-534-41965-8
This text-specific graphing calculator manual guides students in using the TI-83 Plus and TI-86.

For Instructors and Students

InfoTrac® College Edition http://infotrac.thomsonlearning.com
Automatically packaged free with the text!
You and your students receive four months of anytime, anywhere access to InfoTrac College Edition. This vast online library offers the full text of articles from almost 4,000 scholarly and popular publications, updated daily and going back as far as 22 years. It's a great resource for exploring, completing assignments, or simply catching up on the news.

Visit Us on the Web!
http://mathematics.brookscole.com
Here, you'll find information on the full range of mathematics texts available from Brooks/Cole, as well as discipline-related resources for you and your students. For instance, *NewsEdge* offers daily feeds of the latest news in your field. You can also link to the text-specific Web site for all of Waner and Costenoble's books.

For Instructors

Instructor's Suite CD-ROM
ISBN: 0-534-41961-5
This CD-ROM contains Complete Solutions to every exercise in Microsoft Word and PDF format, Microsoft® PowerPoint® slides, and test items in Word format. For qualified adopters only.

Instructor's Solutions Manual *by Waner and Costenoble*
ISBN: 0-534-41959-3
The instructor's solutions manual provides worked-out solutions to the even-numbered problems in the text as well as complete solutions to the case study problems. Solutions have been completely rewritten and expanded by the authors.

Test Bank
ISBN: 0-534-41962-3
The test bank contains 110 test items per chapter with a problem grid breaking down the questions by section.

BCA Testing
ISBN: 0-534-41963-1
BCA Testing is a revolutionary text-specific testing suite that allows you to customize exams and track your students' progress—all in an accessible, browser-based format. The Internet-ready tool offers full algorithmic generation of problems and free response mathematics, with automatic homework grading that flows directly to your gradebook.

MyCourse 2.1
Ask us about our new FREE online course builder!
Brooks/Cole offers you a simple solution for a custom course Web site that allows you to assign, track, and report on student progress; load your syllabus; and more. Contact your Thomson•Brooks/Cole representative for details. In addition to the Web site, Brooks/Cole offers a complete ancillary package.

Acknowledgments

This project would not have been possible without the contributions and suggestions of numerous colleagues, students and friends. We are particularly grateful to our colleagues at Hofstra and elsewhere who used and gave us useful feedback on the Second Edition. We are also grateful to everyone at Brooks/Cole for their encouragement and guidance throughout the project. Specifically, we would like to thank Curt Hinrichs for his unflagging enthusiasm, Cheryll Linthicum for whipping the book into shape, and Karin Sandberg for letting the world know about it.

We would also like to thank the authors who contributed to the ancillaries associated with this text: James Ball, Florence Chambers, Edwin C. Hackleman, Larry J. Stephens, Mark Stevenson, and Patrick C. Ward. Finally, we would like to thank accuracy checker Jerrold W. Grossman for his meticulous reading of the manuscript and the numerous reviewers who provided many helpful suggestions that have shaped the development of this book.

Mark Clark
Palomar University

Jon Cole
St. John's University

Matt Coleman
Fairfield University

Casey Cremins
University of Maryland

James Czachor
Fordham University

Julie Daberkow
University of Maryland

Deborah Denvir
Marshall University

Michael Ecker
Pennsylvania State University

Janice Epstein
Texas A & M University

Joseph Erdeky
Prince George Community College

Candy Giovanni
Michigan State University

Joe Guthrie
University of Texas–El Paso

Bennette Harris
University of Wisconsin–Whitewater

Loek Helminck
North Carolina State University

Victor Kaftal
University of Cincinnati

Thomas Keller
Southwest Texas State University

Thomas Kelley
Metropolitan State College of Denver

Keith Kendig
Cleveland State University

Greg Klein
Texas A & M University

Michael Moses
George Washington University

Terry A. Nyman
University of Wisconsin–Fox Valley

Ralph Oberste-Vorth
University of South Florida

James Osterburg
University of Cincinnati

James Parks
State University of New York–Potsdam

Joanne Peeples
El Paso Community College

Mihaela Poplicher
University of Cincinnati

Timothy Ray
Southeast Missouri State University

E. Arthur Robinson, Jr.
George Washington University

Gordon Savin
University of Utah

Daniel Scanlon
Orange Coast College

Brad Shelton
University of Oregon

Cynthia Siegel
University of Missouri–St. Louis

Stephen Stuckwisch
Auburn University

Stephen Suen
University of South Florida

Doug Ulmer
University of Arizona

Jackie Vogel
Pellissippi State Community College

Will Watkins
University of Texas–Pan American

Denise Widup
University of Wisconsin–Parkside

Janet E. Yi
Ball State University

Stefan Waner
Steven R. Costenoble

Applied Calculus

THIRD EDITION

FUNCTIONS AND LINEAR MODELS

CASE STUDY

Modeling Spending on Internet Advertising

You are the new director of Impact Advertising's Internet division, which has enjoyed a steady 0.25% of the Internet advertising market. You have drawn up an ambitious proposal to expand your division in light of your anticipation that Internet advertising will continue to skyrocket. The vice president in charge of Financial Affairs feels that current projections (based on a linear model) do not warrant the level of expansion you propose. How can you persuade the vice president that those projections do not fit the data convincingly?

COURTESY VALUECLICK.COM

INTERNET RESOURCES FOR THIS CHAPTER

At the Web site, follow the path

> Web Site → Everything for Finite Math → Chapter 1

where you will find a detailed chapter summary you can print out, a true/false quiz, and a collection of review exercises. You will also find downloadable Excel tutorials for each section, an online grapher, an online regression utility, and other resources. In addition, complete interactive text and exercises have been placed on the Web site, covering the following optional topic:

> **New Functions from Old: Scaled and Shifted Functions**

Introduction

To analyze recent trends in spending on Internet advertising and to make reasonable projections, we need a mathematical model of this spending. Where do we start? To apply mathematics to real-world situations like this, we need a good understanding of basic mathematical concepts. Perhaps the most fundamental of these concepts is that of a function: a relationship that shows how one quantity depends on another. Functions may be described numerically and, often, algebraically. They can also be described graphically—a viewpoint that is extremely useful.

The simplest functions—the ones with the simplest formulas and the simplest graphs—are linear functions. Because of their simplicity, they are also among the most useful functions and can often be used to model real-world situations, at least over short periods of time. In discussing linear functions, we will meet the concepts of slope and rate of change, which are the starting point of the mathematics of change.

In the last section of this chapter, we discuss *simple linear regression:* construction of linear functions that best fit given collections of data. Regression is used extensively in applied mathematics, statistics, and quantitative methods in business. The inclusion of regression utilities in computer spreadsheets like Excel® makes this powerful mathematical tool readily available for anyone to use.

Algebra Review

For this chapter you should be familiar with real numbers and intervals. To review this material, see Appendix A.

1.1 Functions from the Numerical and Algebraic Viewpoints

The following table gives the weights of a particular child at various ages in her first year:

Age (mo)	0	2	3	4	5	6	9	12
Weight (lb)	8	9	13	14	16	17	18	19

Let's write $W(0)$ for the child's weight at birth (in pounds), $W(2)$ for her weight at 2 months, and so on [we read $W(0)$ as "W of 0"]. Thus, $W(0) = 8$, $W(2) = 9$, $W(3) = 13, \ldots, W(12) = 19$. More generally, if we write t for the age of the child (in months) at any time during her first year, then we write $W(t)$ for the weight of the child at age t. We call W a **function** of the variable t, meaning that for each value of t between 0 and 12, W gives us a single corresponding number $W(t)$ (the weight of the child at that age).

Age t → W → Weight $W(t)$

Figure 1

In general, we think of a function as a way of producing new objects from old ones. The functions we deal with in this text produce new numbers from old numbers. The numbers we have in mind are the *real* numbers, including not only positive and negative integers and fractions but also numbers like $\sqrt{2}$ or π (see Appendix A for more on real numbers). For this reason, the functions we use are called **real-valued functions of a real variable.** For example, the function W takes the child's age in months and returns her weight in pounds at that age (Figure 1).

If we introduce another variable—y, say—that stands for the weight of the child, then we can write $y = W(t)$. The function W then tells us exactly how y depends on t. We call y the **dependent variable,** since its value depends on the value of t, the **independent variable.**

A function may be specified in several different ways. It may be specified **numerically,** by giving the values of the function for a number of values of the independent variable, as in the preceding table. It may be specified **verbally,** as in "Let $W(t)$ be the weight of the child at age t months in her first year."[1] In some cases we may be able to use an algebraic formula to calculate the function, and we say that the function is specified **algebraically.** In Section 1.2 we will see that a function may also be specified **graphically.**

Question For which values of t does it make sense to ask for $W(t)$? In other words, for which ages t is the function W defined?

Answer Since $W(t)$ refers to the weight of the child at age t months *in her first year,* $W(t)$ is defined when t is any number between 0 and 12—that is, when $0 \leq t \leq 12$. Using interval notation (see Appendix A), we can say that $W(t)$ is defined when t is in the interval $[0, 12]$. The set of values of the independent variable for which a function is defined is called its **domain** and is a necessary part of the definition of the function. Notice that the preceding table gives the value of $W(t)$ at only some of the infinitely many possible values in the domain $[0, 12]$.

Here is a summary of the terms we've just introduced.

Functions
A **real-valued function f of a real-valued variable x** assigns to each real number x in a specified set of numbers, called the **domain** of f, a unique real number $f(x)$, read "f of x."

The variable x is called the **independent variable.** If $y = f(x)$, we call y the **dependent variable.**

Note on Domains
The domain of a function is not always specified explicitly; if no domain is specified for the function f, we take the domain to be the largest set of numbers x for which $f(x)$ makes sense. This "largest possible domain" is sometimes called the **natural domain.**

Quick Examples
1. Let $W(t)$ be the weight (in pounds) at age t months of a particular child during her first year. The independent variable is t. If we write $y = W(t)$, then the dependent variable is y, the child's weight. The domain of W is $[0, 12]$ because it was specified that W gives the child's weight during her first year.

---

[1]Specifying a function verbally in this way is useful for understanding what the function is doing, but it gives no numerical information.

2. Let $f(x) = 1/x$. The function f is specified algebraically. Some specific values of f are

$$f(2) = \frac{1}{2} \qquad f(3) = \frac{1}{3} \qquad f(-1) = \frac{1}{-1} = -1$$

Here, $f(0)$ is not defined because there is no such number as $1/0$. The natural domain of f consists of all real numbers except zero because $f(x)$ makes sense for all values of x other than $x = 0$.

Example 1 • A Numerically Specified Function: Airline Profits

The following table shows the combined third-quarter operating profits of U.S. airlines each year from 1996 to 2001, with $t = 6$ representing 1996:

t (years since 1990)	6	7	8	9	10	11
P ($ billions)	5.3	6.0	7.8	7.2	6.0	−3.5

Profits are for domestic travel only, excluding commuter airlines and small regional carriers.
SOURCE: Bureau of Transportation Statistics, http://www.bts.gov/oai/indicators/yrlyopfinan.html, 2002.

Viewing P as a function of t, give its domain and the values $P(6)$, $P(10)$, and $P(11)$. Estimate and interpret the value $P(8.5)$.

Solution The domain of P is the set of numbers t, with $6 \leq t \leq 11$—that is, $[6, 11]$. From the table, we have

$$P(6) = 5.3 \qquad \text{\$5.3 billion profits in the third quarter of 1996}$$

$$P(10) = 6.0 \qquad \text{\$6.0 billion profits in the third quarter of 2000}$$

$$P(11) = -3.5 \qquad \text{\$3.5 billion loss in the third quarter of 2001}$$

What about $P(8.5)$? Since $P(8) = 7.8$ and $P(9) = 7.2$, we estimate that

$$P(8.5) \approx 7.5 \qquad \text{7.5 is midway between 7.8 and 7.2.}$$

We call the process of estimating values for a function between points where it is already known **interpolation.**

Question How should we interpret $P(8.5)$?

Answer $P(8)$ represents the third-quarter profits for the year beginning January 1998, and $P(9)$ represents the third-quarter profits for the year beginning January 1999. Thus, the most logical interpretation of $P(8.5)$ is that it represents the third-quarter profits for the year beginning July 1998—that is, the first-quarter profits for 1999.[2]

---

[2]Airline profits were actually $7.9 billion in the first quarter of 1999. The fact that this is higher than the interpolated estimate probably results from the peak travel season in January.

✳ *Before we go on . . .*

Question Can we use the table to estimate $P(t)$ for values of t *outside* the domain—say, $t = 15$?

Answer Strictly speaking, $P(t)$ is defined only when $6 \le t \le 11$. However, we could consider a function with a larger domain, like the function that gives the third-quarter profits each year from 1996 to 2005. Estimating values for a function outside a range where it is already known is called **extrapolation.** As a general rule, extrapolation is far less reliable than interpolation: Predicting the future from current data is difficult.

The two functions we have looked at so far were both specified **numerically,** meaning that we were given numerical values of the function evaluated at *certain* values of the independent variable. It would be more useful if we had a formula that would allow us to calculate the value of the function for *any* value of the independent variable we wished. If a function is specified by a formula, we say that it is specified **algebraically.**

Example 2 • An Algebraically Defined Function

Let f be the function specified by

$$f(x) = -0.4x^2 + 7x - 23$$

with domain $(-2, 10]$. This formula gives an approximation of the airline profit function P in Example 1. Use the formula to calculate $f(0), f(10), f(-1), f(a)$, and $f(x + h)$. Is $f(-2)$ defined?

Solution Let's check first that the values we are asked to calculate are all defined. Since the domain is stated to be $(-2, 10]$, the quantities $f(0), f(10)$, and $f(-1)$ are all defined. The quantities $f(a)$ and $f(x + h)$ will also be defined if a and $x + h$ are understood to be in $(-2, 10]$. However, $f(-2)$ is not defined, since -2 is not in the domain $(-2, 10]$.

If we take the formula for $f(x)$ and substitute 0 for x (replace x everywhere it occurs by 0), we get

$$f(0) = -0.4(0)^2 + 7(0) - 23 = -23$$

so $f(0) = -23$. Similarly,

$$f(10) = -0.4(10)^2 + 7(10) - 23 = -40 + 70 - 23 = 7$$

$$f(-1) = -0.4(-1)^2 + 7(-1) - 23 = -0.4 - 7 - 23 = -30.4$$

$$f(a) = -0.4a^2 + 7a - 23 \qquad \text{Substitute } a \text{ for } x.$$

$$f(x + h) = -0.4(x + h)^2 + 7(x + h) - 23 \qquad \text{Substitute } (x + h) \text{ for } x.$$

$$= -0.4x^2 - 0.8xh - 0.4h^2 + 7x + 7h - 23$$

✳ *Before we go on . . .* Had we not specified anything about the domain of f, we would have used the natural domain of f. In this case the natural domain is the set of all real numbers, since $-0.4x^2 + 7x - 23$ is defined for every real number x.

We said that the function f given in this example is an approximation of the profit function P of Example 1. The following table compares some of their values:

x	8	9	10
$P(x)$	7.8	7.2	6.0
$f(x)$	7.4	7.6	7.0

The function f is a "best-fit," or regression quadratic, curve based on the data in Example 1 (coefficients are rounded and the third-quarter 2001 figure is excluded). We will learn more about regression later in this chapter.

We call the algebraic function f an **algebraic model** of U.S. airline third-quarter profits since it models, or represents (approximately), the third-quarter profits, using an algebraic formula. The particular kind of algebraic model we used is called a **quadratic model** (see the end of this section for the names of some commonly used models).

Notes

1. Instead of using *function notation*

$$f(x) = -0.4x^2 + 7x - 23 \qquad \text{Function notation}$$

we could use *equation notation*

$$y = -0.4x^2 + 7x - 23 \qquad \text{Equation notation}$$

(the choice of the letter y is conventional), and we say that "y is a function of x." Note also that there is nothing magical about the letter x. We might just as well say

$$f(t) = -0.4t^2 + 7t - 23$$

which defines *exactly the same function* as $f(x) = -0.4x^2 + 7x - 23$. For example, to calculate $f(10)$ from the formula for $f(t)$, we would substitute 10 for t, getting $f(10) = 7$, just as we did using the formula for $f(x)$.

2. It is important to place parentheses around the number at which you are evaluating a function. For instance, if $g(x) = x^2 + 1$, then

$$g(-2) = (-2)^2 + 1 = 4 + 1 = 5 \quad ✓ \quad \text{Not } -2^2 + 1 = -4 + 1 = -3 \quad ✗$$

$$g(x + h) = (x + h)^2 + 1 \quad ✓ \qquad \text{Not } x^2 + h + 1 \quad \text{or} \quad x + h^2 + 1 \quad ✗$$

T Example 3 • Evaluating a Function with Technology

Evaluate the function $f(x) = -0.4x^2 + 7x - 23$ of Example 2 for $x = 0, 1, 2, \ldots, 10$.

Solution This task is tedious to do by hand, but various technologies can make evaluating a function easier.

Graphing Calculator

There are several ways to evaluate an algebraically defined function on a graphing calculator such as the TI-83. First, enter the function in the Y= screen, as

$$Y_1 = -0.4*X^2+7*X-23$$

Negative (-) and minus (−) are different keys on the TI-83.

or $$Y_1 = -0.4X^2+7X-23$$

(See Appendix A for a discussion of technology formulas.) Then, to evaluate $f(0)$, for example, enter the following in the home screen:

$$Y_1(0)$$ This evaluates the function Y_1 at 0.

Alternatively, you can use the Table feature: After entering the function under Y_1, press 2nd TABLE and set Indpnt to Ask. (You do this once and for all; it will permit you to specify values for x in the table screen.) Then, press 2nd TABLE, and you will be able to evaluate the function at several values of x. Whichever method you use, you should obtain the following set of values:

x	0	1	2	3	4	5	6	7	8	9	10
$f(x)$	−23	−16.4	−10.6	−5.6	−1.4	2	4.6	6.4	7.4	7.6	7

Excel

To create a table of values of f using Excel, first set up two columns—one for the values of x and one for the values of $f(x)$. To enter the sequence of values 0, 1, . . . , 10 in the x column, start by entering the first two values, 0 and 1, highlight both of them, and drag the *fill handle* (the little dot at the lower right-hand corner of the selection) down until you reach row 12. (Why 12?)

Now enter the formula for f in cell B2. The technology formula (see Appendix A) for f is

$$-0.4*x^2+7*x-23$$ Technology formula

To get the formula to use for Excel, replace each occurrence of x by the name of the cell holding the value of x (cell A2 in this case) to obtain

$$=-0.4*A2^2+7*A2-23$$ A2 refers to the cell containing x.

(The formula

$$=-0.4*x^2+7*x-23$$

will also work in many versions of Excel, provided you have entered the heading x in cell A1 as shown. Try it!)

Enter this formula in cell B2 and drag it down (using the fill handle) to cell B12, as shown below (on the left), to obtain the result shown on the right.

	A	B	C
1	x	f(x)	
2		0	=-0.4*A2^2+7*A2-23
3		1	
4		2	
5		3	
6		4	
7		5	
8		6	
9		7	
10		8	
11		9	
12		10	

	A	B
1	x	f(x)
2		0	-23
3		1	-16.4
4		2	-10.6
5		3	-5.6
6		4	-1.4
7		5	2
8		6	4.6
9		7	6.4
10		8	7.4
11		9	7.6
12		10	7

 Web Site At the Web site, follow the path

Web site → Online Utilities → Function Evaluator & Grapher

Enter

$-0.4*x^2+7*x-23$

or $\quad -0.4x^2+7x-23$

in the f(x) = box, enter the values of x in the Values of x boxes (you can use the tab key to move from one box to the next), and press Evaluate.

Question In Example 3 the values of $f(x)$ differ slightly from those of $P(x)$. Is this the best we can do with an algebraic model? Can't we get the airline profit data exactly?

Answer It is possible to find algebraic formulas that give the same exact values as Example 1, but such formulas would be far more complicated than the one given and quite possibly less useful.[3]

Question How do we measure how closely the algebraic model approximates the actual data? For instance, I notice that the approximation is close when $x = 8$ but not so close when $x = 10$.

Answer In the discussion on regression at the end of this chapter, we describe precise ways of measuring this "goodness of fit." These measurements are based on a simple procedure: Take the differences between the actual values and the values given by the model, square them, and then add the results.

---

[3]Methods for obtaining such formulas are described in most numerical analysis textbooks. One reason that more complex formulas are often less realistic than simple ones is that it is often random phenomena in the real world, rather than algebraic relationships, that cause data to fluctuate. Attempting to model these random fluctuations using algebraic formulas amounts to imposing mathematical structure where structure does not exist.

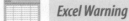

Excel Warning

In interpreting a negative sign at the start of an expression, Excel uses a different convention from the usual mathematical one used by almost all other technology and programming languages:

Excel Formula	Usual Interpretation	Excel Interpretation	
$-x^2$	$-x^2$	$(-x)^2$	Same as x^2 Different
$2-x^2$	$2 - x^2$	$2 - x^2$	The same
$-1*x^2$	$-x^2$	$-x^2$	The same
$-(x^2)$	$-x^2$	$-x^2$	The same

Thus, if a formula begins with $-x^2$, you should enter it in Excel as

$$=-1*x\text{^}2 \quad \text{or} \quad -(x\text{^}2) \qquad \text{With } x \text{ replaced by the cell holding } x$$

Quick Examples

1. To enter $-x^2 + 4x - 3$ in Excel, type $=-1*x\text{^}2+4*x-3$.

2. To enter $-3x^2 + 4x - 3$ in Excel, type $=-3*x\text{^}2+4*x-3$.

3. To enter $4x - x^2$ in Excel, type $=4*x-x\text{^}2$.

In short, you need to be careful only when the expression you want to use begins with a negative sign in front of an x.

Sometimes, as in Example 4, we need to use several formulas to specify a single function.

Example 4 • A Piecewise-Defined Function: Semiconductors

The percentage $U(t)$ of semiconductor equipment manufactured in the United States during the period 1981–1994 can be approximated by the following function of time t in years ($t = 0$ represents 1980):[4]

$$U(t) = \begin{cases} 76 - 3.3t & \text{if } 1 \le t \le 10 \\ 18 + 2.5t & \text{if } 10 < t \le 14 \end{cases}$$

What percentage of semiconductor equipment was manufactured in the United States in the years 1982, 1990, and 1992?

Solution   The years 1982, 1990, and 1992 correspond, respectively, to $t = 2$, 10, and 12.

$t = 2$: $U(2) = 76 - 3.3(2) = 69.4$ We use the first formula since $1 \le t \le 10$.

$t = 10$: $U(10) = 76 - 3.3(10) = 43$ We use the first formula since $1 \le t \le 10$.

$t = 12$: $U(12) = 18 + 2.5(12) = 48$ We use the second formula since $10 < t \le 14$.

Thus, the percentage of semiconductor equipment that was manufactured in the United States was 69.4% in 1982, 43% in 1990, and 48% in 1992.

---

[4]SOURCE for data: VLSI Research/*New York Times*, October 9, 1994, sec. 3, p. 2.

Using Logical Expressions with Technology

The following technology formula defines the function U in graphing calculators, Excel, and the Web site (follow Web Site → Online Utilities → Function Evaluator & Grapher) as well as several other technologies:

$$(x<=10)*(76-3.3*x)+(x>10)*(18+2.5*x) \qquad \text{Excel, Web site}$$

$$(X\leq10)*(76-3.3*X)+(X>10)*(18+2.5*X) \qquad \text{TI-83}$$

In the TI-83 the logical operators ($\leq$ and $>$, for example) can be found by pressing $\boxed{\text{2nd}}$ $\boxed{\text{TEST}}$. When x is less than or equal to 10, the logical expression $(x<=10)$ evaluates to 1 because it is true, and the expression $(x>10)$ evaluates to 0 because it is false. The value of the function is given by the expression $(76-3.3*x)$. When x is greater than 10, the expression $(x<=10)$ evaluates to 0, and the expression $(x>10)$ evaluates to 1, so the value of the function is given by the expression $(18+2.5*x)$.

As in Example 3, you can use the Table feature to compute several values of the function at once.

In Excel we can set up a worksheet as shown:

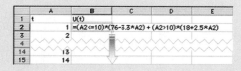

Using the IF Function in Excel

The following worksheet shows how we can get the same result using the IF function in Excel:

The IF function evaluates its first argument, which tests to see if the value of t is in the range $t \leq 10$. If the first argument is true, IF returns the result of evaluating its second argument; if not, it returns the result of evaluating its third argument.

The functions we used in Examples 1–4 are **mathematical models** of real-life situations, since they model, or represent, situations in mathematical terms.

Mathematical Modeling

To *mathematically model* a situation means to represent it in mathematical terms. The particular representation used is called a **mathematical model** of the situation. Mathematical models do not always represent a situation perfectly or completely. Some (like that in Example 2) represent a situation only approximately, whereas others represent only some aspects of the situation.

Quick Examples

Situation	**Model**
1. Albano's bank balance is twice Bravo's.	$a = 2b$ ($a =$ Albano's balance; $b =$ Bravo's)
2. The temperature is now 10° F and increasing by 20° per hour.	$T(t) = 10 + 20t$ ($t =$ time in hours; $T =$ temperature)
3. The volume of a rectangular solid with square base is obtained by multiplying the area of its base by its height.	$V = x^2 h$ ($h =$ height; $x =$ length of a side of the base)
4. U.S. airline profits	The table in Example 1 is a **numerical model** of U.S. airline profits. The function in Example 2 is an **algebraic model** of U.S. airline profits.
5. Semiconductor manufacture in the United States	Example 4 gives a **piecewise algebraic model** of the percentage of semiconductor equipment manufactured in the United States.

Table 1 lists some common types of functions that are often used to model real-world situations.

Table 1 • Common Types of Algebraic Functions

	Type of Function	**Example**
Linear	$f(x) = mx + b$ m, b constant	$f(x) = 3x - 2$ 3*x−2
Quadratic	$f(x) = ax^2 + bx + c$ a, b, c constant $(a \neq 0)$	$f(x) = -3x^2 + x - 1$ −3*x^2+x−1
Cubic	$f(x) = ax^3 + bx^2 + cx + d$ a, b, c, d constant $(a \neq 0)$	$f(x) = 2x^3 - 3x^2 + x - 1$ 2*x^3−3*x^2+x−1
Polynomial	$f(x) = ax^n + bx^{n-1} + \cdots + rx + s$ $a, b, \ldots, r, s$ constant (Includes all of the above functions)	All of the above and $f(x) = x^6 - x^4 + x - 3$ x^6−x^4+x−3
Exponential	$f(x) = Ab^x$ A, b constant $(b$ positive)	$f(x) = 3(2^x)$ 3*2^x
Rational	$f(x) = \dfrac{P(x)}{Q(x)}$ $P(x)$ and $Q(x)$ polynomials	$f(x) = \dfrac{x^2 - 1}{2x + 5}$ (x^2−1)/(2*x+5)

Functions and models other than linear ones are called **nonlinear.**

> ## Guideline: How Many Decimal Places?
>
> **Question** When I use technology to evaluate a function, I often get numbers with many decimals, like 2.034 239 847 2. How many decimal places should I round to?
>
> **Answer** General rules of thumb are these:
>
> 1. When dealing with a mathematical model in which the coefficients are already rounded to a certain number of digits, round your answer to the same number of digits, since the additional digits are meaningless. (See Appendix A for a discussion on significant digits.) For instance, if
>
> $$f(x) = -0.62x^2 + 4.3x - 2.1$$
>
> has rounded coefficients, then $f(2.1) \approx 4.1958$, which we should round to two digits: 4.2.
>
> 2. If the coefficients in the model are exact, then retain as many digits as practically convenient or as called for in the problem.
>
> 3. Never round the answers to intermediate steps in your calculation. *Only round at the end of the calculation.*

1.1 EXERCISES

In Exercises 1–4, evaluate or estimate each expression based on the following table:

x	-3	-2	-1	0	1	2	3
$f(x)$	1	2	4	2	1	0.5	0.25

1. a. $f(0)$ **b.** $f(2)$ **2. a.** $f(-1)$ **b.** $f(1)$

3. a. $f(2) - f(-2)$ **b.** $f(-1)f(-2)$ **c.** $-2f(-1)$

4. a. $f(1) - f(-1)$ **b.** $f(1)f(-2)$ **c.** $3f(-2)$

5. Given $f(x) = 4x - 3$, find **a.** $f(-1)$ **b.** $f(0)$ **c.** $f(1)$
 d. $f(y)$ **e.** $f(a + b)$

6. Given $f(x) = -3x + 4$, find **a.** $f(-1)$ **b.** $f(0)$ **c.** $f(1)$
 d. $f(y)$ **e.** $f(a + b)$

7. Given $f(x) = x^2 + 2x + 3$, find **a.** $f(0)$ **b.** $f(1)$
 c. $f(-1)$ **d.** $f(-3)$ **e.** $f(a)$ **f.** $f(x + h)$

8. Given $g(x) = 2x^2 - x + 1$, find **a.** $g(0)$ **b.** $g(-1)$
 c. $g(r)$ **d.** $g(x+h)$

9. Given $g(s) = s^2 + \dfrac{1}{s}$, find **a.** $g(1)$ **b.** $g(-1)$ **c.** $g(4)$
 d. $g(x)$ **e.** $g(s + h)$ **f.** $g(s + h) - g(s)$

10. Given $h(r) = \dfrac{1}{r + 4}$, find **a.** $h(0)$ **b.** $h(-3)$
 c. $h(-5)$ **d.** $h(x^2)$ **e.** $h(x^2 + 1)$ **f.** $h(x^2) + 1$

11. Given
$$f(t) = \begin{cases} -t & \text{if } t < 0 \\ t^2 & \text{if } 0 \le t < 4 \\ t & \text{if } t \ge 4 \end{cases}$$
find **a.** $f(-1)$ **b.** $f(1)$ **c.** $f(4) - f(2)$ **d.** $f(3)f(-3)$

12. Given
$$f(t) = \begin{cases} t - 1 & \text{if } t \le 1 \\ 2t & \text{if } 1 < t < 5 \\ t^3 & \text{if } t \ge 5 \end{cases}$$
find **a.** $f(0)$ **b.** $f(1)$ **c.** $f(4) - f(2)$ **d.** $f(5) + f(-5)$

In Exercises 13–16, say whether $f(x)$ is defined for the given values of x. If it is defined, give its value.

13. $f(x) = x - \dfrac{1}{x^2}$, with domain $(0, +\infty)$ **a.** $x = 4$ **b.** $x = 0$
 c. $x = -1$

14. $f(x) = \dfrac{2}{x} - x^2$, with domain $[2, +\infty)$ **a.** $x = 4$ **b.** $x = 0$
 c. $x = 1$

15. $f(x) = \sqrt{x + 10}$, with domain $[-10, 0)$ **a.** $x = 0$
 b. $x = 9$ **c.** $x = -10$

16. $f(x) = \sqrt{9 - x^2}$, with domain $(-3, 3)$ **a.** $x = 0$
 b. $x = 3$ **c.** $x = -3$

In Exercises 17–20, find and simplify **a.** $f(x + h) - f(x)$
 b. $\dfrac{f(x + h) - f(x)}{h}$

17. $f(x) = x^2$ **18.** $f(x) = 3x - 1$

19. $f(x) = 2 - x^2$ **20.** $f(x) = x^2 + x$

T In Exercises 21–24, first give the technology formula for the given function and then use technology to evaluate the function for the given values of x (when defined there).

21. $f(x) = 0.1x^2 - 4x + 5; x = 0, 1, \ldots, 10$

22. $g(x) = 0.4x^2 - 6x - 0.1; x = -5, -4, \ldots, 4, 5$

23. $h(x) = \dfrac{x^2 - 1}{x^2 + 1}$; $x = 0.5, 1.5, 2.5, \ldots, 10.5$ (Round all answers to four decimal places.)

24. $r(x) = \dfrac{2x^2 + 1}{2x^2 - 1}$; $x = -1, 0, 1, \ldots, 9$ (Round all answers to four decimal places.)

APPLICATIONS

25. Employment The following table lists the approximate number of people employed in the United States during the period 1995–2001, on July 1 of each year ($t = 5$ represents 1995):

Year, t	5	6	7	8	9	10	11
Employment, $P(t)$ (millions)	117	120	123	125	130	132	132

The given values represent nonfarm employment and are approximate. SOURCE: Bureau of Labor Statistics/*New York Times*, December 17, 2001, p. C3.

  a. Find or estimate $P(5)$, $P(10)$, and $P(9.5)$. Interpret your answers.

  b. What is the domain of P?

26. Cell Phone Sales The following table lists the net sales (after-tax revenue) at the Finnish cell phone company Nokia for each year in the period 1995–2001 ($t = 5$ represents 1995):

Year, t	5	6	7	8	9	10	11
Nokia Net Sales, $P(t)$ ($ billions)	8	8	10	16	20	27	28

SOURCE: Nokia/*New York Times*, February 6, 2002, p. A3.

  a. Find or estimate $P(5)$, $P(10)$, and $P(7.5)$. Interpret your answers.

  b. What is the domain of P?

27. Coffee Shops The number $C(t)$ of coffee shops and related enterprises in the United States can be approximated by the following function of time t in years since 1990:[5]

$$C(t) = \begin{cases} 500t + 800 & \text{if } 0 \le t \le 4 \\ 1300t - 2400 & \text{if } 4 < t \le 10 \end{cases}$$

  a. Evaluate $C(0)$, $C(4)$, and $C(5)$ and interpret the results.

  b. Use the model to estimate when there were 5400 coffee shops in the United States.

 c. Use technology to generate a table of values for $C(t)$ with $t = 0, 1, \ldots, 10$.

28. Semiconductors The percentage $J(t)$ of semiconductor equipment manufactured in Japan during the period 1981–1994 can be approximated by the following function of time t in years since 1980:[6]

$$J(t) = \begin{cases} 17 + 3.1t & \text{if } 1 \le t \le 10 \\ 68 - 2t & \text{if } 10 < t \le 14 \end{cases}$$

  a. Evaluate $J(1)$, $J(10)$, and $J(12)$ and interpret the results.

  b. Use the model to estimate, to the closest year, when Japan manufactured 40% of all semiconductor equipment.

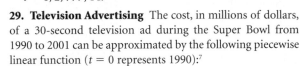 **c.** Use technology to generate a table of values for $J(t)$ with $t = 1, 2, \ldots, 14$.

29. Television Advertising The cost, in millions of dollars, of a 30-second television ad during the Super Bowl from 1990 to 2001 can be approximated by the following piecewise linear function ($t = 0$ represents 1990):[7]

$$C(t) = \begin{cases} 0.08t + 0.6 & \text{if } 0 \le t < 8 \\ 0.355t - 1.6 & \text{if } 8 \le t \le 11 \end{cases}$$

  a. Give the technology formula for C and complete the following table of values of the function C:

t	0	1	2	3	4	5	6	7	8	9	10	11
$C(t)$												

  b. Between 1998 and 2000, the cost of a Super Bowl ad was increasing at a rate of $_____ million per year.

30. Internet Purchases The percentage $p(t)$ of buyers of new cars who used the Internet for research or purchase since 1997 is given by the following function ($t = 0$ represents 1997):[8]

$$p(t) = \begin{cases} 10t + 15 & \text{if } 0 \le t < 1 \\ 15t + 10 & \text{if } 1 \le t \le 4 \end{cases}$$

  a. Give the technology formula for p and complete the following table of values of the function p:

t	0	0.5	1	1.5	2	2.5	3	3.5	4
$p(t)$									

  b. Between 1998 and 2000, the percentage of buyers of new cars who used the Internet for research or purchase was increasing at a rate of _____% per year.

[5]SOURCE: Specialty Coffee Association of America/*New York Times*, August 13, 1995, p. F10.

[6]SOURCE for data: VLSI Research/*New York Times*, October 9, 1994, sec. 3, p. 2.

[7]SOURCE: *New York Times*, January 26, 2001, p. C1.

[8]Model is based on data through 2000 (the 2000 value was estimated). SOURCE: J. D. Power Associates/*New York Times*, January 25, 2000, p. C1.

31. Income Taxes The U.S. federal income tax is a function of taxable income. Write T for the tax owed on a taxable income of I dollars. In 2001 the function T for a single taxpayer was specified as follows:

If your taxable income was		Your tax is	Of the amount over—
Over—	But not over—		
$0	$6,000	----------10%	$0
6,000	27,050	$600.00 + 15%	6,000
27,050	65,550	3,757.50 + 27%	27,050
65,550	136,750	14,152.50 + 30%	65,550
136,750	297,350	35,512.50 + 35%	136,750
297,350	----------	91,722.50 + 38.6%	297,350

These were the tax rates that went into effect as of July 1, 2001. The lowest bracket, 10%, was implemented not as a tax bracket but as a midyear rebate to the taxpayer of (in most cases) 5% of $6000, or $300, representing the difference between the old tax of 15% in this range and the new tax of 10%.

What was the tax owed by a single taxpayer on a taxable income of $26,000? On a taxable income of $65,000?

32. Income Taxes The income tax function T in Exercise 31 can also be written in the following form:

$$T(I) = \begin{cases} 0.10I & \text{if } 0 < I \le 6000 \\ 600 + 0.15(I - 6000) & \text{if } 6000 < I \le 27{,}050 \\ 3757.50 + 0.27(I - 27{,}050) & \text{if } 27{,}050 < I \le 65{,}550 \\ 14{,}152.50 + 0.30(I - 65{,}550) & \text{if } 65{,}550 < I \le 136{,}750 \\ 35{,}512.50 + 0.35(I - 136{,}750) & \text{if } 136{,}750 < I \le 297{,}350 \\ 91{,}722.50 + 0.386(I - 297{,}350) & \text{if } I > 297{,}350 \end{cases}$$

What was the tax owed by a single taxpayer on a taxable income of $25,000? On a taxable income of $125,000?

33. Demand The demand for Sigma Mu Fraternity plastic brownie dishes is

$$q(p) = 361{,}201 - (p + 1)^2$$

where q represents the number of brownie dishes Sigma Mu can sell each month at a price of p¢. Use this function to determine

a. the number of brownie dishes Sigma Mu can sell each month if the price is set at 50¢.

b. the number of brownie dishes they can unload each month if they give them away.

c. the lowest price at which Sigma Mu will be unable to sell any dishes.

34. Revenue The total weekly revenue earned at Royal Ruby Retailers is given by

$$R(p) = -\frac{4}{3}p^2 + 80p$$

where p is the price (in dollars) RRR charges per ruby. Use this function to determine

a. the weekly revenue, to the nearest dollar, when the price is set at $20/ruby.

b. the weekly revenue, to the nearest dollar, when the price is set at $200/ruby (interpret your result).

c. the price RRR should charge in order to obtain a weekly revenue of $1200.

35. Investments in South Africa The number of U.S. companies that invested in South Africa from 1986 through 1994 closely followed the function

$$n(t) = 5t^2 - 49t + 232$$

Here, t is the number of years since 1986, and $n(t)$ is the number of U.S. companies that own at least 50% of their South African subsidiaries and employ 1000 or more people.[9]

a. Find the appropriate domain of n.

b. Is $t \ge 0$ an appropriate domain? Give reasons for your answer.

36. Sony Net Income The annual net income for Sony Corporation from 1989 through 1994 can be approximated by the function

$$I(t) = -77t^2 + 301t + 524$$

Here, t is the number of years since 1989, and $I(t)$ is Sony's net income in millions of dollars for the corresponding fiscal year.[10]

a. Find the appropriate domain of I.

b. Is $[0, +\infty)$ an appropriate domain? Give reasons for your answer.

37. Spending on Corrections The following table shows the annual spending by all states in the United States on corrections ($t = 0$ represents the year 1990):

Year, t	0	1	2	3	4	5	6	7
Spending, S ($ billions)	16	18	18	20	22	26	28	30

SOURCE: National Association of State Budget Officers/*New York Times*, February 28, 1999, p. A1. Data are rounded.

a. Which of the following functions best fits the given data? (*Warning:* None of them fits exactly, but one fits more closely than the others.)

(1) $S(t) = -0.2t^2 + t + 16$

(2) $S(t) = 0.2t^2 + t + 16$

(3) $S(t) = t + 16$

b. Use your answer to part (a) to "predict" spending on corrections in 1998, assuming that the trend continued.

---

[9]The model is the authors' (least-squares quadratic regression with coefficients rounded to nearest integer). SOURCE: Investor Responsibility Research Center Inc., Fleming Martin/*New York Times*, June 7, 1994, p. D1.

[10]The model is the authors' (least-squares quadratic regression with coefficients rounded to nearest integer). SOURCE: Sony Corporation/*New York Times*, May 20, 1994, p. D1.

38. Spending on Corrections Repeat Exercise 37, this time choosing from the following functions:

$$(1)\ S(t) = 16 + 2t$$

$$(2)\ S(t) = 16 + t + 0.5t^2$$

$$(3)\ S(t) = 16 + t - 0.5t^2$$

39. Toxic Waste Treatment The cost of treating waste by removing PCPs rises rapidly as the quantity of PCPs removed increases. Here is a possible model:

$$C(q) = 2000 + 100q^2$$

where q is the reduction in toxicity (in pounds of PCPs removed per day) and $C(q)$ is the daily cost (in dollars) of this reduction.

a. Find the cost of removing 10 pounds/day of PCPs.

b. Government subsidies for toxic waste cleanup amount to

$$S(q) = 500q$$

where q is as above and $S(q)$ is the daily dollar subsidy. Calculate the net cost function $N(q)$ (the cost of removing q pounds of PCPs per day after the subsidy is taken into account) given the cost function and subsidy above, and find the net cost of removing 20 pounds/day of PCPs.

40. Dental Plans A company pays for its employees' dental coverage at an annual cost C given by

$$C(q) = 1000 + 100\sqrt{q}$$

where q is the number of employees covered and $C(q)$ is the annual cost in dollars.

a. If the company has 100 employees, find its annual outlay for dental coverage.

b. Assuming that the government subsidizes coverage by an annual dollar amount of

$$S(q) = 200q$$

calculate the net cost function $N(q)$ to the company and calculate the net cost of subsidizing its 100 employees. Comment on your answer.

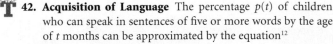

41. Acquisition of Language The percentage $p(t)$ of children who can speak in at least single words by the age of t months can be approximated by the equation[11]

$$p(t) = 100\left(1 - \frac{12,200}{t^{4.48}}\right) \qquad (t \geq 8.5)$$

a. Give a technology formula for p.

b. Create a table of values of p for $t = 9, 10, \ldots, 20$ (rounding answers to one decimal place).

c. What percentage of children can speak in at least single words by the age of 12 months?

d. By what age are 90% or more children speaking in at least single words?

42. Acquisition of Language The percentage $p(t)$ of children who can speak in sentences of five or more words by the age of t months can be approximated by the equation[12]

$$p(t) = 100\left(1 - \frac{5.27 \times 10^{17}}{t^{12}}\right) \qquad (t \geq 30)$$

a. Give a technology formula for p.

b. Create a table of values of p for $t = 30, 31, \ldots, 40$ (rounding answers to one decimal place).

c. What percentage of children can speak in sentences of five or more words by the age of 36 months?

d. By what age are 75% or more children speaking in sentences of five or more words?

COMMUNICATION AND REASONING EXERCISES

43. Complete the following: If the market price m of gold varies with time t, then the independent variable is _____, and the dependent variable is _____.

44. Complete the following: If weekly profit P is specified as a function of selling price s, then the independent variable is _____, and the dependent variable is _____.

45. Complete the following: The function notation for the equation $y = 4x^2 - 2$ is _____.

46. Complete the following: The equation notation for $C(t) = -0.34t^2 + 0.1t$ is _____.

47. You now have 200 sound files on your hard drive, and this number is increasing by 10 sound files each day. Find a mathematical model for this situation.

48. The amount of free space left on your hard drive is now 50 gigabytes (GB) and is decreasing by 5 GB/month. Find a mathematical model for this situation.

49. Why is the following assertion false? "If $f(x) = x^2 - 1$, then $f(x + h) = x^2 + h - 1$."

50. Why is the following assertion false? "If $f(2) = 2$ and $f(4) = 4$, then $f(3) = 3$."

51. True or false: Every function can be specified numerically.

52. Which supplies more information about a situation: a numerical model or an algebraic model?

---

[11]The model is the authors' and is based on data presented in the article "The Emergence of Intelligence" by William H. Calvin, *Scientific American* (October 1994): 101–107.

[12]Ibid.

1.2 *Functions from the Graphical Viewpoint*

Consider again the function W discussed in Section 1.1, giving a child's weight during her first year. If we represent the data given in Section 1.1 graphically by plotting the given pairs of numbers $(t, W(t))$, we get Figure 2. (We have connected successive points by line segments.)

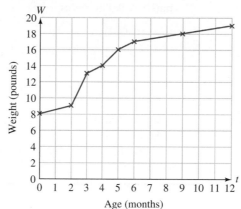

Figure 2

Suppose now that we had only the graph without the table of data given in Section 1.1. We could use the graph to find values of W. For instance, to find $W(9)$ from the graph, we do the following:

1. Find the desired value of t at the bottom of the graph ($t = 9$ in this case).

2. Estimate the height (W coordinate) of the corresponding point on the graph (18 in this case).

Thus, $W(9) \approx 18$ pounds.[13]

We say that Figure 1 specifies the function W **graphically.** The graph is not a very accurate specification of W; the actual weight of the child would follow a smooth curve rather than a jagged line. However, the jagged line is useful in that it permits us to interpolate: For instance, we can estimate that $W(1) \approx 8.5$ pounds.

Example 1 • A Function Specified Graphically: Magazine Circulation

Figure 3 shows the approximate annual circulation of *Outside Magazine* for the period 1993–2000 ($t = 0$ represents December 1993; $C(t)$ represents the circulation in the year ending at time t). Estimate and interpret $C(0)$, $C(2)$, and $C(4.5)$. What is the domain of C?

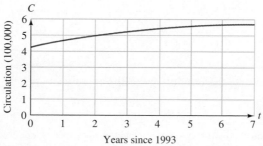

Figure 3

SOURCES: 1993–1997 data: Audit Bureau of Circulations/ Publishers Information Bureau/*New York Times*, October 27, 1997, p. D13. 2000 data: http://writenews.com/ 2000/030600_gearunlimited_outside.htm.

---

[13]In a graphically defined function, we can never know the y coordinates of points exactly; no matter how accurately a graph is drawn, we can only obtain *approximate* values of the coordinates of points. That is why we have been using the word *estimate* rather than *calculate* and why we say "$W(9) \approx 18$" rather than "$W(9) = 18$."

Solution We carefully estimate the C coordinates of the points with t coordinates 0, 2, and 4.5:

$$C(0) \approx 4.2$$

meaning that the circulation in the year ending December 1993 ($t = 0$) was approximately 420,000 copies;

$$C(2) \approx 5.0$$

meaning that the circulation in the year ending December 1995 ($t = 2$) was approximately 500,000 copies;

$$C(4.5) \approx 5.5$$

meaning that the circulation in the year ending June 1998 ($t = 4.5$) was approximately 550,000 copies.

The domain of C is the set of all values of t for which $C(t)$ is defined: $0 \le t \le 7$, or $[0, 7]$.

Sometimes we are interested in drawing the graph of a function that has been specified in some other way—perhaps numerically or algebraically. We do this by plotting points with coordinates $(x, f(x))$.[14] Here is the formal definition of a graph.

Graph of a Function

The **graph of the function f** is the set of all points $(x, f(x))$ in the xy plane, where *we restrict the values of x to lie in the domain of f.*

To obtain the graph of a function, plot points of the form $(x, f(x))$ for several values of x in the domain of f. The shape of the entire graph can usually be inferred from sufficiently many points. (Calculus can give us information that allows us to draw a graph with relatively few points plotted.)

Quick Example

To sketch the graph of the function

$$f(x) = x^2 \qquad \text{Function notation}$$

or $\quad y = x^2 \qquad \text{Equation notation}$

with domain the set of all real numbers, first choose some convenient values of x in the domain and compute the corresponding y-coordinates.

x	-3	-2	-1	0	1	2	3
$y = x^2$	9	4	1	0	1	4	9

Plotting these points gives the picture on the left, suggesting the graph on the right.*

*If you plot more points, you will find that they lie on a smooth curve as shown. That is why we did not use line segments to connect the points.

[14]Graphing utilities typically draw graphs by plotting and joining a large number of points.

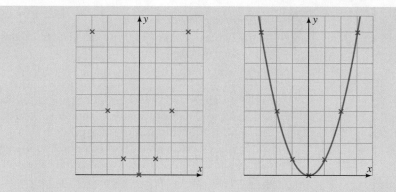

(This particular curve happens to be called a **parabola,** and its lowest point, at the origin, is called its **vertex.**)

Example 2 • Drawing the Graph of a Function: Web-Site Revenue

The monthly revenue[15] R from users logging on to your gaming site depends on the monthly access fee p you charge according to the formula

$$R(p) = -5600p^2 + 14,000p \qquad (0 \leq p \leq 2.5)$$

(R and p are in dollars.) Sketch the graph of R. Find the access fee that will result in the largest monthly revenue.

Solution To sketch the graph of R by hand, we plot points of the form $(x, R(x))$ for several values of x in the domain $[0, 2.5]$ of R. First, we calculate several points:

p	0	0.5	1	1.5	2	2.5
$R(p) = -5600p^2 + 14,000p$	0	5600	8400	8400	5600	0

Graphing these points gives the graph shown in Figure 4(a), suggesting the parabola shown in Figure 4(b).

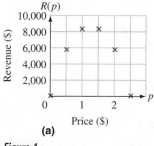

(a)

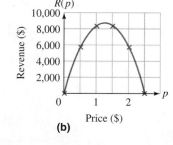

(b)

Figure 4

The revenue graph appears to reach its highest point when $p = 1.25$, so setting the access fee at $1.25 appears to result in the largest monthly revenue.[16]

---

[15]The **revenue** resulting from one or more business transactions is the total payment received, sometimes called the gross proceeds.

[16]We are hedging our language with words like *suggesting* and *appears* because the few points we have plotted don't, by themselves, allow us to draw these conclusions with certainty.

Graphing Calculator

As the name suggests, graphing calculators are designed for graphing functions. If you enter

    Y₁ = -5600*X^2+14000*X

in the Y= screen, you can reproduce the graph shown in Figure 4(b) by setting the window coordinates to match Figure 4(a): $\text{Xmin} = 0$, $\text{Xmax} = 2.5$, $\text{Ymin} = 0$, $\text{Ymax} = 10000$, and then pressing GRAPH.

If you want to plot individual points [as in Figure 4(a)] on the TI-83, enter the data in the STAT EDIT mode with the values of p in L_1 and the value of $R(p)$ in L_2. Then go to the Y= window and turn Plot1 on by selecting it and pressing ENTER. Now select ZoomStat ZOOM 9 to obtain the plot.

Excel

Here's how to plot this graph using Excel. As in Example 3 of Section 1.1, we create a table of values for the function by entering the values of the independent variable p in column A and the formula for the function in cell B2, then copying the formula into the cells beneath it as shown:

	A	B	C
1	p	R	
2	0	=-5600*A2^2+14000*A2	
3	0.5		
4	1		
5	1.5		
6	2		
7	2.5		

To draw the graph shown in Figure 4(a), select (highlight) both columns of data and then ask Excel to insert a chart. When it asks you to specify what type of chart, select the XY (Scatter) option. This tells the program that your data specify the x and y coordinates of a sequence of points. In the same dialog box, select the option that shows points connected by lines. Select Next to bring up a new dialog called Data Type, where you should make sure that the Series in Columns option is selected, telling the program that the x and y coordinates are arranged vertically, down columns. You can then set various other options, add labels, and otherwise fiddle with it until it looks nice.

✻ **Before we go on . . .** To get a smoother curve in Excel, you need to plot many more points. Here is a method of plotting 100 points (in addition to the starting point), similar to the Excel worksheet posted on the Web site at
Web Site → Everything for Finite Math → Chapter 1 → Excel Tutorials →
Section 1.2: Functions from the Graphical Viewpoint
Moreover, if you decide to follow this method, you can save the resulting spreadsheet and use it as an "Excel graphing calculator" to graph any other function (see next page).

Excel Graphing Worksheet
Set up your worksheet as follows:

	A	B	C	D	E	F
1	p	R				
2	=$F$2	= −5600*A2^2 + 14000*A2			Xmin	0
3	=A2+$F$5				Xmax	2.5
4					Points	100
5					Delta X	=(F3−F2)/F4
6						
7						
100						
101						
102						

(Columns C and D are empty in case you want to add additional functions to graph.) The 101 values of the *x* coordinate (the price *p*) will appear in column A. The corresponding values of the *y* coordinate (the revenue *R*) will appear in column B. In column F you see some settings: Xmin = 0 and Xmax = 2.5 (see Figure 4). Delta X (in cell F5) is the amount by which the *x* coordinate is increased as you go from one *x* value in column A to the next, starting with Xmin in A2. Enter the formula for *R*(*p*) in cell B2 and copy it into the other cells in column B, as shown.

When done, graph the data in columns A and B, choosing the scatter-plot option with subtype `Points connected by lines with no markers`. You should see a curve similar to that in Figure 4(b).

Save this worksheet with its graph, and you can use it to graph new functions as follows:

1. Enter the new values for Xmin and Xmax in column F.
2. Enter the new function in cell B2 (using A2 in place of *x*).
3. Copy the contents of cell B2 to cells B3–B102.

The graph will be updated automatically.

Web Site
Follow

Web site → Online Utilities

and then click on any of the following:

Function Evaluator & Grapher	Gives a small graph and also values.
Java Graphing Utility	A high-quality Java grapher
Excel Graphing Utility	An Excel sheet that graphs[17]

Instructions for these graphers are available on the Web pages.

---

[17]Since the grapher requires macros, make sure that macros are *enabled* when Excel prompts you. If macros are disabled, the grapher will not work.

Note: Equation and Function Form

As we discussed after Example 2 in Section 1.1, we could write the function R in the example above in equation form as

$$R = -5600p^2 + 14{,}000p \qquad \text{Equation form}$$

The independent variable is p, and the dependent variable is R. Function notation and equation notation, using the same letter for the function name and the dependent variable, are used interchangeably throughout the literature. It is important to be able to switch back and forth from function notation to equation notation easily.

Vertical-Line Test

Every point in the graph of a function has the form $(x, f(x))$ for some x in the domain of f. Since f assigns a *single* value $f(x)$ to each value of x in the domain, it follows that, in the graph of f, there should be only one y corresponding to any such value of x—namely, $y = f(x)$. In other words, *the graph of a function cannot contain two or more points with the same x coordinate—that is, two or more points on the same vertical line.* On the other hand, a vertical line at a value of x not in the domain will not contain any points in the graph. This gives us the following rule.

Vertical-Line Test

For a graph to be the graph of a function, every vertical line must intersect the graph in *at most* one point.

Quick Examples

As illustrated below, only graph B passes the vertical-line test, so only graph B is the graph of a function.

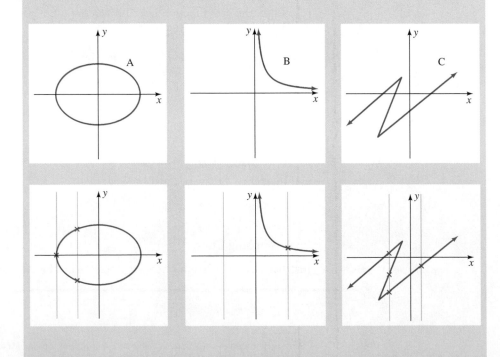

Graphing Piecewise-Defined Functions

Let's revisit the semiconductor example from Section 1.1.

Example 3 • Graphing a Piecewise-Defined Function: Semiconductors

The percentage $U(t)$ of semiconductor equipment manufactured in the United States during the period 1981–1994 can be approximated by the following function of time t in years ($t = 0$ represents 1980):[18]

$$U(t) = \begin{cases} 76 - 3.3t & \text{if } 1 \le t \le 10 \\ 18 + 2.5t & \text{if } 10 < t \le 14 \end{cases}$$

Graph the function U.

Solution As in Example 2, we can sketch the graph of U by hand by computing $U(t)$ for a number of values of t, plotting these points on the graph, and then connecting them.

t	1	4	6	8	10	12	14
$U(t)$	72.7	62.8	56.2	49.6	43	48	53

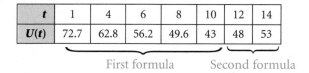

First formula   Second formula

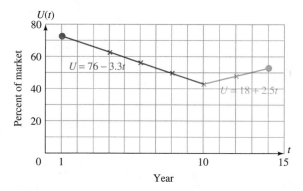

Figure 5

The graph (Figure 5) has the following features:

1. The first formula (the descending line) is used for $1 \le t \le 10$.

2. The second formula (ascending line) is used for $10 < t \le 14$.

3. The domain is $[1, 14]$, so the graph is cut off at $t = 1$ and $t = 14$.

4. The heavy dots at the ends indicate the endpoints of the domain.

T To graph the function U using technology, consult Example 4 of Section 1.1, which shows how to enter this piecewise-defined function and obtain a table of values. Example 2 shows how to then draw the graph (highlight the table of values and use an XY (Scatter) plot).

---

[18]SOURCE for data: VLSI Research/*New York Times*, October 9, 1994, sec. 3, p. 2.

Example 4 • Graphing More Complicated Piecewise-Defined Functions

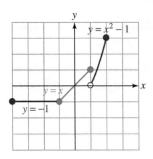

Graph the function f specified by

$$f(x) = \begin{cases} -1 & \text{if } -4 \leq x < -1 \\ x & \text{if } -1 \leq x \leq 1 \\ x^2 - 1 & \text{if } 1 < x \leq 2 \end{cases}$$

Figure 6

Solution The domain of f is $[-4, 2]$, since $f(x)$ is only specified when $-4 \leq x \leq 2$. Further, the function changes formulas when $x = -1$ and $x = 1$.

To sketch the graph by hand, we first sketch the three graphs $y = -1$, $y = x$, and $y = x^2 - 1$ and then use the appropriate portion of each (Figure 6).

Question What are solid dots and open dots doing there?

Answer Solid dots indicate points on the graph, whereas open dots indicate points *not* on the graph. For example, when $x = 1$, the inequalities in the formula tell us that we are to use the middle formula (x) rather than the bottom one ($x^2 - 1$). Thus, $f(1) = 1$, not 0, so we place a solid dot at $(1, 1)$ and an open dot at $(1, 0)$.

Graphing Calculator

On the TI-83 you can enter this function as

$$\underbrace{(-1)*(\text{X}<-1)}_{\text{First formula}} + \underbrace{\text{X}*(-1\leq\text{X and X}\leq1)}_{\text{Second formula}} + \underbrace{(\text{X}^2-1)*(1<\text{X})}_{\text{Third formula}}$$

The logical operator and is found in the TEST LOGIC menu. The following alternative formula will also work:

$$(-1)*(\text{X}<-1) + \text{X}*(-1\leq\text{X})*(\text{X}\leq1) + (\text{X}^2-1)*(1<\text{X})$$

In the graph the TI-83 will not handle the transition at $x = 1$ correctly; it will connect the two parts of the graph with a spurious line segment.

Excel

For Excel you can use the following formula (the same as the alternative TI-83 formula):[19]

$$= \underbrace{(-1)*(\text{x}<-1)}_{\text{First formula}} + \underbrace{\text{x}*(-1<=\text{x})*(\text{x}<=1)}_{\text{Second formula}} + \underbrace{(\text{x}^2-1)*(1<\text{x})}_{\text{Third formula}}$$

Since one of the formulas (the third) specifying the function f is not linear, we need to plot many points in Excel to get a smooth graph. Here is one possible setup (for a smoother curve, plot more points):

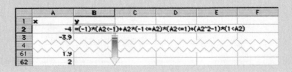

---

[19]We could also use a nested IF statement to accomplish the same result: =IF(x<-1,-1,IF(x<=1,x, x^2-1)).

This results in the graph shown in Figure 7. (Notice that, like the graphing calculator, Excel does not handle the transition at $x = 1$ correctly and connects the two parts of the graph with a spurious line segment.)

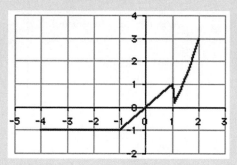

Figure 7

We end this section with a list of some useful types of functions and their graphs (Table 2).

Table 2 • *Functions and Their Graphs*

Type of Function	Examples	
Linear $f(x) = mx + b$ m, b constant Graphs of linear functions are straight lines.	$y = x$	$y = -2x + 2$
Quadratic $f(x) = ax^2 + bx + c$ a, b, c constant $\quad(a \neq 0)$ Graphs of quadratic functions are called **parabolas.**	$y = x^2$	$y = -2x^2 + 2x + 4$
Technology formulas:	x^2	-2*x^2+2*x+4
Cubic $f(x) = ax^3 + bx^2 + cx + d$ a, b, c, d constant $\quad(a \neq 0)$	$y = x^3$	$y = -x^3 + 3x^2 + 1$
Technology formulas:	x^3	-(x^3)+3*x^2+1

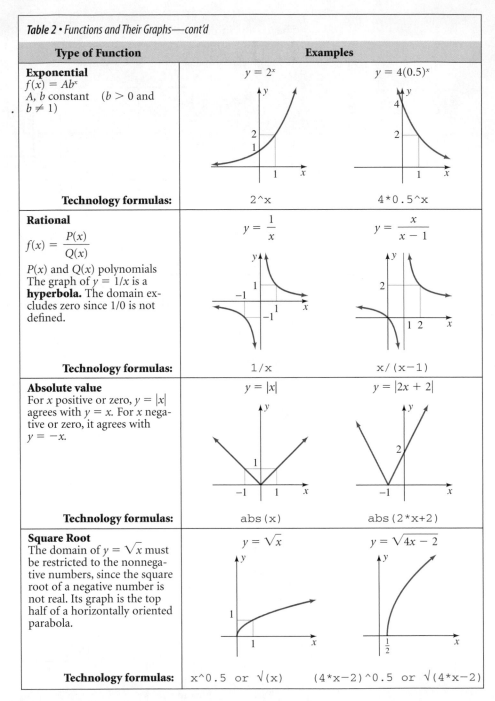

Table 2 • Functions and Their Graphs—cont'd

Type of Function	Examples	
Exponential $f(x) = Ab^x$ A, b constant $(b > 0$ and $b \neq 1)$	$y = 2^x$	$y = 4(0.5)^x$
Technology formulas:	`2^x`	`4*0.5^x`
Rational $f(x) = \dfrac{P(x)}{Q(x)}$ $P(x)$ and $Q(x)$ polynomials The graph of $y = 1/x$ is a **hyperbola.** The domain excludes zero since 1/0 is not defined.	$y = \dfrac{1}{x}$	$y = \dfrac{x}{x-1}$
Technology formulas:	`1/x`	`x/(x-1)`
Absolute value For x positive or zero, $y = \lvert x \rvert$ agrees with $y = x$. For x negative or zero, it agrees with $y = -x$.	$y = \lvert x \rvert$	$y = \lvert 2x + 2 \rvert$
Technology formulas:	`abs(x)`	`abs(2*x+2)`
Square Root The domain of $y = \sqrt{x}$ must be restricted to the nonnegative numbers, since the square root of a negative number is not real. Its graph is the top half of a horizontally oriented parabola.	$y = \sqrt{x}$	$y = \sqrt{4x - 2}$
Technology formulas:	`x^0.5` or `√(x)`	`(4*x-2)^0.5` or `√(4*x-2)`

New Functions from Old: Scaled and Shifted Functions (Online Optional Section)

If you follow the path

Web site $\rightarrow$ Everything for Calculus $\rightarrow$ Chapter 1

you will find a link to this section, with complete online interactive text, examples, and exercises on scaling and translating the graph of a function by changing the formula.

1.2 EXERCISES

In Exercises 1–4, use the graph of the function f to find approximations of the given values.

1.

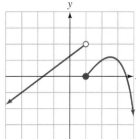

a. $f(1)$ **b.** $f(2)$
c. $f(3)$ **d.** $f(5)$
e. $f(3) - f(2)$

2.

a. $f(1)$ **b.** $f(2)$
c. $f(3)$ **d.** $f(5)$
e. $f(3) - f(2)$

3.

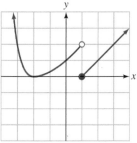

a. $f(-3)$ **b.** $f(0)$
c. $f(1)$ **d.** $f(2)$

e. $\dfrac{f(3) - f(2)}{3 - 2}$

4.

a. $f(-2)$ **b.** $f(0)$
c. $f(1)$ **d.** $f(3)$

e. $\dfrac{f(3) - f(1)}{3 - 1}$

 In Exercises 5 and 6, match the functions to the graphs. Using technology to draw the graphs is suggested but not required.

5. a. $f(x) = x$ $(-1 \le x \le 1)$
   b. $f(x) = -x$ $(-1 \le x \le 1)$
   c. $f(x) = \sqrt{x}$ $(0 < x < 4)$

   d. $f(x) = x + \dfrac{1}{x} - 2$ $(0 < x < 4)$

   e. $f(x) = |x|$ $(-1 \le x \le 1)$
   f. $f(x) = x - 1$ $(-1 \le x \le 1)$

(I)

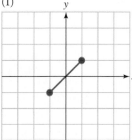

(II)

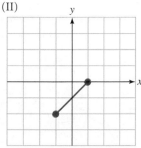

(III)

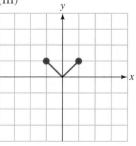

(IV)

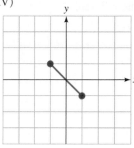

(V)

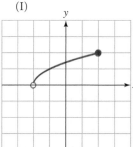

(VI)

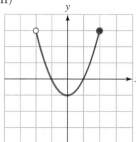

6. a. $f(x) = -x + 4$ $(0 < x \le 4)$
   b. $f(x) = 2 - |x|$ $(-2 < x \le 2)$
   c. $f(x) = \sqrt{x + 2}$ $(-2 < x \le 2)$
   d. $f(x) = -x^2 + 2$ $(-2 < x \le 2)$

   e. $f(x) = \dfrac{1}{x} - 1$ $(0 < x \le 4)$

   f. $f(x) = x^2 - 1$ $(-2 < x \le 2)$

(I)

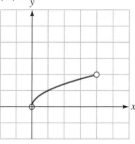

(II)

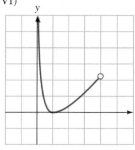

(III)

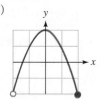

(IV)

(V)

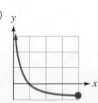

(VI)

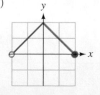

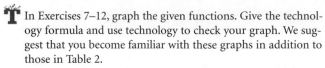

In Exercises 7–12, graph the given functions. Give the technology formula and use technology to check your graph. We suggest that you become familiar with these graphs in addition to those in Table 2.

7. $f(x) = -x^3$ (domain $(-\infty, +\infty)$)

8. $f(x) = x^3$ (domain $[0, +\infty)$)

9. $f(x) = x^4$ (domain $(-\infty, +\infty)$)

10. $f(x) = \sqrt[3]{x}$ (domain $(-\infty, +\infty)$)

11. $f(x) = \dfrac{1}{x^2}$ ($x \neq 0$)

12. $f(x) = x + \dfrac{1}{x}$ ($x \neq 0$)

In Exercises 13–18, sketch the graph of the given function, evaluate the given expressions, and then use technology to duplicate the graphs. Give the technology formula.

13. $f(x) = \begin{cases} x & \text{if } -4 \leq x < 0 \\ 2 & \text{if } 0 \leq x \leq 4 \end{cases}$

 a. $f(-1)$ **b.** $f(0)$ **c.** $f(1)$

14. $f(x) = \begin{cases} -1 & \text{if } -4 \leq x \leq 0 \\ x & \text{if } 0 < x \leq 4 \end{cases}$

 a. $f(-1)$ **b.** $f(0)$ **c.** $f(1)$

15. $f(x) = \begin{cases} x^2 & \text{if } -2 < x \leq 0 \\ \dfrac{1}{x} & \text{if } 0 < x \leq 4 \end{cases}$

 a. $f(-1)$ **b.** $f(0)$ **c.** $f(1)$

16. $f(x) = \begin{cases} -x^2 & \text{if } -2 < x \leq 0 \\ \sqrt{x} & \text{if } 0 < x < 4 \end{cases}$

 a. $f(-1)$ **b.** $f(0)$ **c.** $f(1)$

17. $f(x) = \begin{cases} x & \text{if } -1 < x \leq 0 \\ x + 1 & \text{if } 0 < x \leq 2 \\ x & \text{if } 2 < x \leq 4 \end{cases}$

 a. $f(0)$ **b.** $f(1)$ **c.** $f(2)$ **d.** $f(3)$

18. $f(x) = \begin{cases} -x & \text{if } -1 < x < 0 \\ x - 2 & \text{if } 0 \leq x \leq 2 \\ -x & \text{if } 2 < x \leq 4 \end{cases}$

 a. $f(0)$ **b.** $f(1)$ **c.** $f(2)$ **d.** $f(3)$

APPLICATIONS

Sales of Sport Utility Vehicles Exercises 19–22 refer to the following graph, which shows the number $f(t)$ of sports utility vehicles (SUVs) sold in the United States each year from 1990 through 2003 ($t = 0$ represents 1990, and $f(t)$ represents sales in year t in thousands of vehicles).

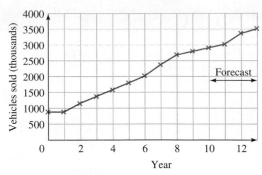

SOURCES: Oak Ridge National Laboratory, Light Vehicle MPG and Market Shares System, AutoPacific, *The US Car and Light Truck Market*, 1999, pp. 24, 120, 121.

19. Estimate $f(6), f(9)$, and $f(7.5)$ and interpret your answers.

20. Estimate $f(5), f(11)$, and $f(1.5)$ and interpret your answers.

21. Which is larger: $f(6) - f(5)$ or $f(10) - f(9)$? Interpret the answer.

22. Which is larger: $f(10) - f(8)$ or $f(13) - f(11)$? Interpret the answer.

23. Employment The following graph shows the number $N(t)$ of people, in millions, employed in the United States (t is time in years, and $t = 0$ represents January 2000).

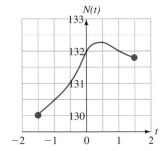

SOURCE: Haver Analytics: The Conference Board/*New York Times*, November 24, 2001.

 a. What is the domain of N?

 b. Estimate $N(-0.5)$, $N(0)$, and $N(1)$ and interpret your answers.

 c. On which interval is $N(t)$ falling? Interpret the result.

24. Productivity The following graph shows an index $P(t)$ of productivity in the United States (t is time in years, and $t = 0$ represents January 2000).

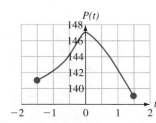

SOURCE: Haver Analytics: The Conference Board/*New York Times*, November 24, 2001.

 a. What is the domain of P?

 b. Estimate $P(-0.5)$, $P(0)$, and $P(1.5)$ and interpret your answers.

 c. On which interval is $P(t) \geq 144$? Interpret the result.

25. Soccer Gear The East Coast College soccer team is planning to buy new gear for its road trip to California. The cost per shirt depends on the number of shirts the team orders, as shown in the following table:

x (shirts ordered)	5	25	40	100	125
$A(x)$ (cost/shirt, $)	22.91	21.81	21.25	21.25	22.31

   a. Which of the following functions best models the data?

     (A) $A(x) = 0.005x + 20.75$

     (B) $A(x) = 0.01x + 20 + \dfrac{25}{x}$

     (C) $A(x) = 0.0005x^2 - 0.07x + 23.25$

     (D) $A(x) = 25.5(1.08)^{(x-5)}$

   b. Graph the model you chose in part (a) for $10 \le x \le 100$. Use your graph to estimate the lowest cost per shirt and the number of shirts the team should order to obtain the lowest price per shirt.

26. Hockey Gear The South Coast College hockey team wants to purchase wool hats for its road trip to Alaska. The cost per hat depends on the number of hats the team orders, as shown in the following table:

x (hats ordered)	5	25	40	100	125
$A(x)$ (cost/hat, $)	25.50	23.50	24.63	30.25	32.70

   a. Which of the following functions best models the data?

     (A) $A(x) = 0.05x + 20.75$

     (B) $A(x) = 0.1x + 20 + \dfrac{25}{x}$

     (C) $A(x) = 0.0008x^2 - 0.07x + 23.25$

     (D) $A(x) = 25.5(1.08)^{(x-5)}$

   b. Graph the model you chose in part (a) with $5 \le x \le 30$. Use your graph to estimate the lowest cost per hat and the number of hats the team should order to obtain the lowest price per hat.

27. Hawaiian Tourism The following table shows the number of visitors to Hawaii during three different years:

Year	1985	1991	1994
Visitors (millions)	4.9	7.0	6.5

Source: Hawaii Visitors Bureau/*New York Times,* September 5, 1995, p. A12.

Which of the following kind of model would best fit the given data? Explain your choice of model. (A, a, b, c, and m are constants.)

(A) Linear: $n(t) = mt + b$

(B) Quadratic: $n(t) = at^2 + bt + c$

(C) Exponential: $n(t) = Ab^t$

28. Magazine Advertising The following table shows the number of advertising pages in *Outside Magazine* for the period 1993–1997 ($t = 0$ represents December 1993):

Year	1993	1994	1995
Advertising Pages	1000	1040	1080

Source: Audit Bureau of Circulations/Publishers Information Bureau/*New York Times,* October 27, 1997, p. D13.

Which of the following kind of model would best fit the given data? Explain your choice of model. (A, a, b, c, and m are constants.)

(A) Linear: $n(t) = mt + b$

(B) Quadratic: $n(t) = at^2 + bt + c$

(C) Exponential: $n(t) = Ab^t$

29. Acquisition of Language The percentage $p(t)$ of children who can speak in at least single words by the age of t months can be approximated by the equation[20]

$$p(t) = 100\left(1 - \frac{12{,}200}{t^{4.48}}\right) \qquad (t \ge 8.5)$$

   a. Give a technology formula for p.

   b. Graph p for $8.5 \le t \le 20$ and $0 \le p \le 100$. Use your graph to answer parts (c) and (d).

   c. What percentage of children can speak in at least single words by the age of 12 months? (Round your answer to the nearest percentage point.)

   d. By what age are 90% of children speaking in at least single words? (Round your answer to the nearest month.)

30. Acquisition of Language The percentage $p(t)$ of children who can speak in sentences of five or more words by the age of t months can be approximated by the equation[21]

$$p(t) = 100\left(1 - \frac{5.27 \times 10^{17}}{t^{12}}\right) \qquad (t \ge 30)$$

   a. Give a technology formula for p.

   b. Graph p for $30 \le t \le 45$ and $0 \le p \le 100$. Use your graph to answer parts (c) and (d).

   c. What percentage of children can speak in sentences of five or more words by the age of 36 months? (Round your answer to the nearest percentage point.)

   d. By what age are 75% of children speaking in sentences of five or more words? (Round your answer to the nearest month.)

[20]The model is the authors' and is based on data presented in the article "The Emergence of Intelligence" by William H. Calvin, *Scientific American* (October 1994): 101–107.

[21]Ibid.

31. Coffee Shops (Compare Exercise 27 in Section 1.1.) The number $C(t)$ of coffee shops and related enterprises in the United States can be approximated by the following function of time t in years since 1990:[22]

$$C(t) = \begin{cases} 500t + 800 & \text{if } 0 \le t \le 4 \\ 1300t - 2400 & \text{if } 4 < t \le 10 \end{cases}$$

Sketch the graph of C and use your graph to estimate, to the nearest year, when there were 9000 coffee shops in the United States.

32. Semiconductors (Compare Exercise 28 in Section 1.1.) The percentage $J(t)$ of semiconductor equipment manufactured in Japan during the period 1981–1994 can be approximated by the following function of time t in years since 1980:[23]

$$J(t) = \begin{cases} 17 + 3.1t & \text{if } 1 \le t \le 10 \\ 68 - 2t & \text{if } 10 < t \le 14 \end{cases}$$

Sketch the graph of J and use your graph to estimate, to the nearest year, when Japan manufactured 40% of all semiconductor equipment.

 33. Television Advertising The cost, in millions of dollars, of a 30-second television ad during the Super Bowl in the years 1990–2001 can be approximated by the following piecewise linear function ($t = 0$ represents 1990):[24]

$$C(t) = \begin{cases} 0.08t + 0.6 & \text{if } 0 \le t < 8 \\ 0.355t - 1.6 & \text{if } 8 \le t \le 11 \end{cases}$$

a. Give a technology formula for C and use technology to graph the function C.
b. Based on the graph, a Super Bowl ad first exceeded $2 million in what year?

[22]SOURCE: Specialty Coffee Association of America/*New York Times*, August 13, 1995, p. F10.

[23]SOURCE: VLSI Research/*New York Times*, October 9, 1994, sec. 3, p. 2.

[24]SOURCE: *New York Times*, January 26, 2001, p. C1.

 34. Internet Purchases The percentage $p(t)$ of buyers of new cars who used the Internet for research or purchase each year since 1997 is given by the following function ($t = 0$ represents 1997):[25]

$$p(t) = \begin{cases} 10t + 15 & \text{if } 0 \le t < 1 \\ 15t + 10 & \text{if } 1 \le t \le 4 \end{cases}$$

a. Give a technology formula for p and use technology to graph the function p.
b. Based on the graph, 50% or more of all new car buyers used the Internet for research or purchase in what years?

COMMUNICATION AND REASONING EXERCISES

35. True or false: Every graphically specified function can also be specified numerically. Explain.

36. True or false: Every algebraically specified function can also be specified graphically. Explain.

37. True or false: Every numerically specified function with domain [0, 10] can also be specified graphically. Explain.

38. True or false: Every graphically specified function can also be specified algebraically. Explain.

39. How do the graphs of two functions differ if they are specified by the same formula but have different domains?

40. How do the graphs of two functions $f(x)$ and $g(x)$ differ if $g(x) = f(x) + 10$? (Try an example.)

41. How do the graphs of two functions $f(x)$ and $g(x)$ differ if $g(x) = f(x - 5)$? (Try an example.)

42. How do the graphs of two functions $f(x)$ and $g(x)$ differ if $g(x) = f(-x)$? (Try an example.)

[25]Model is based on data through 2000 (the 2000 value was estimated). SOURCE: J. D. Power Associates/*New York Times*, January 25, 2000, p. C1.

1.3 Linear Functions

Linear functions are among the simplest functions and are perhaps the most useful of all mathematical functions.

Linear Function

A **linear function** is one that can be written in the form

Quick Examples

$$f(x) = mx + b \qquad \text{Function form} \qquad f(x) = 3x - 1$$

or

$$y = mx + b \qquad \text{Equation form} \qquad y = 3x - 1$$

where m and b are fixed numbers (the names m and b are traditional).*

*Actually, c is sometimes used instead of b. As for m, there has even been some research lately into the question of its origin, but no one knows exactly why the letter m is used.

Linear Functions from the Numerical and Graphical Point of View

Figure 8

The following table shows values of $y = 3x - 1$ ($m = 3$, $b = -1$) for some values of x:

x	-4	-3	-2	-1	0	1	2	3	4
y	-13	-10	-7	-4	-1	2	5	8	11

Its graph is shown in Figure 8.

Looking first at the table, notice that setting $x = 0$ gives $y = -1$, the value of b.

Numerically, b is the value of y when x = 0.

On the graph, the corresponding point $(0, -1)$ is the point where the graph crosses the y axis, and we say that $b = -1$ is the **y intercept** of the graph (Figure 9).

Graphically, b is the y intercept of the graph.

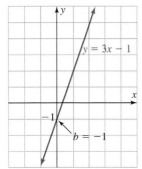

Figure 9
y intercept = $b = -1$

What about m? Looking once again at the table, notice that y increases by $m = 3$ units for every increase of 1 unit in x. This is caused by the term $3x$ in the formula: For every increase of 1 in x, we get an increase of $3 \times 1 = 3$ in y.

Numerically, y increases by m units for every 1-unit increase of x.

Likewise, for every increase of 2 in x, we get an increase of $3 \times 2 = 6$ in y. In general, if x increases by some amount, y will increase by three times that amount. We write

$$\text{change in } y = 3 \times \text{change in } x$$

The Change in a Quantity: Delta Notation

If a quantity q changes from q_1 to q_2, the **change in q** is just the difference:

$$\text{change in } q = \text{second value} - \text{first value}$$

$$= q_2 - q_1$$

Mathematicians traditionally use Δ (delta, the Greek equivalent of the Roman letter D) to stand for change and write the change in q as Δq:

$$\Delta q = \text{change in } q = q_2 - q_1$$

Quick Examples

1. If x is changed from 1 to 3, we write

   $$\Delta x = \text{second value} - \text{first value} = 3 - 1 = 2$$

2. Looking at our linear function, we see that, when x changes from 1 to 3, y changes from 2 to 8. So,

   $$\Delta y = \text{second value} - \text{first value} = 8 - 2 = 6$$

Using delta notation, we can now write, for our linear function,

$$\Delta y = 3\Delta x \qquad \text{change in } y = 3 \times \text{change in } x$$

$$\frac{\Delta y}{\Delta x} = 3$$

Question How does the number $m = 3$ show up in the graph?

Answer Because the value of y increases by exactly 3 units for every increase of 1 unit in x, the graph is a straight line rising by 3 units for every 1 unit we go to the right. We say that we have a **rise** of 3 units for each **run** of 1 unit. Because the value of y changes by $\Delta y = 3\Delta x$ units for every change of Δx units in x, in general we have a rise of $\Delta y = 3\Delta x$ units for each run of Δx units (Figure 10). Thus, we have a rise of 6 for a run of 2, a rise of 9 for a run of 3, and so on. So, $m = 3$ is a measure of the steepness of the line; we call m the **slope of the line:**

$$\text{slope} = m = \frac{\Delta y}{\Delta x} = \frac{\text{rise}}{\text{run}}$$

Graphically, m is the slope of the graph.

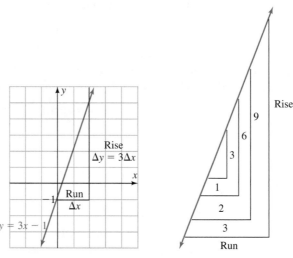

Figure 10 • *Slope* $= m = 3$

In general (replace the number 3 by a general number m), we can say the following.

The Roles of m and b in the Linear Function f(x) = mx + b
Role of m
Numerically If $y = mx + b$, then y changes by m units for every 1-unit change in x. A change of Δx units in x results in a change of $\Delta y = m\Delta x$ units in y. Thus,

$$m = \frac{\Delta y}{\Delta x} = \frac{\text{change in } y}{\text{change in } x} \qquad\qquad m = \frac{\Delta f}{\Delta x} = \frac{\text{change in } f}{\text{change in } x}$$

Equation form ($y = mx + b$)   Function form ($f(x) = mx + b$)

Graphically m is the slope of the line $y = mx + b$:

$$m = \frac{\Delta y}{\Delta x} = \frac{\text{rise}}{\text{run}} = \text{slope}$$

For positive m the graph rises m units for every 1-unit move to the right and rises $\Delta y = m\Delta x$ units for every Δx units moved to the right. For negative m the graph drops $|m|$ units for every 1-unit move to the right and drops $|m|\Delta x$ units for every Δx units moved to the right.

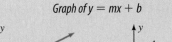

Graph of $y = mx + b$

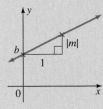

Positive m

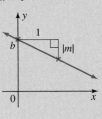

Negative m

Role of b

Numerically When $x = 0$, $y = b$ Equation form

$$f(0) = b$$ Function form

Graphically b is the y intercept of the line $y = mx + b$.

Quick Examples

1. $f(x) = 2x + 1$ has slope $m = 2$ and y intercept $b = 1$. To sketch the graph, we start at the y intercept $b = 1$ on the y axis and then move 1 unit to the right and up $m = 2$ units to arrive at a second point on the graph. Now we connect the two points to obtain the graph on the left.

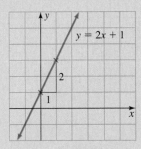

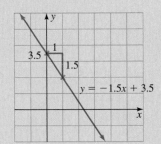

2. The line $y = -1.5x + 3.5$ has slope $m = -1.5$ and y intercept $b = 3.5$. Since the slope is negative, the graph (above right) goes *down* 1.5 units for every 1 unit it moves to the right.

3. Some slopes:

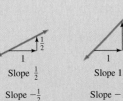

Slope $\frac{1}{2}$ Slope 1   Slope 2   Slope 3

Slope $-\frac{1}{2}$ Slope -1 Slope -2 Slope -3

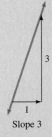

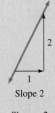

(Notice that the larger the absolute value of the slope, the steeper is the line.)

Example 1 • Recognizing Linear Data Numerically and Graphically

Which of the following two tables gives the values of a linear function? What is the formula for that function?

x	0	2	4	6	8	10	12
$f(x)$	3	-1	-3	-6	-8	-13	-15

x	0	2	4	6	8	10	12
$g(x)$	3	-1	-5	-9	-13	-17	-21

Solution The function f cannot be linear: If it were, we would have $\Delta f = m\Delta x$ for some fixed number m. However, although the change in x between successive entries in the table is $\Delta x = 2$ each time, the change in f is not the same each time. Thus, the ratio $\Delta f/\Delta x$ is not the same for every successive pair of points.

On the other hand, the ratio $\Delta g/\Delta x$ is the same each time, namely,

$$\frac{\Delta g}{\Delta x} = \frac{-4}{2} = -2$$

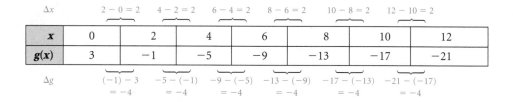

Δx		$2-0=2$		$4-2=2$		$6-4=2$		$8-6=2$		$10-8=2$		$12-10=2$	
x	0		2		4		6		8		10		12
$g(x)$	3		-1		-5		-9		-13		-17		-21
Δg		$(-1)-3$ $=-4$		$-5-(-1)$ $=-4$		$-9-(-5)$ $=-4$		$-13-(-9)$ $=-4$		$-17-(-13)$ $=-4$		$-21-(-17)$ $=-4$	

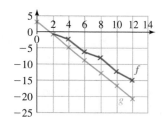

Figure 11

Thus, g is linear with slope $m = -2$. By the table, $g(0) = 3$; hence, $b = 3$. Therefore,

$$g(x) = -2x + 3 \qquad \text{Check that this formula gives the values in the table.}$$

If you graph the points in the tables defining f and g, it becomes easy to see that g is linear and f is not; the points of g lie on a straight line (with slope -2), whereas the points of f do not lie on a straight line (Figure 11).

Excel

The following worksheet shows how you can compute the successive quotients $m = \Delta y/\Delta x$ and then check whether a given set of data shows a linear relationship, in which case all the quotients will be the same. (The shading indicates that the formula is to be copied down only as far as cell C7. [Why not cell C8?])

	A	B	C	D
1	x	f	m	
2	0	3	=(B3-B2)/(A3-A2)	
3	2	-1		
4	4	-3		
5	6	-6		
6	8	-8		
7	10	-13		
8	12	-15		

Example 2 • Graphing a Linear Equation by Hand: Intercepts

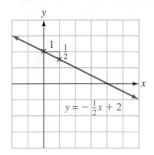

Figure 12

Graph the equation $x + 2y = 4$. Where does the line cross the x and y axes?

Solution We first write y as a linear function of x by solving the equation for y:

$$2y = -x + 4$$

$$y = -\frac{1}{2}x + 2$$

Now we can see that the graph is a straight line with a slope of $-\frac{1}{2}$ and a y intercept of 2. We start at 2 on the y axis and go down $\frac{1}{2}$ unit for every 1 unit we go to the right. The graph is shown in Figure 12.

We already know that the line crosses the y axis at 2. Where does it cross the x axis? Wherever that is, we know that the y coordinate will be 0 at that point. So, we set $y = 0$ and solve for x. It's most convenient to use the equation we were originally given:

$$x + 2(0) = 4$$

$$x = 4$$

The line crosses the x axis at 4.

 Before we go on . . . We could have graphed the equation in another way by first finding the intercepts. Once we know that the line crosses the y axis at 2 and the x-axis at 4, we can draw those two points and then draw the line connecting them.

It's worth noting what we did to find the intercepts in Example 2.

Finding the Intercepts

The **x intercept** of a line is where it crosses the x axis. To find it, set $y = 0$ and solve for x. The **y intercept** is where it crosses the y axis. If the equation of the line is written as $y = mx + b$, then b is the y intercept. Otherwise, set $x = 0$ and solve for y.

Quick Example

Consider the equation $3x - 2y = 6$. To find its x intercept, set $y = 0$ to find $x = \frac{6}{3} = 2$. To find its y intercept, set $x = 0$ to find $y = \frac{6}{-2} = -3$. The line crosses the x axis at 2 and the y axis at -3.

Computing the Slope of a Line

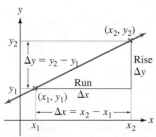

Figure 13

We know that the slope of a line is given by

$$\text{slope} = m = \frac{\text{rise}}{\text{run}} = \frac{\Delta y}{\Delta x}$$

Question Two points—say, (x_1, y_1) and (x_2, y_2)—determine a line in the xy plane. How do we find its slope?

Answer As you can see in Figure 13, the rise is $\Delta y = y_2 - y_1$, the change in the y coordinate from the first point to the second; the run is $\Delta x = x_2 - x_1$, the change in the x coordinate.

Computing the Slope of a Line

We can compute the slope m of the line through the points (x_1, y_1) and (x_2, y_2) using

$$m = \frac{\Delta y}{\Delta x} = \frac{y_2 - y_1}{x_2 - x_1}$$

Quick Examples

1. The slope of the line through $(x_1, y_1) = (1, 3)$ and $(x_2, y_2) = (5, 11)$ is

$$m = \frac{\Delta y}{\Delta x} = \frac{y_2 - y_1}{x_2 - x_1} = \frac{11 - 3}{5 - 1} = \frac{8}{4} = 2$$

2. The slope of the line through $(x_1, y_1) = (1, 2)$ and $(x_2, y_2) = (2, 1)$ is

$$m = \frac{\Delta y}{\Delta x} = \frac{y_2 - y_1}{x_2 - x_1} = \frac{1 - 2}{2 - 1} = \frac{-1}{1} = -1$$

3. Here is an Excel worksheet to compute the slope of the line through two given points (such as those in Quick Example 1):

	A	B	C	D
1	x	y	m	
2	1	3	=(B3-B2)/(A3-A2)	
3	5	11		

4. The line through $(2, 3)$ and $(-1, 3)$ has slope

$$m = \frac{\Delta y}{\Delta x} = \frac{3 - 3}{-1 - 2} = \frac{0}{-3} = 0$$

A line of slope 0 has a 0 rise, so it is a *horizontal* line, as shown below by the graph on the left. On the other hand, the line through $(3, 2)$ and $(3, -1)$ has

$$m = \frac{\Delta y}{\Delta x} = \frac{-1 - 2}{3 - 3} = \frac{-3}{0}$$

which is undefined. If we plot the two points, we see that the line passing through them is *vertical*, as shown by the graph on the right.

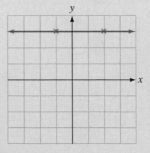

The line through $(2, 3)$ and $(-1, 3)$; $m = 0$.

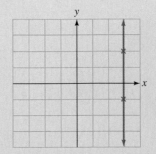

The line through $(3, 2)$ and $(3, -1)$; m is undefined.

Question What if we had chosen to list the two points in Quick Example 1 in reverse order? That is, suppose we had taken $(x_1, y_1) = (5, 11)$ and $(x_2, y_2) = (1, 3)$. What would have happened?

Answer We would have found

$$m = \frac{\Delta y}{\Delta x} = \frac{y_2 - y_1}{x_2 - x_1} = \frac{3 - 11}{1 - 5} = \frac{-8}{-4} = 2$$

the same answer. The order in which we take the points is not important, *as long as we use the same order in the numerator and the denominator.*

Finding a Linear Equation from Data: How to Make a Linear Model

If we happen to know the slope and y intercept of a line, writing down its equation is straightforward. For example, if we know that the slope is 3 and the y intercept is -1, then the equation is $y = 3x - 1$. Sadly, the information we are given is seldom so convenient. For instance, we may know the slope and a point other than the y intercept, two points on the line, or other information.

We describe a straightforward method for finding the equation of a line: the **point-slope** method. As the name suggests, we need two pieces of information:

• The *slope m* (which specifies the direction of the line)

• A *point* (x_1, y_1) on the line (which pins down its location in the plane)

Question What is the equation of the line through the point (x_1, y_1) with slope m?

Answer Since its slope is m, it must have the form

$$y = mx + b$$

for some (unknown) number b. To determine b we use the fact that the line must pass through the point (x_1, y_1), so that

$$y_1 = mx_1 + b \qquad \qquad (x_1, y_1) \text{ satisfies the equation } y = mx + b.$$

Solving for b gives

$$b = y_1 - mx_1$$

The following provides a summary.

The Point-Slope Formula

An equation of the line through the point (x_1, y_1) with slope m is given by

$$y = mx + b \qquad\qquad \text{Equation form}$$

where

$$b = y_1 - mx_1$$

When to Apply the Point-Slope Formula

Apply the point-slope formula to find the equation of a line whenever you are given information about a point and the slope of the line. The formula does not apply if the slope is undefined [as in a vertical line; see Example 3(d)].

1.3 EXERCISES

In Exercises 1–6, a table of values for a linear function is given. Fill in the missing value and calculate m in each case.

1.

x	−1	0	1
y	5	8	

2.

x	−1	0	1
y		−1	−3

3.

x	2	3	5
f(x)	−1	−2	

4.

x	2	4	5
f(x)	−1	−2	

5.

x	−2	0	2
f(x)	4		10

6.

x	0	3	6
f(x)	−1		−5

In Exercises 7–10, first find $f(0)$, if not supplied, and then find an equation of the given linear function.

7.

x	−2	0	2	4
f(x)	−1	−2	−3	−4

8.

x	−6	−3	0	3
f(x)	1	2	3	4

9.

x	−4	−3	−2	−1
f(x)	−1	−2	−3	−4

10.

x	1	2	3	4
f(x)	4	6	8	10

In each of Exercises 11–14, decide which of the two given functions is linear and find its equation.

11.

x	0	1	2	3	4
f(x)	6	10	14	18	22
g(x)	8	10	12	16	22

12.

x	−10	0	10	20	30
f(x)	−1.5	0	1.5	2.5	3.5
g(x)	−9	−4	1	6	11

13.

x	0	3	6	10	15
f(x)	0	3	5	7	9
g(x)	−1	5	11	19	29

14.

x	0	3	5	6	9
f(x)	2	6	9	12	15
g(x)	−1	8	14	17	26

In Exercises 15–24, find the slope of the given line, if it is defined.

15. $y = -\dfrac{3}{2}x - 4$

16. $y = \dfrac{2x}{3} + 4$

17. $y = \dfrac{x + 1}{6}$

18. $y = -\dfrac{2x - 1}{3}$

19. $3x + 1 = 0$

20. $8x - 2y = 1$

21. $3y + 1 = 0$

22. $2x + 3 = 0$

23. $4x + 3y = 7$

24. $2y + 3 = 0$

In Exercises 25–38, graph the given equations.

25. $y = 2x - 1$

26. $y = x - 3$

27. $y = -\dfrac{2}{3}x + 2$

28. $y = -\dfrac{1}{2}x + 3$

29. $y + \dfrac{1}{4}x = -4$

30. $y - \dfrac{1}{4}x = -2$

31. $7x - 2y = 7$

32. $2x - 3y = 1$

33. $3x = 8$

34. $2x = -7$

35. $6y = 9$

36. $3y = 4$

37. $2x = 3y$

38. $3x = -2y$

In Exercises 39–54, calculate the slope, if defined, of the straight line through the given pair of points. Try to do as many as you can without writing anything down except the answer.

39. $(0, 0)$ and $(1, 2)$

40. $(0, 0)$ and $(−1, 2)$

41. $(−1, −2)$ and $(0, 0)$

42. $(2, 1)$ and $(0, 0)$

43. $(4, 3)$ and $(5, 1)$

44. $(4, 3)$ and $(4, 1)$

45. $(1, −1)$ and $(1, −2)$

46. $(−2, 2)$ and $(−1, −1)$

47. $(2, 3.5)$ and $(4, 6.5)$

48. $(10, −3.5)$ and $(0, −1.5)$

49. $(300, 20.2)$ and $(400, 11.2)$

50. $(1, −20.2)$ and $(2, 3.2)$

51. $(0, 1)$ and $\left(-\dfrac{1}{2}, \dfrac{3}{4}\right)$

52. $\left(\dfrac{1}{2}, 1\right)$ and $\left(-\dfrac{1}{2}, \dfrac{3}{4}\right)$

53. (a, b) and (c, d) $(a \neq c)$

54. (a, b) and (c, b) $(a \neq c)$

55. In the following figure, estimate the slopes of all line segments.

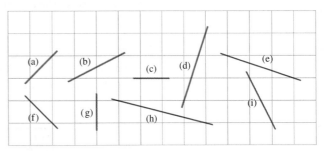

56. In the following figure, estimate the slopes of all line segments.

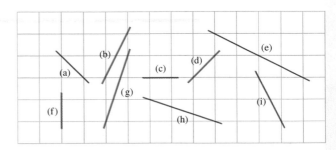

In Exercises 57–70, find linear equations whose graphs are the straight lines.

57. Through $(1, 3)$ with slope 3

58. Through $(2, 1)$ with slope 2

59. Through $(1, -\frac{3}{4})$ with slope $\frac{1}{4}$

60. Through $(0, -\frac{1}{3})$ with slope $\frac{1}{3}$

61. Through $(20, -3.5)$ and increasing at a rate of 10 units of y per unit of x

62. Through $(3.5, -10)$ and increasing at a rate of 1 unit of y per 2 units of x

63. Through $(2, -4)$ and $(1, 1)$

64. Through $(1, -4)$ and $(-1, -1)$

65. Through $(1, -0.75)$ and $(0.5, 0.75)$

66. Through $(0.5, -0.75)$ and $(1, -3.75)$

67. Through $(6, 6)$ and parallel to the line $x + y = 4$

68. Through $(\frac{1}{3}, -1)$ and parallel to the line $3x - 4y = 8$

69. Through $(0.5, 5)$ and parallel to the line $4x - 2y = 11$

70. Through $(\frac{1}{3}, 0)$ and parallel to the line $6x - 2y = 11$

COMMUNICATION AND REASONING EXERCISES

71. How would you test a table of values of x and y to see if it comes from a linear function?

72. You have ascertained that a table of values of x and y corresponds to a linear function. How do you find an equation for that linear function?

73. To what linear function of x does the linear equation $ax + by = c$ $(b \neq 0)$ correspond? Why did we specify $b \neq 0$?

74. Complete the following. The slope of the line with equation $y = mx + b$ is the number of units that _____ increases per unit increase in _____.

75. Complete the following. If, in a straight line, y is increasing three times as fast as x, then its _____ is _____.

76. Suppose that y is decreasing at a rate of 4 units per 3-unit increase of x. What can we say about the slope of the linear relationship between x and y? What can we say about the intercept?

77. If y and x are related by the linear expression $y = mx + b$, how will y change as x changes if m is positive? Negative? Zero?

78. Your friend April tells you that $y = f(x)$ has the property that, whenever x is changed by Δx, the corresponding change in y is $\Delta y = -\Delta x$. What can you tell her about f?

79. Consider the following worksheet [as in Example 3(a)]:

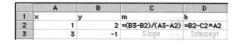

	A	B	C	D
1	x	y	m	b
2		1	2	=(B3-B2)/(A3-A2)	=B2-C2*A2
3		3	-1	Slope	Intercept

What is the effect on the slope of increasing the y co-ordinate of the second point (the point whose coordinates are in row 3)? Explain.

80. Referring to the worksheet in Exercise 79, what is the effect on the slope of increasing the x coordinate of the second point (the point whose coordinates are in row 3)? Explain.

1.4 Linear Models

Using linear functions to describe or approximate relationships in the real world is called **linear modeling.** We start with some examples involving cost, revenue, and profit.

Cost, Revenue, and Profit Functions

Example 1 • Linear Cost Function

As of April 2002, Yellow Cab Chicago's rates were $1.90 on entering the cab plus $1.60 for each mile.[26]

a. Find the cost C of an x-mile trip.

b. Use your answer to calculate the cost of a 40-mile trip.

c. What is the cost of the second mile? The tenth mile?

d. Graph C as a function of x.

---

[26]According to their Web site at http://www.yellowcabchicago.com/.

Solution

a. We are being asked to find how the cost C depends on the length x of the trip, or to find C as a function of x. Here is the cost in a few cases:

Cost of a 1-mile trip: $C = 1.60(1) + 1.90 = \$3.50$ 1 mile @ \$1.60 per mile plus \$1.90

Cost of a 2-mile trip: $C = 1.60(2) + 1.90 = \$5.10$ 2 miles @ \$1.60 per mile plus \$1.90

Cost of a 3-mile trip: $C = 1.60(3) + 1.90 = \$6.70$ 3 miles @ \$1.60 per mile plus \$1.90

Do you see the pattern? The cost of an x-mile trip is given by the linear function

$$C = 1.60x + 1.90 \qquad \text{Equation form}$$

$$C(x) = 1.60x + 1.90 \qquad \text{Function form}$$

Notice that the slope 1.60 is the incremental cost per mile. In this context we call 1.60 the **marginal cost**; the varying quantity $1.60x$ is called the **variable cost.** The C intercept 1.90 is the cost to enter the cab, which we call the **fixed cost.** In general, a linear cost function has the following form.

$$\underset{\substack{\uparrow \\ \text{Marginal cost}}}{C(x) = } \overset{\substack{\text{Variable} \\ \text{cost} \\ \frown}}{m} x + \underset{\substack{\uparrow \\ \text{Fixed cost}}}{b}$$

b. We can use the formula for the cost function to calculate the cost of a 40-mile trip as

$$C(40) = 1.60(40) + 1.90 = \$65.90$$

c. To calculate the cost of the second mile, we *could* proceed as follows:

Find the cost of a 1-mile trip: $C(1) = 1.60(1) + 1.90 = \$3.50$.

Find the cost of a 2-mile trip: $C(2) = 1.60(2) + 1.90 = \$5.10$.

Therefore, the cost of the second mile is $\$5.10 - \$3.50 = \$1.60$.

But notice that this is just the marginal cost. In fact, the marginal cost is the cost of each additional mile, so we could have done this more simply:

cost of second mile = cost of tenth mile = marginal cost = \$1.60

d. Figure 14 shows the graph of the cost function, which we can interpret as a *cost-versus-miles* graph. The fixed cost is the starting height on the left, and the marginal cost is the slope of the line.

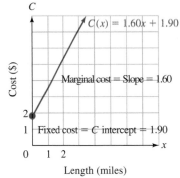

Figure 14

※ *Before we go on . . .* In general, the slope m measures the number of units of change in y per 1-unit change in x, so *we measure m in units of y per unit of x:*

units of slope = units of y per unit of x

In this example, y is the cost C, measured in dollars, and x is the length of a trip, measured in miles. Hence,

units of slope = units of y per unit of x = dollars per mile

The y intercept b, being a value of y, is measured in the same units as y. In this example, b is measured in dollars.

Here is a summary of the terms used in this example, along with an introduction to some new terms.

Cost, Revenue, and Profit Functions

A **cost function** specifies the cost C as a function of the number of items x. Thus, $C(x)$ is the cost of x items. A cost function of the form

$$C(x) = mx + b$$

is called a **linear cost function.** The quantity mx is called the **variable cost,** and the intercept b is called the **fixed cost.** The slope m, the **marginal cost,** measures the incremental cost per item.

The **revenue** resulting from one or more business transactions is the total payment received, sometimes called the gross proceeds. If $R(x)$ is the revenue from selling x items at a price of m each, then R is the linear function $R(x) = mx$, and the selling price m can also be called the **marginal revenue.**

The **profit,** on the other hand, is the *net* proceeds, or what remains of the revenue when costs are subtracted. If the profit depends linearly on the number of items, the slope m is called the **marginal profit.** Profit, revenue, and cost are related by the following formula:

$$\text{profit} = \text{revenue} - \text{cost}$$
$$P = R - C$$

If the profit is negative, say $-\$500$, we refer to a **loss** (of $\$500$ in this case). To **break even** means to make neither a profit nor a loss. Thus, break even occurs when $P = 0$, or

$$R = C \qquad \text{Break even}$$

The **break-even point** is the number of items x at which break even occurs.

Quick Example

If the daily cost (including operating costs) of manufacturing x T-shirts is $C(x) = 8x + 100$ and the revenue obtained by selling x T-shirts is $R(x) = 10x$, then the daily profit resulting from the manufacture and sale of x T-shirts is

$$P(x) = R(x) - C(x) = 10x - (8x + 100) = 2x - 100$$

Break even occurs when $P(x) = 0$, or $x = 50$.

Example 2 • Cost, Revenue, and Profit

The manager of the FrozenAir Refrigerator factory notices that on Monday it cost the company a total of $\$25,000$ to build 30 refrigerators and on Tuesday it cost $\$30,000$ to build 40 refrigerators.

a. Find a linear cost function based on this information. What is the daily fixed cost, and what is the marginal cost?

b. FrozenAir sells its refrigerators for $\$1500$ each. What is the revenue function?

c. What is the profit function? How many refrigerators must FrozenAir sell in a day in order to break even for that day? What will happen if it sells fewer refrigerators? If it sells more?

Solution

a. We are seeking C as a linear function of x, the number of refrigerators sold:

$$C = mx + b$$

We are told that $C = 25{,}000$ when $x = 30$, and this amounts to being told that $(30, 25{,}000)$ is a point on the graph of the cost function. Similarly, $(40, 30{,}000)$ is another point on the line (Figure 15).

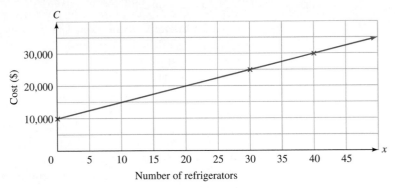

Figure 15

We can now use the point-slope formula to construct a linear cost equation. Recall that we need two items of information: a point on the line and the slope.

- **Point** Let's use the first point: $(x_1, C_1) = (30, 25{,}000)$ *C plays the role of y.*

- **Slope** $m = \dfrac{C_2 - C_1}{x_2 - x_1} = \dfrac{30{,}000 - 25{,}000}{40 - 30} = 500$ *Marginal cost = $500*

The cost function is therefore

$$C = 500x + b$$

where $b = C_1 - mx_1 = 25{,}000 - (500)(30) = 10{,}000$ *Fixed cost = $10,000*

so $C = 500x + 10{,}000$ *Cost function: equation notation*

or $C(x) = 500x + 10{,}000$ *Cost function: function notation*

Since $m = 500$ and $b = 10{,}000$, the factory's fixed cost is $10,000 each day, and its marginal cost is $500 per refrigerator.

Excel
We can also use Excel to compute the cost equation from the given data:

	A	B	C	D
1	x	C	m	b
2	30	25000	=(B3-B2)/(A3-A2)	=B2-C2*A2
3	40	30000	Slope	Intercept

b. The revenue FrozenAir obtains from the sale of a single refrigerator is $1500. So, if it sells x refrigerators, it earns a revenue of

$$R(x) = 1500x$$

c. For the profit, we use the formula

$$\text{profit} = \text{revenue} - \text{cost}$$

For the cost and revenue, we can substitute the answers from parts (a) and (b) and obtain

$$P(x) = R(x) - C(x) \qquad \text{Formula for profit}$$
$$= 1500x - (500x + 10,000) \qquad \text{Substitute } R(x) \text{ and } C(x).$$
$$= 1000x - 10,000$$

Here, $P(x)$ is the daily profit FrozenAir makes by making and selling x refrigerators.

Finally, to break even means to make zero profit. So, we need to find the x such that $P(x) = 0$. All we have to do is set $P(x) = 0$ and solve for x:

$$1000x - 10,000 = 0$$

giving

$$x = \frac{10,000}{1000} = 10$$

To break even, FrozenAir needs to manufacture and sell 10 refrigerators in a day.

For values of x less than the break-even point, 10, $P(x)$ is negative, so the company will have a loss. For values of x greater than the break-even point, $P(x)$ is positive, so the company will make a profit. This is the reason we are interested in the point where $P(x) = 0$. Since $P(x) = R(x) - C(x)$, we can also look at the break-even point as the point where revenue = cost: $R(x) = C(x)$ (see Figure 16).

✳ *Before we go on . . .* We can graph the cost and revenue function and find the break-even point graphically (Figure 16):

$$\text{Cost: } C(x) = 500x + 10,000$$

$$\text{Revenue: } R(x) = 1500x$$

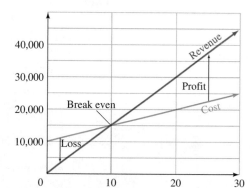

Figure 16 • *Break even occurs at the point of intersection of the graphs of revenue and cost.*

The break-even point is the point where the revenue and cost are equal—that is, where the graphs of cost and revenue cross. Figure 16 confirms that break even occurs when $x = 10$ refrigerators. Or we can use the graph to find the break-even point in the first place. If we use technology, we can zoom in for an accurate estimate of the point of intersection.

Excel

Excel also has an interesting feature called Goal Seek that can be used to find the point of intersection of two lines numerically (rather than graphically). The downloadable Excel tutorial for this section contains detailed instructions on using Goal Seek to find break-even points.

Demand and Supply Functions

The demand for a commodity usually goes down as its price goes up. It is traditional to use the letter q for the (quantity of) demand, as measured, for example, in weekly sales. Consider the following example.

Example 3 • Linear Demand Function

You run a small supermarket and must determine how much to charge for Hot'n'Spicy brand baked beans. The following chart shows weekly sales figures for Hot'n'Spicy at two different prices:

Price/Can, p ($)	0.50	0.75
Demand, q (cans sold/week)	400	350

a. Model the data by expressing the demand q as a linear function of the price p.

b. How do we interpret the slope? The q intercept?

c. How much should you charge for a can of Hot'n'Spicy beans if you want the demand to increase to 410 cans per week?

Solution

a. A **demand equation** or **demand function** expresses demand q (in this case, the number of cans of beans sold per week) as a function of the unit price p (in this case, price per can). We model the demand using the two points we are given: $(0.50, 400)$ and $(0.75, 350)$:

- **Point** $(0.50, 400)$

- **Slope** $m = \dfrac{q_2 - q_1}{p_2 - p_1} = \dfrac{350 - 400}{0.75 - 0.50} = \dfrac{-50}{0.25} = -200$

Thus, the demand equation is

$$q = mp + b$$
$$= -200p + [400 - (-200)(0.50)]$$
$$= -200p + 500$$

Figure 17 shows the data points and the linear model.

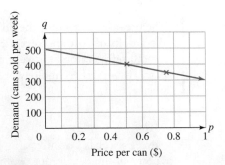

Figure 17

b. The key to interpreting the slope, $m = -200$, is to recall (see Example 1) that we measure the slope in units of y per unit of x. In this example, we mean units of q per unit of p, or the number of cans sold per dollar change in the price. Since m is negative, we see that the number of cans sold decreases as the price increases. We conclude that the weekly sales will drop by 200 cans per $1-increase in the price.

To interpret the q intercept, recall that it gives the q coordinate when $p = 0$. Hence, it is the number of cans you can "sell" every week if you were to give them away.[27]

c. If we want the demand to increase to 410 cans per week, we set $q = 410$ and solve for p:

$$410 = -200p + 500$$

$$200p = 90$$

$$p = \frac{90}{200} = \$0.45/\text{can}$$

Before we go on . . .

Question Just how reliable *is* the linear model?

Answer Look at it this way: The *actual* demand graph could in principle be obtained by tabulating new sales figures for a large number of different prices. If the resulting points were plotted on the pq plane, they would probably suggest a curve and not a straight line. However, if you looked at a small enough portion of this curve, you could closely *approximate* it by a straight line. In other words, *over a small range of values of p, the linear model is accurate.* Linear models of real-world situations are generally reliable only for small ranges of the variables. (This point will come up again in some of the exercises.)

Graphing Calculator

In part (c) we could also find p numerically using the Table feature in the TI-83. First make sure that the table settings permit you to enter values of x: Press [2nd] [TBLSET] and set Indpnt to Ask. Now enter

      Y₁=-200*X+500   Demand equation

and press [2nd] [TABLE]. You will now be able to adjust the price (x) until you find the value at which the demand equals 410.

Excel

In Excel we can use the worksheet from Example 2 to compute m and b for part (a):

	A	B	C	D
1	x	y	m	b
2	0.5	400	=(B3-B2)/(A3-A2)	=B2-C2*A2
3	0.75	350	Formula for Slope	Formula for Intercept

---

[27]Does this seem realistic? Demand is not always unlimited if items were given away. For instance, campus newspapers are sometimes given away, and yet piles of them are often left untaken. Also see the "Before we go on" discussion at the end of this example.

In part (c) we could find p numerically by using a worksheet like the following to compute the demand for many prices starting at $p = \$0.00$, until we find the price at which the demand equals 410.

	A	B	C
1	p	q	
2	0	=-200*A2 + 500	
3	0.01	498	
4	0.02	496	
5			
101	0.43	414	
102	0.44	412	
103	0.45	410	
104	0.46	408	

Demand Function

A **demand equation** or **demand function** expresses demand q (the number of items demanded) as a function of the unit price p (the price per item). A **linear demand function** has the form

$$q(p) = mp + b$$

Interpretation of m

The (usually negative) slope m measures the change in demand per unit change in price. For instance, if p is measured in dollars and q in monthly sales and $m = -400$, then each \$1 increase in the price per item will result in a drop in sales of 400 items per month.

Interpretation of b

The y intercept b gives the demand if the items were given away.

Quick Example

If the demand for T-shirts, measured in daily sales, is given by $q(p) = -4p + 90$, where p is the sale price in dollars, then daily sales drop by four T-shirts for every \$1 increase in price. If the T-shirts were given away, the demand would be 90 T-shirts per day.

We have seen that a demand function gives the number of items consumers are willing to buy at a given price, and a higher price generally results in a lower demand. However, as the price rises, suppliers will be more inclined to produce these items (as opposed to spending their time and money on other products), so supply will generally rise. A **supply function** gives q, the number of items suppliers are willing to make available for sale, as a function of p, the price per item.[28]

---

[28]Although a bit confusing at first, it is traditional to use the same letter q for the quantity of supply and the quantity of demand, particularly when we want to compare them, as in Example 4.

Example 4 • Demand, Supply, and Equilibrium Price

Continuing with Example 3, consider the following chart, which shows weekly sales figures for Hot'n'Spicy at two different prices (the demand), as well as the number of cans per week that you are prepared to place on sale at these prices (the supply).

Price/Can ($)	0.50	0.75
Demand (cans sold/week)	400	350
Supply (cans placed on sale/week)	300	500

a. Model these data with linear demand and supply functions.

b. How much should you charge per can of Hot'n'Spicy beans if you want the demand to equal the supply? How many cans will you sell at that price, known as the **equilibrium price**? What happens if you charge more than the equilibrium price? If you charge less?

Solution

a. We have already modeled the demand function in Example 3:

$$q = -200p + 500$$

To model the supply, we use the first and third rows of the table. We are again given two points: (0.50, 300) and (0.75, 500).

- **Point** (0.50, 300)

- **Slope** $m = \dfrac{q_2 - q_1}{p_2 - p_1} = \dfrac{500 - 300}{0.75 - 0.50} = \dfrac{200}{0.25} = 800$

So, the supply equation is

$$q = mp + b$$
$$= 800p + [300 - (800)(0.50)]$$
$$= 800p - 100$$

b. To find where the demand equals the supply, we equate the two functions:

$$\text{demand} = \text{supply}$$
$$-200p + 500 = 800p - 100$$
$$-1000p = -600$$

so
$$p = \frac{-600}{-1000} = \$0.60$$

This is the equilibrium price. We can find the corresponding demand by substituting 0.60 for p in the demand (or supply) equation.

$$\text{equilibrium demand} = -200(0.60) + 500 = 380 \text{ cans/week}$$

So, to balance supply and demand, you should charge $0.60 per can of Hot'n'Spicy beans, and you should place 380 cans on sale each week.

If we graph supply and demand on the same set of axes, we obtain the graphs shown in Figure 18.

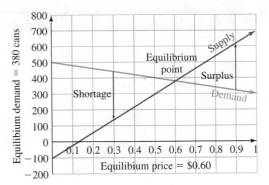

Figure 18 • Equilibrium occurs at the point of intersection of the graphs of supply and demand.

Figure 18 shows what happens if you charge prices other than the equilibrium price. If you charge more—say, $0.90 per can ($p = 0.90$)—then the supply will be larger than demand and there will be a weekly surplus. Similarly, if you charge less—say $0.30 per can—then the supply will be less than the demand and there will be a shortage of Hot'n'Spicy beans.

✳ *Before we go on . . .* We just saw that if you charge less than the equilibrium price there will be a shortage. If you were to raise your price toward the equilibrium, you would sell more items and increase revenue, since it is the supply equation—not the demand equation—that determines what you can sell below the equilibrium price. On the other hand, if you charge more than the equilibrium price, you will be left with a possibly costly surplus of unsold items (and will want to lower prices to reduce the surplus). Prices tend to move toward the equilibrium, so supply tends to equal demand. When supply equals demand, we say that the market **clears.**

Supply Function and Equilibrium Price

A **supply equation** or **supply function** expresses supply q (the number of items a supplier is willing to make available) as a function of the unit price p (the price per item). A **linear supply function** has the form

$$q(p) = mp + b$$

It is usually the case that supply increases as the unit price increases, so m is usually positive.

Demand and supply are said to be in **equilibrium** when demand equals supply. The corresponding values of p and q are called the **equilibrium price** and **equilibrium demand.** To find the equilibrium price, set demand equal to supply and solve for the unit price p. To find the equilibrium demand, evaluate the demand (or supply) function at the equilibrium price.

Change Over Time

Things around us change with time. There are many quantities, such as your income or the temperature in Honolulu, that it is natural to think of as functions of time.

Example 5 • Growth of Sales

The U.S. Air Force's satellite-based Global Positioning System (GPS) allows people with radio receivers to determine their exact location anywhere on Earth. The following table shows the estimated total sales of U.S.-made products that use the GPS:

Year	1994	2000
Sales ($ billions)	0.8	8.3

Data estimated from published graph.
SOURCE: U.S. Global Positioning System Industry Council/*New York Times*, March 5, 1996, p. D1.

a. Use these data to model total sales of GPS-based products as a linear function of time t measured in years since 1994. What is the significance of the slope?

b. Use the model to predict when sales of GPS-based products will reach $13.3 billion, assuming they continue to grow at the same rate.

Solution

a. First, notice that 1994 corresponds to $t = 0$ and 2000 to $t = 6$. Thus, we are given the coordinates of two points on the graph of sales s as a function of time t: $(0, 0.8)$ and $(6, 8.3)$. The slope is

$$m = \frac{s_2 - s_1}{t_2 - t_1} = \frac{8.3 - 0.8}{6 - 0} = \frac{7.5}{6} = 1.25$$

Using the point $(0, 0.8)$, we get

$$s = mt + b$$
$$= 1.25t + 0.8 - (1.25)(0)$$
$$= 1.25t + 0.8$$

Notice that we calculated the slope as the ratio (change in sales)/(change in time). Thus, m is the *rate of change* of sales and is measured in units of sales per unit of time, or billions of dollars per year. In other words, to say that $m = 1.25$ is to say that sales are increasing by $1.25 billion per year.

b. Our model of sales as a function of time is

$$s = 1.25t + 0.8$$

Sales of GPS-based products will reach $13.3 billion when $s = 13.3$, or

$$13.3 = 1.25t + 0.8$$

Solving for t,

$$1.25t = 13.3 - 0.8 = 12.5$$
$$t = \frac{12.5}{1.25} = 10$$

In 2004 ($t = 10$), sales are predicted to reach $13.3 billion.

Example 6 • Velocity

You are driving down the Ohio Turnpike, watching the mileage markers to stay awake. Measuring time in hours after you see the 20-mile marker, you see the following markers each half hour:

Time (h)	0	0.5	1	1.5	2
Marker (mi)	20	47	74	101	128

Find your location *s* as a function of *t*, the number of hours you have been driving. (The number *s* is also called your **position** or **displacement.**)

Solution If we plot the location *s* versus the time *t*, the five markers listed give us the graph in Figure 19. These points appear to lie along a straight line. We can verify this by calculating how far you traveled in each half hour. In the first half hour, you traveled $47 - 20 = 27$ miles. In the second half hour, you traveled $74 - 47 = 27$ miles also. In fact, you traveled exactly 27 miles each half hour. The points we plotted lie on a straight line that rises 27 units for every 0.5 unit we go to the right, for a slope of $27/0.5 = 54$.

To get the equation of that line, notice that we have the *s* intercept, which is the starting marker of 20. From the slope-intercept form (using *s* in place of *y* and *t* in place of *x*), we get

$$s(t) = 54t + 20$$

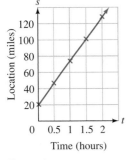

Figure 19

Before we go on . . . Notice the significance of the slope: For every hour you travel, you drive a distance of 54 miles. In other words, you are traveling at a constant velocity of 54 miles per hour. We have uncovered a very important principle:

In the graph of displacement versus time, velocity is given by the slope.

Linear Change Over Time

If a quantity *q* is a linear function of time *t*, so that

$$q(t) = mt + b$$

then the slope *m* measures the **rate of change** of *q*, and *b* is the quantity at time $t = 0$, the **initial quantity.** If *q* represents the position of a moving object, then the rate of change is also called the **velocity.**

Units of m and b

The units of measurement of *m* are units of *q* per unit of time; for instance, if *q* is income in dollars and *t* is time in years, then the rate of change *m* is measured in dollars per year.

The units of *b* are units of *q*; for instance, if *q* is income in dollars and *t* is time in years, then *b* is measured in dollars.

Quick Example

If the accumulated revenue from sales of your video-game software is given by $R(t) = 2000t + 500$ dollars, where *t* is time in years from now, then you have earned $500 in revenue so far, and the accumulated revenue is increasing at a rate of $2000 per year.

Examples 1–6 share the following common theme.

General Linear Models

If $y = mx + b$ is a linear model of changing quantities x and y, then the slope m is the rate at which y is increasing per unit increase in x, and the y intercept b is the value of y that corresponds to $x = 0$.

Units of m and b

The slope m is measured in units of y per unit of x, and the intercept b is measured in units of y.

Quick Example

If the number n of spectators at a soccer game is related to the number g of goals your team has scored so far by the equation $n = 20g + 4$, then you can expect 4 spectators if no goals have been scored and 20 additional spectators per additional goal scored.

Guideline: What to Use as x and y and How to Interpret a Linear Model

Question In a problem where I must find a linear relationship between two quantities, which quantity do I use as x and which do I use as y?

Answer The key is to decide which of the two quantities is the independent variable and which is the dependent variable. Then use the independent variable as x and the dependent variable as y. In other words, *y depends on x.*

Here are examples of phrases that convey this information, usually of the form *"Find y (dependent variable) in terms of x (independent variable)."*

Find the cost in terms of the number of items. $y = $ cost; $x = $ number of items

How does color depend on wavelength? $y = $ color; $x = $ wavelength

If no information is conveyed about which variable is intended to be independent, then you can use whichever is convenient.

Question How do I interpret a general linear model $y = mx + b$?

Answer The key to interpreting a linear model is to remember the units we use to measure m and b:

The slope m is measured in units of y per unit of x; the intercept b is measured in units of y.

For instance, if $y = 4.3x + 8.1$ and you know that x is measured in feet and y in kilograms, then you can already say, "y is 8.1 kilograms when $x = 0$ feet and increases at a rate of 4.3 kilograms per foot" without even knowing anything more about the situation!

1.4 EXERCISES

APPLICATIONS

1. **Cost** A piano manufacturer has a daily fixed cost of $1200 and a marginal cost of $1500 per piano. Find the cost $C(x)$ of manufacturing x pianos in one day. Use your function to answer the following questions:
   a. On a given day, what is the cost of manufacturing 3 pianos?
   b. What is the cost of manufacturing the third piano that day?
   c. What is the cost of manufacturing the eleventh piano that day?

2. **Cost** The cost of renting tuxes for the Choral Society's formal is $20 down, plus $88 per tux. Express the cost C as a function of x, the number of tuxedos rented. Use your function to answer the following questions.
   a. What is the cost of renting 2 tuxes?
   b. What is the cost of the second tux?
   c. What is the cost of the 4098th tux?
   d. What is the marginal cost per tux?

3. **Cost** The RideEm Bicycles factory can produce 100 bicycles in a day at a total cost of $10,500, and it can produce 120 bicycles in a day at a total cost of $11,000. What are the company's daily fixed costs, and what is the marginal cost per bicycle?

4. **Cost** A soft-drink manufacturer can produce 1000 cases of soda in a week at a total cost of $6000 and 1500 cases of soda at a total cost of $8500. Find the manufacturer's weekly fixed costs and marginal cost per case of soda.

5. **Break-Even Analysis** Your college newspaper, *The Collegiate Investigator,* has fixed production costs of $70 per edition and marginal printing and distribution costs of 40¢/copy. *The Collegiate Investigator* sells for 50¢/copy.
   a. Write down the associated cost, revenue, and profit functions.
   b. What profit (or loss) results from the sale of 500 copies of *The Collegiate Investigator*?
   c. How many copies should be sold in order to break even?

6. **Break-Even Analysis** The Audubon Society at Enormous State University (ESU) is planning its annual fund-raising "Eatathon." The society will charge students 50¢/serving of pasta. The only expenses the society will incur are the cost of the pasta, estimated at 15¢/serving, and the $350 cost of renting the facility for the evening.
   a. Write down the associated cost, revenue, and profit functions.
   b. How many servings of pasta must the Audubon Society sell in order to break even?
   c. What profit (or loss) results from the sale of 1500 servings of pasta?

7. **Demand** Sales figures show that your company sold 1960 pen sets each week when they were priced at $1/pen set and 1800 pen sets each week when they were priced at $5/pen set. What is the linear demand function for your pen sets?

8. **Demand** A large department store is prepared to buy 3950 of your neon-colored shower curtains per month for $5 each, but only 3700 shower curtains per month for $10 each. What is the linear demand function for your neon-colored shower curtains?

9. **Demand for Microprocessors** The following table shows the worldwide quarterly sales of microprocessors and average wholesale price:

Year	1997 (2nd quarter)	1997 (3rd quarter)	1998 (1st quarter)
Wholesale Price ($)	235	215	210
Sales (millions)	21	24	23

SOURCE: International Data Corporation/*New York Times,* "Computer Economics 101," June 9, 1998, p. D6. Data are rounded.

Use the 1997 second- and third-quarter data to obtain a linear demand function for microprocessors. Is the 1998 first-quarter data consistent with the demand equation? Explain.

10. **Demand for Microprocessors** Referring to the data in Exercise 9, use the 1997 second-quarter data and the 1998 first-quarter data to obtain a linear demand function for microprocessors. Use your model to give an estimate of worldwide sales if the price changes to $215. Can all three data points be represented by the same linear demand function?

11. **Equilibrium Price** You can sell 90 Chia pets each week if they are marked at $1/Chia but only 30 each week if they are marked at $2/Chia. Your Chia supplier is prepared to sell you 20 Chias each week if they are marked at $1/Chia and 100 each week if they are marked at $2/Chia.
    a. Write down the associated linear demand and supply functions.
    b. At what price should the Chias be marked so that there is neither a surplus nor a shortage of Chias?

12. **Equilibrium Price** The demand for your college newspaper is 2000 copies each week if the paper is given away free of charge and drops to 1000 each week if the charge is 10¢/copy. However, the university is only prepared to supply 600 copies each week free of charge but will supply 1400 each week at 20¢/copy.
    a. Write down the associated linear demand and supply functions.
    b. At what price should the college newspapers be sold so that there is neither a surplus nor a shortage of papers?

13. **Swimming Pool Sales** The following graph shows approximate annual sales of new in-ground swimming pools in the United States:

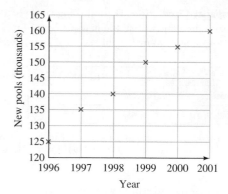

SOURCE: PK Data/*New York Times*, July 5, 2001,
p. C1. 2001 figure was estimated.

a. Find two points on the graph such that the slope of the line segment joining them is the largest possible.

b. What does your answer tell you about swimming pool sales?

14. **Swimming Pool Sales** Repeat Exercise 13, using the following graph for sales of new above-ground swimming pools in the United States.

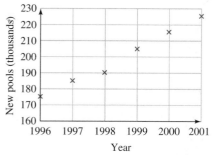

SOURCE: PK Data/*New York Times*, July 5, 2001,
p. C1. 2001 figure was estimated.

15. **Online Shopping** The number of online shopping transactions in the United States increased from 350 million transactions in 1999 to 450 million transactions in 2001.[29] Find a linear model for the number N of online shopping transactions in year t, with $t = 0$ corresponding to 2000. What is the significance of the slope? What are its units?

16. **Online Shopping** The percentage of people in the United States who have ever purchased anything online increased from 20% in January 2000 to 35% in January 2002.[30] Find a linear model for the percentage P of people in the United States who have ever purchased anything online in year t, with $t = 0$ corresponding to January 2000. What is the significance of the slope? What are its units?

17. **Medicare Spending** Annual federal spending on Medicare (in constant 2000 dollars) was projected to increase from $240 billion in 2000 to $600 billion in 2025.[31]

a. Use this information to express s, the annual spending on Medicare (in billions of dollars), as a linear function of t, the number of years since 2000. How fast is Medicare predicted to rise in the coming years?

b. Use your model to predict Medicare spending in 2040, assuming the spending trend continues.

18. **Pasta Imports** During the period 1990–2001, U.S. imports of pasta increased from 290 million pounds in 1990 ($t = 0$) by an average of 40 million pounds/year.[32]

a. Use these data to express q, the annual U.S. imports of pasta (in millions of pounds), as a linear function of t, the number of years since 1990.

b. Use your model to estimate U.S. pasta imports in 2005, assuming the import trend continues.

19. **Velocity** The position of a model train, in feet along a railroad track, is given by

$$s(t) = 2.5t + 10$$

after t seconds.

a. How fast is the train moving?

b. Where is the train after 4 seconds?

c. When will it be 25 feet along the track?

20. **Velocity** The height of a falling sheet of paper, in feet from the ground, is given by

$$s(t) = -1.8t + 9$$

after t seconds.

a. What is the velocity of the sheet of paper?

b. How high is it after 4 seconds?

c. When will it reach the ground?

21. **Fast Cars** A police car was traveling down Ocean Parkway in a high-speed chase from Jones Beach. It was at Jones Beach at exactly 10 P.M. ($t = 10$) and was at Oak Beach, 13 miles from Jones Beach, at exactly 10:06 P.M.

a. How fast was the police car traveling?

b. How far was the police car from Jones Beach at time t?

22. **Fast Cars** The car that was being pursued by the police in Exercise 21 was at Jones Beach at exactly 9:54 P.M. ($t = 9.9$) and passed Oak Beach (13 miles from Jones Beach) at exactly 10:06 P.M., where it was overtaken by the police.

a. How fast was the car traveling?

b. How far was the car from Jones Beach at time t?

[29]Second half of 2001 data was estimated. SOURCE: Odyssey Research/*New York Times*, November 5, 2001, p. C1.

[30]Ibid. January 2002 data was estimated.

[31]Data are rounded. SOURCE: The Urban Institute's Analysis of the 1999 Trustee's Report, http://www.urban.org.

[32]Data are rounded. SOURCES: Department of Commerce/*New York Times*, September 5, 1995, p. D4; International Trade Administration, http://www.ita.doc.gov/, March 31, 2002.

23. Fahrenheit and Celsius In the Fahrenheit temperature scale, water freezes at 32°F and boils at 212°F. In the Celsius scale, water freezes at 0°C and boils at 100°C. Assuming that the Fahrenheit temperature F and the Celsius temperature C are related by a linear equation, find F in terms of C. Use your equation to find the Fahrenheit temperatures corresponding to 30°C, 22°C, –10°C, and –14°C, to the nearest degree.

24. Fahrenheit and Celsius Use the information about Celsius and Fahrenheit given in Exercise 23 to obtain a linear equation for C in terms of F and use your equation to find the Celsius temperatures corresponding to 104°F, 77°F, 14°F, and −40°F, to the nearest degree.

25. Income The well-known romance novelist Celestine A. Lafleur (a.k.a. Bertha Snodgrass) has decided to sell the screen rights to her latest book, *Henrietta's Heaving Heart,* to Boxoffice Success Productions for $50,000. In addition, the contract ensures Ms. Lafleur royalties of 5% of the net profits.[33] Express her income I as a function of the net profit N and determine the net profit necessary to bring her an income of $100,000. What is her marginal income (share of each dollar of net profit)?

26. Income Due to the enormous success of the movie *Henrietta's Heaving Heart* based on a novel by Celestine A. Lafleur (see Exercise 25), Boxoffice Success Productions decides to film the sequel, *Henrietta, Oh Henrietta.* At this point, Bertha Snodgrass (whose novels now top the best-seller lists) feels she is in a position to demand $100,000 for the screen rights and royalties of 8% of the net profits. Express her income I as a function of the net profit N and determine the net profit necessary to bring her an income of $1,000,000. What is her marginal income (share of each dollar of net profit)?

27. Newspaper Circulation The following table gives the average daily circulation of all the newspapers of the Gannett Company and Newhouse Newspapers, Inc., two major newspaper companies.

	Gannett	Newhouse
Number of Newspapers	82	26
Daily Circulation (millions)	5.8	3.0

Circulation figures are rounded to the nearest 0.1 million and reflect average daily circulation as of September 30, 1994. SOURCE: Newspaper Association of America/*New York Times,* July 30, 1995, p. E6.

a. Use these data to express a company's daily circulation in millions c as a linear function of the number n of newspapers it publishes.

b. What information does the slope give about newspaper publishers?

c. Gannett Company was planning to add 11 new newspapers in 1995. According to your model, how would this affect daily circulation?

28. Newspaper Circulation The following table gives the average daily circulation of all the newspapers of Knight Ridder and Dow Jones, two major publishers of newspapers.

	Knight Ridder	Dow Jones
Number of Newspapers	27	22
Daily Circulation (millions)	3.6	2.4

Circulation figures are rounded to the nearest 0.1 million and reflect average daily circulation as of September 30, 1994. SOURCE: Newspaper Association of America/*New York Times,* July 30, 1995, p. E6.

a. Use these data to express a company's daily circulation c as a linear function of the number n of newspapers it publishes.

b. The Times Mirror Company publishes 11 newspapers and has a daily circulation of 2.6 million. To what extent is this consistent with the model?

c. Find a domain for your model that is consistent with a circulation of between 0 and 4.8 million.

Exercises 29–32 are based on the following data, showing the total amount of milk and cheese, in billions of pounds, produced in 13 western and 12 north-central states in 1999 and 2000.

Milk

Year	1999	2000
Western States	56	60
North-Central States	57	59

SOURCE: Department of Agriculture/*New York Times,* June 28, 2001, p. C1. Figures are approximate.

Cheese

Year	1999	2000
Western States	2.7	3.0
North-Central States	3.9	4.0

SOURCE: Department of Agriculture/*New York Times,* June 28, 2001, p. C1. Figures are approximate.

29. Use the data from both years to find a linear relationship giving the amount of milk w produced in the western states as a function of the amount of milk n produced in north-central states. Use your model to predict how much milk western states will produce if north-central states produce 50 billion pounds.

30. Use the data from both years to find a linear relationship giving the amount of cheese w produced in western states as a function of the amount of cheese n produced in north-central states. Use your model to predict how much cheese western states will produce if north-central states produce 3.4 billion pounds.

---

[33]Percentages of net profit are commonly called "monkey points." Few movies ever make a net profit on paper, and anyone with any clout in the business gets a share of the *gross,* not the net.

31. Use the data to model cheese production c in western states as a function of milk production m in western states. According to the model, how many pounds of cheese are produced for every 10 pounds of milk?

32. Repeat Exercise 31 for north-central states.

33. Biology The Snowtree cricket behaves in a rather interesting way: The rate at which it chirps depends linearly on the temperature. One summer evening you hear a cricket chirping at a rate of 140 chirps/minute, and you notice that the temperature is 80°F. Later in the evening, the cricket has slowed to 120 chirps/minute, and you notice that the temperature has dropped to 75°F. Express the temperature T as a linear function of the cricket's rate of chirping r. What is the temperature if the cricket is chirping at a rate of 100 chirps/minute?

34. Muscle Recovery Time Most workout enthusiasts will tell you that muscle recovery time is about 48 hours. But it is not quite as simple as that; the recovery time ought to depend on the number of sets you do involving the muscle group in question. For example, if you do no sets of biceps exercises, then the recovery time for your biceps is (of course) zero. To take a compromise position, let's assume that if you do three sets of exercises on a muscle group then its recovery time is 48 hours. Use these data to write a linear function that gives the recovery time (in hours) in terms of the number of sets affecting a particular muscle. Use this model to calculate how long it would take your biceps to recover if you did 15 sets of curls. Comment on your answer with reference to the usefulness of a linear model.

35. Profit Analysis—Aviation The operating cost of a Boeing 747-100, which seats up to 405 passengers, is estimated to be $5132 per hour.[34] If an airline charges each passenger a fare of $100 per hour of flight, find the hourly profit P it earns operating a 747-100 as a function of the number of passengers x (be sure to specify the domain). What is the least number of passengers it must carry in order to make a profit?

36. Profit Analysis—Aviation The operating cost of a McDonnell Douglas DC 10-10, which seats up to 295 passengers, is estimated to be $3885 per hour.[35] If an airline charges each passenger a fare of $100 per hour of flight, find the hourly profit P it earns operating a DC 10-10 as a function of the number of passengers x (be sure to specify the domain). What is the least number of passengers it must carry in order to make a profit?

37. Break-Even Analysis (based on a question from a CPA exam) The Oliver Company plans to market a new product. Based on its market studies, Oliver estimates that it can sell up to 5500 units in 2005. The selling price will be $2 per unit. Variable costs are estimated to be 40% of total revenue. Fixed costs are estimated to be $6000 for 2005. How many units should the company sell to break even?

38. Break-Even Analysis (based on a question from a CPA exam) The Metropolitan Company sells its latest product at a unit price of $5. Variable costs are estimated to be 30% of the total revenue, and fixed costs amount to $7000 per month. How many units should the company sell each month in order to break even, assuming that it can sell up to 5000 units per month at the planned price?

39. Break-Even Analysis (from a CPA exam) Given the following notations, write a formula for the break-even sales level.

SP = selling price per unit

FC = total fixed cost

VC = variable cost per unit

40. Break-Even Analysis (based on a question from a CPA exam) Given the following notation, give a formula for the total fixed cost.

SP = selling price per unit

VC = variable cost per unit

BE = break-even sales level in units

41. Break-Even Analysis—Organized Crime The organized crime boss and perfume king, Butch (Stinky) Rose, has daily overheads (bribes to corrupt officials, motel photographers, wages for hit men, explosives, etc.) amounting to $20,000/day. On the other hand, he has a substantial income from his counterfeit perfume racket: He buys imitation French perfume (Chanel N° 22.5) at $20/gram, pays an additional $30/100 grams for transportation, and sells it via his street thugs for $600/gram. Specify Stinky's profit function $P(x)$, where x is the quantity (in grams) of perfume he buys and sells, and use your answer to calculate how much perfume should pass through his hands each day so that he breaks even.

42. Break-Even Analysis—Disorganized Crime Lately, Butch (Stinky) Rose's counterfeit Chanel N° 22.5 racket has run into difficulties; it seems that the *authentic* Chanel N° 22.5 perfume is selling at only $500/gram, whereas his street thugs have been selling the counterfeit perfume for $600/gram, and his costs are $400/gram plus $30/gram transportation costs and commission. (The perfume's smell is easily detected by specially trained Chanel Hounds, and this necessitates elaborate packaging measures.) He therefore decides to price it at $420/gram in order to undercut the competition. Specify Stinky's profit function $P(x)$, where x is the quantity (in grams) of perfume he buys and sells, and use your answer to calculate how much perfume should pass through his hands each day so that he breaks even. Interpret your answer.

43. Television Advertising The cost, in millions of dollars, of a 30-second television ad during the Super Bowl in the years 1990 to 2001 can be approximated by the following piecewise linear function ($t = 0$ represents 1990):[36]

$$C(t) = \begin{cases} 0.08t + 0.6 & \text{if } 0 \leq t < 8 \\ 0.355t - 1.6 & \text{if } 8 \leq t \leq 11 \end{cases}$$

How fast and in what direction was the cost of an ad during the Super Bowl changing in 1999?

[34]In 1992. Source: Air Transportation Association of America.

[35]Ibid.

[36]Source: *New York Times*, January 26, 2001, p. C1.

44. Semiconductors The percentage $J(t)$ of semiconductor equipment manufactured in Japan during the period 1981–1994 can be approximated by the following function of time t in years since 1980:[37]

$$J(t) = \begin{cases} 17 + 3.1t & \text{if } 1 \le t \le 10 \\ 68 - 2t & \text{if } 10 < t \le \end{cases}$$

How fast and in what direction was the percentage of Japan's contribution to the semiconductor market changing in 1986?

45. Internet Sales The following table shows the percentage of buyers of new cars who used the Internet for research or purchase between 1997 and 2000 ($t = 0$ represents 1997):

Year, t	0	1	3
Percentage, p	15	25	55

SOURCE: J. D. Power Associates/*New York Times*, January 25, 2000, p. C1. 2000 figure is an estimate.

a. Model the percentage p as a linear function of t from 1997 ($t = 0$) to 1998 ($t = 1$).
b. Model p as a linear function of time t from 1998 to 2000.
c. Use the answers you obtained in parts (a) and (b) to model p as a piecewise-defined function of t from 1997 to 2000.
d. According to the model, what percentage of buyers of new cars used the Internet for research or purchase in 1999?

46. Internet Sales The following table shows the percentage of buyers of new cars who used the Internet for purchase (including referrals to dealers) between 1997 and 2000 ($t = 0$ represents 1997):

Year, t	0	1	3
Percentage, p	0	1	5

SOURCE: J. D. Power Associates/*New York Times*, January 25, 2000, p. C1. 2000 figure was estimated.

a. Model the percentage p as a linear function of t from 1997 ($t = 0$) to 1998.
b. Model p as a linear function of time t from 1998 to 2000.
c. Use the answers you obtained in parts (a) and (b) to model p as a piecewise-defined function of t from 1997 to 2000.
d. According to the model, what percentage of buyers of new cars used the Internet for purchase in 1999?

47. Career Choices In 1989 approximately 30,000 college-bound high school seniors intended to major in computer and information sciences. This number decreased to approximately 23,000 in 1994 and rose to 60,000 in 1999.[38] Model

this number C as a piecewise linear function of the time t in years since 1989 and use your model to estimate the number of college-bound high school seniors who intended to major in computer and information sciences in 1992.

48. Career Choices In 1989 approximately 100,000 college-bound high school seniors intended to major in engineering. This number decreased to approximately 85,000 in 1994 and rose to 88,000 in 1999.[39] Model this number E as a piecewise linear function of the time t in years since 1989 and use your model to estimate the number of college-bound high school seniors who intended to major in engineering in 1995.

49. Divorce Rates A study found that the divorce rate d appears to depend on the ratio r of available men to available women.[40] When the ratio was 1.3 (130 available men per 100 available women) the divorce rate was 22%. It rose to 35% if the ratio grew to 1.6 and rose to 30% if the ratio dropped to 1.1. Model these data by expressing d as a piecewise linear function of r and extrapolate your model to estimate the divorce rate if there are the same number of available men as women.

50. Retirement In 1950 the number N of retirees was approximately 150 per 1000 people aged 20–64. In 1990 this number rose to approximately 200 and is projected to rise to 275 in 2020.[41] Model N as a piecewise linear function of the time t in years since 1950 and use your model to project the number of retirees per 1000 people aged 20–64 in 2010.

COMMUNICATION AND REASONING EXERCISES

51. If y is measured in bootlags[42] and x is measured in Martian yen and $y = mx + b$, then m is measured in _____, and b is measured in _____.

52. If the slope in a linear relationship is measured in miles per dollar, then the independent variable is measured in _____, and the dependent variable is measured in _____.

53. The velocity of an object is given by $v = 0.1t + 20$ m/sec., where t is time in seconds. The object is

(A) moving with fixed speed   **(B)** accelerating

(C) decelerating   **(D)** impossible to say from the given information

---

[38]SOURCE: The College Board; National Science Foundation/*New York Times*, September 2, 1999, p. C1.

[39]Ibid.

[40]The cited study, by Scott J. South and associates, appeared in the *American Sociological Review* (February 1995). Figures are rounded. SOURCE: *New York Times*, February 19, 1995, p. 40.

[41]SOURCE: Social Security Administration/*New York Times*, April 4, 1999, p. WK3.

[42]An ancient Martian unit of length; 1 bootlag is the mean distance from a Martian's foreleg to its hindleg.

[37]SOURCE: VLSI Research/*New York Times*, October 9, 1994, sec. 3, p. 2.

54. The position of an object is given by $x = 0.2t - 4$, where t is time in seconds. The object is

(A) moving with fixed speed   (B) accelerating

(C) decelerating   (D) impossible to say from the given information

55. If a quantity is changing linearly with time and it increases by 10 units in the first day, what can you say about its behavior in the third day?

56. The quantities Q and T are related by a linear equation of the form

$$Q = mT + b$$

When $T = 0$, Q is positive but decreases to a negative quantity when T is 10. What are the signs of m and b? Explain your answers.

57. Suppose the cost function is $C(x) = mx + b$ (with m and b positive), the revenue function is $R(x) = kx$ ($k > m$), and the number of items is increased from the break-even quantity. Does this result in a loss or a profit, or is it impossible to say? Explain your answer.

58. You have been constructing a demand equation and have obtained a (correct) expression of the form $p = mq + b$, whereas you would have preferred one of the form $q = mp + b$. Should you simply switch p and q in the answer; should you start again from scratch, using p in the role of x and q in the role of y; or should you solve your demand equation for q? Give reasons for your decision.

1.5   Linear Regression

We have seen how to find a linear model given two data points: We find the equation of the line that passes through them. However, we often have more than two data points, and they will rarely all lie on a single straight line but may often come close to doing so. The problem is to find the line coming *closest* to passing through all of the points.

Suppose, for example, we are conducting research for a cable TV company interested in expanding into China and we come across the following figures showing the growth of the cable market there:

Year, t ($t = 0$ represents 2000)	−4	−3	−2	−1	0	1	2	3
Households with Cable, y (millions)	50	55	57	60	68	72	80	83

Data are approximate, and the 2001–2003 figures are estimated. SOURCES: HSBC Securities, Bear Stearns/*New York Times*, March 23, 2001, p. C1.

A plot of these data suggests a roughly linear growth of the market (Figure 20).

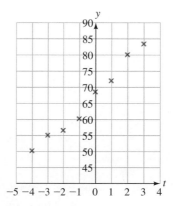

Figure 20

These points suggest a line, although they clearly do not all lie on a single straight line. Figure 21 shows the points together with several lines, some fitting better than others.

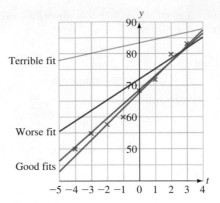

Figure 21

Question Can we precisely measure which lines fit better than others? For instance, which of the two lines labeled as "good fits" in Figure 21 models the data more accurately?

Answer We begin by considering, for each of the years 1996 through 2003, the difference between the actual number of households with cable (the **observed value**) and the number of households with cable predicted by a linear equation (the **predicted value**). The difference between the predicted value and the observed value is called the **residue:**

$$\text{residue} = \text{observed value} - \text{predicted value}$$

On the graph the residues measure the vertical distances between the (observed) data points and the line (Figure 22), and they tell us how far the linear model is from predicting the number of households with cable.

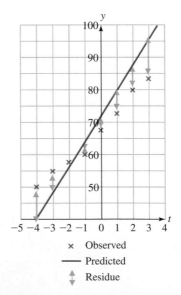

Figure 22

The more accurate our model, the smaller the residues should be. We can combine all the residues into a single measure of accuracy by adding their *squares*. (We square the residues in part to make them all positive.) The sum of the squares of the residues is called the **sum-of-squares error, SSE.**[43] Smaller values of SSE indicate more accurate models.

Here are some definitions and formulas for what we have been discussing.

Observed Values, Predicted Values, Residues, and Sum-of-Squares Error (SSE)

Observed and Predicted Values

Suppose we are given a collection of data points $(x_1, y_1), \ldots, (x_n, y_n)$. The n quantities $y_1, y_2, \ldots, y_n$ are called the **observed y values.** If we model these data with a linear equation

$$\hat{y} = mx + b \qquad \hat{y} \text{ stands for "estimated } y\text{" or "predicted } y\text{."}$$

then the y values we get by substituting the given x values into the equation are called the **predicted y values:**

$$\hat{y}_1 = mx_1 + b \qquad \text{Substitute } x_1 \text{ for } x.$$

$$\hat{y}_2 = mx_2 + b \qquad \text{Substitute } x_2 \text{ for } x.$$

$$\vdots$$

$$\hat{y}_n = mx_n + b \qquad \text{Substitute } x_n \text{ for } x.$$

Quick Example

Consider the three data points $(0, 2)$, $(2, 5)$, and $(3, 6)$. The observed y values are $y_1 = 2$, $y_2 = 5$, and $y_3 = 6$. If we model these data with the equation $\hat{y} = x + 2.5$, then the predicted values are

$$\hat{y}_1 = x_1 + 2.5 = 0 + 2.5 = 2.5$$

$$\hat{y}_2 = x_2 + 2.5 = 2 + 2.5 = 4.5$$

$$\hat{y}_3 = x_3 + 2.5 = 3 + 2.5 = 5.5$$

Residues and Sum-of-Squares Error (SSE)

If we model a collection of data $(x_1, y_1), \ldots, (x_n, y_n)$ with a linear equation $\hat{y} = mx + b$, then the **residues** are the n quantities (actual value – predicted value):

$$(y_1 - \hat{y}_1), (y_2 - \hat{y}_2), \ldots, (y_n - \hat{y}_n)$$

The **sum-of-squares error (SSE)** is the sum of the squares of the residues:

$$\text{SSE} = (y_1 - \hat{y}_1)^2 + (y_2 - \hat{y}_2)^2 + \cdots + (y_n - \hat{y}_n)^2$$

Quick Example

For the data and linear approximation given above, the residues are

$$y_1 - \hat{y}_1 = 2 - 2.5 = -0.5$$

$$y_2 - \hat{y}_2 = 5 - 4.5 = 0.5$$

$$y_3 - \hat{y}_3 = 6 - 5.5 = 0.5$$

$$\text{SSE} = (-0.5)^2 + (0.5)^2 + (0.5)^2 = 0.75$$

---

[43]Why not add the absolute values of the residues instead? Here is one reason: Take any two numbers, say, 1 and 3. Their average, 2, has the property that the sum of the squares of the residues, $(\bar{x} - 1)^2 + (\bar{x} - 3)^2$, is as small as possible when $\bar{x} = 2$; other values for $\bar{x}$ would result in a larger sum. However, all values of $\bar{x}$ between 1 and 3 result in the *same* value for the sum of the absolute values of the residues, $|\bar{x} - 1| + |\bar{x} - 3| = 2$.

Example 1 • Computing the Sum-of-Squares Error

Using the preceding data on the cable TV market in China, compute SSE for the linear models $y = 8t + 72$ and $y = 5t + 68$. Which model is the better fit?

Solution   We begin by creating a table showing the values of t, the observed (given) values of y, and the values predicted by the first model:

Year, t	Observed, y	Predicted, $\hat{y} = 8t + 72$
-4	50	40
-3	55	48
-2	57	56
-1	60	64
0	68	72
1	72	80
2	80	88
3	83	96

We now add two new columns for the residues and their squares:

Year, t	Observed, y	Predicted, $\hat{y} = 8t + 72$	Residue, $y - \hat{y}$	Residue^2, $(y - \hat{y})^2$
-4	50	40	$50 - 40 = 10$	$10^2 = 100$
-3	55	48	$55 - 48 = 7$	$7^2 = 49$
-2	57	56	$57 - 56 = 1$	$1^2 = 1$
-1	60	64	$60 - 64 = -4$	$(-4)^2 = 16$
0	68	72	$68 - 72 = -4$	$(-4)^2 = 16$
1	72	80	$72 - 80 = -8$	$(-8)^2 = 64$
2	80	88	$80 - 88 = -8$	$(-8)^2 = 64$
3	83	96	$83 - 96 = -13$	$(-13)^2 = 169$

SSE, the sum of the squares of the residues, is then the sum of the entries in the last column,

$$\text{SSE} = 479$$

Repeating the process using the second model, $y = 5t + 68$, yields SSE = 23. Thus, the second model is a better fit.

Graphing Calculator

We can use the List feature in the TI-83 to automate the computation of SSE. Press $\boxed{\text{STAT}}$ EDIT to obtain the list screen, where you can enter the given data in the first two columns, called L_1 and L_2. (If there is already data in a column you want to use, you can clear it by highlighting the column heading—for example, L_3—using the arrow key, and pressing $\boxed{\text{CLEAR}}$ $\boxed{\text{ENTER}}$.) To compute the predicted values, highlight the heading L_3 using the arrow keys and press $\boxed{\text{ENTER}}$, followed by the formula for the predicted values:

$8*L_1+72$ $\quad\quad\quad\quad$ L_1 is $\boxed{\text{2nd}}$ $\boxed{1}$.

L_1	L_2	L_3
-4	50	
-3	55	
-2	57	
-1	60	
0	68	
1	72	
2	80	
3	83	

$L_3=8*L_1+72$

Pressing $\boxed{\text{ENTER}}$ again will fill column 3 with the predicted values. Next, highlight the heading L_4, press $\boxed{\text{ENTER}}$, and enter

$(L_2-L_3)\,\hat{}\,2$ $\quad$ Squaring the residues

L_1	L_2	L_3	L_4
-4	50	40	
-3	55	48	
-2	57	56	
-1	60	64	
0	68	72	
1	72	80	
2	80	88	
3	83	96	

$L_4=(L_2-L_3)\,\hat{}\,2$

Pressing $\boxed{\text{ENTER}}$ again will fill column 4 with the squares of the residues. To compute SSE, the sum of the entries in L_4, go to the home screen and enter sum(L_4). The sum function is found by pressing $\boxed{\text{2nd}}$ $\boxed{\text{LIST}}$ and selecting MATH.

Excel

In Excel begin by setting up your worksheet with the observed data in two columns, t and y, and the predicted data for the first model in the third:

	A	B	C	D	E	F
1	t	y (Observed)	y (Predicted)		m	b
2	-4	50	=$E$2*A2+$F$2		8	72
3	-3	55				
4	-2	57				
5	-1	60				
6	0	68				
7	1	72				
8	2	80				
9	3	83				

Notice that, instead of using the numerical equation for the first model in column C, we used absolute references to the cells containing the slope m and the intercept b. This way, we can switch from one linear model to the next by changing only m and b in cells E2 and F2. (We have deliberately left column D empty in anticipation of the next step.)

CHAPTER 1 REVIEW TEST

1. Graph the following functions and equations:
   a. $y = -2x + 5$
   b. $2x - 3y = 12$
   c. $y = \begin{cases} \frac{1}{2}x & \text{if } -1 \le x \le 1 \\ x - 1 & \text{if } 1 < x \le 3 \end{cases}$
   d. $f(x) = 4x - x^2$, with domain $[0, 4]$

2. Decide whether each of the functions specified below is based on a linear, quadratic, exponential, or absolute value model.

x	-2	0	1	2	4
$f(x)$	4	2	1	0	2
$g(x)$	-5	-3	-2	-1	1
$h(x)$	1.5	1	0.75	0.5	0
$k(x)$	0.25	1	2	4	16
$u(x)$	0	4	3	0	-12

3. Find the equation of each line.
   a. Through $(3, 2)$ with slope -3
   b. Through $(-1, 2)$ and $(1, 0)$
   c. Through $(1, 2)$ parallel to $x - 2y = 2$
   d. With slope $\frac{1}{2}$ crossing $3x + y = 6$ at its x intercept

OHaganBooks.com—MODELING SALES, REVENUES, AND DEMAND

4. As your online bookstore, OHaganBooks.com, has grown in popularity, you have been monitoring book sales as a function of the traffic at your site (measured in "hits" per day) and have obtained the following model:

$$n(x) = \begin{cases} 0.02x & \text{if } 0 \le x \le 1000 \\ 0.025x - 5 & \text{if } 1000 < x \le 2000 \end{cases}$$

   where $n(x)$ is the average number of books sold in a day in which there are x hits at the site.
   a. On average, how many books per day does your model predict you will sell when you have 500 hits/day? 1000 hits/day? 1500 hits/day?
   b. What does the coefficient 0.025 tell you about your book sales?
   c. According to the model, how many hits per day will be needed in order to sell an average of 30 books/day?

5. Your monthly books sales had increased quite dramatically over the past few months but now appear to be leveling off. Here are the sales figures for the past 6 months:

Month, t	1	2	3	4	5	6
Daily Book Sales, S	12.5	37.5	62.5	72.0	74.5	75.0

   a. Which of the following models best approximates the data?
   (A) $S(t) = \dfrac{300}{4 + 100(5^{-t})}$ (B) $S(t) = 13.3t + 8.0$
   (C) $S(t) = -2.3t^2 + 30.0t - 3.3$ (D) $S(t) = 7(3^{0.5t})$
   b. What do each of the above models predict for the sales in the next few months: rising, falling, or leveling off?

6. To increase business at OHaganBooks.com, you plan to place more banner ads at well-known Internet portals. So far, you have the following data on the average number of hits per day at OHaganBooks.com versus your monthly advertising expenditures:

Advertising Expenditure ($/month)	2000	5000
Web Site Traffic (hits/day)	1900	2050

   You decide to construct a linear model giving h, the average number of hits/day as a function of the advertising expenditure c.
   a. What is the model you obtain?
   b. Based on your model, how much traffic can you anticipate if you budget $6000/month for banner ads?
   c. Your goal is to eventually increase traffic at your site to an average of 2500 hits/day. Based on your model, how much do you anticipate you will need to spend on banner ads in order to accomplish this?

7. A month ago you increased expenditure on banner ads to $6000/month, and you have noticed that the traffic at OHaganBooks.com has not increased to the level predicted by the linear model in Exercise 6. Fitting a quadratic function to the data you have gives the model

$$h = -0.000005c^2 + 0.085c + 1750$$

   where h is the daily traffic (hits) at your Web site and c is the monthly advertising expenditure.
   a. According to this model, what is the current traffic at your site?
   b. Does this model give a reasonable prediction of traffic at expenditures larger than $8500/month? Why?

8. Besides selling books, you are generating additional revenue at OHaganBooks.com through your new online publishing service. Readers pay a fee to download the entire text of a novel. Author royalties and copyright fees cost you an average of $4/novel, and the monthly cost of operating and maintaining the service amounts to $900/month. You are currently charging readers $5.50/novel.
   a. What are the associated cost, revenue, and profit functions?
   b. How many novels must you sell each month in order to break even?
   c. If you lower the charge to $5.00/novel, how many books will you need to sell in order to break even?

9. To generate a profit from your online publishing service, you need to know how the demand for novels depends on the price you charge. During the first month of the service, you were charging $10.00/novel and sold 350. Lowering the price to $5.50/novel had the effect of increasing demand to 620 novels/month.

   a. Use the given data to construct a linear demand equation.

   b. Use the demand equation you constructed in part (a) to estimate the demand if you raised the price to $15.00/novel.

   c. Using the information on cost given in Exercise 8, determine which of the three prices ($5.50, $10.00, and $15.00) would result in the largest profit and the size of that profit.

10. It is now several months later and you have tried selling your online novels at a variety of prices, with the following results:

Price ($)	5.50	10.00	12.00	15.00
Demand (monthly sales)	620	350	300	100

   a. Use the given data to obtain a linear regression model of demand. (Round coefficients to four decimal places.)

   b. Use the demand model you constructed in part (a) to estimate the demand if you charged $8.00/novel. (Round the answer to the nearest novel.)

 # ADDITIONAL ONLINE REVIEW

If you follow the path

Web site → Everything for Finite Mathematics → Chapter 1

you will find the following additional resources to help you review:

A comprehensive chapter summary (including examples and interactive features)

Additional review exercises (including interactive exercises and many with help)

Interactive section-by-section online tutorials

Section-by-section Excel tutorials

A true/false chapter quiz

Several useful utilities, including graphers and a regression tool

2

NONLINEAR MODELS

CASE STUDY

Checking Up on Malthus

In 1798 Thomas R. Malthus (1766–1834) published an influential pamphlet, later expanded into a book, titled *An Essay on the Principle of Population as It Affects the Future Improvement of Society*. One of his main contentions was that population grows geometrically (exponentially) while the supply of resources such as food grows only arithmetically (linearly). Some 200 years later, you have been asked to check the validity of Malthus's contention. How do you go about doing so?

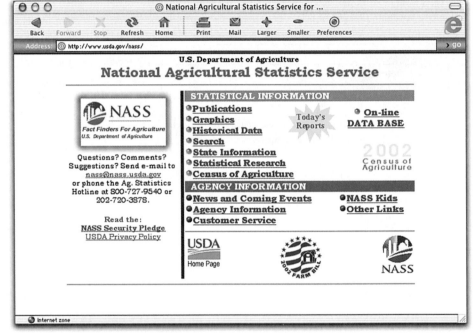

INTERNET RESOURCES FOR THIS CHAPTER

At the Web site, follow the path

    Web Site → Everything for Calculus → Chapter 2

where you will find a detailed chapter summary you can print out, a true/false quiz, and a collection of review questions. You will also find downloadable Excel tutorials for each section, an online regression utility you can use to model with many of the functions we discuss in this chapter, an online grapher, and other resources. In addition, complete text and interactive exercises have been placed on the Web site, covering several optional topics:

Inverse functions

Linear and exponential regression

Using and deriving algebraic properties of logarithms

Introduction

To see if Malthus was right, we need to see if the data fit the models (linear and exponential) he suggested or if other models would be better. We saw in Chapter 1 how to fit a linear model. In this chapter we discuss how to construct models using various *nonlinear* functions.

The nonlinear functions we consider in this chapter are the *quadratic* functions, the simplest nonlinear functions; the *exponential* functions, essential for discussing many kinds of growth and decay, including the growth (and decay) of money in finance and the initial growth of an epidemic; the *logarithmic* functions, needed to fully understand the exponential functions; and the *logistic* functions, used to model growth with an upper limit, like the spread of an epidemic.

A Algebra Review

For this chapter, you should be familiar with the algebra reviewed in Appendix A, Section A.2.

2.1 Quadratic Functions and Models

In Chapter 1 we studied linear functions. Linear functions are useful, but in real-life applications they are often accurate for only a limited range of values of the variables. The relationship between two quantities is often best modeled by a curved line rather than a straight line. The simplest function with a graph that is not straight line is a *quadratic* function.

Quadratic Function

A **quadratic function** of the variable x is a function that can be written in the form

$$f(x) = ax^2 + bx + c \qquad \text{Function form}$$

or

$$y = ax^2 + bx + c \qquad \text{Equation form}$$

where a, b, and c are fixed numbers (with $a \neq 0$).

Quick Examples

1. $f(x) = 3x^2 - 2x + 1$ $a = 3, b = -2, c = 1$

2. $g(x) = -x^2$ $a = -1, b = 0, c = 0$

3. $R(p) = -5600p^2 + 14{,}000p$ $a = -5600, b = 14{,}000, c = 0$

Every quadratic function $f(x) = ax^2 + bx + c$ $(a \neq 0)$ has a **parabola** as its graph. Figure 1 shows the two possible orientations of the graph.

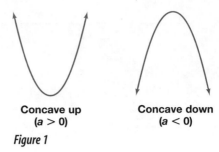

Concave up
$(a > 0)$

Concave down
$(a < 0)$

Figure 1

Following is a summary of some features of parabolas that we can use to sketch the graph of any quadratic function.[1]

Features of a Parabola

The graph of $f(x) = ax^2 + bx + c$ $(a \neq 0)$ is a parabola with the following features.

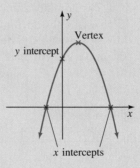

- **Vertex** The x coordinate of the vertex is $-\frac{b}{2a}$. The y coordinate is $f\left(-\frac{b}{2a}\right)$.
- **x Intercepts** If any, these occur when $f(x) = 0$; that is, when

  $$ax^2 + bx + c = 0$$

We can solve this equation for x using the quadratic formula. Thus, the x intercepts are

$$x = \frac{-b \pm \sqrt{b^2 - 4ac}}{2a}$$

If the **discriminant** $b^2 - 4ac$ is positive, there are two x intercepts. If it is zero, there is a single x intercept (at the vertex). If it is negative, there are no x intercepts (so the parabola doesn't touch the x axis at all).

- **y Intercept** This occurs when $x = 0$, so

  $$y = a(0)^2 + b(0) + c = c$$

- **Symmetry** The parabola is symmetric with respect to the vertical line through the vertex, which is the line $x = -\frac{b}{2a}$.

---

[1] We will not fully justify the formula for the vertex and the axis of symmetry until we have studied some calculus, although it is possible to do so with just algebra.

Note that the x intercepts can also be written as

$$x = -\frac{b}{2a} \pm \frac{\sqrt{b^2 - 4ac}}{2a}$$

making it clear that they are located symmetrically on either side of the line $x = -\frac{b}{2a}$. This partially justifies the claim that the whole parabola is symmetric with respect to this line.

Example 1 • Sketching the Graph of a Quadratic Function

Sketch the graph of $f(x) = x^2 + 2x - 8$ by hand.

Figure 2

Solution Here, $a = 1$, $b = 2$, and $c = -8$. Since $a > 0$, the parabola is concave up (Figure 2).

• *Vertex* The x coordinate of the vertex is

$$x = -\frac{b}{2a} = -\frac{2}{2} = -1$$

To get its y coordinate, we substitute the value of x back into $f(x)$ to get

$$y = f(-1) = (-1)^2 + 2(-1) - 8 = 1 - 2 - 8 = -9$$

Thus, the coordinates of the vertex are $(-1, -9)$.

• *x Intercepts* To calculate the x intercepts (if any), we solve the equation

$$x^2 + 2x - 8 = 0$$

Luckily, this equation factors as $(x + 4)(x - 2) = 0$. Thus, the solutions are $x = -4$ and $x = 2$, so these values are the x intercepts. (We could also have used the quadratic formula here.)

• *y Intercept* The y intercept is given by $c = -8$.

• *Symmetry* The graph is symmetric around the vertical line $x = -1$.

Now we can sketch the curve as in Figure 3.

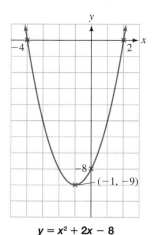

$$y = x^2 + 2x - 8$$

Figure 3

Example 2 • One x Intercept and No x Intercepts

Sketch the graph of each quadratic function, showing the location of the vertex and intercepts.

a. $f(x) = 4x^2 - 12x + 9$ **b.** $g(x) = -\frac{1}{2}x^2 + 4x - 12$

Solution

a. We have $a = 4$, $b = -12$, and $c = 9$. Since $a > 0$, this parabola is concave up.

$$\text{Vertex } x = -\frac{b}{2a} = \frac{12}{8} = \frac{3}{2} \qquad \text{\textit{x coordinate of vertex}}$$

$$y = f\left(\frac{3}{2}\right) = 4\left(\frac{3}{2}\right)^2 - 12\left(\frac{3}{2}\right) + 9 = 0 \qquad \text{\textit{y coordinate of vertex}}$$

Thus, the vertex is at the point $\left(\frac{3}{2}, 0\right)$.

$$\text{\textit{x Intercepts }} 4x^2 - 12x + 9 = 0$$

$$(2x - 3)^2 = 0$$

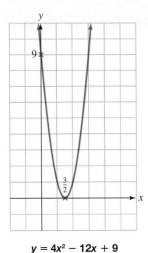

$$y = 4x^2 - 12x + 9$$

Figure 4

The only solution is $2x - 3 = 0$, or $x = \frac{3}{2}$. Note that this coincides with the vertex, which lies on the x axis.

> y Intercept $c = 9$

> *Symmetry* The graph is symmetric around the vertical line $x = \frac{3}{2}$.

The graph is the narrow parabola shown in Figure 4.

b. Here, $a = -\frac{1}{2}$, $b = 4$, and $c = -12$. Since $a < 0$, the parabola is concave down. The vertex has x coordinate $-\frac{b}{2a} = 4$, with corresponding y coordinate $f(4) = -\frac{1}{2}(4)^2 + 4(4) - 12 = -4$. Thus, the vertex is at $(4, -4)$.

For the x intercepts, we must solve $-\frac{1}{2}x^2 + 4x - 12 = 0$. If we try to use the quadratic formula, we discover that the discriminant is $b^2 - 4ac = 16 - 24 = -8$. Since the discriminant is negative, there are no solutions of the equation, so there are no x intercepts.

The y intercept is given by $c = -12$, and the graph is symmetric around the vertical line $x = 4$.

Since there are no x intercepts, the graph lies entirely below the x axis, as shown in Figure 5.

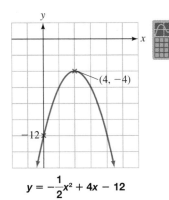

$$y = -\frac{1}{2}x^2 + 4x - 12$$

Figure 5

Graphing Calculator

Question We already know how to graph functions on the TI-83. However, what should we use for Xmin and Xmax?

Answer We would like our graph to show the vertex as well as all the intercepts. To see the vertex, make sure that its x coordinate is between Xmin and Xmax. To see the x intercepts, make sure that they are also between Xmin and Xmax. To see the y intercept, make sure that $x = 0$ is between Xmin and Xmax. Thus, to see everything, choose Xmin and Xmax so that the interval [Xmin, Xmax] contains the x coordinate of the vertex, the x intercepts, and 0.

Once Xmin and Xmax are chosen, you can obtain convenient values of Ymin and Ymax by pressing ZOOM and selecting the option ZoomFit. (Make sure that your quadratic equation is entered in the Y=screen before doing this!)

Excel

We can set up our worksheet so that all we have to enter are the coefficients a, b, and c and a range of x values for the graph. Here is a possible layout that will plot 100 points using the coefficients for part (a) (similar to the Excel Graphing Worksheet we used in Example 2 of Section 1.2):

	A	B	C	D
1	x	y	a	4
2	=$D$4		b	-12
3	=A2+$D$6		c	9
4			Xmin	-10
5			Xmax	10
6			Delta X	=(D5-D4)/100
7				
101				
102				

To add the y coordinates, we use the technology formula

```
a*x^2+b*x+c
```

replacing *a*, *b*, and *c* with (absolute) references to the cells containing their values:

	A	B	C	D
1	x	y	a	4
2		-10	=$D$1*A2^2+$D$2*A2+$D$3	b	-12
3		-9.8		c	9
4		-9.6		Xmin	-10
5		-9.4		Xmax	10
6		-9.2		Delta X	0.2
7					
101					
102		10			

Graphing the data in columns A and B gives the graph shown in Figure 6.

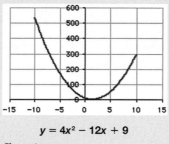

$$y = 4x^2 - 12x + 9$$

Figure 6

We can go further and compute the exact coordinates of the vertex and intercepts:

	A	B	C	D	E	F
1	x	y	a	4	Vertex (x)	=-D2/(2*D1)
2	-10	529	b	-12	Vertex (y)	=D1*F1^2+D2*F1+D3
3	-9.8	510.76	c	9	X Intercept 1	=(-D2-SQRT(D2^2-4*D1*D3))/(2*D1)
4	-9.6	492.84	Xmin	-10	X Intercept 2	=(-D2+SQRT(D2^2-4*D1*D3))/(2*D1)
5	-9.4	475.24	Xmax	10	Y Intercept	=D3
6	-9.2	457.96	Delta X	0.2		
7						

The completed sheet should look like this:

	A	B	C	D	E	F
1	x	y	a	4	Vertex (x)	1.5
2	-10	529	b	-12	Vertex (y)	0
3	-9.8	510.76	c	9	X Intercept 1	1.5
4	-9.6	492.84	Xmin	-10	X Intercept 2	1.5
5	-9.4	475.24	Xmax	10	Y Intercept	9
6	-9.2	457.96	Delta X	0.2		
7						

We can now save this sheet as a template to handle all quadratic functions. For instance, to do part (b), we just change the values of *a*, *b*, and *c* in column D to $a = -1/2$ (we enter it as =-1/2 or Excel might think it's a calendar date!), $b = 4$, and $c = -12$.

APPLICATIONS

Recall that the **revenue** resulting from one or more business transactions is the total payment received. Thus, if q units of some item are sold at p dollars per unit, the revenue resulting from the sale is

revenue = price $\times$ quantity

$$R = pq$$

Example 3 • Demand and Revenue

A publishing company predicts that the demand equation for the sale of its latest sci-fi novel is

$$q = -2000p + 150,000$$

where q is the number of books it can sell each year at a price of $\$p$ per book. What price should it charge to obtain the maximum annual revenue?

Solution The total revenue depends on the price, as follows:

$R = pq$	Formula for revenue
$\quad = p(-2000p + 150,000)$	Substitute for q from demand equation.
$\quad = -2000p^2 + 150,000p$	Simplify.

We are after the price p that gives the maximum possible revenue. Notice that what we have is a quadratic function of the form $R(p) = ap^2 + bp + c$, where $a = -2000$, $b = 150,000$, and $c = 0$. Since a is negative, the graph of the function is a parabola, concave down, so its vertex is its highest point (Figure 7). The p coordinate of the vertex is

$$p = -\frac{b}{2a} = -\frac{150,000}{-4000} = 37.5$$

This value of p gives the highest point on the graph and thus gives the largest value of $R(p)$. We may conclude that the company should charge $\$37.50$ per book to maximize its annual revenue.

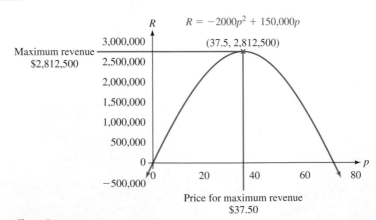

Figure 7

✳ **Before we go on ...** You might ask what the maximum annual revenue is. Since $R(p)$ gives us the revenue at a price of $\$p$, the answer is $R(37.5) = -2000(37.5)^2 + 150,000(37.5) = 2,812,500$. In other words, the company will earn total annual revenues from this book amounting to $\$2,812,500$.

Example 4 • Demand, Revenue, and Profit

As the operator of Workout Fever Health Club, you calculate your demand equation to be

$$q = -0.06p + 46$$

where q is the number of members in the club and p is the annual membership fee you charge.

a. Since you are running a shoestring operation, your annual operating costs amount to only $5000 per year. Find the annual revenue and profit as functions of the membership price p.

b. At what price should you set annual memberships in order to break even? Could you still break even if your operating costs went up to $10,000 per year?

c. At what price should you set annual memberships to obtain the maximum profit?

Solution

a. The annual revenue is given by

$$R = pq \qquad \qquad \text{Formula for revenue}$$

$$= p(-0.06p + 46) \qquad \text{Substitute for } q \text{ from demand equation.}$$

$$= -0.06p^2 + 46p \qquad \text{Simplify.}$$

and the annual cost C is fixed at $5000. Thus, the profit function is

$$P = R - C \qquad \qquad \text{Formula for profit}$$

$$= -0.06p^2 + 46p - 5000 \qquad \text{Substitute for revenue and cost.}$$

b. First recall from Section 1.4 that break-even occurs when the profit equals zero:

$$P = 0 \quad \text{or} \quad -0.06p^2 + 46p - 5000 = 0$$

We now have a quadratic equation in p, and we must solve it for the break-even price p. To do this, we can use the quadratic formula. Alternatively, we can use our Excel worksheet from Example 2 with $a = -0.06$, $b = 46$, and $c = -5000$ to locate the x intercepts (which in this example are the p intercepts). We obtain

$$p \approx 635.55 \text{ or } 131.12$$

Although you may be tempted to choose the larger figure, thinking that you'll get a larger total fee, remember that both these options will result in your operation breaking even—there is no advantage to one or the other. Thus, to break even, you should charge either $635.55 or $131.12 per year.

   If the operating costs went up to $10,000 per year, then the break-even equation would become

$$-0.06p^2 + 46p - 10,000 = 0$$

This equation has no real solutions since the discriminant, $b^2 - 4ac$, is negative. You simply cannot break even with these costs, and you might as well close down the operation because the profit is always negative (the graph of profit versus price lies below the p axis).

c. We saw in part (a) that the profit function is

$$P = -0.06p^2 + 46p - 5000$$

The graph of P is a parabola whose vertex has p coordinate $-b/(2a) = 46/0.12 \approx$ 383.33. Thus, for a maximum profit, you should charge an annual membership fee of $383.33.

✴ **Before we go on ...** Figure 8 shows the revenue function $R = -0.06p^2 + 46p$ together with the cost lines $C = 5000$ and $C = 10,000$.

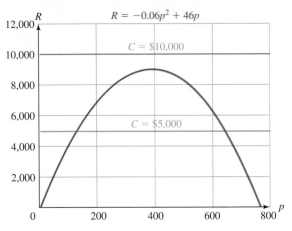

Figure 8

The horizontal lines at heights of $5000 and $10,000 represent the two fixed annual costs, and the parabola is the graph of the revenue function. The two intersection points are the break-even points we calculated in part (b). We can zoom and trace to obtain the coordinates of the break-even point and confirm the calculations. The higher cost line fails to touch the revenue curve, which confirms our conclusion that the health club cannot break even with annual costs of $10,000.

The graph also shows the price you should charge to obtain the largest profit—it is the p coordinate of the vertex of the revenue graph. The highest annual revenue is its R coordinate.

Fitting a Quadratic Function to Data: Quadratic Regression

In Section 1.5 we saw how to fit a regression line to a collection of data points. Here, we see how to use technology to obtain the **quadratic regression curve** associated with a set of points.

Question What exactly *is* the quadratic regression curve?

Answer It is similar to the regression line: Start with a set of "observed" data points and a quadratic model $y = ax^2 + bx + c$ that may or may not pass through these points. Use the quadratic model to generate a set of "predicted" values of y. Then measure the associated sum-of-squares error (SSE) in the same way as in Section 1.5. (Recall that SSE is the sum of the squares of the residues—the differences between the observed y values and the values predicted by the model.) Now adjust the values of a, b, and c to obtain the smallest possible value of SSE, and, *voilá*, you have the regression curve.[2]

---

[2]This can be done in various ways by using technology (see Example 5) or by solving a certain set of equations determined by the data.

ⵟ *Example 5 • Prison Population*

The following table shows the total population in U.S. state and federal prisons in the period 1970–1997 as a function of time in years ($t = 0$ represents 1970):

Year, t	0	3	6	9	12	15	18	21	24	27
Population (millions)	0.18	0.18	0.26	0.30	0.40	0.50	0.65	0.80	1.05	1.25

Sources: Bureau of Justice Statistics, N.Y. State Dept. of Correctional Services/*New York Times,* January 9, 2000, p. WK3.

a. Is a linear model appropriate for these data?

b. Find the quadratic model

$$P(t) = at^2 + bt + c$$

that best fits the data.

Solution

a. To see whether a linear model is appropriate, we plot the data points and the regression line using one of the methods of Example 3 in Section 1.5 (Figure 9).

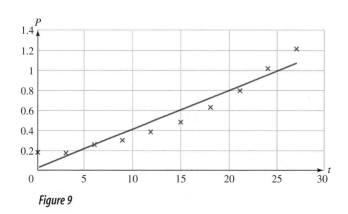

Figure 9

From the graph we can see that the given data suggest a curve and not a straight line: The observed points are above the regression line near the ends but below in the middle. (We would expect the data points from a linear relation to fall randomly above and below the regression line.)

b. The quadratic model that best fits the data is the quadratic regression model. As with linear regression, there are algebraic formulas to compute a, b, and c, but they are rather involved. However, we can exploit the fact that these formulas are built in to graphing calculators and spreadsheets.

Graphing Calculator

Using $\boxed{\text{STAT}}$ EDIT, we enter the data in the TI-83 putting the x coordinates (values of t) in L_1 and the populations in L_2, just as in Section 1.5. Then we press $\boxed{\text{STAT}}$, select CALC, and choose the option QuadReg. Pressing $\boxed{\text{ENTER}}$ gives the quadratic regression curve in the home screen:

$$y \approx 0.0015x^2 - 0.0006x + 0.1837 \quad \text{Coefficients rounded to four decimal places}$$

Thus, the quadratic regression model is

$$P(t) = 0.0015t^2 - 0.0006t + 0.1837$$

Figure 10 shows the points and regression curve. Clearly, this model gives an excellent fit.

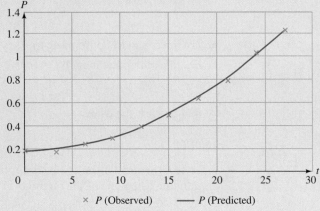

Figure 10

Excel

As in Section 1.5, Example 3, we start with a scatter plot of the original data, click on the chart, and select Add Trendline... from the Chart menu. Then select a Polynomial type (order 2) and, under Options, check the option Display equation on chart.

Web Site

Follow the path

Web Site → Online Utilities → Simple Regression

enter the data as x and y coordinates, and click on the y=ax^2+bx+c button to find the quadratic regression equation. You can then graph the regression curve and data points by clicking on Graph.

2.1 EXERCISES

In Exercises 1–10, sketch the graphs of the quadratic functions, indicating the coordinates of the vertex, the y intercept, and the x intercepts (if any).

1. $f(x) = x^2 + 3x + 2$

2. $f(x) = -x^2 - x$

3. $f(x) = -x^2 + 4x - 4$

4. $f(x) = x^2 + 2x + 1$

5. $f(x) = -x^2 - 40x + 500$

6. $f(x) = x^2 - 10x - 600$

7. $f(x) = x^2 + x - 1$

8. $f(x) = x^2 + \sqrt{2}x + 1$

9. $f(x) = x^2 + 1$

10. $f(x) = -x^2 + 5$

In Exercises 11–14, for each demand equation, express the total revenue R as a function of the price p per item, sketch the graph of the resulting function, and determine the price p that maximizes total revenue in each case.

11. $q = -4p + 100$

12. $q = -3p + 300$

13. $q = -2p + 400$

14. $q = -5p + 1200$

In Exercises 15–18, use technology to find the quadratic regression curve through the given points. (Round all coefficients to four decimal places.)

15. $\{(1, 2), (3, 5), (4, 3), (5, 1)\}$

16. $\{(-1, 2), (-3, 5), (-4, 3), (-5, 1)\}$

17. $\{(-1, 2), (-3, 5), (-4, 3)\}$

18. $\{(2, 5), (3, 5), (5, 3)\}$

APPLICATIONS

19. SUV Sales in Europe Sport utility vehicle (SUV) sales in Europe increased substantially in the last 5 years of the 1990s, as shown in the following chart:

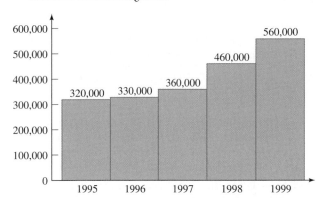

a. If you want to use a function of the form $f(t) = at^2 + bt + c$ to model the sales, with $t =$ time in years since 1995, should the coefficient a be positive or negative for the most accurate model of sales? Why?

b. Which of the following models best approximates the data given?

(A) $f(t) = 18{,}000t^2 - 10{,}000t + 320{,}000$

(B) $f(t) = -18{,}000t^2 + 10{,}000t + 320{,}000$

(C) $f(t) = 18{,}000t^2 + 10{,}000t - 320{,}000$

c. Which year would correspond to the vertex of the graph of the correct model from part (b)? What is the danger of extrapolating the data backward beyond that year?

Data are rounded (the 1999 figure is estimated). SOURCE: Standard & Poor's DRI, company reports/*New York Times,* December 14, 1999, p. C1.

20. Auto Sales in the United States Sales of family vehicles in the United States continued to rise during the last 5 years of the 1990s, as shown in the following chart:

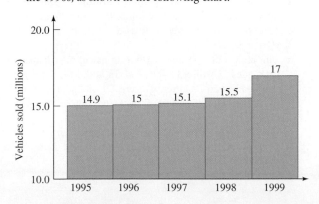

a. If you want to use a function of the form $f(t) = at^2 + bt + c$ to model the sales, with $t =$ time in years since 1995, why should the coefficient a be nonzero for the most accurate model?

b. Which of the following models best approximates the data given?

(A) $f(t) = 0.22t^2 + 0.42t - 15$

(B) $f(t) = -0.22t^2 + 0.42t + 15$

(C) $f(t) = 0.22t^2 - 0.42t + 15$

c. Which year would correspond most closely to the vertex of the graph of the correct model from part (b)? Would you be more comfortable using the model to extrapolate to the left of 1995 or to the right of 1999? Why?

Data are rounded (the 1999 figure is estimated). SOURCE: Wards company reports, Bloomberg Financial/*New York Times,* December 15, 1999, p. C1.

21. Sport Utility Vehicles The average weight of an SUV can be approximated by

$$W = 3t^2 - 90t + 4200 \qquad (5 \le t \le 27)$$

where t is its year of manufacture ($t = 0$ represents 1970) and W is the average weight of an SUV in pounds. Sketch the graph of W as a function of t. According to the model, in what year were SUVs the lightest? What was their average weight in that year?

SOURCE: The quadratic model is based on data published in the *New York Times,* November 30, 1997, p. 43.

22. Sedans The average weight of a sedan can be approximated by

$$W = 6t^2 - 240t + 4800 \qquad (5 \le t \le 27)$$

where t is its year of manufacture ($t = 0$ represents 1970) and W is the average weight of a sedan in pounds. Sketch the graph of W as a function of t. According to the model, in what year were sedans the lightest? What was their average weight in that year?

SOURCE: The quadratic model is based on data published in the *New York Times,* November 30, 1997, p. 43.

23. Fuel Efficiency The fuel efficiency (in miles per gallon) of an SUV depends on its weight according to the formula[3]

$$E = 0.000\,001\,6x^2 - 0.016x + 54 \qquad (1800 \le x \le 5400)$$

where x is the weight of an SUV in pounds. According to the model, what is the weight of the least fuel-efficient SUV? Would you trust the model for weights greater than the answer you obtained? Explain.

SOURCES: Environmental Protection Agency, National Highway Traffic Safety Administration; American Automobile Manufacturers' Association, Ford Motor Company/*New York Times,* November 30, 1997, p. 43.

24. Global Warming The amount of carbon dioxide (in pounds per 15,000 miles) released by a typical SUV depends on its fuel efficiency according to the formula

$$W = 32x^2 - 2080x + 44{,}000 \qquad (12 \le x \le 33)$$

---

[3]Fuel efficiency assumes 50% city driving and 50% highway driving. The model is based on a quadratic regression using data from 18 models of SUVs.

where x is the fuel efficiency of an SUV in miles per gallon. According to the model, what is the fuel efficiency of the SUV with the least carbon dioxide pollution? Comment on the reliability of the model for fuel efficiencies that exceed your answer.

Sources: Environmental Protection Agency, National Highway Traffic Safety Administration; American Automobile Manufacturers' Association, Ford Motor Company/*New York Times*, November 30, 1997, p. 43.

25. Revenue The market research department of Better Baby Buggy Co. notices that when its buggies are priced at $80 each, it can sell 100 each month. However, when the price is raised to $100, it sells only 90 each month. Assuming that the demand is linear, at what price should it sell the buggies to get the largest revenue? What is the largest monthly revenue?

26. Revenue Better Baby Buggy Co. has just come out with a new model, the Turbo. The market research department now estimates that the company can sell 200 Turbos each month at $60, but only 120 each month at $100. Assuming that the demand is linear, at what price should it sell its buggies to get the largest revenue? What is the largest monthly revenue?

27. Revenue Pack-Em-In Real Estate is building a new housing development. The more houses it builds, the less people will be willing to pay, due to the crowding and smaller lot sizes. In fact, if it builds 40 houses in this particular development, it can sell them for $200,000 each, but if it builds 60 houses, it will only be able to get $160,000 each. How many houses should Pack-Em-In build to get the largest revenue? What is the largest possible revenue?

28. Revenue Pack-Em-In has another development in the works. If it builds 50 houses in this development, it will be able to sell them at $190,000 each, but if it builds 70 houses, it will get only $170,000 each. How many houses should they build to get the largest revenue? What is the largest possible revenue?

29. Web-Site Profit Encouraged by the popularity of your gaming Web site, www.mudbeast.net, you have decided to charge users who log on to the site. When you charged a $2 access fee, your web counter showed a demand of 280 "hits" per month. After lowering the price to $1.50, activity increased to 560 hits per month.
   a. Construct a linear demand function for your Web site and hence obtain the monthly revenue R as a function of the access fee x.
   b. Your Internet provider charges you $30 per month to maintain your site. Write your monthly profit P as a function of the access fee x and hence determine the access fee you should charge to obtain the largest possible monthly profit. What is the largest possible monthly profit?

30. T-Shirt Profit The two fraternities Sigma Alpha Mu and Ep Sig plan to raise money jointly to benefit homeless people on Long Island. They will sell *Star Wars* (Episode VII) T-shirts in the student center but are not sure how much to charge. Sigma Alpha Mu treasurer Solo recalls that they once sold 400 shirts in a week at $8 each, but Ep Sig treasurer Justino

claims that, based on past experience, they can sell 600 each week if they charge $4 each.
   a. Based on this anecdotal information, construct a linear demand equation for *Star Wars* T-shirts and hence obtain the weekly revenue R as a function of the unit price x.
   b. The university administration charges the fraternities $500 per week for use of the Student Center. Write down the monthly profit P as a function of the unit price x and hence determine how much the fraternities should charge to obtain the largest possible weekly profit. What is the largest possible weekly profit?

31. Web-Site Profit The latest demand equation for your gaming Web site, www.mudbeast.net, is given by

$$q = -400x + 1200$$

where q is the number of hits per month and x is the access fee you charge. Your Internet provider bills you as follows:

   Site maintenance fee: $20 per month

   High-volume access fee: 50¢ per hit

Find the monthly cost as a function of the access fee x. Hence, find the monthly profit as a function of x and determine the access fee you should charge to obtain the largest possible monthly profit. What is the largest possible monthly profit?

32. T-Shirt Profit The latest demand equation for your *Star Trek* T-shirts is given by

$$q = -40x + 600$$

where q is the number of shirts you can sell in 1 week if you charge $x per shirt. The Student Council charges you $400 per week for use of their facilities, and the T-shirts cost you $5 each. Find the weekly cost as a function of the unit price x. Hence, find the weekly profit as a function of x and determine the unit price you should charge to obtain the largest possible weekly profit. What is the largest possible weekly profit?

33. Nightclub Management You have just opened a new nightclub, Russ' Techno Pitstop but are unsure of how high to set the cover charge (entrance fee). One week you charged $10/per guest and averaged 300 guests each night. The next week you charged $15 per guest and averaged 250 guests each night.
   a. Find a linear demand equation showing the number of guests q per night as a function of the cover charge p.
   b. Find the nightly revenue R as a function of the cover charge p.
   c. The club will provide two free nonalcoholic drinks for each guest, costing $3 per head. In addition, the nightly overheads (rent, salaries, dancers, DJ, and so on) amount to $3000. Find the cost C as a function of the cover charge p.
   d. Now find the profit in terms of the cover charge p and hence determine the cover charge you should charge for a maximum profit.

34. Television Advertising As the sales manager for Montevideo Productions, you are planning to review the prices you charge clients for television advertisement development. You currently charge each client an hourly development fee of $2500. With this pricing structure, the demand, measured by the number of contracts Montevideo signs per month, is 15 contracts. This is down 5 contracts from the figure last year, when your company charged only $2000.

a. Construct a linear demand equation giving the number of contracts q as a function of the hourly fee p Montevideo charges for development.

b. On average, Montevideo bills for 50 hours of production time on each contract. Give a formula for the total revenue obtained by charging p per hour.

c. The costs to Montevideo Productions are estimated as follows:

Fixed costs: $120,000 per month

Variable costs: $80,000 per contract

Express Montevideo's monthly cost (i) as a function of the number q of contracts and (ii) as a function of the hourly production charge p.

d. Express Montevideo's monthly profit as a function of the hourly development fee p and find the price it should charge to maximize the profit.

For Exercises 35–38, use technology to solve.

35. Advertising Revenue The following table shows the annual advertising revenue earned by America Online (AOL) during the last 3 years of the 1990s. ($t = -1$ represents 1999.)

Year, t	-3 (1997)	-2	-1 (1999)
Revenue, R ($ millions)	150	360	760

Figures are rounded to the nearest $10 million. SOURCES: AOL; Forrester Research/*New York Times*, January 31, 2000, p. C1.

Find a quadratic model for these data and use your model to estimate AOL's advertising revenue in 2000.

36. Religion The following table shows the population of Roman Catholic nuns in the United States during the last 25 years of the 1900s ($t = 0$ represents 2000):

Year, t	-25 (1975)	-15	-5 (1995)
Population, P	130,000	120,000	80,000

Figures are rounded. SOURCE: Center for Applied Research in the Apostolate/*New York Times*, January 16, 2000, p. A1.

Find a quadratic model for these data and use your model to estimate the number of Roman Catholic nuns in 2000.

37. Income The following table shows the average after-tax income of the wealthiest 1% of U.S. taxpayers for the years 1990–1998:

Year, t	0 (1990)	1	2	3	4	5	6	7	8 (1998)
Income, I ($ thousands)	400	350	400	370	372	400	450	520	600

Data are approximate. Incomes are in constant 1998 dollars. SOURCE: IRS/*New York Times*, February 26, 2001, p. C2.

Find a quadratic regression model for these data. Graph the model together with the data. Assuming the trend continued, estimate the average after-tax income of the wealthiest 1% of U.S. taxpayers in 1999. Round all coefficients and answers to three significant digits.

38. Income The following table shows the share of all personal after-tax income of the wealthiest 1% of U.S. taxpayers for the years 1990–1998:

Year, t	0 (1990)	1	2	3	4	5	6	7	8 (1998)
Share of Income, P (%)	12	11	12	11	11	12	13.5	14.5	15.5

Data are approximate. Incomes are in constant 1998 dollars. SOURCE: IRS/*New York Times*, February 26, 2001, p. C2.

Find a quadratic regression model for these data. Graph the model together with the data. Assuming the trend continued, estimate the share of all personal after-tax income of the wealthiest 1% of U.S. taxpayers in 1999. Round all coefficients to three significant digits and the 1999 percentage to the nearest 1%.

COMMUNICATION AND REASONING EXERCISES

39. Suppose the graph of revenue as a function of unit price is a parabola that is concave down. What is the significance of the coordinates of the vertex, the x intercepts and the y intercept?

40. Suppose the height of a stone thrown vertically upward is given by a quadratic function of time. What is the significance of the coordinates of the vertex, the (possible) x intercepts, and the y intercept?

41. How might you tell, roughly, whether a set of data should be modeled by a quadratic rather than a linear equation?

42. A member of your study group tells you that, since the following set of data do not suggest a straight line, they are best modeled by a quadratic. Comment on her suggestion.

x	0	2	4	6	8
y	1	2	1	0	1

43. Explain why, if demand is a linear function of unit price p (with negative slope), then there must be a *single value of p* that results in the maximum revenue.

44. Explain why, if the average cost of a commodity is given by $y = 0.1x^2 - 4x - 2$, where x is the number of units sold, there is a single choice of x that results in the lowest possible average cost.

45. If the revenue function for a particular commodity is $R(p) = -50p^2 + 60p$, what is the (linear) demand function? Give a reason for your answer.

46. If the revenue function for a particular commodity is $R(p) = -50p^2 + 60p + 50$, can the demand function be linear? What is the associated demand function?

2.2 Exponential Functions and Models

The quadratic functions we discussed in Section 2.1 can be used to model many nonlinear phenomena. However, exponential functions give better models in some applications, including population growth, radioactive decay, the growth or depreciation of financial investments, and many other phenomena.

To manipulate exponential functions, we need to manipulate exponents. The following list, taken from the algebra review in the appendix, gives the laws of exponents we will be using.

The Laws of Exponents

If b and c are positive and x and y are any real numbers, then the following laws hold:

Law	Quick Example
1. $b^x b^y = b^{x+y}$	$2^3 2^2 = 2^5 = 32$
2. $\dfrac{b^x}{b^y} = b^{x-y}$	$\dfrac{4^3}{4^2} = 4^{3-2} = 4^1 = 4$
3. $b^{-x} = \dfrac{1}{b^x}$	$9^{-0.5} = \dfrac{1}{9^{0.5}} = \dfrac{1}{3}$
4. $b^0 = 1$	$(3.3)^0 = 1$
5. $(b^x)^y = b^{xy}$	$(2^3)^2 = 2^6 = 64$
6. $(bc)^x = b^x c^x$	$(4 \cdot 2)^2 = 4^2 2^2 = 64$
7. $\left(\dfrac{b}{c}\right)^x = \dfrac{b^x}{c^x}$	$\left(\dfrac{4}{3}\right)^2 = \dfrac{4^2}{3^2} = \dfrac{16}{9}$

Here are the functions we will study in this section.

Exponential Function

An **exponential function** has the form

$$f(x) = Ab^x \qquad \text{Technology: } \texttt{A*b\^{}x}$$

where A and b are constants with b positive. We call b the **base** of the exponential function.

Quick Examples

1. $f(x) = 2^x$ $A = 1, b = 2$; technology: $\texttt{2\^{}x}$

  $f(1) = 2^1 = 2$ $\texttt{2\^{}1}$

  $f(-3) = 2^{-3} = \dfrac{1}{8}$ $\texttt{2\^{}(-3)}$

  $f(0) = 2^0 = 1$ $\texttt{2\^{}0}$

2. $g(x) = 20(3^x)$ $A = 20, b = 3$; technology: `20*3^x`

$g(2) = 20(3^2) = 20(9) = 180$ `20*3^2`

$g(-1) = 20(3^{-1}) = 20\left(\dfrac{1}{3}\right) = 6\dfrac{2}{3}$ `20*3^(-1)`

3. $h(x) = 2^{-x} = \left(\dfrac{1}{2}\right)^x$ $A = 1, b = \dfrac{1}{2}$; technology: `2^(-x)` or `(1/2)^x`

$h(1) = 2^{-1} = \dfrac{1}{2}$ `2^(-1)` or `(1/2)^1`

$h(2) = 2^{-2} = \dfrac{1}{4}$ `2^(-2)` or `(1/2)^2`

4. $k(x) = 3 \cdot 2^{-4x} = 3(2^{-4})^x$ $A = 3, b = 2^{-4}$; technology: `3*2^(-4*x)`

$k(-2) = 3 \cdot 2^{-4(-2)}$ `3*2^(-4*(-2))`

$= 3 \cdot 2^8 = 3 \cdot 256 = 768$

Exponential Functions from the Numerical and Graphical Points of View

The following table shows values of $f(x) = 3(2^x)$ for some values of x ($A = 3, b = 2$):

x	-3	-2	-1	0	1	2	3
$f(x)$	$\dfrac{3}{8}$	$\dfrac{3}{4}$	$\dfrac{3}{2}$	3	6	12	24

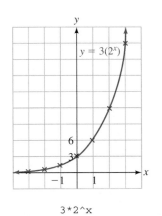

$y = 3(2^x)$

`3*2^x`

Figure 11

Its graph is shown in Figure 11. As we can see in the table, setting $x = 0$ gives $y = 3$, the value of A.

Numerically, A is the value of y when x = 0.

On the graph, the corresponding point $(0, 3)$ is the y intercept.

What about b? Notice from the table that the value of y is multiplied by $b = 2$ for every increase of 1 in x. If we decrease x by 1, the y coordinate gets *divided* by $b = 2$.

Numerically, y is multiplied by b for every 1-unit increase of x.

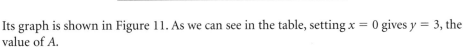

x	-3	-2	-1	0	1	2	3
$f(x)$	$\dfrac{3}{8}$	$\dfrac{3}{4}$	$\dfrac{3}{2}$	3	6	12	24

Multiply by 2.

On the graph, if we move 1 unit to the right from any point on the curve, the y coordinate doubles. Thus, the curve becomes dramatically steeper as the value of x increases. This phenomenon is called **exponential growth.**

Question How do we graph the function $f(x) = 3(2^x)$ using technology?

Answer All you need to know is the technology formula: `3*2^x` for this exponential function. If using Excel, see the procedure we developed in Section 1.2 for drawing smooth curves (or better yet, use the spreadsheet you saved from that section). If using the Web site, you can follow Web Site → Online Utilities and choose the Java Graphing Utility, the Excel Grapher, or the Function Evaluator and Grapher.

Example 1 • Recognizing Exponential Data Numerically and Graphically

Some of the values of two functions, f and g, are given in the following table:

x	-2	-1	0	1	2
$f(x)$	-7	-3	1	5	9
$g(x)$	$\frac{2}{9}$	$\frac{2}{3}$	2	6	18

One of these functions is linear, and the other is exponential. Which is which?

Solution   Remember that a linear function increases (or decreases) by the same amount every time x increases by 1. The values of f behave this way: Every time x increases by 1, the value of $f(x)$ increases by 4. Therefore, f is a linear function with a *slope* of 4. Since $f(0) = 1$, we see that

$$f(x) = 4x + 1$$

is a linear formula that fits the data.

On the other hand, every time x increases by 1, the value of $g(x)$ is *multiplied* by 3. Since $g(0) = 2$, we find that

$$g(x) = 2(3^x)$$

is an exponential function fitting the data.

We can visualize the two functions f and g by plotting the data points (Figure 12).

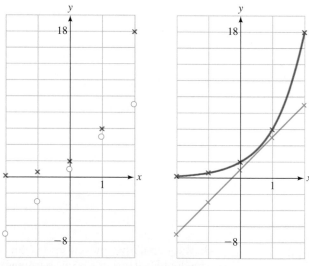

Figure 12

The data points for $f(x)$ clearly lie along a straight line, whereas the points for $g(x)$ lie along a curve, as shown in the graph on the right. The y coordinate of each point for $g(x)$ is three times the y coordinate of the preceding point, demonstrating that the curve is an exponential one.

Exponential Function Numerically and Graphically

For the exponential function $f(x) = Ab^x$,

Role of b

If x increases by 1, $f(x)$ is multiplied by b.

If x increases by 2, $f(x)$ is multiplied by b^2.

⋮

If x increases by Δx, $f(x)$ is multiplied by $b^{\Delta x}$.

Role of A

$f(0) = A$, so A is the y intercept of the graph of f.

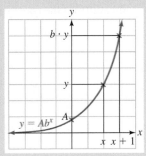

If x increases by 1, y is multiplied by b.

Quick Examples

1. $f_1(x) = 2^x$, $f_2(x) = 2^{-x}$

	A	B	C
1	x	2^x	2^(-x)
2	-3	1/8	8
3	-2	1/4	4
4	-1	1/2	2
5	0	1	1
6	1	2	1/2
7	2	4	1/4
8	3	8	1/8

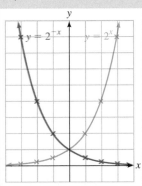

Note that $2^{-x} = \left(\dfrac{1}{2}\right)^x$. So, if x increases by 1, y is multiplied by $\dfrac{1}{2}$. The function $f_1(x) = 2^x$ illustrates exponential growth, and $f_2(x) = \left(\dfrac{1}{2}\right)^x$ illustrates the opposite phenomenon: **exponential decay.**

2. $f_1(x) = 2^x$, $f_2(x) = 3^x$, $f_3(x) = 1^x$

	A	B	C	D
1	x	2^x	3^x	1^x
2	-3	1/8	1/27	1
3	-2	1/4	1/9	1
4	-1	1/2	1/3	1
5	0	1	1	1
6	1	2	3	1
7	2	4	9	1
8	3	8	27	1

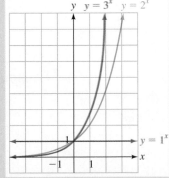

If x increases by 1, 3^x is multiplied by 3. Note also that all the graphs pass through $(0, 1)$. (Why?)

Fitting an Exponential Function to Two Data Points

In Section 1.3 we discussed a method for calculating the equation of the line that passes through two given points. In the following example, we find a similar method for calculating the equation of the exponential curve through two given points.

Example 2 • Finding the Exponential Curve through Two Points

Find an equation of the exponential curve through $(1, 6)$ and $(3, 24)$.

Solution   We want an equation of the form

$$y = Ab^x \qquad (b > 0)$$

Substituting the coordinates of the given points, we get

$$6 = Ab^1 \qquad\qquad \text{Substitute } (1, 6).$$

$$24 = Ab^3 \qquad\qquad \text{Substitute } (3, 24).$$

If we now divide the second equation by the first, we get

$$\frac{24}{6} = \frac{Ab^3}{Ab} = b^2$$

$$b^2 = 4$$

$$b = 2 \qquad\qquad \text{Since } b > 0$$

Now that we have b, we can substitute its value into the first equation to obtain

$$6 = 2A \qquad\qquad \text{Substitute } b = 2 \text{ into the equation } 6 = Ab.$$

$$A = 3$$

We have both constants, $A = 3$ and $b = 2$, so the model is

$$y = 3(2^x)$$

✸ **Before we go on . . .** Example 7 will show how to use technology to fit an exponential function to two or more data points.

APPLICATIONS

Recall some terminology we mentioned earlier: A quantity y experiences **exponential growth** if $y = Ab^t$ with $b > 1$. (Here we use t for the independent variable, thinking of time.) It experiences **exponential decay** if $y = Ab^t$ with $0 < b < 1$. Here are some examples of the numerous applications of exponential growth and decay.

Example 3 • Exponential Growth: Epidemics

In the early stages of the AIDS epidemic during the 1980s, the number of cases in the United States was increasing by about 50% every 6 months. By the start of 1983, there were approximately 1600 AIDS cases in the United States.[4]

a. Assuming an exponential growth model, find a function that predicts the number of people infected t years after the start of 1983.

---

[4]Data based on regression of the 1982–1986 figures. SOURCE: Centers for Disease Control and Prevention, *HIV/AIDS Surveillance Report, 2000;* 12 (no. 2).

b. Use the model to estimate the number of people infected by October 1, 1986, and by December 31, 1986.

Solution

a. We could use the method in Example 2 or reason as follows: At time $t = 0$ (January 1, 1983), the number of people infected was 1600, so $A = 1600$. Every 6 months, the number of cases increased to 150% of the number 6 months earlier—that is, to 1.50 times that number. Each year, it therefore increased to $(1.50)^2 = 2.25$ times the number one year earlier. Hence, after t years, we need to multiply the original 1600 by 2.25^t, so the model is

$$y = 1600(2.25^t) \text{ cases}$$

If instead we wish to use the method from Example 2, we need two data points. We are given one point: $(0, 1600)$. Since y increased by 50% every 6 months, 6 months later it reached $1600 + 800 = 2400$ $(t = 0.5)$. This information gives a second point: $(0.5, 2400)$. We can now apply the method in Example 2 to find the model above.

b. October 1, 1986, corresponds to $t = 3.75$ (since October 1 is 9 months, or $9/12 = 0.75$ of a year, after January 1). Substituting this value of t in the model gives

$$y = 1600(2.25^{3.75}) \approx 33,481 \text{ cases} \qquad \texttt{=1600*2.25\^{}3.75}$$

By the end of 1986, the model predicts that

$$y = 1600(2.25^4) = 41,006 \text{ cases}$$

(The actual number of cases was around 41,700.)

✸ **Before we go on ...** Increasing by 50% every 6 months couldn't continue for very long, and this is borne out by observations. If increasing by 50% every 6 months did continue, then by January, 2003 $(t = 20)$, the number of infected people would have been

$$1600(2.25^{20}) \approx 17,700,000,000$$

a number that is more than 50 times the size of the U.S. population! Thus, the exponential model is unreliable for predicting long-term trends.

Epidemiologists use more sophisticated models to measure the spread of epidemics, and these models predict a leveling-off phenomenon as the number of cases becomes a significant part of the total population. We discuss such a model, the **logistic function,** in Section 2.4. However, the exponential growth model *is* fairly reliable *in the early stages* of an epidemic.

Exponential functions arise in finance and economics mainly through the idea of **compound interest.** Suppose you invest $500 (the **present value**) in an investment account with an annual yield of 15%, and the interest is reinvested at the end of every year (we say that the interest is **compounded** once a year). Let t represent the number of years since you made the initial $500 investment. Each year, the investment is worth 115% (or 1.15 times) its value the previous year. The **future value** A of your investment changes over time t, so we think of A as a function of t. The following table illustrates how we can calculate the future value for several values of t:

t	0	1	2	3
Future Value, $A(t)$	500	575	661.25	760.44

$$\begin{array}{ccccc} A & 500(1.15) & 500(1.15)^2 & 500(1.15)^3 \\ & \times 1.15 & \times 1.15 & \times 1.15 \end{array}$$

Thus, $A(t) = 500(1.15)^t$. A traditional way to write this formula is

$$A(t) = P(1 + r)^t$$

where P is the present value ($P = 500$) and r is the annual interest rate ($r = 0.15$).

If, instead of compounding the interest once a year, we compound it every 3 months (four times a year), we would earn one-quarter of the interest ($r/4$) every 3 months. Since this would happen $4t$ times in t years, the formula for the future value becomes

$$A(t) = P\left(1 + \frac{r}{4}\right)^{4t}$$

Compound Interest

If an amount (**present value**) P is invested for t years at an annual rate of r and if the interest is compounded (reinvested) m times per year, then the **future value** A is

$$A(t) = P\left(1 + \frac{r}{m}\right)^{mt}$$

A special case is **interest compounded once a year:**

$$A(t) = P(1 + r)^t$$

Quick Example

If \$2000 is invested for $2\frac{1}{2}$ years in a mutual fund with an annual yield of 12.6% and the earnings are reinvested each month, then $P = 2000$, $r = 0.126$, $m = 12$, and $t = 2.5$, which gives

$$A(2.5) = 2000\left(1 + \frac{0.126}{12}\right)^{12 \times 2.5} \qquad \text{2000*(1+0.126/12)^(12*2.5)}$$

$$= 2000(1.0105)^{30} = \$2736.02$$

Example 4 • Compound Interest: Investments

Consider the scenario in the preceding Quick Example: You invest \$2000 in a mutual fund with an annual yield of 12.6%, and the interest is reinvested each month.

a. Find the associated exponential model.

b. Use the model to estimate the year when the value of your investment will reach \$5000.

Solution

a. Apply the formula

$$A(t) = P\left(1 + \frac{r}{m}\right)^{mt}$$

with $P = 2000$, $r = 0.126$, and $m = 12$. We get

$$A(t) = 2000\left(1 + \frac{0.126}{12}\right)^{12t}$$

$$= 2000(1.0105)^{12t} \qquad \text{2000*(1+0.126/12)^(12*t)}$$

This is the exponential model. (What would happen if we left out the last set of parentheses in the technology formula?)

b. We need to find the value of t for which $A(t) = \$5000$, so we need to solve the equation

$$5000 = 2000(1.0105)^{12t}$$

In Section 2.3 we will learn how to use logarithms to do this algebraically, but we can answer the question now using a graphing calculator, a spreadsheet, or the Function Evaluator and Grapher tool at the Web site. Just enter the model and compute the balance at the end of several years. Here is a table you might obtain in Excel:

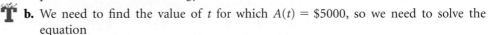

	A	B	C
1	t	A	
2	0	2000*(1+0.126/12)^(12*A2)	
3	1		
4	2		
5	3		
6	4		
7	5		
8	6		
9	7		
10	8		
11	9		

$\rightarrow$

	A	B
1	t	A
2	0	2000
3	1	2267.07459
4	2	2569.81361
5	3	2912.97957
6	4	3301.97098
7	5	3742.90726
8	6	4242.72498
9	7	4809.287
10	8	5451.50619
11	9	6179.48559

Since the balance first exceeds $5000 in year 8, the answer is $t = 8$ years.

Example 5 • Exponential Decay: Carbon Dating

Carbon-14, an unstable isotope of carbon, decays slowly to nitrogen. The amount of carbon-14 remaining in a sample that originally contained A grams is approximately

$$C(t) = A(0.999\ 879)^t$$

where t is time in years. A plant unearthed in an archaeological dig contains 0.50 gram of carbon-14 and is known to be 50,000 years old. How much carbon-14 did the plant originally contain?

Solution We are given the following information: $C = 0.50$, $A =$ the unknown, and $t = 50,000$. Substituting gives

$$0.50 = A(0.999\ 879)^{50,000}$$

Solving for A gives

$$A = \frac{0.5}{0.999\ 879^{50,000}} \approx 212 \text{ grams}$$

Before we go on ... The formula we used for A has the form

$$A(t) = \frac{C}{0.999\ 879^t}$$

which gives the original amount of carbon-14 t years ago in terms of the amount C that is left now. A similar formula can be used in finance to find the present value, given the future value.

The Number e and More Applications

In nature we find examples of growth that occurs *continuously,* as though "interest" is being added more often than every second or fraction of a second. To model this, we need to see what happens to our compound interest formula as we let m (the number of times interest is added per year) become extremely large. Something very interesting does happen: We end up with a more compact and elegant formula than we began with. To begin to see why, let's look at a very simple situation.

Suppose we invest \$1 in the bank for 1 year at 100% interest, compounded m times per year. If $m = 1$ then 100% interest is added every year, and so our money doubles at the end of the year. In general, the accumulated capital at the end of the year is

$$A = 1\left(1 + \frac{1}{m}\right)^m = \left(1 + \frac{1}{m}\right)^m$$

(1+1/m)^m

Now, we are interested in what A becomes for large values of m. We can make a chart that shows how this quantity behaves as m increases:

m	1	10	100	1,000	10,000	100,000	1,000,000	10,000,000
$\left(1 + \dfrac{1}{m}\right)^m$	2	2.593 742 46	2.704 813 83	2.716 923 93	2.718 145 93	2.718 268 24	2.718 280 47	2.718 281 69

Something interesting *does* seem to be happening! The numbers appear to be getting closer and closer to a specific value. In mathematical terminology we say that the numbers **converge** to a fixed number, 2.718 28 . . . , called the **limiting value** of the quantities $\left(1 + \frac{1}{m}\right)^m$. This number, called e, is one of the most important in mathematics. The number e is irrational, just as the more familiar number π is, so we cannot write down its exact numerical value. To 20 decimal places, $e = 2.718\ 281\ 828\ 459\ 045\ 235\ 36. . . .$

We now say that, if \$1 is invested for 1 year at 100% interest **compounded continuously,** the accumulated money at the end of that year will amount to \$$e$ = \$2.72 (to the nearest cent). But what about the following more general question?

Question   Suppose we invest an amount P for t years at an interest rate of r, compounded continuously. What will be the accumulated amount A at the end of that period?

Answer   In the special case above (P, t, and r all equal to 1), we took the compound interest formula and let m get larger and larger. We do the same more generally, after a little preliminary work with the algebra of exponentials:

$$A = P\left(1 + \frac{r}{m}\right)^{mt}$$

$$= P\left(1 + \frac{1}{\frac{m}{r}}\right)^{mt} \qquad \text{Substitute } \frac{r}{m} = \frac{1}{\frac{m}{r}}.$$

$$= P\left(1 + \frac{1}{\frac{m}{r}}\right)^{(m/r)rt} \qquad \text{Substitute } m = \left(\frac{m}{r}\right)r.$$

$$= P\left[\left(1 + \frac{1}{\frac{m}{r}}\right)^{(m/r)}\right]^{rt} \qquad \text{Use the rule } a^{bc} = (a^b)^c.$$

For continuous compounding of interest, we let m, and hence m/r, get very large. This affects only the term in brackets, which converges to e, and we get the formula

$$A = Pe^{rt}$$

Question How do I get powers of e on a graphing calculator or in Excel?

Answer On the TI-83, enter e^x as `e^(x)`, where `e^(` can be obtained by pressing $\boxed{\text{2nd}}$ $\boxed{\text{LN}}$. Excel has a built-in function called EXP; EXP(x) gives the value of e^x.

The Number e, Continuous Compounding, and Continuous Growth

The number e is the limiting value of the quantities $\left(1 + \dfrac{1}{m}\right)^m$ as m gets larger and larger and has the value 2.718 281 828 459 045 235 36. . . . To obtain the number e on the TI-83, enter `e^(1)`. In Excel, enter `=EXP(1)`.

If \$$P$ is invested at an annual interest rate r compounded continuously, the accumulated amount after t years is

$$A(t) = Pe^{rt} \qquad\qquad \texttt{P*e\^(r*t)} \quad \text{or} \quad \texttt{P*EXP(r*t)}$$

Quick Examples

1. If \$100 is invested in an account that bears 15% interest compounded continuously, at the end of 10 years the investment will be worth

$$A(10) = 100e^{(0.15)(10)} = \$448.17 \quad \texttt{100*e\^(0.15*10)} \;\text{or}\; \texttt{100*EXP(0.15*10)}$$

2. If \$1 is invested in an account that bears 100% interest compounded continuously, at the end of x years the investment will be worth

$$A(x) = e^x \text{ dollars.}$$

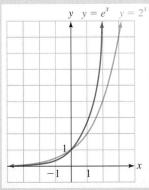

Graph of $A(x) = e^x$

Technology formula: `e^(x)` **or** `EXP(x)`

Note If we write $A(t) = P(e^r)^t$, we see that $A(t)$ is an exponential function of t, where the base is $b = e^r$, so we have really not introduced a new kind of function. As we will see in Chapter 4, the exponential function with base e exhibits some interesting properties when we measure its rate of change, and this is the real mathematical importance of e.

Example 6 • Continuous Compounding

a. You invest \$10,000 at Fastrack Savings & Loan, which pays 6% compounded continuously. Express the balance in your account as a function of the number of years t and calculate the amount of money you will have after 5 years.

b. Your friend has just invested \$10,000 in Constant Growth Funds, whose stocks are continuously declining at a rate of 6% per year. How much will her investment be worth in 5 years?

Solution

a. We use the continuous growth formula with $P = 10,000$, $r = 0.06$, and t variable, getting

$$A(t) = Pe^{rt} = 10,000e^{0.06t}$$

In 5 years,

$$A(5) = 10,000e^{0.06(5)} = 10,000e^{0.3} \approx \$13,498.59$$

b. Since the investment is depreciating, we use a negative value for r and take $P = 10,000$, $r = -0.06$, and $t = 5$, getting

$$A(t) = Pe^{rt} = 10,000e^{-0.06t}$$

$$A(5) = 10,000e^{-0.06(5)} = 10,000e^{-0.3} \approx \$7408.18$$

✳ **Before we go on . . .**

Question How does continuous compounding compare with monthly compounding?

Answer To repeat the calculation in part (a) using monthly compounding instead of continuous compounding, we use the compound interest formula with $P = 10,000$, $r = 0.06$, $m = 12$, and $t = 5$ and find

$$A(5) = 10,000\left(1 + \frac{0.06}{12}\right)^{60} \approx \$13,488.50$$

Thus, continuous compounding earns you approximately \$10 more than monthly compounding. On a 5-year, \$10,000 investment, this is little to get excited about.

Exponential Regression

Starting with a set of data that suggests an exponential curve, we can use technology to compute the exponential regression curve in much the same way as we did for the quadratic regression curve in Example 5 of Section 2.1.

⊤ Example 7 • Exponential Regression: Dot Com Job Cuts

The following table shows monthly job cuts at Internet companies in the period January 2000–January 2001 ($t = 0$ represents January 2000):

Month, t	0	2	4	6	8	10	12
Job Cuts	200	25	2500	2000	4500	8500	12,500

Data are approximate. SOURCE: Challanger, Gray & Christmas/*New York Times*, February 21, 2001, p. C1.

a. Find the exponential regression model

$$N(t) = Ab^t$$

for the number of job cuts.

b. Use this regression model to estimate the number of job cuts in October 2000 ($t = 9$); the actual figure was approximately 5500.

Solution

a. We will use technology to answer this part of the example.

Graphing Calculator

To find the equation of the regression exponential curve using a TI-83, enter the coordinates of the data points in the lists L_1 and L_2 as in Example 5 of Section 2.1, press STAT, select CALC, and choose the option ExpReg. Pressing ENTER gives the regression equation in the home screen:

$$N(t) \approx 101.59(1.5530)^t$$

The data points together with the regression curve are shown in Figure 13.

Excel

Start with a scatter plot of the observed data as in Example 5 of Section 2.1, click on the chart, and select Add Trendline... from the Chart menu. Then select the Exponential type and, under Options, check the option Display equation on chart. The result will appear as in Figure 13.

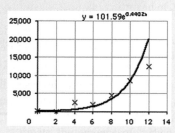

Figure 13

Notice that the regression curve is given in the form Ae^{kt} rather than Ab^t. To transform it, write

$$101.59e^{0.4402t} = 101.59(e^{0.4402})^t = 101.59(1.5530)^t$$

Web Site

Follow the path

Web Site → Online Utilities → Simple Regression

enter the data as x and y coordinates, and press the y=a(b^x) button to find the regression equation. You can then graph the curve and data points by pressing graph.

b. Using the model $N(t) \approx 101.59(1.5530)^t$, we find that

$$N(9) \approx 101.59(1.5530)^9 \approx 5338$$

which is close to the actual number of around 5500.

✴ Before we go on... We said in Section 2.1 that the regression curve gives the smallest value of the sum-of-squares error, SSE (the sum of the squares of the residues). However, exponential regression as computed via technology generally minimizes the sum of the squares of the residues of the *logarithms* (logarithms are discussed in Section 2.3).

69. Aspirin Soon after taking an aspirin, a patient has absorbed 300 milligrams of the drug. If the amount of aspirin in the bloodstream decays exponentially, with half being removed every 2 hours, find the amount of aspirin in the bloodstream after 5 hours.

70. Alcohol After several drinks, a person has a blood alcohol level of 200 mg/dL (milligrams per deciliter). If the amount of alcohol in the blood decays exponentially, with one-fourth being removed every hour, find the person's blood alcohol level after 4 hours.

71. Profit South African Breweries (SAB) reported profits of $360 million in 1997 ($t = 0$) and $480 million in 2000 ($t = 3$). Use this information to find (a) a linear model and (b) an exponential model for SAB's profit P as a function of time t since 1997. (Round all coefficients to four decimal places.) Which, if either, of these models would you judge to be applicable to the data shown below?

t	0 (1997)	1	2	3	4 (2001)
Profit ($ millions)	360	380	320	480	360

SOURCE: South African Breweries corporate Web site, http://www.sab.co.za/investfr.asp, April 2002.

72. Assets South African Breweries (SAB) reported fixed assets of R9500 million in 1997 ($t = 0$) and R29,300 million in 2001 ($t = 4$). Use this information to find (a) a linear model and (b) an exponential model for SAB's fixed assets as a function of time t since 1997. (Round all coefficients to four significant digits.) Which, if either, of these models would you judge to be applicable to the data shown below?

t	0 (1997)	1	2	3	4 (2001)
Fixed Assets (millions of rand, R)	9500	11,100	16,100	22,900	29,300

SOURCE: South African Breweries corporate Web site, http://www.sab.co.za/investfr.asp, April 2002.

73. Frogs Frogs in Nassau County have been breeding like flies! Each year, the pledge class of Epsilon Delta is instructed by the brothers to tag all the frogs residing on the ESU campus (Nassau County Branch) as an educational exercise. Two years ago they managed to tag all 50,000 of them (with little Epsilon Delta Fraternity tags). This year's pledge class discovered that last year's tags had all fallen off, and they wound up tagging a total of 75,000 frogs.
   a. Find an exponential model for the frog population.
   b. Assuming exponential population growth and that all this year's tags have fallen off, how many tags should Epsilon Delta order for next year's pledge class?

74. Flies Flies in Suffolk County have been breeding like frogs! Three years ago the Health Commission caught 4000 flies in 1 hour in a trap. This year it caught 7000 flies in 1 hour.
   a. Find an exponential model for the fly population.

b. Assuming exponential population growth, how many flies should the commission expect to catch next year?

75. U.S. Population The U.S. population was 180 million in 1960 and 281 million in 2000.
   a. Use these data to give an exponential growth model showing the U.S. population P as a function of time t in years since 1960.
   b. Assuming exponential growth, what will be the population in 2020?

Figures are rounded to three significant digits. SOURCE: *Statistical Abstract of the United States,* U.S. Census Bureau, 2001.

76. World Population World population was estimated at 2.56 billion people in 1950 and 6.23 billion people in 2002.
   a. Use these data to give an exponential growth model showing the world population P as a function of time t in years since 1950.
   b. Assuming exponential growth, what was the world population in A.D. 1000? Comment on your answer.

Figures are rounded to three significant digits. SOURCE: U.S. Census Bureau, *International Data Base,* September 2000.

77. Investments Rock Solid Bank & Trust is offering a CD (certificate of deposit) that pays 4% compounded continuously. How much interest would a $1000 deposit earn over 10 years?

78. Savings FlybynightSavings.com is offering a savings account that pays 31% interest compounded continuously. How much interest would a deposit of $2000 earn over 10 years?

79. Global Warming The most abundant greenhouse gas is carbon dioxide. According to a United Nations worst-case scenario prediction, the amount of carbon dioxide in the atmosphere (in parts of volume per million) can be approximated by

$$C(t) \approx 277e^{0.00353t} \qquad (0 \le t \le 350)$$

where t is time in years since 1750.
   a. Use the model to estimate the amount of carbon dioxide in the atmosphere in 1950, 2000, 2050, and 2100.
   b. According to the model, when, to the nearest decade, will the level surpass 700 parts per million?

Exponential regression based on the 1750 figure and the 2100 U.N. prediction. SOURCES: Tom Boden/Oak Ridge National Laboratory, Scripps Institute of Oceanography/University of California, International Panel on Climate Change/*New York Times,* December 1, 1997, p. F1.

80. Global Warming Repeat Exercise 79 using the U.N. midrange scenario prediction:

$$C(t) \approx 277e^{0.00267t} \qquad (0 \le t \le 350)$$

where t is time in years since 1750.

In Exercises 81–84, use technology to solve.

81. Camera Sales The following table shows the number of cameras sold by Polaroid in the years 1994–1998:

t	0 (1994)	1	2	3	4 (1998)
Cameras Sold (millions)	6.3	5.2	5.0	5.0	4.7

SOURCES: Polaroid, Bloomberg Financial Markets/*New York Times,* March 27, 2000, p. C14.

a. Use exponential regression to model the number of cameras sold by Polaroid as a function of time t since 1994. Include a sketch of the points and the regression curve. (Round the coefficients to three decimal places.)

b. Extrapolate your model to "predict" the number of cameras sold by Polaroid in 1999 (to the nearest 0.1 million cameras).

c. Polaroid actually sold 10 million cameras in 1999. The difference between the predicted value and the actual value is the *residual*. What is the 1999 ($t = 5$) residual, and how does it compare with those of the other years? Comment on the result.

82. **Video-Game Sales** The following table gives Nintendo's revenue from Game Boy video-games in the years 1996–1999:

t	0 (1996)	1	2	3 (1999)
Game Boy Revenue ($ billions)	0.18	0.2	0.44	1.2

SOURCE: Gerard Klauer Mattison & Company/*New York Times*, March 25, 2000, p. C1.

a. Use exponential regression to model the revenue from Game Boy sales as a function of time t since 1996. Include a sketch of the points and the regression curve. (Round the coefficients to three decimal places.)

b. Extrapolate your model to "predict" the revenue from Game Boy sales in 2000 (to the nearest $0.1 billion).

c. Revenues from Game Boy sales in 2000 were actually $0.98 billion. Recall that the difference between the predicted value and the actual value is the residual. What is the 2000 ($t = 4$) residual, and how does it compare with those of the other years?

83. **Grants** The following table shows the annual spending on grants by U.S. foundations from 1976 to 2001:

t	0 (1976)	5	10	15	20	25 (2001)
Spending ($ billions)	6	7	10	12	15	29

Figures are rounded and adjusted for inflation. SOURCE: The Foundation Center/*New York Times*, April 2, 2002, p. A21.

a. Use exponential regression to model the annual spending on grants by U.S. foundations as a function of time in years since 1976 and graph the data points and regression curve. (Round coefficients to four decimal places.)

b. According to your model, by what annual percentage has spending on grants by U.S. foundations been increasing over the period shown?

c. Use your model to estimate 1994 spending to the nearest $1 billion.

84. **Foundations** The following table shows the total number of active grant-making foundations in the United States from 1975 to 2000:

t	0 (1975)	5	10	15	20	25 (2000)
Foundations (thousands)	22	22	25	32	40	57

Figures are rounded. SOURCE: The Foundation Center/*New York Times*, April 2, 2002, p. A21.

a. Use exponential regression to model the number of active grant-making foundations in the United States as a function of time in years since 1975 and graph the data points and regression curve. (Round coefficients to four decimal places.)

b. According to your model, by what annual percentage has the number of grant-making foundations been increasing over the period shown?

c. Use your model to estimate, to the nearest 1000, the number of active grant-making foundations in 1994.

COMMUNICATION AND REASONING EXERCISES

85. Which of the following three functions will be largest for large values of x?
    (A) $f(x) = x^2$ **(B)** $r(x) = 2^x$ **(C)** $h(x) = x^{10}$

86. Which of the following three functions will be smallest for large values of x?
    (A) $f(x) = x^{-2}$ **(B)** $r(x) = 2^{-x}$ **(C)** $h(x) = x^{-10}$

87. What limitations apply to using an exponential function to model growth in real-life situations? Illustrate your answer with an example.

88. Describe two real-life situations in which a linear model would be more appropriate than an exponential model and two situations in which an exponential model would be more appropriate than a linear model.

89. Describe a real-life situation in which a quadratic model would be more appropriate than an exponential model and one in which an exponential model would be more appropriate than a quadratic model.

90. Explain in words why 5% per year compounded continuously yields more interest than 5% per year compounded monthly.

91. How would you check whether data points of the form $(1, y_1), (2, y_2), (3, y_3)$ lie on an exponential curve?

92. You are told that the points $(1, y_1), (2, y_2), (3, y_3)$ lie on an exponential curve. Express y_3 in terms of y_1 and y_2.

93. Your local banker tells you that the reason his bank doesn't compound interest continuously is that it would be too demanding of computer resources because the computer would need to spend a great deal of time keeping all accounts updated. Comment on his reasoning.

94. Your other local banker tells you that the reason her bank doesn't offer continuously compounded interest is that it is equivalent to offering a fractionally higher interest rate compounded daily. Comment on her reasoning.

2.3 *Logarithmic Functions and Models*

Logarithms were invented by John Napier (1550–1617) in the late 16th century as a means of aiding calculation. His invention made possible the prodigious hand calculations of astronomer Johannes Kepler (1571–1630), who was the first to describe accurately the orbits and the motions of the planets. Today, computers and electronic calculators have done away with that use of logarithms, but many other uses remain. In particular, the logarithm is a key tool for manipulating exponential models.

From the equation

$$2^3 = 8$$

we can see that the power to which we need to raise 2 in order to get 8 is 3. We abbreviate the phrase "the power to which we need to raise 2 in order to get 8" as "$\log_2 8$." Thus, another way of writing the equation $2^3 = 8$ is

$$\log_2 8 = 3 \qquad \text{The power to which we need to raise 2 in order to get 8 is 3.}$$

This is read "the base 2 logarithm of 8 is 3" or "the log, base 2, of 8 is 3."

Here is the general definition.

Base b Logarithm

The **base b logarithm of x,** $\log_b x$, is the power to which we need to raise b in order to get x. Symbolically,

$$\log_b x = y \qquad \text{means} \qquad b^y = x$$
Logarithmic form $\qquad\qquad$ Exponential form

Note

The number $\log_b x$ is defined only if b and x are both positive and $b \neq 1$. Thus, it is impossible to compute, say, $\log_3(-9)$ (since there is no power of 3 that equals -9) or $\log_1(2)$ (since there is no power of 1 that equals 2).

Quick Examples

1. The following table lists some exponential equations and their equivalent logarithmic forms:

Exponential Form	$10^3 = 1000$	$4^2 = 16$	$3^3 = 27$	$5^1 = 5$	$7^0 = 1$	$4^{-2} = \frac{1}{16}$	$25^{1/2} = 5$
Logarithmic Form	$\log_{10} 1000 = 3$	$\log_4 16 = 2$	$\log_3 27 = 3$	$\log_5 5 = 1$	$\log_7 1 = 0$	$\log_4 \frac{1}{16} = -2$	$\log_{25} 5 = \frac{1}{2}$

2. $\log_3 9 =$ the power to which we need to raise 3 in order to get 9. Since $3^2 = 9$, this power is 2, so $\log_3 9 = 2$.

3. $\log_{10} 10{,}000 =$ the power to which we need to raise 10 in order to get 10,000. Since $10^4 = 10{,}000$, this power is 4, so $\log_{10} 10{,}000 = 4$.

4. $\log_3 \frac{1}{27}$ is the power to which we need to raise 3 in order to get $\frac{1}{27}$. Since $3^{-3} = \frac{1}{27}$, this power is -3, so $\log_3 \frac{1}{27} = -3$.

5. $\log_b 1 = 0$ for every positive number b other than 1 because $b^0 = 1$.

Using Technology

To compute $\log_b x$ using technology, use the following formulas (see "Change-of-Base Formula" on page 112 for an explanation of the first one):

TI-83: `log(x)/log(b)` $\qquad$ Example: $\log_2(16)$ is `log(16)/log(2)`.
Excel: `=LOG(x,b)` $\qquad$ Example: $\log_2(16)$ is `=LOG(16,2)`.

As we mentioned earlier, one important use for logarithms is to solve equations in which the unknown appears in the exponent, as in Example 1.

Example 1 • Solving Equations with Unknowns in the Exponent

Solve the following equations:

a. $5^{-x} = 125$ **b.** $3^{2x-1} = 6$ **c.** $100(1.005)^{3x} = 200$

Solution

a. Write the given equation $5^{-x} = 125$ in logarithmic form:

$$-x = \log_5 125$$

This gives $x = -\log_5 125 = -3$.

b. In logarithmic form, $3^{2x-1} = 6$ becomes

$$2x - 1 = \log_3 6$$

$$2x = 1 + \log_3 6$$

$$x = \frac{1 + \log_3 6}{2} \approx \frac{1 + 1.6309}{2} \approx 1.3155$$

c. We cannot write the given equation, $100(1.005)^{3x} = 200$, directly in exponential form. We must first divide both sides by 100:

$$1.005^{3x} = \frac{200}{100} = 2$$

$$3x = \log_{1.005} 2$$

$$x = \frac{\log_{1.005} 2}{3} \approx \frac{138.9757}{3} \approx 46.3252$$

Logarithms with base 10 and base e are frequently used, so they have special names and notations.

Common Logarithm and Natural Logarithm
The following are standard abbreviations.

		TI-83 and Excel Formula
Base 10: $\log_{10} x = \log x$	Common logarithm	`log(x)`
Base e: $\log_e x = \ln x$	Natural logarithm	`ln(x)`

Quick Examples

Logarithmic Form

1. $\log 10{,}000 = 4$

2. $\log 10 = 1$

3. $\log \dfrac{1}{10{,}000} = -4$

4. $\ln e = 1$

5. $\ln 1 = 0$

6. $\ln 2 = 0.693\,147\,18\ldots$

Exponential Form

$10^4 = 10{,}000$

$10^1 = 10$

$10^{-4} = \dfrac{1}{10{,}000}$

$e^1 = e$

$e^0 = 1$

$e^{0.69314718\ldots} = 2$

Change-of-Base Formula

Some technologies (such as calculators) do not permit direct calculation of logarithms other than common and natural logarithms. To compute logarithms with other bases with these technologies, we can use the following formulas.

$$\log_b a = \frac{\log a}{\log b} = \frac{\ln a}{\ln b} \qquad \text{Change-of-base formula}$$

For example, to find $\log_{3.45} 2.261$, we divide $\log 2.261$ by $\log 3.45$, getting 0.6588 (to four significant digits). We get the same answer by dividing $\ln 2.261$ by $\ln 3.45$. (Try it.)*

Quick Examples

1. $\log_{11} 9 = \dfrac{\log 9}{\log 11} \approx 0.916\ 31$ `log(9)/log(11)` or `ln(9)/ln(11)`

2. $\log_{3.2}\left(\dfrac{1.42}{3.4}\right) \approx -0.750\ 65$ `log(1.42/3.4)/log(3.2)`

---

*Here is a quick explanation of why this formula works: To calculate $\log_b a$, we ask, "To what power must we raise b to get a?" To check the formula, we try using $\log a/\log b$ as the exponent:

$$b^{\log a/\log b} = (10^{\log b})^{\log a/\log b} \qquad \text{Since } b = 10^{\log b}$$
$$= 10^{\log a} = a$$

so this exponent works!

Question The title of this section is "Logarithmic Functions and Models." What exactly is a logarithmic function?

Answer Recall that an exponential function has the form $f(x) = Ab^x$. Analogously, a **logarithmic function** is a function of the form $f(x) = \log_b x + C$.

Logarithmic Function

A **logarithmic function** has the form

$$f(x) = \log_b x + C$$

Quick Examples

1. $f(x) = \log x$ \qquad\qquad $b = 10, C = 0$

2. $g(x) = \ln x - 5$ \qquad\qquad $b = e, C = -5$

Alternative Form

We can write any logarithmic function in the following alternative form:

$$f(x) = A \ln x + C$$

Quick Examples

1. $f(x) = 4.3 \ln x - 6.3$ \qquad\qquad $A = 4.3, C = -6.3$

2. $g(x) = \ln x - 5$ \qquad\qquad $A = 1, C = -5$

Question What is the difference between the two forms of the logarithmic function?

Answer None, really, they're equivalent: We can start with an equation in the first form and use the change-of-base formula to rewrite it:

$$f(x) = \log_b x + C$$

$$= \frac{\ln x}{\ln b} + C \qquad \text{Change-of-base formula}$$

$$= \left(\frac{1}{\ln b}\right) \ln x + C$$

Our function now has the form $f(x) = A \ln x + C$, where $A = \frac{1}{\ln b}$. We can go the other way as well, to rewrite $A \ln x + C$ in the form $\log_b x + C$.

Example 2 • Graphs of Logarithmic Functions

a. Sketch the graph of $f(x) = \log_2 x$ by hand.

b. Use technology to compare the graph in part (a) with the graphs of $\log_b x$ for $b = \frac{1}{4}$, $\frac{1}{2}$, and 4.

Solution

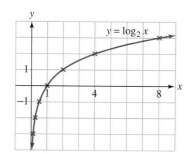

Figure 14

a. To sketch the graph of $f(x) = \log_2 x$ by hand, we begin with a table of values. Since $\log_2 x$ is not defined when $x = 0$, we choose several values of x close to zero and also some larger values, all chosen so that their logarithms are easy to compute:

x	$\frac{1}{8}$	$\frac{1}{4}$	$\frac{1}{2}$	1	2	4	8
$f(x) = \log_2 x$	-3	-2	-1	0	1	2	3

Graphing these points and joining them by a smooth curve gives us Figure 14.

b. We enter the logarithmic functions in graphing utilities as follows (note the use of the change-of-base formula in the TI-83 version):

TI-83
```
Y₁=log(X)/log(0.25)
Y₂=log(X)/log(0.5)
Y₃=log(X)/log(2)
Y₄=log(X)/log(4)
```

Excel
```
=LOG(x,0.25)
=LOG(x,0.5)
=LOG(x,2)
=LOG(x,4)
```

Figure 15 shows the resulting graphs.

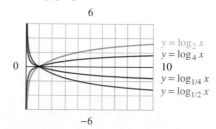

Figure 15

Before we go on... Notice that the graphs all pass through the point $(1, 0)$. (Why?) Notice further that the graphs of the logarithmic functions with bases less than 1 are upside-down versions of the others. Finally, how are these graphs related to the graphs of exponential functions?

The following lists some important algebraic properties of logarithms.

Logarithmic Identities

The following identities hold for all positive bases $a \neq 1$ and $b \neq 1$, all positive numbers x and y, and every real number r. These identities follow from the laws of exponents.

Identity   **Quick Example**

1. $\log_b(xy) = \log_b x + \log_b y$ $\log_2 16 = \log_2 8 + \log_2 2$

2. $\log_b\left(\dfrac{x}{y}\right) = \log_b x - \log_b y$ $\log_2\left(\dfrac{5}{3}\right) = \log_2 5 - \log_2 3$

3. $\log_b(x^r) = r \log_b x$ $\log_2(6^5) = 5 \log_2 6$

4. $\log_b b = 1; \log_b 1 = 0$ $\log_2 2 = 1; \ln e = 1; \log_{11} 1 = 0$

5. $\log_b\left(\dfrac{1}{x}\right) = -\log_b x$ $\log_2\left(\dfrac{1}{3}\right) = -\log_2 3$

6. $\log_b x = \dfrac{\log_a x}{\log_a b}$ $\log_2 5 = \dfrac{\log_{10} 5}{\log_{10} 2} = \dfrac{\log 5}{\log 2}$

Relationship with Exponential Functions

The following two identities demonstrate that the operations of taking the base b logarithm and raising b to a power are "inverse" to each other.*

Identity   **Quick Example**

1. $\log_b(b^x) = x$ $\log_2(2^7) = 7$

    In words: The power to which you raise b in order to get b^x is x (!).

2. $b^{\log_b x} = x$ $5^{\log_5 8} = 8$

    In words: Raising b to the power to which it must be raised to get x, yields x (!).

---

*See the online topic on inverse functions mentioned in "Internet Topics."

 ## (1) Using and Deriving Algebraic Properties of Logarithms and (2) Inverse Functions (Online Optional Section)

At the Web site, you will find a list of the logarithmic identities you can print out, some exercises on using the identities, and a discussion on where they come from. Follow the path

    Web Site → Everything for Calculus → Chapter 2 → Using and Deriving Algebraic Properties of Logarithms

    For a general discussion of inverse functions, including further discussion of the relationship between logarithmic and exponential functions, follow the path

    Web Site → Everything for Calculus → Chapter 2 → Inverse Functions

Fitting a Logarithmic Function to Data

Example 3 • Finding the Logarithmic Curve through Two Points

Find an equation of the logarithmic curve through the points $(3, 5)$ and $(27, 7)$.

Solution We want an equation of the form

$$y = \log_b x + C \quad (b > 0)$$

Substituting the coordinates of the given points, we get

$$5 = \log_b(3) + C$$

$$7 = \log_b(27) + C$$

If we now subtract the first equation from the second, the C's cancel and we get

$$2 = \log_b(27) - \log_b(3)$$

$$2 = \log_b\left(\frac{27}{3}\right) \qquad \text{Logarithmic identity (2)}$$

$$2 = \log_b(9)$$

So,

$$b^2 = 9 \qquad \text{Exponential form}$$

$$b = 3 \qquad \text{Since } b > 0$$

Now that we have b, we can substitute its value into the first equation to obtain

$$5 = \log_3(3) + C \qquad \text{Substitute } b = 3 \text{ in the equation } 5 = \log_b(3) + C.$$

$$5 = 1 + C$$

$$C = 4$$

Now we have both constants: $C = 4$ and $b = 3$, so the model is

$$y = \log_3 x + 4$$

✸ ***Before we go on...*** See Example 6 for a discussion of the use of technology to fit a logarithmic function to two or more data points.

APPLICATIONS

Example 4 • Investments: How Long?

Ten-year government bonds in Italy are yielding an average of 5.2% per year.[5] At that interest rate, how long will it take a €1000 investment to be worth €1500 if the interest is compounded monthly?

Solution Substituting $A = 1500$, $P = 1000$, $r = 0.052$, and $m = 12$ in the compound interest equation gives

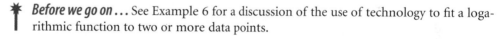

$$1500 = 1000\left(1 + \frac{0.052}{12}\right)^{12t}$$

$$\approx 1000(1.004\ 333)^{12t}$$

---

[5] In 2001. SOURCE: ECB, Reuters, and BE.

and we must solve for t. We first divide both sides by 1000, getting an equation in exponential form:

$$1.5 = 1.004\ 333^{12t}$$

In logarithmic form, this becomes

$$12t = \log_{1.004333}(1.5)$$

We can now solve for t:

$$t = \frac{\log_{1.004333}(1.5)}{12} \approx 7.8 \text{ years} \qquad \texttt{log(1.5)/(log(1.004333)*12)}$$

Thus, it will take approximately 7.8 years for a €1000 investment to be worth €1500.

Example 5 • Half-Life

a. The weight of carbon-14 that remains in a sample that originally contained A grams is given by

$$C(t) = A(0.999\ 879)^t$$

where t is time in years. Find the **half-life**, the time it takes half of the carbon-14 in a sample to decay.

b. Another radioactive material has a half-life of 7000 years. How long will it take for 99.95% of the substance in a sample to decay?

Solution

a. We want to find the value of t for which $C(t) = $ the weight of undecayed carbon-14 left = half the original weight = $0.5A$. Substituting, we get

$$0.5A = A(0.999\ 879)^t$$

Dividing both sides by A gives

$$0.5 = 0.999\ 879^t \qquad\qquad\qquad \text{Exponential form}$$

$$t = \log_{0.999879} 0.5 \approx 5728 \text{ years} \qquad \text{Logarithmic form}$$

b. This time we are given the half-life, which we can use to find the exponential model. Since half of the sample decays in 7000 years, a sample of A grams now ($t = 0$) will decay to $0.5A$ grams in 7000 years ($t = 7000$). This gives two data points: $(0, A)$ and $(7000, 0.5A)$. We can use the method in Example 2 of Section 2.2 to find

$$y = A(0.5)^{t/7000}$$

(Actually, we would get $y = Ab^t$ where $b = 0.5^{1/7000}$, but we can rewrite this formula as above if we do not evaluate $0.5^{1/7000}$.) Now that we have the model, we can answer the question by substituting

$$y = \text{amount of } \textit{undecayed} \text{ material left} = 0.05\% \text{ of the original amount} = 0.0005A$$

We have

$$0.0005A = A(0.5)^{t/7000}$$

Dividing both sides by A gives

$$0.0005 = 0.5^{t/7000}$$

$$\frac{t}{7000} = \log_{0.5}(0.0005)$$

$$t = 7000 \log_{0.5}(0.0005) \approx 76,760 \text{ years}$$

Logarithmic Regression

If we start with a set of data that suggests a logarithmic curve we can, by repeating the methods from previous sections, use technology to find the logarithmic regression curve $y = \log_b x + C$ approximating the data.

Example 6 • Research and Development

The following table shows the total spent in the United States on research and development in areas other than defense for the period 1985–2000 ($t = 5$ represents 1985):

Year, t	5	6	7	8	9	10	11	12
Spending ($ billions)	25	24	26	27	30	32	34	36
Year, t	13	14	15	16	17	18	19	20
Spending ($ billions)	35	37	37	36	37	38	40	41

Data are approximate and are given in constant 2001 dollars. SOURCES: Economic Research Service; U.S. Dept. of Agriculture, American Association for the Advancement of Science, "National Plant Breeding Study 1996" by Dr. Kenneth J. Frey, Iowa State University/*New York Times*, May 15, 2001, p. F1.

Find the best-fit logarithmic model of the form

$$S(t) = A \ln t + C$$

and use the model to project total spending on research in 2005, assuming the trend continues.

Solution

Graphing Calculator
To find the equation of the logarithmic regression curve using a TI-83, enter the coordinates of the data points in the lists L_1 and L_2, press $\boxed{\text{STAT}}$, select CALC, and choose the option LnReg. Pressing $\boxed{\text{ENTER}}$ gives us the regression equation in the home screen:

$$S(t) = 12.457 \ln t + 2.951$$

Plotting the data and regression curve will give us the graph shown in Figure 16.
Since 2005 is represented by $t = 25$, we have

$$S(25) = 12.457 \ln(25) + 2.951 \approx 43 \qquad \text{Why did we round the result to two significant digits?}$$

So, research and development spending is predicted to be around $43 billion in 2005.

Excel

We start, as usual, with a scatter plot of the observed data and add a `Logarithmic` trendline. Figure 16 shows the result.

y = 12.457 ln(x) + 2.9507

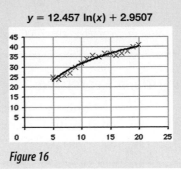

Figure 16

✷ *Before we go on ...* Our model seems to give reasonable estimates when we extrapolate forward, but extrapolating backward is quite another matter: The logarithm curve drops sharply to the left of the given range and becomes negative for small values of t (Figure 17).

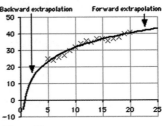

Figure 17

2.3 EXERCISES

In Exercises 1–4, complete the given tables.

1.

Exponential Form	$10^4 = 10,000$	$4^2 = 16$	$3^3 = 27$	$5^1 = 5$	$7^0 = 1$	$4^{-2} = \frac{1}{16}$
Logarithmic Form						

2.

Exponential Form	$4^3 = 64$	$10^{-1} = 0.1$	$2^8 = 256$	$5^0 = 1$	$(0.5)^2 = 0.25$	$6^{-2} = \frac{1}{36}$
Logarithmic Form						

3.

Exponential Form						
Logarithmic Form	$\log_{0.5} 0.25 = 2$	$\log_5 1 = 0$	$\log_{10} 0.1 = -1$	$\log_4 64 = 3$	$\log_2 256 = 8$	$\log_2 \frac{1}{4} = -2$

4.

Exponential Form						
Logarithmic Form	$\log_5 5 = 1$	$\log_4 \frac{1}{16} = -2$	$\log_4 16 = 2$	$\log_{10} 10,000 = 4$	$\log_3 27 = 3$	$\log_7 1 = 0$

In Exercises 5–12, use logarithms to solve the given equation. (Round answers to four decimal places.)

5. $3^x = 5$ **6.** $4^x = 3$ **7.** $5^{-2x} = 40$

8. $6^{3x+1} = 30$ **9.** $4.16e^x = 2$ **10.** $5.3(10^x) = 2$

11. $5(1.06^{2x+1}) = 11$ **12.** $4(1.5^{2x-1}) = 8$

In Exercises 13–18, find the logarithmic curve $y = \log_b x + C$ through the given pair of points. (Round coefficients to four decimal places.)

13. $(1, 2), (9, 4)$ **14.** $(1, 2), (\frac{1}{2}, \frac{3}{2})$

15. $(10, 3), (100, 4)$ **16.** $(1000, 0), (10,000, 1)$

17. $(10, 25), (20, 41)$ **18.** $(10, 20), (30, 40)$

APPLICATIONS

19. Investments How long will it take a $500 investment to be worth $700 if it is continuously compounded at 10% per year? (Give the answer to two decimal places.)

20. Investments How long will it take a $500 investment to be worth $700 if it is continuously compounded at 15% per year? (Give the answer to two decimal places.)

21. Investments How long, to the nearest year, will it take an investment to triple if it is continuously compounded at 10% per year?

22. Investments How long, to the nearest year, will it take me to become a millionaire if I invest $1000 at 10% interest compounded continuously?

23. Carbon Dating The amount of carbon-14 remaining in a sample that originally contained A grams is given by

$$C(t) = A(0.999\ 879)^t$$

where t is time in years. If tests on a fossilized skull reveal that 99.95% of the carbon-14 has decayed, how old is the skull?

24. Carbon Dating Refer to Exercise 23. How old is a fossil in which only 30% of the carbon-14 has decayed?

Government Bonds Exercises 25–32 are based on the following table, which lists annual percent yields on 10-year government bonds in several countries in 2002:

Country	U.S.	Japan	Germany	Britain	Italy	Mexico
Yield (%)	5.3	1.5	5.2	5.3	5.4	6.0

Source: ECB, Reuters, and BE. Source for Mexican rate: http://www.latin-focus.com.

25. Assuming that you invest $10,000 in U.S. government bonds, how long (to the nearest year) must you wait before your investment is worth $15,000 if the interest is compounded annually?

26. Assuming that you invest $10,000 in Japanese government bonds, how long (to the nearest year) must you wait before your investment is worth $15,000 if the interest is compounded annually?

27. If you invest $10,400 in German government bonds and the interest is compounded monthly, how many months will it take for your investment to grow to $20,000?

28. If you invest $10,400 in British government bonds and the interest is compounded monthly, how many months will it take for your investment to grow to $20,000?

29. How long, to the nearest year, will it take an investment in Mexico to double its value if the interest is compounded every 6 months?

30. How long, to the nearest year, will it take an investment in Italy to double its value if the interest is compounded every 6 months?

31. If the interest on U.S. government bonds is compounded continuously, how long will it take the value of an investment to double? (Give the answer correct to two decimal places.)

32. If the interest on Mexican government bonds is compounded continuously, how long will it take the value of an investment to double? (Give the answer correct to two decimal places.)

33. Automobiles The rate of auto thefts triples every 6 months.
a. Determine the base b for an exponential model of the rate of auto thefts as a function of time in months.
b. Find the doubling time.

34. Television Theft The rate of television thefts is doubling every 4 months.
a. Determine the base b for an exponential model of the rate of television thefts as a function of time in months.
b. Find the tripling time.

Exercises 35–40 involve the half-life of various substances. See Example 5 in Section 2.2.

35. Radioactive Decay Uranium-235 is used as fuel for some nuclear reactors. It has a half-life of 710 million years. How long will it take 10 grams of uranium-235 to decay to 1 gram? (Round your answer to three significant digits.)

36. Radioactive Decay Plutonium-239 is used as fuel for some nuclear reactors and as the fissionable material in atomic bombs. It has a half-life of 24,400 years. How long will it take 10 grams of plutonium-239 to decay to 1 gram? (Round your answer to three significant digits.)

37. Radioactive Decay You are trying to determine the half-life of a new radioactive element you have isolated. You start with 1 gram, and 2 days later you determine that it has decayed to 0.7 gram. What is its half-life? (Round your answer to three significant digits.)

38. Radioactive Decay You have just isolated a new radioactive element. If you can determine its half-life, you will win the Nobel Prize in physics. You purify a sample of 2 grams. One of your colleagues steals half of it, and 3 days later you find that 0.1 gram of the radioactive material is still left. What is the half-life? (Round your answer to three significant digits.)

39. Aspirin Soon after taking an aspirin, a patient has absorbed 300 milligrams (mg) of the drug. If the amount of aspirin in the bloodstream decays exponentially, with half being removed every 2 hours, find the time it will take for the amount of aspirin in the bloodstream to decrease to 100 mg.

40. Alcohol After several drinks a person has a blood alcohol level of 200 mg/dL (milligrams per deciliter). If the amount of alcohol in the blood decays exponentially, with one-fourth being removed every hour, find the time it will take for the person's blood alcohol level to decrease to 80 mg/dL.

For Exercises 41–44, use technology to solve.

41. Research Expenditures The following table gives the (approximate) spending on agriculture research in 1976–1996 by the public sector:

Year, t	6 (1976)	8	10	12	14	16	18	20	22	24	26 (1996)
Expenditure, E ($ billions)	2.5	3.0	3.0	3.1	3.0	3.1	3.2	3.3	3.4	3.3	3.1

Data are approximate. SOURCES: Economic Research Service, U.S. Dept. of Agriculture, American Association for the Advancement of Science, "National Plant Breeding Study 1996" by Kenneth J. Frey, Iowa State University/*New York Times*, May 15, 2001, p. F1.

  a. Find the logarithmic regression model of the form $E(t) = A \ln t + C$. (Round the coefficients to four significant digits.)
  b. In 1991 the public sector spent approximately \$3.3 billion on agriculture research. Does the model reflect this figure accurately?
  c. Which of the following is correct? The model predicts that
   (A) Public-sector expenditure on agriculture will increase without bound.
   (B) Public-sector expenditure on agriculture will level off at around \$3.3 billion.
   (C) Public-sector expenditure on agriculture will eventually decrease.

42. Research Expenditures The following table gives the (approximate) spending on agriculture research in 1976–1996 by the private sector:

Year, t	6 (1976)	8	10	12	14	16	18	20	22	24	26 (1996)
Expenditure, E ($ billions)	2.4	2.8	3.2	3.0	3.4	3.5	3.5	3.7	3.7	3.7	4.0

Data are approximate. SOURCES: Economic Research Service, U.S. Dept. of Agriculture, American Association for the Advancement of Science, "National Plant Breeding Study 1996" by Kenneth J. Frey, Iowa State University/*New York Times*, May 15, 2001, p. F1.

  a. Find the logarithmic regression model of the form $E(t) = A \ln t + C$. (Round coefficients to four significant digits).
  b. In 1991 the private sector spent approximately \$3.7 billion on agriculture research. Does the model reflect this figure accurately?
  c. Which of the following is correct? If you increase A by 0.1 and decrease C by 0.1 in the logarithmic model, then
   (A) The new model predicts eventually lower expenditures.
   (B) The long-term prediction is essentially the same.
   (C) The new model predicts eventually higher expenditures.

43. Market Share The following table shows the U.S. market share of sport utility vehicles (SUVs) (including crossover utility vehicles—smaller SUVs based on car designs) for the period 1994–2001:

Year, t	4 (1994)	5	6	7	8	9	10	11 (2001)
Market Share (%)	10	11.5	12	15	17.5	19	19.5	21

Market share is given as a percentage of all vehicles sold in the United States. 2001 figure is based on sales through April. SOURCE: Ward's Auto Infobank/*New York Times*, May 19, 2001, p. C1.

Find the logarithmic regression model of the form $M(t) = A \ln t + C$. Comment on the long-term suitability of the model.

44. Market Share The following table shows the U.S. market share of large-size SUVs for the period 1994–2001:

Year, t	4 (1994)	5	6	7	8	9	10	11 (2001)
Market Share (%)	1.5	2	2.5	3.5	4.5	5	5	5

Market share is given as a percentage of all vehicles sold in the United States. 2001 figure is based on sales through April. SOURCE: Ward's Auto Infobank/*New York Times*, May 19, 2001, p. C1.

Find the logarithmic regression model of the form $M(t) = A \ln t + C$. Comment on the suitability of the model to estimate the market share during the period 1990–1993.

45. Richter Scale The Richter scale is used to measure the intensity of earthquakes. The Richter scale rating of an earthquake is given by the formula

$$R = \frac{2}{3}(\log E - 4.4)$$

where E is the energy released by the earthquake (measured in joules).[6]

  a. The San Francisco earthquake of 1906 registered $R = 8.2$ on the Richter scale. How many joules of energy were released?
  b. In 1989 another San Francisco earthquake registered 7.1 on the Richter scale. Compare the two: The energy released in the 1989 earthquake was what percentage of the energy released in the 1906 quake?
  c. Show that if two earthquakes registering R_1 and R_2 on the Richter scale release E_1 and E_2 joules of energy, respectively, then

$$\frac{E_2}{E_1} = 10^{1.5(R_2 - R_1)}$$

  d. Fill in the blank: If one earthquake registers 2 points more on the Richter scale than another, then it releases ____ times the amount of energy.

---

[6]A joule is a unit of energy; 100 joules of energy would light a 100-watt light bulb for 1 second.

46. Sound Intensity The loudness of a sound is measured in *decibels.* The decibel level of a sound is given by the formula

$$D = 10 \log \frac{I}{I_0}$$

where D is the decibel level (dB), I is its intensity in watts per square meter (W/m²), and $I_0 = 10^{-12}$ W/m² is the intensity of a barely audible "threshold" sound. A sound intensity of 90 dB or greater causes damage to the average human ear.

a. Find the decibel levels of each of the following, rounding to the nearest decibel:

Whisper: 115×10^{-12} W/m²

TV (average volume from 10 feet): 320×10^{-7} W/m²

Loud music: 900×10^{-3} W/m²

Jet aircraft (from 500 feet): 100 W/m²

b. Which of the above sounds damages the average human ear?

c. Show that if two sounds of intensity I_1 and I_2 register decibel levels of D_1 and D_2 respectively, then

$$\frac{I_2}{I_1} = 10^{0.1(D_2 - D_1)}$$

d. Fill in the blank: If one sound registers 1 decibel more than another, then it is _____ times as intense.

47. Sound Intensity The decibel level of a TV set decreases with the distance from the set according to the formula

$$D = 10 \log\left(\frac{320 \times 10^7}{r^2}\right)$$

where D is the decibel level and r is the distance from the TV set in feet.

a. Find the decibel level (to the nearest decibel) at distances of 10, 20, and 50 feet.

b. Express D in the form $D = A + B \log r$ for suitable constants A and B. (Round A and D to two significant digits.)

c. How far must a listener be from a TV so that the decibel level drops to zero? (Round the answer to two significant digits.)

48. Acidity The acidity of a solution is measured by its pH, which is given by the formula

$$\text{pH} = -\log(\text{H}^+)$$

where H^+ measures the concentration of hydrogen ions in moles per liter.[7] The pH of pure water is 7. A solution is referred to as *acidic* if its pH is below 7 and as *basic* if its pH is above 7.

a. Calculate the pH of each of the following substances:

Blood: 3.9×10^{-8} mole/liter

Milk: 4.0×10^{-7} mole/liter

Soap solution: 1.0×10^{-11} mole/liter

Black coffee: 1.2×10^{-7} mole/liter

b. How many moles of hydrogen ions are contained in 1 liter of acid rain that has a pH of 5.0?

c. Complete the following sentence: If the pH of a solution increases by 1.0, then the concentration of hydrogen ions _____.

COMMUNICATION AND REASONING EXERCISES

49. Why is the logarithm of a negative number not defined?

50. Of what use are logarithms, now that they are no longer needed to perform complex calculations?

51. If $y = 4^x$, then $x =$ _____.

52. If $y = \log_6 x$, then $x =$ _____.

53. Your company's market share is undergoing steady growth. Explain why a logarithmic function is *not* appropriate for long-term future prediction of your market share.

54. Your company's market share is undergoing steady growth. Explain why a logarithmic function is *not* appropriate for long-term backward extrapolation of your market share.

55. If a town's population is increasing exponentially with time, how is time increasing with population? Explain.

56. If a town's population is increasing logarithmically with time, how is time increasing with population? Explain.

57. Evaluate $2^{\log_2 8}$.

58. Evaluate $e^{\ln x}$.

59. Evaluate $\ln(e^x)$.

60. Simplify $\ln\sqrt{a}$.

---

[7]A mole corresponds to about 6.0×10^{23} hydrogen ions. (This number is known as Avogadro's number.)

2.4 Logistic Functions and Models

In 2001 there was a growing belief among analysts that the number of Internet users in the United States was "reaching a plateau." This is suggested by the graph in Figure 18, which shows N, the number of U.S. households connected to the Internet, as a function of t, the number of quarters since the start of 1999.

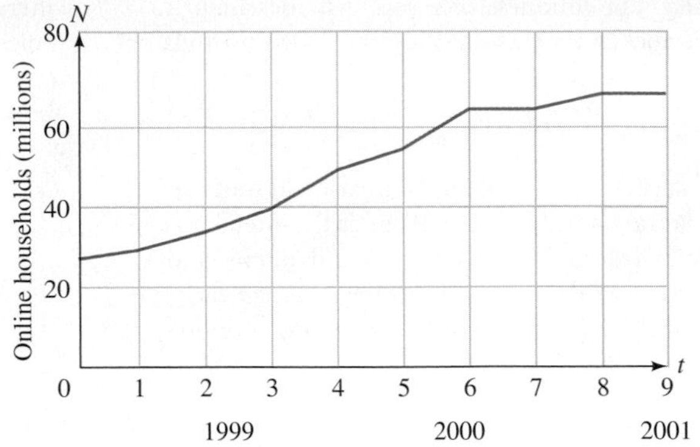

SOURCE: Telecommunications Reports International/*New York Times*, May 21, 2001, p. C9.

Figure 18 • U.S. Households Connected to the Internet

The graph starts leveling off around the middle of 2000, and some Internet consultants suggested that N would reach a plateau of approximately 70 million (about two-thirds of all U.S. households) by 2004.

Question What kind of function can we use to model N?

Answer The left-hand part of the curve, from $t = 0$ to, say, $t = 4$, looks like exponential growth: N behaves (roughly) like an exponential function, growing at a rate of about 15% per quarter. Then, as the market starts to become saturated, the growth of N slows, and its value approaches the saturation point. This kind of behavior is typical in the demand for a new technology or product (in this case, the Internet). The **logistic function** has just this kind of behavior, growing exponentially at first and then leveling off. In addition to modeling the demand for a new technology, logistic functions are often used in epidemic and population modeling. In an epidemic the number of infected people often grows exponentially at first and then slows when a significant proportion of the entire susceptible population is infected and the epidemic has "run its course." Similarly, populations may grow exponentially at first and then slow as they approach the capacity of the available resources.

Logistic Function
A **logistic function** has the form

$$f(x) = \frac{N}{1 + Ab^{-x}}$$

for some constants A, N, and b ($b > 0$ and $b \neq 1$).

Quick Example

$N = 6$, $A = 2$, and $b = 1.1$ gives $f(x) = \dfrac{6}{1 + 2(1.1^{-x})}$.   `6/(1+2*1.1^-x)`

$f(0) = \dfrac{6}{1 + 2} = 2$ The y intercept is $\dfrac{N}{1 + A}$.

$f(1000) = \dfrac{6}{1 + 2(1.1^{-1000})} \approx \dfrac{6}{1 + 0} = 6 = N$ When x is large, $f(x) \approx N$.

Graph of a Logistic Function

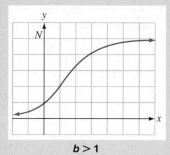

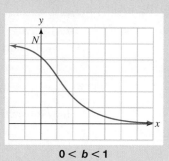

$$b > 1 \qquad\qquad\qquad 0 < b < 1$$

$$y = \frac{N}{1 + Ab^{-x}}$$

Properties of the Logistic Curve $y = \dfrac{N}{1 + Ab^{-x}}$

- The graph is an S-shaped curve sandwiched between the horizontal lines $y = 0$ and $y = N$. N is called the **limiting value** of the logistic curve.
- If $b > 1$, the graph rises; if $b < 1$, the graph falls.
- The y-intercept is $\dfrac{N}{1 + A}$.

Note

If we write b^{-x} as e^{-kx} (where $k = \ln b$), we get the following alternative form of the logistic function:

$$f(x) = \frac{N}{1 + Ae^{-kx}}$$

Question How does the constant b affect the graph?

Answer To understand the role of b, we first rewrite the logistic function by multiplying top and bottom by b^x:

$$f(x) = \frac{N}{1 + Ab^{-x}} = \frac{Nb^x}{(1 + Ab^{-x})\,b^x}$$

$$= \frac{Nb^x}{b^x + A} \qquad\qquad \text{Since } b^{-x}b^x = 1$$

For values of x close to 0, the quantity b^x is close to 1, so the denominator is approximately $1 + A$, giving

$$f(x) \approx \frac{Nb^x}{1 + A} = \left(\frac{N}{1 + A}\right)b^x$$

In other words, $f(x)$ is approximately exponential with base b for values of x close to zero. Put another way, if x represents time, then initially the logistic function behaves like an exponential function.

To summarize:

Logistic Function for Small x and the Role of b

For small values of x, we have

$$\frac{N}{1 + Ab^{-x}} \approx \left(\frac{N}{1 + A}\right)b^x$$

Thus, for small x, the logistic function grows approximately exponentially with base b.

Quick Example

Let

$$f(x) = \frac{50}{1 + 24(3^{-x})} \qquad\qquad N = 50, A = 24, b = 3$$

Then

$$f(x) \approx \left(\frac{50}{1 + 24}\right)(3^x) = 2(3^x)$$

for small values of x. The following figure compares their graphs:

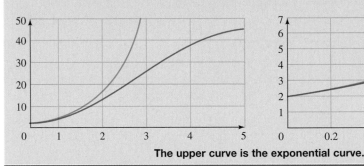 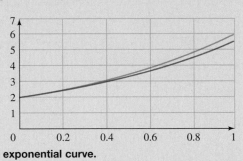

The upper curve is the exponential curve.

Modeling with the Logistic Function

Example 1 • Epidemics

A flu epidemic is spreading through the U.S. population. An estimated 150 million people are susceptible to this particular strain, and it is predicted that all susceptible people will eventually become infected. There are 10,000 people already infected, and the number is doubling every 2 weeks. Use a logistic function to model the number of people infected. Hence, predict when, to the nearest week, 1 million people will be infected.

Solution Let t be time in weeks and let $P(t)$ be the total number of people infected at time t. We want to express P as a logistic function of t, so that

$$P(t) = \frac{N}{1 + Ab^{-t}}$$

We are told that, in the long run, 150 million people will be infected, so that

$$N = 150,000,000 \qquad\qquad \text{Limiting value of } P$$

At the current time ($t = 0$), 10,000 people are infected, so

$$10,000 = \frac{N}{1 + A} = \frac{150,000,000}{1 + A} \qquad \text{Value of } P \text{ when } t = 0$$

Solving for A gives

$$10,000(1 + A) = 150,000,000$$

$$1 + A = 15,000$$

$$A = 14,999$$

What about b? At the beginning of the epidemic (t near 0), P is growing approximately exponentially, doubling every 2 weeks. Using the technique of Section 2.2, we find that the exponential curve passing through the points (0, 10,000) and (2, 20,000) is

$$y = 10,000(\sqrt{2})^t$$

giving us $b = \sqrt{2}$. Now that we have the constants N, A, and b, we can write down the logistic model:

$$P(t) = \frac{150,000,000}{1 + 14,999(\sqrt{2})^{-t}}$$

The graph of this function is shown in Figure 19.

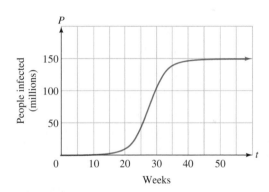

Figure 19

Now we tackle the question of prediction: When will 1 million people be infected? In other words: When is $P(t) = 1,000,000$?

$$1,000,000 = \frac{150,000,000}{1 + 14,999(\sqrt{2})^{-t}}$$

$$1,000,000[1 + 14,999(\sqrt{2})^{-t}] = 150,000,000$$

$$1 + 14,999(\sqrt{2})^{-t} = 150$$

$$14,999(\sqrt{2})^{-t} = 149$$

$$(\sqrt{2})^{-t} = \frac{149}{14,999}$$

$$-t = \log_{\sqrt{2}}\left(\frac{149}{14,999}\right) \approx -13.31 \qquad \text{Logarithmic form}$$

Thus, 1 million people will be infected by about the 13th week.

Logistic Regression

Can we find a logistic model that fits the data on Internet connections we considered at the beginning of this section? We encounter immediate problems: Since we cannot predict the future, it is impossible to say what the limiting value, or plateau, is. The article from which the data were obtained did quote some analysts as suggesting that the plateau is approximately 70 million, but why should we take their word for it? Furthermore, the data already graphed suggest anything but an *exact* logistic curve, so we will need to do some form of regression.

Example 2 • Logistic Regression

Here are the data graphed in Figure 18:

Quarter, t	0	1	2	3	4	5	6	7	8	9
Online Households, H (millions)	28	30	35	40	50	55	65	65	68	69

Find a logistic regression curve of the form

$$H(t) = \frac{N}{1 + Ab^{-t}}$$

Compare the long-term prediction of your model with the analysts' claim that $N \approx 70$.

Solution

Graphing Calculator
As with the other forms of regression in this chapter, start by entering the coordinates of the data points in the lists L_1 and L_2, press $\boxed{\text{STAT}}$, select CALC, and choose the option Logistic. Pressing $\boxed{\text{ENTER}}$ gives the logistic regression equation in the home screen:

$$H(t) \approx \frac{79.317}{1 + 2.2116e^{-0.325957t}}$$

This is not exactly the form we are seeking, but we can convert it to that form by writing

$$e^{-0.325957t} = (e^{0.325957})^{-t} \approx 1.3854^{-t}$$

so,

$$H(t) \approx \frac{79.317}{1 + 2.2116(1.3854)^{-t}}$$

The data points together with the regression curve are shown in Figure 21.

Excel
Excel does not have a built-in logistic regression calculation, so we use an alternative method that works for any type of regression curve. We first use rough estimates for N, A, and b, and compute the sum-of-squares error (SSE; see Section 1.5) directly:

	A	B	C	D	E	F	G
1	Quarter	H (Observed)	H (Predicted)	Residue^2	N	A	b
2	0	28	=$E$2/(1+$F$2*$G$2^(-A2))	=(C2-B2)^2	70	2	1.5
3	1	30					
4	2	35					
5	3	40				SSE	
6	4	50				=SUM(D2:D11)	
7	5	55					
8	6	65					
9	7	65					
10	8	68					
11	9	69					

Cells E2:G2 contain our initial rough estimates of N, A, and b. For N we used the analysts' long-term projection of 70 (notice that the y coordinates do appear to level off around 70). For A we used the fact that the y intercept is $N/(1 + A)$. In other words,

$$28 = \frac{70}{1 + A}$$

Since a very rough estimate is all we are after, using $A = 2$ will do just fine. For b we choose a value a little over 1 (since the graph rises). Cell C2 contains the formula for $H(t)$, and the square of the resulting residue is computed in D2. Cell F6 will contain SSE. The completed spreadsheet should look like this:

	A	B	C	D	E	F	G
1	Quarter	H (Observed)	H (Predicted)	Residue^2	N	A	b
2	0	28	23.33333333	21.77777778	70	2	1.5
3	1	30	30	1.26218E-29			
4	2	35	37.05882353	4.238754325			
5	3	40	43.95348837	15.63007031		SSE	
6	4	50	50.17699115	0.031325867		92.5220071	
7							

The best-fit curve will result from values of N, A, and b that give a minimum value for SSE. We will use Excel's Solver, found in the Tools menu, to find these values for us. (If Solver does not appear in the Tools menu, select Add-Ins in the Tools menu and install it.)

Figure 20 shows the dialog box with the necessary fields completed to solve the problem.

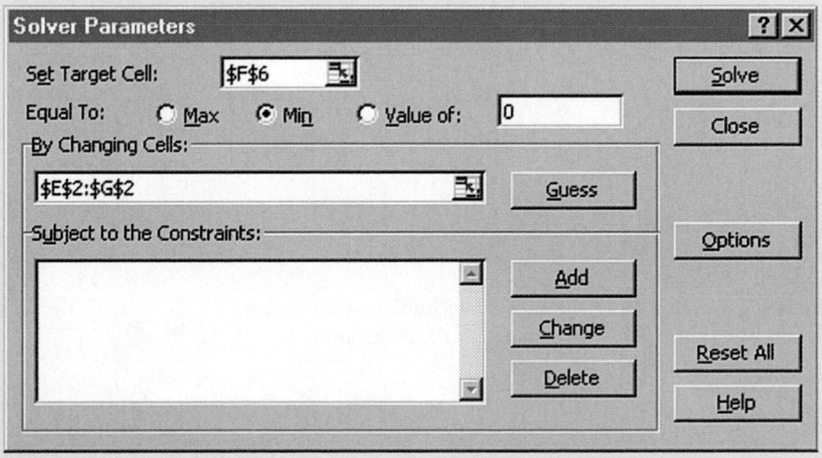

Figure 20

- The `Target Cell` refers to the cell that contains SSE.
- `Min` is selected because we are minimizing SSE.
- `Changing Cells` is obtained by selecting the cells that contain the current values of N, A, and b.

When you are done, click on `Solve`, telling Solver to `Keep Solver Solution` when done, and you will find $N \approx 79.317$, $A \approx 2.2116$, and $b \approx 1.3854$ so that

$$H(t) \approx \frac{79.317}{1 + 2.2116(1.3854)^{-t}}$$

Figure 21 shows the data points together with the regression curve.

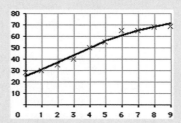

Figure 21

Notice that $N \approx 79$, predicting a plateau of 79 million households, somewhat more than predicted by the analysts.

Note　Logistic regression estimates all three constants N, A, and b for a model $y = N/(1 + Ab^{-x})$. However, there are times, as in Example 1, when we already know the limiting value N and require estimates of only A and b. In such cases, we can use exponential regression to compute these estimates: First, rewrite the logistic equation as

$$\frac{N}{y} = 1 + Ab^{-x}$$

so that

$$\frac{N}{y} - 1 = Ab^{-x} = A(b^{-1})^x$$

This equation gives $\frac{N}{y} - 1$ as an exponential function of x. Thus, if we do exponential regression using the data points $(x, \frac{N}{y} - 1)$, we can obtain estimates for A and b^{-1} (and hence b). This is done in Exercises 34 and 33.

2.4 EXERCISES

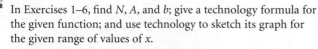

In Exercises 1–6, find N, A, and b; give a technology formula for the given function; and use technology to sketch its graph for the given range of values of x.

1. $f(x) = \dfrac{7}{1 + 6(2^{-x})}$; $[0, 10]$

2. $g(x) = \dfrac{4}{1 + 0.333(4^{-x})}$; $[0, 2]$

3. $f(x) = \dfrac{10}{1 + 4(0.3^{-x})}$; $[-5, 5]$

4. $g(x) = \dfrac{100}{1 + 5(0.5^{-x})}$; $[-5, 5]$

5. $h(x) = \dfrac{2}{0.5 + 3.5(1.5^{-x})}$; $[0, 15]$

(First divide top and bottom by 0.5.)

6. $k(x) = \dfrac{17}{2 + 6.5(1.05^{-x})}$; [0, 100]

(First divide top and bottom by 2.)

In Exercises 7–10, find the logistic function f with the given properties.

7. $f(0) = 10$; f has limiting value 200; and for small values of x, f is approximately exponential and doubles with every increase of 1 in x.

8. $f(0) = 1$; f has limiting value 10; and for small values of x, f is approximately exponential and grows by 50% with every increase of 1 in x.

9. f has limiting value 6 and passes through $(0, 3)$ and $(1, 4)$.

10. f has limiting value 4 and passes through $(0, 1)$ and $(1, 2)$.

In Exercises 11–16, choose the logistic function that best approximates the given curve.

11.

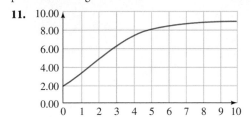

(A) $f(x) = \dfrac{6}{1 + 0.5(3^{-x})}$

(B) $f(x) = \dfrac{9}{1 + 3.5(2^{-x})}$

(C) $f(x) = \dfrac{9}{1 + 0.5(1.01)^{-x}}$

12.

(A) $f(x) = \dfrac{8}{1 + 7(2)^{-x}}$

(B) $f(x) = \dfrac{8}{1 + 3(2)^{-x}}$

(C) $f(x) = \dfrac{6}{1 + 11(5)^{-x}}$

13.

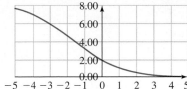

(A) $f(x) = \dfrac{8}{1 + 7(0.5)^{-x}}$

(B) $f(x) = \dfrac{8}{1 + 3(0.5)^{-x}}$

(C) $f(x) = \dfrac{8}{1 + 3(2)^{-x}}$

14.

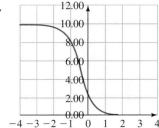

(A) $f(x) = \dfrac{10}{1 + 3(1.01)^{-x}}$

(B) $f(x) = \dfrac{8}{1 + 7(0.1)^{-x}}$

(C) $f(x) = \dfrac{10}{1 + 3(0.1)^{-x}}$

15.

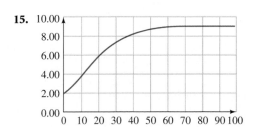

(A) $f(x) = \dfrac{18}{2 + 7(5)^{-x}}$

(B) $f(x) = \dfrac{18}{2 + 3(1.1)^{-x}}$

(C) $f(x) = \dfrac{18}{2 + 7(1.1)^{-x}}$

16.

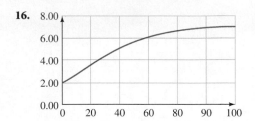

(A) $f(x) = \dfrac{14}{2 + 5(15)^{-x}}$

(B) $f(x) = \dfrac{14}{1 + 13(1.05)^{-x}}$

(C) $f(x) = \dfrac{14}{2 + 5(1.05)^{-x}}$

In Exercises 17–20, use technology to find a logistic regression curve $y = N/(1 + Ab^{-x})$ approximating the given data. Draw a graph showing the data points and regression curve. (Round b to three significant digits and A and N to two significant digits.)

17.

x	0	20	40	60	80	100
y	2.1	3.6	5.0	6.1	6.8	6.9

18.

x	0	30	60	90	120	150
y	2.8	5.8	7.9	9.4	9.7	9.9

19.

x	0	20	40	60	80	100
y	30.1	11.6	3.8	1.2	0.4	0.1

20.

x	0	30	60	90	120	150
y	30.1	20	12	7.2	3.8	2.4

APPLICATIONS

21. Epidemics There are currently 1000 cases of Venusian flu in a total susceptible population of 10,000, and the number of cases is increasing by 25% each day. Find a logistic model for the number of cases of Venusian flu and use your model to predict the number of flu cases 1 week from now.

22. Epidemics Last year's epidemic of Martian flu began with a single case in a total susceptible population of 10,000. The number of cases was increasing initially by 40% each day. Find a logistic model for the number of cases of Martian flu and use your model to predict the number of flu cases 3 weeks into the epidemic.

23. Sales You have sold 100 "I ♥ Calculus" T-shirts, and sales appear to be doubling every 5 days. You estimate the total market for "I ♥ Calculus" T-shirts to be 3000. Give a logistic model for your sales and use it to predict, to the nearest day, when you will have sold 700 T-shirts.

24. Sales In Russia the average consumer drank two servings of Coca-Cola® in 1993. This amount appeared to be increasing exponentially with a doubling time of 2 years. Given a long-range market saturation estimate of 100 servings per year, find a logistic model for the consumption of Coca-Cola in Russia and use your model to predict when, to the nearest year, the average consumption will be 50 servings per year.

The doubling time is based on retail sales of Coca-Cola products in Russia. Sales in 1993 were double those in 1991 and were expected to double again by 1995. Source: *New York Times*, September 26, 1994, p. D2.

25. Big Brother The following chart shows the number of wiretaps authorized by U.S. courts in 1990–2000:

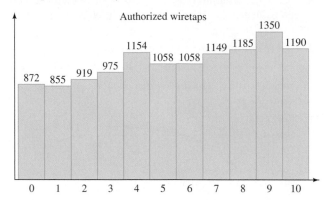

Authorized wiretaps

a. Which of the following logistic functions best models the data? (t is the number of years since 1990.) Try to determine the correct model without actually computing data points.

(A) $W(t) = \dfrac{1500}{1 + 0.77(1.16)^{-t}}$

(B) $W(t) = \dfrac{1500}{1 + 0.77(1.96)^{-t}}$

(C) $W(t) = \dfrac{800}{1 + 0.87(1.16)^{-t}}$

(D) $W(t) = \dfrac{800}{1 + 0.87(0.96)^{-t}}$

b. According to the model you selected, at what percentage was the number of authorized wiretaps growing around 1990?

Source: 2000 Wiretap Report, Administrative Office of the U.S. Courts, http://www.epic.org/privacy/wiretap/stats/2000_report/default.html.

26. Big Brother The following chart shows the number of wiretaps authorized by U.S. federal courts in 1990–2000:

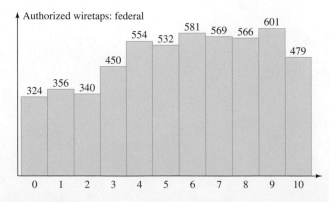

Authorized wiretaps: federal

a. Which of the following logistic functions best models the data? (t is the number of years since 1990.) Try to determine the correct model without actually computing data points.

(A) $W(t) = \dfrac{600}{1 + 1.00(3.52)^{-t}}$

(B) $W(t) = \dfrac{3200}{1 + 1.00(1.62)^{-t}}$

(C) $W(t) = \dfrac{600}{1 + 1.00(0.94)^{-t}}$

(D) $W(t) = \dfrac{600}{1 + 1.00(1.62)^{-t}}$

b. According to the model you selected, at what percentage was the number of authorized wiretaps growing around 1990?

SOURCE: 2000 Wiretap Report, Administrative Office of the U.S. Courts, http://www.epic.org/privacy/wiretap/stats/2000_report/default.html.

27. Computer Use The following graph shows the actual percentage of U.S. households with a computer in 2000 as a function of household income (the data points) and a logistic model of these data (the curve):

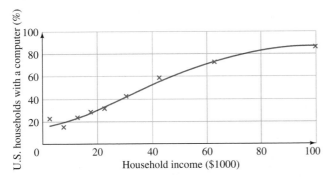

The logistic model is

$$P(x) = \dfrac{91}{1 + 5.35(1.05)^{-x}} \text{ percent}$$

where x is the houshold income in thousands of dollars.

a. According to the model, what percentage of extremely wealthy households had computers?

b. For low incomes the logistic model is approximately exponential. Which exponential model best approximates $P(x)$ for small x?

c. According to the model, 50% of households of what income had computers in 2000? (Round the answer to the nearest $1000.)

Income levels are midpoints of income brackets. (The top income level is an estimate.) SOURCE: NTIA and ESA, U.S. Dept. of Commerce, using U.S. Bureau of the Census *Current Population, 2000.*

28. Internet Use The following graph shows the actual percentage of U.S. residents who used the Internet at home in 2000 as a function of income (the data points) and a logistic model of these data (the curve):

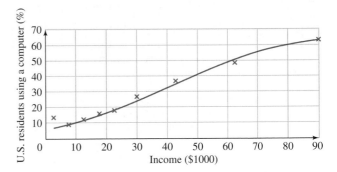

The logistic model is given by

$$P(x) = \dfrac{64.2}{1 + 9.6(1.06)^{-x}} \text{ percent}$$

where x is an individual's income in thousands of dollars.

a. According to the model, what percentage of extremely wealthy people used the Internet at home?

b. For low incomes the logistic model is approximately exponential. Which exponential model best approximates $P(x)$ for small x?

c. According to the model, 50% of individuals with what income used the Internet at home in 2000? (Round the answer to the nearest $1000.)

Income levels are midpoints of income brackets. (The top income level is an estimate.) SOURCE: NTIA and ESA, U.S. Dept. of Commerce, using U.S. Bureau of the Census *Current Population, 2000.*

For Exercises 29–32, use technology to solve.

29. Online Book Sales The following table shows the number of books sold online in the United States in the period 1997–2000 ($t = 0$ represents 1997):

Year, t	0	1	2	3
Book Sales (millions)	4.5	20.0	58.2	78.0

SOURCE: Ipsos-NPD Book Trends/*New York Times*, April 16, 2001, p. C1.

a. What is the logistic regression model for the data? (Round all coefficients to three significant digits.) At what value (to three significant digits) does the model predict that book sales will level off?

b. The 2000 figure ($t = 3$) represents approximately 7% of the total number of books sold. Some analysts predict that book sales will level off at around 15% of the market. Is this prediction consistent with the logistic regression model? Comment on the answer.

c. In what year does the model predict that book sales first exceed 80 million?

30. South African Exports The following table shows the value of South African exports to African nations for the period 1991–1999 ($t = 0$ represents 1990):

Year, t	1	2	3	4	5	6	7	8	9
Exports (billions of rand, R)	5	6	7	9	13	18	20	20	22

Data are approximate. SOURCES: ABSA, *South Africa's Foreign Trade, 2001 Edition/New York Times*, February 17, 2002, p. 14.

a. What is the logistic regression model for the data? (Round all coefficients to three significant digits.) At what value does the model predict that the exports will level off?

b. In 2000 South Africa exported approximately R29 billion to African nations. Comment on this figure in the light of your model.

31. Prison Population The following chart shows the total population in state and federal prisons in 1970–1997 as a function of time in years ($t = 0$ represents 1970). The data are shown beneath the graph:

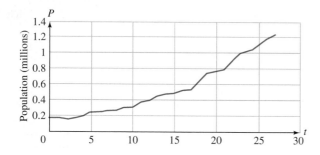

Year, t	0	1	2	3	4	5	6
Population (millions)	0.18	0.18	0.17	0.18	0.2	0.25	0.26
Year, t	7	8	9	10	11	12	13
Population (millions)	0.27	0.27	0.30	0.32	0.38	0.40	0.45
Year, t	14	15	16	17	18	19	20
Population (millions)	0.49	0.50	0.53	0.55	0.65	0.75	0.78
Year, t	21	22	23	24	25	26	27
Population (millions)	0.80	0.92	1.00	1.05	1.12	1.20	1.25

Data are approximate. Download the data at

Web Site → Everything for Calculus → Chapter 2 Exercise Data

rather than attempting to enter it manually. SOURCES: Bureau of Justice Statistics, New York State Dept. of Correctional Services/*New York Times*, January 9, 2000, p. WK3.

a. Use technology to obtain a logistic regression curve (round A and N to two significant digits and b to three significant digits). Graph the data and regression curve. Based on the graph, would you judge the logistic model a good fit?

b. At what percent rate was the prison population growing in 1970?

c. What does the model predict as the long-term prison population (to the nearest 0.1 million)?

32. Prison Population Repeat Exercise 31, using the following data for the population in New York State prisons in 1970–1997:

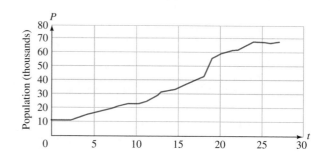

Year, t	0	1	2	3	4	5	6
Population (thousands)	12	12	12	13	15	17	18
Year, t	7	8	9	10	11	12	13
Population (thousands)	20	22	23	23	25	28	32
Year, t	14	15	16	17	18	19	20
Population (thousands)	33	35	38	40	43	55	59
Year, t	21	22	23	24	25	26	27
Population (thousands)	61	62	65	68	68	67	68

Data are approximate. Download the data at

Web Site → Everything for Calculus → Chapter 2 Exercise Data

rather than attempting to enter it manually. SOURCES: Bureau of Justice Statistics, New York State Dept. of Correctional Services/*New York Times*, January 9, 2000, p. WK3.

Exercises 33 and 34 are based on the discussion following Example 2. If the limiting value N is known, then

$$\frac{N}{y} - 1 = A(b^{-1})^x$$

and so $\frac{N}{y} - 1$ is an exponential function of x. In Exercises 33 and 34, use the given value of N and the data points $(x, \frac{N}{y} - 1)$ to obtain A and b and hence a logistic model.

33. Education The following chart shows the number of high school graduates in the United States over the period 1993–1999:

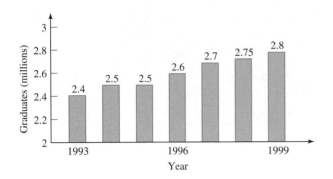

Let t be the number of years since 1993 and find a logistic model based on the assumption that eventually the number of high school graduates will grow to 5 million/year. (Round coefficients.) In what year does your model predict the number of high school graduates will first reach 3.5 million?

Data are rounded. SOURCE: Western Interstate Commission for Higher Education and the College Board/*New York Times*, February 17, 1999, p. B9.

34. College Athletics The percentage of college athletes who are women tended to increase over the preceding 20 years, as shown in the following chart:

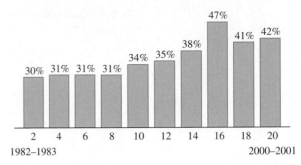

The long-term goal of many athletic policymakers is to attain a parity level of 50% women in college athletics. Using $N = 50$, find a logistic model for the percentage of women college athletes as a function of time t in years ($t = 2$ represents the 1982–1983 academic year). In which year does your model predict that 45% of all college athletes will be women?

COMMUNICATION AND REASONING EXERCISES

35. Logistic functions are commonly used to model the spread of epidemics. Given this fact, explain why a logistic function is also useful to model the spread of a new technology.

36. Why is a logistic function more appropriate than an exponential function for modeling the spread of an epidemic?

37. Give one practical use for logistic regression.

38. What happens to function $P(t) = \dfrac{N}{1 + Ab^{-t}}$ if $A = 0$? If $A < 0$?

CASE STUDY

Reproduced from the U.S. Department of Agriculture Web site

Checking Up on Malthus

In 1798 Thomas R. Malthus (1766–1834) published an influential pamphlet, later expanded into a book, titled *An Essay on the Principle of Population As It Affects the Future Improvement of Society*. One of his main contentions was that population grows geometrically (exponentially) while the supply of resources such as food grows only arithmetically (linearly). This led him to the pessimistic conclusion that population would always reach the limits of subsistence and precipitate famine, war, and ill health, unless population could be checked by other means. He advocated "moral restraint," which includes the pattern of late marriage common in Western Europe at the time and is now common in most developed countries, and which leads to a lower reproduction rate.

Two hundred years later, you have been asked to check the validity of Malthus's contention. That population grows geometrically, at least over short periods of time, is commonly assumed. That resources grow linearly is more questionable. You decide to check the actual production of a common crop, wheat, in the United States. Agricultural statistics like these are available from the U.S. government on the Internet, through the U.S. Department of Agriculture's National Agricultural Statistics Service (NASS). As of 2003, this service was available at http://www.usda.gov/nass/. Looking through this site, you locate data on the U.S. annual production of wheat from 1866 through 2001. (A current link to NASS, as well as the data, is available at the Web site at

Web Site → Applied Calculus→ Chapter 2 Case Study

Year	1866	1867	...	2000	2001
Bushels (thousands)	169,703	210,878	...	2,232,460	1,957,643

Graphing the data from 1900 to 1999 (using Excel, for example), you obtain the graph in Figure 22.

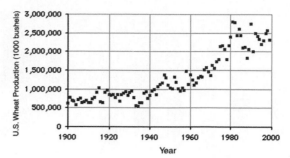

Figure 22

This does not look very linear, particularly in the last half of the 20th century, but you continue checking the mathematics. Using Excel's built-in linear regression capabilities, you find that the line that best fits these data, shown in Figure 23, has $r^2 = 0.8002$. (Recall the discussion of the correlation coefficient r in Section 1.5. A similar statistic is available for other types of regression as well.)

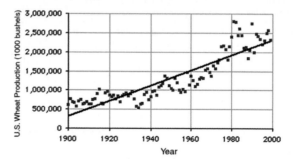

Figure 23

Although that is a fairly high correlation, you notice that the residues are not distributed randomly: The actual wheat production starts out higher than the line, is below the line from about 1920 to about 1970, and then is mostly above the line. This suggests that you would get a better fit from a function whose graph bends upward. But what kind of curve? You decide to compare three different models: quadratic, exponential, and logistic.

Following is a comparison of the results of fitting the three proposed models. (Coefficients are rounded to six significant digits. SSE is the sum-of-squares error, and r^2 is the regression correlation coefficient.)

Model and Graph	SSE	r^2
Quadratic $P(t) \approx 245.067t^2 - 4319.47t + 711{,}703$![quadratic graph]	4.941×10^{12}	0.8807

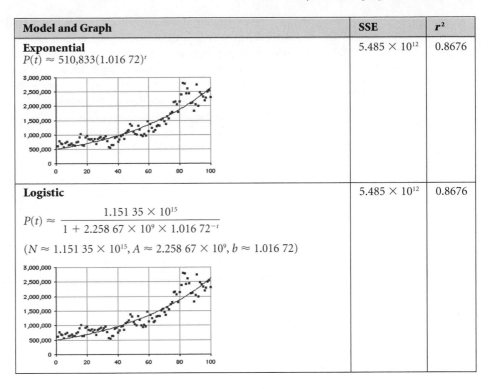

Model and Graph	SSE	r^2
Exponential $P(t) \approx 510{,}833(1.016\ 72)^t$	5.485×10^{12}	0.8676
Logistic $P(t) \approx \dfrac{1.151\ 35 \times 10^{15}}{1 + 2.258\ 67 \times 10^9 \times 1.016\ 72^{-t}}$ $(N \approx 1.151\ 35 \times 10^{15},\ A \approx 2.258\ 67 \times 10^9,\ b \approx 1.016\ 72)$	5.485×10^{12}	0.8676

Question Since the quadratic model gives a slightly higher value for r^2, it follows that the quadratic model is more appropriate than the exponential model, right?

Answer The quadratic model does give a better fit to the data than the exponential model (as evidenced by the values of both SSE and r^2). However, a better fit does not necessarily imply a more appropriate model.

Question Why not?

Answer Suppose, for arguments' sake, that production of wheat *was* growing exponentially according to the above regression model *on average,* but that due to random fluctuations in market conditions and weather patterns, the actual output varied from the predicted output by a random amount. This randomness would explain the fact that the observed values do not lie exactly on the regression curve but are instead scattered about in its vicinity. Moreover, since the scatter is due to random factors, any attempt to obtain a mathematical curve that fits the actual data more closely would be questionable, given our assumption. (For example, we *could* find a degree-99 polynomial whose graph passes through every data point!)

Question Could you not then argue that the linear model might be the most appropriate of all three and that the illusion of curvature is caused by those same random fluctuations?

Answer That could conceivably be happening, but two factors argue against it: First, the positive-negative-positive pattern of the residues in the linear model (Figure 23) strongly suggest that something other than a linear relationship is going on. Second (and we will be vague here), one can perform rigorous statistical tests to support the hypothesis that there is curvature.[8]

Although neither the quadratic nor the exponential model has a clear advantage over the other, the good fit of both models provides reasonable evidence that wheat production, at least, is better described as increasing quadratically or exponentially than linearly, contradicting Malthus.

---

[8]These hypothesis tests are generally discussed in statistics courses.

If you compare the exponential and logistic models, you get additional interesting information. The logistic model seems as though it *ought* to be the most appropriate, since wheat production cannot reasonably be expected to continue increasing exponentially forever; eventually resource limitations must lead to a leveling off of wheat production. Such a leveling off, if it occurred before the population started to level off, would seem to vindicate Malthus's pessimistic predictions.[9]

However, the logistic regression model looks suspect: It predicts a leveling-off value (N) that is orders of magnitude higher than what seems reasonable. Moreover, the logistic model (as well as SSE and r^2) looks almost indistinguishable from the exponential model, suggesting that it is no better. You are forced to conclude that wheat production—even if it is logistic—is still in the early (exponential) stage of growth and thus shows no sign, as yet, of leveling off. In general, for a logistic model to be reliable in its prediction of the leveling-off value N, we would need to see significant evidence of leveling off in the data.

You tentatively conclude that wheat production for the past 100 years is better described as increasing quadratically or exponentially than linearly, contradicting Malthus, and moreover that it shows no sign of leveling off as yet.

EXERCISES

1. Find the best fit line for wheat production for the period 1866–1940. How good a fit is the line?

2. Find the best fit line for wheat production for the period 1940–1997. How good a fit is the line?

3. Find the production figures for another common U.S. crop. Compare the linear, quadratic, exponential, and logistic models. What can you conclude?

4. Below are the census figures for the U.S. population (in thousands) from 1800 to 2000. Compare the linear, quadratic, and exponential models. What can you conclude?

1810	1820	1830	1840	1850	1860	1870	1880	1890	1900
7240	9638	12,861	17,063	23,192	31,443	38,558	50,189	62,980	76,212

1910	1920	1930	1940	1950	1960	1970	1980	1990	2000
92,228	106,022	123,203	132,165	151,326	179,323	203,302	226,542	248,710	281,422

SOURCE: Bureau of the Census, U.S. Dept. of Commerce.

CHAPTER 2 REVIEW TEST

1. Sketch the graph of each of the following quadratic functions, indicating the coordinates of the vertex, the y intercept, and the x intercepts (if any):
   a. $f(x) = x^2 + 2x - 3$ b. $f(x) = -x^2 - x - 1$

2. Find a formula of the form $f(x) = Ab^x$ for each of the following functions:
   a. $f(1) = 2, f(3) = 18$ b. $f(1) = 10, f(3) = 5$

3. Use logarithms to solve the following equations for x:
   a. $3^{-2x} = 4$ b. $2^{2x^2-1} = 2$
   c. $300(10^{3x}) = 315$ d. $P(1 + i)^{mx} = A$

4. In each of the following, find the time required for the investment to reach the desired goal:

   a. $2000 invested at 4% compounded monthly; goal: $3000
   b. $2000 invested at 6.75% compounded daily; goal: $3000
   c. $2000 invested at 3.75% compounded continuously; goal: $3000

5. Find equations for the following logistic functions of x:
   a. Through $(0, 100)$, initially increasing by 50% per unit of x, and limiting value 900
   b. Initially exponential of the form $y = 5(1.1)^x$ with limiting value 25
   c. Passing through $(0, 5)$ and decreasing from a limiting value of 20 to 0 at a rate of 20% per unit of x when x is near 0

---

[9]See the exercise set for data on population.

OHaganBooks.com—NONLINEAR TRENDS

6. The daily traffic ("hits per day") at OHaganBooks.com seems to depend on the monthly expenditure on advertising through banner ads on well-known Internet portals. The following model, based on information you have collected over the past few months, shows the approximate relationship:

$$h = -0.000\,005c^2 + 0.085c + 1750$$

where h is the average number of hits per day at OHaganBooks.com and c is the monthly advertising expenditure.

a. According to the model, what monthly advertising expenditure will result in the largest volume of traffic at OHaganBooks.com? What is that volume?

b. In addition to predicting a maximum volume of traffic, the model predicts that the traffic will eventually drop to zero if the advertising expenditure is increased too far. What expenditure (to the nearest dollar) results in no Web-site traffic?

c. What feature of the formula for this quadratic model indicates that it will predict an eventual decline in traffic as advertising expenditure increases?

7. Some time ago, you formulated the following linear model of demand:

$$q = -60p + 950$$

where q is the monthly demand for OHaganBooks.com's online novels at a price of p dollars per novel.

a. Use this model to express the monthly revenue as a function of the unit price p and hence determine the price you should charge for a maximum monthly revenue.

b. Author royalties and copyright fees cost you an average of $4 per novel, and the monthly cost of operating and maintaining the online publishing service amounts to $900 per month. Express the monthly profit P as a function of the unit price p and hence determine the unit price you should charge for a maximum monthly profit. What is the resulting profit (or loss)?

8. In the period immediately following its initial public offering (IPO), OHaganBooks.com's stock doubled in value every 3 hours.

a. If you bought $10,000 worth of the stock when it was first offered, how much was your stock worth after 8 hours?

b. How long from the initial offering did it take your investment to reach $50,000?

c. After 10 hours of trading, the stock turns around and starts losing one-third of its value every 4 hours. How long (from the initial offering) will it be before your stock is once again worth $10,000?

9. **(Based on a Question from the GRE Economics Exam)** To estimate the rate at which new computer hard drives will have to be retired, OHaganBooks.com uses the "survivor curve":

$$L_x = L_0 e^{-x/t}$$

where L_x = number of surviving hard drives at age x

L_0 = number of hard drives initially

t = average life in years

All the following are implied by the curve *except:*

(A) Some of the equipment is retired during the first year of service.

(B) Some equipment survives three average lives.

(C) More than half the equipment survives the average life.

(D) Increasing the average life of equipment by using more durable materials would increase the number surviving at every age.

(E) The number of survivors never reaches zero.

10. OHaganBooks.com modeled its weekly sales over a period of time with the function

$$s(t) = 6053 + \frac{4474}{1 + e^{-0.55(t-4.8)}}$$

as shown in the following graph (t is measured in weeks):

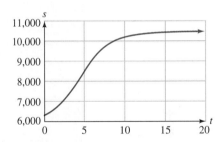

a. As time goes on, it appears that weekly sales are leveling off. At what value are they leveling off?

b. When did weekly sales rise above 10,000?

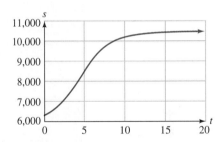 ADDITIONAL ONLINE REVIEW

If you follow the path

Web Site → Everything for Calculus → Chapter 2

you will find the following additional resources to help you review.

A comprehensive chapter summary (including examples and interactive features)

Additional review exercises (including interactive exercises and many with help)

A true/false chapter quiz

Several useful graphing utilities

INTRODUCTION TO THE

CASE STUDY

Reducing Sulfur Emissions

The Environmental Protection Agency (EPA) wants to formulate a policy that will encourage utilities to reduce sulfur emissions. Its goal is to reduce annual emissions of sulfur dioxide by a total of 10 million tons from the current level of 25 million tons by imposing a fixed charge for every ton of sulfur released into the environment per year. The EPA has some data showing the marginal cost to utilities of reducing sulfur emissions. As a consultant to the EPA, you must determine the amount to be charged per ton of sulfur emissions in light of these data.

REPRODUCED FROM THE U.S. ENVIRONMENTAL PROTECTION AGENCY WEB SITE

INTERNET RESOURCES FOR THIS CHAPTER

At the Web site, follow the path

> **Web Site → Everything for Calculus→ Chapter 3**

where you will find links to step-by-step tutorials for the main topics in this chapter, a detailed chapter summary, a true/false quiz, and a collection of sample test questions. You will also find an online grapher and other useful resources. Complete text and interactive exercises have been placed on the Web site covering the following optional topics:

Sketching the graph of the derivative
Proof of the power rule
Continuity and differentiability

DERIVATIVE

Introduction

In the world around us, everything is changing. The mathematics of change is largely about the rate of change: how fast and in which direction the change is occurring. Is the Dow Jones average going up, and if so, how fast? If I raise my prices, how many customers will I lose? If I launch this missile, how fast will it be traveling after 2 seconds, how high will it go, and where will it come down?

We have already discussed the concept of rate of change for linear functions (straight lines), where the slope measures the rate of change. But this works only because a straight line maintains a constant rate of change along its whole length. Other functions rise faster here than there—or rise in one place and fall in another—so that the rate of change varies along the graph. The first achievement of calculus is to provide a systematic and straightforward way of calculating (hence, the name) these rates of change. To describe a changing world, we need a language of change, and that is what calculus is.

The history of calculus is an interesting story of personalities, intellectual movements, and controversy. Credit for its invention is given to two mathematicians: Isaac Newton (1642–1727) and Gottfried Leibniz (1646–1716). Newton, an English mathematician and scientist, developed calculus first, probably in the 1660s. We say "probably" because, for various reasons, he did not publish his ideas until much later. This allowed Leibniz, a German mathematician and philosopher, to publish his own version of calculus first, in 1684. Fifteen years later, stirred up by nationalist fervor in England and on the continent, controversy erupted over who should get the credit for the invention of calculus. The debate got so heated that the Royal Society (of which Newton and Leibniz were both members) set up a commission to investigate the question. The commission decided in favor of Newton, who happened to be president of the society at the time. The consensus today is that both mathematicians deserve credit because they came to the same conclusions working independently. This is not really surprising: Both built on well-known work of other people, and it was almost inevitable that someone would put it all together at about that time.

Algebra Review

For this chapter you should know the algebra reviewed in Appendix A, Section A.2.

3.1 Average Rate of Change

Calculus is the mathematics of change, inspired largely by observation of continuously changing quantities around us in the real world. As an example, the New York metro area consumer confidence index C decreased from 100 points in January 2000 to 80 points in January 2002.[1] As we saw in Chapter 1, the **change** in this index can be measured as the difference:

$$\Delta C = \text{second value} - \text{first value} = 80 - 100 = -20 \text{ points}$$

(The fact that the confidence index decreased is reflected in the negative sign of the change.)

Question How fast was the consumer confidence index dropping?

Answer Since C decreased by 20 points in 2 years, it averaged a $20/2 = 10$-point drop each year. (It actually dropped less than 10 points the first year and more the second, giving an average drop of 10 points each year.) We say that C decreased **at an average rate of 10 points per year.**

Alternatively, we might want to measure this rate in points per month rather than points per year. Since C decreased by 20 points in 24 months, it went down at an average rate of $20/24 \approx 0.833$ point per month.

In both cases we obtained the average rate of change by dividing the change by the corresponding length of time:

$$\text{average rate of change} = \frac{\text{change in } C}{\text{change in time}} = \frac{-20}{2} = -10 \text{ points per year}$$

$$\text{average rate of change} = \frac{\text{change in } C}{\text{change in time}} = \frac{-20}{24} \approx -0.833 \text{ point per month}$$

Average Rate of Change of a Function Using Numerical Data

Example 1 • Average Rate of Change from a Table

The following table lists the approximate value of Standard and Poor's (S&P) 500 stock market index during the 10-year period 1995–2001 ($t = 5$ represents 1995):

Year, t	5	6	7	8	9	10	11
S&P Index, $S(t)$ (points)	500	650	900	1200	1350	1450	1200

The given values are approximate values midway through the given year. SOURCE: Bloomberg Financial Markets/*New York Times*, December 17, 2001, p. C3.

a. What was the average rate of change in the S&P over the 4-year period 1996–2000 (the period $6 \le t \le 10$, or [6, 10] in interval notation); over the 2-year period 1999–2001 (the period $9 \le t \le 11$, or [9, 11]); and over the period [8, 11]?

b. Graph the values shown in the table. How are the rates of change reflected in the graph?

Solution

a. During the 4-year period [6, 10], the S&P changed as follows:

start of the period ($t = 6$): $S(6) = 650$

end of the period ($t = 10$): $S(10) = 1450$

change during the period [6, 10]: $S(10) - S(6) = 800$

---

[1]Figures are approximate. SOURCE: Siena College Research Institute/*New Tork Times,* February 10, 2002, p. L1.

Thus, the S&P increased by 800 points in 4 years, giving an average rate of change of 800/4 = 200 points per year. We can write the calculation this way:

$$\text{average rate of change of } S = \frac{\text{change in } S}{\text{change in } t}$$

$$= \frac{\Delta S}{\Delta t}$$

$$= \frac{S(10) - S(6)}{10 - 6} = \frac{800}{4} = 200 \text{ points per year}$$

Interpreting the result: During the period [6, 10] (that is, 1996–2000), the S&P increased at an average rate of 200 points per year.

Similarly, the average rate of change during the period [9, 11] was

$$\text{average rate of change of } S = \frac{\Delta S}{\Delta t} = \frac{S(11) - S(9)}{11 - 9} = \frac{1200 - 1350}{11 - 9}$$

$$= -\frac{150}{2} = -75 \text{ points per year}$$

Interpreting the result: During the period [9, 11], the S&P *decreased* at an average rate of 75 points per year.

Finally, during the period [8, 11], the average rate of change was

$$\text{average rate of change of } S = \frac{\Delta S}{\Delta t} = \frac{S(11) - S(8)}{11 - 8} = \frac{1200 - 1200}{11 - 8}$$

$$= \frac{0}{3} = 0 \text{ points per year}$$

Interpreting the result: During the period [8, 11], the average rate of change of the S&P was zero points per year (even though its value did fluctuate during that period).

b. In Chapter 1 we saw that the rate of change of a quantity that changes linearly with time is measured by the slope of its graph. However, the S&P index does not change linearly with time. Figure 1 shows the data plotted two different ways: (a) as a bar chart and (b) as a piecewise linear graph. Bar charts are more commonly used in the media, but Figure 1(b) illustrates the changing index price more clearly.

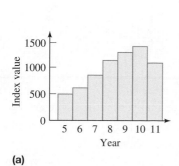

(a)

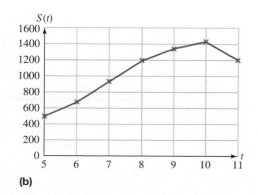

(b)

Figure 1

We saw in part (a) that the average rate of change of S over the interval $[6, 10]$ is the ratio

$$\text{rate of change of } S = \frac{\Delta S}{\Delta t} = \frac{\text{change in } S}{\text{change in } t} = \frac{S(10) - S(6)}{10 - 6}$$

Notice that this rate of change is also the slope of the line through P and Q shown in Figure 2.

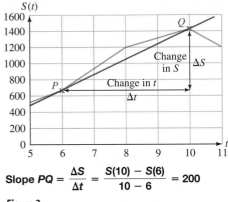

$$\text{Slope } PQ = \frac{\Delta S}{\Delta t} = \frac{S(10) - S(6)}{10 - 6} = 200$$

Figure 2

Interpreting the result: The average rate of change of the S&P over the interval $[6, 10]$ is the slope of the line passing through the points on the graph where $t = 6$ and $t = 10$.

Similarly, the average rates of change of the S&P over the intervals $[9, 11]$ and $[8, 11]$ are the slopes of the lines through pairs of corresponding points.

Excel

We compute average rates of change the same way as we compute slopes. For instance, to compute the rate of change of S over the period $[6, 10]$, you can set up your worksheet as shown:

	A	B	C	D
1	t	S(t)		
2	5	500	=(B7-B3)/(A7-A3)	
3	6	650	Rate of change over [6, 10]	
4	7	900		
5	8	1200		
6	9	1350		
7	10	1450		
8	11	1200		

Here is the formal definition of the average rate of change of a function over an interval.

Change and Average Rate of Change of f over [a, b]: Difference Quotient

The **change** in $f(x)$ over the interval $[a, b]$ is

$$\text{change in } f = \Delta f$$

$$= \text{second value} - \text{first value}$$

$$= f(b) - f(a)$$

The **average rate of change** of $f(x)$ over the interval $[a, b]$ is

$$\text{average rate of change of } f = \frac{\text{change in } f}{\text{change in } x}$$

$$= \frac{\Delta f}{\Delta x} = \frac{f(b) - f(a)}{b - a}$$

$$= \text{slope of line through points } P \text{ and } Q$$
$$\text{(see figure)}$$

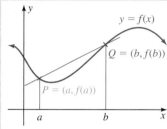

Average rate of change = Slope of PQ

We also call this average rate of change the **difference quotient** of f over the interval $[a, b]$. (It is the *quotient* of the *differences* $f(b) - f(a)$ and $b - a$.) A line through two points of a graph like P and Q is called a **secant line** of the graph.

Units

- The units of the change in f are the units of $f(x)$.
- The units of the average rate of change of f are units of $f(x)$ per unit of x.

Quick Example

If $f(3) = -1$ billion dollars, $f(5) = 0.5$ billion dollars, and x is measured in years, then the change and average rate of change of f over the interval $[3, 5]$ are given by

$$\text{change in } f = f(5) - f(3) = 0.5 - (-1) = 1.5 \text{ billion dollars}$$

$$\text{average rate of change} = \frac{f(5) - f(3)}{5 - 3} = \frac{0.5 - (-1)}{2} = 0.75 \text{ billion dollars per year}$$

Alternative Formula: Average Rate of Change of f over [a, a + h]

(Replace b above by $a + h$.) The average rate of change of f over the interval $[a, a + h]$ is

$$\text{average rate of change of } f = \frac{f(a + h) - f(a)}{h}$$

> ### Guideline: Recognizing When and How to Compute the Average Rate of Change and How to Interpret the Answer
>
> **Question** How do I know, by looking at the wording of a problem, that it is asking for an average rate of change?
>
> **Answer** If a problem does not ask for an average rate of change directly, it might do so indirectly, as in "On average, how fast is quantity q increasing?"
>
> **Question** If I know that a problem calls for computing an average rate of change, how should I compute it? By hand or using technology?
>
> **Answer** All the computations can be done by hand, but when hand calculations are not called for, using technology might save time.
>
> **Question** Lots of problems ask us to "interpret" the answer. How do I do that for questions involving average rates of change?
>
> **Answer** The units of the average rate of change are often the key to interpreting the results:
>
> The units of the average rate of change of $f(x)$ are units of $f(x)$ per unit of x.
>
> Thus, for instance, if $f(x)$ is the cost, in dollars, of a trip of x miles in length, and the average rate of change of f is calculated to be 10, then we can say that the cost of a trip rises an average of $10 for each additional mile.

3.1 EXERCISES

In Exercises 1–18, calculate the average rate of change of the given function over the given interval. Where appropriate, specify the units of measurement.

1.

x	0	1	2	3
f(x)	3	5	2	−1

Interval: $[1, 3]$

2.

x	0	1	2	3
f(x)	−1	3	2	1

Interval: $[0, 2]$

3.

x	−3	−2	−1	0
f(x)	−2.1	0	−1.5	0

Interval: $[-3, -1]$

4.

x	−2	−1	0	1
f(x)	−1.5	−0.5	4	6.5

Interval: $[-1, 1]$

5.

t (mo)	2	4	6
R(t) ($ millions)	20.2	24.3	20.1

Interval: $[2, 6]$

6.

x (kilo)	1	2	3
C(x) (£)	2.20	3.30	4.00

Interval: $[1, 3]$

7.

p ($)	5.00	5.50	6.00
q(p) (items)	400	300	150

Interval: $[5, 5.5]$

8.

t (h)	0	0.1	0.2
D(t) (mi)	0	3	6

Interval: $[0.1, 0.2]$

9.

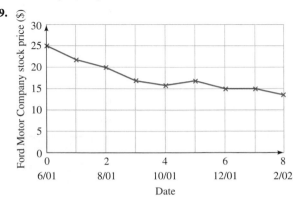

Interval: $[2, 6]$

10.

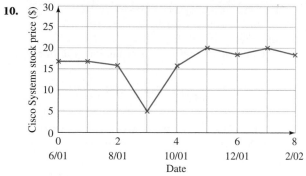

Interval: [3, 7]

11.

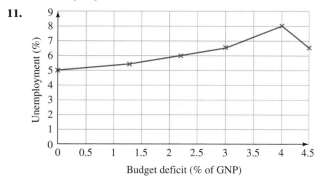

Interval: [0, 4]

12.

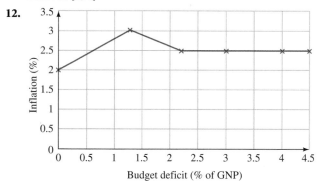

Interval: [0, 4]

13. $f(x) = x^2 - 3$; [1, 3]   **14.** $f(x) = 2x^2 + 4$; [−1, 2]

15. $f(x) = 2x + 4$; [−2, 0]   **16.** $f(x) = \dfrac{1}{x}$; [1, 4]

17. $f(x) = \dfrac{x^2}{2} + \dfrac{1}{x}$; [2, 3]   **18.** $f(x) = 3x^2 - \dfrac{x}{2}$; [3, 4]

In Exercises 19–24, use technology to calculate the average rate of change of the given function f over the intervals $[a, a + h]$ where $h = 1, 0.1, 0.01, 0.001,$ and 0.0001.

19. $f(x) = 2x^2$; $a = 0$ **20.** $f(x) = \dfrac{x^2}{2}$; $a = 1$

21. $f(x) = \dfrac{1}{x}$; $a = 2$ **22.** $f(x) = \dfrac{2}{x}$; $a = 1$

23. $f(x) = x^2 + 2x$; $a = 3$ **24.** $f(x) = 3x^2 - 2x$; $a = 0$

APPLICATIONS

25. Employment The following table lists the approximate number of people employed in the United States during the period 1995–2001, on July 1 of each year ($t = 5$ represents 1995):

Year, t	5	6	7	8	9	10	11
Employment, $P(t)$ (millions)	117	120	123	125	130	132	132

The given (approximate) values represent nonfarm employment.
Source: Bureau of Labor Statistics/*New York Times*, December 17, 2001, p. C3.

Compute and interpret the average rate of change of $P(t)$ (a) over the period 1995–2000 (that is, [5, 10]) and (b) over the period [10, 11]. Be sure to state the units of measurement.

26. Cell Phone Sales The following table lists the net sales (after-tax revenue) at Nokia, the Finnish cell phone company, during the period 1995–2001 ($t = 5$ represents 1995):

Year, t	5	6	7	8	9	10	11
Nokia Net Sales, $P(t)$ ($ billions)	8	8	10	16	20	27	28

Source: Nokia/*New York Times*, February 6, 2002, p. A3.

Compute and interpret the average rate of change of $P(t)$ (a) over the period [5, 6] and (b) over the period [7, 11]. Be sure to state the units of measurement.

27. Venture Capital The following table shows the number of companies that invested in venture capital each year during the period 1995–2001 ($t = 5$ represents 1995):

Year, t	5	6	7	8	9	10	11
Companies, $N(t)$	100	150	300	400	1000	1700	900

2001 figure is a projection based on data through September. Source: Venture Economics/ National Venture Capital Association/*New York Times*, February 3, 2002, p. BU4.

During which 2-year interval(s) was the average rate of change of $N(t)$ (a) greatest and (b) least? Interpret your answers by referring to the rates of change.

28. Venture Capital The following table shows the amount of money that companies invested in venture capital during the period 1995–2001 ($t = 5$ represents 1995):

Year, t	5	6	7	8	9	10	11
Investment, $M(t)$ ($ billions)	0.05	0.5	1	2	8	16	4

2001 figure is a projection based on data through September. Source: Venture Economics/ National Venture Capital Association/*New York Times*, February 3, 2002, p. BU4.

During which 3-year interval(s) was the average rate of change of $M(t)$ (a) greatest and (b) least? Interpret your answers by referring to the rates of change.

29. Collegiate Sports The following chart shows the number of women's college soccer teams in the United States from 1982 to 2001 ($t = 1$ represents the 1981–1982 academic year):

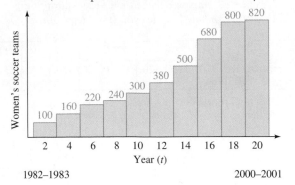

1982–1983   2000–2001

a. On average, how fast was the number of women's college soccer teams growing over the 4-year period starting in the 1992–1993 academic year?

b. By inspecting the graph, determine whether the 4-year average rates of change increased or decreased beginning in 1994–1995.

Source: NCAA/*New York Times*, May 9, 2002, p. D4.

30. Collegiate Sports The following chart shows the number of men's college wrestling teams in the United States from 1982 to 2001 ($t = 1$ represents the 1981–1982 academic year):

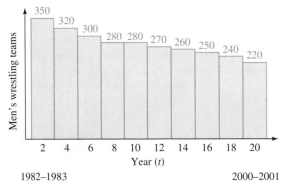

1982–1983   2000–2001

a. On average, how fast was the number of men's college wrestling teams decreasing over the 8-year period starting in the 1984–1985 academic year?

b. By inspecting the graph, determine when the number of men's college wrestling teams were declining the fastest.

Source: NCAA/*New York Times*, May 9, 2002, p. D4.

31. Boat Sales The following table shows U.S. sales of recreational boats during the period 1997–2001 ($t = 7$ represents 1997):

Year, t	7	8	9	10	11
Boats Sold, $N(t)$ (thousands)	620	580	590	580	540

Figures are approximate and represent new recreational boats sold. Source: National Marine Manufacturers Association/*New York Times*, January 10, 2002, p. C1.

a. Find the interval(s) over which the average rate of change of N was the most negative. What was that rate of change? Interpret your answer.

b. The **percent change of N over the interval** $[a, b]$ is defined to be

$$\text{percent change of } N = \frac{\text{change in } N}{\text{first value}} = \frac{N(b) - N(a)}{N(a)}$$

Compute the percent change of N over the interval $[9, 11]$ and the average rate of change. Interpret the answers.

32. Boat Sales The following table shows total revenues from U.S. sales of recreational boats during the period 1997–2001 ($t = 7$ represents 1997):

Year, t	7	8	9	10	11
Revenue, $R(t)$ ($ billions)	7	6	7.5	8	8.5

Figures are approximate and represent the total spent on new and used boats, as well as accessories. Source: National Marine Manufacturers Association/*New York Times*, January 10, 2002, p. C1.

a. Find the interval(s) over which the average rate of change of R was the most negative. What was that rate of change? Interpret your answer.

b. The **percent change of R over the interval** $[a, b]$ is defined to be

$$\text{percent change of } R = \frac{\text{change in } R}{\text{first value}} = \frac{R(b) - R(a)}{R(a)}$$

Compute the percent change of R over the interval $[9, 11]$ and the average rate of change. Interpret the answers.

33. Online Shopping The following graph shows the annual number $N(t)$ of online shopping transactions in the United States for the period January 2000–January 2002 ($t = 0$ represents January 2000):

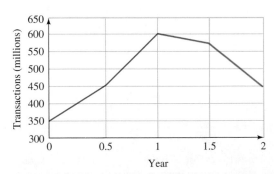

a. Estimate the average rate of change of $N(t)$ over the intervals $[0, 1]$, $[1, 2]$, and $[0, 2]$. Interpret your answers.

b. How can the average rate of change of $N(t)$ over $[0, 2]$ be obtained from the rates over $[0, 1]$ and $[1, 2]$?

Second half of 2001 data is an estimate. Source: Odyssey Research/*New York Times*, November 5, 2001, p. C1.

34. Online Shopping The following graph shows the percentage of people in the United States who have ever purchased anything online for the period January 2000–January 2002 ($t = 0$ represents January 2000):

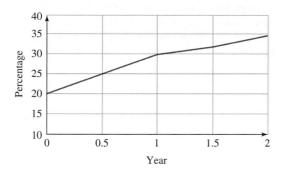

a. Estimate the average rate of change of $P(t)$ over the interval $[0, 2]$ and interpret your answer.

b. Are any of the 1-year average rates of change greater than the 2-year rate? In your answer refer to the slopes of certain lines.

January 2002 data is estimated. SOURCE: Odyssey Research/*New York Times*, November 5, 2001, p. C1.

35. Prison Population The following chart shows the total population in state and federal prisons in 1970–1997 as a function of time in years ($t = 0$ represents 1970), together with the regression line:

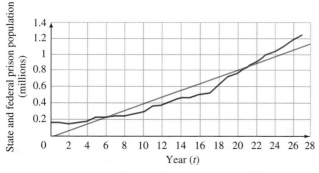

a. Over the period $[0, 6]$, the average rate of change of prison population was
(**A**) less than (**B**) greater than (**C**) approximately equal to the rate predicted by the regression line.

b. Over the period $[6, 21]$, the average rate of change of prison population was
(**A**) less than (**B**) greater than (**C**) approximately equal to the rate predicted by the regression line.

c. Over the period $[21, 26]$, the average rate of change of prison population was
(**A**) less than (**B**) greater than (**C**) approximately equal to the rate predicted by the regression line.

d. Estimate, to one significant digit, the average rate of change of the prison population over the period $[0, 27]$. (Be careful to state the units of measurement.) How does it compare to the slope of the regression line?

SOURCES: Bureau of Justice Statistics; New York State Dept. of Correctional Services/*New York Times*, January 9, 2000, p. WK3.

36. Prison Population The following chart shows the total population in New York State prisons in 1970–1999 as a function of time in years ($t = 0$ represents 1970), together with the regression line:

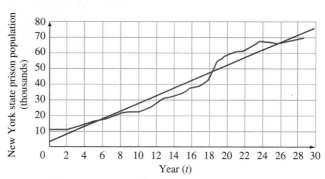

a. Over the period $[6, 26]$, the average rate of change of prison population was
(**A**) less than (**B**) greater than (**C**) approximately equal to the rate predicted by the regression line.

b. Over the period $[10, 26]$, the average rate of change of prison population was
(**A**) less than (**B**) greater than (**C**) approximately equal to the rate predicted by the regression line.

c. Over the period $[0, 6]$, the average rate of change of prison population was
(**A**) less than (**B**) greater than (**C**) approximately equal to the rate predicted by the regression line.

d. Estimate, to the nearest whole number, the average rate of change of the prison population over the period $[0, 29]$. (Be careful to state the units of measurement.) How does it compare to the slope of the regression line?

SOURCES: Bureau of Justice Statistics; New York State Dept. of Correctional Services/*New York Times*, January 9, 2000, p. WK3.

37. Market Volatility A *volatility index* generally measures the extent to which a market undergoes sudden changes in value. The volatility of the S&P 500 (as measured by one such index) was decreasing at an average rate of 0.2 point per year during 1991–1995 and was increasing at an average rate of about 0.3 point per year during 1995–1999. In 1995 the volatility of the S&P was 1.1. Use this information to give a rough sketch of the volatility of the S&P 500 as a function of time, showing its values in 1991 and 1999.

SOURCE: Sanford C. Bernstein Co./*New York Times*, March 24, 2000, p. C1.

38. Market Volatility The volatility (see Exercise 37) of the Nasdaq had an average rate of change of 0 points/year during 1992–1995 and increased at an average rate of 0.2 point per year during 1995–1998. In 1995 the volatility of the Nasdaq was 1.1. Use this information to give a rough sketch of the volatility of the Nasdaq as a function of time.

SOURCE: Sanford C. Bernstein Co./*New York Times*, March 24, 2000, p. C1.

39. Market Index Joe Downs runs a small investment company from his basement. Every week he publishes a report on the success of his investments, including the progress of the "Joe Downs Index." At the end of one particularly memorable week, he reported that the index for that week had the value $I(t) = 1000 + 1500t - 800t^2 + 100t^3$ points, where t represents the number of business days into the week; t ranges from 0 at the beginning of the week to 5 at end of the week. The graph of I is shown below:

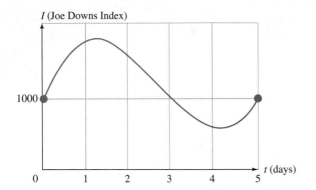

I (Joe Downs Index)

1000

0   1   2   3   4   5 t (days)

On average, how fast and in what direction was the index changing over the first 2 business days (the interval $[0, 2]$)?

40. Market Index Refer to the Joe Downs Index in Exercise 39. On average, how fast and in which direction was the index changing over the last 3 business days (the interval $[2, 5]$)?

41. Swimming Pool Sales The number of in-ground swimming pools built in the United States each year from 1996 to 2001 can be approximated by

$$n(t) = -0.45t^2 + 15t + 54 \text{ thousand pools} \quad (6 \le t \le 11)$$

where t is the year since 1990. Compute the average rate of change of $n(t)$ over the interval $[6, 10]$ and interpret your answer.

Based on a regression model. SOURCE: PK Data/*New York Times,* July 5, 2001, p. C1.

42. Swimming Pool Sales Repeat Exercise 41, using the following model for the number of above-ground swimming pools built in the United States each year:

$$n(t) = 0.36t^2 + 4.1t + 140 \text{ thousand pools} \quad (6 \le t \le 11)$$

Based on a regression model. SOURCE: PK Data/*New York Times,* July 5, 2001, p. C1.

43. Income The average after-tax income of the wealthiest 1% of U.S. taxpayers can be approximated by

$$I(t) = 7.32t^2 - 34.53t + 401.4 \text{ thousand dollars} \quad (0 \le t \le 8)$$

where t is the year since 1990. Compute the average rate of change of I over the period $[4, 8]$ and interpret the result.

Based on a regression of approximate data. Figures are in constant 1998 dollars. SOURCE: IRS/*New York Times,* February 26, 2001, p. C2.

44. Income The share of all personal after-tax income of the wealthiest 1% of U.S. taxpayers can be approximated by

$$P(t) = 0.145t^2 - 0.681t + 11.95 \text{ percent} \quad (0 \le t \le 8)$$

where t is the year since 1990. Compute the average rate of change of P over the period $[4, 8]$ and interpret the result.

Based on a regression of approximate data. Figures are in constant 1998 dollars. SOURCE: IRS/*New York Times,* February 26, 2001, p. C2.

45. Ecology Increasing numbers of manatees ("sea sirens") have been killed by boats off the Florida coast. The following graph shows the relationship between the number of boats registered in Florida and the number of manatees killed each year:

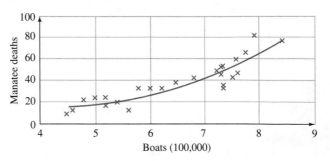

Boats (100,000)

The regression curve shown is given by

$$f(x) = 3.55x^2 - 30.2x + 81 \text{ manatees} \quad (4.5 \le x \le 8.5)$$

where x is the number of boats (in hundreds of thousands) registered in Florida in a particular year and $f(x)$ is the number of manatees killed by boats in Florida that year.

a. Compute the average rate of change of f over the intervals $[5, 6]$ and $[7, 8]$.

b. What does the answer to part (a) tell you about the manatee deaths per boat?

Regression model is based on data from 1976 to 2000. SOURCE: Florida Department of Highway Safety & Motor Vehicles, Florida Marine Institute/*New York Times,* February 12, 2002, p. F4.

46. Ecology Refer to Exercise 45.

a. Compute the average rate of change of f over the intervals $[5, 7]$ and $[6, 8]$.

b. Had we used a linear model instead of a quadratic one, how would the two answers in part (a) be related to each other?

47. Advertising Revenue The following table shows the annual advertising revenue earned by America Online (AOL) during the last 3 years of the 1990s:

Year	1997	1998	1999
Revenue ($ millions)	150	360	760

Figures are rounded to the nearest $10 million.
SOURCES: AOL; Forrester Research/*New York Times,* January 31, 2000, p. C1.

These data can be modeled by

$$R(t) = 95t^2 + 115t + 150 \text{ million dollars} \quad (0 \le t \le 2)$$

where t is time in years since December 1997.

a. What was the average rate of change of R over the period 1997–1999? Interpret the result.

b. Which of the following is true? From 1997 to 1999, annual online advertising revenues
   (A) increased at a faster and faster rate.
   (B) increased at a slower and slower rate.
   (C) decreased at a faster and faster rate.
   (D) decreased at a slower and slower rate.
c. Use the model to project the average rate of change of R over the 1-year period ending December 2000. Interpret the result.

48. Religion The following table shows the U.S. population of Roman Catholic nuns during the last 25 years of the 1900s:

Year	1975	1985	1995
Population	130,000	120,000	80,000

Figures are rounded. SOURCE: Center for Applied Research in the Apostolate/*New York Times,* January 16, 2000, p. A1.

These data can be modeled by

$$P(t) = -0.15t^2 + 0.50t + 130 \text{ thousand nuns} \quad (0 \le t \le 20)$$

where t is time in years since December 1975.
a. What was the average rate of change of P over the period 1975–1995? Interpret the result.
b. Which of the following is true? From 1975 to 1995, the population of nuns
   (A) increased at a faster and faster rate.
   (B) increased at a slower and slower rate.
   (C) decreased at a faster and faster rate.
   (D) decreased at a slower and slower rate.
c. Use the model to project the average rate of change of P over the 10-year period ending December 2005. Interpret the result.

COMMUNICATION AND REASONING EXERCISES

49. Describe three ways of determining the average rate of change of f over an interval $[a, b]$. Which of the three methods is *least* precise? Explain.

50. If f is a linear function of x with slope m, what is its average rate of change over any interval $[a, b]$?

51. Sketch the graph of a function whose average rate of change over $[0, 3]$ is negative but whose average rate of change over $[1, 3]$ is positive.

52. Sketch the graph of a function whose average rate of change over $[0, 2]$ is positive but whose average rate of change over $[0, 1]$ is negative.

53. If the rate of change of quantity A is 2 units of quantity A per unit of quantity B and the rate of change of quantity B is 3 units of quantity B per unit of quantity C, what is the rate of change of quantity A with respect to quantity C?

54. If the rate of change of quantity A is 2 units of quantity A per unit of quantity B, what is the rate of change of quantity B with respect to quantity A?

55. A certain function has the property that its average rate of change over the interval $[1, 1 + h]$ (for positive h) increases as h decreases. Which of the following graphs could be the graph of f?

(A)   **(B)**

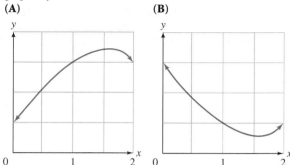

(C)

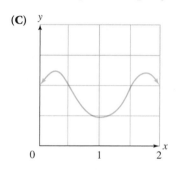

56. A certain function has the property that its average rate of change over the interval $[1, 1 + h]$ (for positive h) decreases as h decreases. Which of the following graphs could be the graph of f?

(A)   **(B)**

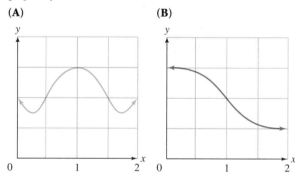

(C)

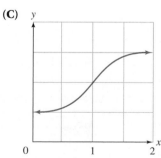

57. Is it possible for a company's revenue to have a negative 3-year average rate of growth but a positive average rate of growth in 2 of the 3 years? (If not, explain; if so, illustrate with an example.)

58. Is it possible for a company's revenue to have a larger 2-year average rate of change than either of the 1-year average rates of change? (If not, explain why with the aid of a graph; if so, illustrate with an example.)

59. The average rate of change of f over $[1, 3]$ is
(A) always equal to   (B) never equal to
(C) sometimes equal to
the average of its average rates of change over $[1, 2]$ and $[2, 3]$.

60. The average rate of change of f over $[1, 4]$ is
(A) always equal to   (B) never equal to
(C) sometimes equal to
the average of its average rates of change over $[1, 2]$, $[2, 3]$, and $[3, 4]$.

3.2 The Derivative: Numerical and Graphical Viewpoints

In Example 4 of Section 3.1, we looked at the average rate of change of the function $R(t) = 7500 + 500t - 100t^2$, representing the value of the U.S. dollar in Indonesian currency (rupiah) during a hypothetical 5-day period, over the intervals $[3, 3 + h]$ for successively smaller values of h. Here are the values we got:

h getting smaller; interval $[3, 3 + h]$ getting smaller $\rightarrow$

h	1	0.1	0.01	0.001	0.0001
Average Rate of Change over $[3, 3 + h]$	-200	-110	-101	-100.1	-100.01

Rate of change approaching -100 points per day $\rightarrow$

The average rate of change of the value of the U.S. dollar over smaller and smaller periods of time, starting at the instant $t = 3$ (noon on Thursday), appears to be getting closer and closer to -100 rupiahs per day. As we look at these shrinking periods of time, we are getting closer to looking at what happens at the *instant $t = 3$*. So it seems reasonable to say that the average rates of change are approaching the **instantaneous rate of change** at $t = 3$, which the table suggests was -100 rupiahs per day. This is how fast the value of the U.S. dollar was changing at $t = 3$ (noon on Thursday).

The process of letting h get smaller and smaller is called taking the **limit** as h approaches 0. We write $h \rightarrow 0$ as shorthand for "h approaches zero." Thus, taking the limit of the average rates of change as $h \rightarrow 0$ gives us the instantaneous rate of change.

Question All these intervals $[3, 3 + h]$ are intervals to the right of 3. What about small intervals to the *left* of 3, such as $[2.9, 3]$?

Answer We can compute the average rate of change of our function for such intervals by choosing h to be negative ($h = -1, -0.1, -0.01$, and so on) and using the same difference quotient formula we used for positive h:

$$\text{average rate of change of } R \text{ over } [3 + h, 3] = \frac{R(3) - R(3 + h)}{3 - (3 + h)}$$

$$= \frac{R(3 + h) - R(3)}{h}$$

Here are the results we get using negative h:

h getting closer to 0; interval $[3 + h, 3]$ getting smaller $\rightarrow$

h	-1	-0.1	-0.01	-0.001	-0.0001
Average Rate of Change over $[3 + h, 3]$	0	-90	-99	-99.9	-99.99

Rate of change approaching -100 points per day $\rightarrow$

Notice that the average rates of change are again getting closer and closer to -100 as h approaches 0, suggesting once again that the instantaneous rate of change is -100 rupiahs per day.

Instantaneous Rate of Change of f(x) at x = a: Derivative
The **instantaneous rate of change** of $f(x)$ at $x = a$ is obtained as follows:

1. Compute the average rate of change of f,

$$\frac{f(a + h) - f(a)}{h}$$

over shorter and shorter intervals $[a, a + h]$ by using smaller and smaller values of h, approaching the value 0. Also allow $h < 0$, in which case the interval is $[a + h, a]$.

2. If these average rates of change approach a fixed number (for h both positive and negative), that number is the instantaneous rate of change. We write

$$\text{instantaneous rate of change} = \lim_{h \to 0} \frac{f(a + h) - f(a)}{h}$$

(We read $\lim_{h \to 0}$ as "the limit as h approaches zero of.") We also call the instantaneous rate of change the **derivative** of $f(x)$ at $x = a$, which we write as $f'(a)$ (read "f prime of a"). Thus,

$$f'(a) = \lim_{h \to 0} \frac{f(a + h) - f(a)}{h}$$

$$= \text{instantaneous rate of change of } f(x) \text{ at } x = a$$

$$= \text{derivative of } f(x) \text{ at } x = a$$

Units
The units of $f'(a)$ are the same as the units of the average rate of change: units of f per unit of x.

Quick Example
If $f(x) = 7500 + 500x - 100x^2$, then the two tables above suggest that

$$f'(3) = \lim_{h \to 0} \frac{f(3 + h) - f(3)}{h} = -100$$

Notes
1. For now, we will trust our intuition when it comes to limits. (We discuss limits in detail in the last sections of this chapter.)

2. In this section we only approximate derivatives. In Section 3.3 we will begin to see how we find the exact values of derivatives.

3. $f'(a)$ is a number we can calculate, or at least approximate, for various values of a, as we have done in the example above. Since $f'(a)$ depends on the value of a, we can think of f' as a *function of a*. (We return to this idea at the end of this section.) An old name for f' is "the function *derived from f*," which has been shortened to the *derivative of f*.

4. It is because f' is a function that we sometimes refer to $f'(a)$ as "the derivative of f *evaluated* at a," or the "derivative of $f(x)$ evaluated at $x = a$."

5. Finding the derivative of f is called **differentiating f.**

Question We found the instantaneous rate of change above by observing that the average rates of change over the smaller and smaller intervals [3, 3.1], [3, 3.01], [3, 3.001], . . . , as well as [2.9, 3], [2.99, 3], [2.999, 3], . . . , approach a fixed number (−100). What happens if the average rates of change do not approach any fixed number at all or if they approach one number on the intervals using positive h and another on those using negative h?

Answer That can certainly happen. When, as in the example above, the average rates of change $[f(a + h) − f(a)]/h$ *do* approach a fixed value, we say that $f'(a)$ **exists,** or that f is **differentiable** at the point $x = a$. Thus, $f(x) = 7500 + 500x − 100x^2$ is differentiable at $x = 3$. If, on the other hand, the average rates of change $[f(a + h) − f(a)]/h$ do *not* approach some fixed number as h approaches 0, then f is **not differentiable** at $x = a$. That is, $f'(a)$ **does not exist.** Examples are $f(x) = |x|$ and $f(x) = x^{1/3}$, which are both not differentiable at $x = 0$ (see Sections 3.3 and 3.4). However, it is comforting to know that all polynomials and exponential functions are differentiable at *every* point.

Example 1 • Instantaneous Rate of Change

The air temperature (in degrees Fahrenheit) one spring morning was given by the function $f(t) = 50 + 0.1t^4$, t hours after 7:00 A.M. ($0 \leq t \leq 4$).

a. How fast was the temperature rising at 9:00 A.M.?

b. How is the instantaneous rate of change of temperature at 9:00 A.M. reflected in the graph of temperature versus time?

Solution

a. We are being asked to find the instantaneous rate of change of the temperature at $t = 2$, so we need to find $f'(2)$. To do this we examine the average rates of change

$$\frac{f(2 + h) − f(2)}{h} \qquad \text{average rate of change = difference quotient}$$

for values of h approaching 0. Calculating the average rate of change over $[2, 2 + h]$ for $h = 1, 0.1, 0.01, 0.001$, and 0.0001, we get the following values (rounded to four decimal places):[3]

h	1	0.1	0.01	0.001	0.0001
Average Rate of Change over $[2, 2 + h]$	6.5	3.4481	3.2241	3.2024	3.2002

Here are the values we get using negative values of h:

h	−1	−0.1	−0.01	−0.001	−0.0001
Average Rate of Change over $[2 + h, 2]$	1.5	2.9679	3.1761	3.1976	3.1998

The average rates of change are clearly approaching the number 3.2, so we can say that $f'(2) = 3.2$. Thus, at 9:00 A.M., the temperature was rising at the rate of 3.2°F per hour.

b. We saw in Section 3.1 that the average rate of change of f over an interval is the slope of the secant line through the corresponding points on the graph of f. Figure 6 illustrates this for the intervals $[2, 2 + h]$ with $h = 1, 0.5$, and 0.1.

---

[3]See Example 4 in Section 3.1 for instructions to quickly compute these values using technology.

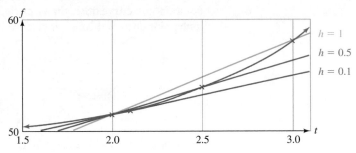

Figure 6

All three secant lines pass though the point $(2, f(2)) = (2, 51.6)$ on the graph of f. Each passes through a second point on the curve, and this second point gets closer and closer to $(2, 51.6)$ as h gets closer to 0. What seems to be happening is that the secant lines are getting closer and closer to a line that just touches the curve at $(2, 51.6)$: the **tangent line** at $(2, 51.6)$, shown in Figure 7.

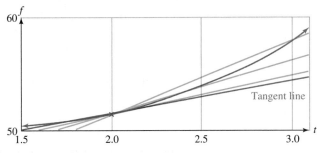

Figure 7

Question What is the slope of this tangent line?

Answer Since the slopes of the secant lines are getting closer and closer to 3.2 and since the secant lines are approaching the tangent line, the tangent line must have slope 3.2. In other words,

Graphically, $f'(2)$ is the slope of the tangent line at the point on the graph where $x = 2$.

✳ *Before we go on . . .*
Question What is the difference between $f(2)$ and $f'(2)$? What does each mean?

Answer Briefly, $f(2)$ is the *value of f* when $t = 2$, whereas $f'(2)$ is the *rate at which f is changing* when $t = 2$. Here,

$$f(2) = 50 + 0.1(2)^4 = 51.6°F$$

Thus, at 9:00 A.M. ($t = 2$), the temperature was 51.6°F. On the other hand,

$$f'(x) = 3.2°F/hour \qquad \text{Units of slope are units of } f \text{ per unit of } t.$$

This means that, at 9:00 A.M. ($t = 2$), the temperature was increasing at a rate of 3.2°F per hour.

Question What exactly *is* the "tangent line to the graph at a point"?

Answer A tangent line to a *circle* is a line that touches the circle in just one point. A tangent line gives the circle "a glancing blow," as shown in Figure 8.

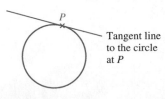

Figure 8

For a smooth curve other than a circle, a tangent line may touch the curve at more than one point or pass through it (Figure 9).

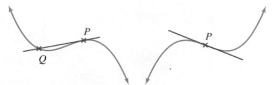

Tangent line at _P_ intersects graph at _Q_.   **Tangent line at _P_ passes through curve at _P_.**

Figure 9

However, all the tangent lines we have sketched have the following property in common: If we focus on a small portion of the curve very close to the point _P_—in other words, if we zoom in on the graph near the point _P_—the curve will appear almost straight and almost indistinguishable from the tangent line (Figure 10).

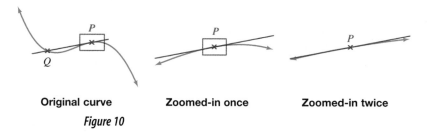

Original curve   **Zoomed-in once**   **Zoomed-in twice**

Figure 10

Try zooming in on the curve shown in Figures 6 and 7 near the point where $x = 2$.

Secant and Tangent Lines

The *slope of the secant line* through the points on the graph of f where $x = a$ and $x = a + h$ is given by the average rate of change, or difference quotient,

$$m_{\text{sec}} = \text{slope of secant} = \text{average rate of change} = \frac{f(a + h) - f(a)}{h}$$

The *slope of the tangent line* through the point on the graph of f where $x = a$ is given by the instantaneous rate of change, or derivative,

$$m_{\text{tan}} = \text{slope of tangent} = \text{derivative} = f'(a) = \lim_{h \to 0} \frac{f(a + h) - f(a)}{h}$$

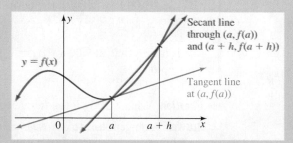

Notes

1. It might happen that the tangent line is vertical at some point or does not exist at all. These are the cases in which f is not differentiable at the given point. (See below.)

2. We can now give a more precise definition of what we mean by the tangent line to a point P on the graph of f at a given point.

Definition of Tangent Line

The **tangent line** to the graph of f at the point $P(a, f(a))$ is the straight line passing through P with slope $f'(a)$.

Quick Approximation of the Derivative

Question Do we always need to make tables of difference quotients as above in order to calculate an approximate value for the derivative?

Answer We can usually *approximate* the value of the derivative by using a single, small value of h. In the example above, the value $h = 0.0001$ would have given a pretty good approximation. The problems with using a fixed value of h are that (1) we do not get an *exact* answer, only an *approximation* of the derivative, and (2) how good an approximation it is depends on the function that we're differentiating.[4] However, with most of the functions we'll be considering, it is a good enough approximation.

Calculating a Quick Approximation of the Derivative

We can calculate an approximate value of $f'(a)$ by using the formula

$$f'(a) \approx \frac{f(a + h) - f(a)}{h} \qquad \text{Rate of change over } [a, a + h]$$

with a small value of h. The value $h = 0.0001$ often works (but see Example 2 for a graphical way of determining a good value to use).

Alternative Formula: The Balanced Difference Quotient

The following alternative formula, which measures the rate of change of f over the interval $[a - h, a + h]$, often gives a more accurate result and is the one used in many calculators:

$$f'(a) \approx \frac{f(a + h) - f(a - h)}{2h} \qquad \text{Rate of change over } [a - h, a + h]$$

Note

For these approximations to be valid, the function f must be differentiable; that is, $f'(a)$ must exist.

Example 2 • Quick Approximation of the Derivative

Calculate an approximate value of $f'(1.5)$ if $f(x) = x^2 - 4x$ and then find the equation of the tangent line at the point on the graph where $x = 1.5$.

---

[4]In fact, no matter how small the value we decide to use for h, it is possible to craft a function f for which the difference quotient at a is not even close to $f'(a)$.

Here is one last comment on Leibniz notation. In Example 3 we could have written the velocity either as s' or as ds/dt, as we chose to do. To write the answer to the question, that the velocity at $t = 2$ seconds was 36 feet per second, we can write either

$$s'(2) = 36 \qquad \text{or} \qquad \left.\frac{ds}{dt}\right|_{t=2} = 36$$

The notation $\big|_{t=2}$ is read "evaluated at $t = 2$." Similarly, if $y = f(x)$, we can write the instantaneous rate of change of f at $x = 5$ either in functional notation as

$$f'(5) \qquad \qquad \text{The derivative of } f, \text{ evaluated at } x = 5$$

or in Leibniz notation as

$$\left.\frac{dy}{dx}\right|_{x=5} \qquad \qquad \text{The derivative of } y, \text{ evaluated at } x = 5$$

The latter notation is obviously more cumbersome than the functional notation $f'(5)$, but the notation dy/dx has compensating advantages. You should feel free to use whichever notation is more convenient at the time. In fact, you should practice using both notations.

The Derivative Function

The derivative $f'(x)$ is a number we can calculate, or at least approximate, for various values of x. Since $f'(x)$ depends on the value of x, we may think of f' as a function of x. This function is the **derivative function.**

Derivative Function

If f is a function, its **derivative function** f' is the function whose value $f'(x)$ is the derivative of f at x. Its domain is the set of all x at which f is differentiable. Equivalently, f' associates to each x the slope of the tangent to the graph of the function f at x, or the rate of change of f at x. The formula for the derivative function is

$$f'(x) = \lim_{h \to 0} \frac{f(x + h) - f(x)}{h} \qquad \qquad \text{Derivative function}$$

Quick Examples

1. Let $f(x) = 3x - 1$. The graph of f is a straight line that has slope 3 everywhere. In other words, $f'(x) = 3$ for every choice of x; that is, f' is a constant function.

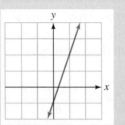

Original function f:
$f(x) = 3x - 1$

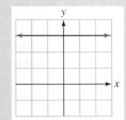

Slope function f':
$f'(x) = 3$

2. Given the graph of a function f, we can get a rough sketch of the graph of f' by estimating the slope of the tangent to the graph of f at several points, as illustrated below.*

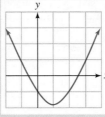
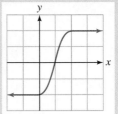

Original function f:	Slope function f':
$y = f(x)$	$y = f'(x)$

For x between -2 and 0, the graph of f is linear with slope -2. As x increases from 0 to 2, the slope increases from -2 to 2. For x larger than 2, the graph of f is linear with slope 2. (Notice that, when $x = 1$, the graph of f has a horizontal tangent, so $f'(1) = 0$.)

---

*This method is illustrated online at Web Site → Online Text → Sketching the Graph of the Derivative.

Example 4 shows how we can use technology to graph the (approximate) derivative of a function, where it exists.

Example 4 • *Tabulating and Graphing the Derivative with Technology*

Use technology to obtain a table of values of and graph the derivative of

$$f(x) = -2x^2 + 6x + 5 \ (-5 \le x \le 5)$$

Solution

Graphing Calculator

On the TI-83 the easiest way to obtain quick approximations of the derivative of a given function is to use the built-in nDeriv function, which calculates balanced difference quotients. On the Y= screen, we first enter the function:

```
Y₁=-2x^2+6x+5
```

Then, we set

```
Y₂=nDeriv(Y₁,X,X)
```

which is the TI-83's approximation of $f'(x)$. Alternatively, we can enter the balanced difference quotient directly:

```
Y₂=(Y₁(X+0.001)-Y₁(X-0.001))/0.002
```

(The TI-83 uses $h = 0.001$ by default in the balanced difference quotient when calculating nDeriv, but this can be changed by giving a value of h as a fourth argument, like nDeriv(Y₁,X,X,0.0001).) To see a table of approximate values of the derivative, press [2nd] [TABLE] and choose a collection of values for x. To graph the function or its derivative, we can graph Y_1 or Y_2 in a window showing the given domain $[-5, 5]$. The graphs of f and f' are shown in Figure 15.

Excel

We start with a table of values for the function *f*, set up in the same way as our graphing worksheet:

	A	B	C	D	E
1	x	f(x)			
2	=$E$2	=-2*A2^2+6*A2+5		Xmin	-5
3	=A2+$E$5			Xmax	5
4				Points	100
5				h	=(E3-E2)/E4
6					
100					
101					
102					

Next, we compute approximate derivatives in column C:

	A	B	C	D	E
1	x	f(x)	f'(x)		
2	-5	-75	=(B3-B2)/(A3-A2)	Xmin	-5
3	-4.9	-72.42		Xmax	5
4	-4.8	-69.88		Points	100
5	-4.7	-67.38		h	0.1
6					
100					
101	4.9	-13.62			
102	5	-15			

We cannot paste the difference quotient formula into cell C102. (Why?) Notice that this worksheet uses the ordinary difference quotients, $[f(x + h) - f(x)]/h$. If you prefer you can use balanced difference quotients $[f(x + h) - f(x - h)]/(2h)$, in which case cells C2 and C102 would both have to be left blank.

We then graph the function and the derivative on different graphs as follows. First, graph the function *f* in the usual way, using columns A and B. Then make a copy of this graph and click on it once. Columns A and B should be outlined, indicating that these are the columns used in the graph. By dragging from the center of the bottom edge (Figure 14), we can move the column B box over to column C as shown. The graph will then show the derivative (columns A and C) as in Figure 15.

99	4.7	-10.98	-13
100	4.8	-12.28	-13.4
101	4.9	-13.62	-13.8
102	5	-15	
103			

99	4.7	-10.98	-13
100	4.8	-12.28	-13.4
101	4.9	-13.62	-13.8
102	5	-15	
103			

Figure 14

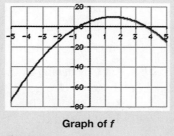

Graph of *f*

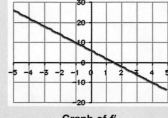

Graph of *f'*

Figure 15

✳ *Before we go on . . .* We said that f' records the slope of (the tangent line to) the function f at each point. Notice that the graph of f' in Figure 15 confirms that the slope of the graph of f is decreasing as x increases from -5 to 5. Note also that the graph of f reaches a high point at $x = 1.5$ (the vertex of the parabola). At that point the slope of the tangent is zero; that is, $f'(1.5) = 0$, as we see in the graph of f'.

📅 Example 5 • Market Growth

The number N of U.S. households connected to the Internet can be modeled by the logistic function

$$N(t) = \frac{79.317}{1 + 2.2116(1.3854)^{-t}} \text{ million households} \qquad (0 \le t \le 9)$$

where t is the number of quarters since the start of 1999.[7] Graph both N and its derivative and hence determine when Internet usage was growing most rapidly.

Solution Using one of the methods in Example 4, we obtain the graphs shown in Figure 16.

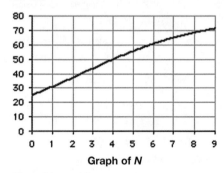
Graph of N

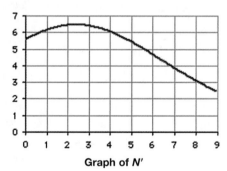
Graph of N'

Figure 16

From the graph on the right, we see that N' reaches a peak somewhere between $t = 2$ and $t = 3$ (sometime during the third quarter of 1999). Recalling that N' measures the *slope* of the graph of N, we can conclude that the graph of N is steepest between $t = 2$ and $t = 3$, indicating that, according to the model, the number of U.S. households connected to the Internet was growing most rapidly during the third quarter of 1999. Notice that this is not so easy to see directly on the graph of N.

To determine the point of maximum growth more accurately, we can zoom in on the graph of N' using the range $2 \le t \le 3$ (Figure 17).

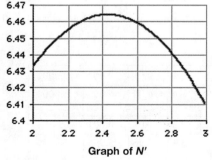
Graph of N'

Figure 17

---

[7]The model was obtained in Section 2.4 by logistic regression. SOURCE: Telecommunications Reports International/*New York Times*, May 21, 2001, p. C9.

We can now see that N' reaches its highest point around $t = 2.4$, so we conclude that the number of U.S. households connected to the Internet was growing most rapidly shortly before the midpoint of the third quarter of 1999.

✸ *Before we go on...*
Question What other information does the graph of N' give us about U.S. Internet usage?

Answer The graph of N' gives us a great deal of quantitative as well as qualitative information. As just one example, in Figure 16 we can see that the maximum value of N' is approximately 6.5, indicating that Internet usage in the United States never grew at a faster rate than about 6.5 million households per quarter.

Guideline: Recognizing When and How to Compute the Instantaneous Rate of Change

Question How do I know, by looking at the wording of a problem, that it is asking for an instantaneous rate of change?

Answer If a problem does not ask for an instantaneous rate of change directly, it might do so indirectly, as in "How fast is quantity q increasing?" or "Find the rate of increase of q."

Question If I know that a problem calls for estimating an instantaneous rate of change, how should I estimate it: with a table showing smaller and smaller values of h or by using a quick approximation?

Answer For most practical purposes, a quick approximation is accurate enough. Use a table showing smaller and smaller values of h when you would like to check the accuracy.

Question Which should I use in computing a quick approximation: the balanced difference quotient or the ordinary difference quotient?

Answer In general, the balanced difference quotient gives a more accurate answer.

3.2 EXERCISES

In Exercises 1–4, estimate the derivative from the table of average rates of change.

1.

h	1	0.1	0.01	0.001	0.0001
Ave. Rate of Change of f over $[5, 5 + h]$	12	6.4	6.04	6.004	6.0004
h	−1	−0.1	−0.01	−0.001	−0.0001
Ave. Rate of Change of f over $[5 + h, 5]$	3	5.6	5.96	5.996	5.9996

Estimate $f'(5)$.

2.

h	1	0.1	0.01	0.001	0.0001
Ave. Rate of Change of g over $[7, 7 + h]$	4	4.8	4.98	4.998	4.9998
h	−1	−0.1	−0.01	−0.001	−0.0001
Ave. Rate of Change of g over $[7 + h, 7]$	5	5.3	5.03	5.003	5.0003

Estimate $g'(7)$.

3.

h	1	0.1	0.01	0.001	0.0001
Ave. Rate of Change of r over $[-6, -6+h]$	-5.4	-5.498	-5.4998	$-5.499\,982$	$-5.499\,998\,22$
h	-1	-0.1	-0.01	-0.001	-0.0001
Ave. Rate of Change of r over $[-6+h, -6]$	-7.52	-6.13	-5.5014	$-5.500\,014\,4$	$-5.500\,001\,444$

Estimate $r'(-6)$.

4.

h	1	0.1	0.01	0.001	0.0001
Ave. Rate of Change of s over $[0, h]$	-2.52	-1.13	-0.6014	$-0.600\,014\,4$	$-0.600\,001\,444$
h	-1	-0.1	-0.01	-0.001	-0.0001
Ave. Rate of Change of s over $[h, 0]$	-0.4	-0.598	-0.5998	$-0.599\,982$	$-0.599\,998\,22$

Estimate $s'(0)$.

In Exercises 5–8, consider the functions as representing the value of the U.S. dollar in Indian rupees as a function of the time t in days.[8] Find the average rates of change of $R(t)$ over the time intervals $[t, t + h]$, where t is as indicated and $h = 1, 0.1$, and 0.01 days. Hence, estimate the instantaneous rate of change of R at time t, specifying the units of measurement. (Use smaller values of h to check your estimates.)

5. $R(t) = 50t - t^2; t = 5$ **6.** $R(t) = 60t - 2t^2; t = 3$

7. $R(t) = 100 + 20t^3; t = 1$ **8.** $R(t) = 1000 + 50t - t^3; t = 2$

In Exercises 9–12, each function gives the cost to manufacture x items. Find the average cost per unit of manufacturing h more items (that is, the average rate of change of the total cost) at a production level of x, where x is as indicated and $h = 10$ and 1. Hence, estimate the instantaneous rate of change of the total cost at the given production level x, specifying the units of measurement. (Use smaller values of h to check your estimates.)

9. $C(x) = 10,000 + 5x - \dfrac{x^2}{10,000}; x = 1000$

10. $C(x) = 20,000 + 7x - \dfrac{x^2}{20,000}; x = 10,000$

11. $C(x) = 15,000 + 100x + \dfrac{1000}{x}; x = 100$

12. $C(x) = 20,000 + 50x + \dfrac{10,000}{x}; x = 100$

In Exercises 13–18, in each graph given, say at which labeled point the slope of the tangent is (a) greatest and (b) least (in the sense that -7 is less than 1).

13.
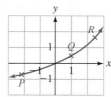

14.
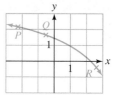

15.
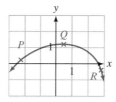

16.
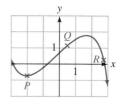

17.
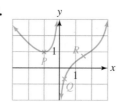

18.
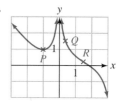

In Exercises 19–22, the graph of a function is shown together with the tangent line at a point P. Estimate the derivative of f at the corresponding x value.

19.
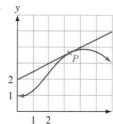

20.
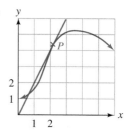

21.
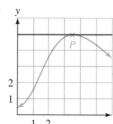

22.
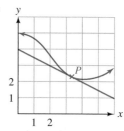

In Exercises 23–26, three slopes are given. For each slope determine at which of the labeled points on the graph the tangent line has that slope.

23. a. 0 **b.** 4 **c.** -1

24. a. 0 **b.** 1 **c.** -1

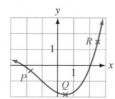

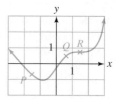

---

[8]The U.S. dollar was trading at around 49 rupees in July 2002.

25. a. 0 **b.** 3 **c.** −3   **26. a.** 0 **b.** 3 **c.** 1

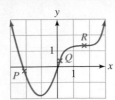

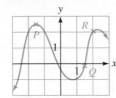

In Exercises 27–30, find the approximate coordinates of all points (if any) where the slope of the tangent is (a) 0, (b) 1, and (c) −1.

27.

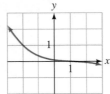

28.

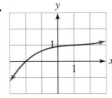

29.

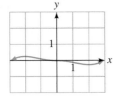

30.

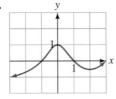

31. Complete the following: The tangent to the graph of the function f at the point where $x = a$ is the line passing through the point _____ with slope _____.

32. Complete the following: The difference quotient for f at the point where $x = a$ gives the slope of the _____ line that passes through _____.

33. Let f have the graph shown.

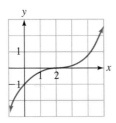

a. The average rate of change of f over the interval $[2, 4]$ is
(**A**) greater than   (**B**) less than
(**C**) approximately equal to
$f'(2)$.

b. The average rate of change of f over the interval $[-1, 1]$ is
(**A**) greater than   (**B**) less than
(**C**) approximately equal to
$f'(0)$.

c. Over the interval $[0, 2]$, the instantaneous rate of change of f is
(**A**) increasing.   (**B**) decreasing.   (**C**) neither.

d. Over the interval $[0, 4]$, the instantaneous rate of change of f is
(**A**) increasing and then decreasing.
(**B**) decreasing and then increasing.
(**C**) always increasing.
(**D**) always decreasing.

e. When $x = 4$, $f(x)$ is
(**A**) approximately 0 and increasing at a rate of about 0.7 unit per unit of x.
(**B**) approximately 0 and decreasing at a rate of about 0.7 unit per unit of x.
(**C**) approximately 0.7 and increasing at a rate of about 1 unit per unit of x.
(**D**) approximately 0.7 and increasing at a rate of about 3 units per unit of x.

34. A function f has the following graph:

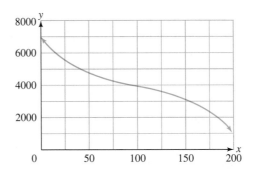

a. The average rate of change of f over $[0, 200]$ is
(**A**) greater than   (**B**) less than
(**C**) approximately equal to
the instantaneous rate of change at $x = 100$.

b. The average rate of change of f over $[0, 200]$ is
(**A**) greater than   (**B**) less than
(**C**) approximately equal to
the instantaneous rate of change at $x = 150$.

c. Over the interval $[0, 50]$, the instantaneous rate of change of f is
(**A**) increasing and then decreasing.
(**B**) decreasing and then increasing.
(**C**) always increasing.   (**D**) always decreasing.

d. On the interval $[0, 200]$, the instantaneous rate of change of f is
(**A**) always positive.   (**B**) always negative.
(**C**) negative, positive, and then negative.

e. $f'(100)$ is
(**A**) greater than   (**B**) less than
(**C**) approximately equal to
$f'(25)$.

In Exercises 35–38, use a quick approximation to estimate the derivative of the given function at the indicated point.

35. $f(x) = 1 - 2x$; $x = 2$ **36.** $f(x) = \dfrac{x}{3} - 1$; $x = -3$

37. $f(x) = \dfrac{x^2}{4} - \dfrac{x^3}{3}$; $x = -1$ **38.** $f(x) = \dfrac{x^2}{2} + \dfrac{x}{4}$; $x = 2$

In Exercises 39–46, estimate the indicated derivative by any method.

39. $g(t) = \dfrac{1}{t^5}$; estimate $g'(1)$

40. $s(t) = \dfrac{1}{t^3}$; estimate $s'(-2)$

41. $y = 4x^2$; estimate $\dfrac{dy}{dx}\Big|_{x=2}$

42. $y = 1 - x^2$; estimate $\dfrac{dy}{dx}\Big|_{x=-1}$

43. $s = 4t + t^2$; estimate $\dfrac{ds}{dt}\Big|_{t=-2}$

44. $s = t - t^2$; estimate $\dfrac{ds}{dt}\Big|_{t=2}$

45. $R = \dfrac{1}{p}$; estimate $\dfrac{dR}{dp}\Big|_{p=20}$

46. $R = \sqrt{p}$; estimate $\dfrac{dR}{dp}\Big|_{p=400}$

In Exercises 47–52, (a) use any method to estimate the slope of the tangent to the graph of the given function at the point with the given x coordinate and (b) find an equation of the tangent line in part (a). In each case, sketch the curve together with the appropriate tangent line.

47. $f(x) = x^3$; $x = -1$

48. $f(x) = x^2$; $x = 0$

49. $f(x) = x + \dfrac{1}{x}$; $x = 2$

50. $f(x) = \dfrac{1}{x^2}$; $x = 1$

51. $f(x) = \sqrt{x}$; $x = 4$

52. $f(x) = 2x + 4$; $x = -1$

Note: Exercises marked with $\boxed{\text{E}}$ are based on logarithmic and exponential functions.

$\boxed{\text{E}}$ In Exercises 53–56, estimate the given quantity.

53. $f(x) = e^x$; estimate $f'(0)$ **54.** $f(x) = 2e^x$; estimate $f'(1)$

55. $f(x) = \ln x$; estimate $f'(1)$ **56.** $f(x) = \ln x$; estimate $f'(2)$

57. Which is correct? The derivative function assigns to each value x
   (A) the average rate of change of f at x.
   (B) the slope of the tangent to the graph of f at $(x, f(x))$.
   (C) the rate at which f is changing over the interval $[x, x + h]$ for $h = 0.0001$.
   (D) the balanced difference quotient $[f(x + h) - f(x - h)]/(2h)$ for $h \approx 0.0001$.

58. Which is correct? The derivative function $f'(x)$ tells us
   (A) the slope of the tangent line at each of the points $(x, f(x))$.
   (B) the approximate slope of the tangent line at each of the points $(x, f(x))$.
   (C) the slope of the secant line through $(x, f(x))$ and $(x + h, f(x + h))$ for $h = 0.0001$.
   (D) the slope of a certain secant line through each of the points $(x, f(x))$.

In Exercises 59–64, match the graph of f to the graph of f' (the graphs of f' are shown after Exercise 64).

59.

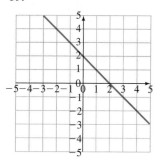

60.

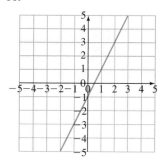

61.

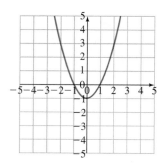

62.

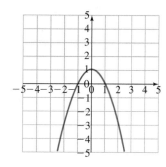

63.

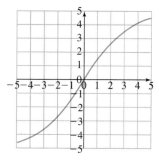

64.

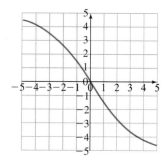

Graphs of derivatives for Exercises 59–64:

(A)

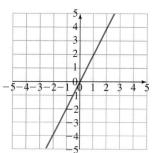

(B)

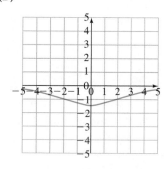

(C)

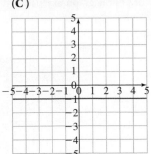

(D)

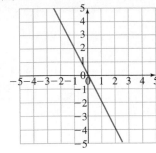

(E)

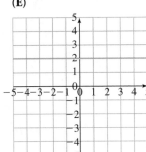

(F)

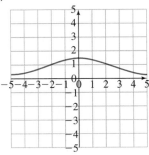

In Exercises 65–70, use technology to graph the derivative of the given function for the given range of values of x. Then use your graph to estimate all values of x (if any) where (a) the given function is not differentiable and (b) the tangent line to the graph of the given function is horizontal. Round answers to one decimal place.

65. $f(x) = x^4 + 2x^3 - 1$ $(-2 \leq x \leq 1)$

66. $f(x) = -x^3 - 3x^2 - 1$ $(-3 \leq x \leq 1)$

67. $h(x) = |x - 3|$ $(-5 \leq x \leq 5)$

68. $h(x) = 2x + (x - 3)^{1/3}$ $(-5 \leq x \leq 5)$

69. $f(x) = x - 5(x - 1)^{2/5}$ $(-4 \leq x \leq 6)$

70. $f(x) = |2x + 5| - x^2$ $(-4 \leq x \leq 4)$

APPLICATIONS

71. Demand Suppose the demand for a new brand of sneakers is given by

$$q = \frac{5,000,000}{p}$$

where p is the price per pair of sneakers, in dollars, and q is the number of pairs of sneakers that can be sold at price p. Find $q(100)$ and estimate $q'(100)$. Interpret your answers.

72. Demand Suppose the demand for an old brand of TV is given by

$$q = \frac{100,000}{p + 10}$$

where p is the price per TV set, in dollars, and q is the number of TV sets that can be sold at price p. Find $q(190)$ and estimate $q'(190)$. Interpret your answers.

73. Swimming Pool Sales The following graph shows the approximate annual U.S. sales of in-ground swimming pools. Also shown is the tangent line (and its slope) at the point corresponding to year 2000.

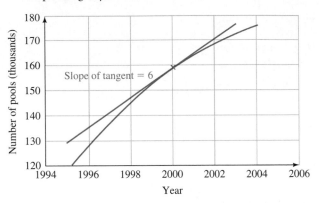

a. What does the graph tell you about swimming pool sales in 2000?

b. According to the graph, is the rate of change of swimming pool sales increasing or decreasing? Why?

Based on a regression model using 1996–2001 data. Source: PK Data/*New York Times*, July 5, 2001, p. C1.

74. Swimming Pool Sales Repeat Exercise 73, using the following graph showing the approximate annual U.S. sales of above-ground swimming pools.

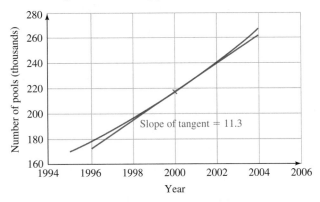

Based on a regression model using 1996–2001 data. Source: PK Data/*New York Times*, July 5, 2001, p. C1.

75. Prison Population The following curve is a model of the total population in state prisons as a function of time in years ($t = 0$ represents 1980):

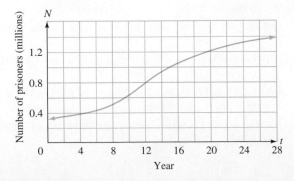

a. Which is correct? Over the period [16, 20], the instantaneous rate of change of N is
(A) increasing.   (B) decreasing.

b. Which is correct? The instantaneous rate of change of the prison population at $t = 12$ is
(A) less than (B) greater than (C) approximately equal to
the average rate of change over the interval [0, 24].

c. Which is correct? Over the period [0, 28], the instantaneous rate of change of N is
(A) increasing and then decreasing.
(B) decreasing and then increasing.
(C) always increasing.   (D) always decreasing.

d. According to the model, the total state prison population was increasing fastest around what year?

e. Roughly estimate the instantaneous rate of change of N at $t = 16$ by using a balanced difference quotient with $h = 4$. Interpret the result.

The prison population represented excludes federal prisons. SOURCE: Bureau of Justice Statistics/*New York Times*, June 9, 2001, p. A10.

76. **Demand for Freon** The demand for chlorofluorocarbon-12 (CFC-12)—the ozone-depleting refrigerant commonly known as Freon[9]—has been declining significantly in response to regulation and concern about the ozone layer. The graph below represents a model for the projected demand for CFC-12 as a function of time in years ($t = 0$ represents 1990):

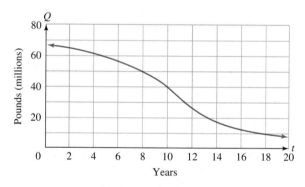

a. Which is correct? Over the period [12, 20], the instantaneous rate of change of Q is
(A) increasing.   (B) decreasing.

b. Which is correct? The instantaneous rate of change of demand for Freon at $t = 10$ is
(A) less than   (B) greater than
(C) approximately equal to
the average rate of change over the interval [0, 20].

c. Which is correct? Over the period [0, 20], the instantaneous rate of change of Q is
(A) increasing and then decreasing.
(B) decreasing and then increasing.
(C) always increasing.   (D) always decreasing.

d. According to the model, the demand for Freon was decreasing most rapidly around what year?

e. Roughly estimate the instantaneous rate of change of Q at $t = 13$ by using a balanced difference quotient with $h = 5$. Interpret the result.

SOURCE: The Automobile Consulting Group/*New York Times*, December 26, 1993, p. F23. The exact figures were not given, and the chart is a reasonable facsimile of the chart that appeared in the *New York Times*.

77. **Velocity** If a stone is dropped from a height of 400 feet, its height after t seconds is given by $s = 400 - 16t^2$.
a. Find its average velocity over the period [2, 4].
b. Estimate its instantaneous velocity at time $t = 4$.

78. **Velocity** If a stone is thrown down at 120 feet per second from a height of 1000 feet, its height after t seconds is given by $s = 1000 - 120t - 16t^2$.
a. Find its average velocity over the period [1, 3].
b. Estimate its instantaneous velocity at time $t = 3$.

79. **Online Shopping** The annual number of online shopping transactions in the United States for the period January 2000–January 2002 can be approximated by

$$N(t) = -180t^2 + 440t + 320 \text{ million transactions} \quad (0 \le t \le 2)$$

where t is time in years since January 2000.
a. Compute the average rate of change of $N(t)$ over the interval [1, 2] and interpret your answer.
b. Estimate the instantaneous rate of change of $N(t)$ at $t = 1$ and interpret your answer.
c. The answers to parts (a) and (b) have opposite signs. What does this indicate about online shopping?

Based on a regression model. (Second half of 2001 data is an estimate.) SOURCE: Odyssey Research/*New York Times*, November 5, 2001, p. C1.

80. **Online Shopping** For the period January 2000–January 2002, the percentage of people in the United States who had ever purchased anything online can be approximated by

$$P(t) = -2.6t^2 + 13t + 19 \text{ percent} \quad (0 \le t \le 2)$$

where t is time in years since January 2000.
a. Compute the average rate of change of $P(t)$ over the interval [1, 2] and interpret your answer.
b. Estimate the instantaneous rate of change of $P(t)$ at $t = 1$ and interpret your answer.
c. The answer to part (b) is greater than the answer to part (a). What does this indicate about online shopping?

Based on a regression model. (Second half of 2001 data is an estimate.) SOURCE: Odyssey Research/*New York Times*, November 5, 2001, p. C1.

81. **Advertising Revenue** The following table shows the annual advertising revenue earned by America Online (AOL) during the last 3 years of the 1990s:

Year	1997	1998	1999
Revenue ($ millions)	150	360	760

Figures are rounded to the nearest $10 million.
SOURCES: AOL; Forrester Research/*New York Times*, January 31, 2000, p. C1.

[9]The name given to it by DuPont.

These data can be modeled by

$$R(t) = 95t^2 + 115t + 150 \text{ million dollars} \qquad (0 \le t \le 2)$$

where t is time in years since December 1997.

a. How fast was AOL's advertising revenue increasing in December 1998?

b. Which of the following is true? During 1997–1999, annual online advertising revenues

  (A) increased at a faster and faster rate.

  (B) increased at a slower and slower rate.

  (C) decreased at a faster and faster rate.

  (D) decreased at a slower and slower rate.

c. Use the model to project the instantaneous rate of change of R in December 2000. Interpret the result.

82. Religion The following table shows the U.S. population of Roman Catholic nuns during the last 25 years of the 1900s:

Year	1975	1985	1995
Population	130,000	120,000	80,000

Figures are rounded. SOURCE: Center for Applied Research in the Apostolate/*New York Times,* January 16, 2000, p. A1.

These data can be modeled by

$$P(t) = -0.15t^2 + 0.50t + 130 \text{ thousand nuns} \quad (0 \le t \le 20)$$

where t is time in years since December 1975.

a. How fast was the number of Roman Catholic nuns decreasing in December 1995?

b. Which of the following is true? During 1980–1995, the population of nuns

  (A) increased at a faster and faster rate.

  (B) increased at a slower and slower rate.

  (C) decreased at a faster and faster rate.

  (D) decreased at a slower and slower rate.

c. Use the model to project the instantaneous rate of change of P in December 2005. Interpret the result.

83. Online Services On January 1, 1996, America Online (AOL) was the biggest online service provider, with 4.5 million subscribers, and was adding new subscribers at a rate of 60,000/week. If $A(t)$ is the number of AOL subscribers t weeks after January 1, 1996, what do the given data tell you about values of the function A and its derivative?

SOURCE: Information and Interactive Services Report/*New York Times,* January 2, 1996, p. C14.

84. Online Services On January 1, 1996, Prodigy was the third-biggest online service provider, with 1.6 million subscribers, but was losing subscribers. If $P(t)$ is the number of Prodigy subscribers t weeks after January 1, 1996, what do the given data tell you about values of the function P and its derivative?

SOURCE: Information and Interactive Services Report/*New York Times,* January 2, 1996, p. C14.

85. Learning to Speak Let $p(t)$ represent the percentage of children who can speak at the age of t months.

a. It is found that $p(10) = 60$ and $\left.\dfrac{dp}{dt}\right|_{t=10} = 18.2$. What does this mean?

b. As t increases, what happens to p and dp/dt?

SOURCE: Based on data presented in William H. Calvin, "The Emergence of Intelligence," *Scientific American* (October 1994): 101–107.

86. Learning to Read Let $p(t)$ represent the number of children in your class who learned to read at the age of t years.

a. Assuming that everyone in your class could read by the age of 7, what does this tell you about $p(7)$ and $\left.\dfrac{dp}{dt}\right|_{t=7}$?

b. Assuming that 25.0% of the people in your class could read by the age of 5 and that 25.3% of them could read by the age of 5 years and 1 month, estimate $\left.\dfrac{dp}{dt}\right|_{t=5}$. Remember to give its units.

E **87. Sales** Weekly sales of a new brand of sneakers are given by

$$S(t) = 200 - 150e^{-t/10}$$

pairs sold per week, where t is the number of weeks since the introduction of the brand. Estimate $S(5)$ and $\left.\dfrac{dS}{dt}\right|_{t=5}$ and interpret your answers.

E **88. Sales** Weekly sales of an old brand of TV are given by

$$S(t) = 100e^{-t/5}$$

sets per week, where t is the number of weeks after the introduction of a competing brand. Estimate $S(5)$ and $\left.\dfrac{dS}{dt}\right|_{t=5}$ and interpret your answers.

E **89. Computer Use** The percentage of U.S. households with a computer in 2000 as a function of household income can be modeled by the logistic function

$$P(x) = \frac{91}{1 + 5.35(1.05)^{-x}} \text{ percent} \qquad (0 \le x \le 100)$$

where x is the household income in thousands of dollars.

a. Estimate $P(50)$ and $P'(50)$. What do the answers tell you about computer use in the United States?

b. Use a graph of the function for $0 \le x \le 100$ to describe how the derivative behaves for values of x approaching 100.

SOURCE: NTIA and ESA, U.S. Dept. of Commerce, using U.S. Bureau of the Census Current Population, 2000.

E **90. Online Book Sales** The number of books sold online in the United States for the period 1997–2000 can be modeled by the logistic function

$$N(t) = \frac{82.8}{1 + 21.8(7.14)^{-t}} \qquad (0 \le t \le 3)$$

where t is time in years since the start of 1997 and $N(t)$ is the number of books sold in the year beginning at time t.

a. Estimate $N(1)$ and $N'(1)$. What do the answers tell you about online book sales?

b. Use a graph of the function for $0 \le t \le 3$ to estimate, to the nearest 6 months, when N' was greatest.

SOURCE: Ipsos-NPD Book Trends/*New York Times,* April 16, 2001, p. C1.

91. Embryo Development The oxygen consumption of a turkey embryo increases from the time the egg is laid through the time the turkey chick hatches. In a brush turkey, the oxygen consumption (in milliliters per hour) can be approximated by

$$c(t) = -0.0012t^3 + 0.12t^2 - 1.83t + 3.97 \quad (20 \le t \le 50)$$

where t is the time (in days) since the egg was laid. (An egg will typically hatch at around $t = 50$.) Use technology to graph $c'(t)$ and use your graph to answer the following questions.

a. Over the interval [20, 32], the derivative c' is
  (A) increasing and then decreasing.
  (B) decreasing and then increasing.
  (C) decreasing.   **(D)** increasing.
b. When, to the nearest day, is the oxygen consumption increasing at the fastest rate?
c. When, to the nearest day, is the oxygen consumption increasing at the slowest rate?

Source: The model approximates graphical data published in Roger S. Seymour, "The Brush Turkey," *Scientific American* (December 1991): 108–114.

92. Embryo Development The oxygen consumption of a bird embryo increases from the time the egg is laid through the time the chick hatches. In a typical galliform bird, the oxygen consumption (in milliliters per hour) can be approximated by

$$c(t) = -0.0027t^3 + 0.14t^2 - 0.89t + 0.15 \quad (8 \le t \le 30)$$

where t is the time (in days) since the egg was laid. (An egg will typically hatch at around $t = 28$.) Use technology to graph $c'(t)$ and use your graph to answer the following questions.

a. Over the interval [8, 30], the derivative c' is

  (A) increasing and then decreasing.

  (B) decreasing and then increasing.

  (C) decreasing.   **(D)** increasing.

b. When, to the nearest day, is the oxygen consumption increasing the fastest?
c. When, to the nearest day, is the oxygen consumption increasing at the slowest rate?

Source: The model approximates graphical data published in Roger S. Seymour, "The Brush Turkey," *Scientific American* (December 1991): 108–114.

Exercises 93 and 94 are applications of Einstein's special theory of relativity and relate to objects that are moving extremely fast. In science fiction terminology, a speed of *warp* 1 is the speed of light—about 3×10^8 meters per second. (Thus, for instance, a speed of warp 0.8 corresponds to 80% of the speed of light—about 2.4×10^8 meters per second.)

93. Lorentz Contraction According to Einstein's special theory of relativity, a moving object appears to get shorter to a stationary observer as its speed approaches the speed of light. If a spaceship that has a length of 100 meters at rest travels at a speed of warp p, its length in meters, as measured by a stationary observer, is given by

$$L(p) = 100\sqrt{1 - p^2}$$

with domain [0, 1). Estimate $L(0.95)$ and $L'(0.95)$. What do these figures tell you?

94. Time Dilation Another prediction of Einstein's special theory of relativity is that, to a stationary observer, clocks (as well as all biological processes) in a moving object appear to go more and more slowly as the speed of the object approaches that of light. If a spaceship travels at a speed of warp p, the time it takes for an onboard clock to register 1 second, as measured by a stationary observer, will be given by

$$T(p) = \frac{1}{\sqrt{1 - p^2}}$$

with domain [0, 1). Estimate $T(0.95)$ and $T'(0.95)$. What do these figures tell you?

COMMUNICATION AND REASONING EXERCISES

95. Company A's profits are given by $P(0) = \$1$ million and $P'(0) = -\$1$ million. Company B's profits are given by $P(0) = -\$1$ million and $P'(0) = \$1$ million. In which company would you rather invest? Why?

96. Company C's profits are given by $P(0) = \$1$ million and $P'(0) = \$0.5$ million. Company D's profits are given by $P(0) = \$0.5$ million and $P'(0) = \$1$ million. In which company would you rather invest? Why?

97. During the 1-month period starting last January 1, your company's profits increased at an average rate of change of $4 million per month. On January 1 profits were increasing at an instantaneous rate of $5 million per month. Which of the following graphs could represent your company's profits? Why?

(A)

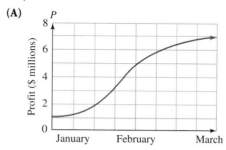

(B)

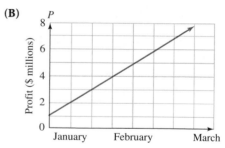

Solution Substituting $a = 3$ into the definition of the derivative, we get

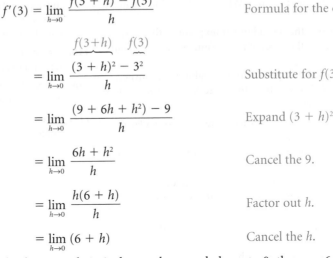

$$f'(3) = \lim_{h \to 0} \frac{f(3 + h) - f(3)}{h} \qquad \text{Formula for the derivative}$$

$$= \lim_{h \to 0} \frac{\overbrace{(3 + h)^2}^{f(3+h)} - \overbrace{3^2}^{f(3)}}{h} \qquad \text{Substitute for } f(3) \text{ and } f(3 + h).$$

$$= \lim_{h \to 0} \frac{(9 + 6h + h^2) - 9}{h} \qquad \text{Expand } (3 + h)^2.$$

$$= \lim_{h \to 0} \frac{6h + h^2}{h} \qquad \text{Cancel the 9.}$$

$$= \lim_{h \to 0} \frac{h(6 + h)}{h} \qquad \text{Factor out } h.$$

$$= \lim_{h \to 0} (6 + h) \qquad \text{Cancel the } h.$$

Now we let h approach 0. As h gets closer and closer to 0, the sum $6 + h$ clearly gets closer and closer to $6 + 0 = 6$. Thus,

$$f'(3) = \lim_{h \to 0} (6 + h) = 6 \qquad \text{As } h \to 0, (6 + h) \to 6.$$

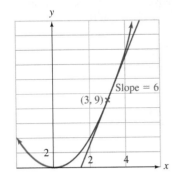

Figure 18

✴ **Before we go on . . .** We did the following calculation: If $f(x) = x^2$, then $f'(3) = 6$. In other words, the tangent to the graph of $y = x^2$ at the point $(3, 9)$ has slope 6 (Figure 18).

There is nothing very special about $a = 3$ in Example 1. Let's try to compute $f'(x)$ for general x.

Example 2 • Calculating the Derivative Function Algebraically

Let $f(x) = x^2$. Use the definition of the derivative to compute $f'(x)$ algebraically.

Solution Once again, our starting point is the definition of the derivative in terms of the difference quotient:

$$f'(x) = \lim_{h \to 0} \frac{f(x + h) - f(x)}{h} \qquad \text{Formula for the derivative}$$

$$= \lim_{h \to 0} \frac{\overbrace{(x + h)^2}^{f(x + h)} - \overbrace{x^2}^{f(x)}}{h} \qquad \text{Substitute for } f(x) \text{ and } f(x + h).$$

$$= \lim_{h \to 0} \frac{(x^2 + 2xh + h^2) - x^2}{h} \qquad \text{Expand } (x + h)^2.$$

$$= \lim_{h \to 0} \frac{2xh + h^2}{h} \qquad \text{Cancel the } x^2.$$

$$= \lim_{h \to 0} \frac{h(2x + h)}{h} \qquad \text{Factor out } h.$$

$$= \lim_{h \to 0} (2x + h) \qquad \text{Cancel the } h.$$

Now we let h approach 0. As h gets closer and closer to 0, the sum $2x + h$ clearly gets closer and closer to $2x + 0 = 2x$. Thus,

$$f'(x) = \lim_{h \to 0} (2x + h) = 2x$$

This is the derivative function. Now that we have a *formula* for the derivative of f, we can obtain $f'(a)$ for any value of a we choose by simply evaluating f' there. For instance,

$$f'(3) = 2(3) = 6$$

as we saw in Example 1.

✳ ***Before we go on . . .*** The graphs of f and of f' are familiar: The graph of f is a parabola opening upward with its vertex at the origin, and the graph of f' is a straight line through the origin with slope 2. When $x < 0$, the parabola slopes downward, which is reflected in the fact that the derivative $2x$ is negative there. When $x > 0$, the parabola slopes upward, which is reflected in the fact that the derivative is positive there. The parabola has a horizontal tangent line at $x = 0$, reflected in the fact that $2x = 0$ there.

Example 3 • More Computations of Derivative Functions

Compute the derivative $f'(x)$ for each of the following functions:

a. $f(x) = x^3$ **b.** $f(x) = 2x^2 - x$ **c.** $f(x) = \dfrac{1}{x}$

Solution

a. $f'(x) = \displaystyle\lim_{h \to 0} \frac{f(x + h) - f(x)}{h}$ Derivative formula

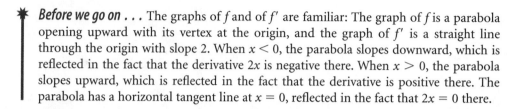

$$= \lim_{h \to 0} \frac{(x + h)^3 - x^3}{h}$$ Substitute for $f(x)$ and $f(x + h)$.

$$= \lim_{h \to 0} \frac{(x^3 + 3x^2h + 3xh^2 + h^3) - x^3}{h}$$ Expand $(x + h)^3$.

$$= \lim_{h \to 0} \frac{3x^2h + 3xh^2 + h^3}{h}$$ Cancel the x^3.

$$= \lim_{h \to 0} \frac{h(3x^2 + 3xh + h^2)}{h}$$ Factor out h.

$$= \lim_{h \to 0} (3x^2 + 3xh + h^2)$$ Cancel the h.

$$= 3x^2$$ Let h approach zero.

b. $f'(x) = \lim\limits_{h \to 0} \dfrac{f(x + h) - f(x)}{h}$ Derivative formula

$$= \lim\limits_{h \to 0} \frac{\overbrace{[2(x + h)^2 - (x + h)]}^{f(x+h)} - \overbrace{(2x^2 - x)}^{f(x)}}{h}$$ Substitute for $f(x)$ and $f(x + h)$.

$$= \lim\limits_{h \to 0} \frac{(2x^2 + 4xh + 2h^2 - x - h) - (2x^2 - x)}{h}$$ Expand.

$$= \lim\limits_{h \to 0} \frac{4xh + 2h^2 - h}{h}$$ Cancel the $2x^2$ and x.

$$= \lim\limits_{h \to 0} \frac{h(4x + 2h - 1)}{h}$$ Factor out h.

$$= \lim\limits_{h \to 0} (4x + 2h - 1)$$ Cancel the h.

$$= 4x - 1$$ Let h approach zero.

c. $f'(x) = \lim\limits_{h \to 0} \dfrac{f(x + h) - f(x)}{h}$ Derivative formula

$$= \lim\limits_{h \to 0} \frac{\left(\overbrace{\dfrac{1}{x + h}}^{f(x+h)} - \overbrace{\dfrac{1}{x}}^{f(x)}\right)}{h}$$ Substitute for $f(x)$ and $f(x + h)$.

$$= \lim\limits_{h \to 0} \frac{\left[\dfrac{x - (x + h)}{(x + h)x}\right]}{h}$$ Subtract the fractions.

$$= \lim\limits_{h \to 0} \frac{1}{h}\left[\frac{x - (x + h)}{(x + h)x}\right]$$ Dividing by h = multiplying by $\dfrac{1}{h}$.

$$= \lim\limits_{h \to 0} \frac{-h}{h(x + h)x}$$ Simplify.

$$= \lim\limits_{h \to 0} \frac{-1}{(x + h)x}$$ Cancel the h.

$$= \frac{-1}{x^2}$$ Let h approach zero.

In Example 4 we redo Example 3 of Section 3.2, this time getting an exact, rather than approximate, answer.

Example 4 • Velocity

My friend Eric, an enthusiastic baseball player, claims he can "probably" throw a ball upward at a speed of 100 feet per second (ft/s). Our physicist friends tell us that its height s (in feet) t seconds later would be $s(t) = 100t - 16t^2$. Find the ball's instantaneous velocity function and its velocity exactly 2 seconds after Eric throws it.

Solution The instantaneous velocity function is the derivative ds/dt, which we calculate as follows:

$$\frac{ds}{dt} = \lim_{h \to 0} \frac{s(t + h) - s(t)}{h}$$

Let's compute $s(t + h)$ and $s(t)$ separately:

$$s(t) = 100t - 16t^2 \qquad s(t + h) = 100(t + h) - 16(t + h)^2$$
$$= 100t + 100h - 16(t^2 + 2th + h^2)$$
$$= 100t + 100h - 16t^2 - 32th - 16h^2$$

Therefore,

$$\frac{ds}{dt} = \lim_{h \to 0} \frac{s(t + h) - s(t)}{h}$$

$$= \lim_{h \to 0} \frac{100t + 100h - 16t^2 - 32th - 16h^2 - (100t - 16t^2)}{h}$$

$$= \lim_{h \to 0} \frac{100h - 32th - 16h^2}{h}$$

$$= \lim_{h \to 0} \frac{h(100 - 32t - 16h)}{h}$$

$$= \lim_{h \to 0} 100 - 32t - 16h$$

$$= (100 - 32t) \text{ ft/s}$$

Thus, the velocity exactly 2 seconds after Eric throws it is

$$\left.\frac{ds}{dt}\right|_{t=2} = 100 - 32(2) = 36 \text{ ft/s}$$

This verifies the accuracy of the approximation we made in Section 3.2.

✸ **Before we go on . . .** From the derivative function we can now describe the behavior of the velocity of the ball: Immediately on release ($t = 0$), the ball is traveling at 100 feet per second upward. The ball then slows down; precisely, it loses 32 feet per second of speed every second. When, exactly, does the velocity become 0 and what happens after that?

Question So, when we have to compute the limit as $h \to 0$, we just substitute $h = 0$?

Answer No, it's really a little more complicated than that. If you examine these calculations carefully, you'll see that we can't substitute $h = 0$ until the very end. And, there are several subtle things going on that allow us to do it even then. We'll address this issue when we look at limits more carefully in the last sections of this chapter.

Question It looks like it is really useful to be able to find a formula for the derivative function given a formula for the original function. But, do we have to calculate the limit of the difference quotient every time?

Answer As it turns out, no. In Section 3.4 we will start to look at shortcuts for finding derivatives that allow us to bypass the definition of the derivative in many cases.

Recall from Section 3.2 that a function is **differentiable** at a point a if $f'(a)$ exists—that is, if the difference quotient $[f(a + h) - f(a)]/h$ approaches a fixed value as h approaches 0. In that section we mentioned that the function $f(x) = |x|$ is not differentiable at $x = 0$. We find out why in Example 5.

Example 5 • Nondifferentiable Function

Numerically, graphically, and algebraically investigate the differentiability of the function $f(x) = |x|$ at the points (a) $x = 1$ and (b) $x = 0$.

Solution

a. We compute

$$f'(1) = \lim_{h \to 0} \frac{f(1 + h) - f(1)}{h}$$

$$= \lim_{h \to 0} \frac{|1 + h| - 1}{h}$$

Numerically, we can make tables of the values of the average rate of change $(|1 + h| - 1)/h$ for h positive or negative and approaching 0:

h	1	0.1	0.01	0.001	0.0001
Average Rate of Change over $[1, 1 + h]$	1	1	1	1	1

h	−1	−0.1	−0.01	−0.001	−0.0001
Average Rate of Change over $[1 + h, 1]$	1	1	1	1	1

From these tables it appears that $f'(1)$ is equal to 1. We can verify that algebraically: For h sufficiently small, $1 + h$ is positive (even if h is negative), and so

$$f'(1) = \lim_{h \to 0} \frac{1 + h - 1}{h}$$

$$= \lim_{h \to 0} \frac{h}{h} \qquad \text{Cancel the 1s.}$$

$$= \lim_{h \to 0} 1 \qquad \text{Cancel the } h.$$

$$= 1$$

Graphically, we are seeing the fact that the tangent line at the point $(1, 1)$ has slope 1 because the graph is a straight line with slope 1 near that point (Figure 19).

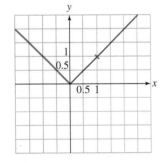

Figure 19

b. $f'(0) = \lim_{h \to 0} \frac{f(0 + h) - f(0)}{h}$

$$= \lim_{h \to 0} \frac{|0 + h| - 0}{h}$$

$$= \lim_{h \to 0} \frac{|h|}{h}$$

If we make tables of values in this case, we get the following:

h	1	0.1	0.01	0.001	0.0001
Average Rate of Change over $[0, 0 + h]$	1	1	1	1	1

h	-1	-0.1	-0.01	-0.001	-0.0001
Average Rate of Change over $[0 + h, 0]$	-1	-1	-1	-1	-1

Now we have a problem. What is $f'(1)$? Is it $+1$ or -1? For the limit and hence the derivative to exist, the average rates of change should approach the same number for both positive and negative h. Since they do not, f is not differentiable at $x = 0$. We can verify this conclusion algebraically: If h is positive, then $|h| = h$, and so the ratio $|h|/h$ is 1, regardless of how small h is. Thus, according to the values of the difference quotients with $h > 0$, the limit should be 1. On the other hand, if h is negative, then $|h| = -h$ (positive), and so $|h|/h = -1$, meaning that the limit should be -1. Since the limit cannot be both -1 and 1 (it must be a single number for the derivative to exist), we conclude that $f'(0)$ does not exist.

To see what is happening graphically, take a look at Figure 20, which shows zoomed-in views of the graph of f near $x = 0$.

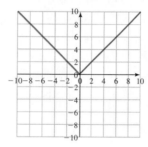

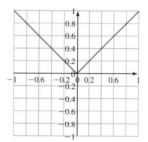

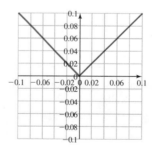

Figure 20

No matter what scale we use to view the graph, it has a sharp corner at $x = 0$ and hence has no tangent line there. Since there is no tangent line at $x = 0$, the function is not differentiable there.

✱ ***Before we go on . . .*** If we repeat the computation of part (a) using any nonzero value for a in place of 1, we see that f is differentiable there as well. If a is positive, we find that $f'(a) = 1$, and if a is negative, $f'(a) = -1$. In other words, the derivative function is

$$f'(x) = \begin{cases} -1 & \text{if } x < 0 \\ 1 & \text{if } x > 0 \end{cases}$$

Of course, $f'(x)$ is not defined when $x = 0$.

Guideline: Computing Derivatives Algebraically and Recognizing When a Function Is Not Differentiable

Question The algebraic computation of $f'(x)$ seems to require a number of steps. How do I remember what to do and when?

Answer If you examine the computations in the examples above, you will find the following pattern:

1. Write out the formula for $f'(x)$ and then substitute $f(x + h)$ and $f(x)$.

2. Expand and simplify the *numerator* of the expression, but not the denominator.

3. After simplifying the numerator, factor out an h to cancel with the h in the denominator. If h does not factor out of the numerator, you might have made an error. (A frequent error is a wrong sign.)

4. After canceling the h, you should be able to see what the limit is by letting $h \to 0$.

Question Is there a way of telling from its formula whether a function f is not differentiable at a point?

Answer Here are some indicators to look for in the formula for f:

• The absolute value of some expression; f may not be differentiable at points where that expression is zero.

   Example: $f(x) = 3x^2 - |x - 4|$ is not differentiable at $x = 4$.

• A fractional power smaller than 1 of some expression; f may not be differentiable at points where that expression is zero. (Such functions will be discussed further in Section 3.4.)

   Example: $f(x) = (x^2 - 16)^{2/3}$ is not differentiable at $x = \pm 4$.

3.3 EXERCISES

In Exercises 1–14, compute $f'(a)$ algebraically. Then compute the derivative function $f'(x)$ algebraically.

1. $f(x) = x^2 + 1; a = 2$
2. $f(x) = x^2 - 3; a = 1$
3. $f(x) = 3x - 4; a = -1$
4. $f(x) = -2x + 4; a = -1$
5. $f(x) = 3x^2 + x; a = 1$
6. $f(x) = 2x^2 + x; a = -2$
7. $f(x) = 2x - x^2; a = -1$
8. $f(x) = -x - x^2; a = 0$
9. $f(x) = x^3 + 2x; a = 2$
10. $f(x) = x - 2x^3; a = 1$
11. $f(x) = mx + b; a = 43$
12. $f(x) = \dfrac{x}{k} - b \ (k \neq 0); a = 12$
13. $f(x) = \dfrac{-1}{x}; a = 1$
14. $f(x) = \dfrac{2}{x}; a = 5$

In Exercises 15–24, compute the indicated derivative.

15. $R(t) = -0.3t^2; R'(2)$
16. $S(t) = 1.4t^2; S'(-1)$
17. $U(t) = 5.1t^2 + 5.1; U'(3)$
18. $U(t) = -1.3t^2 + 1.1; U'(4)$
19. $U(t) = -1.3t^2 - 4.5t; U'(1)$
20. $U(t) = 5.1t^2 - 1.1t; U'(1)$
21. $L(r) = 4.25r - 5.01; L'(1.2)$
22. $L(r) = -1.02r + 5.7; L'(3.1)$
23. $q(p) = \dfrac{2.4}{p} + 3.1; q'(2)$
24. $q(p) = \dfrac{1}{0.5p} - 3.1; q'(2)$

In Exercises 25–30, find the equation of the tangent to the graph at the indicated point.

25. $f(x) = x^2 - 3; a = 2$
26. $f(x) = x^2 + 1; a = 2$
27. $f(x) = -2x - 4; a = 3$
28. $f(x) = 3x + 1; a = 1$
29. $f(x) = x^2 - x; a = -1$
30. $f(x) = x^2 + x; a = -1$

In Exercises 31–36, investigate the differentiability of the given function at the given points numerically (that is, use a table of values). If $f'(a)$ exists, give its approximate value.

31. $f(x) = x^{1/3}$
   a. $a = 1$ b. $a = 0$

32. $f(x) = x + |1 - x|$
   a. $a = 1$ **b.** $a = 0$

33. $f(x) = [x(1 - x)]^{1/3}$
   a. $a = 1$ **b.** $a = 0$

34. $f(x) = (1 - x)^{2/3}$
   a. $a = -1$ **b.** $a = 1$

35. $f(x) = x + (x + 1)^{1/5}$
   a. $a = -1$ **b.** $a = 1$

36. $f(x) = |x(x - 1)|$
   a. $a = 0$ **b.** $a = 1$

In Exercises 37–42, investigate the differentiability of the given function at the given points algebraically. If $f'(a)$ exists, give its value.

37. $f(x) = |x| + x$
   a. $a = 1$ **b.** $a = 0$

38. $f(x) = |x| - 2x$
   a. $a = 1$ **b.** $a = 0$

39. $f(x) = |x + 3|$
   a. $a = -3$ **b.** $a = 3$

40. $f(x) = |x - 2|$
   a. $a = -2$ **b.** $a = 2$

41. $f(x) = |x| + |x - 1|$
   a. $a = 1$ **b.** $a = 0$

42. $f(x) = |x| - |x - 1|$
   a. $a = 1$ **b.** $a = 0$

APPLICATIONS

43. Swimming Pool Sales The number of in-ground swimming pools built in the United States each year from 1996 to 2001 can be approximated by

$$n(t) = -0.45t^2 + 15t + 54 \text{ thousand pools} \quad (6 \le t \le 11)$$

where t is the time in years since 1990. At what rate was this number increasing in 2000?

Source: PK Data/*New York Times*, July 5, 2001, p. C1.

44. Swimming Pool Sales The number of above-ground swimming pools built in the United States each year from 1996 to 2001 can be approximated by

$$n(t) = 0.36t^2 + 4.1t + 140 \text{ thousand pools} \quad (6 \le t \le 11)$$

where t is the time in years since 1990. At what rate was this number increasing in 2000?

Source: PK Data/*New York Times*, July 5, 2001, p. C1.

45. Velocity If a stone is dropped from a height of 400 feet, its height after t seconds is given by $s = 400 - 16t^2$. Find its instantaneous velocity function and its velocity at time $t = 4$.

46. Velocity If a stone is thrown down at 120 feet per second from a height of 1000 feet, its height after t seconds is given by $s = 1000 - 120t - 16t^2$. Find its instantaneous velocity function and its velocity at time $t = 3$.

47. SUV Sales in Europe Sport utility vehicle (SUV) sales in Europe increased substantially during the last 5 years of the 1990s, as shown in the following chart:

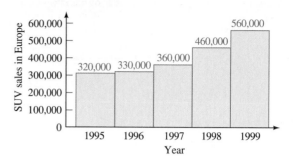

The function

$$N(t) = 18,000t^2 - 10,000t + 320,000 \text{ vehicles per year}$$
$$(0 \le t \le 5)$$

gives a good approximation to the data, where t is time in years since January 1995. According to the model, how fast were annual SUV sales increasing in January 1997?

Data are rounded (the 1999 figure is estimated). Source: Standard & Poor's DRI, company reports/*New York Times*, December 14, 1999, p. C1.

48. Auto Sales in the United States Sales of family vehicles in the United States continued to rise during the last 5 years of the 1990s, as shown in the following chart:

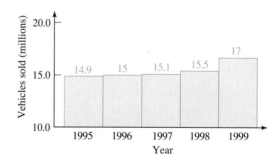

We can approximate the data with the function

$$N(t) = 0.22t^2 - 0.42t + 15 \text{ million vehicles per year}$$
$$(0 \le t \le 5)$$

where t is time in years since January 1995. According to the model, how fast were annual sales of family vehicles increasing in January 1997?

Data are rounded (the 1999 figure is estimated). Source: Wards company reports, Bloomberg Financial/*New York Times*, December 15, 1999, p. C1.

49. **Ecology** Increasing numbers of manatees ("sea sirens") have been killed by boats off the Florida coast. The following graph shows the relationship between the number of boats registered in Florida and the number of manatees killed each year:

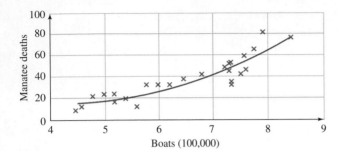

Boats (100,000)

The regression curve shown is given by

$$f(x) = 3.55x^2 - 30.2x + 81 \text{ manatee deaths } (4.5 \le x \le 8.5)$$

where x is the number of boats (hundreds of thousands) registered in Florida in a particular year and $f(x)$ is the number of manatees killed by boats in Florida that year. Compute and interpret $f'(8)$.

Regression model is based on data from 1976 to 2000. SOURCES: Florida Dept. of Highway Safety and Motor Vehicles, Florida Marine Institute/*New York Times*, February 12, 2002, p. F4.

50. **SAT Scores by Income** The following graph shows U.S. verbal SAT scores as a function of parents' income level:

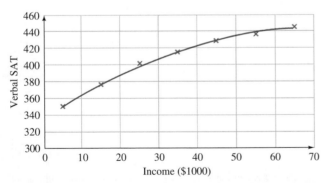

Income ($1000)

Based on 1994 data. SOURCE: The College Board/*New York Times*, March 5, 1995, p. E16.

The regression curve shown is given by

$$f(x) = -0.021x^2 + 3.0x + 336 \qquad (5 \le x \le 65)$$

where $f(x)$ is the average SAT verbal score of a student whose parents earn x thousand dollars per year. Compute and interpret $f'(30)$.

Regression model is based on 1994 data. SOURCE: The College Board/*New York Times*, March 5, 1995, p. E16.

COMMUNICATION AND REASONING EXERCISES

51. Of the three methods (numerical, graphical, algebraic) we can use to estimate the derivative of a function at a given value of x, which is always the most accurate? Explain.

52. Explain why we cannot put $h = 0$ in the formula

$$f'(a) = \lim_{h \to 0} \frac{f(a + h) - f(a)}{h}$$

for the derivative of f.

53. Your friend Muffy claims that, since the balanced difference quotient is more accurate, it would be better to use that instead of the usual difference quotient when computing the derivative algebraically. Comment on this advice.

54. Use the balanced difference quotient formula,

$$f'(a) = \lim_{h \to 0} \frac{f(a + h) - f(a - h)}{2h}$$

to compute $f'(3)$ when $f(x) = x^2$. What do you find?

55. A certain function f has the property that $f'(a)$ does not exist. How is that reflected in the attempt to compute $f'(a)$ algebraically?

56. One cannot put $h = 0$ in the formula

$$f'(a) = \lim_{h \to 0} \frac{f(a + h) - f(a)}{h}$$

for the derivative of f. (See Exercise 52). However, in the last step of each of the computations in the text, we are effectively setting $h = 0$ when taking the limit. What is going on here?

3.4 *Derivatives of Powers, Sums, and Constant Multiples*

So far in this chapter we have approximated derivatives using difference quotients, and we have done exact calculations using the definition of the derivative as the limit of a difference quotient. In general, we would prefer to have an exact calculation, and it is also very useful to have a formula for the derivative function when we can find one. However, the calculation of a derivative as a limit is often tedious, so it would be nice to have a quicker method. We discuss the first of the shortcut rules in this section. By the end of Chapter 4, we will be able to find fairly quickly the derivative of almost any function we can write.

Shortcut Formula: The Power Rule

If you look again at Examples 2 and 3 in Section 3.3, you may notice a pattern:

$$f(x) = x^2 \to f'(x) = 2x$$

$$f(x) = x^3 \to f'(x) = 3x^2$$

Shortcut: The Power Rule

If n is any constant and $f(x) = x^n$, then

$$f'(x) = nx^{n-1}$$

Quick Examples

1. If $f(x) = x^2$, then $f'(x) = 2x^1 = 2x$.
2. If $f(x) = x^3$, then $f'(x) = 3x^2$.
3. If $f(x) = x$, rewrite as $f(x) = x^1$, so $f'(x) = 1x^0 = 1$.
4. If $f(x) = 1$, rewrite as $f(x) = x^0$, so $f'(x) = 0x^{-1} = 0$.

Proof of the Power Rule
(Online Optional Section)

At the Web site, follow the path

    Web Site → Everything for Calculus → Chapter 3 → Proof of the Power Rule

for a proof of the power rule.

Example 1 • Using the Power Rule for Negative and Fractional Exponents

Calculate the derivatives of the following:

a. $f(x) = \dfrac{1}{x}$ **b.** $f(x) = \dfrac{1}{x^2}$ **c.** $f(x) = \sqrt{x}$

Solution

a. Rewrite[10] as $f(x) = x^{-1}$. Then $f'(x) = (-1)x^{-2} = -\dfrac{1}{x^2}$.

b. Rewrite as $f(x) = x^{-2}$. Then $f'(x) = (-2)x^{-3} = -\dfrac{2}{x^3}$.

c. Rewrite as $f(x) = x^{0.5}$. Then $f'(x) = 0.5x^{-0.5} = \dfrac{0.5}{x^{0.5}}$. Alternatively, rewrite $f(x)$ as $x^{1/2}$,

so that $f'(x) = \dfrac{1}{2}x^{-1/2} = \dfrac{1}{2x^{1/2}} = \dfrac{1}{2\sqrt{x}}$.

Before we go on . . . By rewriting the given functions before taking derivatives, we are converting them from **rational,** or **radical, form** (as in, say, $1/x^2$ or $\sqrt{x}$) to **exponent form** (as in x^{-2} and $x^{0.5}$) to enable us to use the power rule (see the following caution).

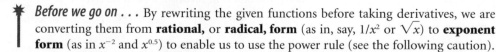

[10]To brush up on negative and fractional exponents, see Section A.2 in Appendix A.

Caution

We cannot apply the power rule to terms in the denominators or under square roots. For example:

1. The derivative of $\dfrac{1}{x^2}$ is *not* $\dfrac{1}{2x}$ but is $-\dfrac{2}{x^3}$.   See Example 1(b).

2. The derivative of $\sqrt{x^3}$ is *not* $\sqrt{3x^2}$ but is $1.5x^{0.5}$.   Rewrite $\sqrt{x^3}$ as $x^{3/2}$ or $x^{1.5}$ and apply the power rule.

Some of the derivatives in Example 1 are very useful to remember, so we summarize them in Table 1. We suggest that you add to this table as you learn more derivatives. It is *extremely* helpful to remember the derivatives of common functions such as $1/x$ and $\sqrt{x}$, even though they can be obtained using the power rule as in Example 1.

Another Notation: Differential Notation

Here is a useful notation based on the d notation we discussed in Section 3.2. **Differential notation** is based on an abbreviation for the phrase "the derivative with respect to x." For example, we learned that if $f(x) = x^3$, then $f'(x) = 3x^2$. When we say "$f'(x) = 3x^2$," we mean the following:

The derivative of x^3 with respect to x equals $3x^2$.

You may wonder why we sneaked in the words "with respect to x." All this means is that the variable of the function is x and not any other variable.[11] Since we use the phrase "the derivative with respect to x" often, we use the following abbreviation.

Differential Notation; Differentiation

$\dfrac{d}{dx}$ means "the derivative with respect to x."

Thus, $\dfrac{d}{dx}[f(x)]$ is the same thing as $f'(x)$, the derivative of $f(x)$ with respect to x. If y is a function of x, then the derivative of y with respect to x is

$$\frac{d}{dx}(y) \quad \text{or, more compactly,} \quad \frac{dy}{dx}$$

To **differentiate** a function $f(x)$ with respect to x means to take its derivative with respect to x.

Notes

1. $\dfrac{dy}{dx}$ is the Leibniz notation for the derivative we discussed in Section 3.2 (see the discussion before Example 3 there).

2. Leibniz notation illustrates units nicely: Units of $\dfrac{dy}{dx}$ are units of y per unit of x.

Table 1 • *Table of Derivative Formulas*

$f(x)$	$f'(x)$
1	0
x	1
x^2	$2x$
x^3	$3x^2$
x^n	nx^{n-1}
$\dfrac{1}{x}$	$-\dfrac{1}{x^2}$
$\dfrac{1}{x^2}$	$-\dfrac{2}{x^3}$
$\sqrt{x}$	$\dfrac{1}{2\sqrt{x}}$

[11]This may seem odd in the case of $f(x) = x^3$, since there are no other variables to worry about. But in expressions like st^3 that involve variables other than x, it is necessary to specify just what the variable of the function is. This is the same reason that we write "$f(x) = x^3$" rather than just "$f = x^3$."

Quick Examples

Differential Notation

In Words	Formula
1. The derivative with respect to x of x^3 is $3x^2$.	$\dfrac{d}{dx}[x^3] = 3x^2$
2. The derivative with respect to t of $\dfrac{1}{t}$ is $-\dfrac{1}{t^2}$.	$\dfrac{d}{dt}\left[\dfrac{1}{t}\right] = -\dfrac{1}{t^2}$

Leibniz Notation

1. If $y = x^4$, then $\dfrac{dy}{dx} = 4x^3$.

2. If $u = \dfrac{1}{t^2}$, then $\dfrac{du}{dt} = -\dfrac{2}{t^3}$.

The Rules for Sums and Constant Multiples

We can now find the derivatives of more complicated functions, such as polynomials, using the following rules.

Derivatives of Sums, Differences, and Constant Multiples

If $f(x)$ and $g(x)$ are any two functions with derivatives $f'(x)$ and $g'(x)$, respectively, and if c is any constant, then

$$[f(x) \pm g(x)]' = f'(x) \pm g'(x)$$

$$[cf(x)]' = cf'(x)$$

In Words

- The derivative of a sum is the sum of the derivatives, and the derivative of a difference is the difference of the derivatives.
- The derivative of c times a function is c times the derivative of the function.

Differential Notation

$$\frac{d}{dx}[f(x) \pm g(x)] = \frac{d}{dx}f(x) \pm \frac{d}{dx}g(x)$$

$$\frac{d}{dx}[cf(x)] = c\frac{d}{dx}f(x)$$

Quick Examples

1. $\dfrac{d}{dx}[x^2 - x^4] = \dfrac{d}{dx}[x^2] - \dfrac{d}{dx}[x^4] = 2x - 4x^3$

2. $\dfrac{d}{dx}[7x^3] = 7\dfrac{d}{dx}[x^3] = 7(3x^2) = 21x^3$

In other words, we multiply the coefficient (7) by the exponent (3) and then decrease the exponent by 1.

3. $\dfrac{d}{dx}[12x] = 12\dfrac{d}{dx}[x] = 12(1) = 12$

In other words, the derivative of a constant times x is that constant.

4. $\dfrac{d}{dx}[-x^{0.5}] = \dfrac{d}{dx}[(-1)x^{0.5}] = (-1)\dfrac{d}{dx}[x^{0.5}] = (-1)(0.5)x^{-0.5} = -0.5x^{-0.5}$

5. $\dfrac{d}{dx}[12] = \dfrac{d}{dx}[12(1)] = 12\dfrac{d}{dx}[1] = 12(0) = 0$

In other words, the derivative of a constant is zero.

6. If my company earns twice as much (annual) revenue as yours and the derivative of your revenue function is the curve on the left, then the derivative of my revenue function is the curve on the right.

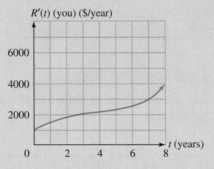
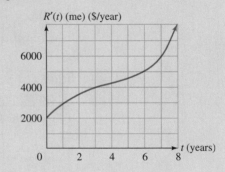

7. Suppose a company's revenue R and cost C are changing with time. Then so is the profit, $P(t) = R(t) - C(t)$, and the rate of change of the profit is

$$P'(t) = R'(t) - C'(t)$$

In words: The derivative of the profit is the derivative of revenue minus the derivative of cost.

A proof of the sum rule appears at the end of this section. The proof of the rule for constant multiples is similar.

Example 2 • Combining the Sum and Constant Multiple Rules and Dealing with x in the Denominator

Find the derivatives of the following:

a. $f(x) = 3x^2 + 2x - 4$ **b.** $f(x) = \dfrac{2x}{3} - \dfrac{6}{x} + \dfrac{2}{3x^{0.2}} - \dfrac{x^4}{2}$

Solution

a. $\dfrac{d}{dx}[3x^2 + 2x - 4] = \dfrac{d}{dx}[3x^2] + \dfrac{d}{dx}[2x - 4]$ Rule for sums

$\qquad\qquad = \dfrac{d}{dx}[3x^2] + \dfrac{d}{dx}[2x] - \dfrac{d}{dx}[4]$ Rule for differences

$\qquad\qquad = 3(2x) + 2(1) - 0$ See Quick Example 2.

$\qquad\qquad = 6x + 2$

b. Notice that f has x and powers of x in the denominator. We deal with these terms the same way we did in Example 1, by rewriting them in exponent form:

$$f(x) = \frac{2x}{3} - \frac{6}{x} + \frac{2}{3x^{0.2}} - \frac{x^4}{2} \qquad \text{Rational form}$$

$$= \frac{2}{3}x - 6x^{-1} + \frac{2}{3}x^{-0.2} - \frac{1}{2}x^4 \qquad \text{Exponent form}$$

We are now ready to take the derivative:

$$f'(x) = \frac{2}{3}(1) - 6(-1)x^{-2} + \frac{2}{3}(-0.2)x^{-1.2} - \frac{1}{2}(4x^3)$$

$$= \frac{2}{3} + 6x^{-2} - \frac{0.4}{3}x^{-1.2} - 2x^3 \qquad \text{Exponent form}$$

$$= \frac{2}{3} + \frac{6}{x^2} - \frac{0.4}{3x^{1.2}} - 2x^3 \qquad \text{Rational form}$$

✱ **Before we go on . . .**
Notice that in part (a) we had three terms in the expression for $f(x)$, not just two. By applying the rule for sums and differences twice, we saw that the derivative of a sum or difference of three terms is the sum or difference of the derivatives of the terms. (One of those terms had zero derivative, so the final answer had only two terms.) In fact, the derivative of a sum or difference of any number of terms is the sum or difference of the derivatives of the terms. Put another way, to take the derivative of a sum or difference of any number of terms, we take derivatives term by term.

Note
Nothing forces us to use only x as the independent variable when taking derivatives (although it is traditional to give x preference). For instance, Example 2(a) can be rewritten as

$$\frac{d}{dt}[3t^2 + 2t - 4] = 6t + 2 \qquad \frac{d}{dt} \text{ means "derivative with respect to } t."$$

or $$\frac{d}{du}[3u^2 + 2u - 4] = 6u + 2 \qquad \frac{d}{du} \text{ means "derivative with respect to } u."$$

In the examples above, we saw instances of the following important facts. (Think about these graphically to see why they must be true.)

The Derivative of a Constant Times x and the Derivative of a Constant
If c is any constant, then

Quick Examples

$$\frac{d}{dx}[cx] = c \qquad \frac{d}{dx}[6x] = 6 \qquad \frac{d}{dx}[-x] = -1$$

$$\frac{d}{dx}[c] = 0 \qquad \frac{d}{dx}[5] = 0 \qquad \frac{d}{dx}[\pi] = 0$$

3.5 A First Application: Marginal Analysis

In Chapter 1 we considered linear *cost functions* of the form $C(x) = mx + b$, where C is the total cost, x is the number of items, and m and b are constants. The slope m is the *marginal cost*. It measures the *cost of one more item*. Notice that the derivative of $C(x) = mx + b$ is $C'(x) = m$. In other words, for a linear cost function, *the marginal cost is the derivative of the cost function.*

In general, we make the following definition.

Marginal Cost

A **cost function** specifies the total cost C as a function of the number of items x. In other words, $C(x)$ is the total cost of x items. The **marginal cost function** is the derivative $C'(x)$ of the cost function $C(x)$. It measures the rate of change of cost with respect to x.

Units

The units of marginal cost are units of cost (dollars, say) per item.

Interpretation

We interpret $C'(x)$ as the approximate cost of one more item.*

Quick Example

If $C(x) = 400x + 1000$ dollars, then the marginal cost function is $C'(x) = \$400$ per item (a constant).

---

*See Example 1.

Example 1 • Marginal Cost

Suppose the cost in dollars to manufacture portable CD players is given by

$$C(x) = 150,000 + 20x - 0.0001x^2$$

where x is the number of CD players manufactured.[12] Find the marginal cost function $C'(x)$ and use it to estimate the cost of manufacturing the 50,001st CD player.

Solution Since

$$C(x) = 150,000 + 20x - 0.0001x^2$$

the marginal cost function is

$$C'(x) = 20 - 0.0002x$$

---

[12]You might well ask where on Earth this formula came from. There are two approaches to obtaining cost functions in real life: analytical and empirical. The analytical approach is to calculate the cost function from scratch. For example, in this situation, we might have fixed costs of $150,000, plus a production cost of $20 per CD player. The term $0.0001x^2$ may reflect a cost saving for high levels of production, such as a bulk discount in the cost of electronic components. In the empirical approach, we first obtain the cost at several different production levels by direct observation. This gives several points on the (as yet unknown) cost versus production-level graph. Then find the equation of a curve that best fits these points, usually using regression.

The units of $C'(x)$ are units of C (dollars) per unit of x (CD players). Thus, $C'(x)$ is measured in dollars per CD player.

The cost of the 50,001st CD player is the amount by which the total cost would rise if we increased production from 50,000 CD players to 50,001. Thus, we need to know the rate at which the total cost rises as we increase production. This rate of change is measured by the derivative, or marginal cost, which we just computed. At $x = 50,000$, we get

$$C'(50,000) = 20 - 0.0002(50,000) = \$10 \text{ per CD player}$$

In other words, we estimate that the 50,001st CD player will cost approximately $10.

✱ **Before we go on . . .** The marginal cost is really only an approximation to the cost of the 50,001st CD player:

$$C'(50,000) \approx \frac{C(50,001) - C(50,000)}{1} \qquad \text{Set } h = 1 \text{ in the definition of the derivative.}$$

$$= C(50,001) - C(50,000)$$

$$= \text{cost of the 50,001st CD player}$$

The exact cost of the 50,001st CD player is

$$C(50,001) - C(50,000) = [150,000 + 20(50,001) - 0.0001(50,001)^2]$$

$$- [150,000 + 20(50,000) - 0.0001(50,000)^2]$$

$$= \$9.9999$$

So, the marginal cost is a good approximation to the actual cost.

Graphically, we are using the tangent line to approximate the cost function near a production level of 50,000. Figure 22 shows the graph of the cost function together with the tangent line at $x = 50,000$. Notice that the tangent line is essentially indistinguishable from the graph of the function for some distance on either side of 50,000.

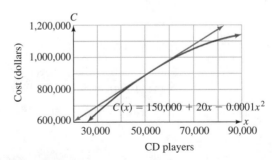

Figure 22

Notes

1. In general, the difference quotient $[C(x + h) - C(x)]/h$ gives the **average cost per item** to produce h more items at a current production level of x items. (Why?)

2. Notice that $C'(x)$ is much easier to calculate than $[C(x + h) - C(x)]/h$. (Try it.)

Example 2 • Marginal Revenue and Profit

As the operator of Workout Fever Health Club, you calculate your demand equation to be

$$q = -0.06p + 46 \qquad (200 \le p \le 700)$$

where q is the number of members in the club and p is the annual membership fee you charge. Annual operating costs amount to only $5000 per year, and each member costs the club $100 per year for medical evaluation and personal training.

a. Calculate and graph the annual revenue and profit as functions of q.

b. Compute the marginal (annual) revenue and profit functions, $R'(q)$ and $P'(q)$. How do we interpret them?

c. Calculate the marginal (annual) profit for $q = 15$, $q = 20$, and $q = 25$ and interpret the results.

Solution

a. The annual revenue is given by $R = pq$. Since we need to express R as a function of q only, we must replace p in this equation by a function of q. The relationship between p and q is given in the above demand equation, and we want p as a function of q, so we first need to rewrite the demand equation by solving for p. This gives

$$p = \frac{46 - q}{0.06} \qquad \text{Solve the demand equation for } p.$$

We substitute this expression for p in $R = pq$ to obtain

$$R(q) = \frac{(46 - q)q}{0.06} \qquad \text{Substitute for } p \text{ in the revenue equation.}$$

$$= \frac{46q - q^2}{0.06}$$

This is the annual revenue as a function of q. For the profit function, recall that

$$P = R - C \qquad \text{revenue} - \text{cost}$$

Since it costs Workout Fever $100 per member each year, the annual cost for q members is $100q$. Adding in the fixed cost gives an annual cost of

$$C(q) = 100q + 5000$$

so

$$P(q) = R(q) - C(q) = \frac{46q - q^2}{0.06} - 100q - 5000$$

Question　We know that the domain of the demand function is $200 \le p \le 700$. What is the corresponding domain for the profit and revenue functions?

Answer　We need to find the range of values of q corresponding to $200 \le p \le 700$. Now $p = \$200$ corresponds to a demand of $-0.06(200) + 46 = 34$, and $p = \$700$ to a demand of $-0.06(700) + 46 = 4$. Further, since the demand equation is linear, values of p between these two extremes result in values of q between 4 and 34. This gives us the domain $4 \le q \le 34$ for the profit and revenue as functions of q.

Graphs of the revenue and profit functions are shown in Figure 23.

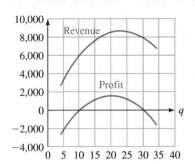

Figure 23

Notice that the revenue peaks at around 23 members and the profit peaks at around 20 members.

b. Taking the derivatives of $R(q)$ and $P(q)$ from part (a) gives

$$R'(q) = \frac{46 - 2q}{0.06} \quad \text{and} \quad P'(q) = \frac{46 - 2q}{0.06} - 100 = R'(q) - 100$$

The units for both are dollars per member. Think of these functions as approximating the *additional* revenue and profit the club receives for each member.

c. Substituting the given values of q gives

$$P'(15) = \frac{46 - 2(15)}{0.06} - 100 = \$166.67 \text{ per member}$$

This means that, at a membership level of 15 members, the profit would increase by about $166.67 with the addition of one more member. It would therefore pay Workout Fever to increase its membership by lowering the membership fee. This is confirmed by looking at the graph in Figure 23.

$$P'(20) = \frac{46 - 2(20)}{0.06} - 100 = \$0 \text{ per member}$$

This means that at a membership level of 20 members, the profit would remain approximately the same with the addition of one more member. In Figure 23 we can see that a membership level of 20 corresponds to the maximum possible profit.

$$P'(25) = \frac{46 - 2(25)}{0.06} - 100 = -\$166.67 \text{ per member}$$

This means that at a membership level of 25 members, the profit would decrease by about $166.67 with the addition of one more member. Workout Fever should therefore decrease its membership by raising the membership fee (see Figure 23).

✳ *Before we go on . . .* This analysis shows that, for a maximum annual profit, Workout Fever should maintain a membership of more than 15 but less than 25. The fact that $P'(20) = 0$ tells us that the health club should adjust the price to maintain a membership of exactly 20 members. (This is confirmed graphically in Figure 23.) In general, *for maximum profit, $P'(q)$ must be* 0.

Notice that computation of the derivatives makes it easy to find out exactly where the revenue and profit graphs reach a maximum. These exact locations are not so clear on the original graphs of revenue and profit.

Example 3 • Marginal Product

A consultant determines that Precision Manufacturer's annual profit (in dollars) is given by

$$P(n) = -200,000 + 400,000n - 4600n^2 - 10n^3 \quad (10 \le n \le 50)$$

where n is the number of assembly-line workers it employs.

a. Compute $P'(n)$. $P'(n)$ is called the **marginal product** at the employment level of n assembly-line workers. What are its units?

b. Calculate $P(20)$ and $P'(20)$ and interpret the results.

c. Precision Manufacturer currently employs 20 assembly-line workers and is considering laying off some of them. What advice would you give the company's management?

Solution

a. Taking the derivative gives

$$P'(n) = 400,000 - 9200n - 30n^2$$

The units of $P'(n)$ are profit (in dollars) per worker.

b. Substituting into the formula for $P(n)$, we get

$$P(20) = -200,000 + 400,000(20) - 4600(20)^2 - 10(20)^3 = \$5,880,000$$

Thus, Precision Manufacturer will make an annual profit of $5,880,000 if it employs 20 assembly-line workers. On the other hand,

$$P'(20) = 400,000 - 9200(20) - 30(20)^2 = \$204,000 \text{ per worker}$$

Thus, at an employment level of 20 assembly-line workers, annual profit is increasing at a rate of $204,000 per additional worker. In other words, if the company were to employ one more assembly-line worker, its annual profit would increase by approximately $204,000.

c. Since the marginal product is positive, profits will increase if the company increases the number of workers and will decrease if it decreases the number of workers, so your advice would be to hire additional assembly-line workers. Downsizing their assembly-line workforce would reduce their annual profits.

Before we go on . . . The following question might have occurred to you.

Question How many additional assembly-line workers should they hire to obtain the maximum annual profit?

Answer Figure 24 shows the graphs of P and P'. (Notice that the graph of the derivative makes it easy to see exactly where P reaches a maximum.)

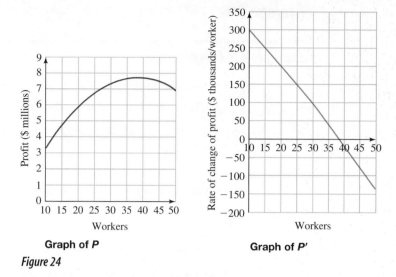

Graph of P

Graph of P'

Figure 24

At the point on the graph where $n = 20$, the slope is positive [as confirmed by our calculation of $P'(20)$]. At approximately $n = 39$, the slope is zero (so the marginal product is zero), and the profit is largest. Figure 25 shows a zoomed-in portion of the graph of P' in which we can see this value of n more accurately.

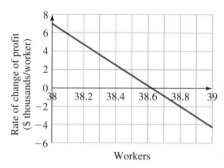

Figure 25

We conclude that the company should employ between 38 and 39 assembly-line workers for a maximum profit. (How could we obtain this answer algebraically?) To see which gives the larger profit, 38 or 39, we check:

$$P(38) = \$7,808,880$$

$$P(39) = \$7,810,210$$

This tells us that the company should employ 39 assembly-line workers for a maximum profit. Thus, instead of laying off any of its 20 assembly-line workers, the company should hire 19 additional assembly-line workers for a total of 39.

Average Cost

Example 4 • Average Cost

Suppose the cost in dollars to manufacture portable CD players is given by

$$C(x) = 150{,}000 + 20x - 0.0001x^2$$

where x is the number of CD players manufactured.

a. Find the average cost per CD player if 50,000 CD players are manufactured.

b. Find a formula for the average cost per CD player if x CD players are manufactured. This function of x is called the **average cost function, $\overline{C}(x)$.**

Solution

a. The total cost of manufacturing 50,000 CD players is given by

$$C(50{,}000) = 150{,}000 + 20(50{,}000) - 0.0001(50{,}000)^2$$

$$= \$900{,}000$$

Since 50,000 CD players cost a total of $900,000 to manufacture, the average cost of manufacturing 1 CD player is this total cost divided by 50,000:

$$\overline{C}(50{,}000) = \frac{900{,}000}{50{,}000} = \$18 \text{ per CD player}$$

Thus, if 50,000 CD players are manufactured, each CD player costs the manufacturer an average of $18 to manufacture.

b. If we replace 50,000 by x, we get the general formula for the average cost of manufacturing x CD players:

$$\overline{C}(x) = \frac{C(x)}{x} = \frac{1}{x}\,(150{,}000 + 20x - 0.0001x^2)$$

$$= \frac{150{,}000}{x} + 20 - 0.0001x \qquad \text{Average cost function}$$

✳ **Before we go on . . .**

Question In Example 1 we calculated the *marginal cost* at a production level of 50,000 CD players as

$$C'(50{,}000) = \$10 \text{ per CD player}$$

What, if anything, does this have to do with the average cost?

Answer Average cost and marginal cost convey different but related information. The average cost $\overline{C}(50{,}000)$ is the cost per item of manufacturing the first 50,000 CD players, whereas the marginal cost $C'(50{,}000)$ gives the (approximate) cost of manufacturing the *next* CD player. Thus, according to our calculations, the first 50,000 CD players cost an average of $18, but it costs only about $10 to manufacture the next one.

Question The marginal cost at a production level of 50,000 CD players is lower than the average cost. Does this mean that the average cost to manufacture CDs is going up or down with increasing volume?

Answer Down. Think about why.

Figure 26 shows the graphs of average and marginal cost. Notice how the decreasing marginal cost seems to pull the average cost down with it.

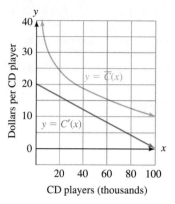

Figure 26

To summarize:

Average Cost

Given a cost function C, the **average cost** of the first x items is given by

$$\overline{C}(x) = \frac{C(x)}{x}$$

The average cost is distinct from the **marginal cost** $C'(x)$, which tells us the approximate cost of the *next* item.

3.5 EXERCISES

In Exercises 1–4, for each cost function find the marginal cost at the given production level x and state the units of measurement. (All costs are in dollars.)

1. $C(x) = 10{,}000 + 5x - 0.0001x^2$; $x = 1000$

2. $C(x) = 20{,}000 + 7x - 0.00005x^2$; $x = 10{,}000$

3. $C(x) = 15{,}000 + 100x + \dfrac{1000}{x}$; $x = 100$

4. $C(x) = 20{,}000 + 50x + \dfrac{10{,}000}{x}$; $x = 100$

In Exercises 5 and 6, find the marginal cost, marginal revenue, and marginal profit functions and find all values of x for which the marginal profit is zero. Interpret your answer.

5. $C(x) = 4x$; $R(x) = 8x - 0.001x^2$

6. $C(x) = 5x^2$; $R(x) = x^3 + 7x + 10$

7. A certain cost function has the following graph:

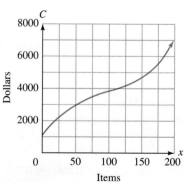

a. The associated marginal cost is
   (A) increasing and then decreasing.
   (B) decreasing and then increasing.
   (C) always increasing.
   (D) always decreasing.

b. The marginal cost is least at approximately
   (A) $x = 0$. **(B)** $x = 50$. **(C)** $x = 100$. **(D)** $x = 150$.

c. The cost of 50 items is
  (A) approximately $20 and increasing at a rate of about $3000 per item.
  (B) approximately $0.50 and increasing at a rate of about $3000 per item.
  (C) approximately $3000 and increasing at a rate of about $20 per item.
  (D) approximately $3000 and increasing at a rate of about $0.50 per item.

8. A certain cost function has the following graph:

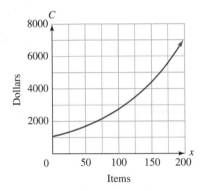

a. The associated marginal cost is
  (A) increasing and then decreasing.
  (B) decreasing and then increasing.
  (C) always increasing.
  (D) always decreasing.
b. When $x = 100$, the marginal cost is
  (A) greater than **(B)** less than **(C)** approximately equal to the average cost.
c. The cost of 150 items is
  (A) approximately $4400 and increasing at a rate of about $40 per item.
  (B) approximately $40 and increasing at a rate of about $4400 per item.
  (C) approximately $4400 and increasing at a rate of about $1 per item.
  (D) approximately $1 and increasing at a rate of about $4400 per item.

APPLICATIONS

9. Advertising Costs The cost, in thousands of dollars, of airing x television commercials during a Super Bowl game is given by

$$C(x) = 150 + 1200x - 0.002x^2$$

a. Find the marginal cost function and use it to estimate how fast the cost is increasing when $x = 4$. Compare this with the exact cost of airing the fifth commercial.
b. Find the average cost function $\overline{C}$ and evaluate $\overline{C}(4)$. What does the answer tell you?

The NBC television network planned to charge approximately $1.2 million per 30-second television spot during the 1996 Super Bowl XXX game. This explains the coefficient of x in the cost function. Source: *New York Times,* May 5, 1995, p. D5.

10. Marginal Cost and Average Cost The cost of producing x teddy bears per day at the Cuddly Companion Co. is calculated by their marketing staff to be given by the formula

$$C(x) = 100 + 40x - 0.001x^2$$

a. Find the marginal cost function and use it to estimate how fast the cost is increasing at a production level of 100 teddy bears. Compare this with the exact cost of producing the 101st teddy bear.
b. Find the average cost function $\overline{C}$ and evaluate $\overline{C}(100)$. What does the answer tell you?

11. Marginal Profit Suppose $P(x)$ represents the profit on the sale of x DVDs. If $P(1000) = 3000$ and $P'(1000) = -3$, what do these values tell you about the profit?

12. Marginal Loss An automobile retailer calculates that its loss on the sale of type M cars is given by $L(50) = 5000$ and $L'(50) = -200$, where $L(x)$ represents the loss on the sale of x type M cars. What do these values tell you about losses?

13. Marginal Profit Your monthly profit (in dollars) from selling magazines is given by

$$P = 5x + \sqrt{x}$$

where x is the number of magazines you sell in a month. If you are currently selling $x = 50$ magazines per month, find your profit and your marginal profit. Interpret your answers.

14. Marginal Profit Your monthly profit (in dollars) from your newspaper route is given by

$$P = 2n - \sqrt{n}$$

where n is the number of subscribers on your route. If you currently have 100 subscribers, find your profit and your marginal profit. Interpret your answers.

15. Marginal Product A car wash firm calculates that its daily profit (in dollars) depends on the number n of workers it employs according to the formula

$$P = 400n - 0.5n^2$$

Calculate the marginal product at an employment level of 50 workers and interpret the result.

16. Marginal Product Repeat Exercise 15, using the formula

$$P = -100n + 25n^2 - 0.005n^4$$

17. Average and Marginal Cost The daily cost to manufacture generic trinkets for gullible tourists is given by the cost function

$$C(x) = -0.001x^2 + 0.3x + 500 \text{ dollars}$$

where x is the number of trinkets.
a. As x increases, the marginal cost
  (A) increases.   **(B)** decreases.
  (C) increases and then decreases.
  (D) decreases and then increases.

b. As x increases, the average cost

  (A) increases.   (B) decreases.
  (C) increases and then decreases.
  (D) decreases and then increases.

c. The marginal cost is

  (A) greater than   (B) equal to   (C) less than
  the average cost when $x = 100$.

18. **Average and Marginal Cost** Repeat Exercise 17 using the following cost function for imitation oil paintings (x is the number of "oil paintings" manufactured):

$$C(x) = 0.1x^2 - 3.5x + 500 \text{ dollars}$$

19. **Marginal Revenue: Pricing Tuna** Assume that the demand function for tuna in a small coastal town is given by

$$p = \frac{20,000}{q^{1.5}} \qquad (200 \le q \le 800)$$

where p is the price (in dollars) per pound of tuna and q is the number of pounds of tuna that can be sold at the price p in 1 month.

a. Calculate the price that the town's fishery should charge for tuna in order to produce a demand of 400 pounds of tuna per month.

b. Calculate the monthly revenue R as a function of the number of pounds of tuna q.

c. Calculate the revenue and marginal revenue (derivative of the revenue with respect to q) at a demand level of 400 pounds per month and interpret the results.

d. If the town fishery's monthly tuna catch amounted to 400 pounds of tuna and the price is at the level in part (a), would you recommend that the fishery raise or lower the price of tuna in order to increase its revenue?

20. **Marginal Revenue: Pricing Tuna** Repeat Exercise 19, assuming a demand equation of

$$p = \frac{60}{q^{0.5}} \qquad (200 \le q \le 800)$$

21. **Advertising Cost** Your company is planning to air a number of television commercials during the ABC television network's presentation of the Academy Awards. ABC is charging your company $685,000 per 30-second spot. Additional fixed costs (development and personnel costs) amount to $500,000, and the network has agreed to provide a discount of

$$D(x) = \$10,000\sqrt{x}$$

for x television spots.

a. Write down the cost function C, marginal cost function C', and average cost function $\overline{C}$.

b. Compute $C'(3)$ and $\overline{C}(3)$. (Round all answers to three significant digits.) Use these two answers to say whether the average cost is increasing or decreasing as x increases.

This is what ABC actually charged for a 30-second spot during the 1995 Academy Awards presentation. Source: *New York Times*, May 5, 1995, p. D5.

22. **Housing Costs** The cost C of building a house is related to the number k of carpenters used and the number e of electricians used by the formula

$$C = 15,000 + 50k^2 + 60e^2$$

a. Assuming that ten carpenters are currently being used, find the marginal cost as a function of e.

b. If ten carpenters and ten electricians are currently being used, use your answer to part (a) to estimate the cost of hiring an additional electrician.

c. If ten carpenters and ten electricians are currently being used, what is the cost of hiring an additional carpenter?

Based on an exercise in A. L. Ostrosky, Jr., and J. V. Koch, *Introduction to Mathematical Economics* (Prospect Heights, Ill.: Waveland Press, 1979).

23. **Emission Control** The cost of controlling emissions at a firm rises rapidly as the amount of emissions reduced increases. Here is a possible model:

$$C(q) = 4000 + 100q^2$$

where q is the reduction in emissions (in pounds of pollutant per day) and C is the daily cost (in dollars) of this reduction.

a. If a firm is currently reducing its emissions by 10 pounds each day, what is the marginal cost of reducing emissions further?

b. Government clean-air subsidies to the firm are based on the formula

$$S(q) = 500q$$

where q is again the reduction in emissions (in pounds per day) and S is the subsidy (in dollars). At what reduction level does the marginal cost surpass the marginal subsidy?

c. Calculate the net cost function, $N(q) = C(q) - S(q)$, given the cost function and subsidy above, and find the value of q that gives the lowest net cost. What is this lowest net cost? Compare your answer to that for part (b) and comment on what you find.

24. **Taxation Schemes** Here is a curious proposal for taxation rates based on income:

$$T(i) = 0.001i^{0.5}$$

where i represents total annual income in dollars and $T(i)$ is the income tax rate as a percentage of total annual income. (Thus, for example, an income of $50,000 per year would be taxed at about 22%, whereas an income of double that amount would be taxed at about 32%.)[13]

a. Calculate the after-tax (net) income $N(i)$ an individual can expect to earn as a function of income i.

b. Calculate an individual's marginal after-tax income at income levels of $100,000 and $500,000.

c. At what income does an individual's marginal after-tax income become negative? What is the after-tax income at that level, and what happens at higher income levels?

d. What do you suspect is the most anyone can earn after taxes? (See note 13.)

---

[13]This model has the following interesting feature: An income of $1 million/year would be taxed at 100%, leaving the individual penniless!

25. Fuel Economy Your Porsche's gas mileage (in miles/gallon) is given as a function $M(x)$ of speed x in miles/hour. It is found that

$$M'(x) = \frac{3600x^{-2} - 1}{(3600x^{-1} + x)^2}$$

Find $M'(10)$, $M'(60)$, and $M'(70)$. What do the answers tell you about your car?

E **26. Marginal Revenue** The estimated marginal revenue for sales of ESU soccer team T-shirts is given by

$$R'(p) = \frac{(8 - 2p)e^{-p^2 + 8p}}{10,000,000}$$

where p is the price (in dollars) the soccer players charge for each shirt. Find $R'(3)$, $R'(4)$, and $R'(5)$. What do the answers tell you?

27. Marginal Cost (from the GRE Economics Test) In a multiple-plant firm in which the different plants have different and continuous cost schedules, if costs of production for a given output level are to be minimized, which of the following is essential?

(A) Marginal costs must equal marginal revenue.

(B) Average variable costs must be the same in all plants.

(C) Marginal costs must be the same in all plants.

(D) Total costs must be the same in all plants.

(E) Output per man-hour must be the same in all plants.

28. Study Time (from the GRE Economics Test) A student has a fixed number of hours to devote to study and is certain of the relationship between hours of study and the final grade for each course. Grades are given on a numerical scale (for example, 0 to 100), and each course is counted equally in computing the grade average. In order to maximize his or her grade average, the student should allocate these hours to different courses so that

(A) the grade in each course is the same.

(B) the marginal product of an hour's study (in terms of final grade) in each course is zero.

(C) the marginal product of an hour's study (in terms of final grade) in each course is equal, although not necessarily equal to zero.

(D) the average product of an hour's study (in terms of final grade) in each course is equal.

(E) the number of hours spent in study for each course are equal.

29. Marginal Product (from the GRE Economics Test) Assume that the marginal product of an additional senior professor is 50% higher than the marginal product of an additional junior professor and that junior professors are paid one-half the amount that senior professors receive. With a fixed overall budget, a university that wishes to maximize its quantity of output from professors should do which of the following?

(A) Hire equal numbers of senior and junior professors.

(B) Hire more senior professors and junior professors.

(C) Hire more senior professors and discharge junior professors.

(D) Discharge senior professors and hire more junior professors.

(E) Discharge all senior professors and half of the junior professors.

30. Marginal Product (Based on a Question from the GRE Economics Test) Assume that the marginal product of an additional senior professor is twice the marginal product of an additional junior professor and that junior professors are paid two-thirds the amount that senior professors receive. With a fixed overall budget, a university that wishes to maximize its quantity of output from professors should do which of the following?

(A) Hire equal numbers of senior and junior professors.

(B) Hire more senior professors and junior professors.

(C) Hire more senior professors and discharge junior professors.

(D) Discharge senior professors and hire more junior professors.

(E) Discharge all senior professors and half of the junior professors.

COMMUNICATION AND REASONING EXERCISES

31. The marginal cost of producing the 1001st item is

(A) equal to   (B) approximately equal to

(C) always slightly greater than

the actual cost of producing the 1001st item.

32. For the cost function $C(x) = mx + b$, the marginal cost of producing the 1001st item is

(A) equal to   (B) approximately equal to

(C) always slightly greater than

the actual cost of producing the 1001st item.

33. What is a cost function? Carefully explain the difference between *average cost* and *marginal cost* in terms of (a) their mathematical definition, (b) graphs, and (c) interpretation.

34. The cost function for your grand piano manufacturing plant has the property that $\overline{C}(1000) = \$3000$/unit and $C'(1000) = \$2500$ per unit. Will the average cost increase or decrease if your company manufactures a slightly larger number of pianos? Explain your reasoning.

35. If the average cost to manufacture one grand piano increases as the production level increases, which is greater, the marginal cost or the average cost?

36. If your analysis of a manufacturing company yielded positive marginal profit but negative profit at the company's current production levels, what would you advise the company to do?

37. If the marginal cost is decreasing, is the average cost necessarily decreasing? Explain.

38. If the average cost is decreasing, is the marginal cost necessarily decreasing? Explain.

39. If a company's marginal average cost is zero at the current production level, positive for a slightly higher production level, and negative for a slightly lower production level, what should you advise the company to do?

40. The **acceleration** of cost is defined as the derivative of the marginal cost function: that is, the derivative of the derivative—or *second derivative*—of the cost function. What are the units of acceleration of cost, and how does one interpret this measure?

3.6 Limits: Numerical and Graphical Approaches (Optional)

The derivative is defined using a limit, and it is now time to say more precisely what that means. It is possible to speak of limits by themselves, rather than in the context of the derivative. The story of limits is a long one that we will try to make concise.

Evaluating Limits Numerically

Start with a simple example: Look at the function $f(x) = 2 + x$ and ask: What happens to $f(x)$ as x approaches 3? The following table shows the value of $f(x)$ for values of x close to and on either side of 3:

x approaching 3 from the left →   ← x approaching 3 from the right

x	2.9	2.99	2.999	2.9999	3	3.0001	3.001	3.01	3.1
$f(x) = 2 + x$	4.9	4.99	4.999	4.9999		5.0001	5.001	5.01	5.1

We have left the entry under 3 blank to emphasize that, when calculating the limit of $f(x)$ as x *approaches* 3, we are not interested in its value when x *equals* 3.

Notice from the table that the closer x gets to 3 from either side, the closer $f(x)$ gets to 5. We write this as

$$\lim_{x \to 3} f(x) = 5 \qquad \text{The limit of } f(x), \text{ as } x \text{ approaches 3, equals 5.}$$

Question Why all the fuss? Can't we simply put $x = 3$ and avoid having to use a table?

Answer This happens to work for *some* functions, but not for *all* functions. Example 1 illustrates this point.

Example 1 • Estimating a Limit Numerically

Use a table to estimate the following limits:

a. $\lim\limits_{x \to 2} \dfrac{x^3 - 8}{x - 2}$ **b.** $\lim\limits_{x \to 0} \dfrac{e^{2x} - 1}{x}$

Solution

a. We cannot simply substitute $x = 2$, because the function $f(x) = (x^3 - 8)/(x - 2)$ is not defined at $x = 2$. (Why?)[14] As above, we can use a table of values, with x approaching 2 from both sides:

---

[14]However, if you factor $x^3 - 8$, you will find that $f(x)$ can be simplified to a function that *is* defined at $x = 2$. This point will be discussed (and this example redone) in Section 3.8. The function in part (b) cannot be simplified in this way.

x approaching 2 from the left → ← x approaching 2 from the right

x	1.9	1.99	1.999	1.9999	2	2.0001	2.001	2.01	2.1
$f(x) = \dfrac{x^3 - 8}{x - 2}$	11.41	11.9401	11.9940	11.9994		12.0006	12.0060	12.0601	12.61

We notice that as x approaches 2 from either side, $f(x)$ approaches 12. This suggests that the limit is 12, and we write

$$\lim_{x \to 2} \frac{x^3 - 8}{x - 2} = 12$$

b. The function $g(x) = (e^{2x} - 1)/x$ is not defined at $x = 0$ (nor can it even be simplified to one which *is* defined at $x = 0$). In the following table, we allow x to approach zero from both sides:

x approaching 0 from the left → ← x approaching 0 from the right

x	−0.1	−0.01	−0.001	−0.0001	0	0.0001	0.001	0.01	0.1
$g(x) = \dfrac{e^{2x} - 1}{x}$	1.8127	1.9801	1.9980	1.9998		2.0002	2.0020	2.0201	2.2140

The table suggests that $\lim\limits_{x \to 0} \dfrac{e^{2x} - 1}{x} = 2$.

Graphing Calculator

On the TI-83 use the Table feature to automate these computations. First, define $Y_1 = (x^3 - 8)/(x - 2)$ for part (a) or $Y_1 = (e^{\wedge}(2x) - 1)/x$ for part (b) and then press 2nd TABLE to list its values for the given values of x. (If the calculator does not allow you to enter values of x, press 2nd TBLSET and set Indpnt to Ask.)

Excel

Spreadsheets are ideal for calculating limits numerically. Set up your spreadsheet to duplicate the table in part (a) as follows:

	A	B	C	D
1	x	f(x)	x	f(x)
2	1.9	=(A2^3-8)/(A2-2)	2.1	
3	1.99		2.01	
4	1.999		2.001	
5	1.9999		2.0001	

(The formula in cell B2 is copied to columns B and D as indicated by the shading.) The values of $f(x)$ will be calculated in columns B and D.

For part (b), use the formula =(EXP(2*A2)-1)/A2 in cell B2 and, in columns A and C, use the values of x shown in the table for part (b).

✳ **Before we go on . . .** Although the table *suggests* that the limit in part (b) is 2, it by no means establishes that fact conclusively. It is *conceivable* (though not in fact the case here) that putting $x = 0.000\,000\,087$ will result in $g(x) = 426$. Using a table can only suggest a value for the limit. In the next two sections, we discuss algebraic techniques for finding limits.

Before we continue, let's make a more formal definition.

Definition of a Limit

If $f(x)$ approaches the number L as x approaches (but is not equal to) a from both sides, then we say that $f(x)$ **approaches L as $x \to a$** ("x approaches a") or that the **limit** of $f(x)$ as $x \to a$ is L. We write

$$\lim_{x \to a} f(x) = L \quad \text{or} \quad f(x) \to L \text{ as } x \to a$$

If $f(x)$ *fails* to approach *a single fixed number* as x approaches a from both sides, then we say that $f(x)$ *has no limit* as $x \to a$ or

$$\lim_{x \to a} f(x) \text{ does not exist.}$$

Quick Examples

1. $\displaystyle\lim_{x \to 3} (2 + x) = 5$ See discussion before Example 1.

2. $\displaystyle\lim_{x \to -2} (3x) = -6$ As x approaches -2, $3x$ approaches -6.

3. $\displaystyle\lim_{x \to 0} (x^2 - 2x + 1)$ exists.   In fact, the limit is 1.

4. $\displaystyle\lim_{x \to 5} \frac{1}{x} = \frac{1}{5}$ As x approaches 5, $\dfrac{1}{x}$ approaches $\dfrac{1}{5}$.

5. $\displaystyle\lim_{x \to 2} \frac{x^3 - 8}{x - 2} = 12$ See Example 1. (We cannot just put $x = 2$ here.)

(For examples of nonexistent limits, see Example 2.)

Notes

1. It is important that $f(x)$ approach the same number as x approaches a from either side. For instance, if $f(x)$ approaches 5 for $x = 1.9, 1.99, 1.999, \ldots$ but approaches 4 for $x = 2.1, 2.01, 2.001, \ldots$ then the limit as $x \to 2$ does not exist. (See Example 2 for such a situation.)

2. It may happen that $f(x)$ does not approach any fixed number at all as $x \to a$ from either side. In this case, we also say that the limit does not exist.

3. We are deliberately suppressing the exact definition of "approaches"; instead we trust your intuition. The following phrasing of the definition of the limit is closer to the one used by mathematicians: *We can make $f(x)$ be as close to L as we like by making x be sufficiently close to a.*

Earlier in this chapter, we examined functions that were not differentiable, and one of them, $|x|$, had the property that its difference quotient at $x = 0$ did not approach a limit as $h \to 0$. The difference quotient in question simplified to $|h|/h$. We now reexamine this limit and some others.

Example 2 • Nonexistent Limits

Do the following limits exist?

a. $\lim\limits_{x \to 0} \dfrac{1}{x^2}$　　**b.** $\lim\limits_{x \to 0} \dfrac{|x|}{x}$　　**c.** $\lim\limits_{x \to 2} \dfrac{1}{x-2}$

Solution

a. Here is a table of values for $f(x) = 1/x^2$, with x approaching 0 from both sides.

x approaching 0 from the left → 　 ← x approaching 0 from the right

x	−0.1	−0.01	−0.001	−0.0001	0	0.0001	0.001	0.01	0.1
$f(x) = \dfrac{1}{x^2}$	100	10,000	1,000,000	100,000,000		100,000,000	1,000,000	10,000	100

The table shows that as x gets closer to zero on either side, $f(x)$ gets larger and larger **without bound**—that is, if you name any number, no matter how large, $f(x)$ will be even larger than that if x is sufficiently close to 0. Since $f(x)$ is not approaching any real number, we conclude that $\lim_{x \to 0} (1/x^2)$ does not exist. Since $f(x)$ is becoming arbitrarily large, we also say that $\lim_{x \to 0} (1/x^2)$ **diverges to $+\infty$,** or just

$$\lim_{x \to 0} \frac{1}{x^2} = +\infty$$

Note: This not meant to imply that the limit exists, since the symbol $+\infty$ does not represent any real number. We write $\lim_{x \to a} f(x) = +\infty$ to indicate two things: (1) The limit does not exist, and (2) the function gets large without bound as x approaches a.

b. Here is a table of values for $f(x) = |x|/x$, with x approaching 0 from both sides.

x approaching 0 from the left → 　 ← x approaching 0 from the right

x	−0.1	−0.01	−0.001	−0.0001	0	0.0001	0.001	0.01	0.1
$f(x) = \dfrac{|x|}{x}$	−1	−1	−1	−1		1	1	1	1

The table shows that $f(x)$ does not approach the same limit as x approaches 0 from both sides. There appear to be two *different* limits: the limit as we approach 0 from the left and the limit as we approach from the right. We write

$$\lim_{x \to 0^-} f(x) = -1$$

read as "the limit as x approaches zero from the left (or from below) is -1," and

$$\lim_{x \to 0^+} f(x) = 1$$

read as "the limit as x approaches 0 from the right (or from above) is 1." These are called the **one-sided limits** of $f(x)$. For f to have a **two-sided limit,** the two one-sided limits must be equal. Since they are not, we conclude that $\lim_{x \to 0} f(x)$ does not exist.

c. Near $x = 2$, we have the following table of values for $f(x) = \dfrac{1}{x-2}$:

x approaching 2 from the left →   ← x approaching 2 from the right

x	1.9	1.99	1.999	1.9999	2	2.0001	2.001	2.01	2.1
$f(x) = \dfrac{1}{x-2}$	-10	-100	-1000	$-10{,}000$		10,000	1000	100	10

Since $1/(x-2)$ is approaching no (single) real number as $x \to 2$, we can see that $\lim_{x \to 2} 1/(x-2)$ does not exist. Notice also that $1/(x-2)$ diverges to $+\infty$ as $x \to 2$ from the positive side (right half of the table) and to $-\infty$ as $x \to 2$ from the left (left half of the table). In other words,

$$\lim_{x \to 2^-} \frac{1}{x-2} = -\infty$$

$$\lim_{x \to 2^+} \frac{1}{x-2} = +\infty$$

$$\lim_{x \to 2} \frac{1}{x-2} \text{ does not exist.}$$

In another useful kind of limit, we let x approach either $+\infty$ or $-\infty$, by which we mean that we let x get arbitrarily large or let x become an arbitrarily large negative number. Example 3 illustrates this.

Example 3 • Limits at Infinity

Use a table to estimate **a.** $\displaystyle\lim_{x \to +\infty} \frac{2x^2 - 4x}{x^2 - 1}$ and **b.** $\displaystyle\lim_{x \to -\infty} \frac{2x^2 - 4x}{x^2 - 1}$.

Solution

a. By saying that x is "approaching $+\infty$," we mean that x is getting larger and larger without bound, so we make the following table:

x approaching $+\infty$ →

x	10	100	1000	10,000	100,000
$f(x) = \dfrac{2x^2 - 4x}{x^2 - 1}$	1.6162	1.9602	1.9960	1.9996	2.0000

(Note that we are only approaching $+\infty$ from the left because we can hardly approach it from the right!) What seems to be happening is that $f(x)$ is approaching 2. Thus, we write

$$\lim_{x \to +\infty} f(x) = 2$$

b. Here, x is approaching $-\infty$, so we make a similar table, this time with x assuming negative values of greater and greater magnitude (read this table from right to left):

← x approaches $-\infty$

x	$-100{,}000$	$-10{,}000$	-1000	-100	-10
$f(x) = \dfrac{2x^2 - 4x}{x^2 - 1}$	2.0000	2.0004	2.0040	2.0402	2.4242

Once again, $f(x)$ is approaching 2. Thus, $\lim_{x \to -\infty} f(x) = 2$.

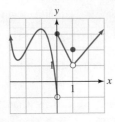

Figure 27

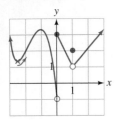

Figure 28

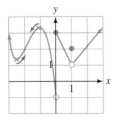

Figure 29

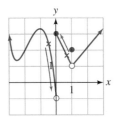

Figure 30

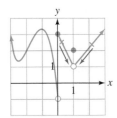

Figure 31

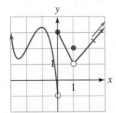

Figure 32

Estimating Limits Graphically

Example 4 • Estimating Limits Graphically

The graph of a function f is shown in Figure 27. (Recall that the solid dots indicate points on the graph, and the hollow dots indicate points *not* on the graph.) From the graph, analyze the following limits:

a. $\lim\limits_{x \to -2} f(x)$ **b.** $\lim\limits_{x \to 0} f(x)$ **c.** $\lim\limits_{x \to 1} f(x)$ **d.** $\lim\limits_{x \to +\infty} f(x)$

Solution Since we are given only a graph of f, we must analyze these limits graphically.

a. Imagine that Figure 27 was drawn on a graphing calculator equipped with a Trace feature that allows us to move a cursor along the graph and see the coordinates as we go. To simulate this, place a pencil point on the graph to the left of $x = -2$ and move it along the curve so that the x coordinate approaches -2 (Figure 28). We evaluate the limit numerically by noting the behavior of the y coordinates.[15] However, we can see directly from the graph that the y coordinate approaches 2. Similarly, if we place our pencil point to the right of $x = -2$ and move it to the left, the y coordinate will approach 2 from that side as well (Figure 29). Therefore, as x approaches -2 from either side, $f(x)$ approaches 2, so

$$\lim_{x \to -2} f(x) = 2$$

b. This time we move our pencil point toward $x = 0$. If we start from the left of $x = 0$ and approach 0 (by moving right), the y coordinate approaches -1 (Figure 30). However, if we start from the right of $x = 0$ and approach 0 (by moving left), the y coordinate approaches 3. Thus (see Example 2),

$$\lim_{x \to 0^-} f(x) = -1 \quad \text{and} \quad \lim_{x \to 0^+} f(x) = 3$$

Since these limits are not equal, we conclude that

$$\lim_{x \to 0} f(x) \text{ does not exist.}$$

In this case there is a "break" in the graph at $x = 0$, and we say that the function is **discontinuous** at $x = 0$ (see Section 3.7).

c. Once more we think about a pencil point moving along the graph with the x coordinate this time approaching $x = 1$ from the left and from the right (Figure 31). As the x coordinate of the point approaches 1 from either side, the y coordinate approaches 1 also. Therefore,

$$\lim_{x \to 1} f(x) = 1$$

d. For this limit, x is supposed to approach infinity. We think about a pencil point moving along the graph farther and farther to the right, as shown in Figure 32. As the x coordinate gets larger, the y coordinate also gets larger and larger without bound. Thus, $f(x)$ diverges to $+\infty$:

$$\lim_{x \to +\infty} f(x) = +\infty$$

Similarly,

$$\lim_{x \to -\infty} f(x) = +\infty$$

---

[15]For a visual animation of this process, look at the online tutorial for this section at the Web site.

✸ *Before we go on . . .* In part (c), $\lim_{x \to 1} f(x) = 1$, but $f(1) = 2$. (Why?) Thus, $\lim_{x \to 1} f(x) \neq f(1)$. In other words, the limit of $f(x)$ as x *approaches* 1 is not the same as the value of f *at* $x = 1$. Always keep in mind that when we evaluate a limit as $x \to a$, *we do not care about the value of the function at* $x = a$. We only care about the value of $f(x)$ as x *approaches a*. In other words, $f(a)$ may or may not equal $\lim_{x \to a} f(x)$.

We can summarize the graphical method we used in Example 4 as follows.

Evaluating Limits Graphically

To decide whether $\lim_{x \to a} f(x)$ exists and to find its value if it does:

1. Draw the graph of $f(x)$ by hand or with graphing technology.

2. Position your pencil point (or the Trace cursor) on a point of the graph to the right of $x = a$.

3. Move the point *along the graph* toward $x = a$ from the right and read the y coordinate as you go. The value the y coordinate approaches (if any) is the limit $\lim_{x \to a^+} f(x)$.

4. Repeat steps 2 and 3, this time starting from a point on the graph to the left of $x = a$ and approaching $x = a$ along the graph from the left. The value the y coordinate approaches (if any) is $\lim_{x \to a^-} f(x)$.

5. If the left and right limits both exist and have the same value L, then $\lim_{x \to a} f(x) = L$. Otherwise, the limit does not exist. The value $f(a)$ is of no relevance whatsoever.

6. To evaluate $\lim_{x \to +\infty} f(x)$, move the pencil point toward the far right of the graph and estimate the value the y coordinate approaches (if any). For $\lim_{x \to -\infty} f(x)$, move the pencil point toward the far left.

In Example 5 we use both the numerical and graphical approaches.

Example 5 • Infinite Limit

Does $\lim\limits_{x \to 0^+} \dfrac{1}{x}$ exist?

Solution

Numerical Method Since we are asked for only the right-hand limit, we need only list values of x approaching 0 from the right:

$\leftarrow x$ approaching 0 from the right

x	0	0.0001	0.001	0.01	0.1
$f(x) = \dfrac{1}{x}$		10,000	1000	100	10

What seems to be happening as x approaches 0 from the right is that $f(x)$ is increasing without bound, as in Example 4(d). That is, if you name any number, no matter how large, $f(x)$ will be even larger than that if x is sufficiently close to zero. Thus, the limit diverges to $+\infty$, so

$$\lim_{x \to 0^+} \frac{1}{x} = +\infty$$

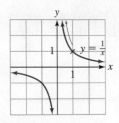

Figure 33

Graphical Method Recall that the graph of $f(x) = \dfrac{1}{x}$ is the standard hyperbola shown in Figure 33. The figure also shows the pencil point moving so that its x coordinate approaches 0 from the right. Since the point moves along the graph, it is forced to go higher and higher. In other words, its y coordinate becomes larger and larger, approaching $+\infty$. Thus, we conclude that

$$\lim_{x \to 0^+} \frac{1}{x} = +\infty$$

✴ ***Before we go on . . .*** You should also check that

$$\lim_{x \to 0^-} \frac{1}{x} = -\infty$$

We say that as x approaches 0 from the left, $\dfrac{1}{x}$ diverges to $-\infty$. Also, check that

$$\lim_{x \to +\infty} \frac{1}{x} = \lim_{x \to -\infty} \frac{1}{x} = 0$$

Application

Example 6 • Demand Equation

Economist Henry Schultz calculated the following demand function for corn:

$$q = \frac{130{,}000}{p^{0.75}}$$

where p is the price in dollars per bushel and q is the number of bushels of corn that could be sold at the price p in 1 year.[16]

a. Estimate $\lim\limits_{p \to +\infty} q(p)$ and interpret the answer.

b. Estimate $\lim\limits_{p \to 0^+} q(p)$ and interpret the answer.

Solution

a. Figure 34 shows a plot of $q(p)$ for $0 < p < 100$ and $0 \le q \le 40{,}000$.

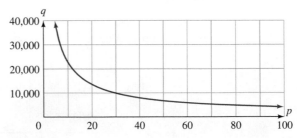

Figure 34

---

[16]This demand function is based on data for the period 1915–1929. SOURCE: Henry Schultz, *The Theory and Measurement of Demand,* as cited in A. L. Ostrosky, Jr., and J. V. Koch, *Introduction to Mathematical Economics* (Prospect Heights, Ill.: Waveland Press, 1979).

Using either the numerical or graphical approach, we find

$$\lim_{p \to +\infty} q(p) = \lim_{p \to +\infty} \frac{130{,}000}{p^{0.75}} = 0$$

Thus, as the price per bushel becomes larger and larger, the demand drops toward zero. This is what we reasonably expect of a demand equation.

b. The limit here is

$$\lim_{p \to 0^+} q(p) = \lim_{p \to 0^+} \frac{130{,}000}{p^{0.75}} = +\infty$$

So, as the price per bushel decreases toward $0, the demand soars without bound. We do not reasonably expect this of a demand equation; even if corn was given away free, there would be a finite demand. (Think of what happens when food or other items are freely available at campus events.) We conclude that this demand equation cannot possibly be valid near $p = 0$, and so we need to restrict its domain. We would need further data to determine the smallest price for which the equation is valid.

3.6 EXERCISES

In Exercises 1–18, estimate the limits numerically.

1. $\lim_{x \to 0} \dfrac{x^2}{x + 1}$

2. $\lim_{x \to 0} \dfrac{x - 3}{x - 1}$

3. $\lim_{x \to 2} \dfrac{x^2 - 4}{x - 2}$

4. $\lim_{x \to 2} \dfrac{x^2 - 1}{x - 2}$

5. $\lim_{x \to -1} \dfrac{x^2 + 1}{x + 1}$

6. $\lim_{x \to -1} \dfrac{x^2 + 2x + 1}{x + 1}$

7. $\lim_{x \to +\infty} \dfrac{3x^2 + 10x - 1}{2x^2 - 5x}$

8. $\lim_{x \to +\infty} \dfrac{6x^2 + 5x + 100}{3x^2 - 9}$

9. $\lim_{x \to -\infty} \dfrac{x^5 - 1000x^4}{2x^5 + 10{,}000}$

10. $\lim_{x \to -\infty} \dfrac{x^6 + 3000x^3 + 1{,}000{,}000}{2x^6 + 1000x^3}$

11. $\lim_{x \to +\infty} \dfrac{10x^2 + 300x + 1}{5x + 2}$

12. $\lim_{x \to +\infty} \dfrac{2x^4 + 20x^3}{1000x^6 + 6}$

13. $\lim_{x \to +\infty} \dfrac{10x^2 + 300x + 1}{5x^3 + 2}$

14. $\lim_{x \to +\infty} \dfrac{2x^4 + 20x^3}{1000x^3 + 6}$

[E] **15.** $\lim_{x \to 2} e^{x-2}$

[E] **16.** $\lim_{x \to +\infty} e^{-x}$

[E] **17.** $\lim_{x \to +\infty} xe^{-x}$

[E] **18.** $\lim_{x \to -\infty} xe^{x}$

In Exercises 19–30, the graph of f is given. Use the graph to compute the quantities asked for.

19. a. $\lim_{x \to 1} f(x)$ **b.** $\lim_{x \to -1} f(x)$

20. a. $\lim_{x \to -1} f(x)$ **b.** $\lim_{x \to 1} f(x)$

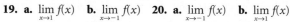

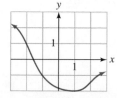

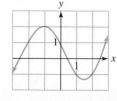

21. a. $\lim_{x \to 0} f(x)$ **b.** $\lim_{x \to 2} f(x)$
c. $\lim_{x \to -\infty} f(x)$ **d.** $\lim_{x \to +\infty} f(x)$

22. a. $\lim_{x \to -1} f(x)$ **b.** $\lim_{x \to 1} f(x)$
c. $\lim_{x \to +\infty} f(x)$ **d.** $\lim_{x \to -\infty} f(x)$

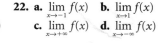

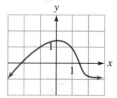

23. a. $\lim_{x \to 2} f(x)$ **b.** $\lim_{x \to 0^+} f(x)$
c. $\lim_{x \to 0^-} f(x)$ **d.** $\lim_{x \to 0} f(x)$
e. $f(0)$ **f.** $\lim_{x \to -\infty} f(x)$

24. a. $\lim_{x \to 3} f(x)$ **b.** $\lim_{x \to 1^+} f(x)$
c. $\lim_{x \to 1^-} f(x)$ **d.** $\lim_{x \to 1} f(x)$
e. $f(1)$ **f.** $\lim_{x \to +\infty} f(x)$

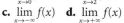

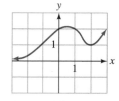

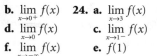

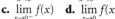

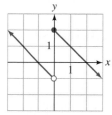

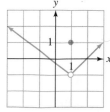

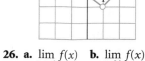

25. a. $\lim_{x \to -2} f(x)$ **b.** $\lim_{x \to -1^+} f(x)$
c. $\lim_{x \to -1^-} f(x)$ **d.** $\lim_{x \to -1} f(x)$
e. $f(-1)$ **f.** $\lim_{x \to +\infty} f(x)$

26. a. $\lim_{x \to -1} f(x)$ **b.** $\lim_{x \to 0^+} f(x)$
c. $\lim_{x \to 0^-} f(x)$ **d.** $\lim_{x \to 0} f(x)$
e. $f(0)$ **f.** $\lim_{x \to -\infty} f(x)$

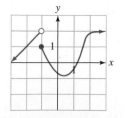

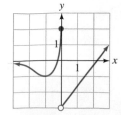

27. a. $\lim\limits_{x\to-1} f(x)$ **b.** $\lim\limits_{x\to0^+} f(x)$ **28. a.** $\lim\limits_{x\to1} f(x)$ **b.** $\lim\limits_{x\to0^+} f(x)$
   c. $\lim\limits_{x\to0^-} f(x)$ **d.** $\lim\limits_{x\to0} f(x)$ **c.** $\lim\limits_{x\to0^-} f(x)$ **d.** $\lim\limits_{x\to0} f(x)$
   e. $f(0)$ **f.** $\lim\limits_{x\to+\infty} f(x)$ **e.** $f(0)$ **f.** $\lim\limits_{x\to-\infty} f(x)$

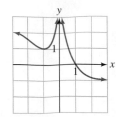

29. a. $\lim\limits_{x\to-1} f(x)$ **b.** $\lim\limits_{x\to0^+} f(x)$ **30. a.** $\lim\limits_{x\to0^-} f(x)$ **b.** $\lim\limits_{x\to1^+} f(x)$
   c. $\lim\limits_{x\to0^-} f(x)$ **d.** $\lim\limits_{x\to0} f(x)$ **c.** $\lim\limits_{x\to0} f(x)$ **d.** $\lim\limits_{x\to1} f(x)$
   e. $f(0)$ **f.** $f(-1)$ **e.** $f(0)$ **f.** $f(1)$

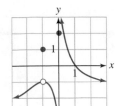
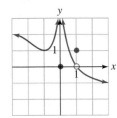

APPLICATIONS

31. Electric Rates The cost of electricity in Portland, Oregon, for residential customers increased suddenly on October 1, 2001, from around $0.06 to around $0.08 per kilowatt-hour. Let $C(t)$ be this cost at time t and let $t = 1$ represent October 1, 2001. What does the given information tell you about $\lim_{t\to1} C(t)$?

SOURCE: Portland General Electric/*New York Times,* February 2, 2002, p. C1.

32. Airline Stocks Prior to the September 11, 2001, attacks, United Airlines stock was trading at around $35/share. Immediately following the attacks, the share price dropped by $15.[17] Let $U(t)$ be this cost at time t and let $t = 11$ represent September 11, 2001. What does the given information tell you about $\lim_{t\to11} U(t)$?

Bottled-Water Sales Annual U.S. sales of domestic bottled water for the period 1990–2000 could be approximated by

$$D(t) = 0.0166t^2 - 0.050t + 0.27 \text{ billion gallons} \quad (0 \le t \le 10)$$

where t is time in years since 1990. Sales of imported bottled water over the same period could be approximated by

$$I(t) = 0.0004t^2 + 0.013t + 0.06 \text{ billion gallons} \quad (0 \le t \le 10)$$

Exercises 33 and 34 are based on these models.

Models are based on regression of approximate data. SOURCE: Beverage Marketing Corporation of New York/*New York Times,* June 21, 2001, p. C1.

33. Assuming the trends shown in the above models continue indefinitely, numerically estimate the following, interpret your answers, and comment on the results.

$$\lim_{t\to+\infty} I(t) \quad \text{and} \quad \lim_{t\to+\infty} \frac{I(t)}{D(t)}$$

34. Repeat Exercise 33, this time calculating

$$\lim_{t\to+\infty} D(t) \quad \text{and} \quad \lim_{t\to+\infty} \frac{D(t)}{I(t)}$$

35. Online Sales The following graph shows the approximate annual U.S. sales of books online for the period 1997–2004:

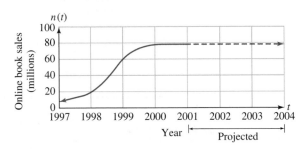

SOURCE: For 1997–2000 data, Ipsos-NPD Book Trends/*New York Times,* April 16, 2001, p. C1.

   a. Estimate $\lim\limits_{t\to+\infty} n(t)$ and interpret your answer.
   b. Estimate $\lim\limits_{t\to+\infty} n'(t)$ and interpret your answer.

36. Employment The following graph shows the number of new employees per year at Amerada Hess Corp. from 1984 ($t = 0$):

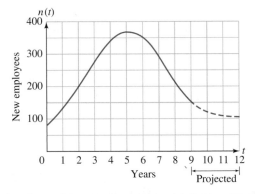

The projected part of the curve (from $t = 9$ on) is fictitious. The model is based on a best-fit logistic curve. SOURCE: *Hoover's Handbook Database* (World Wide Web site), The Reference Press, Inc., Austin, Texas, 1995.

   a. Estimate $\lim\limits_{t\to+\infty} n(t)$ and interpret your answer.
   b. Estimate $\lim\limits_{t\to+\infty} n'(t)$ and interpret your answer.

---

[17]Stock prices are approximate.

⊟ 37. **SAT Scores by Income** The following bar graph shows U.S. verbal SAT scores as a function of parents' income level:

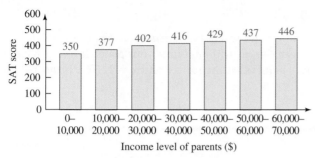

Based on 1994 data. SOURCE: The College Board/*New York Times*, March 5, 1995, p. E16.

These data can be modeled by

$$S(x) = 470 - 136e^{-0.00002645x}$$

where $S(x)$ is the average SAT verbal score of a student whose parents earn $\$x$ per year. Evaluate $\lim_{t \to +\infty} S(x)$ and interpret the result.

⊟ 38. **SAT Scores by Income** The following bar graph shows U.S. math SAT scores as a function of parents' income level:

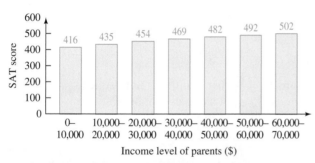

Based on 1994 data. SOURCE: The College Board/*New York Times*, March 5, 1995, p. E16.

These data can be modeled by

$$S(x) = 535 - 136e^{-0.000021286x}$$

where $S(x)$ is the average math SAT score of a student whose parents earn $\$x$ per year. Evaluate $\lim_{x \to +\infty} S(x)$ and interpret the result.

COMMUNICATION AND REASONING EXERCISES

39. Describe the method of evaluating limits numerically. Give at least one disadvantage of this method.

40. Describe the method of evaluating limits graphically. Give at least one disadvantage of this method.

41. What is wrong with the following statement? "Since $f(a)$ is not defined, $\lim_{x \to a} f(x)$ does not exist."

42. What is wrong with the following statement? "If $f(a)$ is defined, then $\lim_{x \to a} f(x)$ exists, and equals $f(a)$."

43. Your friend Dion, a business student, claims that the study of limits that do not exist is completely unrealistic and has nothing to do with the world of business. Give two examples from the world of business that might convince him that he is wrong.

44. If $D(t)$ is the Dow Jones Average at time t and $\lim_{t \to +\infty} D(t) = +\infty$, is it possible that the Dow will fluctuate indefinitely into the future?

45. Give an example of a function f with $\lim_{x \to 1} f(x) = f(2)$.

46. If $S(t)$ represents the size of the universe in billions of light-years at time t years since the Big Bang and $\lim_{t \to +\infty} S(t) = 130,000$, is it possible that the universe will continue to expand forever?

3.7 Limits and Continuity (Optional)

Figure 35

In Section 3.6 we saw examples of graphs that had various kinds of "breaks" or "jumps." For instance, in Example 4 we looked at the graph in Figure 35. This graph appears to have breaks, or **discontinuities**, at $x = 0$ and at $x = 1$. What goes wrong at these points? At $x = 0$ we saw that $\lim_{x \to 0} f(x)$ does not exist because the left- and right-hand limits are not the same. Thus, the discontinuity at $x = 0$ seems to be due to the fact that the limit does not exist there. On the other hand, at $x = 1$, $\lim_{x \to 1} f(x)$ *does* exist (it is equal to 1), but is not equal to $f(1) = 2$.

Thus, we have identified two kinds of discontinuity:

1. Points where the limit of the function does not exist.

$x = 0$ in Figure 35 because $\lim_{x \to 0} f(x)$ does not exist.

2. Points where the limit exists but does not equal the value of the function.

$x = 1$ in Figure 35 because $\lim_{x \to 1} f(x) = 1 \neq f(1)$.

On the other hand, there is no discontinuity at, say, $x = -2$, where we find that $\lim_{x \to -2} f(x)$ exists and equals 2 and $f(-2)$ is also equal to 2. In other words,

$$\lim_{x \to -2} f(x) = 2 = f(-2)$$

The point $x = 2$ is an example of a point where f is **continuous.** (Notice that you can draw the portion of the graph near $x = -2$ without lifting your pencil from the paper.) Similarly, f is continuous at *every* point other than $x = 0$ and $x = 1$. Here is the mathematical definition.

Continuous Function

Let f be a function and let a be a number in the domain of f. Then f is **continuous at *a*** if

1. $\lim_{x \to a} f(x)$ exists, and

2. $\lim_{x \to a} f(x) = f(a)$.

The function f is said to be **continuous on its domain** if it is continuous at each point in its domain.

If f is not continuous at a particular a in its domain, we say that f is **discontinuous** at a or that f has a **discontinuity** at a. Thus, a discontinuity can occur at $x = a$ if either

1. $\lim_{x \to a} f(x)$ does not exist, or

2. $\lim_{x \to a} f(x)$ exists but is not equal to $f(a)$.

Note

If the number a is not in the domain of f—that is, if $f(a)$ is not defined—we will not consider the question of continuity at a. A function cannot be continuous at a point not in its domain, and it cannot be discontinuous there either.

Quick Examples

1. The function shown in Figure 35 is continuous at $x = -1$ and $x = 2$. It is discontinuous at $x = 0$ and $x = 1$ and so is not continuous on its domain.

2. The function $f(x) = x^2$ is continuous on its domain (think of its graph, which contains no breaks).

3. The function f whose graph is shown on the left in the following figure is continuous on its domain. (Although the graph breaks at $x = 2$, that is not a point of its domain.) The function g whose graph is shown on the right is not continuous on its domain because it has a discontinuity at $x = 2$. (Here, $x = 2$ is a point of the domain of g.)

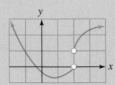

$y = f(x)$: Continuous on its domain

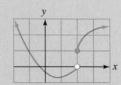

$y = g(x)$: Not continuous on its domain

Example 1 • Continuous and Discontinuous Functions

Which of the following functions are continuous on their domains?

a. $f(x) = \dfrac{1}{x}$

b. $g(x) = \begin{cases} \dfrac{1}{x} & \text{if } x \neq 0 \\ 0 & \text{if } x = 0 \end{cases}$

c. $h(x) = \begin{cases} x + 3 & \text{if } x \leq 1 \\ 5 - x & \text{if } x > 1 \end{cases}$

d. $k(x) = \begin{cases} x + 3 & \text{if } x \leq 1 \\ 1 - x & \text{if } x > 1 \end{cases}$

Solution

a. and **b.** The graphs of f and g are shown in Figure 36.

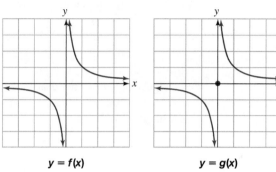

$y = f(x)$ $y = g(x)$

Figure 36

The domain of f consists of all real numbers except 0, and f is continuous at all such numbers. (Notice that 0 is not in the domain of f, so the question of continuity at 0 does not arise.) Thus, f is continuous on its domain.

    The function g, on the other hand, has its domain expanded to include 0, so we now need to check whether g is continuous at 0. From the graph, it is easy to see that g is discontinuous there because $\lim_{x \to 0} g(x)$ does not exist. Thus, g is not continuous on its domain because it is discontinuous at 0.

c. and **d.** The graphs of h and k are shown in Figure 37.

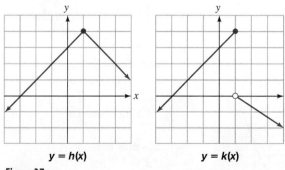

$y = h(x)$ $y = k(x)$

Figure 37

Even though the graph of h is made up of two different line segments, it is continuous at every point of its domain, including $x = 1$ because

$$\lim_{x \to 1} h(x) = 4 = h(1)$$

On the other hand, $x = 1$ is also in the domain of k, but $\lim_{x \to 1} k(x)$ does not exist. Thus, k is discontinuous at $x = 1$ and thus not continuous on its domain.

✳ *Before we go on . . .*
Question What a minute! How can a function like $f(x) = 1/x$ be continuous when its graph has a break in it?

Answer We are not claiming that f is continuous *at every real number*. What we are saying is that f is continuous *on its domain*; the break in the graph occurs at a point not in the domain of f. In other words, f is continuous on the set of all nonzero real numbers; it is not continuous on the set of *all* real numbers, since it is not even defined on the set.

Example 2 • Continuous Except at a Point

In each case, say what, if any, value of $f(a)$ would make f continuous at a.

a. $f(x) = \dfrac{x^3 - 8}{x - 2}$; $a = 2$ **b.** $f(x) = \dfrac{e^{2x} - 1}{x}$; $a = 0$ **c.** $f(x) = \dfrac{|x|}{x}$; $a = 0$

Solution

a. In Figure 38 we see the graph of $f(x) = (x^3 - 8)/(x - 2)$. The point corresponding to $x = 2$ is missing because f is not (yet) defined there. (Your graphing utility will probably miss this subtlety and render a continuous curve.)

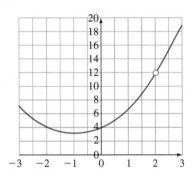

Figure 38

To turn f into a function that is continuous at $x = 2$, we need to "fill in the gap" to obtain a continuous curve. Since the graph suggests that the missing point is $(2, 12)$, let's define $f(2) = 12$.

Does f now become continuous if we take $f(2) = 12$? From the graph or Example 1(a) of Section 3.6,

$$\lim_{x \to 2} f(x) = \lim_{x \to 2} \frac{x^3 - 8}{x - 2} = 12$$

which is now equal to $f(2)$. Thus, $\lim_{x \to 2} f(x) = f(2)$, showing that f is now continuous at $x = 2$.

b. In Example 1(b) of Section 3.6, we saw that

$$\lim_{x \to 0} f(x) = \lim_{x \to 0} \frac{e^{2x} - 1}{x} = 2$$

and so, as in part (a), we must define $f(0) = 2$. This is confirmed by the graph, shown in Figure 39.

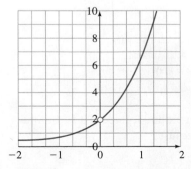

Figure 39

c. We considered the function $f(x) = |x|/x$ in Example 2 in Section 3.6. Its graph is shown in Figure 40. Now we encounter a problem: No matter how we try to fill in the gap at $x = 0$, the result will be a discontinuous function. For example, setting $f(0) = 0$ will result in the discontinuous function shown in Figure 41.

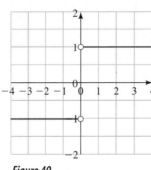
Figure 40

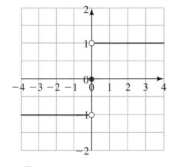
Figure 41

We conclude that it is impossible to assign any value to $f(0)$ to turn f into a function that is continuous at $x = 0$.

We can also see this result algebraically: In Example 2 of Section 3.6, we saw that $\lim_{x \to 0} \dfrac{|x|}{x}$ does not exist. Thus, the resulting function will fail to be continuous at 0, no matter how we define $f(0)$.

✴ *Before we go on . . .* A function not defined at an isolated point is said to have a **singularity** at that point. The function in part (a) has a singularity at $x = 2$, and the functions in parts (b) and (c) have singularities at $x = 0$. The functions in parts (a) and (b) have **removable singularities** because we can make these functions continuous at $x = a$ by properly defining $f(a)$. The function in part (c) has an **essential singularity** because we cannot make f continuous at $x = a$ just by defining $f(a)$ properly.

Continuity and Differentiability (Online Optional Section)

If you follow the path
    Web Site → Online Text → Continuity and Differentiability
you will find online text, interactive examples, and exercises on the relationship between continuity and differentiability.

3.7 EXERCISES

In Exercises 1–12, the graph of a function f is given. Determine whether f is continuous on its domain. If it is not continuous on its domain, say why.

1.

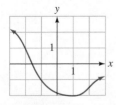

2.

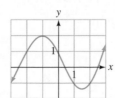

3.

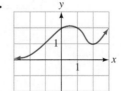

4.

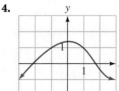

5.

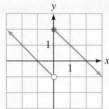

6.

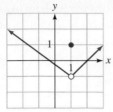

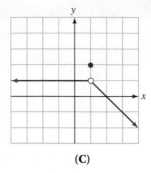

(C)

(D)

7.

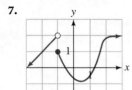

8.

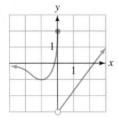

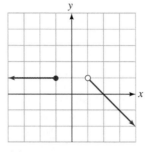

(E)

14.

9.

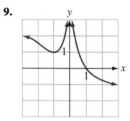

10.

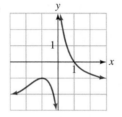

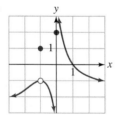

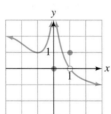

(A)

(B)

11.

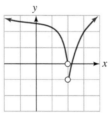

12.

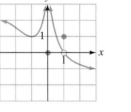

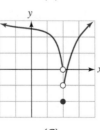

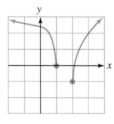

(C)

(D)

In Exercises 13 and 14, identify which (if any) of the given graphs represent functions continuous on their domains.

13.

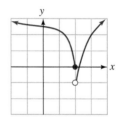

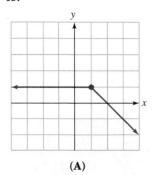

(A)

(B)

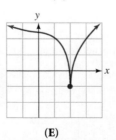

(E)

In Exercises 15–22, use a graph of f or some other method to determine what, if any, value to assign to $f(a)$ to make f continuous at $x = a$.

15. $f(x) = \dfrac{x^2 - 2x + 1}{x - 1}; a = 1$

16. $f(x) = \dfrac{x^2 + 3x + 2}{x + 1}; a = -1$

17. $f(x) = \dfrac{x}{3x^2 - x}; a = 0$

18. $f(x) = \dfrac{x^2 - 3x}{x + 4}; a = -4$

19. $f(x) = \dfrac{3}{3x^2 - x}; a = 0$

20. $f(x) = \dfrac{x - 1}{x^3 - 1}; a = 1$

21. $f(x) = \dfrac{1 - e^x}{x}; a = 0$

22. $f(x) = \dfrac{1 + e^x}{1 - e^x}; a = 0$

In Exercises 23–32, use a graph to determine whether the given function is continuous on its domain. If it is not continuous on its domain, list the points of discontinuity.

23. $f(x) = |x|$

24. $f(x) = \dfrac{|x|}{x}$

25. $g(x) = \dfrac{1}{x^2 - 1}$

26. $g(x) = \dfrac{x - 1}{x + 2}$

27. $f(x) = \begin{cases} x + 2 & \text{if } x < 0 \\ 2x - 1 & \text{if } x \geq 0 \end{cases}$

28. $f(x) = \begin{cases} 1 - x & \text{if } x \leq 1 \\ x - 1 & \text{if } x > 1 \end{cases}$

29. $h(x) = \begin{cases} \dfrac{|x|}{x} & \text{if } x \neq 0 \\ 0 & \text{if } x = 0 \end{cases}$

30. $h(x) = \begin{cases} \dfrac{1}{x^2} & \text{if } x \neq 0 \\ 2 & \text{if } x = 0 \end{cases}$

31. $g(x) = \begin{cases} x + 2 & \text{if } x < 0 \\ 2x + 2 & \text{if } x \geq 0 \end{cases}$

32. $g(x) = \begin{cases} 1 - x & \text{if } x \leq 1 \\ x + 1 & \text{if } x > 1 \end{cases}$

COMMUNICATION AND REASONING EXERCISES

33. If a function is continuous on its domain, is it continuous at every real number? Explain.

34. True or false: The graph of a function that is continuous on its domain is a continuous curve with no breaks in it.

35. True or false: The graph of a function that is continuous at every real number is a continuous curve with no breaks in it.

36. Give an example of a function that is not continuous at $x = -1$ but is not discontinuous there either.

37. Draw the graph of a function that is discontinuous at every integer.

38. Draw the graph of a function that is continuous on its domain but not continuous at any integer.

39. Describe a real-life scenario in the stock market that can be modeled by a discontinuous function.

40. Describe a real-life scenario in your room that can be modeled by a discontinuous function.

3.8   Limits and Continuity: Algebraic Approach (Optional)

Although numerical and graphical estimation of limits is effective, the estimates these methods yield may not be perfectly accurate. The algebraic method, when it can be used, will always yield an exact answer. Moreover, algebraic analysis of a function often enables us to take a function apart and see "what makes it tick."

Let's start with the same example as the one with which we began Section 3.6. Consider the function $f(x) = 2 + x$ and ask: What happens to $f(x)$ as x approaches 3? To answer this algebraically, notice that as x gets closer and closer to 3 the quantity $2 + x$ must get closer and closer to $2 + 3 = 5$. Hence,

$$\lim_{x \to 3} f(x) = \lim_{x \to 3} (2 + x) = 2 + 3 = 5$$

Question Is that all there is to the algebraic method? Just substitute $x = a$?

Answer Notice that by substituting $x = 3$ we *evaluated the function at $x = 3$.* In other words, we relied on the fact that

$$\lim_{x \to 3} f(x) = f(3)$$

In Section 3.7 we said that a function satisfying this equation is said to be *continuous* at $x = 3$. Thus,

If we know that the function f is continuous at a point a, we can compute $\lim_{x \to a} f(x)$ by simply substituting $x = a$ into $f(x)$.

To use this fact, we need to know how to recognize continuous functions when we see them. Geometrically, they are easy to spot: A function is continuous at $x = a$ if its graph has no break at $x = a$. Algebraically, a large class of functions are known to be continuous on their domains—those, roughly speaking, that are *specified by a single formula*.

We can be more precise. A **closed-form function** is any function that can be obtained by combining constants, powers of x, exponential functions, radicals, logarithms, and trigonometric functions (and some other functions we do not encounter in this text) into a *single* mathematical formula by means of the usual arithmetic operations and composition of functions. Examples of closed-form functions are

$$3x^2 - x + 1 \qquad \frac{\sqrt{x^2 - 1}}{6x - 1} \qquad e^{-(4x^2-1)/x} \qquad \sqrt{\log_3(x^2 - 1)}$$

They can be as complicated as we like. The following is *not* a closed-form function:

$$f(x) = \begin{cases} -1 & \text{if } x \leq -1 \\ x^2 + x & \text{if } -1 < x \leq 1 \\ 2 - x & \text{if } 1 < x \leq 2 \end{cases}$$

The reason is that $f(x)$ is not specified by a *single* mathematical formula. What is useful about closed-form functions is the following general rule.

Continuity of Closed-Form Functions

Every closed-form function is continuous on its domain. Thus, if f is a closed-form function and $f(a)$ is defined, we have $\lim_{x \to a} f(x) = f(a)$.

Quick Example

$f(x) = 1/x$ is a closed-form function, and its natural domain consists of all real numbers except 0. Thus, f is continuous at every nonzero real number. That is,

$$\lim_{x \to a} \frac{1}{x} = \frac{1}{a}$$

provided $a \neq 0$.

Mathematics majors spend a great deal of time studying the proof of this result. We ask you to accept it without proof.

Example 1 • Limit of a Closed-Form Function

Evaluate $\lim_{x \to 1} \dfrac{x^3 - 8}{x - 2}$ algebraically.

Solution   First, notice that $(x^3 - 8)/(x - 2)$ is a closed-form function because it is specified by a single algebraic formula. Also, $x = 1$ is in the domain of this function. Therefore,

$$\lim_{x \to 1} \frac{x^3 - 8}{x - 2} = \frac{1^3 - 8}{1 - 2} = 7$$

Question What about $\lim\limits_{x \to 2} \dfrac{x^3 - 8}{x - 2}$?

Answer The point $x = 2$ is not in the domain of the function $(x^3 - 8)/(x - 2)$, so we cannot evaluate this limit by simply substituting $x = 2$. However—and this is the key to finding such limits—some preliminary algebraic simplification will allow us to obtain a closed-form function with $x = 2$ in its domain, as we see in Example 2.

Example 2 • Simplifying to Obtain the Limit

Evaluate $\lim\limits_{x \to 2} \dfrac{x^3 - 8}{x - 2}$ algebraically.

Solution   Again, although $(x^3 - 8)/(x - 2)$ is a closed-form function, $x = 2$ is not in its domain. Thus, we cannot obtain the limit by substitution. Instead, we first simplify $f(x)$ to obtain a new function with $x = 2$ in its domain. To do this, notice first that the numerator can be factored as

$$x^3 - 8 = (x - 2)(x^2 + 2x + 4)$$

Thus,

$$\frac{x^3 - 8}{x - 2} = \frac{(x - 2)(x^2 + 2x + 4)}{x - 2} = x^2 + 2x + 4$$

Once we have canceled the offending $(x - 2)$ in the denominator, we are left with a closed-form function *with 2 in its domain.* Thus,

$$\lim\limits_{x \to 2} \frac{x^3 - 8}{x - 2} = \lim\limits_{x \to 2} (x^2 + 2x + 4)$$

$$= 2^2 + 2(2) + 4 = 12 \qquad \text{Substitute } x = 2.$$

This confirms the answer we found numerically in Example 1 in Section 3.6.

✳ **Before we go on . . .** If the given function fails to simplify, we can always approximate the limit numerically. It may very well be that the limit does not exist in that case.

  The procedure we used may remind you of the "canceling the h" trick we used when computing derivatives algebraically: In the definition of the derivative, we must take the limit of the difference quotient $[f(x + h) - f(x)]/h$ as $h \to 0$. Although the difference quotients we encounter are usually closed-form functions, we cannot evaluate them by substitution: Since h appears in the denominator, $h = 0$ is not in the domain of the difference quotient. However, as in Examples 2 and 3 in Section 3.3, some preliminary algebraic simplification may allow us to obtain a closed-form function with $h = 0$ in its domain. We can then find the limit by substituting $h = 0$ in the new function.

Question There is something suspicious about the last example. If 2 was not in the domain before simplifying but was in the domain after simplifying, we must have changed the function, right?

Answer   Correct. In fact, when we said that

$$\frac{x^3 - 8}{x - 2} = x^2 + 2x + 4$$

Domain excludes 2.   Domain includes 2.

we were lying a little bit. What we really meant is that these two expressions are equal *where both are defined.* The functions $(x^3 - 8)/(x - 2)$ and $x^2 + 2x + 4$ are different functions. The difference is that $x = 2$ is not in the domain of $(x^3 - 8)/(x - 2)$ and is in the domain of $x^2 + 2x + 4$. Since $\lim_{x \to 2} f(x)$ explicitly *ignores* any value that f may have at 2, this does not matter. From the point of view of the limit at 2, these functions *are* equal. In general, we have the following rule.

Functions with Equal Limits

If $f(x) = g(x)$ for all x except possibly $x = a$, then

$$\lim_{x \to a} f(x) = \lim_{x \to a} g(x)$$

We can also use algebraic techniques to analyze functions that are not given in closed form.

Example 3 • Nonclosed-Form Function

For which values of x are the following piecewise-defined functions continuous?

a. $f(x) = \begin{cases} x^2 + 2 & \text{if } x < 1 \\ 2x - 1 & \text{if } x \geq 1 \end{cases}$ **b.** $g(x) = \begin{cases} x^2 - x + 1 & \text{if } x \leq 0 \\ 1 - x & \text{if } 0 < x \leq 1 \\ x - 3 & \text{if } x > 1 \end{cases}$

Solution

a. The function $f(x)$ is given in closed form over the intervals $(-\infty, 1)$ and $[1, +\infty)$. At $x = 1$, $f(x)$ suddenly switches from one closed-form formula to another, so $x = 1$ is the only place where there is a potential problem with continuity. To investigate the continuity of $f(x)$ at $x = 1$, let's calculate the limit there:

$$\lim_{x \to 1^-} f(x) = \lim_{x \to 1^-} (x^2 + 2) \qquad \text{Since } f(x) = x^2 + 2 \text{ for } x < 1$$
$$= (1)^2 + 2 = 3 \qquad x^2 + 2 \text{ is closed form.}$$

$$\lim_{x \to 1^+} f(x) = \lim_{x \to 1^+} (2x - 1) \qquad \text{Since } f(x) = 2x - 1 \text{ for } x > 1$$
$$= 2(1) - 1 = 1 \qquad 2x - 1 \text{ is closed form.}$$

Since the left and right limits are different, $\lim_{x \to 1} f(x)$ does not exist, and so $f(x)$ is discontinuous at $x = 1$.

b. The only potential points of discontinuity for $g(x)$ occur at $x = 0$ and $x = 1$:

$$\lim_{x \to 0^-} g(x) = \lim_{x \to 0^-} x^2 - x + 1 = 1$$
$$\lim_{x \to 0^+} g(x) = \lim_{x \to 0^+} 1 - x = 1$$

Thus, $\lim_{x \to 0} g(x) = 1$. Further, $g(0) = 0^2 - 0 + 1 = 1$ from the formula, and so

$$\lim_{x \to 0} g(x) = g(0)$$

which shows that $g(x)$ is continuous at $x = 0$. At $x = 1$, we have

$$\lim_{x \to 1^-} g(x) = \lim_{x \to 1^-} 1 - x = 0$$
$$\lim_{x \to 1^+} g(x) = \lim_{x \to 1^+} x - 3 = -2$$

so that $\lim_{x \to 1} g(x)$ does not exist. Thus, $g(x)$ is discontinuous at $x = 1$. We conclude that $g(x)$ is continuous at every real number x except $x = 1$.

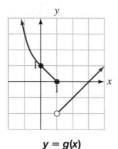

y = g(x)

Figure 42

✴ *Before we go on . . .* Figure 42 shows the graph of *g*. Notice how the discontinuity at $x = 1$ shows up as a break in the graph, whereas at $x = 0$ the two pieces "fit together" at the point $(0, 1)$.

Limits at Infinity

Let's look once again at Example 3 in Section 3.6 and some similar limits.

Example 4 • Limits at Infinity

Compute the following limits, if they exist:

a. $\lim\limits_{x \to +\infty} \dfrac{2x^2 - 4x}{x^2 - 1}$ **b.** $\lim\limits_{x \to -\infty} \dfrac{2x^2 - 4x}{x^2 - 1}$

c. $\lim\limits_{x \to +\infty} \dfrac{-x^3 - 4x}{2x^2 - 1}$ **d.** $\lim\limits_{x \to +\infty} \dfrac{2x^2 - 4x}{5x^3 - 3x + 5}$

Solution

a. and **b.** While calculating the values for the tables used in Example 3 in Section 3.6, you might have noticed that the highest power of *x* in both the numerator and denominator dominated the calculations. For instance, when $x = 100{,}000$, the term $2x^2$ in the numerator has the value of 20,000,000,000, whereas the term $4x$ has the comparatively insignificant value of 400,000. Similarly, the term x^2 in the denominator overwhelms the term -1. In other words, for large values of *x* (or negative values with large magnitude),

$$\frac{2x^2 - 4x}{x^2 - 1} \approx \frac{2x^2}{x^2} \qquad \text{Use only the highest powers top and bottom.}$$

$$= 2$$

Therefore,

$$\lim_{x \to \pm\infty} \frac{2x^2 - 4x}{x^2 - 1} = 2$$

c. Applying the above technique of looking only at highest powers gives

$$\frac{-x^3 - 4x}{2x^2 - 1} \approx \frac{-x^3}{2x^2} \qquad \text{Use only the highest powers top and bottom.}$$

$$= \frac{-x}{2} \qquad \text{Simplify.}$$

As *x* gets large, $-x/2$ gets large and negative, so

$$\lim_{x \to +\infty} \frac{-x^3 - 4x}{2x^2 - 1} = -\infty$$

d. $\dfrac{2x^2 - 4x}{5x^3 - 3x + 5} \approx \dfrac{2x^2}{5x^3} = \dfrac{2}{5x}$. As *x* gets large, $\dfrac{2}{5x}$ gets close to zero, so

$$\lim_{x \to +\infty} \frac{2x^2 - 4x}{5x^3 - 3x + 5} = 0$$

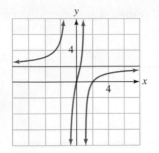

Figure 43

✴ *Before we go on . . .* Let's look again at the limits (a) and (b). We say that the graph of f has a **horizontal asymptote** at $y = 2$ because of the limits we have just calculated. This means that the graph approaches the horizontal line $y = 2$ far to the right or left (in this case, to both the right and left). Figure 43 shows the graph of f together with the line $y = 2$. The graph reveals some additional interesting information: As $x \to 1^+$, $f(x) \to -\infty$, and as $x \to 1^-$, $f(x) \to +\infty$. Thus,

$$\lim_{x \to 1} f(x) \text{ does not exist.}$$

See if you can determine what happens as $x \to -1$.

If you graph the function in part (d), you will again see a horizontal asymptote at $y = 0$. Does the limit in part (c) show a horizontal asymptote?

In Example 4, $f(x)$ was a **rational function**: a quotient of polynomial functions. We calculated the limit of $f(x)$ at $\pm\infty$ by ignoring all powers of x in both the numerator and denominator except for the largest. One can prove that this procedure is valid for any rational function (by dividing both top and bottom by the highest power of x present).

Evaluating the Limit of a Rational Function at $\pm\infty$

If $f(x)$ has the form

$$f(x) = \frac{c_n x^n + \cdots + c_2 x^2 + c_1 x + c_0}{d_m x^m + \cdots + d_2 x^2 + d_1 x + d_0}$$

with the c_i and d_i constants ($c_n \neq 0$ and $d_m \neq 0$), then we can calculate the limit of $f(x)$ as $x \to \pm\infty$ by ignoring all powers of x except the highest in both the numerator and denominator. Thus,

$$\lim_{x \to \pm\infty} f(x) = \lim_{x \to \pm\infty} \frac{c_n x^n}{d_m x^m}$$

L'Hospital's Rule

The limits that cause trouble are those in which we cannot just substitute $x = a$, such as

$$\lim_{x \to 2} \frac{x^3 - 8}{x - 2} \qquad \text{Substituting } x = 2 \text{ yields } \frac{0}{0}.$$

$$\lim_{x \to +\infty} \frac{2x - 4}{x - 1} \qquad \text{Substituting } x = +\infty \text{ yields } \frac{\infty}{\infty}.$$

L'Hospital's rule gives us an alternate way of computing limits such as these without the need to do any preliminary simplification. It will also allow us to compute some limits for which algebraic simplification does not work.[18]

L'Hospital's Rule

If f and g are two differentiable functions such that substituting $x = a$ in the expression $f(x)/g(x)$ gives either $0/0$ or ∞/∞, then

$$\lim_{x \to a} \frac{f(x)}{g(x)} = \lim_{x \to a} \frac{f'(x)}{g'(x)}$$

---

[18]Guillame François Antoine Marquis de L'Hospital (1661–1704) wrote the first textbook on calculus, *Analyse des infiniment petits pour l'intelligence des lignes courbes*, in 1692. The rule now known as L'Hospital's rule appeared first in this book.

That is, we can replace $f(x)$ and $g(x)$ by their *derivatives* and try again to take the limit.

Quick Examples

1. Substituting $x = 2$ in $\dfrac{x^3 - 8}{x - 2}$ yields $\dfrac{0}{0}$. Therefore, L'Hospital's rule applies and

$$\lim_{x \to 2} \frac{x^3 - 8}{x - 2} = \lim_{x \to 2} \frac{3x^2}{1} = \frac{3(2)^2}{1} = 12$$

2. Substituting $x = +\infty$ in $\dfrac{2x - 4}{x - 1}$ yields $\dfrac{\infty}{\infty}$. Therefore, L'Hospital's rule applies and

$$\lim_{x \to +\infty} \frac{2x - 4}{x - 1} = \lim_{x \to +\infty} \frac{2}{1} = 2$$

Example 5 • Applying L'Hospital's Rule

Check whether L'Hospital's rule applies to each of the following limits. If it does, use it to evaluate the limit. Otherwise, use some other method to evaluate the limit.

a. $\displaystyle\lim_{x \to 1} \frac{x^2 - 2x + 1}{4x^3 - 3x^2 - 6x + 5}$ **b.** $\displaystyle\lim_{x \to +\infty} \frac{2x^2 - 4x}{5x^3 - 3x + 5}$

c. $\displaystyle\lim_{x \to 1} \frac{x - 1}{x^3 - 3x^2 + 3x - 1}$ **d.** $\displaystyle\lim_{x \to 1} \frac{x}{x^3 - 3x^2 + 3x - 1}$

Solution
a. Setting $x = 1$ yields

$$\frac{1 - 2 + 1}{4 - 3 - 6 + 5} = \frac{0}{0}$$

Therefore, L'Hospital's rule applies and

$$\lim_{x \to 1} \frac{x^2 - 2x + 1}{4x^3 - 3x^2 - 6x + 5} = \lim_{x \to 1} \frac{2x - 2}{12x^2 - 6x - 6}$$

We are left with a closed-form function. However, we cannot substitute $x = 1$ to find the limit because the function $(2x - 2)/(12x^2 - 6x - 6)$ is still not defined at $x = 1$. In fact, if we set $x = 1$, we again get 0/0. Thus, L'Hospital's rule applies again, and

$$\lim_{x \to 1} \frac{2x - 2}{12x^2 - 6x - 6} = \lim_{x \to 1} \frac{2}{24x - 6}$$

Once again we have a closed-form function, but this time it is defined when $x = 1$, giving

$$\frac{2}{24 - 6} = \frac{1}{9}$$

Thus,

$$\lim_{x \to 1} \frac{x^2 - 2x + 1}{4x^3 - 3x^2 - 6x + 5} = \frac{1}{9}$$

b. Setting $x = +\infty$ yields $\dfrac{\infty}{\infty}$, so

$$\lim_{x \to +\infty} \frac{2x^2 - 4x}{5x^3 - 3x + 5} = \lim_{x \to +\infty} \frac{4x - 4}{15x^2 - 3}$$

Setting $x = +\infty$ again yields $\dfrac{\infty}{\infty}$, so we can apply the rule again to obtain

$$\lim_{x \to +\infty} \frac{4x - 4}{15x^2 - 3} = \lim_{x \to +\infty} \frac{4}{30x}$$

Note that we cannot apply L'Hospital's rule a third time, since setting $x = +\infty$ yields $4/\infty$. However, we can easily see now that the limit is zero. [Refer to Example 4(d).]

c. Setting $x = 1$ yields $0/0$ so, by L'Hospital's rule,

$$\lim_{x \to 1} \frac{x - 1}{x^3 - 3x^2 + 3x - 1} = \lim_{x \to 1} \frac{1}{3x^2 - 6x + 3}$$

We are left with a closed-form function that is still not defined at $x = 1$. Further, L'Hospital's rule no longer applies, since putting $c = 0$ yields $1/0$. To investigate this limit, we must either graph it or create a table of values. If we use either method, we find that

$$\lim_{x \to 1} \frac{x - 1}{x^3 - 3x^2 + 3x - 1} = +\infty$$

d. Setting $x = 1$ in the expression yields $1/0$, so that L'Hospital's rule does not apply here. If we graph the function or create a table of values, we find that the limit does not exist.

Guideline: Strategy for Evaluating Limits Algebraically

Question Is there a systematic way to evaluate a limit $\lim_{x \to a} f(x)$ algebraically?

Answer The following approach is often successful:

Case 1: a Is a Finite Number (*Not* $\pm\infty$)

1. Decide whether f is a closed-form function. If it is not, then check values of x where the function changes from one formula to another.
2. If f is a closed-form function, try substituting $x = a$ in the formula for $f(x)$. Then one of three things will happen:
   - $f(a)$ is defined. Then $\lim_{x \to a} f(x) = f(a)$.
   - $f(a)$ is not defined and has the form $0/0$. Either simplify the expression for f to cancel one of the terms that gives 0 or apply L'Hospital's rule and then start again.
   - $f(a)$ is not defined and has the form $k/0$, where k is not zero. Then the function diverges to $\pm\infty$ as x approaches a from each side. Check this graphically, as in Example 4 (see "Before we go on").

Case 2: $a = \pm\infty$

If the given function is a polynomial or ratio of polynomials, use the technique of Example 4: Focus only on the highest powers of x or use L'Hospital's rule.

3.8 EXERCISES

In Exercises 1–38, calculate the limits algebraically. If a limit does not exist, say why.

1. $\lim_{x \to 0} (x + 1)$

2. $\lim_{x \to 0} (2x - 4)$

3. $\lim_{x \to 2} \dfrac{2 + x}{x}$

4. $\lim_{x \to -1} \dfrac{4x^2 + 1}{x}$

5. $\lim_{x \to -1} \dfrac{x + 1}{x}$

6. $\lim_{x \to 4} (x + \sqrt{x})$

7. $\lim_{x \to 8} (x - \sqrt[3]{x})$

8. $\lim_{x \to 1} \dfrac{x - 2}{x + 1}$

9. $\lim_{h \to 1} (h^2 + 2h + 1)$

10. $\lim_{h \to 0} (h^3 - 4)$

11. $\lim_{h \to 3} 2$

12. $\lim_{h \to 0} -5$

13. $\lim_{h \to 0} \dfrac{h^2}{h + h^2}$

14. $\lim_{h \to 0} \dfrac{h^2 + h}{h^2 + 2h}$

15. $\lim_{x \to 1} \dfrac{x^2 - 2x + 1}{x^2 - x}$

16. $\lim_{x \to -1} \dfrac{x^2 + 3x + 2}{x^2 + x}$

17. $\lim_{x \to 2} \dfrac{x^3 - 8}{x - 2}$

18. $\lim_{x \to -2} \dfrac{x^3 + 8}{x^2 + 3x + 2}$

19. $\lim_{x \to 0^+} \dfrac{1}{x^2}$

20. $\lim_{x \to 0^+} \dfrac{1}{x^2 - x}$

21. $\lim_{x \to -1} \dfrac{x^2 + 1}{x + 1}$

22. $\lim_{x \to -1^-} \dfrac{x^2 + 1}{x + 1}$

23. $\lim_{x \to +\infty} \dfrac{3x^2 + 10x - 1}{2x^2 - 5x}$

24. $\lim_{x \to +\infty} \dfrac{6x^2 + 5x + 100}{3x^2 - 9}$

25. $\lim_{x \to +\infty} \dfrac{x^5 - 1000x^4}{2x^5 + 10,000}$

26. $\lim_{x \to +\infty} \dfrac{x^6 + 3000x^3 + 1,000,000}{2x^6 + 1000x^3}$

27. $\lim_{x \to +\infty} \dfrac{10x^2 + 300x + 1}{5x + 2}$

28. $\lim_{x \to +\infty} \dfrac{2x^4 + 20x^3}{1000x^3 + 6}$

29. $\lim_{x \to +\infty} \dfrac{10x^2 + 300x + 1}{5x^3 + 2}$

30. $\lim_{x \to +\infty} \dfrac{2x^4 + 20x^3}{1000x^6 + 6}$

31. $\lim_{x \to -\infty} \dfrac{3x^2 + 10x - 1}{2x^2 - 5x}$

32. $\lim_{x \to -\infty} \dfrac{6x^2 + 5x + 100}{3x^2 - 9}$

33. $\lim_{x \to -\infty} \dfrac{x^5 - 1000x^4}{2x^5 + 10,000}$

34. $\lim_{x \to -\infty} \dfrac{x^6 + 3000x^3 + 1,000,000}{2x^6 + 1000x^3}$

35. $\lim_{x \to -\infty} \dfrac{10x^2 + 300x + 1}{5x + 2}$

36. $\lim_{x \to -\infty} \dfrac{2x^4 + 20x^3}{1000x^3 + 6}$

37. $\lim_{x \to -\infty} \dfrac{10x^2 + 300x + 1}{5x^3 + 2}$

38. $\lim_{x \to -\infty} \dfrac{2x^4 + 20x^3}{1000x^6 + 6}$

In Exercises 39–46, find all points of discontinuity of the given function.

39. $f(x) = \begin{cases} x + 2 & \text{if } x < 0 \\ 2x - 1 & \text{if } x \geq 0 \end{cases}$

40. $g(x) = \begin{cases} 1 - x & \text{if } x \leq 1 \\ x - 1 & \text{if } x > 1 \end{cases}$

41. $g(x) = \begin{cases} x + 2 & \text{if } x < 0 \\ 2x + 2 & \text{if } 0 \leq x < 2 \\ x^2 + 2 & \text{if } x \geq 2 \end{cases}$

42. $f(x) = \begin{cases} 1 - x & \text{if } x \leq 1 \\ x + 2 & \text{if } 1 < x < 3 \\ x^2 - 4 & \text{if } x \geq 3 \end{cases}$

43. $h(x) = \begin{cases} x + 2 & \text{if } x < 0 \\ 0 & \text{if } x = 0 \\ 2x + 2 & \text{if } x > 0 \end{cases}$

44. $h(x) = \begin{cases} 1 - x & \text{if } x < 1 \\ 1 & \text{if } x = 1 \\ x + 2 & \text{if } x > 1 \end{cases}$

45. $f(x) = \begin{cases} \dfrac{1}{x} & \text{if } x < 0 \\ x & \text{if } 0 \leq x \leq 2 \\ 2^{x-1} & \text{if } x > 2 \end{cases}$

46. $f(x) = \begin{cases} x^3 + 2 & \text{if } x \leq -1 \\ x^2 & \text{if } -1 < x < 0 \\ x & \text{if } x \geq 0 \end{cases}$

APPLICATIONS

47. Law Enforcement Costs of fighting crime in the United States increased steadily in the period 1982–1999. Total spending on police and courts can be approximated, respectively, by

$$P(t) = 1.745t + 29.84 \text{ billion dollars} \qquad (2 \leq t \leq 19)$$

$$C(t) = 1.097t + 10.65 \text{ billion dollars} \qquad (2 \leq t \leq 19)$$

where t is time in years since 1980. Compute

$$\lim_{t \to +\infty} P(t)/C(t)$$

to two decimal places and interpret the result.

Spending is adjusted for inflation and shown in 1999 dollars. Models are based on a linear regression. SOURCE: Bureau of Justice Statistics/*New York Times*, February 11, 2002, p. A14.

48. Law Enforcement Refer to Exercise 47. Total spending on police, courts, and prisons can be approximated, respectively, by

$$P(t) = 1.745t + 29.84 \text{ billion dollars} \qquad (2 \leq t \leq 19)$$

$$C(t) = 1.097t + 10.65 \text{ billion dollars} \qquad (2 \leq t \leq 19)$$

$$J(t) = 1.919x + 12.36 \text{ billion dollars} \qquad (2 \leq t \leq 19)$$

where t is time in years since 1980. Compute

$$\lim_{t \to +\infty} P(t)/[P(t) + C(t) + J(t)]$$

to two decimal places and interpret the result.

Spending is adjusted for inflation and shown in 1999 dollars. Models are based on a linear regression. SOURCE: Bureau of Justice Statistics/*New York Times*, February 11, 2002, p. A14.

Bottled-Water Sales Annual U.S. sales of domestic bottled water for the period 1990–2000 could be approximated by

$$D(t) = 0.0166t^2 - 0.050t + 0.27 \text{ billion gallons} \quad (0 \le t \le 10)$$

where t is time in years since 1990. Sales of imported bottled water over the same period could be approximated by

$$I(t) = 0.0004t^2 + 0.013t + 0.06 \text{ billion gallons} \quad (1 \le t \le 10)$$

Exercises 49 and 50 are based on these models.

Models are based on regression of approximate data. Source: Beverage Marketing Corporation of New York/*New York Times*, June 21, 2001, p. C1.

49. Assuming that the trends shown in the above models continue indefinitely, calculate the limits

$$\lim_{t \to +\infty} I(t) \quad \text{and} \quad \lim_{t \to +\infty} \frac{I(t)}{D(t)}$$

algebraically, interpret your answers, and comment on the results.

50. Repeat Exercise 49, this time calculating

$$\lim_{t \to +\infty} D(t) \quad \text{and} \quad \lim_{t \to +\infty} \frac{D(t)}{I(t)}$$

51. Acquisition of Language The percentage $p(t)$ of children who can speak in at least single words by the age of t months can be approximated by the equation

$$p(t) = 100\left(1 - \frac{12,200}{t^{4.48}}\right) \quad (t \ge 8.5)$$

Calculate $\lim_{t \to +\infty} p(t)$ and $\lim_{t \to +\infty} p'(t)$ and interpret the results.

The model is the authors' and is based on data presented in William H. Calvin, "The Emergence of Intelligence," *Scientific American* (October 1994): 101–107.

52. Acquisition of Language The percentage $q(t)$ of children who can speak in sentences of five or more words by the age of t months can be approximated by the equation

$$q(t) = 100\left(1 - \frac{5.27 \times 10^{17}}{t^{12}}\right) \quad (t \ge 30)$$

If p is the function referred to in Exercise 51, calculate $\lim_{t \to +\infty} [p(t) - q(t)]$ and interpret the result.

The model is the authors' and is based on data presented in William H. Calvin, "The Emergence of Intelligence," *Scientific American* (October 1994): 101–107.

53. Television Advertising The cost, in millions of dollars, of a 30-second television ad during Super Bowls from 1990 to 2001 can be approximated by the following piecewise linear function ($t = 0$ represents 1990):

$$C(t) = \begin{cases} 0.08t + 0.6 & \text{if } 0 \le t < 8 \\ 0.355t - 1.6 & \text{if } 8 \le t \le 11 \end{cases}$$

a. Is C a continuous function of t? Why?

b. Is C a differentiable function of t? Compute $\lim_{t \to 8^-} C'(t)$ and $\lim_{t \to 8^+} C'(t)$ and interpret the results.

Source: *New York Times*, January 26, 2001, p. C1.

54. Internet Purchases The percentage $p(t)$ of buyers of new cars who used the Internet for research or purchase since 1997 is given by the following function ($t = 0$ represents 1997):

$$p(t) = \begin{cases} 10t + 15 & \text{if } 0 \le t < 1 \\ 15t + 10 & \text{if } 1 \le t \le 4 \end{cases}$$

a. Is p a continuous function of t? Why?

b. Is p a differentiable function of t? Compute $\lim_{t \to 1^-} p'(t)$ and $\lim_{t \to 1^+} p'(t)$ and interpret the results.

The model is based on data through 2000 (the 2000 value is estimated). Source: J. D. Power Associates/*New York Times*, January 25, 2000, p. C1.

COMMUNICATION AND REASONING EXERCISES

55. Describe the algebraic method of evaluating limits as discussed in this section and give at least one disadvantage of this method.

56. What is a closed-form function? What can we say about such functions?

57. Your friend Karin tells you that $f(x) = 1/(x - 2)^2$ cannot be a closed-form function because it is not continuous at $x = 2$. Comment on her assertion.

58. Give an example of a function f specified by means of algebraic formulas such that the domain of f consists of all real numbers and f is not continuous at $x = 2$. Is f a closed-form function?

59. What is wrong with the following statement? If $f(x)$ is specified algebraically and $f(a)$ is defined, then $\lim_{x \to a} f(x)$ exists and equals $f(a)$.

60. What is wrong with the following statement? $\lim_{x \to -2} (x^2 - 4)/(x + 2)$ does not exist because substituting $x = -2$ yields 0/0, which is undefined.

61. Find a function that is continuous everywhere except at two points.

62. Find a function that is continuous everywhere except at three points.

CASE STUDY

Reproduced from the U.S. Environmental Protection Agency Web site

Reducing Sulfur Emissions

The Environmental Protection Agency (EPA) wishes to formulate a policy that will encourage utilities to reduce sulfur emissions. Its goal is to reduce annual emissions of sulfur dioxide by a total of 10 million tons from the current level of 25 million tons by imposing a fixed charge for every ton of sulfur released into the environment per year. As a consultant to the EPA, you must determine the amount to be charged per ton of sulfur emissions.

You have the following data, which show the marginal cost to the utility industry of reducing sulfur emissions at several levels of reduction:

Reduction (millions of tons)	8	10	12
Marginal Cost ($/ton)	270	360	779

These figures were produced in a computerized study of reducing sulfur emissions from the 1980 level by the given amounts. SOURCE: U.S. Congress, Congressional Budget Office, *Curbing Acid Rain: Cost, Budget and Coal Market Effects* (Washington, D.C.: Government Printing Office, 1986): xx, xxii, 23, 80.

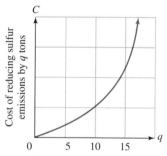

Figure 44

If $C(q)$ is the cost of removing q tons of sulfur dioxide, the table tells you that $C'(8,000,000) = \$270$ per ton, $C'(10,000,000) = \$360$ per ton, and $C'(12,000,000) = \$779$ per ton. Recalling that $C'(q)$ is the slope of the tangent to the graph of the cost function, you can see from the table that this slope is positive and increasing as q increases, so the graph of this cost function has the general shape shown in Figure 44. (Notice that the slope is increasing as you move to the right.) Thus, the utility industry has no cost incentive to reduce emissions.

What you would like to do—if the goal of reducing total emissions by 10 million tons is to be reached—is to alter this cost curve so that it has the general shape shown in Figure 45. In this curve, the cost D to utilities is lowest at a reduction level of 10 million tons, so if the utilities act to minimize cost, they can be expected to reduce emissions by 10 million tons, which is the EPA goal. From the graph, you can see that $D'(10,000,000) = \$0$ per ton, whereas $D'(q)$ is negative for $q < 10,000,000$ and positive for $q > 10,000,000$.

At first you are bothered by the fact that you were not given a cost function. Only the marginal costs were supplied, but you decide to work as best you can without knowing the original cost function $C(q)$.

You now assume that the EPA will impose an annual emission charge of $\$k$ per ton of sulfur released into the environment. It is your job to calculate k. Since you are working with q as the independent variable, you decide that it would be best to formulate the emission charge as a function of q, where q represents the amount by which sulfur emissions are *reduced*. The relationship between the annual sulfur emissions and the amount q by which emissions are reduced from the original 25 million tons is given by

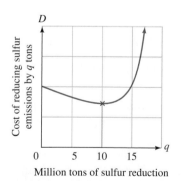

Figure 45

$$\text{annual sulfur emissions} = \text{original emissions} - \text{amount of reduction}$$

$$= 25,000,000 - q$$

Thus, the total annual emission charge to the utilities is

$$k(25,000,000 - q) = 25,000,000k - kq$$

This results in a total cost to the utilities of

total cost = cost of reducing emissions + emission charge

$$D(q) = C(q) + 25,000,000k - kq$$

Even though you have no idea of the form of $C(q)$, you remember that the derivative of a sum is the sum of the derivatives, so you differentiate both sides and obtain

$$D'(q) = C'(q) + 0 - k \qquad \text{The derivative of } kq \text{ is } k \text{ (the slope).}$$

$$= C'(q) - k$$

Remember that you want

$$D'(10,000,000) = 0$$

Thus,

$$C'(10,000,000) - k = 0$$

Referring to the table, you see that

$$360 - k = 0$$

$$k = \$360/\text{ton}$$

In other words, all you need to do is set the emission charge at $k = \$360$ per ton of sulfur emitted. Further, to ensure that the resulting curve will have the general shape shown in Figure 45, you would like to have $D'(q)$ negative for $q < 10,000,000$ and positive for $q > 10,000,000$. To check this, write

$$D'(q) = C'(q) - k$$

$$= C'(q) - 360$$

and refer to the table to obtain

$$D'(8,000,000) = 270 - 360 = -90 < 0 \ ✔$$

$$D'(12,000,000) = 779 - 360 = 419 > 0 \ ✔$$

Thus, based on the given data, the resulting curve will have the shape you require. You therefore inform the EPA that an annual emission charge of \$360 per ton of sulfur released into the environment will create the desired incentive: to reduce sulfur emissions by 10 million tons per year.

One week later, you are informed that this charge would be unrealistic because the utilities cannot possibly afford such a cost. You are asked whether there is an alternative plan that accomplishes the 10-million-ton reduction goal and yet is cheaper to the utilities by \$5 billion per year. You then look at your expression for the emission charge,

$$25,000,000k - kq$$

and notice that, if you decrease this amount by \$5 billion, the derivative will not change at all because the derivative of a constant is zero. Thus, you propose the following revised formula for the emission charge:

$$25,000,000k - kq - 5,000,000,000 = 25,000,000(360) - 360q - 5,000,000,000$$

$$= 4,000,000,000 - 360q$$

At the expected reduction level of 10 million tons, the total amount paid by the utilities will then be

$$4,000,000,000 - 360(10,000,000) = \$400,000,000$$

Thus, your revised proposal is the following: Impose an annual emission charge of $360 per ton of sulfur released into the environment and hand back $5 billion in the form of subsidies. The effect of this policy will be to cause the utilities industry to reduce sulfur emissions by 10 million tons per year and will result in $400 million in annual revenues to the government.

Notice that this policy also provides an incentive for the utilities to search for cheaper ways to reduce emissions. For instance, if they lowered costs to the point where they could achieve a reduction level of 12 million tons, they would have a total emission charge of

$$4,000,000,000 - 360(12,000,000) = -\$320,000,000$$

The fact that this is negative means that the government would be paying the utilities industry $320 million more in annual subsidies than the industry would be paying in per ton emission charges.

EXERCISES

1. Excluding subsidies, what should the annual emission charge be if the goal is to reduce sulfur emissions by 8 million tons?

2. Excluding subsidies, what should the annual emission charge be if the goal is to reduce sulfur emissions by 12 million tons?

3. What is the *marginal emission charge* in your revised proposal (as stated before the exercise set)? What is the relationship between the marginal cost of reducing sulfur emissions before emission charges are implemented and the marginal emission charge, at the optimal reduction under your revised proposal?

4. We said that the revised policy provided an incentive for utilities to find cheaper ways to reduce emissions. How would $C(q)$ have to change to make 12 million tons the optimum reduction?

5. What change in $C(q)$ would make 8 million tons the optimum reduction?

6. If the scenario in Exercise 5 took place, what would the EPA have to do in order to make 10 million tons the optimal reduction once again?

7. Due to intense lobbying by the utility industry, you are asked to revise the proposed policy so that the utility industry will pay no charge if sulfur emissions are reduced by the desired 10 million tons. How can you accomplish this?

8. Suppose that instead of imposing a fixed charge per ton of emission, you decide to use a sliding scale so that the total charge to the industry for annual emissions of x tons will be $\$kx^2$ for some k. What must k be to again make 10 million tons the optimum reduction? (The derivative of kx^2 is $2kx$.)

CHAPTER 3 REVIEW TEST

1. For each of the following functions, find the average rate of change over the interval $[a, a + h]$ for $h = 1, 0.01$, and 0.001. (Round answers to four decimal places.) Then estimate the slope of the tangent line to the graph of the function at a.

a. $f(x) = \dfrac{1}{x + 1}$; $a = 0$ **b.** $f(x) = x^x$; $a = 2$

[E] **c.** $f(x) = e^{2x}$; $a = 0$ [E] **d.** $f(x) = \ln(2x)$; $a = 1$

2. Below are the graphs of several functions with four points marked. Determine at which (if any) of these points the derivative of the function is: (i) −1, (ii) 0, (iii) 1, and (iv) 2.

a. 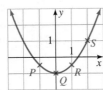   **b.**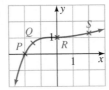

c. 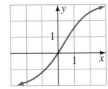   **d.**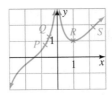

3. Let f have the graph shown:

Select the correct answer.
a. The average rate of change of f over the interval $[0, 2]$ is
   (A) greater than   **(B)** less than
   (C) approximately equal to
   $f'(0)$.
b. The average rate of change of f over the interval $[−1, 1]$ is
   (A) greater than   **(B)** less than
   (C) approximately equal to
   $f'(0)$.
c. Over the interval $[0, 2]$, the instantaneous rate of change of f is
   (A) increasing.
   (B) decreasing.
   (C) neither increasing nor decreasing.
d. Over the interval $[−2, 2]$, the instantaneous rate of change of f is
   (A) increasing and then decreasing.
   (B) decreasing and then increasing.
   (C) approximately constant.

e. When $x = 2$, $f(x)$ is
   (A) approximately 1 and increasing at a rate of about 2.5 units per unit of x.
   (B) approximately 1.2 and increasing at a rate of about 1 unit per unit of x.
   (C) approximately 2.5 and increasing at a rate of about 0.5 units per unit of x.
   (D) approximately 2.5 and increasing at a rate of about 2.5 unit per unit of x.

4. Use the definition of the derivative to calculate the derivative of each of the following functions algebraically.

a. $f(x) = x^2 + x$ **b.** $f(x) = \dfrac{1}{x} + 1$

5. Find the derivative of each of the following functions.

a. $f(x) = 10x^5 + \dfrac{1}{2}x^4 - x + 2$

b. $f(x) = \dfrac{10}{x^5} + \dfrac{1}{2x^4} - \dfrac{1}{x} + 2$

c. $f(x) = 3x^3 + 3\sqrt[3]{x}$ **d.** $f(x) = \dfrac{2}{x^{2.1}} - \dfrac{x^{0.1}}{2}$

6. Evaluate the given expressions.

a. $\dfrac{d}{dx}\left[x + \dfrac{1}{x^2}\right]$ **b.** $\dfrac{d}{dx}\left[2x - \dfrac{1}{x}\right]$

c. $\dfrac{d}{dx}\left[\dfrac{4}{3x} - \dfrac{2}{x^{0.1}} + \dfrac{x^{1.1}}{3.2} - 4\right]$

d. $\dfrac{d}{dx}\left[\dfrac{4}{x} + \dfrac{x}{4} - |x|\right]$

 7. Use any method to graph the derivative of each of the following functions. In each case, choose a range of x values and y values that shows the interesting features of the graph.

a. $f(x) = 10x^5 + \dfrac{1}{2}x^4 - x + 2$

b. $f(x) = \dfrac{10}{x^5} + \dfrac{1}{2x^4} - \dfrac{1}{x} + 2$

c. $f(x) = 3x^3 + 3\sqrt[3]{x}$ **d.** $f(x) = \dfrac{2}{x^{2.1}} - \dfrac{x^{0.1}}{2}$

OHaganBooks.com—
MONITORING CHANGE

8. Since the start of July, OHaganBooks.com has seen its weekly sales increase, as shown in the following table:

Week	1	2	3	4	5	6
Sales (books)	6500	7000	7200	7800	8500	9000

    a. What was the average rate of increase of weekly sales over this time period?

    b. During which 1-week interval(s) did the rate of increase of sales exceed the average rate?

    c. During which 2-week interval(s) did the weekly sales rise at the highest average rate, and what was that average rate?

9. OHaganBooks.com fit the curve

$$w(t) = -3.7t^3 + 74.6t^2 + 135.5t + 6333.3$$

to its weekly sales figures from Question 8, as shown in the following graph:

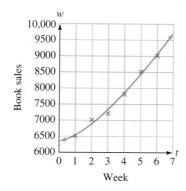

    a. According to the model, what was the rate of increase of sales at the beginning of the second week ($t = 1$)? (Round your answer to the nearest unit.)

    b. If we extrapolate the model, what would be the rate of increase of weekly sales at the beginning of week 8 ($t = 7$)?

    c. Graph the function w for $0 \le t \le 20$. Would it be realistic to use the function to predict sales through week 20? Why?

E **10.** OHaganBooks.com decided that the curve above was not suitable for extrapolation, so instead it tried

$$s(t) = 6053 + \frac{4474}{1 + e^{-0.55(t-4.8)}}$$

which is shown in the following graph:

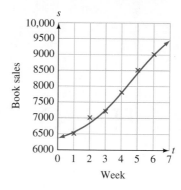

    a. Using this function, estimate the rate of increase of weekly sales at the beginning of week 7 ($t = 6$). (Round your answer to the nearest unit.)

    b. If we extrapolate the model, what would be the rate of increase of weekly sales at the beginning of week 15 ($t = 14$)?

    c. Graph the function s for $0 \le t \le 20$. What is the long-term prediction for weekly sales? What is the long-term prediction for the rate of change of weekly sales?

11. As OHaganBooks.com's sales increase, so do its costs. If we take into account volume discounts from suppliers and shippers, the weekly cost of selling x books is

$$C(x) = -0.00002x^2 + 3.2x + 5400 \text{ dollars}$$

    a. What is the marginal cost at a sales level of 8000 books per week?

    b. What is the average cost per book at a sales level of 8000 books per week?

    c. What is the marginal average cost at a sales level of 8000 books per week?

    d. Interpret the results of parts (a)–(c).

🌐 ADDITIONAL ONLINE REVIEW

If you follow the path

    Web Site → Everything for Calculus → Chapter 3

you will find the following additional resources to help you review:

A comprehensive chapter summary (including examples and interactive features)

Additional review exercises (including interactive exercises and many with help)

A true/false chapter quiz

Several useful graphing utilities

TECHNIQUES OF DIFFERENTIATION

CASE STUDY

Projecting Market Growth

You are on the board of directors at Fullcourt Academic Press. The sales director of the high school division has just burst into your office with a proposal for a major expansion strategy based on the assumption that the number of high school seniors in the United States will be growing at a rate of at least 20,000 per year through the year 2005. Since the figures actually appear to be leveling off, you are suspicious about this estimate. You would like to devise a model that predicts this trend before tomorrow's scheduled board meeting. How do you go about doing this?

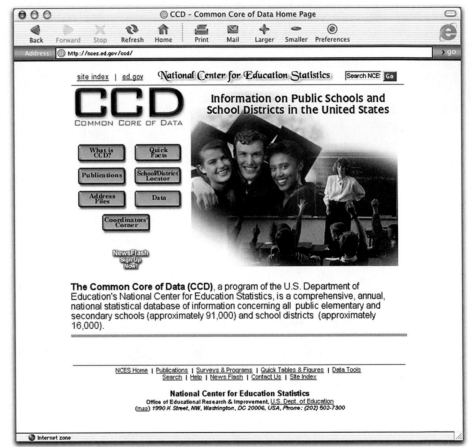

REPRODUCED FROM THE U.S. DEPARTMENT OF EDUCATION, NATIONAL CENTER FOR EDUCATION STATISTICS, WEB SITE

INTERNET RESOURCES FOR THIS CHAPTER

At the Web site, follow the path

Web Site → Everything for Calculus → Chapter 4

where you will find links to step-by-step tutorials for the main topics in this chapter, a detailed chapter summary you can print out, a true/false quiz, and a collection of sample test questions. You will also find an online grapher and other useful resources. Complete text and interactive exercises have been placed on the Web site, covering the following optional topic:

Optional Internet Topic: Linear Approximation and Error Estimation

Introduction

In Chapter 3 we studied the concept of the derivative of a function, and we saw some of the applications for which derivatives are useful. However, the only functions we could differentiate easily were sums of terms of the form ax^n, where a and n are constants.

In this chapter we develop techniques that help us differentiate any closed-form function—that is, any function, no matter how complicated, that can be specified by a formula involving powers, radicals, exponents, and logarithms. (In a later chapter, we will discuss how to add trigonometric functions to this list.) We also show how to find the derivatives of functions that are only specified *implicitly*—that is, functions for which we are not given an explicit formula for y in terms of x but only an equation relating x and y.

 ## Algebra Review

For this chapter you should be familiar with the algebra reviewed in Appendix A, Sections A.3 and A.4.

4.1 The Product and Quotient Rules

We know how to find the derivatives of functions that are sums of powers, like polynomials. In general, if a function is a sum or difference of functions whose derivatives we know, then we know how to find its derivative. But what about *products* and *quotients* of functions whose derivatives we know? For instance, how do we calculate the derivative of something like $x^2/(x + 1)$?

Question I don't see what the problem is. Can't we just say that the derivative of $x^2/(x + 1)$ is $2x/1 = 2x$?

Answer No. Your calculation is based on an assumption that the derivative of a quotient is the quotient of the derivatives. But it is easy to see that this assumption is false. For instance, we know that

$$\frac{d}{dx}\left[\frac{x^3}{x}\right] = \frac{d}{dx}[x^2] = 2x$$

However, if we used the method suggested in your question, we would get

$$\frac{d}{dx}\left[\frac{x^3}{x}\right] = \frac{3x^2}{1} = 3x^2 \qquad \text{✗ WRONG}$$

Thus, the derivative of a quotient is *not* the quotient of the derivatives. Similarly, the following calculation is also wrong:

$$\frac{d}{dx}[x^3 \cdot x] = 3x^2 \cdot 1 = 3x^2 \qquad \text{✗ WRONG}$$

since $x^3 \cdot x = x^4$ and its derivative is $4x^3$, not $3x^2$. Thus, the derivative of a product is *not* the product of the derivatives.

Question If that is not how we find the derivatives of products and quotients, how *do* we?

Answer To motivate the correct method, let's ask *you* a question: We know that the daily revenue resulting from the sale of q items per day at a price of p dollars per item is given by the product, $R = pq$ dollars. Suppose you are currently selling wall posters on campus. At this time your daily sales are 50 posters, and sales are increasing at a rate of 4 per day. Furthermore, you are currently charging $10 per poster, and you are raising the price at a rate of $2 per day. Can you estimate how fast your daily revenue is increasing? In other words, can you estimate the rate of change, dR/dt, of the revenue R?

Question Let me see . . . there are two contributions to the rate of change in daily revenue: the increase in daily sales and the increase in the unit price. We have

$$\frac{dR}{dt} \text{ due to increasing price: } \$2 \text{ per day} \times 50 \text{ posters} = \$100 \text{ per day}$$

$$\frac{dR}{dt} \text{ due to increasing sales: } \$10 \text{ per poster} \times 4 \text{ posters per day} = \$40 \text{ per day}$$

Thus, I estimate my daily revenue to be increasing at a rate of $100 + $40 = $140 per day.

Answer Correct. Now you have just answered your original question about the rate of change of a product. Let's translate what you have said into symbols:

$$\frac{dR}{dt} \text{ due to increasing price: } \frac{dp}{dt} \times q$$

$$\frac{dR}{dt} \text{ due to increasing sales: } p \times \frac{dq}{dt}$$

Thus, the rate of change of revenue is given by

$$\frac{dR}{dt} = \frac{dp}{dt} q + p \frac{dq}{dt}$$

Since $R = pq$, we have discovered the following rule for differentiating a product:

$$\frac{d}{dt}[pq] = \frac{dp}{dt} q + p \frac{dq}{dt}$$

The derivative of a product is the derivative of the first times the second, plus the first times the derivative of the second.

This rule and a similar rule for differentiating quotients are given next. After a little practice using them, we discuss how these results are proved.

Product Rule

$$\frac{d}{dx}[f(x)g(x)] = f'(x)g(x) + f(x)g'(x)$$

In Words The derivative of a product is the derivative of the first times the second, plus the first times the derivative of the second.

Quick Example

$$\frac{d}{dx}[x^2(3x-1)] = \underset{\uparrow}{2x} \cdot \underset{\uparrow}{(3x-1)} + \underset{\uparrow}{x^2} \cdot \underset{\uparrow}{(3)}$$

Derivative of first   Second   First   Derivative of second

Quotient Rule

$$\frac{d}{dx}\left[\frac{f(x)}{g(x)}\right] = \frac{f'(x)g(x) - f(x)g'(x)}{[g(x)]^2}$$

In Words The derivative of a quotient is the derivative of the top times the bottom, minus the top times the derivative of the bottom, all over the bottom squared.

Quick Example

Derivative of top   Bottom   Top   Derivative of bottom

$$\frac{d}{dx}\left[\frac{x^3}{x^2+1}\right] = \frac{3x^2(x^2+1) - x^3 \cdot 2x}{(x^2+1)^2}$$

Bottom squared

Notes

- Don't try to remember the rules by the symbols we have used, but remember them in words. (The slogans are easy to remember, even if the terms are not precise.)
- One more time: *The derivative of a product is* not *the product of the derivatives, and the derivative of a quotient is* not *the quotient of the derivatives.* To find the derivative of a product, you must use the product rule, and to find the derivative of a quotient, you must use the quotient rule. Forgetting this is a mistake everyone makes from time to time.[1]

Question Wait a minute! The expression $2x^3$ is a product, and we already know that its derivative is $6x^2$. Where did we use the product rule?

Answer To differentiate functions such as $2x^3$, we have used the rule from Section 3.4: *The derivative of c times a function is c times the derivative of the function.* The product rule gives us the same result:

Derivative of first   Second   First   Derivative of second

$$\frac{d}{dx}[2x^3] = (0)(x^3) + (2)(3x^2) = 6x^2 \quad \text{Product rule}$$

$$\frac{d}{dx}[2x^3] = (2)(3x^2) = 6x^2 \qquad \text{Derivative of a constant times a function}$$

We do not recommend that you use the product rule to differentiate functions like $2x^3$; continue to use the simpler rule when one of the factors is a constant.

---

[1] Leibniz made this mistake at first, too, so you are in good company.

Example 1 • Using the Product Rule

Compute the following derivatives:

a. $\dfrac{d}{dx}[(x^{3.2} + 1)(1 - x)]$ Simplify the answer.

b. $\dfrac{d}{dx}[(x + 1)(x^2 + 1)(x^3 + 1)]$ Do not expand the answer.

Solution

a. We can do the calculation in two ways.

$$\underset{\substack{\text{Derivative of first} \\ \downarrow}}{}\quad \underset{\substack{\text{Second}\\\downarrow}}{}\quad \underset{\substack{\text{First}\\\downarrow}}{}\quad \underset{\substack{\text{Derivative}\\\text{of second}\\\downarrow}}{}$$

Product Rule $\dfrac{d}{dx}[(x^{3.2} + 1)(1 - x)] = (3.2x^{2.2})(1 - x) + (x^{3.2} + 1)(-1)$

$$= 3.2x^{2.2} - 3.2x^{3.2} - x^{3.2} - 1 \qquad \begin{array}{l}\text{Expand the}\\\text{answer.}\end{array}$$

$$= -4.2x^{3.2} + 3.2x^{2.2} - 1$$

Avoiding the Product Rule First, expand the given expression:

$$(x^{3.2} + 1)(1 - x) = -x^{4.2} + x^{3.2} - x + 1$$

$$\frac{d}{dx}[(x^{3.2} + 1)(1 - x)] = \frac{d}{dx}[-x^{4.2} + x^{3.2} - x + 1]$$

$$= -4.2x^{3.2} + 3.2x^{2.2} - 1$$

In this example the product rule saves us little or no work, but in later sections we will see examples that can be done in no other way. Learn how to use the product rule now!

b. Here we have a product of *three* functions, not just two. We can find the derivative by using the product rule twice:

$$\frac{d}{dx}[(x + 1)(x^2 + 1)(x^3 + 1)] = \frac{d}{dx}[x + 1] \cdot [(x^2 + 1)(x^3 + 1)] + (x + 1) \cdot \frac{d}{dx}[(x^2 + 1)(x^3 + 1)]$$

$$= (1)(x^2 + 1)(x^3 + 1) + (x + 1)[(2x)(x^3 + 1) + (x^2 + 1)(3x^2)]$$

$$= (x^2 + 1)(x^3 + 1) + (x + 1)(2x)(x^3 + 1) + (x + 1)(x^2 + 1)(3x^2)$$

We can see here a more general product rule:

$$(fgh)' = f'gh + fg'h + fgh'$$

Notice that every factor has a chance to contribute to the rate of change of the product. There are similar formulas for products of four or more functions.

Example 2 • Using the Quotient Rule

Compute the derivatives   (a) $\dfrac{d}{dx}\left[\dfrac{1 - 3.2x^{-0.1}}{x + 1}\right]$ and (b) $\dfrac{d}{dx}\left[\dfrac{(x + 1)(x + 2)}{x - 1}\right]$.

Solution

Derivative of top   Bottom   Top   Derivative of bottom

a. $\dfrac{d}{dx}\left[\dfrac{1 - 3.2x^{-0.1}}{x + 1}\right] = \dfrac{(0.32x^{-1.1})(x + 1) - (1 - 3.2x^{-0.1})(1)}{(x + 1)^2}$

Bottom squared

$$= \frac{0.32x^{-0.1} + 0.32x^{-1.1} - 1 + 3.2x^{-0.1}}{(x + 1)^2} \quad \text{Expand the numerator.}$$

$$= \frac{3.52x^{-0.1} + 0.32x^{-1.1} - 1}{(x + 1)^2}$$

b. Here we have both a product and a quotient. Which rule do we use, the product or the quotient rule? Here is a way to decide. Think about how we would calculate, step by step, the value of $(x + 1)(x + 2)/(x - 1)$ for a specific value of x—say, $x = 11$. Here is how we would probably do it:

1. Calculate $(x + 1)(x + 2) = (11 + 1)(11 + 2) = 156$.

2. Calculate $x - 1 = 11 - 1 = 10$.

3. Divide 156 by 10 to get 15.6.

Now ask: What was the last operation we performed? The last operation we performed was division, so we can regard the whole expression as a *quotient*—that is, as $(x + 1)(x + 2)$ *divided by* $(x - 1)$. Therefore, we should use the quotient rule.

The first thing the quotient rule tells us to do is take the derivative of the numerator. Now, the numerator is a product, so we must use the product rule to take its derivative. Here is the calculation:

Derivative of top   Bottom   Top   Derivative of bottom

$$\frac{d}{dx}\left[\frac{(x + 1)(x + 2)}{x - 1}\right] = \frac{[(1)(x + 2) + (x + 1)(1)](x - 1) - [(x + 1)(x + 2)](1)}{(x - 1)^2}$$

Bottom squared

$$= \frac{(2x + 3)(x - 1) - (x + 1)(x + 2)}{(x - 1)^2}$$

$$= \frac{x^2 - 2x - 5}{(x - 1)^2}$$

What is important is to determine the *order of operations* and, in particular, to determine the last operation to be performed. Pretending to do an actual calculation reminds us of the order of operations; we call this technique the **calculation thought experiment.**

✷ *Before we go on . . .* We used the quotient rule because the function was a quotient; we used the product rule to calculate the derivative of the numerator because the numerator was a product. Get used to this. Differentiation rules usually must be used in combination; once you have determined one rule to use, do not assume that you can forget the others.

Here is another way we could have done this problem. Our calculation thought experiment could have taken the following form:

1. Calculate $(x + 1)/(x - 1) = (11 + 1)/(11 - 1) = 1.2$.

2. Calculate $x + 2 = 11 + 2 = 13$.

3. Multiply 1.2 by 13 to get 15.6.

We would have then regarded the expression as a *product*—the product of the factors $(x + 1)/(x - 1)$ and $(x + 2)$—and used the product rule instead. We can't escape the quotient rule, however: We need to use it to take the derivative of the first factor, $(x + 1)/(x - 1)$. Try this approach for practice and check that you get the same answer.

Calculation Thought Experiment

The **calculation thought experiment** is a technique to determine whether to treat an algebraic expression as a product, quotient, sum, or difference. Given an expression, consider the steps you would use in computing its value. If the last operation is multiplication, treat the expression as a product; if the last operation is division, treat the expression as a quotient; and so on.

Quick Examples

1. $(3x^2 - 4)(2x + 1)$ can be computed by first calculating the expressions in parentheses and then multiplying. Since the last step is multiplication, we can treat the expression as a product.

2. $(2x - 1)/x$ can be computed by first calculating the numerator and denominator and then dividing one by the other. Since the last step is division, we can treat the expression as a quotient.

3. $x^2 + (4x - 1)(x + 2)$ can be computed by first calculating x^2, then calculating the product $(4x - 1)(x + 2)$, and finally adding the two answers. Thus, we can treat the expression as a sum.

4. $(3x^2 - 1)^5$ can be computed by first calculating the expression in parentheses and then raising the answer to the fifth power. Thus, we can treat the expression as a power. (We will see how to differentiate powers of expressions in Section 4.2.)

It often happens that the same expression can be calculated in different ways; for example, $(x + 1)(x + 2)/(x - 1)$ can be treated as either a quotient or a product [see Example 2(b)].

Example 3 • Using the Calculation Thought Experiment

Find $\dfrac{d}{dx}\left[6x^2 + 5\left(\dfrac{x}{x - 1}\right)\right].$

Solution The calculation thought experiment tells us that the expression we are asked to differentiate can be treated as a *sum*. Since the derivative of a sum is the sum of the derivatives, we get

$$\frac{d}{dx}\left[6x^2 + 5\left(\frac{x}{x-1}\right)\right] = \frac{d}{dx}[6x^2] + \frac{d}{dx}\left[5\left(\frac{x}{x-1}\right)\right]$$

In other words, we must take the derivatives of $6x^2$ and $5\left(\frac{x}{x-1}\right)$ separately and then add the answers. The derivative of $6x^2$ is $12x$. There are two ways of taking the derivative of $5\left(\frac{x}{x-1}\right)$: We could either first multiply the expression $\left(\frac{x}{x-1}\right)$ by 5 to get $\left(\frac{5x}{x-1}\right)$ and then take its derivative using the quotient rule, or we could pull the 5 out, as we do next:

$$\frac{d}{dx}\left[6x^2 + 5\left(\frac{x}{x-1}\right)\right] = \frac{d}{dx}[6x^2] + \frac{d}{dx}\left[5\left(\frac{x}{x-1}\right)\right] \qquad \text{Derivative of sum}$$

$$= 12x + 5\frac{d}{dx}\left[\frac{x}{x-1}\right] \qquad \text{Constant} \times \text{function}$$

$$= 12x + 5\left[\frac{(1)(x-1) - (x)(1)}{(x-1)^2}\right] \qquad \text{Quotient rule}$$

$$= 12x + 5\left[\frac{-1}{(x-1)^2}\right]$$

$$= 12x - \frac{5}{(x-1)^2}$$

Derivation of the Product Rule

To calculate the derivative of the function $f(x)g(x)$, we use the definition of the derivative:

$$\frac{d}{dx}[f(x)g(x)] = \lim_{h\to 0}\frac{f(x+h)g(x+h) - f(x)g(x)}{h}$$

Question How can we rewrite this expression so that we can evaluate the limit?

Answer There are several ways to do this. Here is one approach: Notice that the numerator reflects a simultaneous change in f [from $f(x)$ to $f(x+h)$] and g [from $g(x)$ to $g(x+h)$]. To separate the two effects, we add and subtract a quantity in the numerator that reflects a change in only one of the functions:

$$\frac{d}{dx}[f(x)g(x)] = \lim_{h\to 0}\frac{f(x+h)g(x+h) - f(x)g(x)}{h}$$

$$= \lim_{h\to 0}\frac{f(x+h)g(x+h) - f(x)g(x+h) + f(x)g(x+h) - f(x)g(x)}{h} \qquad \begin{array}{l}\text{We subtracted and added}\\\text{the quantity}^2 \; f(x)g(x+h).\end{array}$$

$$= \lim_{h\to 0}\frac{[f(x+h) - f(x)]g(x+h) + f(x)[g(x+h) - g(x)]}{h} \qquad \text{Common factors}$$

$$= \lim_{h\to 0}\frac{f(x+h) - f(x)}{h}\,g(x+h) + \lim_{h\to 0}f(x)\left[\frac{g(x+h) - g(x)}{h}\right] \qquad \text{Limit of sum}$$

$$= \lim_{h\to 0}\frac{f(x+h) - f(x)}{h}\,\lim_{h\to 0}g(x+h) + \lim_{h\to 0}f(x)\lim_{h\to 0}\left[\frac{g(x+h) - g(x)}{h}\right] \qquad \text{Limit of product}$$

---

²Adding an appropriate form of zero is an age-old mathematical ploy.

Now we already know the following four limits:

$$\lim_{h \to 0} \frac{f(x + h) - f(x)}{h} = f'(x) \qquad \text{Definition of derivative of } f$$

$$\lim_{h \to 0} \frac{g(x + h) - g(x)}{h} = g'(x) \qquad \text{Definition of derivative of } g$$

$$\lim_{h \to 0} g(x + h) = g(x) \qquad \text{If } g \text{ is differentiable, it must be continuous.}[3]$$

$$\lim_{h \to 0} f(x) = f(x) \qquad \text{Limit of a constant}$$

Putting these limits into the one we're calculating, we get

$$\frac{d}{dx}[f(x)g(x)] = f'(x)g(x) + f(x)g'(x)$$

which is the product rule.

 The quotient rule can be proved in a very similar way. You can find a proof at the Web site by following the path

Web Site → Everything for Calculus → Chapter 4→ Proof of Quotient Rule

4.1 EXERCISES

In Exercises 1–12, (a) calculate the derivative of the given function mentally without using either the product or quotient rule and then (b) use the product or quotient rule to find the derivative. Check that you obtain the same answer.

1. $f(x) = 3x$

2. $f(x) = 2x^2$

3. $g(x) = x \cdot x^2$

4. $g(x) = x \cdot x$

5. $h(x) = x(x + 3)$

6. $h(x) = x(1 + 2x)$

7. $r(x) = 100x^{2.1}$

8. $r(x) = 0.2x^{-1}$

9. $s(x) = \dfrac{2}{x}$

10. $t(x) = \dfrac{x}{3}$

11. $u(x) = \dfrac{x^2}{3}$

12. $s(x) = \dfrac{3}{x^2}$

In Exercises 13–20, calculate dy/dx. Simplify your answer.

13. $y = 3x(4x^2 - 1)$

14. $y = 3x^2(2x + 1)$

15. $y = x^3(1 - x^2)$

16. $y = x^5(1 - x)$

17. $y = (2x + 3)^2$

18. $y = (4x - 1)^2$

19. $y = x\sqrt{x}$

20. $y = x^2\sqrt{x}$

In Exercises 21–50, calculate dy/dx. You need not expand your answers.

21. $y = (x + 1)(x^2 - 1)$

22. $y = (4x^2 + x)(x - x^2)$

23. $y = (2x^{0.5} + 4x - 5)(x - x^{-1})$

24. $y = (x^{0.7} - 4x - 5)(x^{-1} + x^{-2})$

25. $y = (2x^2 - 4x + 1)^2$

26. $y = (2x^{0.5} - x^2)^2$

27. $y = \left(\dfrac{x}{3.2} + \dfrac{3.2}{x} \right)(x^2 + 1)$

28. $y = \left(\dfrac{x^{2.1}}{7} + \dfrac{2}{x^{2.1}} \right)(7x - 1)$

29. $y = x^2(2x + 3)(7x + 2)$

30. $y = x(x^2 - 3)(2x^2 + 1)$

31. $y = (5.3x - 1)(1 - x^{2.1})(x^{-2.3} - 3.4)$

32. $y = (1.1x + 4)(x^{2.1} - x)(3.4 - x^{-2.1})$

33. $y = (\sqrt{x} + 1)\left(\sqrt{x} + \dfrac{1}{x^2} \right)$

34. $y = (4x^2 - \sqrt{x})\left(\sqrt{x} - \dfrac{2}{x^2} \right)$

35. $y = \dfrac{2x + 4}{3x - 1}$

36. $y = \dfrac{3x - 9}{2x + 4}$

37. $y = \dfrac{2x^2 + 4x + 1}{3x - 1}$

38. $y = \dfrac{3x^2 - 9x + 11}{2x + 4}$

39. $y = \dfrac{x^2 - 4x + 1}{x^2 + x + 1}$

40. $y = \dfrac{x^2 + 9x - 1}{x^2 + 2x - 1}$

41. $y = \dfrac{x^{0.23} - 5.7x}{1 - x^{-2.9}}$

42. $y = \dfrac{8.43x^{-0.1} - 0.5x^{-1}}{3.2 + x^{2.9}}$

43. $y = \dfrac{\sqrt{x} + 1}{\sqrt{x} - 1}$

44. $y = \dfrac{\sqrt{x} - 1}{\sqrt{x} + 1}$

---

[3]For a proof of this fact, see the online text "Continuity and Differentiability" available at Web Site → Online Text.

45. $y = \dfrac{\left(\dfrac{1}{x} + \dfrac{1}{x^2}\right)}{x + x^2}$

46. $y = \dfrac{\left(1 - \dfrac{1}{x^2}\right)}{x^2 - 1}$

47. $y = \dfrac{(x + 3)(x + 1)}{3x - 1}$

48. $y = \dfrac{x}{(x - 5)(x - 4)}$

49. $y = \dfrac{(x + 3)(x + 1)(x + 2)}{3x - 1}$

50. $y = \dfrac{3x - 1}{(x - 5)(x - 4)(x - 1)}$

In Exercises 51–56, compute the derivatives.

51. $\dfrac{d}{dx}\left[(x^2 + x)(x^2 - x)\right]$

52. $\dfrac{d}{dx}\left[(x^2 + x^3)(x + 1)\right]$

53. $\dfrac{d}{dx}\left[(x^3 + 2x)(x^2 - x)\right]\Big|_{x=2}$

54. $\dfrac{d}{dx}\left[(x^2 + x)(x^2 - x)\right]\Big|_{x=1}$

55. $\dfrac{d}{dt}\left[(t^2 - t^{0.5})(t^{0.5} + t^{-0.5})\right]\Big|_{t=1}$

56. $\dfrac{d}{dt}\left[(t^2 + t^{0.5})(t^{0.5} - t^{-0.5})\right]\Big|_{t=1}$

In Exercises 57–62, find the equation of the line tangent to the graph of the given function at the point with the indicated x coordinate.

57. $f(x) = (x^2 + 1)(x^3 + x); x = 1$

58. $f(x) = (x^{0.5} + 1)(x^2 + x); x = 1$

59. $f(x) = \dfrac{x + 1}{x + 2}; x = 0$

60. $f(x) = \dfrac{\sqrt{x} + 1}{\sqrt{x} + 2}; x = 4$

61. $f(x) = \dfrac{x^2 + 1}{x}; x = -1$

62. $f(x) = \dfrac{x}{x^2 + 1}; x = 1$

APPLICATIONS

63. Revenue The monthly sales of Sunny Electronics' new sound system are given by $S(x) = 20x - x^2$ hundred units per month, x months after its introduction. The price Sunny charges is $p(x) = 1000 - x^2$ dollars per sound system, x months after introduction. The revenue Sunny earns must then be $R(x) = 100S(x)p(x)$. Find, 5 months after the introduction, the rate of change of monthly sales, the rate of change of the price, and the rate of change of revenue. Interpret your answers.

64. Revenue The monthly sales of Sunny Electronics' new portable CD player is given by $S(x) = 20x - x^2$ hundred units per month, x months after its introduction. The price Sunny charges is $p(x) = 100 - x^2$ dollars per CD player, x months after introduction. The revenue Sunny earns must then be $R(x) = 100S(x)p(x)$. Find, 6 months after the introduction, the rate of change of monthly sales, the rate of change of the price, and the rate of change of revenue. Interpret your answers.

65. Revenue Dorothy Wagner is currently selling 20 "I ♥ Calculus" T-shirts each day, but sales are dropping at a rate of 3 per day. She is currently charging $7 per T-shirt, but to compensate for dwindling sales, she is increasing the unit price by $1 per day. How fast and in what direction is her daily revenue currently changing?

66. Pricing Policy Let's turn Exercise 65 around a little: Dorothy Wagner is currently selling 20 "I ♥ Calculus" T-shirts each day, but sales are dropping at a rate of 3 per day. She is currently charging $7 per T-shirt, and she wishes to increase her daily revenue by $10 per day. At what rate should she increase the unit price to accomplish this (assuming that the price increase does not affect sales)?

67. Bus Travel Thoroughbred Bus Company finds that its monthly costs for one particular year were given by $C(t) = 10,000 + t^2$ dollars after t months. After t months the company had $P(t) = 1000 + t^2$ passengers per month. How fast is its cost per passenger changing after 6 months?

68. Bus Travel Thoroughbred Bus Company finds that its monthly costs for one particular year were given by $C(t) = 100 + t^2$ dollars after t months. After t months the company had $P(t) = 1000 + t^2$ passengers per month. How fast is its cost per passenger changing after 6 months?

69. Fuel Economy Your muscle car's gas mileage (in miles per gallon) is given as a function $M(x)$ of speed x in miles per hour, where

$$M(x) = \dfrac{3000}{x + 3600x^{-1}}$$

Calculate $M'(x)$ and then $M'(10)$, $M'(60)$, and $M'(70)$. What do the answers tell you about your car?

70. Fuel Economy Your used Chevy's gas mileage (in miles per gallon) is given as a function $M(x)$ of speed x in miles per hour, where

$$M(x) = \dfrac{4000}{x + 3025x^{-1}}$$

Calculate $M'(x)$ and hence determine the *sign* of each of the following: $M'(40)$, $M'(55)$, and $M'(60)$. Interpret your results.

71. Saudi Oil Revenues The spot price of crude oil during the period 1999–2001 can be approximated by

$$P(t) = -10t^2 + 32 \text{ dollars per barrel} \qquad (-1 \le t \le 1)$$

in year t, where $t = 0$ represents 2000. Saudi Arabia's crude oil production over the same period can be approximated by

$$Q(t) = -70.8t^2 + 78 \text{ million barrels per day} \quad (-1 \le t \le 1)$$

Use these models to estimate Saudi Arabia's daily oil revenue and its rate of change in 2001 (that is, when $t = 1$).

SOURCE: Deutsche Bank/*New York Times,* November 21, 2001, p. A3.

72. Russian Oil Revenues Russia's crude oil production during the period 1999–2001 can be approximated by

$$Q(t) = 4 + 0.2t \text{ million barrels per day} \qquad (-1 \le t \le 1)$$

in year t, where $t = 0$ represents 2000. Use the model for the spot price in Exercise 71 to estimate Russia's daily oil revenue and its rate of change in 2001.

SOURCE: Deutsche Bank/*New York Times,* November 21, 2001, p. A3.

73. Military Spending The annual cost per active-duty armed service member in the United States increased from $80,000 in 1995 to an estimated $120,000 in 2007. In 1995 there were 1.5 million armed service personnel, and this number was projected to decrease to 1.4 million in 2003. Use linear models for annual cost and personnel to estimate to the nearest $10 million the rate of change of total military personnel costs in 2002.

Annual costs are adjusted for inflation. SOURCES: Department of Defense; Stephen Daggett, military analyst; Congressional Research Service/*New York Times,* April 19, 2002, p. A21.

74. Military Spending in the 1990s The annual cost per active-duty armed service member in the United States increased from $80,000 in 1995 to $90,000 in 2000. In 1990 there were 2 million armed service personnel, and this number decreased to 1.5 million in 2000. Use linear models for annual cost and personnel to estimate to the nearest $10 million the rate of change of total military personnel costs in 1995.

Annual costs are adjusted for inflation. SOURCES: Department of Defense; Stephen Daggett, military analyst; Congressional Research Service/*New York Times,* April 19, 2002, p. A21.

75. Biology—Reproduction The Verhulst model for population growth specifies the reproductive rate of an organism as a function of the total population according to the following formula:

$$R(p) = \frac{r}{1 + kp}$$

where p is the total population in thousands of organisms, r and k are constants that depend on the particular circumstances and the organism being studied, and $R(p)$ is the reproduction rate in thousands of organisms per hour. If $k = 0.125$ and $r = 45$, find $R'(p)$ and then $R'(4)$. Interpret the result.

SOURCE: F. C. Hoppensteadt and C. S. Peskin, *Mathematics in Medicine and the Life Sciences* (New York: Springer-Verlag, 1992): 20–22.

76. Biology—Reproduction Another model, the predator satiation model for population growth, specifies that the reproductive rate of an organism as a function of the total population varies according to the following formula:

$$R(p) = \frac{rp}{1 + kp}$$

where p is the total population in thousands of organisms, r and k are constants that depend on the particular circumstances and the organism being studied, and $R(p)$ is the reproduction rate in new organisms per hour. Given that $k = 0.2$ and $r = 0.08$, find $R'(p)$ and $R'(2)$. Interpret the result.

SOURCE: F. C. Hoppensteadt and C. S. Peskin, *Mathematics in Medicine and the Life Sciences* (New York: Springer-Verlag, 1992): 20–22.

77. Embryo Development Bird embryos consume oxygen from the time the egg is laid through the time the chick hatches. For a typical galliform bird, the total oxygen consumption (in milliliters) t days after the egg was laid can be approximated by

$$C(t) = -0.0163t^4 + 1.096t^3 - 10.704t^2 + 3.576t \quad (t \le 30)$$

(An egg will usually hatch at around $t = 28$.) Suppose at time $t = 0$ you have a collection of 30 newly laid eggs and the number of eggs decreases linearly to zero at time $t = 30$ days. How fast is the total oxygen consumption of your collection of embryos changing after 25 days? (Answer to the nearest whole number.) Interpret the result.

SOURCE: The model is derived from graphical data published in Roger S. Seymour, "The Brush Turkey," *Scientific American* (December 1991): 108–114.

78. Embryo Development Turkey embryos consume oxygen from the time the egg is laid through the time the chick hatches. For a brush turkey, the total oxygen consumption (in milliliters) t days after the egg was laid can be approximated by

$$C(t) = -0.00708t^4 + 0.952t^3 - 21.96t^2 + 95.328t \quad (t \le 50)$$

(An egg will typically hatch at around $t = 50$.) Suppose at time $t = 0$ you have a collection of 100 newly laid eggs and the number of eggs decreases linearly to zero at time $t = 50$ days. How fast is the total oxygen consumption of your collection of embryos changing after 40 days? (Answer to the nearest whole number.) Interpret the result.

SOURCE: The model is derived from graphical data published in Roger S. Seymour, "The Brush Turkey," *Scientific American* (December 1991): 108–114.

79. Bottled-Water Sales Annual U.S. sales of bottled water (including sparkling water) from 1984 to 2000 can be approximated by

$$W(t) = 0.0010t^3 - 0.028t^2 + 0.28t - 0.70 \text{ billion gallons}$$
$$(4 \le t \le 20)$$

where t is time in years since 1980. Sales of sparkling water over the same period can be approximated by

$$S(t) = -0.0013t^2 + 0.030t - 0.0069 \text{ billion gallons}$$
$$(4 \le t \le 20)$$

a. What are represented by the functions $W(t) - S(t)$ and $S(t)/W(t)$? What does the derivative of $S(t)/W(t)$ tell you?

b. Compute $\dfrac{d}{dt}\left[\dfrac{S(t)}{W(t)}\right]\Big|_{t=10}$ and interpret the result.

Models are based on regression of approximate data. Source: Beverage Marketing Corporation of New York/*New York Times*, June 21, 2001, p. C1.

80. Bottled-Water Sales Annual U.S. sales of bottled water (including imported water) can be approximated by

$$W(t) = 0.0010t^3 - 0.028t^2 + 0.28t - 0.70 \text{ billion gallons}$$
$$(4 \le t \le 20)$$

where t is time in years since 1980. Sales of imported bottled water over the same period can be approximated by

$$I(t) = 0.00040t^2 + 0.0052t - 0.028 \text{ billion gallons}$$
$$(4 \le t \le 20)$$

a. What are represented by the functions $W(t) - I(t)$ and $I(t)/W(t)$? What does the derivative of $I(t)/W(t)$ tell you?

b. Compute $\dfrac{d}{dt}\left[\dfrac{I(t)}{W(t)}\right]\Big|_{t=10}$ and interpret the result.

Models are based on regression of approximate data. Source: Beverage Marketing Corporation of New York/*New York Times*, June 21, 2001, p. C1.

COMMUNICATION AND REASONING EXERCISES

81. You have come across the following in a newspaper article: "Revenues of HAL Home Heating Oil Inc. are rising by $4.2 million per year. This is due to an annual increase of 70¢ per gallon in the price HAL charges for heating oil and an increase in sales of 6 million gallons of oil per year." Comment on this analysis.

82. Your friend says that since average cost is obtained by dividing the cost function by the number of units x, it follows that the derivative of average cost is the same as marginal cost because the derivative of x is 1. Comment on this analysis.

83. Find a demand function $q(p)$ such that, at a price per item of $p = \$100$, revenue will rise if the price per item is increased.

84. What must be true about a demand function $q(p)$ so that, at a price per item of $p = \$100$, revenue will decrease if the price per item is increased?

85. You and I are both selling a steady 20 T-shirts per day. The price I am getting for my T-shirts is increasing twice as fast as yours, but your T-shirts are currently selling for twice the price of mine. Whose revenue is increasing fastest: yours, mine, or neither? Explain.

86. You and I are both selling T-shirts for a steady $20 per shirt. Sales of my T-shirts are increasing at twice the rate of yours, but you are currently selling twice as many as I am. Whose revenue is increasing fastest: yours, mine, or neither? Explain.

87. Marginal Product (from the GRE Economics Test) Which of the following statements about average product and marginal product is correct?

(A) If average product is decreasing, marginal product must be less than average product.

(B) If average product is increasing, marginal product must be increasing.

(C) If marginal product is decreasing, average product must be less than marginal product.

(D) If marginal product is increasing, average product must be decreasing.

(E) If marginal product is constant over some range, average product must be constant over that range.

88. Marginal Cost (Based on a Question from the GRE Economics Test) Which of the following statements about average cost and marginal cost is correct?

(A) If average cost is increasing, marginal cost must be increasing.

(B) If average cost is increasing, marginal cost must be decreasing.

(C) If average cost is increasing, marginal cost must be larger than average cost.

(D) If marginal cost is increasing, average cost must be increasing.

(E) If marginal cost is increasing, average cost must be larger than marginal cost.

4.2 *The Chain Rule*

We can now find the derivatives of expressions involving powers of x combined using addition, subtraction, multiplication, and division, but we still cannot take the derivative of an expression like $(3x + 1)^{0.5}$. For this we need one more rule. The function $h(x) = (3x + 1)^{0.5}$ is not a sum, difference, product, or quotient. We can use the calculation thought experiment to find the last operation we would perform in calculating $h(x)$:

1. Calculate $3x + 1$.

2. Take the 0.5 power (square root) of the answer.

Thus, the last operation is "take the 0.5 power." We do not yet have a rule for finding the derivative of the 0.5 power of a quantity other than x.

There is a way to build $h(x)$ out of two simpler functions: $f(x) = x^{0.5}$ (the function that corresponds to the second step in the calculation above) and $u(x) = 3x + 1$ (the first step):

$$h(x) = (3x + 1)^{0.5}$$
$$= f(3x + 1) \qquad\qquad f(x) = x^{0.5}$$
$$= f(u(x)) \qquad\qquad u(x) = 3x + 1$$

We say that h is the **composite** of f and u. We read $f(u(x))$ as "f of u of x."

To compute $h(1)$, say, we first compute $3 \cdot 1 + 1 = 4$ and then take the square root of 4, giving $h(1) = 2$. To compute $f(u(1))$, we follow exactly the same steps: First compute $u(1) = 4$ and then $f(u(1)) = f(4) = 2$. We always compute $f(u(x))$ numerically from the inside out: Given x, first compute $u(x)$ and then $f(u(x))$.

Now, f and u are functions *whose derivatives we know*. The *chain rule* allows us to use our knowledge of the derivatives of f and u to find the derivative of $f(u(x))$. For the purposes of stating the rule, let's avoid some of the nested parentheses by abbreviating $u(x)$ as u. Thus, we write $f(u)$ instead of $f(u(x))$ and remember that u is a function of x.

Chain Rule

If f is a differentiable function of u and u is a differentiable function of x, then the composite $f(u)$ is a differentiable function of x, and

$$\frac{d}{dx}[f(u)] = f'(u)\frac{du}{dx} \qquad\qquad \text{Chain rule}$$

In Words The derivative of f(quantity) is the derivative of f, evaluated at that quantity, times the derivative of the quantity.

Quick Examples

1. Let $f(u) = u^2$. Then

$$\frac{d}{dx}[u^2] = 2u\frac{du}{dx} \qquad\qquad \text{Since } f'(u) = 2u$$

The derivative of a quantity squared is 2 times the quantity, times the derivative of the quantity.

2. Let $f(u) = u^{0.5}$. Then

$$\frac{d}{dx}[u^{0.5}] = 0.5u^{-0.5}\frac{du}{dx} \qquad\qquad \text{Since } f'(u) = 0.5u^{-0.5}$$

The derivative of a quantity raised to the 0.5 is 0.5 times the quantity raised to the -0.5, times the derivative of the quantity.

As the Quick Examples illustrate, for every power of a function u whose derivative we know, we now get a "generalized" differentiation rule. The table gives more examples.

Original Rule	Generalized Rule	In Words
$\dfrac{d}{dx}[x^2] = 2x$	$\dfrac{d}{dx}[u^2] = 2u\,\dfrac{du}{dx}$	The derivative of a quantity squared is twice the quantity, times the derivative of the quantity.
$\dfrac{d}{dx}[x^3] = 3x^2$	$\dfrac{d}{dx}[u^3] = 3u^2\,\dfrac{du}{dx}$	The derivative of a quantity cubed is three times the quantity squared, times the derivative of the quantity.
$\dfrac{d}{dx}\left[\dfrac{1}{x}\right] = -\dfrac{1}{x^2}$	$\dfrac{d}{dx}\left[\dfrac{1}{u}\right] = -\dfrac{1}{u^2}\,\dfrac{du}{dx}$	The derivative of 1 over a quantity is negative 1 over the quantity squared, times the derivative of the quantity.

Power Rule	Generalized Power Rule	In Words
$\dfrac{d}{dx}[x^n] = nx^{n-1}$	$\dfrac{d}{dx}[u^n] = nu^{n-1}\,\dfrac{du}{dx}$	The derivative of a quantity raised to the n is n times the quantity raised to the $n - 1$, times the derivative of the quantity.

Question Why should I accept the chain rule?

Answer To motivate it, let's see why it is true in a few special cases: when $f(u) = u^n$, where $n = \pm 1, \pm 2, \pm 3, \ldots$. In these cases, the chain rule tells us, for example, that

$$\frac{d}{dx}[u^2] = 2u\,\frac{du}{dx}$$

$$\frac{d}{dx}[u^3] = 3u^2\,\frac{du}{dx}$$

and so on. But we could have done the first of these another way, using the product rule:

$$\frac{d}{dx}[u^2] = \frac{d}{dx}[u \cdot u] = \frac{du}{dx}u + u\,\frac{du}{dx} = 2u\,\frac{du}{dx}$$

which gives us the same result. Similarly, we can use the product rule to verify that

$$\frac{d}{dx}[u^3] = 3u^2\,\frac{du}{dx}$$

and so on for higher positive powers of u. We can now use the quotient rule and the chain rule for positive powers to verify the chain rule for *negative* powers.

Question The argument that the chain rule works in this special case does not compel me to accept that it works in general. How can I convince myself of the general case?

Answer Although a proof of the chain rule is beyond the scope of this book, you can find one on the Web site by following the path

Web Site → Everything for Calculus → Chapter 4 → Proof of Chain Rule

Example 1 • Using the Chain Rule

Compute the following derivatives:

a. $\dfrac{d}{dx}[(2x^2 + x)^3]$ **b.** $\dfrac{d}{dx}[(x^3 + x)^{100}]$ **c.** $\dfrac{d}{dx}[\sqrt{3x + 1}]$

Solution

a. Using the calculation thought experiment, we see that the last operation we would perform in calculating $(2x^2 + x)^3$ is that of *cubing*. Thus, we think of $(2x^2 + x)^3$ as a *quantity cubed*. We can use two similar methods to calculate its derivative.

Method 1: Using the Formula We think of $(2x^2 + x)^3$ as u^3, where $u = 2x^2 + x$. By the formula,

$$\frac{d}{dx}[u^3] = 3u^2\frac{du}{dx} \qquad \text{Generalized power rule}$$

Now substitute for u:

$$\frac{d}{dx}[(2x^2 + x)^3] = 3(2x^2 + x)^2\frac{d}{dx}[2x^2 + x]$$

$$= 3(2x^2 + x)^2(4x + 1)$$

Method 2: Using the Verbal Form If we prefer to use the verbal form, we get

The derivative of $(2x^2 + x)$ cubed is three times $(2x^2 + x)$ squared, times the derivative of $(2x^2 + x)$.

In symbols,

$$\frac{d}{dx}[(2x^2 + x)^3] = 3(2x^2 + x)^2(4x + 1)$$

as we obtained above.

b. First, the calculation thought experiment: If we were computing $(x^3 + x)^{100}$, the last operation we would perform is *raising a quantity to the power* 100. Thus, we are dealing with *a quantity raised to the power* 100, and so we must again use the generalized power rule. According to the verbal form of the generalized power rule, the derivative of a quantity raised to the power 100 is 100 times that quantity to the power 99, times the derivative of that quantity. In symbols,

$$\frac{d}{dx}[(x^3 + x)^{100}] = 100(x^3 + x)^{99}(3x^2 + 1)$$

c. We first rewrite the expression $\sqrt{3x + 1}$ as $(3x + 1)^{0.5}$ and then use the generalized power rule as in parts (a) and (b):

The derivative of a quantity raised to the 0.5 power is 0.5 times the quantity raised to the -0.5 power, times the derivative of the quantity.

Thus,

$$\frac{d}{dx}[(3x + 1)^{0.5}] = 0.5(3x + 1)^{-0.5} \cdot 3 = 1.5(3x + 1)^{-0.5}$$

✳ ***Before we go on . . .*** The following are examples of common errors:

$$\frac{d}{dx}[(x^3 + x)^{100}] = 100(3x^2 + 1)^{99} \qquad ✗\ WRONG$$

$$\frac{d}{dx}[(x^3 + x)^{100}] = 100(x^3 + x)^{99} \qquad ✗\ WRONG$$

Remember that the generalized power rule says that the derivative of a quantity to the power 100 is 100 times *that same quantity* raised to the power 99, *times the derivative of that quantity.*

Question It seems that there are now two formulas for the derivative of an *n*th power:

1. $\dfrac{d}{dx}[x^n] = nx^{n-1}$

2. $\dfrac{d}{dx}[u^n] = nu^{n-1}\dfrac{du}{dx}$

Which one do I use?

Answer Formula 1 is the original power rule, which applies only to a power of *x*. For instance, it applies to x^{10}, but it does not apply to $(2x + 1)^{10}$ because the quantity that is being raised to a power is not *x*. Formula 2 applies to a power of any *function of x*, such as $(2x + 1)^{10}$. It can even be used in place of the original power rule. For example, if we take $u = x$ in formula 2, we obtain

$$\frac{d}{dx}[x^n] = nx^{n-1}\frac{dx}{dx}$$

$$= nx^{n-1} \qquad \text{The derivative of } x \text{ with respect to } x \text{ is 1.}$$

Thus, the generalized power rule really *is* a generalization of the original power rule, as its name suggests.

Example 2 • More Examples Using the Chain Rule

Find (a) $\dfrac{d}{dx}[(2x^5 + x^2 - 20)^{-2/3}]$, (b) $\dfrac{d}{dx}\left[\dfrac{1}{\sqrt{x+2}}\right]$, and (c) $\dfrac{d}{dx}\left[\dfrac{1}{x^2 + x}\right]$.

Solution
Each of the given functions is or can be rewritten as a power of a function whose derivative we know. Thus, we can use the method of Example 1.

a. $\dfrac{d}{dx}[(2x^5 + x^2 - 20)^{-2/3}] = -\dfrac{2}{3}(2x^5 + x^2 - 20)^{-5/3}(10x^4 + 2x)$

b. $\dfrac{d}{dx}\left[\dfrac{1}{\sqrt{x+2}}\right] = \dfrac{d}{dx}[x+2]^{-1/2} = -\dfrac{1}{2}(x+2)^{-3/2} \cdot 1 = -\dfrac{1}{2(x+2)^{3/2}}$

c. $\dfrac{d}{dx}\left[\dfrac{1}{x^2 + x}\right] = \dfrac{d}{dx}[x^2 + x]^{-1} = -(x^2 + x)^{-2}(2x + 1) = -\dfrac{2x + 1}{(x^2 + x)^2}$

✻ **Before we go on . . .** In part (c), we could have used the quotient rule instead of the generalized power rule. We can think of the quantity $1/(x^2 + x)$ in two different ways using the calculation thought experiment:

1. As 1 divided by something—in other words, as a quotient

2. As something raised to the -1 power

Of course, we get the same derivative using either approach.

We now look at some more complicated examples.

Example 3 • Harder Examples Using the Chain Rule

Find $\dfrac{dy}{dx}$ in each case.

a. $y = [(x + 1)^{-2.5} + 3x]^{-3}$ **b.** $y = (x + 10)^3\sqrt{1 - x^2}$

Solution

a. The calculation thought experiment tells us that the last operation we would perform in calculating y is raising the quantity $[(x + 1)^{-2.5} + 3x]$ to the power -3. We use the generalized power rule:

$$\frac{dy}{dx} = -3[(x + 1)^{-2.5} + 3x]^{-4}\frac{d}{dx}[(x + 1)^{-2.5} + 3x]$$

We are not yet done; we must still find the derivative of $(x + 1)^{-2.5} + 3x$. Finding the derivative of a complicated function in several steps helps keep the problem manageable. Continuing, we have

$$\frac{dy}{dx} = -3[(x + 1)^{-2.5} + 3x]^{-4}\frac{d}{dx}[(x + 1)^{-2.5} + 3x]$$

$$= -3[(x + 1)^{-2.5} + 3x]^{-4}\left(\frac{d}{dx}[(x + 1)^{-2.5}] + \frac{d}{dx}[3x]\right) \quad \text{Derivative of a sum}$$

Now we have two derivatives left to calculate. The second of these we know to be 3, and the first is the derivative of a quantity raised to the -2.5 power. Thus,

$$\frac{dy}{dx} = -3[(x + 1)^{-2.5} + 3x]^{-4}[-2.5(x + 1)^{-3.5} \cdot 1 + 3]$$

b. The expression $(x + 10)^3\sqrt{1 - x^2}$ is a product, so we use the product rule:

$$\frac{d}{dx}[(x + 10)^3\sqrt{1 - x^2}] = \left(\frac{d}{dx}[(x + 10)^3]\right)\sqrt{1 - x^2} + (x + 10)^3\left(\frac{d}{dx}\sqrt{1 - x^2}\right)$$

$$= 3(x + 10)^2\sqrt{1 - x^2} + (x + 10)^3\frac{1}{2\sqrt{1 - x^2}}(-2x)$$

$$= 3(x + 10)^2\sqrt{1 - x^2} - \frac{x(x + 10)^3}{\sqrt{1 - x^2}}$$

APPLICATIONS

The next example is a new treatment of Example 3 from Section 3.5.

Example 4 • Marginal Product

Precision Manufacturer is informed by a consultant that its annual profit is given by

$$P = -200{,}000 + 4000q - 0.46q^2 - 0.000\,01q^3$$

where q is the number of surgical lasers it sells each year. The consultant also informs Precision that the number of surgical lasers it can manufacture each year depends on the number n of assembly-line workers it employs according to the equation

$$q = 100n \qquad \text{Each worker contributes 100 lasers per year.}$$

Use the chain rule to find the marginal product dP/dn.

Solution We could calculate the marginal product by substituting the expression for q in the expression for P to obtain P as a function of n (as given in Chapter 3) and then finding dP/dn. Alternatively—and this will simplify the calculation—we can use the chain rule. To see how the chain rule applies, notice that P is a function of q, where q in turn is given as a function of n. By the chain rule,

$$\frac{dP}{dn} = P'(q)\frac{dq}{dn} \qquad \text{Chain rule}$$

$$= \frac{dP}{dq}\frac{dq}{dn} \qquad \text{Notice how the "quantities" } dq \text{ appear to cancel.}$$

Now we compute

$$\frac{dP}{dq} = 4000 - 0.92q - 0.000\,03q^2 \quad \text{and} \quad \frac{dq}{dn} = 100$$

Substituting into the equation for dP/dn gives

$$\frac{dP}{dn} = (4000 - 0.92q - 0.000\,03q^2)(100)$$

$$= 400{,}000 - 92q - 0.003q^2$$

Notice that the answer has q as a variable. We can express dP/dn as a function of n by substituting $100n$ for q:

$$\frac{dP}{dn} = 400{,}000 - 92(100n) - 0.003(100n)^2$$

$$= 400{,}000 - 9200n - 30n^2$$

The equation

$$\frac{dP}{dn} = \frac{dP}{dq}\frac{dq}{dn}$$

in Example 4 is an appealing way of writing the chain rule because it suggests that the "quantities" dq cancel. In general, we can write the chain rule as follows.

Chain Rule in Differential Notation

If y is a differentiable function of u and u is a differentiable function of x, then

$$\frac{dy}{dx} = \frac{dy}{du}\frac{du}{dx}$$

Notice how the units cancel:

$$\frac{\text{units of } y}{\text{units of } x} = \frac{\text{units of } y}{\cancel{\text{units of } u}}\frac{\cancel{\text{units of } u}}{\text{units of } x}$$

Quick Example

If $y = u^3$, where $u = 4x + 1$, then

$$\frac{dy}{dx} = \frac{dy}{du}\frac{du}{dx} = 3u^2 \cdot 4 = 12u^2 = 12(4x + 1)^2$$

You can see one of the reasons we still use Leibniz differential notation: The chain rule looks like a simple "cancellation" of *du* terms.

Example 5 • Marginal Revenue

Suppose a company's weekly revenue R is given as a function of the unit price p, and p in turn is given as a function of weekly sales q (by means of a demand equation). If

$$\left.\frac{dR}{dp}\right|_{q=1000} = \$40 \text{ per } \$1 \text{ increase in price}$$

and

$$\left.\frac{dp}{dq}\right|_{q=1000} = -\$20 \text{ per additional item sold per week}$$

find the marginal revenue when sales are 1000 items per week.

Solution The marginal revenue is dR/dq. By the chain rule, we have

$$\frac{dR}{dq} = \frac{dR}{dp}\frac{dp}{dq}$$

Units: revenue per item
= revenue per \$1 price increase × price increase per additional item

Since we are interested in the marginal revenue at a demand level of 1000 items per week, we have

$$\left.\frac{dR}{dq}\right|_{q=1000} = (40)(-20) = -\$800 \text{ per additional item sold}$$

Thus, if the price is lowered to increase the demand from 1000 to 1001 items per week, the weekly revenue will drop by approximately \$800.

4.2 EXERCISES

In Exercises 1–12, mentally calculate the derivatives of the functions.

1. $f(x) = (2x + 1)^2$

2. $f(x) = (3x - 1)^2$

3. $f(x) = (x - 1)^{-1}$

4. $f(x) = (2x - 1)^{-2}$

5. $f(x) = (2 - x)^{-2}$

6. $f(x) = (1 - x)^{-1}$

7. $f(x) = (2x + 1)^{0.5}$

8. $f(x) = (-x + 2)^{1.5}$

9. $f(x) = (4x - 1)^{-1}$

10. $f(x) = (x + 7)^{-2}$

11. $f(x) = \dfrac{1}{3x - 1}$

12. $f(x) = \dfrac{1}{(x + 1)^2}$

In Exercises 13–46, calculate the derivatives of the functions.

13. $f(x) = (x^2 + 2x)^4$

14. $f(x) = (x^3 - x)^3$

15. $f(x) = (2x^2 - 2)^{-1}$

16. $f(x) = (2x^3 + x)^{-2}$

17. $g(x) = (x^2 - 3x - 1)^{-5}$

18. $g(x) = (2x^2 + x + 1)^{-3}$

19. $h(x) = \dfrac{1}{(x^2 + 1)^3}$

20. $h(x) = \dfrac{1}{(x^2 + x + 1)^2}$

21. $r(x) = (0.1x^2 - 4.2x + 9.5)^{1.5}$

22. $r(x) = (0.1x - 4.2x^{-1})^{0.5}$

23. $r(s) = (s^2 - s^{0.5})^4$

24. $r(s) = (2s + s^{0.5})^{-1}$

25. $f(x) = \sqrt{1 - x^2}$

26. $f(x) = \sqrt{x + x^2}$

27. $h(x) = 2[(x + 1)(x^2 - 1)]^{-1/2}$

28. $h(x) = 3[(2x - 1)(x - 1)]^{-1/3}$

29. $h(x) = (3.1x - 2)^2 - \dfrac{1}{(3.1x - 2)^2}$

30. $h(x) = \left(3.1x^2 - 2 - \dfrac{1}{3.1x - 2}\right)^2$

31. $f(x) = [(6.4x - 1)^2 + (5.4x - 2)^3]^2$

32. $f(x) = (6.4x - 3)^{-2} + (4.3x - 1)^{-2}$

33. $f(x) = (x^2 - 3x)^{-2}(1 - x^2)^{0.5}$

34. $f(x) = (3x^2 + x)(1 - x^2)^{0.5}$

35. $s(x) = \left(\dfrac{2x + 4}{3x - 1}\right)^2$

36. $s(x) = \left(\dfrac{3x - 9}{2x + 4}\right)^3$

37. $g(z) = \left(\dfrac{z}{1 + z^2}\right)^3$ **38.** $g(z) = \left(\dfrac{z^2}{1 + z}\right)^2$

39. $f(x) = [(1 + 2x)^4 - (1 - x)^2]^3$

40. $f(x) = [(3x - 1)^2 + (1 - x)^5]^2$

41. $t(x) = [2 + (x + 1)^{-0.1}]^{4.3}$

42. $t(x) = [(x + 1)^{0.1} - 4x]^{-5.1}$

43. $r(x) = (\sqrt{2x + 1} - x^2)^{-1}$

44. $r(x) = (\sqrt{x + 1} + \sqrt{x})^3$

45. $f(x) = \{1 + [1 + (1 + 2x)^3]^3\}^3$

46. $f(x) = 2x + [2x + (2x + 1)^3]^3$

In Exercises 47–54, find the indicated derivatives. In each case the independent variable is a (unspecified) function of t.

47. $y = x^{100} + 99x^{-1};\ \dfrac{dy}{dt}$ **48.** $y = x^{0.5}(1 + x);\ \dfrac{dy}{dt}$

49. $s = \dfrac{1}{r^3} + r^{0.5};\ \dfrac{ds}{dt}$ **50.** $s = r + r^{-1};\ \dfrac{ds}{dt}$

51. $V = \dfrac{4}{3}\pi r^3;\ \dfrac{dV}{dt}$ **52.** $A = 4\pi r^2;\ \dfrac{dA}{dt}$

53. $y = x^3 + \dfrac{1}{x}, x = 2$ when $t = 1,\ \left.\dfrac{dx}{dt}\right|_{t=1} = -1;\ \left.\dfrac{dy}{dt}\right|_{t=1}$

54. $y = \sqrt{x} + \dfrac{1}{\sqrt{x}}, x = 9$ when $t = 1,\ \left.\dfrac{dx}{dt}\right|_{t=1} = -1;\ \left.\dfrac{dy}{dt}\right|_{t=1}$

APPLICATIONS

55. Online Trading The average commission c (in dollars) per trade earned by the Charles Schwab Corporation decreased as more of its customers traded online, according to the formula

$$c(u) = 100u^2 - 160u + 110 \text{ dollars/trade}$$
$$(u = \text{fraction of trades done online})$$

During that time, the fraction of online trades increased according to

$$u(t) = 0.42 + 0.02t \qquad (t = \text{months since January 1, 1998})$$

Use direct substitution to express the average commission per trade c as a function of time (do not simplify the expression) and then use the chain rule to estimate the rate of change of average commission per trade at the beginning of September 1998. Be sure to specify the units.

The models in this exercise are based on data for the period Sept. 1997–Sept. 1998. The function c is reliable only in the range $0.4 \le u \le 0.6$. Source: Charles Schwab/*New York Times*, February 10, 1999, p. C1.

56. Online Trading The profitability p (measured in quarterly net income) of the Charles Schwab Corporation increased as more of its customers traded online according to the formula

$$p(u) = 520u^2 - 300u + 100 \text{ million dollars}$$
$$(u = \text{fraction of trades done online})$$

During that time, the fraction of online trades increased according to

$$u(t) = 0.42 + 0.02t \qquad (t = \text{months since January 1, 1998})$$

Use direct substitution to express the quarterly profits p as a function of time (do not simplify the expression) and then use the chain rule to estimate the rate of change of profitability at the beginning of September 1998. Be sure to specify the units.

The models in this exercise are based on data for the period Sept. 1997–Sept. 1998. The function c is reliable only in the range $0.4 \le u \le 0.6$. Source: Charles Schwab/*New York Times*, February 10, 1999, p. C1.

57. Crime Statistics The murder rate in large cities (over 1 million residents) can be related to that in smaller cities (500,000–1,000,000 residents) by the following linear model:

$$y = 1.5x - 1.9 \qquad (15 \le x \le 25)$$

where y is the murder rate (in murders per 100,000 residents each year) in large cities and x is the murder rate in smaller cities. During the period 1991–1998, the murder rate in large cities was decreasing at an average rate of 3 murders per 100,000 residents each year. Use the chain rule to estimate how fast the murder rate was changing in smaller cities during that period. (Show how you used the chain rule in your answer.)

The model is a linear regression model. Source: Federal Bureau of Investigation, Supplementary Homicide Reports/*New York Times*, May 29, 2000, p. A12.

58. Crime Statistics Following is a quadratic model relating the murder rates described in Exercise 57:

$$y = 0.1x^2 - 3x + 39 \qquad (15 \le x \le 25)$$

In 1996 the murder rate in smaller cities was approximately 22 murders per 100,000 residents each year and was decreasing at a rate of approximately 2.5 murders per 100,000 residents each year. Use the chain rule to estimate how fast the murder rate was changing for large cities. (Show how you used the chain rule in your answer.)

59. Marginal Revenue Weekly sales of rubies by Royal Ruby Retailers (RRR) is given by

$$q = -\dfrac{4p}{3} + 80$$

where p is the price per ruby (in dollars).

a. Express RRR's weekly revenue as a function of p and then calculate $\left.\dfrac{dR}{dp}\right|_{q=60}$.

b. Use the demand equation to calculate $\left.\dfrac{dp}{dq}\right|_{q=60}$.

c. Use the answers to parts (a) and (b) to find the marginal revenue at a demand level of 60 rubies/week and interpret the result.

60. Marginal Revenue Repeat Exercise 59 for a demand level of $q = 52$ rubies per week.

61. Marginal Product Paramount Electronics has an annual profit given by

$$P = -100,000 + 5000q - 0.25q^2$$

where q is the number of laptop computers it sells each year. The number of laptop computers it can make and sell each year depends on the number n of electrical engineers Paramount employs, according to the equation

$$q = 30n + 0.01n^2$$

Use the chain rule to find $\left.\dfrac{dP}{dn}\right|_{n=10}$ and interpret the result.

62. Marginal Product Refer to Exercise 61. Give a formula for the average profit per computer

$$\bar{P} = \frac{P}{q}$$

as a function of q and then determine the **marginal average product**, $d\bar{P}/dn$ at an employee level of ten engineers. Interpret the result.

63. Ecology Manatees are grazing sea mammals sometimes referred to as "sea sirens." Increasing numbers of manatees have been killed by boats off the Florida coast, as shown in the following chart:

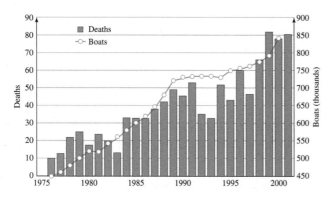

Since 1976 the number M of manatees killed by boats each year is roughly linear, with

$$M(t) = 2.48t + 6.87 \text{ manatees} \quad (1 \le t \le 26)$$

where t is the number of years since 1975. Over the same period, the total number B of boats registered in Florida has also been increasing at a roughly linear rate, given by

$$B(t) = 15,700t + 444,000 \text{ boats} \quad (1 \le t \le 25)$$

Use the chain rule to give an estimate of dM/dB. What does the answer tell you about manatee deaths?

Rounded regression model. Sources: Florida Department of Highway Safety & Motor Vehicles, Florida Marine Institute/*New York Times,* February 12, 2002, p. F4.

64. Ecology Refer to Exercise 63. If we use only the data from 1990 on, we obtain the following linear models:

$$M(t) = 3.87t + 35.0 \text{ manatees} \quad (0 \le t \le 11)$$

$$B(t) = 9020t + 712,000 \text{ boats} \quad (0 \le t \le 10)$$

where t is the number of years since 1990. Repeat Exercise 63 using these models.

Rounded regression model. Sources: Florida Department of Highway Safety & Motor Vehicles, Florida Marine Institute/*New York Times,* February 12, 2002, p. F4.

65. Pollution An offshore oil well is leaking oil and creating a circular oil slick. If the radius of the slick is growing at a rate of 2 miles/hour, find the rate at which the area is increasing when the radius is 3 miles. (The area of a disc of radius r is $A = \pi r^2$.)

66. Mold A mold culture in a dorm refrigerator is circular and growing. The radius is growing at a rate of 0.3 centimeter/day. How fast is the area growing when the culture is 4 centimeters in radius? (The area of a disc of radius r is $A = \pi r^2$.)

67. Budget Overruns The Pentagon is planning to build a new, spherical satellite. As is typical in these cases, the specifications keep changing so that the size of the satellite keeps growing. In fact, the radius of the planned satellite is growing 0.5 foot/week. Its cost will be \$1000/cubic foot. At the point when the plans call for a satellite 10 feet in radius, how fast is the cost growing? (The volume of a solid sphere of radius r is $V = \frac{4}{3}\pi r^3$.)

68. Soap Bubbles The soap bubble I am blowing has a radius that is growing at a rate of 4 centimeters/second. How fast is the surface area growing when the radius is 10 centimeters? (The surface area of a sphere of radius r is $S = 4\pi r^2$.)

69. Revenue Growth The demand for the Cyberpunk II arcade video game is modeled by the logistic curve

$$q(t) = \frac{10,000}{1 + 0.5e^{-0.4t}}$$

where $q(t)$ is the total number of units sold t months after its introduction.

a. Use technology to estimate $q'(4)$.

b. Assume that the manufacturers of Cyberpunk II sell each unit for \$800. What is the company's marginal revenue dR/dq?

c. Use the chain rule to estimate the rate at which revenue is growing 4 months after the introduction of the video game.

70. Information Highway The amount of information transmitted each month on the Internet for the years from 1988 to 1994 can be modeled by the equation

$$q(t) = \frac{2e^{0.69t}}{3 + 1.5e^{-0.4t}}$$

where q is the amount of information transmitted each month in billions of data packets and t is the number of years since the start of 1988.

a. Use technology to estimate $q'(2)$.

b. Assume that it costs \$5 to transmit 1 million packets of data. What is the marginal cost $C'(q)$?

c. How fast was the cost increasing at the start of 1990?

Source: This is the authors' model, based on figures published in the *New York Times,* November 3, 1993.

Money Stock Exercises 71–74 are based on the following demand function for money (taken from a question on the GRE economics test):

$$M_d = 2 \times y^{0.6} \times r^{-0.3} \times p$$

where M_d = demand for nominal money balances (money stock)

y = real income

r = an index of interest rates

p = an index of prices

These exercises also use the idea of **percentage rate of growth**:

$$\text{percentage rate of growth of } M = \frac{\text{rate of growth of } M}{M} = \frac{\dfrac{dM}{dt}}{M}.$$

71. If the interest rate and price level are to remain constant while real income grows at 5 percent per year, the money stock must grow at what percent per year?

72. If real income and price level are to remain constant while the interest rate grows at 5 percent per year, the money stock must change by what percent per year?

73. If the interest rate is to remain constant while real income grows at 5 percent per year and the price level rises at 5 percent per year, the money stock must grow at what percent per year?

74. If real income grows by 5 percent per year, the interest rate grows by 2 percent per year, and the price level drops by 3 percent per year, the money stock must change by what percent per year?

COMMUNICATION AND REASONING EXERCISES

75. Complete the following: The derivative of 1 over a glob is -1 over _____.

76. Complete the following: The derivative of the square root of a glob is 1 over _____.

77. What is wrong with the following?

$$\frac{d}{dx}\left[(3x^3 - x)^3\right] = 3(9x^2 - 1)^2 \qquad \text{✗ WRONG}$$

78. What is wrong with the following?

$$\frac{d}{dx}\left[\left(\frac{3x^2 - 1}{2x - 2}\right)^3\right] = 3\left(\frac{3x^2 - 1}{2x - 2}\right)^2 \frac{6x}{2} \qquad \text{✗ WRONG}$$

79. Formulate a simple procedure for deciding whether to apply first the chain rule, the product rule, or the quotient rule when finding the derivative of a function.

80. Give an example of a function f with the property that calculating $f'(x)$ requires use of the following rules in the given order: (1) the chain rule, (2) the quotient rule, and (3) the chain rule.

81. Give an example of a function f with the property that calculating $f'(x)$ requires use of the chain rule five times in succession.

82. What can you say about composites of linear functions?

4.3   *Derivatives of Logarithmic and Exponential Functions*

At this point, we know how to take the derivative of any algebraic expression in x (involving powers, radicals, and so on). We now turn to the derivatives of logarithmic and exponential functions.

Derivative of the Natural Logarithm

$$\frac{d}{dx}[\ln x] = \frac{1}{x}$$

Recall that $\ln x = \log_e x$.

Quick Examples

1. $\dfrac{d}{dx}[3 \ln x] = 3 \cdot \dfrac{1}{x} = \dfrac{3}{x}$ Derivative of a constant times a function

2. $\dfrac{d}{dx}[x \ln x] = 1 \cdot \ln x + x \cdot \dfrac{1}{x}$ Product rule, since $x \ln x$ is a product

$\qquad\qquad = \ln x + 1$

The above simple formula works only for the natural logarithm (the logarithm with base e). For logarithms with bases other than e, we have the following.

Derivative of the Logarithm with Base b

$$\frac{d}{dx}[\log_b x] = \frac{1}{x \ln b}$$ Notice that if $b = e$ we get the same formula as above.

Quick Examples

1. $\dfrac{d}{dx}[\log_3 x] = \dfrac{1}{x \ln 3} \approx \dfrac{1}{1.0986x}$

2. $\dfrac{d}{dx}[\log_2(x^4)] = \dfrac{d}{dx}[4 \log_2 x]$ We used the logarithm identity $\log_b(x^r) = r \log_b x.$

    $= 4 \cdot \dfrac{1}{x \ln 2} \approx \dfrac{4}{0.6931x}$

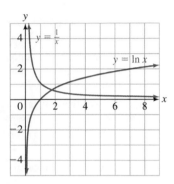

Figure 1

Question Where do these formulas come from?

Answer We will see at the end of this section. For now, let's look at the graphs of $y = \ln x$ and $y = 1/x$ to see whether it is *reasonable* that the derivative of $\ln x$ is $1/x$.

We refer to Figure 1 and notice that if x is close to zero, then the graph of $\ln x$ is rising steeply, so the derivative should be a large positive number. And indeed, $1/x$ is positive and large. As x increases, the graph of $\ln x$ becomes less steep, although it continues to rise. Therefore, the derivative (which gives the slope of the tangent line) should become smaller but remain positive, and this is exactly what happens to $1/x$. So $1/x$ at least behaves the way the derivative of $\ln x$ is supposed to behave.

If we were to take the derivative of the natural logarithm of a *quantity* (a function of x), rather than just x, we would need to use the chain rule, as follows.

Derivatives of Logarithms of Functions

Original Rule	Generalized Rule	In Words
$\dfrac{d}{dx}[\ln x] = \dfrac{1}{x}$	$\dfrac{d}{dx}[\ln u] = \dfrac{1}{u}\dfrac{du}{dx}$	The derivative of the natural logarithm of a quantity is 1 over that quantity, times the derivative of that quantity.
$\dfrac{d}{dx}[\log_b x] = \dfrac{1}{x \ln b}$	$\dfrac{d}{dx}[\log_b u] = \dfrac{1}{u \ln b}\dfrac{du}{dx}$	The derivative of the log to base b of a quantity is 1 over the product of $\ln b$ and that quantity, times the derivative of that quantity.

Quick Examples

1. $\dfrac{d}{dx}[\ln(x^2 + 1)] = \dfrac{1}{x^2 + 1}\dfrac{d}{dx}[x^2 + 1]$ $u = x^2 + 1$ (see note*)

    $= \dfrac{1}{x^2 + 1}(2x) = \dfrac{2x}{x^2 + 1}$

2. $\dfrac{d}{dx}[\log_2(x^3 + x)] = \dfrac{1}{(x^3 + x)\ln 2}\dfrac{d}{dx}[x^3 + x]$ $u = x^3 + x$

    $= \dfrac{1}{(x^3 + x)\ln 2}(3x^2 + 1) = \dfrac{3x^2 + 1}{(x^3 + x)\ln 2}$

---

*If we were to evaluate $\ln(x^2 + 1)$, the last operation we would perform is to take the natural logarithm of a quantity. Thus, the calculation thought experiment tells us that we are dealing with $\ln$ *of a quantity*, and so we need the generalized logarithm rule as stated above.

Example 1 • Derivative of a Logarithmic Function

Find $\dfrac{d}{dx}[\ln\sqrt{x+1}]$.

Solution The calculation thought experiment tells us that we have the natural logarithm of a quantity, so

$$\frac{d}{dx}[\ln\sqrt{x+1}] = \frac{1}{\sqrt{x+1}}\frac{d}{dx}[\sqrt{x+1}] \qquad \frac{d}{dx}[\ln u] = \frac{1}{u}\frac{du}{dx}$$

$$= \frac{1}{\sqrt{x+1}} \cdot \frac{1}{2\sqrt{x+1}} \qquad \frac{d}{dx}[\sqrt{u}] = \frac{1}{2\sqrt{u}}\frac{du}{dx}$$

$$= \frac{1}{2(x+1)}$$

✴ **Before we go on . . .** What happened to the square root? As with many problems involving logarithms, we could have done this one differently and with less bother if we had simplified the expression $\ln\sqrt{x+1}$ using the properties of logarithms *before* differentiating. Doing this, we get

$$\ln\sqrt{x+1} = \ln(x+1)^{1/2} = \frac{1}{2}\ln(x+1) \qquad \text{Simplify the logarithm first.}$$

Thus,

$$\frac{d}{dx}[\ln\sqrt{x+1}] = \frac{d}{dx}\left[\frac{1}{2}\ln(x+1)\right]$$

$$= \frac{1}{2}\left(\frac{1}{x+1}\right) \cdot 1 = \frac{1}{2(x+1)}$$

the same answer as above.

Example 2 • Derivative of a Logarithmic Function

Find $\dfrac{d}{dx}(\ln[(1+x)(2-x)])$.

Solution This time we simplify the expression $\ln[(1+x)(2-x)]$ before taking the derivative:

$$\ln[(1+x)(2-x)] = \ln(1+x) + \ln(2-x) \qquad \text{Simplify the logarithm first.}$$

Thus,

$$\frac{d}{dx}(\ln[(1+x)(2-x)]) = \frac{d}{dx}[\ln(1+x)] + \frac{d}{dx}[\ln(2-x)]$$

$$= \frac{1}{1+x} - \frac{1}{2-x} \qquad \text{Because } \frac{d}{dx}[\ln(2-x)] = -\frac{1}{2-x}$$

$$= \frac{1-2x}{(1+x)(2-x)}$$

✴ **Before we go on . . .** For practice, try doing this example without simplifying first. What other differentiation rule do you need to use?

Example 3 • Logarithm of an Absolute Value

Find $\dfrac{d}{dx}[\ln |x|]$.

Solution Before we start, you might ask why we are considering the natural log of the absolute value of x to begin with. The reason is this: $\ln x$ is defined only for positive values of x, so its domain is the set of positive real numbers. The domain of $\ln |x|$, on the other hand, is the set of *all* nonzero real numbers. For example, $\ln |-2| = \ln 2 \approx 0.6931$. For this reason, it often turns out to be more useful than the ordinary logarithmic function.

Now we'll get to work. The calculation thought experiment tells us that $\ln |x|$ is the natural logarithm of a quantity, so we use the chain rule:

$$\frac{d}{dx}[\ln |x|] = \frac{1}{|x|}\frac{d}{dx}|x| \qquad\qquad u = |x|$$

$$= \frac{1}{|x|}\frac{|x|}{x} \qquad\qquad \text{Recall that } \frac{d}{dx}[|x|] = \frac{|x|}{x}.$$

$$= \frac{1}{x}$$

✸ **Before we go on . . .** Figure 2a shows the graphs of $y = \ln |x|$ and $y = 1/x$. Figure 2(b) shows the graphs of $y = \ln |x|$ and $y = 1/|x|$. You should be able to see from these graphs why the derivative of $\ln |x|$ is $1/x$ and not $1/|x|$.

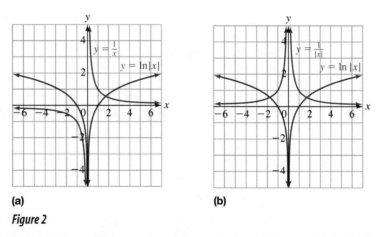

(a)   (b)

Figure 2

This last example, in conjunction with the chain rule, gives us the following formulas.

Derivative of Logarithms of Absolute Values

Original Rule	Generalized Rule	In Words
$\dfrac{d}{dx}[\ln |x|] = \dfrac{1}{x}$	$\dfrac{d}{dx}[\ln |u|] = \dfrac{1}{u}\dfrac{du}{dx}$	The derivative of the natural logarithm of the absolute value of a quantity is 1 over that quantity, times the derivative of that quantity.
$\dfrac{d}{dx}[\log_b |x|] = \dfrac{1}{x \ln b}$	$\dfrac{d}{dx}[\log_b |u|] = \dfrac{1}{u \ln b}\dfrac{du}{dx}$	The derivative of the log to base b of the absolute value of a quantity is 1 over the product of $\ln b$ and that quantity, times the derivative of that quantity.

Quick Examples

1. $\dfrac{d}{dx}[\ln |x^2 + 1|] = \dfrac{1}{x^2 + 1}\dfrac{d}{dx}[x^2 + 1]$ $\qquad u = x^2 + 1$

   $= \dfrac{1}{x^2 + 1}(2x) = \dfrac{2x}{x^2 + 1}$

2. $\dfrac{d}{dx}[\log_2 |x^3 + x|] = \dfrac{1}{(x^3 + x)\ln 2}\dfrac{d}{dx}[x^3 + x]$ $\qquad u = x^3 + x$

   $= \dfrac{1}{(x^3 + x)\ln 2}(3x^2 + 1) = \dfrac{3x^2 + 1}{(x^3 + x)\ln 2}$

In other words, when taking the derivative of the logarithm of the absolute value of a quantity, we can simply ignore the absolute value!

We now turn to the derivatives of *exponential* functions—that is, functions of the form $f(x) = a^x$. We begin by showing how *not* to differentiate them.

Caution The derivative of a^x is *not* xa^{x-1}. The power rule applies only to *constant* exponents. In this case the exponent is decidedly *not* constant, and so the power rule does not apply.

The following shows the correct way of differentiating a^x. (We justify the formula, as well as the formula for the derivative of $\log_b x$, at the end of the section.)

Derivatives of Exponential Functions
Derivative of e^x

$$\frac{d}{dx}[e^x] = e^x$$

Thus, e^x has the amazing property that its derivative is itself!* For bases other than e, we have the following generalization.

Derivative of b^x
If b is any positive number, then

$$\frac{d}{dx}[b^x] = b^x \ln b$$

Note that if $b = e$, we obtain the previous formula.

*There is another—very simple—function that is its own derivative. What is it?

Quick Examples

1. $\dfrac{d}{dx}[3^x] = 3^x \ln 3$

2. $\dfrac{d}{dx}[3e^x] = 3\dfrac{d}{dx}[e^x] = 3e^x$

3. $\dfrac{d}{dx}\left[\dfrac{e^x}{x}\right] = \dfrac{e^x x - e^x(1)}{x^2}$ Quotient rule

   $= \dfrac{e^x(x-1)}{x^2}$

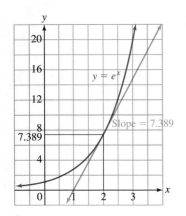

Figure 3

Question   Why should I believe this? How can the derivative of the function e^x be the same as the original function?

Answer   At the end of this section, we will give a derivation of this formula. For now let's consider whether it is *plausible* that e^x equals its own derivative. For a function f to equal its own derivative, its value at every point x must be the same as the slope of the tangent at that point. For instance, if $f(2) = 5$, then the slope of the tangent at the point where $x = 2$ should also be 5. To check whether $f(x) = e^x$ has this property, let's take a look at the graph of $f(x) = e^x$ (Figure 3). Notice that as x increases, both the y coordinate [the value of $f(x)$] and the slope [the value of $f'(x)$] are increasing. More specifically, if we look at any particular value of x—say, $x = 2$—we notice that $f(x)$ and the slope of the tangent at x appear to be the same. [For instance, $f(2) \approx 7.4$, and the slope at $x = 2$ is also approximately 7.4.]

If we were to take the derivative of e raised to a *quantity*, not just x, we would need to use the chain rule, as follows.

Derivatives of Exponentials of Functions

Original Rule	Generalized Rule	In Words
$\dfrac{d}{dx}[e^x] = e^x$	$\dfrac{d}{dx}[e^u] = e^u \dfrac{du}{dx}$	The derivative of e raised to a quantity is e raised to that quantity, times the derivative of that quantity.
$\dfrac{d}{dx}[b^x] = b^x \ln b$	$\dfrac{d}{dx}[b^u] = b^u \ln b \dfrac{du}{dx}$	The derivative of b raised to a quantity is b raised to that quantity, times $\ln b$, times the derivative of that quantity.

Quick Examples

1. $\dfrac{d}{dx}[e^{x^2+1}] = e^{x^2+1}\dfrac{d}{dx}[x^2+1]$ $u = x^2 + 1$ (see the note*)

   $= e^{x^2+1}(2x) = 2x\, e^{x^2+1}$

2. $\dfrac{d}{dx}[2^{3x}] = 2^{3x} \ln 2 \dfrac{d}{dx}[3x]$ $u = 3x$

   $= 2^{3x}(\ln 2)(3) = (3 \ln 2)2^{3x}$

*The calculation thought experiment tells us that we have e raised to a quantity.

APPLICATIONS

Example 4 • Epidemics

In the early stages of the AIDS epidemic during the 1980s, the number of cases in the United States was increasing by about 50% every 6 months. By the start of 1983, there were approximately 1600 AIDS cases in the United States.[4] Had this trend continued, how many new cases per year would have been occurring by the start of 1993?

Solution To find the answer, we must first model this exponential growth using the methods of Chapter 2. Referring to Example 3 in Section 2.2, we find that t years after the start of 1983 the number of cases is

$$A = 1600(2.25^t)$$

We are asking for the number of new cases each year. In other words, we want the rate of change, dA/dt:

$$\frac{dA}{dt} = 1600(2.25)^t \ln 2.25 \text{ cases per year}$$

At the start of 1993, $t = 10$, so the number of new cases each year is

$$\left.\frac{dA}{dt}\right|_{t=10} = 1600(2.25)^{10} \ln 2.25 \approx 4{,}300{,}000 \text{ cases per year}$$

✳ **Before we go on . . .** The reason this figure is so large is that we assumed that exponential growth—the 50% increase every 6 months—would continue. A more realistic model for the spread of a disease is the logistic model. (See Section 2.4, as well as Example 5.)

Example 5 • Sales Growth

The sales of the Cyberpunk II arcade video game can be modeled by the logistic curve

$$q(t) = \frac{10{,}000}{1 + 0.5e^{-0.4t}}$$

where $q(t)$ is the total number of units sold t months after its introduction. How fast is the game selling 2 years after its introduction?

Solution We are asked for $q'(24)$. We can find the derivative of $q(t)$ using the quotient rule, or we can first write

$$q(t) = 10{,}000(1 + 0.5e^{-0.4t})^{-1}$$

and then use the generalized power rule:

$$q'(t) = -10{,}000(1 + 0.5e^{-0.4t})^{-2}(0.5e^{-0.4t})(-0.4)$$

$$= \frac{2000e^{-0.4t}}{(1 + 0.5e^{-0.4t})^2}$$

---

[4]Data based on regression of 1982–1986 figures. SOURCE: Centers for Disease Control and Prevention, *HIV/AIDS Surveillance Report, 2000,* 12 (no. 2).

Thus,

$$q'(24) = \frac{2000e^{-0.4(24)}}{(1 + 0.5e^{-0.4(24)})^2} \approx 0.135 \text{ units per month}$$

So, after 2 years, sales are quite slow.

Graphing Calculator
We can check this answer graphically. If we plot the total sales curve for $0 \le t \le 30$ and $6000 \le q \le 10{,}000$, we get the graph shown in Figure 4. Notice that total sales level off at about 10,000 units.* We computed $q'(24)$, which is the slope of the curve at the point with t coordinate 24. If we zoom in to the portion of the curve near $t = 24$, we obtain the graph shown in Figure 5.

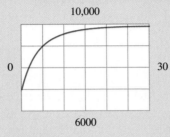

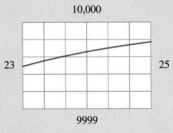

Figure 4   *Figure 5*

The curve is almost linear in this range. If we use the two endpoints of this segment of the curve, $(23, 9999.4948)$ and $(25, 9999.7730)$, we can approximate the derivative as

$$\frac{9999.7730 - 9999.4948}{25 - 23} = 0.1391$$

which is accurate to two decimal places.

*We can also say this using limits: $\lim_{t \to +\infty} q(t) = 10{,}000$.

Let's now explain where the formulas for the derivatives of $\ln x$ and e^x come from. We start with $\ln x$.

Question Why does $\dfrac{d}{dx} [\ln x] = \dfrac{1}{x}$?

Answer To compute $d/dx [\ln x]$, we need to use the definition of the derivative. We also use properties of the logarithm to help evaluate the limit.

$$\frac{d}{dx} [\ln x] = \lim_{h \to 0} \frac{\ln(x + h) - \ln x}{h} \qquad \text{Definition of the derivative}$$

$$= \lim_{h \to 0} \frac{1}{h} [\ln(x + h) - \ln x] \qquad \text{Algebra}$$

$$= \lim_{h \to 0} \frac{1}{h} \ln\left(\frac{x + h}{x}\right) \qquad \text{Properties of the logarithm}$$

$$= \lim_{h \to 0} \frac{1}{h} \ln\left(1 + \frac{h}{x}\right) \qquad \text{Algebra}$$

$$= \lim_{h \to 0} \ln\left(1 + \frac{h}{x}\right)^{1/h} \qquad \text{Properties of the logarithm}$$

which we rewrite as

$$\lim_{h \to 0} \ln\left[\left(1 + \frac{1}{(x/h)}\right)^{x/h}\right]^{1/x}$$

As $h \to 0^+$, the quantity x/h is getting large and positive, and so the quantity in brackets is approaching e (see the definition of e in Section 2.2), which leaves us with

$$\ln[e]^{1/x} = \frac{1}{x}\ln e = \frac{1}{x}$$

which is the derivative we are after.[5] What about the limit as $h \to 0^-$? We will glide over that case and leave it for the interested reader to pursue.[6]

The rule for the derivative of $\log_b x$ follows from the fact that $\log_b x = \ln x/\ln b$.

Question Why does $\dfrac{d}{dx}[e^x] = e^x$?

Answer To find the derivative of e^x, we use a shortcut.[7] Write $g(x) = e^x$. Then

$$\ln g(x) = x$$

Take the derivative of both sides of this equation to get

$$\frac{g'(x)}{g(x)} = 1 \quad \text{or} \quad g'(x) = g(x) = e^x$$

In other words, the exponential function with base e is its own derivative. The rule for exponential functions with other bases follows from the equality $a^x = e^{x \ln a}$ (Why?) and the chain rule. (Try it.)

4.3 EXERCISES

In Exercises 1–14, mentally find the derivatives of the functions.

1. $f(x) = \ln(x - 1)$
2. $f(x) = \ln(x + 3)$
3. $f(x) = \log_2 x$
4. $f(x) = \log_3 x$
5. $g(x) = \ln|x^2 + 3|$
6. $g(x) = \ln|2x - 4|$
7. $h(x) = e^{x+3}$
8. $h(x) = e^{x^2}$
9. $f(x) = e^{-x}$
10. $f(x) = e^{1-x}$
11. $g(x) = 4^x$
12. $g(x) = 5^x$
13. $h(x) = 2^{x^2-1}$
14. $h(x) = 3^{x^2-x}$

In Exercises 15–70, find the derivatives of the functions.

15. $f(x) = x \ln x$
16. $f(x) = 3 \ln x$
17. $f(x) = (x^2 + 1)\ln x$
18. $f(x) = (4x^2 - x)\ln x$
19. $f(x) = (x^2 + 1)^5 \ln x$
20. $f(x) = (x + 1)^{0.5} \ln x$
21. $g(x) = \ln|3x - 1|$
22. $g(x) = \ln|5 - 9x|$

23. $g(x) = \ln|2x^2 + 1|$
24. $g(x) = \ln|x^2 - x|$
25. $g(x) = \ln(x^2 - 2.1x^{0.3})$
26. $g(x) = \ln(x - 3.1x^{-1})$
27. $h(x) = \ln[(-2x + 1)(x + 1)]$
28. $h(x) = \ln[(3x + 1)(-x + 1)]$
29. $h(x) = \ln\left[\dfrac{3x + 1}{4x - 2}\right]$
30. $h(x) = \ln\left[\dfrac{9x}{4x - 2}\right]$
31. $r(x) = \ln\left|\dfrac{(x + 1)(x - 3)}{-2x - 9}\right|$
32. $r(x) = \ln\left|\dfrac{-x + 1}{(3x - 4)(x - 9)}\right|$
33. $s(x) = \ln(4x - 2)^{1.3}$
34. $s(x) = \ln(x - 8)^{-2}$
35. $s(x) = \ln\left|\dfrac{(x + 1)^2}{(3x - 4)^3(x - 9)}\right|$

---

[5]We actually used the fact that the logarithmic function is continuous when we took the limit.

[6]Here is an outline of the argument for negative h. Since x must be positive for $\ln x$ to be defined, we find that $x/h \to -\infty$ as $h \to 0^-$, and so we must consider the quantity $(1 + 1/m)^m$ for large *negative m*. It turns out that the limit is still e (check it numerically!), and so the computation above still works.

[7]This shortcut is an example of a technique called *logarithmic differentiation*, which is occasionally useful. We will see it again in Section 4.4.

36. $s(x) = \ln \left| \dfrac{(x+1)^2(x-3)^4}{2x+9} \right|$

37. $h(x) = \log_2(x+1)$

38. $h(x) = \log_3(x^2 + x)$

39. $r(t) = \log_3\left(t + \dfrac{1}{t}\right)$

40. $r(t) = \log_3(t + \sqrt{t})$

41. $f(x) = (\ln |x|)^2$

42. $f(x) = \dfrac{1}{\ln |x|}$

43. $r(x) = \ln(x^2) - [\ln(x-1)]^2$

44. $r(x) = [\ln(x^2)]^2$

45. $f(x) = xe^x$

46. $f(x) = 2e^x - x^2 e^x$

47. $r(x) = \ln(x+1) + 3x^3 e^x$

48. $r(x) = \ln |x + e^x|$

49. $f(x) = e^x \ln |x|$

50. $f(x) = e^x \log_2 |x|$

51. $f(x) = e^{2x+1}$

52. $f(x) = e^{4x-5}$

53. $h(x) = e^{x^2 - x + 1}$

54. $h(x) = e^{2x^2 - x + 1/x}$

55. $s(x) = x^2 e^{2x-1}$

56. $s(x) = \dfrac{e^{4x-1}}{x^3 - 1}$

57. $r(x) = (e^{2x-1})^2$

58. $r(x) = (e^{2x^2})^3$

59. $g(x) = \dfrac{e^x + e^{-x}}{e^x - e^{-x}}$

60. $g(x) = \dfrac{1}{e^x + e^{-x}}$

61. $g(x) = e^{3x-1}e^{x-2}e^x$

62. $g(x) = e^{-x+3}e^{2x-1}e^{-x+11}$

63. $f(x) = \dfrac{1}{x \ln x}$

64. $f(x) = \dfrac{e^{-x}}{xe^x}$

65. $f(x) = [\ln(e^x)]^2 - \ln[(e^x)^2]$

66. $f(x) = e^{\ln x} - e^{2\ln(x^2)}$

67. $f(x) = \ln|\ln x|$

68. $f(x) = \ln|\ln|\ln x||$

69. $s(x) = \ln\sqrt{\ln x}$

70. $s(x) = \sqrt{\ln(\ln x)}$

In Exercises 71–76, find the equations of the straight lines.

Use graphing technology to check your answers by plotting the given curve together with the tangent line.

71. Tangent to $y = e^x \log_2 x$ at the point $(1, 0)$

72. Tangent to $y = e^x + e^{-x}$ at the point $(0, 2)$

73. Tangent to $y = \ln\sqrt{2x + 1}$ at the point where $x = 0$

74. Tangent to $y = \ln\sqrt{2x^2 + 1}$ at the point where $x = 1$

75. At right angles to $y = e^{x^2}$ at the point where $x = 1$

76. At right angles to $y = \log_2(3x + 1)$ at the point where $x = 1$

APPLICATIONS

77. Investments If \$10,000 is invested in a savings account offering 4% per year, compounded continuously, how fast is the balance growing after 3 years?

78. Investments If \$20,000 is invested in a savings account offering 3.5% per year, compounded continuously, how fast is the balance growing after 3 years?

79. Investments If \$10,000 is invested in a savings account offering 4% per year, compounded semiannually, how fast is the balance growing after 3 years?

80. Investments If \$20,000 is invested in a savings account offering 3.5% per year, compounded semiannually, how fast is the balance growing after 3 years?

81. Population Growth The population of Lower Anchovia was 4 million at the start of 1995 and was doubling every 10 years. How fast was it growing each year at the start of 1995? (Round your answer to three significant digits.)

82. Population Growth The population of Upper Anchovia was 3 million at the start of 1996 and doubling every 7 years. How fast was it growing each year at the start of 1996? (Round your answer to three significant digits.)

83. Radioactive Decay Plutonium-239 has a half-life of 24,400 years. How fast is a lump of 10 grams decaying after 100 years?

84. Radioactive Decay Carbon-14 has a half-life of 5730 years. How fast is a lump of 20 grams decaying after 100 years?

85. Life Span in Ancient Rome The percentage $P(t)$ of people surviving to age t years in ancient Rome can be approximated by

$$P(t) = 92e^{-0.0277t}$$

Calculate $P'(22)$ and explain what the result indicates.

Source: Based on graphical data in Marvin Minsky, "Will Robots Inherit the Earth?" *Scientific American* (October 1994): 109–113.

86. Communication among Bees The audible signals honeybees use to communicate are in the frequency range 0–500 hertz and are generated by their wings. The speed with which a honeybee's wings must move the air to generate these signals depends on the frequency of the signal, and this relationship can be approximated by

$$V(f) = 95.6e^{0.0049f}$$

where $V(f)$ is the speed of the air near the honeybee's wings in millimeters per second and f is the frequency of the communication signal in hertz. Calculate $V'(200)$ and explain what the result indicates.

Source: Based on graphical data in the article Wolfgang H. Kirchner and William F. Towne, "The Sensory Basis of the Honeybee's Dance Language," *Scientific American* (June 1994): 74–80.

87. SAT Scores by Income The following chart shows U.S. average verbal SAT scores as a function of parents' income level:

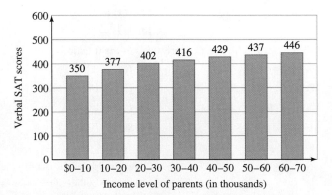

Source: The College Board/*New York Times*, March 5, 1995, p. E16.

a. The data can best be modeled by which of the following?

(A) $S(x) = 470 - 136e^{-0.000\ 026\ 4x}$

(B) $S(x) = 136e^{-0.000\ 026\ 4x}$

(C) $S(x) = 355(1.000\ 004^x)$

(D) $S(x) = 470 - 355(1.000\ 004^x)$

[$S(x)$ is the average verbal SAT score of students whose parents earn x per year.]

b. Use $S'(x)$ to predict how a student's verbal SAT score is affected by a $1000 increase in parents' income for a student whose parents earn $45,000.

c. Does $S'(x)$ increase or decrease as x increases? Interpret your answer.

88. SAT Scores by Income The following chart shows U.S. average math SAT scores as a function of parents' income level:

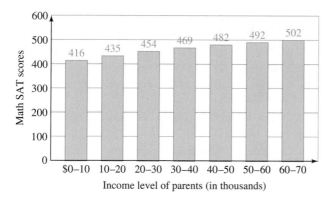

SOURCE: The College Board/*New York Times*, March 5, 1995, p. E16.

a. The data can best be modeled by which of the following?

(A) $S(x) = 535 - 415(1.000\ 003^x)$

(B) $S(x) = 535 - 136e^{0.000\ 021\ 3x}$

(C) $S(x) = 535 - 136e^{-0.000\ 021\ 3x}$

(D) $S(x) = 415(1.000\ 003^x)$

[$S(x)$ is the average math SAT score of students whose parents earn x per year.]

b. Use $S'(x)$ to predict how a student's math SAT score is affected by a $1000 increase in parents' income for a student whose parents earn $45,000.

c. Does $S'(x)$ increase or decrease as x increases? Interpret your answer.

89. Cell Phone Revenues The number of cell phone subscribers in China for the period 2000–2005 was projected to follow the equation

$$N(t) = 39t + 68 \text{ million subscribers}$$

in year t ($t = 0$ represents 2000). The average annual revenue per cell phone user was $350 in 2000. Assuming that, due to competition, the revenue per cell phone user decreases continuously at an annual rate of 10%, give a formula for the annual revenue in year t. Hence, project the annual revenue and its rate of change in 2002. Round all answers to the nearest billion dollars or billion dollars per year.

Based on a regression of projected figures (coefficients are rounded). SOURCE: Intrinsic Technology/*New York Times*, November 24, 2000, p. C1.

90. Cell Phone Revenues The annual revenue for cell phone use in China for the period 2000–2005 was projected to follow the equation

$$R(t) = 14t + 24 \text{ billion dollars}$$

in year t ($t = 0$ represents 2000). At the same time, there were approximately 68 million subscribers in 2000. Assuming that the number of subscribers increases continuously at an annual rate of 10%, give a formula for the annual revenue per subscriber in year t. Hence, project to the nearest dollar the annual revenue per subscriber and its rate of change in 2002. (Be careful with units!)

Not allowing for discounting due to increased competition. SOURCE: Intrinsic Technology/*New York Times*, November 24, 2000, p. C1.

91. Epidemics The flu epidemic described in Example 1 in Section 2.4 followed the curve

$$P = \frac{150,000,000}{1 + 14,999e^{-0.3466t}}$$

where P is the number of people infected and t is the number of weeks after the start of the epidemic. How fast is the epidemic growing (that is, how many new cases are there each week) after 20 weeks? After 30 weeks? After 40 weeks? (Round your answers to three significant digits.)

92. Epidemics Another epidemic follows the curve

$$P = \frac{200,000,000}{1 + 20,000e^{-0.549t}}$$

where t is in years. How fast is the epidemic growing after 10 years? After 20 years? After 30 years? (Round your answers to three significant digits.)

93. Big Brother The following chart shows the number of wiretaps authorized each year by U.S. courts from 1990 to 2000 ($t = 0$ represents 1990):

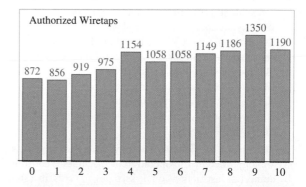

SOURCE: *2000 Wiretap Report*, Administrative Office of the U.S. Courts, http://www.epic.org/privacy/wiretap/stats/2000_report/default.html

These data can be approximated with the logistic model

$$W(t) = \frac{1500}{1 + 0.77(1.16)^{-t}} \qquad (0 \le t \le 10)$$

where $W(t)$ is the number of authorized wiretaps in year $1990 + t$.

a. Calculate $W'(t)$ and use it to approximate $W'(6)$. To how many significant digits should we round the answer? Why? What does the answer tell you?

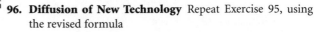

b. Graph the function $W'(t)$. Based on the graph, the number of wiretaps authorized each year (choose one)
   (A) increased at a decreasing rate
   (B) decreased at an increasing rate
   (C) increased at an increasing rate
   (D) decreased at a decreasing rate
   from 1990 to 2000.

94. Big Brother The following chart shows the number of wiretaps authorized each year by U.S. federal courts from 1990 to 2000 ($t = 0$ represents 1990).

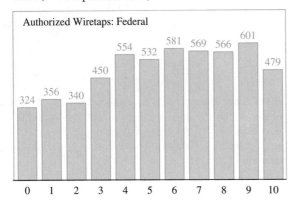

Authorized Wiretaps: Federal

324	0
356	1
340	2
450	3
554	4
532	5
581	6
569	7
566	8
601	9
479	10

SOURCE: *2000 Wiretap Report,* Administrative Office of the U.S. Courts, http://www.epic.org/privacy/wiretap/stats/2000_report/default.html

These data can be approximated with the logistic model

$$W(t) = \frac{600}{1 + 1.00(1.62)^{-t}} \qquad (0 \le t \le 10)$$

where $W(t)$ is the number of authorized wiretaps in year $1990 + t$.

a. Calculate $W'(t)$ and use it to approximate $W'(6)$. To how many significant digits should we round the answer? Why? What does the answer tell you?

b. Graph the function $W'(t)$. Based on the graph, the number of wiretaps authorized each year (choose one)
   (A) increased at an increasing rate
   (B) decreased at a decreasing rate
   (C) increased at a decreasing rate
   (D) decreased at an increasing rate
   from 1990 to 2000.

95. Diffusion of New Technology Numeric control is a technology whereby the operation of machines is controlled by numerical instructions on disks, tapes, or cards. In a study, E. Mansfield and associates modeled the growth of this technology using the equation

$$p(t) = \frac{0.80}{1 + e^{4.46 - 0.477t}}$$

where $p(t)$ is the fraction of firms using numeric control in year t.

a. Graph this function for $0 \le t \le 20$ and estimate $p'(10)$ graphically. Interpret the result.

b. Use your graph to estimate $\lim_{t \to +\infty} p(t)$ and interpret the result.

c. Compute $p'(t)$, graph it, and again find $p'(10)$.

d. Use your graph to estimate $\lim_{t \to +\infty} p'(t)$ and interpret the result.

SOURCE: "The Diffusion of a Major Manufacturing Innovation," in *Research and Innovation in the Modern Corporation* (New York: Norton, 1971): 186–205.

96. Diffusion of New Technology Repeat Exercise 95, using the revised formula

$$p(t) = \frac{0.90e^{-0.1t}}{1 + e^{4.50 - 0.477t}}$$

which takes into account that in the long run this new technology will eventually become outmoded and will be replaced by a newer technology. Draw your graphs using the range $0 \le t \le 40$.

COMMUNICATION AND REASONING EXERCISES

97. Complete the following: The derivative of e raised to a glob is _____.

98. Complete the following: The derivative of the natural logarithm of a glob is _____.

99. Complete the following: The derivative of 2 raised to a glob is _____.

100. Complete the following: The derivative of the base 2 logarithm of a glob is _____.

101. What is wrong with the following?
$$\frac{d}{dx}[3^{2x}] = (2x)3^{2x-1} \qquad \text{X \ WRONG}$$

102. What is wrong with the following?
$$\frac{d}{dx}[\ln(3x^2 - 1)] = \frac{1}{6x} \qquad \text{X \ WRONG}$$

103. The number N of music downloads on campus is growing exponentially with time. Can $N'(t)$ grow linearly with time? Explain.

104. The number N of graphing calculators sold on campus is decaying exponentially with time. Can $N'(t)$ grow with time? Explain.

Exercises 105–108 are based on the following definition: The **percent rate of change,** or **fractional rate of change,** of a function is defined to be the ratio $f'(x)/f(x)$. (It is customary to express this as a percentage when speaking about percent rate of change.)

105. Show that the fractional rate of change of the exponential function e^{kx} is equal to k, which is often called its **fractional growth rate.**

106. Show that the fractional rate of change of $f(x)$ is the rate of change of $\ln[f(x)]$.

107. Let $A(t)$ represent a quantity growing exponentially. Show that the percent rate of change, $A'(t)/A(t)$, is constant.

108. Let $A(t)$ be the amount of money in an account paying interest compounded some number of times per year. Show that the percent rate of growth, $A'(t)/A(t)$, is constant. What might this constant represent?

4.4 Implicit Differentiation

Consider the equation $y^5 + y + x = 0$, whose graph is shown in Figure 6.

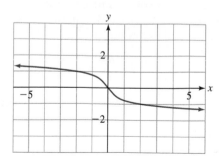

Figure 6

How did we find this graph? We did not solve for y as a function of x; that is impossible. In fact, we solved for x in terms of y to find points to plot. Nonetheless, the graph in Figure 6 is the graph of a function because it passes the vertical-line test: Every vertical line crosses the graph no more than once, so for each value of x there is no more than one corresponding value of y. Since we cannot solve for y explicitly in terms of x, we say that the equation determines y as an **implicit function** of x.

Now, suppose we want to find the slope of the tangent line to this curve at, say, the point $(2, -1)$ (which, you should check, is a point on the curve).

Question How would we find dy/dx without first solving for y?

Answer We use the chain rule and a little cleverness. We think of y as a function of x and take the derivative with respect to x of both sides of the equation:

$$y^5 + y + x = 0 \qquad \text{Original equation}$$

$$\frac{d}{dx}[y^5 + y + x] = \frac{d}{dx}[0] \qquad \text{Derivative with respect to } x \text{ of both sides}$$

$$\frac{d}{dx}[y^5] + \frac{d}{dx}[y] + \frac{d}{dx}[x] = 0 \qquad \text{Derivative rules}$$

Now we must be careful. The derivative *with respect to x* of y^5 is *not* $5y^4$. Rather, because y is a function of x, we must use the chain rule, which tells us that

$$\frac{d}{dx}[y^5] = 5y^4 \frac{dy}{dx}$$

Thus, we get

$$5y^4 \frac{dy}{dx} + \frac{dy}{dx} + 1 = 0$$

We want to find dy/dx, so we solve for it:

$$(5y^4 + 1)\frac{dy}{dx} = -1 \qquad \text{Isolate } \frac{dy}{dx} \text{ on one side.}$$

$$\frac{dy}{dx} = -\frac{1}{5y^4 + 1} \qquad \text{Divide both sides by } 5y^4 + 1.$$

Question The formula we just found for dy/dx is not a function of x because there is a y in it. Is this okay?

Answer We should not expect to obtain dy/dx as an explicit function of x if y was not an explicit function of x to begin with. The result is still useful because we can evaluate the derivative at any point on the graph. For instance, at the point $(2, -1)$ on the graph, we get

$$\frac{dy}{dx} = -\frac{1}{5y^4 + 1} = -\frac{1}{5(-1)^4 + 1} = -\frac{1}{6}$$

Thus, the slope of the tangent line to the curve $y^5 + y + x = 0$ at the point $(2, -1)$ is $-1/6$. Figure 7 shows the graph and this tangent line.

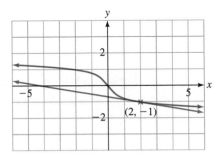

Figure 7

This procedure of differentiating an equation to find dy/dx without first solving the equation for y is called **implicit differentiation.**

An equation in x and y need not determine y as a function of x. Consider, for example, the equation

$$2x^2 + y^2 = 2$$

Solving for y yields $y = \pm\sqrt{2 - 2x^2}$. The $\pm$ sign reminds us that for some values of x there are two corresponding values for y. We can graph this equation by superimposing the graphs of

$$y = \sqrt{2 - 2x^2} \quad \text{and} \quad y = -\sqrt{2 - 2x^2}$$

The graph, an *ellipse*, is shown in Figure 8.

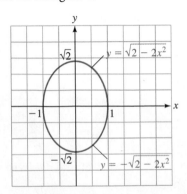

Figure 8

The graph of $y = \sqrt{2 - 2x^2}$ constitutes the top half of the ellipse, and the graph of $y = -\sqrt{2 - 2x^2}$ constitutes the bottom half.

Example 1 • Implicit Differentiation

Find the slope of the tangent line to the ellipse $2x^2 + y^2 = 2$ at the point $(1/\sqrt{2}, 1)$.

Solution Because $(1/\sqrt{2}, 1)$ is on the top half of the ellipse in Figure 8, we *could* differentiate the function $y = \sqrt{2 - 2x^2}$ to obtain the result, but it is actually easier to apply implicit differentiation to the original equation:

$$2x^2 + y^2 = 2 \qquad \text{Original equation}$$

$$\frac{d}{dx}[2x^2 + y^2] = \frac{d}{dx}[2] \qquad \text{Derivative with respect to } x \text{ of both sides}$$

$$4x + 2y\frac{dy}{dx} = 0$$

$$2y\frac{dy}{dx} = -4x$$

$$\frac{dy}{dx} = -\frac{4x}{2y} = -\frac{2x}{y}$$

To find the slope at $(1/\sqrt{2}, 1)$ we now substitute for x and y:

$$\left.\frac{dy}{dx}\right|_{(1/\sqrt{2},1)} = -\frac{2/\sqrt{2}}{1} = -\sqrt{2}$$

Thus, the slope of the tangent to the ellipse at the point $(1/\sqrt{2}, 1)$ is $-\sqrt{2} \approx -1.414$.

Example 2 • Tangent Line to an Implicit Curve

Find the equation of the tangent line to the curve $\ln y = xy$ at the point where $y = 1$.

Solution First, we use implicit differentiation to find dy/dx:

$$\frac{d}{dx}[\ln y] = \frac{d}{dx}[xy] \qquad \text{Take } \frac{d}{dx} \text{ of both sides.}$$

$$\frac{1}{y}\boxed{\frac{dy}{dx}} = (1)y + x\boxed{\frac{dy}{dx}} \qquad \text{Chain rule on left, product rule on right}$$

We have boxed the quantity dy/dx to remind us that it is the unknown. To solve for it, we bring all the terms containing dy/dx to the left-hand side and all terms not containing it to the right-hand side:

$$\frac{1}{y}\boxed{\frac{dy}{dx}} - x\boxed{\frac{dy}{dx}} = y \qquad \text{Bring the terms with } \frac{dy}{dx} \text{ to the left.}$$

$$\boxed{\frac{dy}{dx}}\left(\frac{1}{y} - x\right) = y \qquad \text{Factor out } \frac{dy}{dx}.$$

$$\boxed{\frac{dy}{dx}}\left(\frac{1 - xy}{y}\right) = y$$

$$\frac{dy}{dx} = y\left(\frac{y}{1 - xy}\right) = \frac{y^2}{1 - xy} \qquad \text{Solve for } \frac{dy}{dx}.$$

The derivative gives the slope of the tangent line, so we want to evaluate the derivative at the point where $y = 1$. However, the formula for dy/dx requires values for both x and y. We get the value of x by substituting $y = 1$ in the original equation:

$$\ln y = xy$$

$$\ln(1) = x \cdot 1$$

But $\ln(1) = 0$, and so $x = 0$ for this point. Thus,

$$\left.\frac{dy}{dx}\right|_{(0,1)} = \frac{1^2}{1 - (0)(1)} = 1$$

Therefore, the tangent line is the line through $(x, y) = (0, 1)$ with slope 1, which is

$$y = x + 1$$

 Before we go on . . . This is an example of an implicit function in which it is simply not possible to solve for y. Try it.

Sometimes, it is easiest to differentiate a complicated function of x by first taking the logarithm and then using implicit differentiation—a technique called **logarithmic differentiation.**

Example 3 • Logarithmic Differentiation

Find $\dfrac{d}{dx}\left[\dfrac{(x + 1)^{10}(x^2 + 1)^{11}}{(x^3 + 1)^{12}}\right]$ without using the product or quotient rules.

Solution Write

$$y = \frac{(x + 1)^{10}(x^2 + 1)^{11}}{(x^3 + 1)^{12}}$$

and then take the natural logarithm of both sides:

$$\ln y = \ln\left[\frac{(x + 1)^{10}(x^2 + 1)^{11}}{(x^3 + 1)^{12}}\right]$$

We can use properties of the logarithm to simplify the right-hand side:

$$\ln y = \ln(x + 1)^{10} + \ln(x^2 + 1)^{11} - \ln(x^3 + 1)^{12}$$

$$= 10\ln(x + 1) + 11\ln(x^2 + 1) - 12\ln(x^3 + 1)$$

Now we can find $\dfrac{dy}{dx}$ using implicit differentiation:

$$\frac{1}{y}\frac{dy}{dx} = \frac{10}{x + 1} + \frac{22x}{x^2 + 1} - \frac{36x^2}{x^3 + 1} \qquad \text{Take } \frac{d}{dx} \text{ of both sides.}$$

$$\frac{dy}{dx} = y\left(\frac{10}{x + 1} + \frac{22x}{x^2 + 1} - \frac{36x^2}{x^3 + 1}\right) \qquad \text{Solve for } \frac{dy}{dx}.$$

$$= \frac{(x + 1)^{10}(x^2 + 1)^{11}}{(x^3 + 1)^{12}}\left(\frac{10}{x + 1} + \frac{22x}{x^2 + 1} - \frac{36x^2}{x^3 + 1}\right) \qquad \text{Substitute for } y.$$

✷ **Before we go on . . .** Redo this example using the product and quotient rules (and the chain rule) instead of logarithmic differentiation and compare the answers. Compare also the amount of work involved in both methods.

APPLICATION

Productivity usually depends on both labor and capital. Suppose, for example, you are managing an automobile assembly plant. You can measure its productivity by counting the number of automobiles the plant produces each year. As a measure of labor, you can use the number of employees, and as a measure of capital, you can use its operating budget. The so-called *Cobb–Douglas model* uses a function of the form

$$P = Kx^a y^{1-a} \qquad \text{Cobb–Douglas model for productivity}$$

where P stands for the number of automobiles produced each year, x is the number of employees, and y is the operating budget. The numbers K and a are constants that depend on the particular factory studied, with a between 0 and 1.

Example 4 • Cobb–Douglas Production Function

The automobile assembly plant you manage has the Cobb–Douglas production function

$$P = x^{0.3}y^{0.7}$$

where P is the number of automobiles it produces each year, x is the number of employees, and y is the daily operating budget (in dollars). Assume a production level of 1000 automobiles per year.

a. Find $\dfrac{dy}{dx}$.

b. Evaluate this derivative at $x = 80$ and interpret the answer.

Solution   Since the production level is 1000 automobiles per year, we have $P = 1000$, so the equation is

$$1000 = x^{0.3}y^{0.7}$$

a. We find $\dfrac{dy}{dx}$ by implicit differentiation:

$$0 = \frac{d}{dx}[x^{0.3}y^{0.7}] \qquad\qquad \text{Take } \frac{d}{dx} \text{ of both sides.}$$

$$0 = 0.3x^{-0.7}y^{0.7} + x^{0.3}(0.7)y^{-0.3}\frac{dy}{dx} \qquad \text{Product and chain rules}$$

$$-0.7x^{0.3}y^{-0.3}\frac{dy}{dx} = 0.3x^{-0.7}y^{0.7} \qquad\qquad \text{Bring term with } \frac{dy}{dx} \text{ to left.}$$

$$\frac{dy}{dx} = -\frac{0.3x^{-0.7}y^{0.7}}{0.7x^{0.3}y^{-0.3}} \qquad\qquad \text{Solve for } \frac{dy}{dx}.$$

$$= -\frac{3y}{7x}$$

b. To evaluate this derivative at $x = 80$, we must first find the corresponding value of y. To obtain y, we substitute $x = 80$ in the original equation $1000 = x^{0.3}y^{0.7}$ and solve for y:

$$1000 = 80^{0.3}y^{0.7}$$

$$y^{0.7} = \frac{1000}{80^{0.3}}$$

To obtain y on its own, raise both sides to the power $\frac{1}{0.7}$ to obtain

$$y = (y^{0.7})^{1/0.7}$$

$$= \left(\frac{1000}{80^{0.3}}\right)^{1/0.7} \approx 2951.92$$

Now that we have the corresponding value for y, we evaluate the derivative at $x = 80$ and $y = 2951.92$:

$$\left.\frac{dy}{dx}\right|_{x=80} \approx -\frac{3(2951.92)}{7(80)} \approx -15.81 \qquad \text{Using } \frac{dy}{dx} = -\frac{3y}{7x}$$

How do we interpret this result? The first clue is to look at the units of the derivative: We recall that the units of dy/dx are units of y per unit of x. Since y is the daily budget, its units are dollars; since x is the number of employees, its units are employees. Thus,

$$\left.\frac{dy}{dx}\right|_{x=80} \approx -\$15.81 \text{ per employee}$$

Next, recall that dy/dx measures the rate of change of y as x changes. Since the answer is negative, the daily budget to maintain production of 1000 automobiles is decreasing by approximately $15.81 per additional employee at an employment level of 80 employees. In other words, increasing the workforce by one worker will result in a saving of approximately $15.81 per day. Roughly speaking, *a new employee is worth $15.81 per day* at the current levels of employment and production.

4.4 EXERCISES

In Exercises 1–10, find dy/dx using implicit differentiation. In each case compare your answer with the result obtained by first solving for y as a function of x and then taking the derivative.

1. $2x + 3y = 7$

2. $4x - 5y = 9$

3. $x^2 - 2y = 6$

4. $3y + x^2 = 5$

5. $2x + 3y = xy$

6. $x - y = xy$

7. $e^x y = 1$

8. $e^x y - y = 2$

9. $y \ln x + y = 2$

10. $\dfrac{\ln x}{y} = 2 - x$

In Exercises 11–30, find the indicated derivative using implicit differentiation.

11. $x^2 + y^2 = 5; \dfrac{dy}{dx}$

12. $2x^2 - y^2 = 4; \dfrac{dy}{dx}$

13. $x^2 y - y^2 = 4; \dfrac{dy}{dx}$

14. $xy^2 - y = x; \dfrac{dy}{dx}$

15. $3xy - \dfrac{y}{3} = \dfrac{2}{x}; \dfrac{dy}{dx}$

16. $\dfrac{xy}{2} - y^2 = 3; \dfrac{dy}{dx}$

17. $x^2 - 3y^2 = 8; \dfrac{dx}{dy}$

18. $(xy)^2 + y^2 = 8; \dfrac{dx}{dy}$

19. $p^2 - pq = 5p^2q^2; \dfrac{dp}{dq}$

20. $q^2 - pq = 5p^2q^2; \dfrac{dp}{dq}$

21. $xe^y - ye^x = 1; \dfrac{dy}{dx}$

22. $x^2e^y - y^2 = e^x; \dfrac{dy}{dx}$

23. $e^{st} = s^2; \dfrac{ds}{dt}$

24. $e^{s^2t} - st = 1; \dfrac{ds}{dt}$

25. $\dfrac{e^x}{y^2} = 1 + e^y$; $\dfrac{dy}{dx}$ **26.** $\dfrac{x}{e^y} + xy = 9y$; $\dfrac{dy}{dx}$

27. $\ln(y^2 - y) + x = y$; $\dfrac{dy}{dx}$ **28.** $\ln(xy) - x \ln y = y$; $\dfrac{dy}{dx}$

29. $\ln(xy + y^2) = e^y$; $\dfrac{dy}{dx}$ **30.** $\ln(1 + e^{xy}) = y$; $\dfrac{dy}{dx}$

In Exercises 31–34, use logarithmic differentiation to find dy/dx.

31. $y = (x^3 + x)\sqrt{x^3 + 2}$ **32.** $y = \sqrt{\dfrac{x-1}{x^2+2}}$

33. $y = x^x$ **34.** $y = x^{-x}$

In Exercises 35–46, use implicit differentiation to evaluate dy/dx at the indicated point on the graph. (If only the x coordinate is given, you must also find the y coordinate.)

35. $4x^2 + 2y^2 = 12$; $(1, -2)$ **36.** $3x^2 - y^2 = 11$; $(-2, 1)$

37. $2x^2 - y^2 = xy$; $(-1, 2)$ **38.** $2x^2 + xy = 3y^2$; $(-1, -1)$

39. $3x^{0.3}y^{0.7} = 10$; $x = 20$ **40.** $2x^{0.4}y^{0.6} = 10$; $x = 50$

41. $x^{0.4}y^{0.6} - 0.2x^2 = 100$; $x = 20$

42. $x^{0.4}y^{0.6} - 0.3x^2 = 10$; $x = 10$

43. $e^{xy} - x = 4x$; $x = 3$

44. $e^{-xy} + 2x = 1$; $x = -1$

45. $\ln(x + y) - x = 3x^2$; $x = 0$

46. $\ln(x - y) + 1 = 3x^2$; $x = 0$

APPLICATIONS

47. Demand The demand equation for soccer tournament T-shirts is

$$pq - 2000 = q$$

where q is the number of T-shirts the Enormous State University soccer team can sell for $\$p$ each.

a. How many T-shirts can the team sell at $\$5$ each?

b. Find $\dfrac{dq}{dp}\bigg|_{p=5}$ and interpret the result.

48. Cost Equations The cost c (in cents) of producing x gallons of Ectoplasm hair gel is given by the cost equation

$$c^2 - 10cx = 200$$

a. Find the cost of producing 1 gallon and 3.5 gallons.

b. Evaluate $\dfrac{dc}{dx}$ at $x = 1$ and $x = 3.5$ and interpret the results.

49. Housing Costs The cost C (in dollars) of building a house is related to the number k of carpenters used and the number e of electricians used by the formula

$$C = 15,000 + 50k^2 + 60e^2$$

If the cost of the house comes to $\$200,000$, find $\dfrac{dk}{de}\bigg|_{e=15}$ and interpret your result.

Based on an exercise in A. L. Ostrosky, Jr., and J. V. Koch, *Introduction to Mathematical Economics* (Springfield, Ill.: Waveland Press, 1979).

50. Employment An employment research company estimates that the value of a recent MBA graduate to an accounting company is

$$V = 3e^2 + 5g^3$$

where V is the value of the graduate, e is the number of years of prior business experience, and g is the graduate school grade-point average. If $V = 200$, find de/dg when $g = 3.0$ and interpret the result.

51. Grades A production formula for a student's performance on a difficult English examination is

$$g = 4tx - 0.2t^2 - 10x^2 \quad (t < 30)$$

where g is the score the student can expect to obtain, t is the number of hours of study for the examination, and x is the student's grade-point average.

a. How long should a student with a 3.0 grade-point average study in order to score 80 on the examination?

b. Find dt/dx for a student who earns a score of 80, evaluate it when $x = 3.0$, and interpret the result.

Based on an exercise in A. L. Ostrosky, Jr., and J. V. Koch, *Introduction to Mathematical Economics* (Springfield, Ill.: Waveland Press, 1979).

52. Grades Repeat Exercise 51, using the following production formula for a basket-weaving examination:

$$g = 10tx - 0.2t^2 - 10x^2 \quad (t < 10)$$

Comment on the result.

53. Productivity The number of CDs that Snappy Hardware can manufacture at its plant in 1 day is given by

$$P = x^{0.6}y^{0.4}$$

where x is the number of workers at the plant and y is the annual expenditure at the plant (in dollars). Compute $\dfrac{dy}{dx}\bigg|_{x=100}$ at a production level of 20,000 CDs per day and interpret the result.

54. Productivity Repeat Exercise 53, using a productivity formula of

$$P = x^{0.5}y^{0.5}$$

Exercises 55 and 56 are based on the following demand function for money (taken from a question on the GRE economics test):

$$M_d = 2 \times y^{0.6} \times r^{-0.3} \times p$$

where M_d = demand for nominal money balances (money stock)

      y = real income

      r = index of interest rates

      p = index of prices

55. Money Stock If real income grows while the money stock and the price level remain constant, the interest rate must change at what rate? (First find dr/dy and then dr/dt; your answers will be expressed in terms of r, y, and $\dfrac{dy}{dt}$.)

56. Money Stock If real income grows while the money stock and the interest rate remain constant, the price level must change at what rate? (See the hint in Exercise 55.)

COMMUNICATION AND REASONING EXERCISES

57. Use logarithmic differentiation to give another proof of the product rule.

58. Use logarithmic differentiation to give a proof of the quotient rule.

59. If y is given explicitly as a function of x by an equation $y = f(x)$, compare finding dy/dx by implicit differentiation to finding it explicitly in the usual way.

60. Explain why one should not expect dy/dx to be a function of x if y is not a function of x.

61. True or false? If y is a function of x and $dy/dx \neq 0$ at some point, then, regarding x as an implicit function of y, we have

$$\frac{dx}{dy} = \frac{1}{dy/dx}$$

Explain your answer.

62. If you are given an equation in x and y such that dy/dx is a function of x only, what can you say about the graph of the equation?

CASE STUDY

Projecting Market Growth

Reproduced from the U.S. Department of Education Web site

You are on the board of directors at Fullcourt Academic Press, and TJM, the sales director of the high school division, has just burst into your office with data showing the number of high school graduates each year over the past decade (Figure 9).

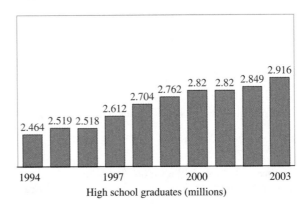

Data starting in 2000 are projections. SOURCE: U.S. Department of Education, http://nces.ed.gov/pubs2001/proj01/tables/table23.asp

Figure 9

TJM is pleased that the figures appear to support a basic premise of his recent proposal for a major expansion strategy: The number of high school seniors in the United States will be growing at a rate of at least 20,000 per year through the year 2005. The rate of increase, as he points out, has averaged around 50,000 per year since 1994, so it would not be overly optimistic to assume that the trend will continue—at least for the next 5 years.

Although you are tempted to support TJM's proposal at the next board meeting, you would like to estimate first whether the 20,000 figure is a realistic expectation, especially since the graph suggests that the number of graduates began to "level off" (in the language of calculus, the *derivative appears to be decreasing*) during the second half of the period. Moreover, you recall reading somewhere that the numbers of students in the lower grades have also begun to level off, so it is safe to predict that the slowing of growth in the senior class will continue over the next few years. You really need precise data about numbers in the lower grades in order to make a meaningful prediction, but TJM's report is scheduled to be presented tomorrow and you would like a quick and easy way of "extending the curve to the right" by then.

It would certainly be helpful if you had a mathematical model of the data in Figure 9 that you could use to project the current trend. But what kind of model should you use? A linear model would be no good because it would not show any change in the derivative (the derivative of a linear function is constant). In addition, best-fit polynomial and exponential functions do not accurately reflect the leveling off, as you realize after trying to fit a few of them (Figure 10).

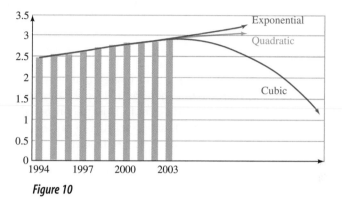

Figure 10

You then recall that a logistic curve can model the leveling-off property you desire, and so you try fitting a curve of the form

$$y = \frac{N}{1 + Ab^{-t}}$$

Figure 11 shows the best-fit logistic curve, which eventually levels off at around $N = 3.3$.

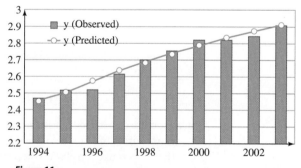

Figure 11

The leveling-off prediction certainly seems reasonable, but you are slightly troubled by the shape of the regression curve: It doesn't seem to follow the S shape of the data very convincingly. Moreover, the curve doesn't appear to fit the data significantly more snugly than the quadratic or cubic models.[8] To reassure yourself, you decide to look for another kind of S-shaped model as a backup.

---

[8]There is another, more mathematical reason for not using logistic regression to predict long-term leveling off. The regression value of the long-term level N is extremely sensitive to the values of the other coefficients. As a result, there can be good fits to the same set of data with a wide variety of values of N.

After flipping through a calculus book, you stumble across a function whose graph looks rather like the one you have (Figure 12).

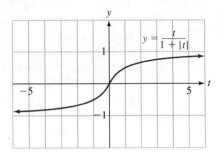

$$y = \frac{t}{1 + |t|}$$

Figure 12

Curves of this form are sometimes called *predator satiation curves,* and they are used to model the population of predators in an environment with limited prey. Although the curve does seem to have the proper shape, you realize that you will need to shift and scale the function in order to fit the actual data. The most general scaled and shifted version of this curve has the form

$$y = c + b\frac{a(t - m)}{1 + a|t - m|} \qquad (a, b, c, m \text{ constant})$$

and you decide to try a model of this form, where y represents the number of high school seniors (in millions) and t represents years since 1994.[9]

For the moment you postpone the question of finding the best values for the constants a, b, c, and m and decide first to calculate the derivative of the model in terms of the given constants. The derivative dy/dt will represent the rate of increase of high school graduates, which is exactly what you wish to estimate.

The main part of the function is a quotient, so you start with the quotient rule:

$$\frac{dy}{dt} = b\frac{a(1 + a|t - m|) - a(t - m)\frac{d}{dt}(1 + a|t - m|)}{(1 + a|t - m|)^2}$$

You now recall the formula for the derivative of an absolute value:

$$\frac{d}{dt}[|t|] = \frac{|t|}{t} = \begin{cases} 1 & \text{if } t > 0 \\ -1 & \text{if } t < 0 \end{cases}$$

which, you are disturbed to notice, is not defined at $t = 0$. Undaunted, you make a mental note and press on. Since you need the derivative of the absolute value of a quantity other than t, you use the chain rule, which tells you that

$$\frac{d}{dt}[|u|] = \frac{|u|}{u}\frac{du}{dt}$$

Thus,

$$\frac{d}{dt}[|t - m|] = \frac{|t - m|}{t - m} \cdot 1 \qquad u = t - m \quad (m = \text{constant})$$

---

[9]The scaled and shifted function is obtained from the original $t/(1 + |t|)$ by scaling by a factor of b in the y direction, a factor of a in the t direction, and then shifting m units in the t direction and c units in the y direction. For a detailed online treatment of scaled and shifted functions, follow the path

Web Site $\rightarrow$ Everything for Calculus $\rightarrow$ Chapter 1 $\rightarrow$ New Functions from Old: Scaled and Shifted Functions

(This is not defined when $t = m$.) Substituting into the formula for dy/dt, you find

$$\frac{dy}{dt} = b\frac{a(1 + a|t - m|) - a(t - m) \cdot a\frac{|t - m|}{(t - m)}}{(1 + a|t - m|)^2}$$

$$= b\frac{a(1 + a|t - m|) - a^2|t - m|}{(1 + a|t - m|)^2} \qquad \text{Cancel } (t - m).$$

$$= a\frac{b}{(1 + a|t - m|)^2} \qquad a^2|t - m| - a^2|t - m| = 0$$

It is interesting that, although the derivative of $|t - m|$ is not defined when $t = m$, the offending term $t - m$ was canceled so that dy/dt seems to be defined[10] at $t = m$.

Now you have a simple-looking expression for dy/dt, which will give you an estimate of the rate of change of the high school senior population. However, you still need values for the constants a, b, c, and m. (You don't really need the value of c to compute the derivative—where has it gone?—but it is a part of the model.) How do you find the values of a, b, c, and m that result in the curve that best fits the given data?

Turning once again to your calculus book (see the discussion of logistic regression in Section 2.4), you see that a best-fit curve is one that minimizes the sum-of-squares error (SSE). Here is an Excel spreadsheet showing the errors for $a = 1$, $b = 1$, $c = 1$, and $m = 1$:

	A	B	C	D	E	F
1	t (Year)	y (Observed)	y (Predicted)	Residue^2	Constants	
2	0	2.464	0.5	3.857296	a	1
3	1	2.519	1	2.307361	b	1
4	2	2.518	1.5	1.036324	c	1
5	3	2.612	1.666666667	0.89365511	m	1
6	4	2.704	1.75	0.910116		
7	5	2.762	1.8	0.925444	SSE:	13.8344343
8	6	2.82	1.833333333	0.97351111		
9	7	2.82	1.857142857	0.92709388		
10	8	2.849	1.875	0.948676		
11	9	2.916	1.888888889	1.05495723		

The first two columns show the observed data ($t =$ years since 1994, and $y =$ number of high school graduates in millions). The formula for y (Predicted) is our model

$$y = c + b\frac{a(t - m)}{1 + a|t - m|}$$

entered in cell C2 as

```
=$F$4+$F$3*$F$2*(A2-$F$5)/(1+$F$2*ABS(A2-$F$5))
    c + b * a *(t - m) /(1+  a *   |t - m|)
```

and then copied into the cells below it. Since the square error (Residue^2) is defined as the square of the difference between y (Observed) and y (Predicted), we enter

```
=(C2-B2)^2
```

in cell D2 and then copy into the cells below it. The SSE (sum of the entries in D2–D11) is then placed in cell F7.

---

[10] In fact, it is defined and has the value given by the formula just derived: ab. To show this takes a bit more work. How might you do it?

The values of a, b, c, and m shown are initial values and don't matter too much (but see below); you will have Excel change these values in order to improve your model. The smaller the SSE is, the better your model. [For a perfect fit, the `y (Observed)` column would equal the `y (Predicted)` column, and SSE would be zero.] Hence, the goal is now to find values of a, b, c, and m that make the value of SSE as small as possible. (See the discussion in Section 1.4.) Finding these values analytically is an extremely difficult mathematical problem. However, software, such as Excel's built-in Solver routine, can be used to find *numerical* solutions.[11]

Figure 13 shows how to set up Solver to find the best values for a, b, c, and m for the setup used in this spreadsheet.

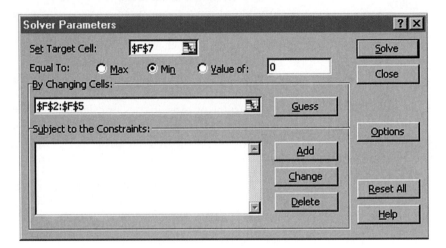

Figure 13

The `Target Cell`, `$F$7`, contains the value of SSE, which is to be minimized. The `Changing Cells` are the cells containing the values of the constants a, b, c, and m that we want to change. That's it.

Now click on `Solve`. After thinking about it for a few seconds, Excel gives the optimal values of a, b, c, and m in cells F2–F5 and the minimum value of SSE in cell D12.[12] You find

$$a = 0.242\,501\,75 \qquad b = 0.410\,271\,1 \qquad c = 2.654\,478\,73 \qquad m = 3.521\,195\,6$$

with SSE $\approx 0.002\,88$, which is a better fit than the logistic regression curve (SSE $\approx 0.006\,53$)

Figure 14 shows that not only does this choice of model and constants give an excellent fit, but also the curve seems to follow the S shape more convincingly than the logistic curve.

---

[11]See the similar discussion in Section 2.4. If `Solver` does not appear in the `Tools` menu, you should first install it using your Excel installation software. (Solver is one of the Excel Add-Ins.)

[12]Depending on the settings in Solver, you may need to run the utility twice in succession to reach the minimum value of SSE.

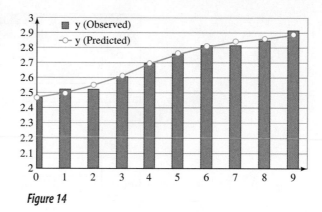

Figure 14

Figure 15 shows how the model predicts the long-term leveling-off phenomenon you were looking for.

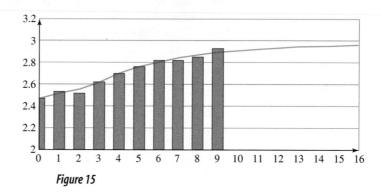

Figure 15

You turn back to the problem at hand: projecting the rate of increase of the number of high school graduates in 2005. You have the formula

$$\frac{dy}{dt} = \frac{ab}{(1 + a|t - m|)^2}$$

and values for the constants. So you compute

$$\frac{dy}{dt} = \frac{(0.242\ 501\ 75)(0.410\ 271\ 1)}{[1 + 0.242\ 501\ 75(11 - 3.521\ 195\ 6)]^2} \qquad t = 11 \text{ in } 2005$$

$$\approx 0.0126 \text{ million students/year}$$

or 12,600 students per year—far less than the optimistic estimate of 20,000 in the proposal!

You now conclude that TJM's prediction is suspect and that further research will have to be done before the board can support the proposal.

Question How accurately does the model predict the number of high school graduates?

Answer Using a regression curve–fitting model to make long-term predictions is always risky. A more accurate model would have to take into account such factors as the birthrate and current school populations at all levels. The U.S. Department of Education has used more sophisticated models to make the projections shown below, which we compare with those predicted by our model.

Year	U.S. Dept. of Ed. Projections (millions)	Model Predictions (millions)
2004	2.921	2.91
2005	2.929	2.92
2006	2.986	2.93
2007	3.054	2.94
2008	3.132	2.95
2009	3.127	2.96
2010	3.103	2.96
2011	3.063	2.97

Question Which values of the constants should I use as starting values when using Excel to find the best-fit curve?

Answer If the starting values of the constants are far from the optimal values, Solver may find a nonoptimal solution. Thus, you need to obtain some rough initial estimate of the constants by examining the graph. Figure 16 shows some important features of the curve that you can use to obtain estimates of a, b, c, and m by inspecting the graph.

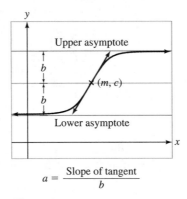

$$a = \frac{\text{Slope of tangent}}{b}$$

Figure 16

From the graph, m and c are the coordinates of the point on the curve where it is steepest, and b is the vertical distance from that point to the upper or lower asymptote (where the curve levels off). To estimate a, first estimate the slope of the tangent at the point of steepest inclination and then divide by b. If b is negative (and a is positive), we obtain an upside-down version of the curve (Figure 17).

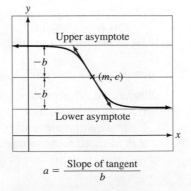

$$a = \frac{\text{Slope of tangent}}{b}$$

Figure 17

EXERCISES

1. In 1993 there were 2.49 million high school graduates. What does the regression model "predict" for 1993? What is the residue ($y_{\text{predicted}} - y_{\text{observed}}$)? (Round answers to the nearest 0.01 million.)

2. What is the long-term prediction of the model?

3. Find $\lim\limits_{t \to 0} \dfrac{dy}{dt}$ and interpret the result.

 For Exercises 4–6, use Excel to solve.

4. You receive a memo to the effect that the 1994 and 1995 figures are not accurate. Use Solver to reestimate the best-fit constants a, b, c, and m in the absence of this data and obtain new estimates for the 1994 and 1995 data. What does the new model predict the rate of change in the number of high school seniors will be in 2005?

5. Shifted Logistic Model Using the original data, find the best-fit shifted logistic curve of the form

$$f(t) = c + \frac{N}{1 + Ab^{-t}}$$

(Start with the following values: $c = 0, N = 3$, and $A = b = 1$. You might have to run Solver twice in succession to minimize SSE.) Graph the data together with the model. What is SSE? Is the model as accurate a fit as the model used in the text? How do the long-term predictions of the two models compare with the U.S. Department of Education projections? What does this model predict will be the growth rate of the number of high school graduates in 2005? Round the coefficients in the model and all answers to four decimal places.

6. Demand for Freon The demand for chloroflurocarbon-12 (CFC-12)—the ozone-depleting refrigerant commonly known as Freon[13]—has been declining significantly in response to regulation and concern about the ozone layer. The chart below shows the projected demand for CFC-12 for the period 1994–2005.

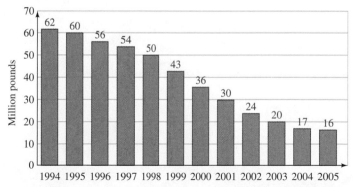

SOURCE: The Automobile Consulting Group/*New York Times*, December 26, 1993, p. F23. The exact figures were not given, and the chart is a reasonable facsimile of the chart that appeared in the *New York Times*.

---

[13]The name given to it by DuPont.

a. Use Solver to obtain the best-fit equation of the form

$$f(t) = c + b\frac{a(t - m)}{1 + a|t - m|}$$

where t = years since 1990. Use your function to estimate the total demand for CFC-12 from the start of the year 2000 to the start of 2010. (Start with the following values: $a = 1$, $b = -25$, $c = 35$, and $m = 10$, and round your answers to four decimal places.)

b. According to your model, how fast is the demand for Freon declining in 2000?

CHAPTER 4 REVIEW TEST

1. Find the derivative of each function.

a. $f(x) = e^x(x^2 - 1)$

b. $f(x) = \dfrac{x^2 + 1}{x^2 - 1}$

c. $f(x) = (x^2 - 1)^{10}$

d. $f(x) = \dfrac{1}{(x^2 - 1)^{10}}$

e. $f(x) = e^x(x^2 + 1)^{10}$

f. $f(x) = \left(\dfrac{x - 1}{3x + 1}\right)^3$

g. $f(x) = e^{x^2 - 1}$

h. $f(x) = (x^2 + 1)e^{x^2 - 1}$

i. $f(x) = \ln(x^2 - 1)$

j. $f(x) = \dfrac{\ln(x^2 - 1)}{x^2 - 1}$

2. Find dy/dx for each of the following equations.

a. $x^2 - y^2 = x$

b. $2xy + y^2 = y$

c. $e^{xy} + xy = 1$

d. $\ln\left(\dfrac{y}{x}\right) = y$

3. Find all values of x (if any) where the tangent line to the graph of the given equation is horizontal.

a. $y = x - e^{2x-1}$

b. $y = e^{x^2}$

c. $y = \dfrac{x}{x + 1}$

d. $y = \sqrt{x}(x - 1)$

OHaganBooks.com—COMPOUNDING RATES OF CHANGE

4. At the moment, OHaganBooks.com is selling 1000 books each week, and its sales are rising at a rate of 200 books per week. Also, it is now selling all its books for $20 each, but its price is dropping at a rate of $1 per week.

a. At what rate is OHaganBooks.com's revenue rising or falling?

b. John O'Hagan would have preferred to see the company's revenue increase at a rate of $5000 per week. At what rate would sales have to have been increasing to accomplish that goal, assuming that all the other information is as given?

c. The percentage rate of change of a quantity Q is Q'/Q. Why is the percentage rate of change of revenue always equal to the sum of the percentage rates of change of unit price and weekly sales?

5. At the beginning of last week, OHaganBooks.com stock was selling for $100 per share, rising at a rate of $50 per year. Its earnings amounted to $1 per share, rising at a rate of $0.10 per year.

a. At what rate was its price-to-earnings (P/E) ratio, the ratio of its stock price to its earnings per share, rising or falling?

b. Curt Hinrichs, who recently invested in OHaganBooks.com stock, would have liked to see the P/E ratio increase at a rate of 100 points per year. How fast would the stock have to have been rising, assuming that all the other information is as given?

c. The percentage rate of change of a quantity Q is Q'/Q. Why is the percentage rate of change of P/E always equal to the percentage rate of change of unit price minus the percentage rate of change of earnings?

6. OHaganBooks.com modeled its weekly sales over a period of time with the function

$$s(t) = 6053 + \frac{4474}{1 + e^{-0.55(t-4.8)}}$$

as shown in the following graph:

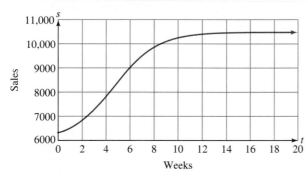

a. Compute $s'(t)$.

b. Use the answer to part (a) to compute the rate of increase of weekly sales at the beginning of the 7th week ($t = 6$). (Round your answer to the nearest unit.)

c. Find the rate of increase of weekly sales at the beginning of the 15th week ($t = 14$).

7. The number of "hits" on OHaganBooks.com's Web site was 1000 per day at the beginning of the year, growing at a rate of 5% per week. If this growth rate continued for the whole year (52 weeks), find the rate of increase (in hits per day per week) at the end of the year.

8. The price p that OHaganBooks.com charges for its latest leather-bound gift edition of *The Lord of the Rings* is related to the demand q in weekly sales by the equation

$$100pq + q^2 = 5,000,000$$

Suppose the price is set at $40, which would make the demand 1000 copies per week.

a. Using implicit differentiation, compute the rate of change of demand with respect to price and interpret the result. (Round the answer to two decimal places.)

b. Use the result of part (a) to compute the rate of change of revenue with respect to price. Should the price be raised or lowered to increase revenue?

![www] ADDITIONAL ONLINE REVIEW

If you follow the path
    Web Site → Everything for Calculus → Chapter 4
you will find the following additional resources to help you review:

A comprehensive chapter summary (including examples and interactive features)

Additional review exercises (including interactive exercises and many with help)

A true/false chapter quiz

Online tutorials for almost all sections of the chapter

5

APPLICATIONS OF THE

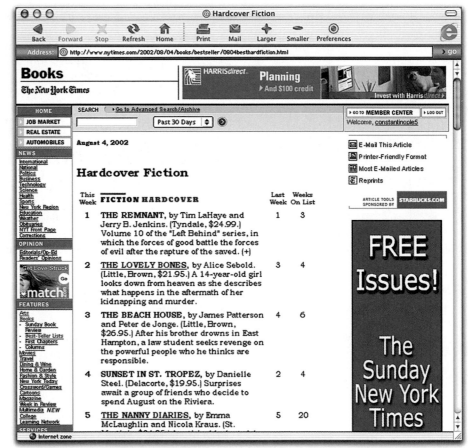

© 2003 The New York Times Company. Reprinted by Permission.

INTERNET RESOURCES FOR THIS CHAPTER

At the Web site, follow the path

   Web Site → Everything for Calculus → Chapter 5

where you will find links to step-by-step tutorials for the main topics in this chapter, a detailed chapter summary you can print out, a true/false quiz, and a collection of sample test questions. You will also find downloadable Excel tutorials for each section, several online graphers, and other resources.

DERIVATIVE

Introduction

In this chapter we begin to see the power of calculus as an optimization tool. In Chapter 2 we saw how to price an item in order to get the largest revenue when the demand functions is linear. Using calculus, we can handle much more general, nonlinear functions. In Section 5.1 we show how calculus can be used to solve the problem of finding the values of a variable that lead to a maximum or minimum value of a given function. In Section 5.2 we show how this helps us in various real-world applications.

Another theme in this chapter is that calculus can help us draw and understand the graph of a function. By the time you have completed the material in Section 5.1, you will be able to locate and sketch some of the important features of a graph. In Section 5.3 we discuss further how to explain what you see in a graph (drawn, for example, using graphing technology) and to locate its most important points.

We also include sections on related rates and elasticity of demand. The first of these (Section 5.4) examines further the concept of the derivative as a rate of change. The second (Section 5.5) returns to the problem of optimizing revenue based on the demand equation, looking at it in a new way that leads to an important idea in economics—elasticity.

 Algebra Review

For this chapter you should be familiar with the algebra reviewed in Appendix A, Sections A.5 and A.6.

5.1 Maxima and Minima

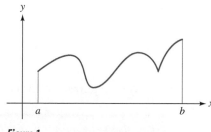

Figure 1

Figure 1 shows the graph of a function f whose domain is the closed interval $[a, b]$. A mathematician sees lots of interesting things going on here. There are hills and valleys, and even a small chasm (called a *cusp*) toward the right. For many purposes the important features of this curve are the highs and lows. Suppose, for example, you know that the price of the stock of a certain company will follow this graph during the course of a week. Although you would certainly make a handsome profit if you bought at time a and sold at time b, your best strategy would be to follow the old adage to "buy low and sell high," buying at all the lows and selling at all the highs.

Figure 2 shows the graph once again with the highs and lows marked.

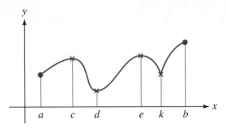

Figure 2

Mathematicians have names for these points: The highs (at the x values c, e, and b) are referred to as **relative maxima,** and the lows (at the x values a, d, and k) are referred to as **relative minima.** Collectively, these highs and lows are referred to as **relative extrema.** (A point of language: The singular forms of the plurals *minima, maxima,* and *extrema* are *minimum, maximum,* and *extremum.*)

Why do we refer to these points as *relative* extrema? Take a look at the point corresponding to $x = c$. It is the highest point of the graph *compared to other points nearby.* If you were an extremely nearsighted mountaineer standing at point c, you would *think* that you were at the highest point of the graph, not being able to see the distant peaks at $x = e$ and $x = b$.

Let's translate into mathematical terms. We are talking about the heights of various points on the curve. The height of the curve at $x = c$ is $f(c)$, so we are saying that $f(c)$ is greater than $f(x)$ for every x near c. For instance, *f(c) is the greatest value that f(x) has for all choices of x between a and d* (see Figure 3).

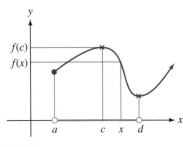

Figure 3

We can phrase the formal definition as follows.

Relative Extremum

- f has a **relative maximum** at c if there is some interval (r, s) (even a very small one) containing c for which $f(c) \geq f(x)$ for all x between r and s for which $f(x)$ is defined.
- f has a **relative minimum** at c if there is an interval (r, s) (even a very small one) containing c for which $f(c) \leq f(x)$ for all x between r and s for which $f(x)$ is defined.

Relative maxima and minima are sometimes also referred to as **local** maxima and minima.

Quick Examples

In Figure 2 f has the following relative extrema:

1. A relative maximum at c, as shown by the interval (a, d)

2. A relative maximum at e, as shown by the interval (d, k)

3. A relative maximum at b, as shown by the interval $(k, b + 1)$
   Note that $f(x)$ is not defined for $x > b$. However, $f(b) \geq f(x)$ for every x in the interval $(k, b + 1)$ *for which $f(x)$ is defined*—that is, for every x in $(k, b]$.

4. A relative minimum at d, as shown by the interval (c, e)

5. A relative minimum at k, as shown by the interval (e, b)

6. A relative minimum at a, as shown by the interval $(a - 1, c)$ (See Quick Example 3.)

Note

Our definition of relative extremum allows f to have a relative extremum at an endpoint of its domain; the definitions used in some books do not. In view of examples like our stock market-investing strategy, we see no good reason not to count endpoints as extrema.

Looking carefully at Figure 2, we can see that the lowest point on the whole graph is where $x = d$ and the highest point is where $x = b$. This means that $f(d)$ is the smallest value of f on the whole domain of f (the interval $[a, b]$) and $f(b)$ is the largest value. We call these the *absolute* minimum and maximum.

Absolute Extremum

- f has an **absolute maximum** at c if $f(c) \geq f(x)$ for every x in the domain of f.
- f has an **absolute minimum** at c if $f(c) \leq f(x)$ for every x in the domain of f.

Absolute maxima and minima are sometimes also referred to as **global** maxima and minima.

Quick Examples

1. In Figure 2 f has an absolute maximum at b and an absolute minimum at d.

2. If $f(x) = x^2$, then $f(x) \geq f(0)$ for every real number x. Therefore, $f(x) = x^2$ has an absolute minimum at $x = 0$.

Question Is there only one absolute maximum of f?

Answer Although there is only one absolute maximum *value* of f, this value may occur at many different values of x. An extreme case is that of a constant function; since we use $\geq$ in the definition, a constant function has an absolute maximum (and minimum) at every point in its domain.

Now, how do we go about locating extrema? In many cases we can get a good idea by using graphing technology to zoom in on a maximum or minimum and approximate its coordinates. However, calculus gives us a way to find the exact locations of the extrema and at the same time to understand why the graph of a function behaves the way it does. In fact, it is often best to combine the powers of graphing technology with those of calculus, as we will see.

In Figure 4 we see the graph from Figure 1 once more, but we have labeled each extreme point as one of three types.

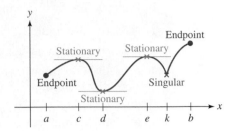

Figure 4

At the points labeled Stationary, the tangent lines to the graph are horizontal, so have slope zero. Remember that the slope of the tangent line is given by the derivative. Thus, the derivative of f is zero at each of the stationary points. In other words,

$$f'(c) = 0 \qquad f'(d) = 0 \qquad f'(e) = 0$$

To find the exact location of each of these extrema, we solve the equation $f'(x) = 0$. Any time $f'(x) = 0$, we say that f has a **stationary point** at x because the rate of change of f is zero there. We shall call an *extremum* that occurs at a stationary point a **stationary extremum.**

There is a relative minimum in Figure 4 at $x = k$, but there is no horizontal tangent there. In fact, there is no tangent line at all; $f'(k)$ is not defined. [Recall a similar situation with the graph of $f(x) = |x|$ at $x = 0$.] When $f'(x)$ does not exist, we say that f has a **singular point** at x. We call an extremum that occurs at a singular point a **singular extremum.** The points that are either stationary or singular we call collectively the **critical points** of f.

The remaining two extrema are at the **endpoints** of the domain.[1] As we see in the figure, they are (almost) always either relative maxima or relative minima.

Question Are there any other types of relative extrema?

Answer No; relative extrema of a function always occur at critical points or endpoints (a rigorous proof is beyond the scope of this book).[2]

Locating Candidates for Relative Extrema

If f is a real-valued function, then its relative extrema occur among the following types of points:

1. **Stationary Points** f has a stationary point at x if x is in the domain and $f'(x) = 0$. To locate stationary points, set $f'(x) = 0$ and solve for x.

2. **Singular Points** f has a singular point at x if x is in the domain and $f'(x)$ is not defined. To locate singular points, find values of x where $f'(x)$ is *not* defined but $f(x)$ *is* defined.

---

[1] Remember that we do allow local extrema at endpoints.

[2] Here is an outline of the argument. Suppose f has a local maximum (say, at $x = a$) at a point other than an endpoint of the domain. Then either f is differentiable there or it is not. If it is not, then we have a singular point. If f is differentiable at $x = a$, then consider the slope of the secant line through the points where $x = a$ and $x = a + h$ for small positive h. Since f has a local maximum at $x = a$, it is falling (or level) to the right of $x = a$, and so the slope of this secant line must be ≤ 0. Thus, we must have $f'(a) \leq 0$ in the limit as $h \to 0$. On the other hand, if h is small and *negative*, then the corresponding secant line must have slope ≥ 0 because f is also falling (or level) as we move left from $x = a$, and so $f'(a) \geq 0$. Since $f'(a)$ is both ≥ 0 and ≤ 0, it must be zero, and so we have a stationary point at $x = a$.

3. Endpoints The x coordinates of endpoints are endpoints of the domain, if any. Recall that closed intervals contain endpoints, but open intervals do not.

Once we have the x coordinates of a candidate for a relative extremum, we find the corresponding y coordinate using $y = f(x)$.

Quick Examples

1. **Stationary Points** Let $f(x) = x^3 - 12x$. Then to locate the stationary points, set $f'(x) = 0$ and solve for x. This gives $3x^2 - 12 = 0$, so f has stationary points at $x = \pm 2$. Thus, the stationary points are $(-2, f(-2)) = (-2, 16)$ and $(2, f(2)) = (2, -16)$.

2. **Singular Points** Let $f(x) = 3(x - 1)^{1/3}$. Then $f'(x) = (x - 1)^{-2/3} = 1/(x - 1)^{2/3}$. $f'(1)$ is not defined, although $f(1)$ *is* defined. Thus, the (only) singular point occurs at $x = 1$. Its coordinates are $(1, f(1)) = (1, 0)$.

3. **Endpoints** Let $f(x) = 1/x$, with domain $(-\infty, 0) \cup [1, +\infty)$. Then the only endpoint in the domain of f occurs when $x = 1$ and has coordinates $(1, 1)$. The natural domain of $1/x$ has no endpoints.

Remember, though, that these are only *candidates* for relative extrema. It is quite possible, as we will see, to have a stationary point (or singular point) that is neither a relative maximum nor a relative minimum.

Now let's look at some examples of finding maxima and minima. In all these examples, we use the following procedure: First, find the derivative, which we examine to find the stationary points and singular points. Next, make a table listing the x coordinates of the critical points and endpoints, together with their y coordinates. Use this table to make a rough sketch of the graph. From the table and rough sketch, we usually have enough data to be able to say where the extreme points are and what kind they are.

Example 1 • Maxima and Minima

Find the relative and absolute maxima and minima of

$$f(x) = x^2 - 2x$$

on the interval $[0, 4]$.

Solution We first calculate $f'(x) = 2x - 2$. We use this derivative to locate the stationary and singular points.

- *Stationary Points* To locate the stationary points, we solve the equation $f'(x) = 0$, or

$$2x - 2 = 0$$

getting $x = 1$. The domain of the function is $[0, 4]$, so $x = 1$ is in the domain. Thus, the only candidate for a stationary relative extremum occurs when $x = 1$.

- *Singular Points* We look for points where the derivative is not defined. However, the derivative is $2x - 2$, which is defined for every x. Thus, there are no singular points and hence no candidates for singular relative extrema.

- *Endpoints* The domain is $[0, 4]$, so the endpoints occur when $x = 0$ and $x = 4$.

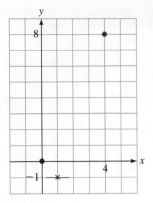

Figure 5

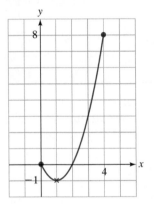

Figure 6

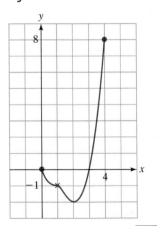

Figure 7

We record these values of x in a table, together with the corresponding y coordinates (values of f):

x	0	1	4
$f(x) = x^2 - 2x$	0	-1	8

This gives us three points on the graph, $(0, 0)$, $(1, -1)$, and $(4, 8)$, which we plot in Figure 5. We remind ourselves that the point $(1, -1)$ is a stationary point of the graph by drawing in a part of the horizontal tangent line. Connecting these points must give us a graph something like that in Figure 6. Notice that the graph has a horizontal tangent line at $x = 1$ but not at either of the endpoints because the endpoints are not stationary points.

From Figure 6 we can see that f has the following extrema:

x	$y = x^2 - 2x$	Classification
0	0	Relative maximum (endpoint)
1	-1	Absolute minimum (stationary point)
4	8	Absolute maximum (endpoint)

✳ **Before we go on . . .** How can we be sure that the graph doesn't look like Figure 7? If it did, there would be another critical point somewhere between $x = 1$ and $x = 4$. But we already know that there aren't any other critical points. The table we made listed all of the possible extrema; there can be no more.

It's also useful to consider the following. We found that $f'(1) = 0$; f has a stationary point at $x = 1$. What are the values of $f'(x)$ like to the left and right of 1? Let's make a table with some values:

x	0	1	2
$f'(x) = 2x - 2$	-2	0	2
	↘		↗

At $x = 0$, $f'(0) = -2 < 0$, so the graph has negative slope and f is **decreasing**; its values are going down as x increases. We note this with the downward pointing arrow in the chart. At $x = 2$, $f'(2) = 2 > 0$, so the graph has positive slope and f is **increasing**; its values are going up as x increases. In fact, since $f'(x) = 0$ only at $x = 1$, we know that $f'(x) < 0$ for all x in $[0, 1)$, and we can say that f is decreasing on the interval $[0, 1]$. Similarly, f is increasing on $[1, 4]$. So, starting at $x = 0$, the graph of f goes down until we reach $x = 1$, and then it goes back up. This is another way of checking that the stationary point at $x = 1$ is a relative minimum (using the derivative in this way to check what happens at a critical point is called using the **first derivative test**).[3]

Note

Here is some terminology: If the point (a, b) is a maximum (or minimum) of f, we sometimes say that ***f* has a maximum (or minimum) value of *b* at *x* = *a*.** Thus, in Example 1 we could have said the following:

- f has a relative maximum value of 0 at $x = 0$.
- f has a absolute minimum value of -1 at $x = 1$.
- f has a absolute maximum value of 8 at $x = 4$.

---

[3] The **second derivative** of a function (which we discuss in Section 5.3) is the derivative of the ("first") derivative—hence the name "first derivative test."

Example 2 • Unbounded Interval

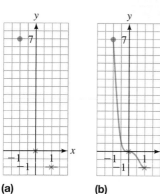

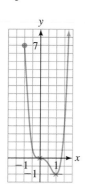

(a) **(b)**

Figure 8

Figure 9

Find all extrema of $f(x) = 3x^4 - 4x^3$ on $[-1, \infty)$.

Solution We first calculate $f'(x) = 12x^3 - 12x^2$.

- *Stationary Points* We solve the equation $f'(x) = 0$, which is

$$12x^3 - 12x^2 = 0 \quad \text{or} \quad 12x^2(x - 1) = 0$$

There are two solutions, $x = 0$ and $x = 1$, and both are in the domain. These are our candidates for the x coordinates of stationary relative extrema.

- *Singular Points* There are no points where $f'(x)$ is not defined, so there are no singular points.

- *Endpoints* The domain is $[-1, \infty)$, so there is one endpoint at $x = -1$.

We record these points in a table with the corresponding y coordinates:

x	-1	0	1
$f(x) = 3x^4 - 4x^3$	7	0	-1

We can either plot these points and sketch the graph by hand or we can turn to technology to help us.

Hand Plot If we plot these points by hand, we obtain Figure 8(a), which suggests Figure 8(b). We can't be sure what happens to the right of $x = 1$. Does the curve go up, or does it go down? To find out let's plot a test point to the right of $x = 1$. Choosing $x = 2$, we obtain $y = 3(2)^4 - 4(2)^3 = 16$, so $(2, 16)$ is another point on the graph. Thus, it must turn upward to the right of $x = 1$, as shown in Figure 9.

T **Technology**

If we use technology to show the graph, we need to choose the viewing window ourselves and should choose it so that it contains the three interesting points we've found so far. Again, we can't be sure yet what happens to the right of $x = 1$; does the graph go up or down from that point? If we set the viewing window to an interval of $[-1, 2]$ for x and $[-2, 8]$ for y, we will leave enough room to the right of $x = 1$ and below $y = -1$ to see what the graph will do. The result will be something like Figure 10.

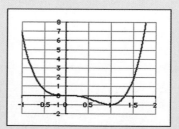

Figure 10

Now we can tell what happens to the right of $x = 1$: The function increases. We know that it cannot later decrease again, because if it did there would have to be another critical point where it turns around and we found that there are no other critical points.

From the graph (drawn either way), we find that f has the following extrema:

x	$y = 3x^4 - 4x^3$	Classification
-1	7	Relative maximum (endpoint)
1	-1	Absolute minimum (stationary point)

✳ **Before we go on . . .** Notice that the stationary point at $x = 0$ is neither a relative maximum nor a relative minimum. It is simply a place where the graph of f flattens out for a moment before it continues to fall. Notice also that f has no absolute maximum because $f(x)$ increases without bound as x gets large.

Example 3 • Singular Point

Find all extrema of $f(t) = t^{2/3}$ on $[-1, 1]$.

Solution First, $f'(t) = \dfrac{2}{3} t^{-1/3}$.

• *Stationary Points* We need to solve

$$\frac{2}{3} t^{-1/3} = 0$$

We can rewrite this equation without the negative exponent:

$$\frac{2}{3t^{1/3}} = 0$$

Now, the only way that a fraction can equal 0 is if the numerator is 0. But the numerator here is 2, so the fraction can never equal 0. Thus, there are no stationary points.

• *Singular Points* Are there any points where

$$f'(t) = \frac{2}{3t^{1/3}}$$

is not defined? Yes, where $t = 0$. Notice that f itself is defined at $t = 0$, so 0 is in the domain. Thus, f has a singular point at $t = 0$.

• *Endpoints* There are two endpoints, -1 and 1.

We now put these three points in a table with the corresponding y coordinates:

t	-1	0	1
$f(t)$	1	0	1

Technology
Since there is only one critical point, at $t = 0$, it is clear from this table that f must decrease from $t = -1$ to $t = 0$ and then increase to $t = 1$. To graph f using technology, choose a viewing window with an interval of $[-1, 1]$ for t and $[0, 1]$ for y. The result will be something like Figure 11.[4]

---

[4] Many graphing calculators will give you only the right-hand half of the graph shown in Figure 11 because fractional powers of negative numbers are not, in general, real numbers. To obtain the whole curve, enter the formula as $Y = (x^2)^{(1/3)}$, a fractional power of the nonnegative function x^2.

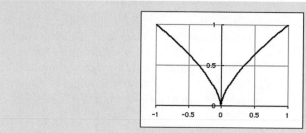

Figure 11

From this graph we find the following extrema for f:

t	$y = t^{2/3}$	Classification
-1	1	Absolute maximum (endpoint)
0	0	Absolute minimum (singular point)
1	1	Absolute maximum (endpoint)

Notice that the absolute maximum value of f is achieved at two values of t: $t = -1$ and $t = 1$.

Before we go on . . . What exactly happens at the singular point at $x = 0$? The derivative is not defined at this point. Let's use limits to investigate what happens to the derivative as we approach 0 from either side:

$$\lim_{t \to 0^-} f'(t) = \lim_{t \to 0^-} \frac{2}{3t^{1/3}} = -\infty$$

$$\lim_{t \to 0^+} f'(t) = \lim_{t \to 0^+} \frac{2}{3t^{1/3}} = +\infty$$

Thus, the graph decreases very steeply approaching $t = 0$ from the left and then rises very steeply as it leaves to the right. It would make sense to say that the tangent line at $x = 0$ is vertical.

In Examples 1 and 3, we could have found the absolute maxima and minima without doing any graphing. In Example 1, after finding the critical points and endpoints, we created the following table:

x	0	1	4
$f(x)$	0	-1	8

From this table we can see that f must decrease from its value of zero at $x = 0$ to -1 at $x = 1$ and then increase to 8 at $x = 4$. The value of 8 must be the largest value it takes on, and the value of -1 must be the smallest, on the interval $[0, 4]$. Similarly, in Example 3 we created the following table:

t	-1	0	1
$f(t)$	1	0	1

From this table we can see that the largest value of f on the interval $[-1, 1]$ is 1 and the smallest value is 0. We are taking advantage of the following fact, whose proof uses some deep and beautiful mathematics (alas, beyond the scope of this book).

Absolute Extrema on a Closed Interval

If f is *continuous* on a closed interval $[a, b]$, then it will have an absolute maximum and an absolute minimum value on that interval. Each absolute extremum must occur either at an endpoint or a critical point. Therefore, the absolute maximum is the largest value in a table of the values of f at the endpoints and critical points, and the absolute minimum is the smallest value.

Quick Example

The function $f(x) = 3x - x^3$ on the interval $[0, 2]$ has one critical point at $x = 1$. The values of f at the critical point and the endpoints of the interval are given in the following table:

x	0	1	2
$f(x)$	0	2	-2

From this table we can say that the absolute maximum value of f on $[0, 2]$ is 2, which occurs at $x = 1$, and the absolute minimum value of f is -2, which occurs at $x = 2$.

As we can see in Example 2 and the following examples, if the domain is not a closed interval, then f may not even have an absolute maximum and minimum, and a table of values is of little help in determining whether it does.

Example 4 • Domain Not a Closed Interval

Find all extrema of $f(x) = x + \dfrac{1}{x}$.

Solution Since no domain is specified, we take the domain to be as large as possible. The function is not defined at $x = 0$ but is at all other points, so we take its domain to be $(-\infty, 0) \cup (0, +\infty)$. We calculate

$$f'(x) = 1 - \frac{1}{x^2}$$

- *Stationary Points* Setting $f'(x) = 0$, we solve

$$1 - \frac{1}{x^2} = 0$$

$$1 = \frac{1}{x^2}$$

$$x^2 = 1 \qquad \text{Multiply both sides by } x^2.$$

$$x = \pm 1$$

Calculating the corresponding values of f, we get the two stationary points $(1, 2)$ and $(-1, -2)$.

- *Singular Points* The only value of x for which $f'(x)$ is not defined is $x = 0$, but then f is not defined there either, so there are no singular points in the domain.

- *Endpoints* The domain, $(-\infty, 0) \cup (0, +\infty)$, has no endpoints.

From this scant information, it is hard to tell what f does. If we are sketching the graph by hand, we will need to plot additional test points to the left and right of the stationary points $x = \pm 1$.

T *Technology*

For the technology approach, let's choose a viewing window with an interval of $[-3, 3]$ for x and $[-4, 4]$ for y, which should leave plenty of room to see how f behaves near the stationary points. The result is something like Figure 12.

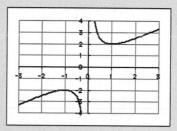

Figure 12

From this graph we can see that f has
- A relative maximum at the point $(-1, -2)$
- A relative minimum at the point $(1, 2)$

Curiously, the relative maximum is lower than the relative minimum! Notice also that, because of the break in the graph at $x = 0$, the graph did not need to rise to get from $(-1, -2)$ to $(1, 2)$.

So far we have been solving the equation $f'(x) = 0$ to obtain our candidates for stationary extrema. However, it is often not easy—or even possible—to solve equations analytically. In Example 5 we show a way around this problem by using graphing technology.

T **Example 5 • Using Technology**

Graph the function $f(x) = (x - 1)^{2/3} - x^2/2$ with domain $[-2, +\infty)$. Also graph its derivative and hence locate and classify all extrema of f, with coordinates accurate to two decimal places.

Solution In Example 1 of Section 3.5, we saw how to draw the graphs of f and f' using technology.[5] Note that the technology formula to use is

```
((x-1)^2)^(1/3)-0.5*x^2
```

instead of

```
(x-1)^(2/3)-0.5*x^2
```

(Why?)

---

[5] Alternatively, download and use the Excel Derivative Grapher worksheet by following the path
Web Site → Online Utilities → Excel Derivative Grapher

Figure 13 shows the resulting graphs.

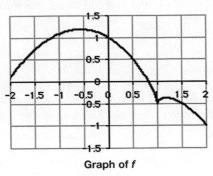

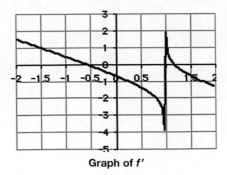

Graph of *f*   Graph of *f′*

Figure 13

If we extend Xmax beyond $x = 2$, we find that the graph continues downward, apparently without any further interesting behavior.

- *Stationary Points* The graph of *f* shows two stationary points, both maxima, at around $x = -0.6$ and $x = 1.2$. Notice that the graph of *f′* is zero at precisely these points. Moreover, it is easier to locate these values accurately on the graph of *f′* because it is easier to pinpoint where a graph crosses the *x* axis than to locate a stationary point. Zooming in to the stationary point at $x \approx -0.6$ results in Figure 14.

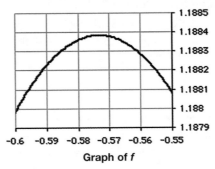

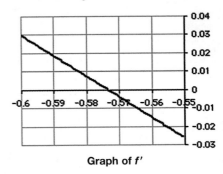

Graph of *f*   Graph of *f′*

Figure 14

From the graph of *f*, we can see that the stationary point is somewhere between -0.58 and -0.57. The graph of *f′* shows more clearly that the zero of *f′*, hence the stationary point of *f*, lies somewhat closer to -0.57 than -0.58. Thus, the stationary point occurs at $x \approx -0.57$, rounded to two decimal places.

In a similar way, we find the second stationary point at $x \approx 1.18$.

- *Singular Points* Going back to Figure 13, we notice what appears to be a cusp (singular point) at the relative minimum around $x = 1$, and this is confirmed by a glance at the graph of *f′*, which seems to take a sudden jump at what value. Zooming in closer suggests that the singular point occurs at exactly $x = 1$. In fact, we can calculate

$$f'(x) = \frac{2}{3(x-1)^{1/3}} - x$$

From this formula we see clearly that $f'(x)$ is defined everywhere except at $x = 1$.

- *Endpoints* The only endpoint in the domain is $x = -2$, which gives a relative minimum.

Therefore, we have found the following approximate extrema for f:

x	$f(x)$	Classification
-2	0.08	Relative minimum (endpoint)
-0.57	1.19	Absolute maximum (stationary point)
1	-0.5	Relative minimum (singular point)
1.18	-0.38	Relative maximum (stationary point)

5.1 EXERCISES

In Exercises 1–12, locate and classify all extrema in each graph. (By *classifying* the extrema, we mean listing whether each extremum is a relative or absolute maximum or minimum.) Also, locate any stationary points or singular points that are not relative extrema.

1.

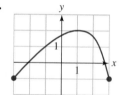

2.

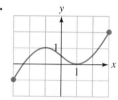

3.

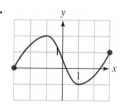

4.

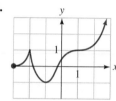

5.

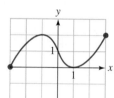

6.

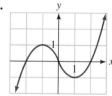

7.

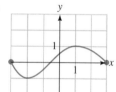

8.

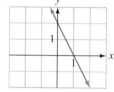

9.

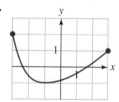

10.

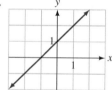

11.

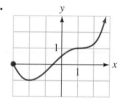

12.

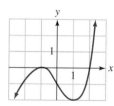

In Exercises 13–20, the graph of the *derivative* of a function f is shown. Determine the x coordinates of all stationary and singular points of f and classify each as a relative maximum, relative minimum, or neither. [Assume that $f(x)$ is defined and continuous everywhere in $[-3, 3]$.]

13.

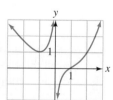

14.

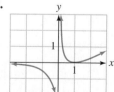

15.

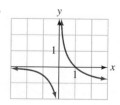

16.

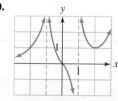

17.

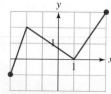

18.

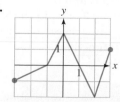

19.

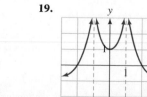

20.

In Exercises 21–48, find the exact location of all the relative and absolute extrema of each function.

21. $f(x) = x^2 - 4x + 1$ with domain $[0, 3]$

22. $f(x) = 2x^2 - 2x + 3$ with domain $[0, 3]$

23. $g(x) = x^3 - 12x$ with domain $[-4, 4]$

24. $g(x) = 2x^3 - 6x + 3$ with domain $[-2, 2]$

25. $f(t) = t^3 + t$ with domain $[-2, 2]$

26. $f(t) = -2t^3 - 3t$ with domain $[-1, 1]$

27. $h(t) = 2t^3 + 3t^2$ with domain $[-2, +\infty)$

28. $h(t) = t^3 - 3t^2$ with domain $[-1, +\infty)$

29. $f(x) = x^4 - 4x^3$ with domain $[-1, +\infty)$

30. $f(x) = 3x^4 - 2x^3$ with domain $[-1, +\infty)$

31. $g(t) = \dfrac{1}{4}t^4 - \dfrac{2}{3}t^3 + \dfrac{1}{2}t^2$ with domain $(-\infty, +\infty)$

32. $g(t) = 3t^4 - 16t^3 + 24t^2 + 1$ with domain $(-\infty, +\infty)$

33. $f(t) = \dfrac{t^2 + 1}{t^2 - 1}; -2 \le t \le 2, t \ne \pm 1$

34. $f(t) = \dfrac{t^2 - 1}{t^2 + 1}$ with domain $[-2, 2]$

35. $f(x) = \sqrt{x}(x - 1); x \ge 0$

36. $f(x) = \sqrt{x}(x + 1); x \ge 0$

37. $g(x) = x^2 - 4\sqrt{x}$

38. $g(x) = \dfrac{1}{x} - \dfrac{1}{x^2}$

39. $g(x) = \dfrac{x^3}{x^2 + 3}$

40. $g(x) = \dfrac{x^3}{x^2 - 3}$

41. $f(x) = x - \ln x$ with domain $(0, +\infty)$

42. $f(x) = x - \ln x^2$ with domain $(0, +\infty)$

43. $g(t) = e^t - t$ with domain $[-1, 1]$

44. $g(t) = e^{-t^2}$ with domain $(-\infty, +\infty)$

45. $f(x) = \dfrac{2x^2 - 24}{x + 4}$

46. $f(x) = \dfrac{x - 4}{x^2 + 20}$

47. $f(x) = xe^{1 - x^2}$

48. $f(x) = x \ln x$ with domain $(0, +\infty)$

T In Exercises 49–52, use graphing technology and the method in Example 5 to find the x coordinates of the critical points, accurate to two decimal places. Find all relative and absolute maxima and minima.

49. $y = x^2 + \dfrac{1}{x - 2}$ with domain $(-3, 2) \cup (2, 6)$

50. $y = x^2 - 10(x - 1)^{2/3}$ with domain $(-4, 4)$

51. $f(x) = (x - 5)^2(x + 4)(x - 2)$ with domain $[-5, 6]$

52. $f(x) = (x + 3)^2(x - 2)^2$ with domain $[-5, 5]$

COMMUNICATION AND REASONING EXERCISES

53. Draw the graph of a function f with domain the set of all real numbers, such that f is not linear and has no relative extrema.

54. Draw the graph of a function g with domain the set of all real numbers, such that g has a relative maximum and minimum but no absolute extrema.

55. Draw the graph of a function that has stationary and singular points but no relative extrema.

56. Draw the graph of a function that has relative, not absolute, maxima and minima but no stationary or singular points.

57. If a stationary point is not a relative maximum, then must it be a relative minimum? Explain your answer.

58. If one endpoint is a relative maximum, must the other be a relative minimum? Explain your answer.

59. We said that if f is continuous on a closed interval $[a, b]$, then it will have an absolute maximum and an absolute minimum. Draw the graph of a function with domain $[0, 1]$ having an absolute maximum but no absolute minimum.

60. Refer to Exercise 59. Draw the graph of a function with domain $[0, 1]$ having no absolute extrema.

5.2 Applications of Maxima and Minima

In many applications we would like to find the largest or smallest possible value of some quantity—for instance, the greatest possible profit or the lowest cost. We call this the *optimal* (best) value. In this section we consider several such examples and use calculus to find the optimal value in each.

In all applications the first step is to translate a written description into a mathematical problem. In the problems we look at in this section, we are asked to find *unknowns;* there is an expression involving those unknowns that must be made as large or as small as possible—the **objective function**; and there may be **constraints**—equations or inequalities relating the variables.[6]

---

[6]If you have studied linear programming, you will notice a similarity here, but unlike the situation in linear programming, neither the objective function nor the constraints need be linear.

Example 1 • Minimizing Average Cost

Gymnast Clothing manufactures expensive hockey jerseys for sale to college bookstores in runs of up to 500. Its cost (in dollars) for a run of x hockey jerseys is

$$C(x) = 2000 + 10x + 0.2x^2$$

How many jerseys should Gymnast produce per run in order to minimize average cost?[7]

Solution Here is the procedure we will follow to solve this problem.

1. *Identify the unknown(s).* There is one unknown: x, the number of hockey jerseys Gymnast should produce per run. (We know this because the question is, How many jerseys . . . ?)

2. *Identify the objective function.* The objective function is the quantity that must be made as small (in this case) as possible. In this example it is the average cost, which is given by

$$\overline{C}(x) = \frac{C(x)}{x} = \frac{2000 + 10x + 0.2x^2}{x} = \frac{2000}{x} + 10 + 0.2x \text{ dollars per jersey}$$

3. *Identify the constraints (if any).* At most 500 jerseys can be manufactured in a run. Also, $\overline{C}(0)$ is not defined. Thus, x is constrained by

$$0 < x \le 500$$

Put another way, the domain of the objective function $\overline{C}(x)$ is (0, 500].

4. *State and solve the resulting optimization problem.* Our optimization problem is to

$$\text{minimize } \overline{C}(x) = \frac{2000}{x} + 10 + 0.2x \qquad \text{Objective function}$$

$$\text{subject to } 0 < x \le 500 \qquad\qquad\qquad \text{Constraint}$$

We now solve this problem as in Section 5.1. We first calculate

$$\overline{C}'(x) = -\frac{2000}{x^2} + 0.2$$

• *Stationary Points* We set $\overline{C}'(x) = 0$ and solve for x:

$$-\frac{2000}{x^2} + 0.2 = 0$$

$$0.2 = \frac{2000}{x^2}$$

$$x^2 = \frac{2000}{0.2} = 10,000$$

$$x = \pm 100$$

We reject $x = -100$ because -100 is not in the domain of $\overline{C}$ (and makes no sense), so we have one stationary point, at $x = 100$. There, the average cost is $\overline{C}(100) = \$50$ per jersey.

---

[7]Why don't we seek to minimize total cost? The answer would be uninteresting; to minimize total cost, we would make *no* jerseys at all. Minimizing the average cost is a more practical objective.

- *Singular Points* The only point at which the formula for $\overline{C}'$ is not defined is $x = 0$, but that is not in the domain of $\overline{C}$, so we have no singular points.
- *Endpoints* We have one endpoint in the domain, at $x = 500$. There, the average cost is $\overline{C}(500) = \$114$.

Technology

Let's plot $\overline{C}$ in a viewing window with the intervals $[0, 500]$ for x and $[0, 150]$ for y, which will show the whole domain and the two interesting points we've found so far. The result is Figure 15.

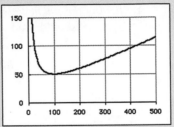

Figure 15

From the graph of $\overline{C}$, we can see that the stationary point at $x = 100$ gives the absolute minimum. We can therefore say that Gymnast Clothing should produce 100 jerseys per run, for a lowest possible average cost of $50 per jersey.

Example 2 • Maximizing Area

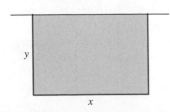

Figure 16

Figure 17

Slim wants to build a rectangular enclosure for his pet rabbit, Killer, against the side of his house, as shown in Figure 16. He has bought 100 feet of fencing. What are the dimensions of the largest area that he can enclose?

Solution

1. *Identify the unknown(s).* To identify the unknown(s), we look at the question: What are the *dimensions* of the largest area he can enclose? Thus, the unknowns are the dimensions of the fence. We call these x and y, as shown in Figure 17.

2. *Identify the objective function.* We look for what it is that we are trying to maximize (or minimize). The phrase "largest area" tells us that our object is to *maximize the area*, which is the product of length and width, so our objective function is

$$A = xy$$

3. *Identify the constraints (if any).* What stops Slim from making the area as large as he wants? He has only 100 feet of fencing to work with. Looking again at Figure 17, we see that the sum of the lengths of the three sides must equal 100, so

$$x + 2y = 100$$

One more point: Because x and y represent the lengths of the sides of the enclosure, neither can be a negative number.

4. *State and solve the resulting optimization problem.* Our mathematical problem is

$$\text{maximize } A = xy \qquad\qquad \text{Objective function}$$

$$\text{subject to } x + 2y = 100, x \geq 0, \text{ and } y \geq 0 \qquad \text{Constraints}$$

We know how to find maxima and minima of a function of one variable, but A appears to depend on two variables. We can remedy this by using a constraint to express one variable in terms of the other. Let's take the constraint $x + 2y = 100$ and solve for x in terms of y:

$$x = 100 - 2y$$

Substituting into the objective function gives

$$A = xy = (100 - 2y)y = 100y - 2y^2$$

and we have eliminated x from the objective function. What about the inequalities? One says that $x \geq 0$, but we want to eliminate x from this as well. We substitute for x again, getting

$$100 - 2y \geq 0$$

Solving this inequality for y gives $y \leq 50$. The second inequality says that $y \geq 0$. Now we can restate our problem with x eliminated:

$$\text{maximize } A(y) = 100y - 2y^2 \text{ subject to } 0 \leq y \leq 50$$

We now proceed with our usual method of solving such problems. We calculate $A'(y) = 100 - 4y$.

- *Stationary Points* Solving $100 - 4y = 0$, we get one stationary point at $y = 25$. There, $A(25) = 1250$.
- *Singular Points* There are no points at which $A'(y)$ is not defined.
- *Endpoints* We have two endpoints, at $y = 0$ and $y = 50$. The corresponding areas are $A(0) = 0$ and $A(50) = 0$.

We record the three points we found in a table:

y	0	25	50
$A(y)$	0	1250	0

It's clear now how A must behave: It increases from zero at $y = 0$ to 1250 at $y = 25$ and then decreases back to zero at $y = 50$. Thus, the largest possible value of A is 1250 square feet, which occurs when $y = 25$. To completely answer the question that was asked, we need to know the corresponding value of x. We have $x = 100 - 2y$, so $x = 50$ when $y = 25$. Thus, Slim should build his enclosure 50 feet across and 25 feet deep (with the "missing" 50-foot side being formed by part of the house).

Before we go on . . . Notice that the problem here came down to finding the absolute maximum value of A on the closed and bounded interval $[0, 50]$. As we noted in Section 5.1, the table of values of A at its critical points and the endpoints of the interval gives us enough information to find the absolute maximum.

Let's stop for a moment and summarize the steps we've taken in these two examples.

Solving an Optimization Problem

1. *Identify the unknown(s), possibly with the aid of a diagram.* These are usually the quantities asked for in the problem.

2. *Identify the objective function.* This is the quantity you are asked to maximize or minimize. You should name it explicitly, as in "let S = surface area."

3. *Identify the constraint(s).* These can be equations relating variables or inequalities expressing limitations on the values of variables.

4. *State the optimization problem.* This will have the form "maximize [minimize] the objective function subject to the constraint(s)."

5. *Eliminate extra variables.* If the objective function depends on several variables, solve the constraint equations to express all variables in terms of one particular variable. Substitute these expressions into the objective function to rewrite it as a function of a single variable. Substitute the expressions into any inequality constraints to help determine the domain of the objective function.

6. *Find the absolute maximum (or minimum) of the objective function.* Use the techniques of Section 5.1.

Example 3 • Maximizing Revenue

Cozy Carriage Company builds baby strollers. Using market research, the company estimates that if it sets the price of a stroller at p dollars, then it can sell $q = 300{,}000 - 10p^2$ strollers per year. What price will bring in the greatest annual revenue?

Solution The question we are asked identifies our main unknown, the price p. However, there is another quantity that we do not know, q, the number of strollers the company will sell each year. The question also identifies the objective function, revenue, which is

$$R = pq$$

Including the equality constraint given to us, that $q = 300{,}000 - 10p^2$, and the "reality" inequality constraints $p \geq 0$ and $q \geq 0$, we can write our problem as

$$\text{maximize } R = pq \text{ subject to } q = 300{,}000 - 10p^2, p \geq 0, \text{ and } q \geq 0$$

We need to eliminate one of our unknowns so that R is expressed as a function of one variable alone. But we are given q in terms of p, so let's substitute to eliminate q:

$$R = pq = p(300{,}000 - 10p^2) = 300{,}000p - 10p^3$$

Substituting in the inequality $q \geq 0$, we get

$$300{,}000 - 10p^2 \geq 0$$

Thus, $p^2 \leq 30{,}000$, which gives $-100\sqrt{3} \leq p \leq 100\sqrt{3}$. When we combine this with $p \geq 0$, we get the following restatement of our problem:

$$\text{maximize } R(p) = 300{,}000p - 10p^3 \text{ such that } 0 \leq p \leq 100\sqrt{3}$$

We solve this problem in much the same way we did the preceding one. We calculate $R'(p) = 300,000 - 30p^2$. Setting $300,000 - 30p^2 = 0$, we find one stationary point at $p = 100$. There are no singular points, and we have the endpoints $p = 0$ and $p = 100\sqrt{3}$. Putting these points in a table and computing the corresponding values of R, we get the following:

p	0	100	$100\sqrt{3}$
$R(p)$	0	20,000,000	0

Thus, Cozy Carriage should price its strollers at $100 each, which will bring in the largest possible revenue of $20,000,000.

Example 4 • Maximizing Resources

Figure 18

Metal Can Company has an order to make cans with a volume of 250 cubic centimeters. What should be the dimensions of the cans in order to use the least amount of metal in their production?

Solution We are asked to find the dimensions of the cans. It is traditional to take as the dimensions of a cylinder the height h and the radius of the base r, as in Figure 18. We are also asked to minimize the total amount of metal used in the can, which is the area of the surface of the cylinder. We can look up the formula or figure it out ourselves: Imagine removing the circular top and bottom and then cutting vertically and flattening out the hollow cylinder to get a rectangle, as shown in Figure 19.

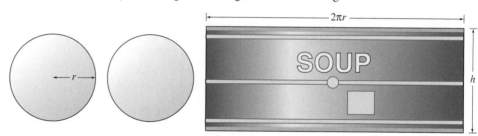

Figure 19

Our objective function is the (total) surface area S of the can. The area of each disc is πr^2, and the area of the rectangular piece is $2\pi rh$. Thus, our objective function is

$$S = 2\pi r^2 + 2\pi rh$$

As usual, there is a constraint: The volume must be exactly 250 cubic centimeters. The formula for the volume of a cylinder is $V = \pi r^2 h$, so

$$\pi r^2 h = 250$$

It is easiest to solve this constraint for h in terms of r:

$$h = \frac{250}{\pi r^2}$$

Substituting in the objective function, we get

$$S = 2\pi r^2 + 2\pi r\frac{250}{\pi r^2} = 2\pi r^2 + \frac{500}{r}$$

Now r cannot be negative or 0, but it can become very large (a very wide but very short can could have the right volume). We therefore take the domain of $S(r)$ to be $(0, +\infty)$, so our mathematical problem is as follows:

$$\text{minimize } S(r) = 2\pi r^2 + \frac{500}{r} \text{ subject to } r > 0$$

Now we calculate

$$S'(r) = 4\pi r - \frac{500}{r^2}$$

To find stationary points, we set this equal to 0 and solve:

$$4\pi r - \frac{500}{r^2} = 0$$

$$4\pi r = \frac{500}{r^2}$$

$$4\pi r^3 = 500$$

$$r^3 = \frac{125}{\pi}$$

$$r = \sqrt[3]{\frac{125}{\pi}} = \frac{5}{\sqrt[3]{\pi}} \approx 3.41$$

The corresponding surface area is approximately $S(3.41) \approx 220$. There are no singular points or endpoints in the domain.

Technology

To see how S behaves near the one stationary point, let's graph it in a viewing window with interval $[0, 5]$ for r and $[0, 300]$ for S. The result is Figure 20. From the graph we can clearly see that the smallest surface area occurs at the stationary point at $r \approx 3.41$. The height of the can will be

$$h = \frac{250}{\pi r^2} \approx 6.83$$

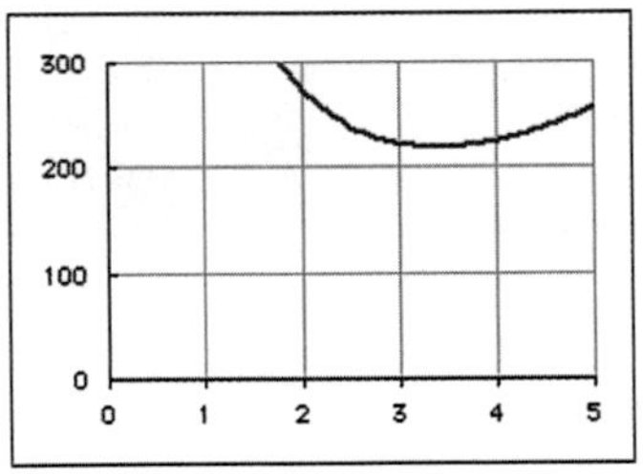

Figure 20

Thus, the can that uses the least amount of metal has a height of approximately 6.83 centimeters and a radius of approximately 3.41 centimeters. Such a can will use approximately 220 square centimeters of metal.

✱ **Before we go on . . .** If we substitute the exact expression

$$r = \sqrt[3]{\frac{125}{\pi}} = \frac{5}{\sqrt[3]{\pi}}$$

into the formula for h, we get

$$h = \frac{10}{\sqrt[3]{\pi}}$$

which is exactly twice r. Put another way, the height is exactly equal to the diameter so that the can looks square when viewed from the side. Have you ever seen a can with that shape? Why do you think most cans do not have this shape?

Example 5 • Allocation of Labor

Gym Sock Company manufactures cotton athletic socks. Production is partially auto-mated through the use of robots. Daily operating costs amount to $50 per laborer and $30 per robot. The number of pairs of socks the company can manufacture in a day is given by a Cobb–Douglas[8] production formula

$$q = 50n^{0.6}r^{0.4}$$

where q is the number of pairs of socks that can be manufactured by n laborers and r robots. Assuming that the company wishes to produce 1000 pairs of socks per day at a minimum cost, how many laborers and how many robots should it use?

Solution The unknowns are the number of laborers n and the number of robots r. The objective is to minimize the daily cost:

$$C = 50n + 30r$$

The constraints are given by the daily quota

$$1000 = 50n^{0.6}r^{0.4}$$

and the fact that n and r are nonnegative. We solve the constraint equation for one of the variables; let's solve for n:

$$n^{0.6} = \frac{1000}{50r^{0.4}} = \frac{20}{r^{0.4}}$$

Taking the $1/0.6$ power of both sides gives

$$n = \left(\frac{20}{r^{0.4}}\right)^{1/0.6} = \frac{20^{1/0.6}}{r^{0.4/0.6}} = \frac{20^{5/3}}{r^{2/3}} \approx \frac{147.36}{r^{2/3}}$$

Substituting in the objective equation gives us the cost as a function of r:

$$C(r) \approx 50\left(\frac{147.36}{r^{2/3}}\right) + 30r$$

$$= 7368r^{-2/3} + 30r$$

The only remaining constraint on r is that $r > 0$. To find the minimum value of $C(r)$, we first take the derivative:

$$C'(r) \approx -4912r^{-5/3} + 30$$

Setting this equal to zero, we solve for r:

$$-4912r^{-5/3} + 30 = 0$$

$$-4912r^{-5/3} = -30$$

$$r^{-5/3} \approx 0.006\,107$$

$$r \approx (0.006\,107)^{-3/5} \approx 21.3$$

The corresponding cost is $C(21.3) \approx \$1600$. There are no singular points or endpoints in the domain of C.

---

[8]Cobb–Douglas production formulas were discussed in Section 4.4.

Technology
To see how C behaves near its stationary point, let's draw its graph in a viewing window with an interval of $[0, 40]$ for r and $[0, 2000]$ for C. The result is Figure 21. From the graph we can see that C does have its minimum at the stationary point. The corresponding value of n is

$$n \approx \frac{147.36}{r^{2/3}} \approx 19.2$$

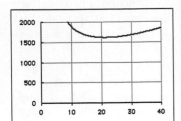

Figure 21

At this point, our solution appears to be this: Use (approximately) 19.2 laborers and (approximately) 21.3 robots to meet the manufacturing quota at a minimum cost. However, we are not interested in fractions of robots or people, so we need to find integer solutions for n and r. If we round these numbers, we get the solution $(n, r) = (19, 21)$. However, a quick calculation shows that

$$q = 50(19)^{0.6}(21)^{0.4} \approx 989 \text{ pairs of socks}$$

which fails to meet the quota of 1000. Thus, we need to round at least one of the quantities n and r *upward* in order to meet the quota. The three possibilities, with corresponding values of q and C, are as follows:

$(n, r) = (20, 21)$, with $q \approx 1020$ and $C = \$1630$

$(n, r) = (19, 22)$, with $q \approx 1007$ and $C = \$1610$

$(n, r) = (20, 22)$, with $q \approx 1039$ and $C = \$1660$

Of these, the solution that meets the quota at a minimum cost is $(n, r) = (19, 22)$. Thus, Gym Sock should use 19 laborers and 22 robots, at a cost of $50 \times 19 + 30 \times 22 = \$1610$, to manufacture $50 \times 19^{0.6} \times 22^{0.4} \approx 1007$ pairs of socks.

5.2 EXERCISES

In Exercises 1–8, solve the optimization problems.

1. Maximize $P = xy$, with $x + y = 10$.

2. Maximize $P = xy$, with $x + 2y = 40$.

3. Minimize $S = x + y$, with $xy = 9$ and both x and $y > 0$.

4. Minimize $S = x + 2y$, with $xy = 2$ and both x and $y > 0$.

5. Minimize $F = x^2 + y^2$, with $x + 2y = 10$.

6. Minimize $F = x^2 + y^2$, with $xy^2 = 16$.

7. Maximize $P = xyz$, with $x + y = 30$ and $y + z = 30$ and $x, y,$ and $z \geq 0$.

8. Maximize $P = xyz$, with $x + z = 12$ and $y + z = 12$ and $x, y,$ and $z \geq 0$.

9. For a rectangle with perimeter 20 to have the largest area, what dimensions should it have?

10. For a rectangle with area 100 to have the smallest perimeter, what dimensions should it have?

APPLICATIONS

11. Fences I want to fence in a rectangular vegetable patch. The fencing for the east and west sides costs $4/foot, and the fencing for the north and south sides costs only $2/foot. I have a budget of $80 for the project. What is the largest area I can enclose?

12. Fences My orchid garden abuts my house so that the house itself forms the northern boundary. The fencing for the southern boundary costs $4/foot, and the fencing for the east and west sides costs $2/foot. If I have a budget of $80 for the project, what is the largest area I can enclose this time?

13. Revenue Hercules Films is deciding on the price of the video release of its film *Son of Frankenstein*. Its marketing people estimate that at a price of p dollars it can sell a total of $q = 200,000 - 10,000p$ copies. What price will bring in the greatest revenue?

14. Profit Hercules Films is also deciding on the price of the video release of its film *Bride of the Son of Frankenstein.* Again, marketing estimates that at a price of p dollars it can sell $q = 200,000 - 10,000p$ copies, but each copy costs $4 to make. What price will give the greatest profit?

15. Revenue The demand for rubies at Royal Ruby Retailers (RRR) is given by the equation

$$q = -\frac{4}{3}p + 80$$

where p is the price RRR charges (in dollars) and q is the number of rubies RRR sells each week. At what price should RRR sell its rubies in order to maximize its weekly revenue?

16. Revenue The consumer demand curve for tissues is given by

$$q = (100 - p)^2 \qquad (0 \le p \le 100)$$

where p is the price per case of tissues and q is the demand in weekly sales. At what price should tissues be sold in order to maximize revenue?

17. Revenue Assume that the demand for tuna in a small coastal town is given by

$$p = \frac{500,000}{q^{1.5}}$$

where q is the number of pounds of tuna that can be sold in a month at p dollars per pound. Assume that the town's fishery wishes to sell at least 5000 pounds of tuna each month.

a. How much should the town's fishery charge for tuna in order to maximize monthly revenue?

b. How much tuna will it sell each month at that price?

c. What will be its resulting revenue?

18. Revenue Economist Henry Schultz devised the following demand function for corn:

$$p = \frac{6,570,000}{q^{1.3}}$$

where q is the number of bushels of corn that could be sold at p dollars per bushel in one year. Assume that at least 10,000 bushels of corn must be sold each year.

a. How much should farmers charge per bushel of corn to maximize annual revenue?

b. How much corn can farmers sell each year at that price?

c. What will be the farmers' resulting revenue?

Based on data for the period 1915–1929. Source: Henry Schultz, *The Theory and Measurement of Demand,* as cited in A. L. Ostrosky, Jr., and J. V. Koch, *Introduction to Mathematical Economics* (Prospect Heights, Ill.: Waveland Press, 1979).

19. Revenue The wholesale price for chicken in the United States fell from 25¢ per pound to 14¢ per pound while per capita chicken consumption rose from 22 pounds per year to 27.5 pounds per year. Assuming that the demand for chicken depends linearly on the price, what wholesale price for chicken maximizes revenues for poultry farmers, and what does that revenue amount to?

Data are provided for the years 1951–1958. Source: U.S. Dept. of Agriculture, *Agricultural Statistics.*

20. Revenue Your underground used-book business is booming. Your policy is to sell all used versions of *Calculus and You* at the same price (regardless of condition). When you set the price at $10, sales amounted to 120 volumes during the first week of classes. The following semester, you set the price at $30 and sold not a single book. Assuming that the demand for books depends linearly on the price, what price gives you the maximum revenue, and what does that revenue amount to?

21. Profit The demand for rubies at Royal Ruby Retailers (RRR) is given by the equation

$$q = -\frac{4}{3}p + 80$$

where p is the price RRR charges (in dollars) and q is the number of rubies RRR sells each week. Assuming that due to extraordinary market conditions, RRR can obtain rubies for $25 each, how much should it charge per ruby to make the greatest possible weekly profit, and what will that profit be?

22. Profit The consumer demand curve for tissues is given by

$$q = (100 - p)^2 \qquad (0 \le p \le 100)$$

where p is the price per case of tissues and q is the demand in weekly sales. If tissues cost $30 per case to produce, at what price should tissues be sold for the largest possible weekly profit? What will that profit be?

23. Profit The demand equation for your company's virtual reality video headsets is

$$p = \frac{1000}{q^{0.3}}$$

where q is the total number of headsets that your company can sell in a week at a price of p dollars. The total manufacturing and shipping cost amounts to $100 per headset.

a. What is the greatest profit your company can make in a week, and how many headsets will your company sell at this level of profit? (Give answers to the nearest whole number.)

b. How much to the nearest $1 should your company charge per headset for the maximum profit?

24. Profit Due to sales by a competing company, your company's sales of virtual reality video headsets have dropped, and your financial consultant revises the demand equation to

$$p = \frac{800}{q^{0.35}}$$

where q is the total number of headsets that your company can sell in a week at a price of p dollars. The total manufacturing and shipping cost still amounts to $100 per headset.

a. What is the greatest profit your company can make in a week, and how many headsets will your company sell at this level of profit? (Give answers to the nearest whole number.)

b. How much to the nearest $1 should your company charge per headset for the maximum profit?

25. Average Cost The cost function for the manufacture of portable CD players is given by

$$C(x) = 150{,}000 + 20x + \frac{x^2}{10{,}000} \text{ dollars}$$

where x is the number of CD players manufactured. How many CD players should be manufactured in order to minimize average cost? What is the resulting average cost of a CD player? (Give your answer to the nearest dollar.)

26. Average Cost Repeat Exercise 25, using the revised cost function

$$C(x) = 150{,}000 + 20x + \frac{x^2}{100} \text{ dollars}$$

27. Net Cost The cost of controlling emissions at a firm rises rapidly as the amount of emissions reduced increases. Here is a possible model:

$$C(q) = 4000 + 100q^2$$

where q is the reduction in emissions (in pounds of pollutant per day) and C is the daily cost to the firm (in dollars) of this reduction. Government clean-air subsidies amount to $500 per pound of pollutant removed. How many pounds of pollutant should the firm remove each day in order to minimize *net* cost (cost minus subsidy)?

28. Net Cost Repeat Exercise 27, using the following cost function:

$$C(q) = 2000 + 200q^2$$

with government subsidies amounting to $100/pound of pollutant removed each day.

29. Box Design Chocolate Box Company is going to make open-topped boxes out of 6-inch × 16-inch rectangles of cardboard by cutting squares out of the corners and folding up the sides. What is the largest volume box it can make this way?

30. Box Design A packaging company is going to make open-topped boxes, with square bases, that hold 108 cubic centimeters. What are the dimensions of the box that can be built with the least material?

31. Luggage Dimensions American Airlines requires that the total outside dimensions (length + width + height) of a checked bag not exceed 62 inches. Suppose you want to check a bag whose height equals its width. What is the largest volume bag of this shape that you can check on an American flight?

Source: According to information on its Web site (http://www.aa.com/) as of August 2002.

32. Luggage Dimensions American Airlines requires that the total outside dimensions (length + width + height) of a carry-on bag not exceed 45 inches. Suppose you want to carry on a bag whose length is twice its height. What is the largest volume bag of this shape that you can carry on an American flight?

Source: According to information on its Web site (http://www.aa.com/) as of August 2002.

33. Luggage Dimensions Fly-by-Night Airlines has a peculiar rule about luggage: The length and width of a bag must add to 45 inches, and the width and height must also add to 45 inches. What are the dimensions of the bag with largest volume that Fly-by-Night will accept?

34. Luggage Dimensions Fair Weather Airlines has a similar rule. It will accept only bags for which the sum of the length and width is 36 inches and the sum of length, height, and twice the width is 72 inches. What are the dimensions of the bag with the largest volume that Fair Weather will accept?

35. Package Dimensions The U.S. Postal Service (USPS) will accept packages only if the length plus girth is no more than 108 inches:[9]

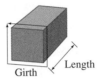
Girth   Length

Assuming that the front face of the package (as shown in the figure) is square, what is the largest volume package that the USPS will accept?

36. Package Dimensions United Parcel Service (UPS) will only accept packages with a length no more than 108 inches and length plus girth no more than 130 inches.[10] (See figure for Exercise 35.) Assuming that the front face of the package (as shown in the figure) is square, what is the largest volume package that UPS will accept?

37. Cell Phone Revenues The number of cell phone subscribers in China in the years 2000–2005 was projected to follow the equation $N(t) = 39t + 68$ million subscribers in year t ($t = 0$ represents January 2000). The average annual revenue per cell phone user was $350 in 2000. If we assume that due to competition the revenue per cell phone user decreases continuously at an annual rate of 30%, we can model the annual revenue as

$$R(t) = 350(39t + 68)e^{-0.3t} \text{ million dollars}$$

Determine (a) when to the nearest 0.1 year the revenue is projected to peak and (b) the revenue to the nearest $1 million at that time.

Based on a regression of projected figures (coefficients are rounded). Source: Intrinsic Technology/*New York Times*, November 24, 2000, p. C1.

38. Cell Phone Revenues Refer to Exercise 37. If we assume instead that the revenue per cell phone user decreases continuously at an annual rate of 20%, we obtain the revenue model

$$R(t) = 350(39t + 68)e^{-0.2t} \text{ million dollars}$$

Determine (a) when to the nearest 0.1 year the revenue is projected to peak and (b) the revenue to the nearest $1 million at that time.

---

[9]The requirement for priority mail as of July 2002.

[10]The requirement as of July 2002.

39. Research and Development Spending on research and development by U.S. drug companies t years after 1970 can be modeled by

$$S(t) = 2.5e^{0.08t} \text{ billion dollars} \qquad (0 \le t \le 31)$$

The number of new drugs approved by the Federal Drug Administration (FDA) over the same period can be modeled by

$$D(t) = 10 + t \text{ drugs per year} \qquad (0 \le t \le 31)$$

When was the function $D(t)/S(t)$ a maximum? What is the maximum value of $D(t)/S(t)$? What does the answer tell you about the cost of developing new drugs?

The exponential model for research and development is based on the 1970 and 2001 spending in constant 2001 dollars, and the linear model for new drugs approved is based on the 6-year moving average from data from 1970 to 2000. Source: Pharmaceutical Research and Manufacturers of America, FDA/*New York Times*, April 19, 2002, p. C1.

40. Research and Development Refer to Exercise 39. If the number of new drugs approved by the FDA had been $10 + 2t$ new drugs each year, when would the function $D(t)/S(t)$ have reached a maximum? What does the answer tell you about the cost of developing new drugs?

41. Asset Appreciation As the financial consultant to a classic auto dealership, you estimate that the total value (in dollars) of its collection of 1959 Chevrolets and Fords is given by the formula

$$v = 300,000 + 1000t^2$$

where t is the number of years from now. You anticipate a continuous inflation rate of 5% per year so that the discounted (present) value of an item that will be worth $\$v$ in t years' time is

$$p = ve^{-0.05t}$$

When would you advise the dealership to sell the vehicles to maximize their discounted value?

42. Plantation Management The value of a fir tree in your plantation increases with the age of the tree according to the formula

$$v = \frac{20t}{1 + 0.05t}$$

where t is the age of the tree in years. Given a continuous inflation rate of 5% per year, the discounted (present) value of a newly planted seedling is

$$p = ve^{-0.05t}$$

At what age (to the nearest year) should you harvest your trees in order to ensure the greatest possible discounted value?

43. Marketing Strategy FeatureRich Software Company has a dilemma. Its new program, Doors-X 2005, is almost ready to go on the market. However, the longer the company works on it, the better it can make the program and the more it can

charge for it. The company's marketing analysts estimate that if it delays t days, it can set the price at $100 + 2t$ dollars. On the other hand, the longer it delays, the more market share they will lose to their main competitor (see Exercise 44) so that if it delays t days it will be able to sell $400,000 - 2500t$ copies of the program. How many days should FeatureRich delay the release in order to get the greatest revenue?

44. Marketing Strategy FeatureRich's main competitor (see Exercise 43) is Moon Systems, and Moon is in a similar predicament. Its product, Walls-Y 2005, could be sold now for $200, but for each day Moon delays, it could increase the price by $4. On the other hand, it could sell 300,000 copies now, but each day it waits will cut sales by 1500. How many days should Moon delay the release in order to get the greatest revenue?

45. Average Profit FeatureRich Software Company sells its graphing program, Dogwood, with a volume discount.[11] If a customer buys x copies, then he pays $\$500\sqrt{x}$. It cost the company $10,000 to develop the program and $2 to manufacture each copy. If a single customer were to buy all copies of Dogwood, how many copies would the customer have to buy for FeatureRich's average profit per copy to be maximized? How are average profit and marginal profit related at this number of copies?

46. Average Profit Repeat Exercise 45 with the charge to the customer $\$600\sqrt{x}$ and the cost to develop the program $9000.

47. Prison Population The U.S. prison population followed the curve

$$N(t) = 0.028234t^3 - 1.0922t^2 + 13.029t + 146.88$$
$$(0 \le t \le 39)$$

in the years 1950–1989. Here t is the number of years since 1950, and N is the number of prisoners in thousands. When to the nearest year was the prison population decreasing most rapidly? When was it increasing most rapidly?

Source: The model is the authors' from data in *Sourcebook of Criminal Justice Statistics, 1990,* U.S. Department of Justice, Bureau of Justice Statistics (Washington D.C.: Government Printing Office): 604.

48. Test Scores Combined SAT scores in the United States in the years 1967–1989 could be approximated by

$$T(t) = -0.01085t^3 + 0.5804t^2 - 10.12t + 962.4$$
$$(0 \le t \le 22)$$

where t is the number of years since 1967 and T is the combined SAT score average. Based on this model, when (to the nearest year) was the average SAT score decreasing most rapidly? When was it increasing most rapidly?

The model is the authors'. Source: Educational Testing Service.

----

[11]This is similar to the way site licenses have been structured for the program Maple®.

49. Embryo Development The oxygen consumption of a bird embryo increases from the time the egg is laid through the time the chick hatches. In the case of a typical galliform bird, the oxygen consumption (in milliliters per hour) can be approximated by

$$c(t) = -0.00271t^3 + 0.137t^2 - 0.892t + 0.149 \quad (8 \le t \le 30)$$

where t is the time (in days) since the egg was laid. (An egg will typically hatch at around $t = 28$.) When to the nearest day is $c'(t)$ a maximum? What does the answer tell you?

The model approximates graphical data published in Roger S. Seymour, "The Brush Turkey," *Scientific American* (December 1991): 108–114.

50. Embryo Development The oxygen consumption of a turkey embryo increases from the time the egg is laid through the time the chick hatches. In the case of a brush turkey, the oxygen consumption (in milliliters per hour) can be approximated by

$$c(t) = -0.00118t^3 + 0.119t^2 - 1.83t + 3.972 \quad (20 \le t \le 50)$$

where t is the time (in days) since the egg was laid. (An egg will typically hatch at around $t = 50$.) When to the nearest day is $c'(t)$ a maximum? What does the answer tell you?

The model approximates graphical data published in Roger S. Seymour, "The Brush Turkey," *Scientific American* (December 1991): 108–114.

51. Minimizing Resources Basic Buckets has an order for plastic buckets holding 5000 cubic centimeters. The buckets are open-topped cylinders, and the company wants to know what dimensions will use the least plastic per bucket. (The volume of an open-topped cylinder with height h and radius r is $\pi r^2 h$, and the surface area is $\pi r^2 + 2\pi rh$.)

52. Optimizing Capacity Basic Buckets would like to build a bucket with a surface area of 1000 square centimeters. What is the volume of the largest bucket they can build? (See Exercise 51.)

53. Agriculture The fruit yield per tree in an orchard containing 50 trees is 100 pounds per tree each year. Due to crowding, the yield decreases by 1 pound per season for every additional tree planted. How may additional trees should be planted for a maximum total annual yield?

54. Agriculture Two years ago your orange orchard contained 50 trees, and the total yield was 75 bags of oranges. Last year you removed 10 of the trees and noticed that the total yield increased to 80 bags. Assuming that the yield per tree depends linearly on the number of trees in the orchard, what should you do this year to maximize your total yield?

For Exercises 55–62, the use of technology is required.

55. Online Book Sales The number of books sold online each year in the United States for the period 1997–2000 can be modeled by

$$N(t) = \frac{82.8}{1 + 21.8(7.14)^{-t}} \text{ million books} \quad (0 \le t \le 3)$$

($t = 0$ represents 1997). During which quarter were online book sales increasing most rapidly? How fast to the nearest 0.1 million books per year were book sales increasing at that time?

The model is a logistic regression. Source: Ipsos-NPD Book Trends/*New York Times*, April 16, 2001, p. C1.

56. Prison Population The prison population in New York State for the period 1970–1997 can be modeled by

$$P(t) = \frac{100,600}{1 + 10.17(1.13)^{-t}} \text{ prisoners} \quad (0 \le t \le 27)$$

($t = 0$ represents 1970). When to the nearest year was the New York State prison population increasing most rapidly? How fast was it increasing at that time?

The model is a logistic regression. Sources: Bureau of Justice Statistics, New York State Dept. of Correctional Services/*New York Times*, January 9, 2000, p. WK3.

57. Bottled-Water Sales Annual U.S. sales of bottled water (including sparkling water) can be approximated by

$$W(t) = 0.001t^3 - 0.028t^2 + 0.28t - 0.70 \text{ billion gallons}$$
$$(4 \le t \le 20)$$

where t is time in years since June 1980. Sales of sparkling water over the same period could be approximated by

$$S(t) = -0.0013t^2 + 0.030t - 0.0069 \text{ billion gallons}$$
$$(4 \le t \le 20)$$

Graph the derivative of $S(t)/W(t)$ for $10 \le t \le 20$. Determine when to the nearest year $d/dt\,[S(t)/W(t)]$ had a relative minimum and find its approximate value at that time. What does the answer tell you?

Models are based on regression of approximate data. Source: Beverage Marketing Corporation of New York/*New York Times*, June 21, 2001, p. C1.

58. Bottled-Water Sales Annual U.S. sales of bottled water (including imported water) can be approximated by

$$W(t) = 0.001t^3 - 0.028t^2 + 0.28t - 0.70 \text{ billion gallons}$$
$$(4 \le t \le 20)$$

where t is time in years since June 1980. Sales of imported bottled water over the same period could be approximated by

$$I(t) = 0.0004t^2 + 0.0052t - 0.028 \text{ billion gallons}$$
$$(4 \le t \le 20)$$

Graph the derivative of $W(t) - I(t)$ for $5 \le t \le 20$. Determine when to the nearest 6 months $d/dt\,[W(t) - I(t)]$ was a minimum and find its approximate value at that time. What does the answer tell you?

Models are based on regression of approximate data. Source: Beverage Marketing Corporation of New York/*New York Times*, June 21, 2001, p. C1.

59. Asset Appreciation You manage a small antique company, which owns a collection of Louis XVI jewelry boxes. Their value v is increasing according to the formula

$$v = \frac{10,000}{1 + 500e^{-0.5t}}$$

where t is the number of years from now. You anticipate an inflation rate of 5% per year so that the present value of an item that will be worth $\$v$ in t years' time is given by

$$p = v(1.05)^{-t}$$

When to the nearest year should you sell the jewelry boxes to maximize their present value? How much to the nearest constant dollar will they be worth at that time?

60. **Harvesting Forests** The following equation models the approximate volume in cubic feet of a typical Douglas fir tree of age t years:

$$V = \frac{22,514}{1 + 22,514t^{-2.55}}$$

The lumber will be sold at $10 per cubic foot, and you do not expect the price of lumber to appreciate in the foreseeable future. On the other hand, you anticipate a general inflation rate of 5% per year so that the present value of an item that will be worth $\$v$ in t years' time is given by

$$p = v(1.05)^{-t}$$

At what age to the nearest year should you harvest a Douglas fir tree in order to maximize its present value? How much to the nearest constant dollar will a Douglas fir tree be worth at that time?

The model is the authors' and is based on data in Tom Tietenberg, *Environmental and Natural Resource Economics,* 3rd ed. (New York: HarperCollins, 1992): 282.

61. **Resource Allocation** Your automobile assembly plant has a Cobb–Douglas production function given by

$$q = x^{0.4}y^{0.6}$$

where q is the number of automobiles it produces each year, x is the number of employees, and y is the daily operating budget (in dollars). Annual operating costs amount to an average of $20,000 per employee plus the operating budget of $365y$. Assume that you wish to produce 1000 automobiles each year at a minimum cost. How many employees should you hire?

62. **Resource Allocation** Repeat Exercise 61, using the production formula

$$q = x^{0.5}y^{0.5}$$

63. **Revenue (Based on a Question in the GRE Economics Test)** If total revenue (TR) is specified by $TR = a + bQ - cQ^2$, where Q is quantity of output and a, b, and c are positive parameters, then TR is maximized for this firm when it produces Q equal to

(A) $b/2ac$.   (B) $b/4c$.   (C) $(a + b)/c$.   (D) $b/2c$.   (E) $c/2b$.

64. **Revenue (Based on a Question in the GRE Economics Test)** If total demand (Q) is specified by $Q = -aP + b$, where P is unit price and a and b are positive parameters, then total revenue is maximized for this firm when it charges P equal to

(A) $b/2a$.   (B) $b/4a$.   (C) a/b.   (D) $a/2b$.   (E) $-b/2a$.

COMMUNICATION AND REASONING EXERCISES

65. Explain why the following problem is uninteresting: A packaging company wishes to make cardboard boxes with open tops by cutting square pieces from the corners of a square sheet of cardboard and folding up the sides. What is the box with the least surface area it can make this way?

66. Explain why finding the production level that minimizes a cost function is frequently uninteresting. What would be a more interesting objective?

67. Your friend Margo claims that all you have to do to find the absolute maxima and minima in applications is set the derivative equal to zero and solve. "All that other stuff about endpoints and so on is a waste of time just to make life hard for us," according to Margo. Explain why she is wrong and find at least one exercise in this exercise set to illustrate your point.

68. You are having a hard time persuading your friend Marco that maximizing revenue is not the same as maximizing profit. He says, "How can you expect to obtain the largest profit if you are not taking in the largest revenue?" Explain why he is wrong and find at least one exercise in this exercise set to illustrate your point.

69. If demand q decreases as price p increases, what does the minimum value of dq/dp measure?

70. Explain how you would solve an optimization problem of the following form: Maximize $P = f(x, y, z)$ subject to $z = g(x, y)$ and $y = h(x)$.

5.3   *The Second Derivative and Analyzing Graphs*

Acceleration

Recall that if $s(t)$ represents the position of a car at time t, then its velocity is given by the derivative: $v(t) = s'(t)$. But one rarely drives a car at a constant speed; the velocity itself is changing. The rate at which the velocity is changing is the **acceleration.** Since the derivative measures the rate of change, acceleration is the derivative of velocity: $a(t) = v'(t)$. Since v is the derivative of s, we can express the acceleration in terms of s:

$$a(t) = v'(t) = (s')'(t) = s''(t)$$

That is, a is the derivative of the derivative of s, which we call the **second derivative** of s and write as s''. (In this context you will often hear the derivative s' referred to as the **first derivative.**)

Acceleration and the Second Derivative

The **acceleration** of a moving object is the derivative of its velocity—that is, the second derivative of the position function.

Quick Example

If t is time in hours and the position of a car at time t is $s(t) = t^3 + 2t^2$ miles, then the car's velocity is $v(t) = s'(t) = 3t^2 + 4t$ miles per hour and its acceleration is $a(t) = v'(t) = s''(t) = 6t + 4$ miles per hour per hour.

Example 1 • Acceleration Due to Gravity

According to the laws of physics, an object near the surface of the earth falling in a vacuum from an initial rest position under the influence of gravity will travel a distance of approximately

$$s(t) = 16t^2 \text{ ft}$$

in t seconds. Find its acceleration.

Solution The velocity of the object is

$$v(t) = s'(t) = 32t \text{ ft/s}$$

Hence, the acceleration is

$$a(t) = s''(t) = 32 \text{ ft/s}^2$$

(We write ft/s² as an abbreviation for feet/second/second—that is, feet per second per second. It is often read "feet per second squared.") Thus, the velocity is increasing by 32 feet per second every second. We say that 32 feet per second per second is the **acceleration due to gravity.** If we ignore air resistance, all falling bodies near the surface of the earth, no matter what their weight, will fall with this acceleration.[12]

✸ *Before we go on . . .* This was one of Galileo's most important discoveries. In very careful experiments using balls rolling down inclined planes, he discovered that the acceleration due to gravity is constant and that it does not depend on the weight or composition of the object falling.[13] A famous, though probably apocryphal, story has him dropping cannonballs of different weights off the leaning tower of Pisa to prove his point.[14]

---

[12]There is nothing unique about the number 32. On other planets the number is different. For example, on Jupiter, the acceleration due to gravity is about three times as large.

[13]An interesting aside: Galileo's experiments depended on getting extremely accurate timings. Since the timepieces of his day were very inaccurate, he used the most accurate time measurement he could: He sang and used the beat as his stopwatch.

[14]A true story: The point was made again during the Apollo 15 mission to the Moon (July 1971) when astronaut David Scott dropped a feather and a hammer from the same height. The Moon has no atmosphere, so the two hit the surface of the Moon simultaneously.

We have written the second derivative of $f(x)$ as $f''(x)$. We could also use differential notation:

$$f''(x) = \frac{d^2f}{dx^2}$$

This notation comes from writing the second derivative as the derivative of the derivative in differential notation:

$$f''(x) = \frac{d}{dx}\left[\frac{df}{dx}\right] = \frac{d^2f}{dx^2}$$

Example 2 • Acceleration of Sales

For the first 15 months after the introduction of a new video game, the total number of units sold (total sales) can be modeled by the curve

$$S(t) = 20e^{0.4t}$$

where t is the time in months since the game was introduced. After about 25 months, total sales follow more closely the curve

$$S(t) = 100{,}000 - 20e^{17-0.4t}$$

How fast are total sales accelerating after 10 months? How fast are they accelerating after 30 months? What do these numbers mean?

Solution   By acceleration we mean the rate of change of the rate of change, which is the second derivative. During the first 15 months, the first derivative of sales is

$$\frac{dS}{dt} = 8e^{0.4t}$$

and so the second derivative is

$$\frac{d^2S}{dt^2} = 3.2e^{0.4t}$$

Thus, after 10 months the acceleration of sales is

$$\left.\frac{d^2S}{dt^2}\right|_{t=10} = 3.2e^4 \approx 175 \text{ units/month/month, or units/month}^2$$

We can also compute total sales

$$S(10) = 20e^4 \approx 1092 \text{ units}$$

and the rate of change of sales

$$\left.\frac{dS}{dt}\right|_{t=10} = 8e^4 \approx 437 \text{ units/month}$$

What do these numbers mean? By the end of the tenth month, a total of 1092 video games have been sold. At that time the game is selling at the rate of 437 units per month. This rate of sales is increasing by 175 units per month per month. More games will be sold each month than the month before.

To analyze the sales after 30 months is similar, using the formula

$$S(t) = 100{,}000 - 20e^{17-0.4t}$$

The derivative is

$$\frac{dS}{dt} = 8e^{17-0.4t}$$

and the second derivative is

$$\frac{d^2S}{dt^2} = -3.2e^{17-0.4t}$$

After 30 months,

$$S(30) = 100{,}000 - 20e^{17-12} \approx 97{,}032 \text{ units}$$

$$\left.\frac{dS}{dt}\right|_{t=30} = 8e^{17-12} \approx 1187 \text{ units/month}$$

$$\left.\frac{d^2S}{dt^2}\right|_{t=30} = -3.2e^{17-12} \approx -475 \text{ units/month}^2$$

By the end of the 30th month, 97,032 video games have been sold, the game is selling at a rate of 1187 units per month, and the rate of sales is *decreasing* by 475 units per month per month. Fewer games are sold each month than the month before.

Concavity

Question I know that the first derivative of f tells where the graph of f is rising [where $f'(x) > 0$] and where it is falling [where $f'(x) < 0$]. What, if anything, does the second derivative tell about the graph of f?

Answer The second derivative tells in what direction the graph of f *curves* or *bends*. Consider the graphs in Figures 22 and 23.

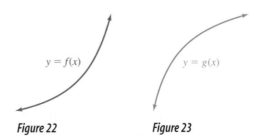

$y = f(x)$ $y = g(x)$

Figure 22   **Figure 23**

Think of a car driving from left to right along each of the roads shown in the two figures. A car driving along the graph of f in Figure 22 will turn to the left (upward); a car driving along the graph of g in Figure 23 will turn to the right (downward). We say that the graph of f is **concave up** and the graph of g is **concave down.** Now think about the derivatives of f and g. The derivative $f'(x)$ starts small but *increases* as the graph gets steeper. Since $f'(x)$ is increasing, its derivative $f''(x)$ must be positive. On the other hand, $g'(x)$ *decreases* as we go to the right. Since $g'(x)$ is decreasing, its derivative $g''(x)$ must be negative. Summarizing, we have the following.

Concavity and the Second Derivative

A curve is **concave up** if its slope is increasing, in which case the second derivative is positive. A curve is **concave down** if its slope is decreasing, in which case the second derivative is negative. A point where the graph of f changes concavity, from concave up to concave down or vice versa, is called a **point of inflection.** At a point of inflection, the second derivative is either zero or undefined.

Quick Example

Consider $f(x) = x^3 - 3x$, whose graph is shown in Figure 24.

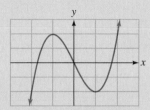

Figure 24

$f''(x) = 6x$ is negative when $x < 0$ and positive when $x > 0$. The graph of f is concave down when $x < 0$ and concave up when $x > 0$. f has a point of inflection at $x = 0$, where the second derivative is 0.

Example 3 • The Point of Diminishing Returns

After the introduction of a new video game, the total number of units sold (total sales) is modeled by the curve

$$S(t) = \frac{100,000}{1 + 5000e^{-0.4t}}$$

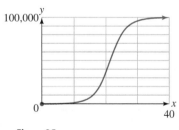

Figure 25

where t is the time in months since the game was introduced (compare Example 2). Where is the graph of S concave up, and where is it concave down? Where are any points of inflection? What does this all mean?

Solution   The graph of S is shown in Figure 25. We see that the graph of S is concave up in the early months and then becomes concave down later. The point of inflection, where the concavity changes, is somewhere around 20 months. With some calculus we can determine exactly where the changeover occurs.

Using the quotient rule, we compute

$$\frac{dS}{dt} = \frac{200,000,000e^{-0.4t}}{(1 + 5000e^{-0.4t})^2}$$

Using the quotient rule again (and simplifying),

$$\frac{d^2S}{dt^2} = \frac{80,000,000e^{-0.4t}(5000e^{-0.4t} - 1)}{(1 + 5000e^{-0.4t})^3}$$

Now, at the point of inflection, this second derivative is either undefined or 0. The denominator is never 0, so there is no point at which the second derivative is not defined. For the second derivative to be 0, the numerator has to be 0, so we set

$$80,000,000e^{-0.4t}(5000e^{-0.4t} - 1) = 0$$

The factor $80{,}000{,}000e^{-0.4t}$ can never be 0, so we must have

$$5000e^{-0.4t} - 1 = 0$$

We can now solve for t:

$$5000e^{-0.4t} = 1$$

$$e^{-0.4t} = \frac{1}{5000} = 0.0002 \qquad\qquad \text{Exponential form}$$

$$-0.4t = \ln(0.0002) \qquad\qquad \text{Logarithmic form}$$

$$t = \frac{\ln(0.0002)}{-0.4} \approx 21.3$$

Thus, the graph of S is concave up for the first 21 months and is concave down after the 22nd month, with the point of inflection being at approximately $t = 21.3$ months.

What does this mean? In Example 2 we noted that where the graph of S is concave up monthly sales are increasing; more units are being sold each month than the month before. Where the graph of S is concave down, monthly sales are decreasing; fewer units are being sold each month than the month before. The point of inflection is the time when sales stop increasing and start to fall off. This is sometimes called the **point of diminishing returns.** Although the total sales figure continues to rise (game units continue to be sold), the rate at which units are sold starts to drop.

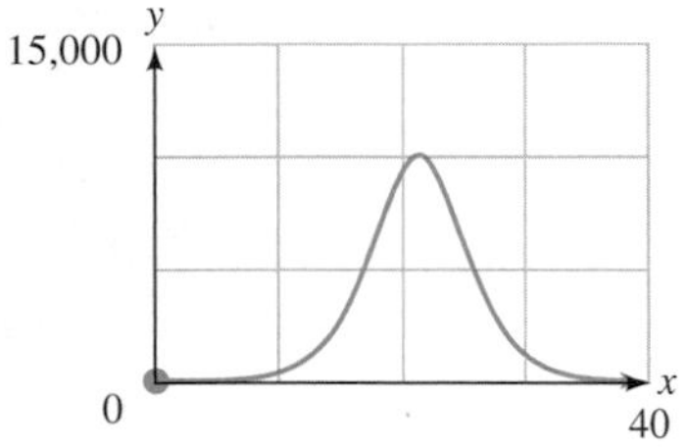

Figure 26

✴ *Before we go on . . .* Since a point of inflection is the point at which the derivative changes from increasing to decreasing or vice versa, a point of inflection is always a relative maximum or minimum of the derivative if the derivative is defined there. The graph of the derivative in this example is shown in Figure 26. We can see that the derivative has a relative (and absolute) maximum of approximately 10,000 at $t = 21.3$ months. That is, the monthly sales are highest in the 22nd month, when the game is selling at a rate of about 10,000 units per month.

Analyzing Graphs

We now have the tools we need to find the most interesting points on the graph of a function. It is easy to use graphing technology to draw the graph, but we need to use calculus to understand what we are seeing. The most interesting features of a graph are the following.

Features of a Graph

1. *The x and y Intercepts* If $y = f(x)$, find the x intercept(s) by setting $y = 0$ and solving for x; find the y intercept by setting $x = 0$.

2. *Relative Extrema* Use the technique of Section 5.1 to locate the relative extrema.

3. *Points of Inflection* Use the technique of this section to find the points of inflection.

4. *Behavior Near Points Where the Function Is Not Defined* If $f(x)$ is not defined at $x = a$, consider $\lim_{x \to a^-} f(x)$ and $\lim_{x \to a^+} f(x)$ to see how the graph of f behaves as x approaches this point.

5. *Behavior at Infinity* Consider $\lim_{x \to -\infty} f(x)$ and $\lim_{x \to +\infty} f(x)$ if appropriate, to see how the graph of f behaves far to the left and right.

⊤ Example 4 • Analyzing a Graph

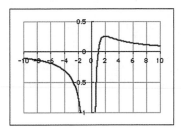

Figure 27

Analyze the graph of $f(x) = \dfrac{1}{x} - \dfrac{1}{x^2}$.

Solution The graph, as drawn using graphing technology, is shown in Figure 27. (Note that $x = 0$ is not in the domain of f.) In the process of locating the interesting features of the graph, we can check that we've included them all in the viewing window.

1. *The x and y Intercepts* We consider $y = 1/x - 1/x^2$. To find the x intercept(s), we set $y = 0$ and solve for x:

$$0 = \frac{1}{x} - \frac{1}{x^2}$$

$$\frac{1}{x} = \frac{1}{x^2}$$

Multiplying both sides by x^2 (we know that x cannot be zero, so we are not multiplying both sides by 0) gives

$$x = 1$$

Thus, there is one x intercept (which we can see in Figure 27) at $x = 1$. We cannot substitute $x = 0$; since $f(0)$ is not defined, the graph does not meet the y axis.

2. *Relative Extrema* We calculate $f'(x) = -1/x^2 + 2/x^3$. To find any stationary points, we set the derivative equal to 0 and solve for x:

$$-\frac{1}{x^2} + \frac{2}{x^3} = 0$$

$$\frac{1}{x^2} = \frac{2}{x^3}$$

$$x = 2$$

Thus, there is one stationary point, at $x = 2$. The graph in Figure 27 shows that f has a relative maximum at this point. The only possible singular point is given at $x = 0$ because $f'(0)$ is not defined. However, $f(0)$ is not defined either, so there are no singular points.

3. *Points of Inflection* We calculate $f''(x) = 2/x^3 - 6/x^4$. To find points of inflection, we set the second derivative equal to 0 and solve for x:

$$\frac{2}{x^3} - \frac{6}{x^4} = 0$$

$$\frac{2}{x^3} = \frac{6}{x^4}$$

$$2x = 6$$

$$x = 3$$

Figure 27 confirms that the graph of f changes from being concave down to being concave up at $x = 3$, so this is a point of inflection. $f''(x)$ is not defined at $x = 0$, but that is not in the domain, so there are no other points of inflection. For example, the graph must be concave down in the whole region $(-\infty, 0)$.

4. *Behavior Near Points Where f Is Not Defined* The only point where $f(x)$ is not defined is $x = 0$. From the graph $f(x)$ appears to go to $-\infty$ as x approaches 0 from either side. To calculate these limits, we rewrite $f(x)$:

$$f(x) = \frac{1}{x} - \frac{1}{x^2} = \frac{x - 1}{x^2}$$

Now, if x is close to 0 (on either side), the numerator $x - 1$ is close to -1, and the denominator is a very small but positive number. The quotient is therefore a negative number of very large magnitude. Therefore,

$$\lim_{x \to 0^-} f(x) = -\infty \quad \text{and} \quad \lim_{x \to 0^+} f(x) = -\infty$$

as appears to happen in Figure 27. We say that f has a **vertical asymptote** at $x = 0$, meaning that the graph approaches the line $x = 0$ without touching it.

5. *Behavior at Infinity* Both $1/x$ and $1/x^2$ go to 0 as x goes to $-\infty$ or $+\infty$; that is,

$$\lim_{x \to -\infty} f(x) = 0 \quad \text{and} \quad \lim_{x \to +\infty} f(x) = 0$$

as we can see happening in Figure 27. We say that f has a **horizontal asymptote** at $y = 0$.

In summary, there is one x intercept at $x = 1$; there is one relative maximum (which, we can now see, is also an absolute maximum) at $x = 2$; there is one point of inflection at $x = 3$, where the graph changes from being concave down to concave up. The graph approaches 0 to the left and to the right, and it goes to $-\infty$ approaching $x = 0$. All these features can be seen in Figure 27.

5.3 EXERCISES

In Exercises 1–10, calculate d^2y/dx^2.

1. $y = 3x^2 - 6$

2. $y = -x^2 + x$

3. $y = \dfrac{2}{x}$

4. $y = -\dfrac{2}{x^2}$

5. $y = 4x^{0.4} - x$

6. $y = 0.2x^{-0.1}$

7. $y = e^{-(x-1)} - x$

8. $y = e^{-x} + e^x$

9. $y = \dfrac{1}{x} - \ln x$

10. $y = x^{-2} + \ln x$

In Exercises 11–16, the position s of a point (in feet) is given as a function of time t (in seconds). Find (a) its acceleration as a function of t and (b) its acceleration at the specified time.

11. $s = 12 + 3t - 16t^2$; $t = 2$

12. $s = -12 + t - 16t^2$; $t = 2$

13. $s = \dfrac{1}{t} + \dfrac{1}{t^2}$; $t = 1$

14. $s = \dfrac{1}{t} - \dfrac{1}{t^2}$; $t = 2$

15. $s = \sqrt{t} + t^2$; $t = 4$

16. $s = 2\sqrt{t} + t^3$; $t = 1$

In Exercises 17–24, the graph of a function is given. Find the approximate coordinates of all points of inflection of each function (if any).

17.

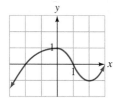

18.

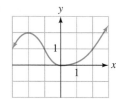

19.

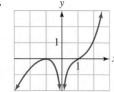

20.

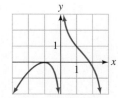

21.

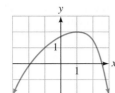

22.

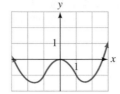

33.

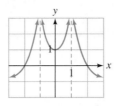

34.

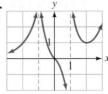

23.

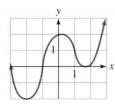

24.

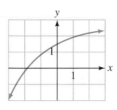

In Exercises 35–60, use technology to draw the graph of the given function, labeling all relative and absolute extrema and points of inflection. Find the coordinates of these points exactly, where possible. Also, find any vertical or horizontal asymptotes. In Exercises 57–60, use graphing technology to approximate the coordinates of the extrema and points of inflection to two decimal places.

In Exercises 25–28, the graph of the *derivative, f'(x)*, is given. Determine the *x* coordinates of all points of inflection of $f(x)$, if any. [Assume that $f(x)$ is defined and continuous everywhere in $[-3, 3]$.]

35. $f(x) = x^2 + 2x + 1$

36. $f(x) = -x^2 - 2x - 1$

37. $f(x) = 2x^3 + 3x^2 - 12x + 1$

38. $f(x) = 4x^3 + 3x^2 + 2$

25.

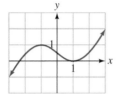

26.

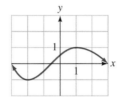

39. $g(x) = x^3 - 12x$, domain $[-4, 4]$

40. $g(x) = 2x^3 - 6x$, domain $[-4, 4]$

41. $g(t) = \frac{1}{4}t^4 - \frac{2}{3}t^3 + \frac{1}{2}t^2$

42. $g(t) = 3t^4 - 16t^3 + 24t^2 + 1$

27.

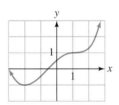

28.

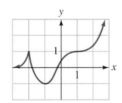

43. $f(t) = \frac{t^2 + 1}{t^2 - 1}$, domain $[-2, 2]$, $t \neq \pm 1$

44. $f(t) = \frac{t^2 - 1}{t^2 + 1}$, domain $[-2, 2]$

45. $g(x) = \frac{x^3}{x^2 + 3}$

In Exercises 29–34, the graph of the *second derivative, f''(x)*, is given. Determine the *x* coordinates of all points of inflection of $f(x)$, if any. [Assume that $f(x)$ is defined and continuous everywhere in $[-3, 3]$.]

46. $g(x) = \frac{x^3}{x^2 - 3}$

47. $f(x) = x + \frac{1}{x}$

29.

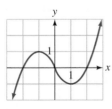

30.

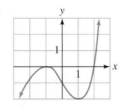

48. $f(x) = x^2 + \frac{1}{x^2}$

49. $g(x) = (x - 3)\sqrt{x}$

50. $g(x) = (x + 3)\sqrt{x}$

51. $f(x) = x - \ln x$, domain $(0, +\infty)$

52. $f(x) = x - \ln x^2$, domain $(0, +\infty)$

53. $f(x) = x^2 + \ln x^2$

54. $f(x) = 2x^2 + \ln x$

55. $g(t) = e^t - t$, domain $[-1, 1]$

31.

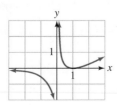

32.

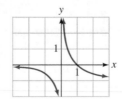

56. $g(t) = e^{-t^2}$

57. $f(x) = x^4 - 2x^3 + x^2 - 2x + 1$

58. $f(x) = x^4 + x^3 + x^2 + x + 1$

59. $f(x) = e^x - x^3$

60. $f(x) = e^x - \frac{x^4}{4}$

APPLICATIONS

61. Epidemics The following graph shows the total number n of people (in millions) infected in an epidemic as a function of time t (in years):

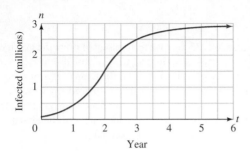

Year

a. When to the nearest year was the rate of new infection largest?

b. When could the Centers for Disease Control announce that the rate of new infection was beginning to drop?

62. Sales The following graph shows the total number of Pomegranate Q4 computers sold since their release (t is in years):

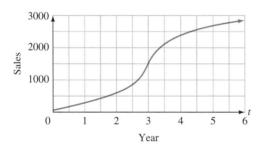

Year

a. When were the computers selling fastest?

b. Explain why this graph might look as it does.

63. Industrial Output The following graph shows the yearly industrial output (measured in billions of dollars) of the Republic of Mars over a 7-year period:

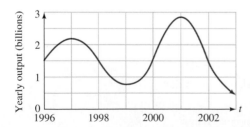

a. When to the nearest year did the rate of change of yearly industrial output reach a maximum?

b. When to the nearest year did the rate of change of yearly industrial output reach a minimum?

c. When to the nearest year did the rate of change of yearly industrial output first start to increase?

64. Profits The following graph shows the yearly profits of Gigantic Conglomerate Inc. (GCI) from 1990 to 2005:

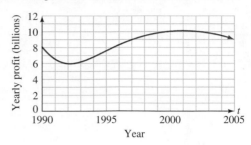

Year

a. When were the profits rising most rapidly?

b. When were the profits falling most rapidly?

c. When could GCI's board of directors legitimately tell stockholders that they had "turned the company around"?

65. Prison Population The total population in state and federal prisons in 1970–1997 can be modeled by

$$p(t) = 0.0015t^2 - 0.0035t + 0.18 \text{ million prisoners}$$
$$(0 \le t \le 27)$$

where t is time in years ($t = 0$ represents 1970). Use the model to estimate the following (give the units of measurement for each answer and round all answers to three significant digits):

a. The prison population in 1985

b. The rate at which the prison population was changing in 1985

c. The rate at which the prison population was accelerating in 1985

Regression model based on the actual data. Sources: Bureau of Justice Statistics, New York State Dept. of Correctional Services/*New York Times,* January 9, 2000, p. WK3.

66. Prison Population Repeat Exercise 65, using the following model for the total prison population of New York State in 1970–1999. Round all answers to three significant digits.

$$p(t) = 0.027t^2 + 1.61t + 8.02 \text{ thousand prisoners}$$
$$(0 \le t \le 29)$$

67. Bottled-Water Sales Annual U.S. sales of bottled water can be approximated by

$$W(t) = 0.001t^3 - 0.028t^2 + 0.28t - 0.70 \text{ billion gallons}$$
$$(4 \le t \le 20)$$

where t is time in years since June 1980. Locate all points of inflection on the graph of W and interpret the result.

Model based on regression of approximate data. Source: Beverage Marketing Corporation of New York/*New York Times,* June 21, 2001, p. C1.

68. Bottled-Water Sales Annual U.S. sales of nonsparkling, nonimported bottled water can be approximated by

$$N(t) = 0.001t^3 - 0.026t^2 + 0.24t - 0.66 \text{ billion gallons}$$
$$(4 \le t \le 20)$$

where t is time in years since June 1980. Locate all points of inflection on the graph of N, and interpret the result.

Model based on regression of approximate data. Source: Beverage Marketing Corporation of New York/*New York Times,* June 21, 2001, p. C1.

69. Education and Crime The following graph shows a striking relationship between the total prison population and the average combined SAT score in the United States:

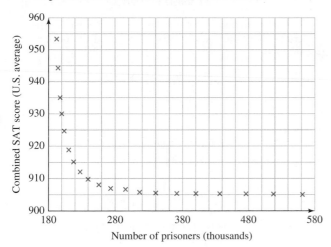

Number of prisoners (thousands)

These data can be accurately modeled by

$$S(n) = 904 + \frac{1326}{(n - 180)^{1.325}} \qquad (192 \le n \le 563)$$

Here, $S(n)$ is the combined U.S. average SAT score at a time when the total U.S. prison population was n thousand.

a. Are there any points of inflection on the graph of S?

b. What does the concavity of the graph of S tell you about prison population and SAT scores?

SOURCE: The model is the authors' from data in *Sourcebook of Criminal Justice Statistics, 1990,* U.S. Department of Justice, Bureau of Justice Statistics (Washington D.C.: Government Printing Office): 604.

70. Education and Crime Refer to the model in Exercise 69.

a. Are there any points of inflection on the graph of S'?

b. When is S'' a maximum? Interpret your answer in terms of prisoners and SAT scores.

71. Patents In 1965 the economist F. M. Scherer modeled the number n of patents produced by a firm as a function of the size s of the firm (measured in annual sales in millions of dollars). He came up with the following equation based on a study of 448 large firms:

$$n = -3.79 + 144.42s - 23.86s^2 + 1.457s^3$$

a. Find $\left. \dfrac{d^2n}{ds^2} \right|_{s=3}$. Is the rate at which patents are produced as the size of a firm gets larger increasing or decreasing with size when $s = 3$? Comment on Scherer's words: "We find diminishing returns dominating."

b. Find $\left. \dfrac{d^2n}{ds^2} \right|_{s=7}$ and interpret the answer.

c. Find the s coordinate of any points of inflection and interpret the result.

SOURCE: F. M. Scherer, "Firm Size, Market Structure, Opportunity, and the Output of Patented Inventions," *American Economic Review* 55 (December 1965): 1097–1125.

72. Returns on Investments A company finds that the number of new products it develops each year depends on the size of its annual research and development budget, x (in thousands of dollars), according to the formula

$$n(x) = -1 + 8x + 2x^2 - 0.4x^3$$

a. Find $n''(1)$ and $n''(3)$ and interpret the results.

b. Find the size of the budget that gives the largest rate of return as measured in new products per dollar (again, called the point of diminishing returns).

 For Exercises 73–78, use technology to solve.

73. Asset Appreciation You manage a small antique store, which owns a collection of Louis XVI jewelry boxes. Their value v is increasing according to the formula

$$v = \frac{10,000}{1 + 500e^{-0.5t}}$$

where t is the number of years from now. You anticipate an inflation rate of 5% per year so that the present value of an item that will be worth $\$v$ in t years' time is given by

$$p = v(1.05)^{-t}$$

What is the greatest rate of increase of the value of your antiques, and when is this rate attained?

74. Harvesting Forests The following equation models the approximate volume in cubic feet of a typical Douglas fir tree of age t years:

$$V = \frac{22,514}{1 + 22,514t^{-2.55}}$$

The lumber will be sold at $10 per cubic foot, and you do not expect the price of lumber to appreciate in the foreseeable future. On the other hand, you anticipate a general inflation rate of 5% per year so that the present value of an item that will be worth $\$v$ in t years' time is given by

$$p = v(1.05)^{-t}$$

What is the largest rate of increase of the value of a fir tree, and when is this rate attained?

SOURCE: The authors' model is based on data in Tom Tietenberg, *Environmental and Natural Resource Economics,* 3rd ed. (New York: HarperCollins, 1992): 282.

75. Asset Appreciation As the financial consultant to a classic auto dealership, you estimate that the total value of its collection of 1959 Chevrolets and Fords is given by the formula

$$v = 300,000 + 1000t^2$$

where t is the number of years from now. You anticipate a continuous inflation rate of 5% per year so that the discounted (present) value of an item that will be worth $\$v$ in t years' time is given by

$$p = ve^{-0.05t}$$

When is the value of the collection of classic cars increasing most rapidly? Decreasing most rapidly?

76. Plantation Management The value of a fir tree in your plantation increases with the age of the tree according to the formula

$$v = \frac{20t}{1 + 0.05t}$$

where t is the age of the tree in years. Given a continuous inflation rate of 5% per year, the discounted (present) value of a newly planted seedling is

$$p = ve^{-0.05t}$$

When is the discounted value of a tree increasing most rapidly? Decreasing most rapidly?

77. Ecology Manatees are grazing sea mammals sometimes referred to as sea sirens. Increasing numbers of manatees have been killed by boats off the Florida coast, as shown in the following chart:

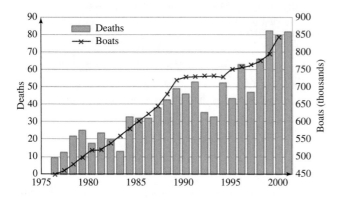

Let $M(t)$ be the number of manatees killed and let $B(t)$ be the number of boats registered in Florida in year t since 1975. These functions can be approximated by the following linear functions:

$$M(t) = 2.48t + 6.87 \text{ manatees}$$

$$B(t) = 15,700t + 444,000 \text{ boats}$$

a. Graph the function $M(t)/B(t)$ and estimate $\lim_{t \to +\infty} M(t)/B(t)$.

b. What does $M(t)/B(t)$ represent? What does $\lim_{t \to +\infty} M(t)/B(t)$ represent?

c. Is the graph of $M(t)/B(t)$ concave up or concave down? The concavity of $M(t)/B(t)$ tells you that

  (A) Fewer manatees are killed per boat each year.

  (B) More manatees are killed per boat each year.

  (C) The number of manatees killed per boat is increasing at a decreasing rate.

  (D) The number of manatees killed per boat is decreasing at an increasing rate.

Rounded regression model. Sources: Florida Department of Highway Safety & Motor Vehicles, Florida Marine Institute/*New York Times,* February 12, 2002, p. F4.

78. Saudi Oil Revenues The spot price of crude oil during the period 1999–2001 can be approximated by

$$P(t) = -10t^2 + 32 \text{ dollars per barrel} \qquad (-1 \le t \le 1)$$

in year t, where $t = 0$ represents June 2000. Saudi Arabia's crude oil production over the same period can be approximated by

$$Q(t) = -70.8t^2 + 78 \text{ million barrels per day} \quad (-1 \le t \le 1)$$

a. Graph the revenue function $R(t) = P(t)Q(t)$ and its derivative for $-1 \le t \le 1$.

b. Locate the approximate t coordinate (to within ± 0.1) of any point(s) of inflection and interpret the result.

Source: Deutsche Bank/*New York Times,* November 21, 2001, p. A3.

COMMUNICATION AND REASONING EXERCISES

79. Complete the following: If the graph of a function is concave up on its entire domain, then its second derivative is _____ on the domain.

80. Complete the following: If the graph of a function is concave up on its entire domain, then its first derivative is _____ on the domain.

81. Company A's profits satisfy $P(0) = \$1$ million, $P'(0) = \$1$ million per year, and $P''(0) = -\$1$ million per year per year. Company B's profits satisfy $P(0) = \$1$ million, $P'(0) = -\$1$ million per year, and $P''(0) = \$1$ million per year per year. There are no points of inflection in either company's profit-curve graph. Sketch two pairs of profit curves: one in which company A ultimately outperforms company B and another in which company B ultimately outperforms company A.

82. Company C's profits satisfy $P(0) = \$1$ million, $P'(0) = \$1$ million per year, and $P''(0) = -\$1$ million per year per year. Company D's profits satisfy $P(0) = \$0$ million, $P'(0) = \$0$ million per year, and $P''(0) = \$1$ million per year per year. There are no points of inflection in either company's profit curve. Sketch two pairs of profit curves: one in which company C ultimately outperforms company D and another in which company D ultimately outperforms company C.

83. Sales of Kent's Tents reached a maximum in January 2002 and declined to a minimum in January 2003 before starting to climb again. The graph of sales shows a point of inflection at June 2002. What is the significance of the point of inflection?

84. The graph of sales of Luddington's Wellington boots is concave down, although sales continue to increase. What properties of the graph of sales versus time are reflected in the following behaviors: (a) a point of inflection next year and (b) a horizontal asymptote?

85. Explain geometrically why the derivative of a function has a relative extremum at a point of inflection, if it is defined there. Which points of inflection give rise to relative maxima in the derivative?

86. If we regard position s as a function of time t, what is the significance of the *third* derivative $s'''(t)$? Describe an everyday scenario in which it arises.

5.4   Related Rates

We know that if Q is a quantity changing over time, then the derivative dQ/dt is the rate at which Q changes over time. In this section we are concerned with what are called **related rates** problems. In such a problem, we have two (sometimes more) related quantities, we know the rate at which one is changing, and we wish to find the rate at which another is changing. A typical example is the following.

Example 1 • The Expanding Circle

The radius of a circle is increasing at a rate of 10 cm/s. How fast is the area increasing at the instant when the radius has reached 5 cm?

Solution   We have two related quantities: the radius of the circle r and its area A. The first sentence of the problem tells us that r is increasing at a certain rate. When we see a sentence referring to speed or change, it is very helpful to rephrase the sentence using the phrase "the rate of change of." Here, we can say

The rate of change of r is 10 cm/s.

Since the rate of change is the derivative, we can rewrite this sentence as the equation

$$\frac{dr}{dt} = 10$$

Similarly, the second sentence of the problem asks how A is changing. We can rewrite that question:

What is the rate of change of A when the radius is 5 cm?

Using mathematical notation, the question is:

What is $\dfrac{dA}{dt}$ *when $r = 5$?*

Thus, knowing one rate of change dr/dt, we wish to find a related rate of change dA/dt. To find exactly how these derivatives are related, we need the equation relating the variables, which is

$$A = \pi r^2 \qquad \text{Formula for area of a circle}$$

To find the relationship between the derivatives, we take the derivative of both sides of this equation *with respect to t.* On the left we get dA/dt. On the right we need to be a little careful. We do not get $2\pi r$ because that would be the derivative with respect to r, not t. We need to remember that r is a function of t and use the chain rule. We get

$$\frac{dA}{dt} = 2\pi r \, \frac{dr}{dt}$$

Now we substitute the given values $r = 5$ and $dr/dt = 10$. This gives

$$\left.\frac{dA}{dt}\right|_{r=5} = 2\pi(5)(10) = 100\pi \approx 314 \text{ cm}^2/\text{s}$$

We can organize our work as follows.

Solving a Related Rates Problem

A. The Problem

1. List the related, changing quantities.

2. Restate the problem in terms of rates of change. Rewrite the problem using mathematical notation for the changing quantities and their derivatives.

B. The Relationship

1. Draw a diagram, if appropriate, showing the changing quantities.

2. Find an equation or equations relating the changing quantities.

3. Take the derivative with respect to time of the equation(s) relating the quantities to get the **derived equation(s),** relating the rates of change of the quantities.

C. The Solution

1. Substitute into the derived equation(s) the given values of the quantities and their derivatives.

2. Solve for the derivative required.

We can illustrate the procedure with the "ladder problem" found in almost every calculus textbook.

Example 2 • The Falling Ladder

Jane is at the top of a 5-foot ladder when it starts to slide down the wall at a rate of 3 feet per minute. Jack is standing on the ground behind her. How fast is the base of the ladder moving when it hits him if Jane is 4 feet from the ground at that instant?

Solution   The first sentence talks about (the top of) the ladder sliding down the wall. Thus, one of the changing quantities is the height of the top of the ladder. The question asked refers to the motion of the base of the ladder, so another changing quantity is the distance of the base of the ladder from the wall. Let's record these variables and follow the outline above to obtain the solution.

A. The Problem

1. The changing quantities are

h = height of the top of the ladder

b = distance of the base of the ladder from the wall

2. We rephrase the problem in words, using the phrase "rate of change":

The rate of change of the height of the top of the ladder is −3 *feet per minute. What is the rate of change of the distance of the base from the wall when the top of the ladder is 4 feet from the ground?*

We can now rewrite the problem mathematically:

$$\frac{dh}{dt} = -3. \text{ Find } \frac{db}{dt} \text{ when } h = 4.$$

B. The Relationship

Figure 28

1. Figure 28 shows the ladder and the variables h and b. Notice that we put in the figure the fixed length, 5, of the ladder, but any changing quantities, like h and b we leave as variables. We do not use any specific values for h or b until the very end.

2. From the figure we can see that h and b are related by the Pythagorean theorem:

$$h^2 + b^2 = 25$$

3. Taking the derivative with respect to time of the equation above gives us the derived equation:

$$2h\,\frac{dh}{dt} + 2b\,\frac{db}{dt} = 0$$

C. The Solution

1. We substitute the known values $dh/dt = -3$ and $h = 4$ into the derived equation:

$$2(4)(-3) + 2b\,\frac{db}{dt} = 0$$

We would like to solve for db/dt, but what do we do with the b in the equation? We can determine b from the equation $h^2 + b^2 = 25$, using the value $h = 4$:

$$16 + b^2 = 25$$
$$b^2 = 9$$
$$b = 3$$

Substituting into the derived equation, we get

$$-24 + 2(3)\frac{db}{dt} = 0$$

2. Solving for db/dt gives

$$\frac{db}{dt} = \frac{24}{6} = 4$$

Thus, the base of the ladder is sliding away from the wall at 4 feet per minute when it hits Jack.

Example 3 • Average Cost

The cost to manufacture x cell phones in a day is

$$C(x) = 10{,}000 + 20x + \frac{x^2}{10{,}000} \text{ dollars}$$

The daily production level is currently $x = 5000$ cell phones and is increasing at a rate of 100 units per day. How fast is the average cost changing?

Solution

A. The Problem

1. The changing quantities are the production level x and the average cost $\overline{C}$.

2. We rephrase the problem as follows:

   The daily production level is $x = 5000$ units, and the rate of change of x is 100 units per day. What is the rate of change of the average cost $\overline{C}$?

   In mathematical notation,

   $$x = 5000 \text{ and } \frac{dx}{dt} = 100. \text{ Find } \frac{d\overline{C}}{dt}.$$

B. The Relationship

1. In this example the changing quantities cannot easily be depicted geometrically.

2. We are given a formula for the *total* cost. We get the *average* cost by dividing the total cost by x:

   $$\overline{C} = \frac{C}{x}$$

   $$= \frac{10,000}{x} + 20 + \frac{x}{10,000}$$

3. Taking derivatives with respect to t of both sides, we get the derived equation:

   $$\frac{d\overline{C}}{dt} = \left(-\frac{10,000}{x^2} + \frac{1}{10,000} \right) \frac{dx}{dt}$$

C. The Solution

Substituting the values from part A into the derived equation, we get

$$\frac{d\overline{C}}{dt} = \left(-\frac{10,000}{5000^2} + \frac{1}{10,000} \right) 100$$

$$= -0.03 \text{ dollars per day}$$

Thus, the average cost is decreasing by 3¢ per day.

The scenario in the following example is similar to Example 5 in Section 5.2.

Example 4 • Automation

The Gym Sock Company manufactures cotton athletic socks. Production is partially automated through the use of robots. The number of pairs of socks the company can manufacture in a day is given by a Cobb–Douglas production formula

$$q = 50n^{0.6}r^{0.4}$$

where q is the number of pairs of socks that can be manufactured by n laborers and r robots. The company currently produces 1000 pairs of socks each day and employs 20 laborers. It is bringing one new robot on line every month. At what rate are laborers being laid off, assuming that the number of socks produced remains constant?

Solution

A. The Problem

1. The changing quantities are the number of laborers n and the number of robots r.

2. $\dfrac{dr}{dt} = 1$. Find $\dfrac{dn}{dt}$ when $n = 20$.

B. The Relationship

1. No diagram is appropriate here.

2. The equation relating the changing quantities:

$$1000 = 50n^{0.6}r^{0.4}$$

$$20 = n^{0.6}r^{0.4}$$

(Productivity is constant at 1000 pairs of socks each day.)

3. The derived equation is

$$0 = 0.6n^{-0.4}\left(\frac{dn}{dt}\right)r^{0.4} + 0.4n^{0.6}r^{-0.6}\left(\frac{dr}{dt}\right)$$

$$= 0.6\left(\frac{r}{n}\right)^{0.4}\left(\frac{dn}{dt}\right) + 0.4\left(\frac{n}{r}\right)^{0.6}\left(\frac{dr}{dt}\right)$$

We solve this equation for dn/dt because we want to find dn/dt below and because the equation becomes simpler when we do this:

$$0.6\left(\frac{r}{n}\right)^{0.4}\left(\frac{dn}{dt}\right) = -0.4\left(\frac{n}{r}\right)^{0.6}\left(\frac{dr}{dt}\right)$$

$$\frac{dn}{dt} = -\frac{0.4}{0.6}\left(\frac{n}{r}\right)^{0.6}\left(\frac{n}{r}\right)^{0.4}\left(\frac{dr}{dt}\right)$$

$$= -\frac{2}{3}\left(\frac{n}{r}\right)\left(\frac{dr}{dt}\right)$$

C. The Solution

Substituting the numbers in part A into the last equation in part B, we get

$$\frac{dn}{dt} = -\frac{2}{3}\left(\frac{20}{r}\right)(1)$$

We need to compute r by substituting the known value of n in the original formula:

$$20 = n^{0.6}r^{0.4}$$

$$20 = 20^{0.6}r^{0.4}$$

$$r^{0.4} = \frac{20}{20^{0.6}} = 20^{0.4}$$

$$r = 20$$

Thus,

$$\frac{dn}{dt} = -\frac{2}{3}\left(\frac{20}{20}\right)(1) = -\frac{2}{3} \text{ laborer per month}$$

The company is laying off laborers at a rate of 2/3 per month, or two every 3 months.

✦ *Before we go on . . .* We can interpret this result as saying that, at the current level of production and number of laborers, one robot is as productive as $\frac{2}{3}$ of a laborer, or three robots are as productive as two laborers.

5.4 EXERCISES

In Exercises 1–8, rewrite the statements and questions in mathematical notation.

1. The population P is currently 10,000 and growing at a rate of 1000 per year.

2. There are presently 400 cases of Bangkok flu, and the number is growing by 30 new cases every month.

3. The annual revenue of your tie-dyed T-shirt operation is currently $7000 but is decreasing by $700 each year. How fast are annual sales changing?

4. A ladder is sliding down a wall so that the distance between the top of the ladder and the floor is decreasing at a rate of 3 feet per second. How fast is the base of the ladder receding from the wall?

5. The price of shoes is rising $5 per year. How fast is the demand changing?

6. Stock prices are rising $1000 per year. How fast is the value of your portfolio increasing?

7. The average global temperature is 60°F and rising by 0.1°F per decade. How fast are annual sales of Bermuda shorts increasing?

8. The country's population is now 2.6 million and is increasing by 1 million people per year. How fast is the annual demand for diapers increasing?

APPLICATIONS

9. **Sunspots** The area of a circular sunspot is growing at a rate of 1200 km²/s.
   a. How fast is the radius growing at the instant when it equals 10,000 km?
   b. How fast is the radius growing at the instant when the sunspot has an area of 640,000 km²?

10. **Puddles** The radius of a circular puddle is growing at a rate of 5 cm/s.
    a. How fast is its area growing at the instant when the radius is 10 cm?
    b. How fast is the area growing at the instant when it equals 36 cm²?

11. **Sliding Ladders** The base of a 50-foot ladder is being pulled away from a wall at a rate of 10 feet per second. How fast is the top of the ladder sliding down the wall at the instant when the base of the ladder is 30 feet from the wall?

12. **Sliding Ladders** The top of a 5-foot ladder is sliding down a wall at a rate of 10 feet per second. How fast is the base of the ladder sliding away from the wall at the instant when the top of the ladder is 3 feet from the ground?

13. **Average Cost** The average cost function for the weekly manufacture of portable CD players is given by

    $$\overline{C}(x) = 150{,}000x^{-1} + 20 + 0.0001x \text{ dollars per player}$$

    where x is the number of CD players manufactured that week. Weekly production is currently 3000 players and is increasing at a rate of 100 per week. What is happening to the average cost?

14. **Average Cost** Repeat Exercise 13, using the revised average cost function

    $$C(x) = 150{,}000x^{-1} + 20 + 0.01x \text{ dollars per player}$$

15. **Demand** Demand for your tie-dyed T-shirts is given by the formula

    $$q = 500 - 100p^{0.5}$$

    where q is the number of T-shirts you can sell each month at a price of p dollars. If you currently sell T-shirts for $15 each and you raise your price by $2 per month, how fast will the demand drop? (Round your answer to a whole number.)

16. **Supply** The number of portable CD players you are prepared to supply to a retail outlet every week is given by the formula

    $$q = 0.1p^2 + 3p$$

    where p is the price it offers you. The retail outlet is currently offering you $40 per CD player. If the price it offers decreases at a rate of $2 per week, how will this affect the number you supply?

17. **Revenue** You can now sell 50 cups of lemonade each week at 30¢ per cup, but demand is dropping at a rate of 5 cups per week each week. Assuming that raising the price does not affect demand, how fast do you have to raise your price if you want to keep your weekly revenue constant?

18. **Revenue** You can now sell 40 cars each month at $20,000 per car, and demand is increasing at a rate of 3 cars per month each month. What is the fastest you could drop your price before your monthly revenue starts to drop?

19. **Demand** Assume that the demand equation for tuna in a small coastal town is

    $$pq^{1.5} = 50{,}000$$

    where q is the number of pounds of tuna that can be sold in 1 month at the price of p dollars per pound. The town's fishery finds that the demand for tuna is currently 900 pounds per month and is increasing at a rate of 100 pounds per month each month. How fast is the price changing?

20. Demand The demand equation for rubies at Royal Ruby Retailers (RRR) is

$$q + \frac{4}{3}p = 80$$

where q is the number of rubies RRR can sell each week at p dollars per ruby. RRR finds that the demand for its rubies is currently 20 rubies per week and is dropping at a rate of 1 per week. How fast is the price changing?

21. Balloons A spherical party balloon is being inflated with helium pumped in at a rate of 3 cubic feet per minute. How fast is the radius growing at the instant when the radius has reached 1 foot? (The volume of a sphere of radius r is $V = \frac{4}{3}\pi r^3$.)

22. More Balloons A rather flimsy spherical balloon is designed to pop at the instant its radius has reached 10 cm. Assuming the balloon is filled with helium at a rate of 10 cm³ per second, calculate how fast the diameter is growing at the instant it pops. (The volume of a sphere of radius r is $V = \frac{4}{3}\pi r^3$.)

23. Ships Sailing Apart The HMS *Dreadnaught* is 40 miles north of Montauk and steaming due north at 20 miles/hour, while the USS *Mona Lisa* is 50 miles east of Montauk and steaming due east at an even 30 miles/hour. How fast is their distance apart increasing?

24. Near Miss My aunt and I were approaching the same intersection, she from the south and I from the west. She was traveling at a steady speed of 10 miles/hour, while I was approaching the intersection at 60 miles/hour. At a certain instant in time, I was one-tenth of a mile from the intersection, while she was one-twentieth of a mile from it. How fast were we approaching each other at that instant?

25. Movement along a Graph A point on the graph of $y = 1/x$ is moving along the curve in such a way that its x coordinate is increasing at a rate of 4 units per second. What is happening to the y coordinate at the instant the y coordinate is equal to 2?

26. Motion around a Circle A point is moving along the circle $x^2 + (y - 1)^2 = 8$ in such a way that its x coordinate is decreasing at a rate of 1 unit per second. What is happening to the y coordinate at the instant when the point has reached $(-2, 3)$?

27. Production The automobile assembly plant you manage has a Cobb–Douglas production function given by

$$P = 10x^{0.3}y^{0.7}$$

where P is the number of automobiles it produces each year, x is the number of employees, and y is the daily operating budget (in dollars). You maintain a production level of 1000 automobiles per year. If you currently employ 150 workers and are hiring new workers at a rate of 10 per year, how fast is your daily operating budget changing?

28. Production Refer to the Cobb–Douglas production formula in Exercise 27. Assume that you maintain a constant workforce of 200 workers and wish to increase production in order to meet a demand that is increasing by 100 automobiles per year. The current demand is 1000 automobiles per year. How fast should your daily operating budget be increasing?

29. Education In 1991 the expected income of an individual depended on his or her educational level according to the following formula:

$$I(n) = 2928.8n^3 - 115,860n^2 + 1,532,900n - 6,760,800$$
$$(12 \le n \le 15)$$

Here, n is the number of school years completed, and $I(n)$ is the individual's expected income. You have completed 13 years of school and are currently a part-time student. Your schedule is such that you will complete the equivalent of 1 year of college every 3 years. Assuming that your salary is linked to the above model, how fast is your income rising? (Round your answer to the nearest \$1.)

Source: The model is a best-fit cubic based on Table 358, U.S. Department of Education, *Digest of Education Statistics, 1991* (Washington, D.C.: Government Printing Office, 1991).

30. Education Refer to the model in Exercise 29. Assume that someone has completed 14 years of school and that her income is increasing by \$10,000/year. How much schooling per year is this rate of increase equivalent to?

31. Employment An employment research company estimates that the value of a recent MBA graduate to an accounting company is

$$V = 3e^2 + 5g^3$$

where V is the value of the graduate, e is the number of years of prior business experience, and g is the graduate school grade-point average (GPA). A company that currently employs graduates with a 3.0 GPA wishes to maintain a constant employee value of $V = 200$ but finds that the GPA of its new employees is dropping at a rate of 0.2 per year. How fast must the experience of its new employees be growing in order to compensate for the decline in GPA?

32. Grades A production formula for a student's performance on a difficult English examination is given by

$$g = 4hx - 0.2h^2 - 10x^2$$

where g is the grade the student can expect to obtain, h is the number of hours of study for the examination, and x is the student's GPA. The instructor finds that students' GPAs have remained constant at 3.0 over the years and that students currently spend an average of 15 hours studying for the examination. However, scores on the examination are dropping at a rate of 10 points per year. At what rate is the average study time decreasing?

Source: Based on an exercise in A. L. Ostrosky, Jr., and J. V. Koch, *Introduction to Mathematical Economics* (Prospect Heights, Ill.: Waveland Press, 1979).

33. Cones A right circular conical vessel is being filled with green industrial waste at a rate of 100 m³/s. How fast is the level rising after 200π cubic meters have been poured in? The cone has height 50 m and radius 30 m at its brim. (The volume of a cone of height h and cross-sectional radius r at its brim is given by $V = \frac{1}{3}\pi r^2 h$.)

34. More Cones A circular conical vessel is being filled with ink at a rate of 10 cm³/s. How fast is the level rising after 20 cm³ have been poured in? The cone has height 50 cm and radius 20 cm at its brim. (The volume of a cone of height h and cross-sectional radius r at its brim is given by $V = \frac{1}{3}\pi r^2 h$.)

35. Cylinders The volume of paint in a right cylindrical can is given by $V = 4t^2 - t$ where t is time in seconds and V is the volume in cm³. How fast is the level rising when the height is 2 cm? The can has a height of 4 cm and a radius of 2 cm. (*Hint:* To get h as a function of t, first solve the volume $V = \pi r^2 h$ for h.)

36. Cylinders A cylindrical bucket is being filled with paint at a rate of 6 cm³ per minute. How fast is the level rising when the bucket starts to overflow? The bucket has radius 30 cm and height 60 cm.

37. Computers versus Income The demand for personal computers in the home rises with household income. For a given community, we can approximate the average number of computers in a home as

$$q = 0.3454 \ln x - 3.047 \qquad (10{,}000 \le x \le 125{,}000)$$

where x is mean household income. Your community has a mean income of $30,000, increasing at a rate of $2000 per year. How many computers per household are there, and how fast is the number of computers in a home increasing? (Round your answer to four decimal places.)

The model is a regression model. Source: For income distribution, Luxembourg Income Study/*New York Times*, August 14, 1995, p. A9. For computer data, Forrester Research/*New York Times*, August 8, 1999, p. BU4.

38. Computers versus Income Refer to the model in Exercise 37. The average number of computers per household in your town is 0.5 and is increasing at a rate of 0.02 computer per household per year. What is the average household income in your town, and how fast is it increasing? (Round your answers to the nearest $10).

Education and Crime The following graph shows a striking relationship between the total prison population and the average combined SAT score in the United States:

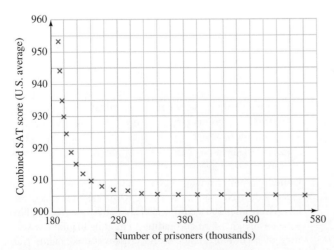

Exercises 39 and 40 are based on the following model for these data:

$$S(n) = 904 + \frac{1326}{(n - 180)^{1.325}} \qquad (192 \le n \le 563)$$

Here, $S(n)$ is the combined average SAT score at a time when the total prison population is n thousand.

The authors' model based on data for the years 1967–1989. Sources: *Sourcebook of Criminal Justice Statistics, 1990*, p. 604/Educational Testing Service.

39. In 1985 the U.S. prison population was 475,000 and increasing at a rate of 35,000 per year. What was the average SAT score, and how fast and in what direction was it changing? (Round your answers to two decimal places.)

40. In 1970 the U.S. combined SAT average was 940 and dropping by 10 points per year. What was the U.S. prison population, and how fast and in what direction was it changing? (Round your answers to the nearest 100.)

Divorce Rates A study found that the divorce rate d (given as a percentage) appears to depend on the ratio r of available men to available women. This function can be approximated by

$$d(r) = \begin{cases} -40r + 74 & \text{if } r \le 1.3 \\ \frac{130r}{3} - \frac{103}{3} & \text{if } r > 1.3 \end{cases}$$

Exercises 41 and 42 are based on this model.

The cited study, by Scott J. South and associates, appeared in the *American Sociological Review* (February, 1995). Figures are rounded. Source: *New York Times*, February 19, 1995, p. 40.

41. There are currently 1.1 available men per available woman in Littleville, and this ratio is increasing by 0.05 per year. What is happening to the divorce rate?

42. There are currently 1.5 available men per available woman in Largeville, and this ratio is decreasing by 0.03 per year. What is happening to the divorce rate?

COMMUNICATION AND REASONING EXERCISES

43. Why is this section titled "Related Rates"?

44. If you know how fast one quantity is changing and need to compute how fast a second quantity is changing, what kind of information do you need?

45. In a related rates problem, there is no limit to the number of changing quantities we can consider. Illustrate this by creating a related rates problem with four changing quantities.

46. If three quantities are related by a single equation, how would you go about computing how fast one of them is changing based on a knowledge of the other two?

47. The demand and unit price for your store's checkered T-shirts are changing with time. Show that the percentage rate of change of revenue equals the sum of the percentage rates of change of price and demand. [The percentage rate of change of a quantity Q is $Q'(t)/Q(t)$.]

48. The number N of employees and the total floor space S of your company are both changing with time. Show that the percentage rate of change of square footage per employee equals the percentage rate of change of S minus the percentage rate of change of N. [The percentage rate of change of a quantity Q is $Q'(t)/Q(t)$.]

49. In solving a related rates problem, a key step is solving the derived equation for the unknown rate of change (once we have substituted the other values into the equation). Call the unknown rate of change X. The derived equation is what kind of equation in X?

50. On a recent exam, you were given a related rates problem based on an algebraic equation relating two variables x and y. Your friend told you that the correct relationship between dx/dt and dy/dt was given by

$$\frac{dx}{dt} = \left(\frac{dy}{dt}\right)^2$$

Could he be correct?

51. Transform the following into a mathematical statement about derivatives: If my grades are improving at twice the speed of yours, then your grades are improving at half the speed of mine.

52. If two quantities x and y are related by a linear equation, how are their rates of change related?

5.5 Elasticity

You manufacture an extremely popular brand of sneakers and want to know what will happen if you increase the selling price. Common sense tells you that demand will drop as you raise the price. But, will the drop in demand be enough to cause your revenue to fall? Or will it be small enough that your revenue will rise because of the higher selling price? For example, if you raise the price by 1%, you might suffer only a 0.5% loss in sales. In this case the loss in sales will be more than offset by the increase in price, and your revenue will rise. In such a case, we say that the demand is **inelastic** because it is not very sensitive to the increase in price. On the other hand, if your 1% price increase results in a 2% drop in demand, then raising the price will cause a drop in revenues. We then say that the demand is **elastic** because it reacts strongly to a price change.

We can use calculus to measure the response of demand to price changes if we have a demand equation for the item we are selling.[15] We need to know by what percent demand will drop for each percent increase in price. In other words, we want the *percentage drop in demand per percentage increase in price.* This ratio is called the **elasticity of demand,** or **price elasticity of demand,** and is usually denoted by E. Let's derive a formula for E in terms of the demand equation.

Assume that we have a demand equation

$$q = f(p)$$

where q stands for the number of items we would sell (per week, per month, or what have you) if we set the price per item at p. Now suppose we increase the price p by a very small amount, Δp. Then our percentage increase in price is $(\Delta p/p) \times 100\%$ (we multiply by 100 to get a percentage). This increase in p will presumably result in a decrease in the demand q. Let's denote this corresponding decrease in q by $-\Delta q$ (we use the minus sign because, by convention, Δq stands for the *increase* in demand). Thus, the percentage decrease in demand is $(-\Delta q/q) \times 100\%$.

---

[15]Coming up with a good demand equation is not always easy. We saw in Chapter 1 that it is possible to find a linear demand equation if we know the sales figures at two different prices. However, such an equation is only a first approximation. To come up with a more accurate demand equation, we might need to gather data corresponding to sales at several different prices and use curve-fitting techniques like regression. Another approach would be an analytic one, based on mathematical modeling techniques an economist might use.

Now E is the ratio

$$E = \frac{\text{percentage decrease in demand}}{\text{percentage increase in price}}$$

$$= \frac{-\frac{\Delta q}{q} \times 100\%}{\frac{\Delta p}{p} \times 100\%}$$

Canceling the 100%s and reorganizing, we get

$$E = -\frac{\Delta q}{\Delta p} \cdot \frac{p}{q}$$

Question Fine, but what small change in price will we use for Δp?

Answer It should probably be pretty small. If, say, we increased the price of sneakers to $1 million per pair, the sales would likely drop to zero. But knowing this tells us nothing about how the market would respond to a modest increase in price. In fact, we'll do the usual thing we do in calculus and let Δp approach 0.

In the expression for E, if we let Δp go to 0, then the ratio $\Delta q / \Delta p$ goes to the derivative dq/dp. This gives us our final and most useful definition of the elasticity.

Price Elasticity of Demand

The **price elasticity of demand E** is the percentage rate of decrease of demand per percentage increase in price. E is given by the formula

$$E = -\frac{dq}{dp} \cdot \frac{p}{q}$$

We say that the demand is **elastic** if $E > 1$, is **inelastic** if $E < 1$, and has **unit elasticity** if $E = 1$.

Quick Example

Suppose the demand equation is $q = 20{,}000 - 2p$, where p is the price in dollars. Then

$$E = -(-2)\frac{p}{20{,}000 - 2p} = \frac{p}{10{,}000 - p}$$

- If $p = \$2000$, then $E = 1/4$, and demand is inelastic at this price.
- If $p = \$8000$, then $E = 4$, and demand is elastic at this price.
- If $p = \$5000$, then $E = 1$, and the demand has unit elasticity at this price.

Question We are generally interested in the price that maximizes revenue. Can elasticity help us with that?

Answer Yes. In ordinary cases, the price that maximizes revenue must give unit elasticity. One way of seeing this is as follows[16]: If the demand is inelastic (which ordinarily occurs at a low unit price) then raising the price by a small percentage—1%, say—results in a smaller percentage drop in demand. For example, in the Quick Example

---

[16]For another—more rigorous—argument, see Exercise 25.

above, if $p = \$2000$, then the demand would drop by only $\frac{1}{4}\%$ for every 1% increase in price. To see the effect on revenue, we use the fact[17] that, for small changes in price,

$$\text{percentage change in revenue} \approx \text{percentage change in price}$$
$$+ \text{percentage change in demand}$$
$$= 1 + \left(-\frac{1}{4}\right) = \frac{3}{4}\%$$

Thus, the revenue will increase by about $\frac{3}{4}\%$. Put another way,

If the demand is inelastic, raising the price increases revenue.

On the other hand, if the price is elastic (which ordinarily occurs at a high unit price), then increasing the price slightly will lower the revenue, so

If the demand is elastic, lowering the price increases revenue.

The price that results in the largest revenue must therefore be at unit elasticity.

Example 1 • Price Elasticity of Demand: Dolls

Suppose the demand equation for Bobby Dolls is given by $q = 216 - p^2$, where p is the price per doll in dollars and q is the number of dolls sold per week.

a. Compute the price elasticity of demand when $p = \$5$ and $p = \$10$ and interpret the results.

b. Find the range of prices for which the demand is elastic and the range for which the demand is inelastic.

c. Find the price at which the weekly revenue is maximized. What is the maximum weekly revenue?

Solution

a. The price elasticity of demand is

$$E = -\frac{dq}{dp} \cdot \frac{p}{q}$$

Taking the derivative and substituting for q gives

$$E = 2p \cdot \frac{p}{216 - p^2} = \frac{2p^2}{216 - p^2}$$

When $p = \$5$,

$$E = \frac{2(5)^2}{216 - 5^2} = \frac{50}{191} \approx 0.26$$

Thus, when the price is set at $5, the demand is dropping at a rate of 0.26% per 1% increase in the price. Since $E < 1$, the demand is inelastic at this price, so raising the price will increase revenue.
    When $p = \$10$,

$$E = \frac{2(10)^2}{216 - 10^2} = \frac{200}{116} \approx 1.72$$

---

[17]See, for example, Exercise 47 in Section 5.4.

Thus, when the price is set at $10, the demand is dropping at a rate of 1.72% per 1% increase in the price. Since $E > 1$, demand is elastic at this price, so raising the price will decrease revenue; lowering the price will increase revenue.

b. and **c.** We answer part (c) first. Setting $E = 1$, we get

$$\frac{2p^2}{216 - p^2} = 1$$

$$2p^2 = 216 - p^2$$

$$3p^2 = 216$$

$$p^2 = 72$$

Thus, we conclude that the maximum revenue occurs when $p = \sqrt{72} \approx \$8.49$. We can now answer part (b): The demand is elastic when $p > \$8.49$ (the price is too high), and the demand is inelastic when $p < \$8.49$ (the price is too low). Finally, we calculate the maximum weekly revenue, which equals the revenue corresponding to the price of $8.49:

$$R = qp = (216 - p^2)p = (216 - 72)\sqrt{72} = 144\sqrt{72} \approx \$1222$$

Graphing Calculator

The TI-83 function `nDeriv` can be used to compute approximations of E at various prices. We set

$\texttt{Y}_1\texttt{=216-X}^2$ Demand equation

$\texttt{Y}_2\texttt{=-nDeriv(Y}_1\texttt{,X,X)*X/Y}_1$ Formula for E

$\texttt{Y}_3\texttt{=X*Y}_1$ Revenue $R = pq$

We can then use the Table feature to list the values of elasticity ($\texttt{Y}_2$) and revenue ($\texttt{Y}_3$) for a range of prices, as shown:

X	Y1	Y2	Y3
4.8	192.96	.23881	926.21
4.9	191.99	.25012	940.75
5	191	.26178	955
5.1	189.99	.2738	968.95
5.2	188.96	.2862	982.59

Excel

To approximate E we can use the following approximation of E:

$$E \approx \frac{\text{percentage decrease in demand}}{\text{percentage increase in price}} \approx -\frac{\Delta q/q}{\Delta p/p}$$

The smaller Δp is, the better the approximation. Let's use $\Delta p = 1\text{¢}$, or 0.01 (which is small compared with the typical prices we consider—around $5 to $10). We start by setting up our worksheet to list a range of prices, in increments of Δp, on either side of a price in which we are interested, such as $p_0 = \$5$:

	A	B
1	p0	5
2	Delta p	0.01
3		
4	p	
5	=B1-10*B2	p0-10*(Delta p)
6	=A5+$B$2	
7		
14		
15		p0
16		
24		
25		p0+10*(Delta p)

	A	B
1	p0	5
2	Delta p	0.01
3		
4	p	
5	4.9	p0-10*(Delta p)
6	4.91	
7		
14		
15	5	p0
16		
24		
25	5.1	p0+10*(Delta p)

We start in cell A5 with the formula for $p_0 - 10\Delta p$ and then successively add Δp going down column A. Notice that the value $p_0 = 5$ appears midway down the list. Next, we compute the corresponding values for the demand q in column B:

	A	B
1	p0	5
2	Delta p	0.01
3		
4	p	q
5	4.9	=216-A5^2
6	4.91	
7		
14		
15	5	
16		
24		
25	5.1	

	A	B
1	p0	5
2	Delta p	0.01
3		
4	p	q
5	4.9	191.99
6	4.91	191.8919
7		
14		
15	5	191
16		
24		
25	5.1	189.99

We add two new columns for the percentage changes in p and q. The formula shown in cell C5 is copied down columns C and D, to row 24. (Why not row 25?)

	A	B	C	D
1	p0	5		
2	Delta p	0.01		
3				
4	p	q	Δp/p	Δq/q
5	4.9	191.99	=(A6-A5)/A5	
6	4.91	191.8919		
7				
14				
15	5	191		
16				
23				
24	5.09	190.0919		
25	5.1	189.99		

The elasticity can now be computed in column E as shown:

	A	B	C	D	E
1	p0	5			
2	Delta p	0.01			
3					
4	p	q	Δp/p	Δq/q	E
5	4.9	191.99	0.002040816	-0.000511	=-D5/C5
6	4.91	191.8919	0.00203666	-0.0005123	
7					
14					
15	5	191	0.002	-0.0005241	
16					
23					
24	5.09	190.0919	0.001964637	-0.0005361	
25	5.1	189.99			

The concept of elasticity can be applied in other situations. In Example 2 we consider *income* elasticity of demand—the percentage increase in demand for a particular item per percentage increase in personal income.

Example 2 • Income Elasticity of Demand: Porsches

You are the sales director at Suburban Porsche and have noticed that demand for Porsches depends on income according to

$$q = 0.005e^{-0.05x^2+x} \qquad (1 \le x \le 10)$$

Here, x is the income of a potential customer in hundreds of thousands of dollars, and q is the probability that the person will actually purchase a Porsche.[18] The **income elasticity of demand** is

$$E = \frac{dq}{dx} \cdot \frac{x}{q}$$

Compute and interpret E for $x = 2$ and 9.

Solution
Question Why is there no negative sign in the formula?

Answer Since we anticipate that the demand will increase as income increases, the ratio

$$\frac{\text{percentage increase in demand}}{\text{percentage increase in income}}$$

will be positive, so there is no need to introduce a negative sign.

Turning to the calculation, since $q = 0.005e^{-0.05x^2+x}$,

$$\frac{dq}{dx} = 0.005e^{-0.05x^2+x}(-0.1x + 1)$$

and so

$$E = \frac{dq}{dx} \cdot \frac{x}{q}$$

$$= 0.005e^{-0.05x^2+x}(-0.1x + 1)\frac{x}{0.005e^{-0.05x^2+x}}$$

$$= x(-0.1x + 1)$$

When $x = 2$, $E = 2[-0.1(2) + 1)] = 1.6$. Thus, at an income level of \$200,000, the probability that a customer will purchase a Porsche increases at a rate of 1.6% per 1% increase in income.

When $x = 9$, $E = 9[-0.1(9) + 1)] = 0.9$. Thus, at an income level of \$900,000, the probability that a customer will purchase a Porsche increases at a rate of 0.9% per 1% increase in income.

---

[18]In other words, q is the fraction of visitors to your showroom having income x who actually purchase a Porsche.

5.5 EXERCISES

APPLICATIONS

1. **Demand for Oranges** The weekly sales of Honolulu Red Oranges is given by $q = 1000 - 20p$. Calculate the price elasticity of demand when the price is $30/orange (yes, $30 per orange!).[19] Interpret your answer. Also, calculate the price that gives a maximum weekly revenue and find this maximum revenue.

2. **Demand for Oranges** Repeat Exercise 1 for weekly sales of $1000 - 10p$.

3. **Tissues** The consumer demand equation for tissues is given by $q = (100 - p)^2$, where p is the price per case of tissues and q is the demand in weekly sales.
   a. Determine the price elasticity of demand E when the price is set at $30 and interpret your answer.
   b. At what price should tissues be sold in order to maximize the revenue?
   c. Approximately how many cases of tissues would be demanded at that price?

4. **Bodybuilding** The consumer demand curve for Professor Stefan Schwartzenegger dumbbells is given by $q = (100 - 2p)^2$, where p is the price per dumbbell and q is the demand in weekly sales. Find the price Professor Schwartzenegger should charge for his dumbbells in order to maximize revenue.

5. **College Tuition** A study of about 1800 U.S. colleges and universities resulted in the demand equation $q = 9859.39 - 2.17p$, where q is the enrollment at a college or university and p is the average annual tuition (plus fees) it charges.
   a. The study also found that the average tuition charged by universities and colleges was $2867. What is the corresponding price elasticity of demand? Interpret your answer.
   b. Based on the study, what would you advise a college to charge its students in order to maximize total revenue, and what would the revenue be?

   SOURCE: Based on a study by A. L. Ostrosky, Jr., and J. V. Koch, as cited in their book *Introduction to Mathematical Economics* (Prospect Heights, Ill.: Waveland Press, 1979): 133.

6. **Demand for Fried Chicken** A fried-chicken franchise finds that the demand equation for its new roasted chicken product, "Roasted Rooster," is given by

$$p = \frac{40}{q^{1.5}}$$

   where p is the price (in dollars) per quarter-chicken serving and q is the number of quarter-chicken servings that can be sold each hour at this price. Express q as a function of p and find the price elasticity of demand when the price is set at $4 per serving. Interpret the result.

7. **Linear Demand Functions** A general linear demand function has the form $q = mp + b$ (m and b constants, with $m \neq 0$).
   a. Obtain a formula for the price elasticity of demand at a unit price of p.
   b. Obtain a formula for the price that maximizes revenue.

8. **Exponential Demand Functions** A general exponential demand function has the form $q = Ae^{-bp}$ (A and b nonzero constants).
   a. Obtain a formula for the price elasticity of demand at a unit price of p.
   b. Obtain a formula for the price that maximizes revenue.

9. **Hyperbolic Demand Functions** A general hyperbolic demand function has the form $q = k/p^r$ (r and k nonzero constants).
   a. Obtain a formula for the price elasticity of demand at unit price p.
   b. How does E vary with p?
   c. What does the answer to part (b) say about the model?

10. **Quadratic Demand Functions** A general quadratic demand function has the form $q = ap^2 + bp + c$ (a, b, and c constants, with $a \neq 0$).
    a. Obtain a formula for the price elasticity of demand at a unit price p.
    b. Obtain a formula for the price or prices that could maximize revenue.

11. **Exponential Demand Functions** The estimated monthly sales of *Mona Lisa* paint-by-number sets is given by the formula $q = 100e^{-3p^2+p}$, where q is the demand in monthly sales and p is the retail price in yen.
    a. Determine the price elasticity of demand E when the retail price is set at ¥3 and interpret your answer.
    b. At what price will revenue be a maximum?
    c. Approximately how many paint-by-number sets will be sold each month at the price in part (b)?

12. **Exponential Demand Functions** Repeat Exercise 12, using the demand equation $q = 100e^{p-3p^2/2}$.

13. **Modeling Linear Demand** You have been hired as a marketing consultant to Johannesburg Burger Supply, and you wish to come up with a unit price for its hamburgers in order to maximize its weekly revenue. To make life as simple as possible, you assume that the demand equation for Johannesburg hamburgers has the linear form $q = mp + b$, where p is the price per hamburger, q is the demand in weekly sales, and m and b are certain constants you must determine.
    a. Your market studies reveal the following sales figures: When the price is set at $2 per hamburger, the sales amount to 3000 per week, but when the price is set at $4 per hamburger, the sales drop to zero. Use these data to calculate the demand equation.
    b. Now estimate the unit price that maximizes weekly revenue and predict what the weekly revenue will be at that price.

---

[19]They are very hard to find, and their possession confers considerable social status.

14. **Modeling Linear Demand** You have been hired as a marketing consultant to Big Book Publishing, and you have been approached to determine the best selling price for the hit calculus text by Whiner and Istanbul entitled *Fun with Derivatives*. You decide to make life easy and assume that the demand equation for *Fun with Derivatives* has the linear form $q = mp + b$, where p is the price per book, q is the demand in annual sales, and m and b are certain constants you'll have to figure out.

   a. Your market studies reveal the following sales figures: When the price is set at $50 per book, the sales amount to 10,000 per year; when the price is set at $80 per book, the sales drop to 1000 per year. Use these data to calculate the demand equation.

   b. Now estimate the unit price that maximizes annual revenue and predict what Big Book Publishing's annual revenue will be at that price.

15. **Income Elasticity of Demand: Live Drama** The likelihood that a child will attend a live theatrical performance can be modeled by

   $$q = 0.01(-0.0078x^2 + 1.5x + 4.1) \qquad (15 \le x \le 100)$$

   Here, q is the fraction of children with annual household income x thousand dollars who will attend a live dramatic performance at a theater during the year. Compute the income elasticity of demand at an income level of $20,000 and interpret the result. (Round your answer to two significant digits.)

   Based on a quadratic regression of data from a 2001 survey. Source: New York Foundation of the Arts http://www.nyfa.org/culturalblueprint/

16. **Income Elasticity of Demand: Live Concerts** The likelihood that a child will attend a live musical performance can be modeled by

   $$q = 0.01(0.0006x^2 + 0.38x + 35) \qquad (15 \le x \le 100)$$

   Here, q is the fraction of children with annual household income x who will attend a live musical performance during the year. Compute the income elasticity of demand at an income level of $30,000 and interpret the result.

   Based on a quadratic regression of data from a 2001 survey. Source: New York Foundation of the Arts http://www.nyfa.org/culturalblueprint/

17. **Income Elasticity of Demand: Computer Usage** The demand for personal computers in the home goes up with household income. The following graph shows some data on computer usage, together with the logarithmic model $q = 0.3454 \ln(x) - 3.047$, where q is the probability that a household with annual income x will have a computer.

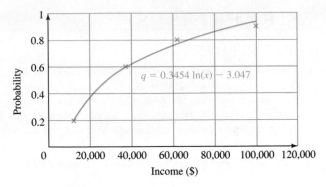

   a. Compute the income elasticity of demand for computers to two decimal places for a household income of $60,000 and interpret the result.

   b. As household income increases, how is income elasticity of demand affected?

   c. How reliable is the given model of demand for incomes well above $120,000? Explain.

   d. What can you say about E for incomes much larger than those shown?

   All figures are approximate. The model is a regression model, and x measures the probability that a given household will have one or more computers. Source: For income distribution, Luxembourg Income Study/New York Times, August 14, 1995, p. A9. For computer data, Forrester Research/New York Times, August 8, 1999, p. BU4.

18. **Income Elasticity of Demand: Internet Usage** The demand for Internet connectivity also goes up with household income. The following graph shows some data on Internet usage, together with the logarithmic model $q = 0.2802 \ln(x) - 2.505$, where q is the probability that a home with annual household income x will have an Internet connection.

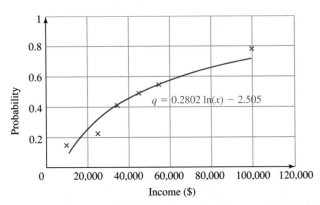

   a. Compute the income elasticity of demand to two decimal places for a household income of $60,000 and interpret the result.

b. As household income increases, how is income elasticity of demand affected?

c. The logarithmic model shown above is not appropriate for incomes well above $100,000. Suggest a model that might be more appropriate.

d. In the model you propose, how does E behave for very large incomes?

All figures are approximate, and the model is a regression model. The Internet connection figures were actually quoted as "share of consumers who use the Internet, by household income." SOURCES: Luxembourg Income Study/*New York Times,* August 14, 1995, p. A9; Commerce Department, Deloitte & Touche Survey/*New York Times,* November 24, 1999, p. C1.

19. Income Elasticity of Demand (Based on a Question on the GRE Economics Test) If $Q = aP^\alpha Y^\beta$ is the individual's demand function for a commodity, where P is the (fixed) price of the commodity, Y is the individual's income, and a, α, and β are parameters, explain why β can be interpreted as the income elasticity of demand.

20. College Tuition (from the GRE Economics Test) A time-series study of the demand for higher education, using tuition charges as a price variable, yields the following result:

$$\frac{dq}{dp} \cdot \frac{p}{q} = -0.4$$

where p is tuition and q is the quantity of higher education. Which of the following is suggested by the result?

(A) As tuition rises, students want to buy a greater quantity of education.

(B) As a determinant of the demand for higher education, income is more important than price.

(C) If colleges lowered tuition slightly, their total tuition receipts would increase.

(D) If colleges raised tuition slightly, their total tuition receipts would increase.

(E) Colleges cannot increase enrollments by offering larger scholarships.

21. Modeling Exponential Demand As the new owner of a supermarket, you have inherited a large inventory of unsold imported Limburger cheese, and you would like to set the price so that your revenue from selling it is as large as possible. Previous sales figures of the cheese are shown in the following table:

Price/Pound, p ($)	3.00	4.00	5.00
Monthly Sales, q (in lb)	407	287	223

a. Use the sales figures for the prices $3 and $5 per pound to construct a demand function of the form $q = Ae^{-bp}$, where A and b are constants you must determine. (Round A and b to two significant digits.)

b. Use your demand function to find the price elasticity of demand at each of the prices listed.

c. At what price should you sell the cheese in order to maximize monthly revenue?

d. If your total inventory of cheese amounts to only 200 pounds and it will spoil 1 month from now, how should you price it in order to receive the greatest revenue? Is this the same answer you got in part (c)? If not, give a brief explanation.

22. Modeling Exponential Demand Repeat Exercise 21, but this time use the sales figures for $4 and $5 per pound to construct the demand function.

COMMUNICATION AND REASONING EXERCISES

23. Complete the following: When demand is inelastic, revenue will decrease if _____.

24. Complete the following: When demand has unit elasticity, revenue will decrease if _____.

25. Given that the demand q is a differentiable function of the unit price p, show that the revenue $R = pq$ has a stationary point when

$$q + p\frac{dq}{dp} = 0$$

Deduce that the stationary points of R are the same as the points of unit price elasticity of demand. (Ordinarily, there is only one such stationary point, corresponding to the absolute maximum of R.) (*Hint:* Differentiate R with respect to p.)

26. Given that the demand q is a differentiable function of income x, show that the quantity $R = q/x$ has a stationary point when

$$q - x\frac{dq}{dx} = 0$$

Deduce that stationary points of R are the same as the points of unit income elasticity of demand. (*Hint:* Differentiate R with respect to x.)

27. Your calculus study group is discussing price elasticity of demand, and a member of the group asks the following question: "Since elasticity of demand measures the response of demand to change in unit price, what is the difference between elasticity of demand and the quantity $-dq/dp$?" How would you respond?

28. Another member of your study group claims that unit price elasticity of demand need not always correspond to maximum revenue. Is he correct? Explain your answer.

CASE STUDY

Production Lot-Size Management

Your publishing company, Knockem Dead Paperbacks, is about to release its next best-seller, *Henrietta's Heaving Heart* by Celestine A. Lafleur. The company expects to sell 100,000 books each month in the next year. You have been given the job of scheduling print runs to meet the anticipated demand and minimize total costs to the company. Each print run has a setup cost of $5000, each book costs $1 to produce, and monthly storage costs for books awaiting shipment average 1¢ per book. What will you do?

If you decide to print all 1.2 million books (the total demand for the year, 100,000 books per month for 12 months) in a single run at the start of the year and sales run as predicted, then the number of books in stock would begin at 1.2 million and decrease to zero by the end of the year, as shown in Figure 29.

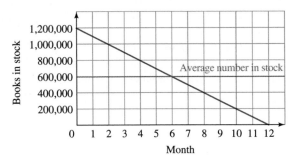

Figure 29

On average you would be storing 600,000 books for 12 months at 1¢ per book, giving a total storage cost of 600,000 × 12 × .01 = $72,000. The setup cost for the single print run would be $5000. When you add to these the total cost of producing 1.2 million books at $1 per book, your total cost would be $1,277,000.

If, on the other hand, you decide to cut down on storage costs by printing the book in two runs of 600,000 each, you would get the picture shown in Figure 30.

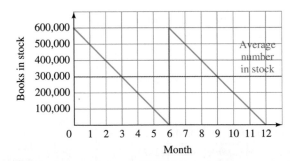

Figure 30

Now, the storage cost would be cut in half because on average there would be only 300,000 books in stock. Thus, the total storage cost would be $36,000, and the setup cost would double to $10,000 (there would now be two runs). The production costs would be the same: 1.2 million books at $1 per book. The total cost would therefore be reduced to $1,246,000, a savings of $31,000 compared to your first scenario.

"Aha!" you say to yourself, after doing these calculations. "Why not drastically cut costs by setting up a run every month?" You calculate that the setup costs alone would be 12 × $5000 = $60,000, which is already more than the setup plus storage costs for two runs, so a run every month will cost too much. Perhaps, then, you should investigate three runs, four runs, and so on, until you find the lowest cost. This strikes you as too laborious a process, especially considering that you will have to do it all over again when planning for Lafleur's sequel, *Lorenzo's Lost Love,* due to be released next year. Realizing that this is an optimization problem, you decide to use some calculus to help you come up with a *formula* that you can use for all future plans. So you get to work.

Instead of working with the number 1.2 million, you use the letter N so that you can be as flexible as possible. (What if *Lorenzo's Lost Love* sells more copies?) Thus, you have a total of N books to be produced for the year. You now calculate the total cost of using x print runs per year. Because you are to produce a total of N books in x print runs, you will have to produce N/x books in each print run. N/x is called the **lot size.** As you can see from Figures 29 and 30, the average number of books in storage will be half that amount, $N/(2x)$.

Now you can calculate the total cost for a year. Write P for the setup cost of a single print run ($P = \$5000$ in your case) and c for the *annual* cost of storing a book (to convert all of the time measurements to years; $c = \$0.12$ here). Finally, write b for the cost of producing a single book ($b = \$1$ here). The costs break down as follows:

Setup costs: x print runs at P dollars per run: $\quad Px$

Storage costs: $N/(2x)$ books stored at c dollars per year: $\quad cN/(2x)$

Production costs: N books at b dollars per book: $\quad \underline{Nb}$

$$\textbf{Total cost:}\quad Px + \frac{cN}{2x} + Nb$$

Remember that P, N, c, and b are all constants and x is the only variable. Thus, your cost function is

$$C(x) = Px + \frac{cN}{2x} + Nb$$

and you need to find the value of x that will minimize $C(x)$. But that's easy! All you need to do is find the relative extrema and select the absolute minimum (if any).

The domain of $C(x)$ is $(0, +\infty)$ because there is an x in the denominator and x can't be negative. To locate the extrema, you start by locating the critical points:

$$C'(x) = P - \frac{cN}{2x^2}$$

The only singular point would be at $x = 0$, but 0 is not in the domain. To find stationary points, you set $C'(x) = 0$ and solve for x:

$$P - \frac{cN}{2x^2} = 0$$

$$2x^2 = \frac{cN}{P}$$

$$x = \sqrt{\frac{cN}{2P}}$$

There is only one stationary point, and there are no singular points or endpoints. To graph the function, you will need to put in numbers for the various constants. Substituting $N = 1{,}200{,}000$, $P = 5000$, $c = 0.12$, and $b = 1$, you get

$$C(x) = 5000x + \frac{72{,}000}{x} + 1{,}200{,}000$$

with the stationary point at

$$x = \sqrt{\frac{(0.12)(1,200,000)}{2(5000)}} \approx 3.79$$

The total cost at the stationary point is

$$C(3.79) \approx 1,237,900$$

You now graph $C(x)$ in a window that includes the stationary point, getting Figure 31.

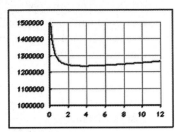

Figure 31

From the graph you can see that the stationary point is an absolute minimum. In the graph it appears that the graph is always concave up, which also tells you that your stationary point is a minimum. You can check the concavity by computing the second derivative:

$$C''(x) = \frac{cN}{x^3} > 0$$

The second derivative is always positive because c, N, and x are all positive numbers, so indeed the graph is always concave up. Now you also know that it works regardless of the particular values of the constants.

So now you are practically done! You know that the absolute minimum cost occurs when you have $x \approx 3.79$ print runs per year. Don't be disappointed that the answer is not a whole number; whole number solutions are rarely found in real scenarios. What the answer (and the graph) do indicate is that either three or four print runs per year will cost the least money. If you take $x = 3$, you get a total cost of

$$C(3) = \$1,239,000$$

If you take $x = 4$, you get a total cost of

$$C(4) = \$1,238,000$$

So, four print runs per year will allow you to minimize your total costs.

EXERCISES

1. *Lorenzo's Lost Love* will sell 2 million copies in a year. The remaining costs are the same. How many print runs should you use now?

2. In general, what happens to the number of runs that minimizes cost if both the setup cost and the total number of books are doubled?

3. In general, what happens to the number of runs that minimizes cost if the setup cost increases by a factor of 4?

4. Assuming that the total number of copies and storage costs are as originally stated, find the setup cost that would result in a single print run.

5. Assuming that the total number of copies and setup cost are as originally stated, find the storage cost that would result in a print run each month.

6. In Figure 30 we assumed that all the books in each run were manufactured in a very short time; otherwise, the figure might have looked more like the following figure, which shows the inventory assuming a slower rate of production.

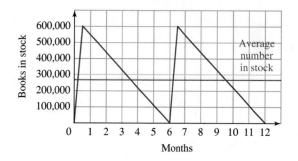

How would this affect the answer?

7. Referring to the general situation discussed in the text, find the cost as a function of the total number of books produced, assuming that the number of runs is chosen to minimize total cost. Also find the average cost per book.

8. Let $\overline{C}$ be the average cost function found in Exercise 7. Calculate $\lim_{N \to +\infty} \overline{C}(N)$ and interpret the result.

CHAPTER 5 REVIEW TEST

1. In each of the following, find all the relative and absolute extrema of the given functions on the given domain (if supplied) or on the largest possible domain (if no domain is supplied).

   a. $f(x) = 2x^3 - 6x + 1$ on $[-2, +\infty)$

   b. $f(x) = \dfrac{x+1}{(x-1)^2}$ on $[-2, 1) \cup (1, 2]$

   c. $g(x) = (x-1)^{2/3}$

   d. $g(x) = x^2 + \ln x$ on $(0, +\infty)$

   e. $h(x) = \dfrac{1}{x} + \dfrac{1}{x^2}$

   f. $h(x) = e^{x^2} + 1$

2. In each of the following, the graph of the function f or its derivative is given. Find the approximate x coordinates of all relative extrema and points of inflection of the original function f (if any).

   a. Graph of f:

   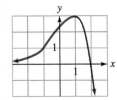

   b. Graph of f:

   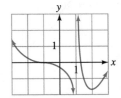

   c. Graph of f':

   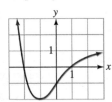

   d. Graph of f':

   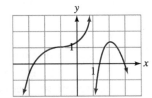

3. In each of the following, use technology to draw the graph of the given function, indicating all relative and absolute extrema and points of inflection. Find the coordinates of these points exactly, where possible. Also indicate any horizontal and vertical asymptotes.

   a. $f(x) = x^3 - 12x$ on $[-2, +\infty)$ **b.** $f(x) = \dfrac{x^2 - 3}{x^3}$

   c. $f(x) = (x-1)^{2/3} + \dfrac{2x}{3}$

OHaganBooks.com—TAKING ADVANTAGE OF CHANGE

4. Demand for the latest best-seller at OHaganBooks.com, *A River Burns Through It*, is given by

$$q = -p^2 + 33p + 9 \qquad (18 \le p \le 28)$$

   copies sold per week when the price is p dollars.

   a. Find the price elasticity of demand as a function of p.

   b. Find the elasticity of demand for this book at a price of \$20 and at a price of \$25. (Round your answers to two decimal places.) Interpret the answers.

   c. What price should the company charge to obtain the largest revenue?

5. Taking into account storage and shipping, it costs OHaganBooks.com

$$C = 9q + 100$$

   dollars to sell q copies of *A River Burns Through It* in 1 week.

   a. If demand is as in Question 4, express the weekly profit earned by OHaganBooks.com from the sale of *A River Burns Through It* as a function of unit price p.

   b. What price should the company charge to get the largest weekly profit?

   c. What is the maximum possible weekly profit?

   d. Compare your answer to part (b) with the price the company should charge to obtain the largest revenue. Explain any difference.

6. OHaganBooks.com modeled its weekly sales over a period of time with the function

$$s(t) = 6053 + \frac{4474}{1 + e^{-0.55(t-4.8)}}$$

   as shown in the following graph:

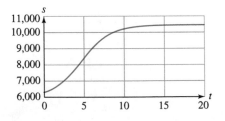

a. Compute and graph $s'(t)$ and use your graph (or some other method) to estimate when to the nearest week the weekly sales were growing fastest.

b. To what feature on the graph of s does your answer to part (a) correspond?

c. The graph of s has a horizontal asymptote. What is the value (s coordinate) of this asymptote, and what is its significance in terms of weekly sales at OHaganBooks.com?

d. The graph of s' has a horizontal asymptote. What is the value (s' coordinate) of this asymptote, and what is its significance in terms of weekly sales at OHaganBooks.com?

7. OHaganBooks.com's Web site has an animated graphic with its name in a rectangle whose height and width change; on either side of the rectangle are semicircles, as in the figure, whose diameters are the same as the height of the rectangle.

For reasons too complicated to explain, the designer wants the combined area of the rectangle and semicircles to remain constant. At one point during the animation, the width of the rectangle is 1 inch, growing at a rate of 0.5 inch per second, and the height is 3 inches. How fast is the height changing?

 ADDITIONAL ONLINE REVIEW

If you follow the path

    Web Site → Everything for Calculus → Chapter 5

you will find the following additional resources to help you review:

A comprehensive chapter summary (including examples and interactive features)

Additional review exercises (including interactive exercises and many with help)

A true/false chapter quiz

Section-by-section online tutorials

6

THE INTEGRAL

CASE STUDY

Wage Inflation

As assistant personnel manager for a large corporation, you have been asked to estimate the average annual wage earned by a worker in your company, from the time the worker is hired to the time the worker retires. You have data about wage increases. How will you estimate this average?

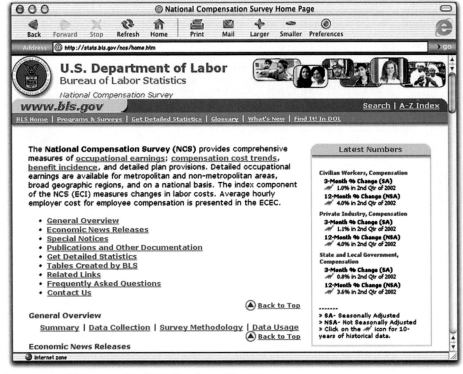

REPRODUCED FROM THE U.S. DEPARTMENT OF LABOR, BUREAU OF LABOR STATISTICS, WEB SITE

INTERNET RESOURCES FOR THIS CHAPTER

At the Web site, follow the path

> Web Site → Everything for Calculus → Chapter 6

where you will find links to step-by-step tutorials for the main topics in this chapter, a detailed chapter summary you can print out, a true/false quiz, and a collection of sample test questions. You will also find downloadable Excel tutorials for each section, an online numerical integration utility, and other resources. Complete text and interactive exercises have been placed on the Web site covering the following optional topic: numerical integration.

Introduction

Roughly speaking, calculus is divided into two parts: **differential calculus** (the calculus of derivatives) and **integral calculus,** which is the subject of this chapter and the next. Integral calculus is concerned with problems that are in some sense the reverse of the problems seen in differential calculus. For example, whereas differential calculus shows how to compute the rate of change of a quantity, integral calculus shows how to find the quantity if we know its rate of change. This idea is made precise in the **Fundamental Theorem of Calculus.** Integral calculus and the Fundamental Theorem of Calculus allow us to solve many problems in economics, physics, and geometry, including one of the oldest problems in mathematics—computing areas of regions with curved boundaries.

6.1 The Indefinite Integral

Having studied differentiation in the preceding chapters, we now discuss how to *reverse* the process.

Question If the derivative of $F(x)$ is $4x^3$, what is $F(x)$?

Answer After a moment's thought, we recognize $4x^3$ as the derivative of x^4. So, we might have $F(x) = x^4$. However, on thinking further, we realize that, for example, $F(x) = x^4 + 7$ works just as well. In fact, $F(x) = x^4 + C$ works for any number C. Thus, there are *infinitely many* possible answers to this question.

In fact, we will see shortly that the formula $F(x) = x^4 + C$ covers *all* possible answers to the question. Let's give a name to what we are doing.

Antiderivative

An **antiderivative** of a function f is a function F such that $F' = f$.

Quick Examples

1. An antiderivative of $4x^3$ is x^4.   Because the derivative of x^4 is $4x^3$

2. Another antiderivative of $4x^3$ is $x^4 + 7$.   Because the derivative of $x^4 + 7$ is $4x^3$

3. An antiderivative of $2x$ is $x^2 + 12$.   Because the derivative of $x^2 + 12$ is $2x$

Thus,

If the derivative of $A(x)$ is $B(x)$, then an antiderivative of $B(x)$ is $A(x)$.

We call the set of *all* antiderivatives of a function the **indefinite integral** of the function.

Indefinite Integral

$$\int f(x)\, dx$$

is read "the **indefinite integral** of $f(x)$ with respect to x" and stands for the set of all antiderivatives of f. Thus, $\int f(x)\, dx$ is a *collection of functions*; it is not a single function nor a number. The function f that is being **integrated** is called the **integrand,** and the variable x is called the **variable of integration.**

Quick Examples

1. $\int 4x^3\, dx = x^4 + C$ Every possible antiderivative of $4x^3$ has the form $x^4 + C$.

2. $\int 2x\, dx = x^2 + C$ Every possible antiderivative of $2x$ has the form $x^2 + C$.

The **constant of integration** C reminds us that we can add any constant and get a different antiderivative.

Question If $F(x)$ is one antiderivative of $f(x)$, why must all other antiderivatives have the form $F(x) + C$?

Answer Suppose $F(x)$ and $G(x)$ are both antiderivatives of $f(x)$, so that $F'(x) = G'(x)$. Consider what this means by looking at Figure 1. If $F'(x) = G'(x)$ for all x, then F and G have the *same slope* at each value of x. This means that their graphs must be *parallel* and hence remain exactly the same vertical distance apart. But that is the same as saying that the functions differ by a constant—that is, that $G(x) = F(x) + C$ for some constant C.[1]

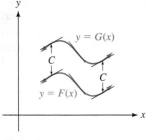

Figure 1

Example 1 • Indefinite Integral

Check that $\int x\, dx = \dfrac{x^2}{2} + C$.

Solution We check by taking the derivative of the right-hand side:

$$\frac{d}{dx}\left[\frac{x^2}{2} + C\right] = \frac{2x}{2} + 0 = x \quad \checkmark$$

Now, we would like to make the process of finding indefinite integrals (antiderivatives) more mechanical. For example, it would be nice to have a power rule for indefinite integrals similar to the one we already have for derivatives. Two cases suggested by the examples above are

$$\int x\, dx = \frac{x^2}{2} + C \quad \text{and} \quad \int x^3\, dx = \frac{x^4}{4} + C$$

---

[1] This argument can be turned into a more rigorous proof—that is, a proof that does not rely on geometric concepts such as parallel graphs. We should also say that the result requires that F and G have the same derivative *on an interval* $[a, b]$.

You should check the last equation by taking the derivative of its right-hand side. These cases suggest the following general statement.

Power Rule for the Indefinite Integral, Part I

$$\int x^n \, dx = \frac{x^{n+1}}{n+1} + C \qquad (\text{if } n \neq -1)$$

In Words To find the integral of x^n, add 1 to the exponent and then divide by the new exponent. This rule works provided that n is not -1.

Quick Examples

1. $\displaystyle\int x^{55} \, dx = \frac{x^{56}}{56} + C$

2. $\displaystyle\int \frac{1}{x^{55}} \, dx = \int x^{-55} \, dx \qquad\qquad$ Exponent form

$$= \frac{x^{-54}}{-54} + C \qquad\qquad \text{When we add 1 to } -55, \text{ we get } -54, \textit{not } -56.$$

$$= -\frac{1}{54x^{54}} + C \qquad\qquad \text{See note.}^*$$

3. $\displaystyle\int 1 \, dx = x + C \qquad\qquad$ Since $1 = x^0$. This is an important special case.

Notes

1. The integral $\displaystyle\int 1 \, dx$ is commonly written as $\displaystyle\int dx$.

2. Similarly, the integral $\displaystyle\int \frac{1}{x^{55}} \, dx$ may be written as $\displaystyle\int \frac{dx}{x^{55}}$.

---

*We are glossing over a subtlety here: The constant of integration C can be different for $x < 0$ and $x > 0$ because the graph breaks at $x = 0$. In general, our understanding will be that the constant of integration may be different on disconnected intervals of the domain.

We can easily check the power rule formula by taking the derivative of the right-hand side:

$$\frac{d}{dx}\left[\frac{x^{n+1}}{n+1} + C\right] = \frac{(n+1)x^n}{n+1} = x^n \quad \checkmark$$

Question What is the reason for the restriction $n \neq -1$?

Answer Let's answer a question with a question: Does the right-hand side of the power rule formula make sense if $n = -1$?

Question Well, no, so what is

$$\int x^{-1} \, dx = \int \frac{1}{x} \, dx$$

Answer Think before reading on: Have you ever seen a function whose derivative is $1/x$? Prodding our memories a little, we recall that $\ln x$ has derivative $1/x$. In fact, as we pointed out when we first discussed it, $\ln |x|$ also has derivative $1/x$, but it has the advantage that its domain is the same as that of $1/x$. Thus, we can fill in the missing case as follows.

Power Rule for the Indefinite Integral, Part II

$$\int x^{-1} \, dx = \ln |x| + C \qquad\qquad \text{Equivalently,} \int \frac{1}{x} \, dx = \ln |x| + C.$$

Here are two other indefinite integrals that come from corresponding formulas for differentiation.

Indefinite Integral of e^x and b^x

$$\int e^x \, dx = e^x + C \qquad\qquad \text{Because } \frac{d}{dx}[e^x] = e^x$$

If b is any positive number other than 1, then

$$\int b^x \, dx = \frac{b^x}{\ln b} + C \qquad\qquad \text{Because } \frac{d}{dx}\left[\frac{b^x}{\ln b}\right] = \frac{b^x \ln b}{\ln b} = b^x$$

Quick Example

$$\int 2^x \, dx = \frac{2^x}{\ln 2} + C$$

Question What about more complicated functions, such as $2x^3 + 6x^5 - 1$?

Answer We need the following rules for integrating sums, differences, and constant multiples.

Rules for the Indefinite Integral

Sum and Difference Rules

$$\int [f(x) \pm g(x)] \, dx = \int f(x) \, dx \pm \int g(x) \, dx$$

In Words The integral of a sum is the sum of the integrals, and the integral of a difference is the difference of the integrals.

Constant Multiple Rule

$$\int kf(x) \, dx = k \int f(x) \, dx \qquad (k \text{ constant})$$

In Words The integral of a constant times a function is the constant times the integral of the function. (In other words, the constant "goes along for the ride.")

Quick Examples

Sum rule: $\int (x^3 + 1) \, dx = \int x^3 \, dx + \int 1 \, dx = \frac{x^4}{4} + x + C \qquad f(x) = x^3; g(x) = 1$

Constant multiple rule: $\int 5x^3 \, dx = 5 \int x^3 \, dx = 5 \frac{x^4}{4} + C \qquad k = 5; f(x) = x^3$

Constant multiple rule: $\int 4 \, dx = 4 \int 1 \, dx = 4x + C \qquad k = 4; f(x) = 1$

Constant multiple rule: $\int 4e^x \, dx = 4 \int e^x \, dx = 4e^x + C \qquad k = 4; f(x) = e^x$

Why are these rules true? Because the derivative of a sum is the sum of the derivatives, and similarly for differences and constant multiples.

Example 2 • Using the Sum and Difference Rules

Find the integrals.

a. $\displaystyle\int (x^3 + x^5 - 1)\, dx$ **b.** $\displaystyle\int \left(x^{2.1} + \frac{1}{x^{1.1}} + \frac{1}{x} + e^x \right) dx$ **c.** $\displaystyle\int (e^x + 3^x - 1)\, dx$

Solution

a. $\displaystyle\int (x^3 + x^5 - 1)\, dx = \int x^3\, dx + \int x^5\, dx - \int 1\, dx$ Sum/difference rule

$$= \frac{x^4}{4} + \frac{x^6}{6} - x + C$$ Power rule

b. $\displaystyle\int \left(x^{2.1} + \frac{1}{x^{1.1}} + \frac{1}{x} + e^x \right) dx = \int (x^{2.1} + x^{-1.1} + x^{-1} + e^x)\, dx$ Exponent form

$$= \int x^{2.1}\, dx + \int x^{-1.1}\, dx + \int x^{-1}\, dx + \int e^x\, dx$$ Sum rule

$$= \frac{x^{3.1}}{3.1} + \frac{x^{-0.1}}{-0.1} + \ln|x| + e^x + C$$ Power rule and exponential rule

$$= \frac{x^{3.1}}{3.1} - \frac{10}{x^{0.1}} + \ln|x| + e^x + C$$ Back to fraction form

c. $\displaystyle\int (e^x + 3^x - 1)\, dx = \int e^x\, dx + \int 3^x\, dx - \int 1\, dx$ Sum/difference rule

$$= e^x + \frac{3^x}{\ln 3} - x + C$$ Power rule and exponential rule

---

✷ **Before we go on . . .** As usual, you should check each of the answers by differentiating.

Question Why is there only a single arbitrary constant C in each of the answers?

Answer We could have written the answer to part (a) as

$$\frac{x^4}{4} + D + \frac{x^6}{6} + E - x + F$$

where D, E, and F are all arbitrary constants. Now suppose, for example, we set $D = 1$, $E = -2$, and $F = 6$. Then the particular antiderivative we get is $x^4/4 + x^6/6 - x + 5$, which has the form $x^4/4 + x^6/6 - x + C$. Thus, we could have chosen the single constant C to be 5 and obtained the same answer. In other words, the answer $x^4/4 + x^6/6 - x + C$ is just as general as the answer $x^4/4 + D + x^6/6 + E - x + F$, but simpler.

In practice we do not explicitly write the integral of a sum as a sum of integrals but just "integrate term by term," much as we learned to differentiate term by term.

Example 3 • Combining the Rules

Find the integrals.

a. $\int (10x^4 + 2x^2 - 3e^x)\, dx$ **b.** $\int \left(\dfrac{2}{x^{0.1}} + \dfrac{x^{0.1}}{2} - \dfrac{3}{4x} \right) dx$ **c.** $\int [3e^x - 2(1.2^x)]\, dx$

Solution

a. We need to integrate separately each of the terms $10x^4$, $2x^2$, and $3e^x$. To integrate $10x^4$, we use the rules for constant multiples and powers:

$$\int 10x^4\, dx = 10 \int x^4\, dx = 10\frac{x^5}{5} + C = 2x^5 + C$$

The other two terms are similar. We get

$$\int (10x^4 + 2x^2 - 3e^x)\, dx = 10\frac{x^5}{5} + 2\frac{x^3}{3} - 3e^x + C = 2x^5 + \frac{2}{3}x^3 - 3e^x + C$$

b. We first convert to exponent form and then integrate term by term:

$$\int \left(\frac{2}{x^{0.1}} + \frac{x^{0.1}}{2} - \frac{3}{4x} \right) dx = \int \left(2x^{-0.1} + \frac{1}{2}x^{0.1} - \frac{3}{4}x^{-1} \right) dx \quad \text{Exponent form}$$

$$= 2\frac{x^{0.9}}{0.9} + \frac{1}{2}\frac{x^{1.1}}{1.1} - \frac{3}{4} \ln |x| + C \quad \begin{array}{l}\text{Integrate term by}\\ \text{term.}\end{array}$$

$$= \frac{20x^{0.9}}{9} + \frac{x^{1.1}}{2.2} - \frac{3}{4} \ln |x| + C$$

c. $\int [3e^x - 2(1.2^x)]\, dx = 3e^x - 2\dfrac{1.2^x}{\ln (1.2)} + C$

Example 4 • Different Variable Name

Find $\int \left(\dfrac{1}{u} + \dfrac{1}{u^2} \right) du$.

Solution   This integral may look a little strange because we are using the letter u instead of x, but there is really nothing special about x. Using u as the variable of integration, we get

$$\int \left(\frac{1}{u} + \frac{1}{u^2} \right) du = \int (u^{-1} + u^{-2})\, du \quad \text{Exponent form}$$

$$= \ln |u| + \frac{u^{-1}}{-1} + C \quad \text{Integrate term by term.}$$

$$= \ln |u| - \frac{1}{u} + C \quad \text{Simplify the result.}$$

✳ *Before we go on . . .* When we compute an indefinite integral, we want the independent variable in the answer to be the same as the variable of integration. Thus, if the integral had been written in terms of x rather than u, we would have written

$$\int \left(\frac{1}{x} + \frac{1}{x^2} \right) dx = \ln |x| - \frac{1}{x} + C$$

APPLICATION: COST AND MARGINAL COST

Example 5 • Finding Cost from Marginal Cost

The marginal cost to manufacture baseball caps at a production level of x caps is $3.20 - 0.001x$ dollars per cap, and the cost of producing 50 caps is $200. Find the cost function.

Solution   We are asked to find the cost function $C(x)$, given that the *marginal* cost function is $3.20 - 0.001x$. Recalling that the marginal cost function is the derivative of the cost function, we have

$$C'(x) = 3.20 - 0.001x$$

and must find $C(x)$. Now $C(x)$ must be an antiderivative of $C'(x)$, so we write

$$C(x) = \int (3.20 - 0.001x) \, dx$$

$$= 3.20x - 0.001 \frac{x^2}{2} + K \qquad\qquad K \text{ is the constant of integration.}$$

$$= 3.20x - 0.0005x^2 + K$$

(Why did we use K and not C for the constant of integration?) Now, unless we know a value for K, we don't really know what the cost function is. However, there is another piece of information we have ignored: The cost of producing 50 baseball caps is $200. In symbols

$$C(50) = 200$$

Substituting in our formula for $C(x)$, we have

$$C(50) = 3.20(50) - 0.0005(50)^2 + K$$

$$200 = 158.75 + K$$

$$K = 41.25$$

Now that we know what K is, we can write down the cost function:

$$C(x) = 3.20x - 0.0005x^2 + 41.25$$

✳ *Before we go on . . .*

Question   What is the significance of the term 41.25?

Answer   If we substitute $x = 0$, we get

$$C(0) = 3.20(0) - 0.0005(0)^2 + 41.25 = 41.25$$

Thus, $41.25 is the cost of producing zero items; in other words, it is the **fixed cost.**

APPLICATION: MOTION IN A STRAIGHT LINE

An important application of the indefinite integral is to the study of motion. The application of calculus to problems about motion is an example of the intertwining of mathematics and physics that is an important part of both. We begin by bringing together some facts, scattered through the last several chapters, that have to do with an object moving in a straight line.

Position, Velocity, and Acceleration: Derivative Form

If $s = s(t)$ is the **position** of an object at time t, then its **velocity** is given by the derivative

$$v = \frac{ds}{dt}$$

In Words   *Velocity is the derivative of position.*

The **acceleration** of an object is given by the derivative

$$a = \frac{dv}{dt}$$

In Words   *Acceleration is the derivative of velocity.*

On Earth a free-falling body experiencing no air resistance accelerates at approximately 32 feet per second per second, or 32 ft/s² (or 9.8 m/s²).

We may rewrite the derivative formulas above as integral formulas.

Position, Velocity, and Acceleration: Integral Form

$$s(t) = \int v(t)\, dt \qquad \text{Because } v = \frac{ds}{dt}$$

$$v(t) = \int a(t)\, dt \qquad \text{Because } a = \frac{dv}{dt}$$

Example 6 • Motion in a Straight Line

You toss a stone upward at a speed of 30 feet per second.

a. Find the stone's velocity as a function of time. How fast and in what direction is it going after 5 seconds?

b. Find the position of the stone as a function of time. Where will it be after 5 seconds?

c. When and where will the stone reach its *zenith*, its highest point?

Solution

a. Let's measure heights above the ground as positive, so that a rising object has positive velocity and the acceleration downward due to gravity is negative. Thus, the acceleration of the stone is given by

$$a(t) = -32 \text{ feet per second per second}$$

We wish to know the velocity, which is an antiderivative of acceleration, so we compute

$$v(t) = \int (-32)\, dt = -32t + C$$

This is the velocity as a function of time t. But what is the value of C? We are told that you tossed the stone upward at 30 feet per second, so when $t = 0$, $v = 30$; that is, $v(0) = 30$. Thus,

$$30 = v(0) = -32(0) + C$$

so $C = 30$ and the formula for velocity is $v(t) = -32t + 30$. In particular, after 5 seconds the velocity will be

$$v(5) = -32(5) + 30 = -130 \text{ feet per second}$$

After 5 seconds the stone is *falling* with a speed of 130 feet per second.

b. We wish to know the position, but position is an antiderivative of velocity. Thus,

$$s(t) = \int v(t)\, dt = \int (-32t + 30)\, dt = -16t^2 + 30t + C$$

Now, to find C we need to know the initial position $s(0)$. We are not told this, so let's measure heights so that the initial position is zero. Then

$$0 = s(0) = C$$

and $s(t) = -16t^2 + 30t$. In particular, after 5 seconds the stone has a height of

$$s(5) = -16(5)^2 + 30(5) = -250 \text{ ft}$$

In other words, the stone is now 250 feet *below* where it was when you first threw it, as shown in Figure 2.

c. The stone reaches its zenith when its height $s(t)$ is at its maximum value, which occurs when $v(t) = s'(t)$ is zero. So we solve

$$v(t) = -32t + 30 = 0$$

getting $t = 30/32 = 15/16 = 0.9375$ second. This is the time when the stone reaches its zenith. The height of the stone at that time is

$$s\left(\frac{15}{16}\right) = -16\left(\frac{15}{16}\right)^2 + 30\left(\frac{15}{16}\right) = 14.0625 \text{ feet}$$

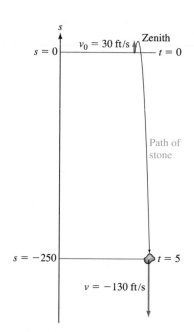

Figure 2

6.1 EXERCISES

In Exercises 1–10, mentally evaluate the integrals.

1. $\int x^5\, dx$

2. $\int x^7\, dx$

3. $\int 6\, dx$

4. $\int (-5)\, dx$

5. $\int x\, dx$

6. $\int (-x)\, dx$

7. $\int (x^2 - x)\, dx$

8. $\int (x + x^3)\, dx$

9. $\int (1 + x)\, dx$

10. $\int (4 - x)\, dx$

In Exercises 11–36, evaluate the integrals.

11. $\int x^{-5}\, dx$

12. $\int x^{-7}\, dx$

13. $\int (x^{2.3} + x^{-1.3})\, dx$

14. $\int (x^{-0.2} - x^{0.2})\, dx$

15. $\int \left(u^2 - \frac{1}{u}\right) du$

16. $\int \left(\frac{1}{v^2} + \frac{2}{v}\right) dv$

17. $\int \sqrt{x}\, dx$

18. $\int \sqrt[3]{x}\, dx$

19. $\int (3x^4 - 2x^{-2} + x^{-5} + 4)\, dx$ **20.** $\int (4x^7 - x^{-3} + 1)\, dx$

21. $\int \left(\dfrac{1}{x} + \dfrac{2}{x^2} - \dfrac{1}{x^3} \right) dx$ **22.** $\int \left(\dfrac{3}{x} - \dfrac{1}{x^5} + \dfrac{1}{x^7} \right) dx$

23. $\int (3x^{0.1} - x^{4.3} - 4.1)\, dx$ **24.** $\int \left(\dfrac{x^{2.1}}{2} - 2.3 \right) dx$

25. $\int \left(\dfrac{3}{x^{0.1}} - \dfrac{4}{x^{1.1}} \right) dx$ **26.** $\int \left(\dfrac{1}{x^{1.1}} - \dfrac{1}{x} \right) dx$

27. $\int \left(2e^x + \dfrac{5}{x} + \dfrac{1}{4} \right) dx$ **28.** $\int \left(-e^x + x^{-2} - \dfrac{1}{8} \right) dx$

29. $\int \left(\dfrac{6.1}{x^{0.5}} + \dfrac{x^{0.5}}{6} - e^x \right) dx$ **30.** $\int \left(\dfrac{4.2}{x^{0.4}} + \dfrac{x^{0.4}}{3} - 2e^x \right) dx$

31. $\int (2^x - 3^x)\, dx$ **32.** $\int (1.1^x + 2^x)\, dx$

33. $\int 100(1.1^x)\, dx$ **34.** $\int 1000(0.9^x)\, dx$

35. $\int \dfrac{x+2}{x^3}\, dx$ **36.** $\int \dfrac{x^2-2}{x}\, dx$

37. Find $f(x)$ if $f(0) = 1$ and the tangent line at $(x, f(x))$ has slope x.

38. Find $f(x)$ if $f(1) = 1$ and the tangent line at $(x, f(x))$ has slope $1/x$.

39. Find $f(x)$ if $f(0) = 0$ and the tangent line at $(x, f(x))$ has slope $e^x - 1$.

40. Find $f(x)$ if $f(1) = -1$ and the tangent line at $(x, f(x))$ has slope $2e^x + 1$.

APPLICATIONS

41. Marginal Cost The marginal cost of producing the xth box of light bulbs is $5 - x/10{,}000$, and the fixed cost is $20{,}000$. Find the cost function $C(x)$.

42. Marginal Cost The marginal cost of producing the xth box of Zip™ disks is $10 + x^2/100{,}000$, and the fixed cost is $100{,}000$. Find the cost function $C(x)$.

43. Marginal Cost The marginal cost of producing the xth roll of film is $5 + 2x + 1/x$. The total cost to produce one roll is 1000. Find the cost function $C(x)$.

44. Marginal Cost The marginal cost of producing the xth box of CDs is $10 + x + 1/x^2$. The total cost to produce 100 boxes is $10{,}000$. Find the cost function $C(x)$.

45. Interest Rates Between 1990 and 1998, the discount interest rate in Japan declined at a rate of 0.7 percentage point per year. Given that the discount interest rate was 6% in 1992, use an indefinite integral to find a formula for the interest rate I (in percentage points) as a function of time t since 1990 ($t = 0$ represents 1990) and use your formula to calculate the interest rate in 1998.

This was the average rate of change over the given period. Sources: Bloomberg Financial Markets/Japan External Trade Organization/*New York Times*, September 20, 1998, p. WK5.

46. Real Estate Between 1990 and 1997, the price of a square foot of land in Tokyo declined at a rate of $25 per year. Given that the price in 1990 was $375, use an indefinite integral to find a formula for the price p as a function of time t since 1990 ($t = 0$ represents 1990) and use your formula to calculate the price in 1997.

This was the average rate of change over the given period. Sources: Bloomberg Financial Markets/Japan External Trade Organization/*New York Times*, September 20, 1998, p. WK5.

47. Bottled-Water Sales The rate of U.S. sales of domestic bottled water for the period 1990–2000 can be approximated by

$$D(t) = 0.0166t^2 - 0.050t + 0.27 \text{ billion gallons per year}$$
$$(0 \le t \le 10)$$

where t is time in years since 1990. Use an indefinite integral to compute the total sales $S(t)$ of domestic bottled water since 1990.

The model is based on regression of approximate data. Source: Beverage Marketing Corporation of New York/*New York Times*, June 21, 2001, p. C1.

48. Bottled-Water Sales The rate of U.S. sales of imported bottled water for the period 1990–2000 can be approximated by

$$I(t) = 0.0004t^2 + 0.013t + 0.06 \text{ billion gallons per year}$$
$$(0 \le t \le 10)$$

where t is time in years since 1990. Use an indefinite integral to compute the total sales $M(t)$ of imported bottled water since 1990.

The model is based on regression of approximate data. Source: Beverage Marketing Corporation of New York/*New York Times*, June 21, 2001, p. C1.

49. Military Spending Write $C(t)$ for the amount the U.S. military spent on personnel in year t, where t is measured in years since 2000. The rate of increase of $C(t)$ was about $3.7 billion per year in 2000 and was projected to decrease to $3 billion per year in 2007.
a. Find a linear model for the rate of change $C'(t)$.
b. Given that the U.S. military spent $135 billion on personnel in 2000, find the annual personnel cost function $C(t)$.

While the cost per active-duty armed service member has been increasing, the number of armed service personnel has been decreasing. Annual costs are adjusted for inflation, and data are based on linear models for annual cost and personnel. Sources: U.S. Dept. of Defense; Stephen Daggett, military analyst, Congressional Research Service/*New York Times*, April 19, 2002, p. A21.

50. Military Spending in the 1990s Recall the function $C(t)$ from Exercise 49. The rate of change of $C(t)$ was about $0.5 billion per year in 1990 and decreased to $-$0.5 billion per year in 1995.
a. Find a linear model for the rate of change $C'(t)$, with t now measured in years since 1990.
b. Given that the U.S. military spent $115 billion on personnel in 1990, find the annual personnel cost function $C(t)$.

While the cost per active-duty armed service member has been increasing, the number of armed service personnel has been decreasing. Annual costs are adjusted for inflation, and data are based on linear models for annual cost and personnel. Sources: U.S. Dept. of Defense; Stephen Daggett, military analyst, Congressional Research Service/*New York Times*, April 19, 2002, p. A21.

51. Motion in a Straight Line The velocity of a particle moving in a straight line is given by $v(t) = t^2 + 1$.
   a. Find an expression for the position s after a time t.
   b. Given that $s = 1$ at time $t = 0$, find the constant of integration C and hence an expression for s in terms of t without any unknown constants.

52. Motion in a Straight Line The velocity of a particle moving in a straight line is given by $v = 3e^t + t$.
   a. Find an expression for the position s after a time t.
   b. Given that $s = 3$ at time $t = 0$, find the constant of integration C and hence an expression for s in terms of t without any unknown constants.

Vertical Motion In Exercises 53–62, neglect the effects of air resistance.

53. If a stone is dropped from a rest position above the ground, how fast and in what direction will it be traveling after 10 seconds?

54. If a stone is thrown upward at 10 feet per second, how fast and in what direction will it be traveling after 10 seconds?

55. Show that if a projectile is thrown upward with a velocity of v_0 feet per second, then it will reach its highest point after $v_0/32$ seconds. (*Hint*: The projectile has zero velocity at its highest point.)

56. Use the result of Exercise 55 to show that if a projectile is thrown upward with a velocity of v_0 feet per second, its highest point will be $v_0^2/64$ feet above the starting point.

Exercises 57–62 use the results of Exercises 55 and 56.

57. I threw a ball up in the air to a height of 20 feet. How fast was the ball traveling when it left my hand?

58. I threw a ball up in the air to a height of 40 feet. How fast was the ball traveling when it left my hand?

59. A piece of chalk is tossed vertically upward by Professor Schwartzenegger and hits the ceiling 100 feet above, with a *BANG*.
   a. What is the minimum speed the piece of chalk must have been traveling to enable it to hit the ceiling?
   b. Assuming that Prof. Schwartzenegger in fact tossed the piece of chalk up at 100 feet per second, how fast was it moving when it struck the ceiling?
   c. Assuming that Prof. Schwartzenegger tossed the chalk up at 100 feet per second and that it recoils from the ceiling with the same speed it had at the instant it hit, how long will it take the chalk to make the return journey and hit the ground?

60. A projectile is fired vertically upward from ground level at 16,000 feet per second.
   a. How high does the projectile go?
   b. How long does it take to reach its zenith (highest point)?
   c. How fast is it traveling when it hits the ground?

61. Strength Professor Strong can throw a 10-pound dumbbell twice as high as Professor Weak can. How much faster can Prof. Strong throw it?

62. Weakness Professor Weak can throw a computer disk three times as high as Professor Strong can. How much faster can Prof. Weak throw it?

COMMUNICATION AND REASONING EXERCISES

63. If $F(x)$ and $G(x)$ are both antiderivatives of $f(x)$, how are $F(x)$ and $G(x)$ related?

64. Your friend Marco claims that once you have one antiderivative of $f(x)$ you have all of them. Explain what he means.

65. Complete the following: The total cost function is a(n) _____ of the _____ cost function.

66. Complete the following: The distance covered is an antiderivative of the _____ function, and the velocity is an antiderivative of the _____ function.

67. If x represents the number of items manufactured and $f(x)$ represents dollars per item, what does $\int f(x)\, dx$ represent? In general, how are the units of $f(x)$ and the units of $\int f(x)\, dx$ related?

68. Complete the following: $-1/x$ is a(n) _____ of $1/x^2$, whereas $\ln x^2$ is not. Also, $-1/x + C$ is the _____ of $1/x^2$, because the _____ of $-1/x + C$ is _____.

69. Give an argument for the rule that the integral of a sum is the sum of the integrals.

70. Is it true that $\int (1/x^3)\, dx = \ln(x^3) + C$? Give a reason for your answer.

71. Give an example to show that the integral of a product is not the product of the integrals.

72. Give an example to show that the integral of a quotient is not the quotient of the integrals.

73. Complete the following: If you take the _____ of the _____ of $f(x)$, you obtain $f(x)$ back. On the other hand, if you take the _____ of the _____ of $f(x)$, you obtain $f(x) + C$.

74. If a Martian told you that the Institute of Alien Mathematics, after a long and difficult search, has announced the discovery of a new antiderivative of $x - 1$ called $M(x)$ [the formula for $M(x)$ is classified information and cannot be revealed here], how would you respond?

6.2   *Substitution*

The chain rule for derivatives gives us an extremely useful technique for finding anti-derivatives. This technique is called **change of variables** or **substitution.**

Recall that to differentiate a function like $(x^2 + 1)^{4.4}$, we first think of the function as $g(u)$ where $u = x^2 + 1$ and $g(u) = u^{4.4}$. We then compute the derivative, using the chain rule, as

$$\frac{d}{dx} g(u) = g'(u) \frac{du}{dx}$$

Any rule for derivatives can be turned into a technique for finding antiderivatives. The chain rule turns into the following formula:

$$\int g'(u) \frac{du}{dx} dx = g(u) + C$$

But, if we write $g(u) + C = \int g'(u) \, du$, we get the following interesting equation:

$$\int g'(u) \frac{du}{dx} dx = \int g'(u) \, du$$

This equation is the one usually called the change of variables formula. We can turn it into a more useful integration technique as follows. Let $f = g'(u)(du/dx)$. We can rewrite the above change of variables formula using f:

$$\int f \, dx = \int \left(\frac{f}{du/dx} \right) du$$

In essence, we are making the formal substitution

$$dx = \frac{1}{du/dx} du$$

Here's the technique:

Substitution Rule
If u is a function of x, then we can use the following formula to evaluate an integral:

$$\int f \, dx = \int \left(\frac{f}{du/dx} \right) du$$

Rather than use the formula directly, we use the following procedure:

1. Write u as a function of x.
2. Take the derivative du/dx and solve for the quantity dx in terms of du.
3. Use the expression you obtain in step 2 to substitute for dx in the given integral and substitute u for its defining expression.

Now let's see how this procedure works in practice.

Example 1 • *Substitution*

Find $\int 4x(x^2 + 1)^{4.4}\, dx$.

Solution To use substitution we need to choose an expression to be u. There is no hard and fast rule, but here is one hint that often works:

Let u be an expression that is being raised to a power.

In this case let's set $u = x^2 + 1$. Continuing the procedure above, we place the calculations for step 2 in a box:

$$u = x^2 + 1$$

Write u as a function of x.

$$\frac{du}{dx} = 2x$$

Take the derivative of u with respect to x.

$$dx = \frac{1}{2x}\, du$$

Solve for dx: $dx = \dfrac{1}{du/dx}\, du$.

Now we *substitute u for its defining expression and substitute for dx* in the original integral:

$$\int 4x(x^2 + 1)^{4.4}\, dx = \int 4x\, u^{4.4}\, \frac{1}{2x}\, du$$

Substitute[2] for u and[3] dx.

$$= \int 2u^{4.4}\, du$$

Cancel the x's and simplify.

We have boiled the given integral down to the much simpler integral $\int 2u^{4.4}\, du$, and we can now write down the solution:

$$2\,\frac{u^{5.4}}{5.4} + C = \frac{2(x^2 + 1)^{5.4}}{5.4} + C$$

Substitute $(x^2 + 1)$ for u in the answer.

Before we go on . . . There are two points to notice here. First, before we can actually integrate with respect to u, *we must eliminate all x's from the integrand.* If we cannot, we may have chosen the wrong expression for u. Second, after integrating, we must substitute back to obtain an expression involving x.

It is easy to check our answer. We differentiate:

$$\frac{d}{dx}\left[\frac{2(x^2 + 1)^{5.4}}{5.4}\right] = \frac{2(5.4)(x^2 + 1)^{4.4}(2x)}{5.4} = 4x(x^2 + 1)^{4.4}$$

Notice how we used the chain rule to check the result obtained by substitution.

When we use substitution, the first step is always to decide what to take as u. Again, there are no set rules, but we see some common cases in the examples.

---

[2] Can you see how this step is equivalent to using the formula $\int f\, dx = \int [f/(du/dx)]\, du$?

[3] If it should bother you that the integral contains both x and u, note that x is now a function of u.

Example 2 • More Substitution

Calculate $\int x^2(x^3 + 1)^2\, dx$.

Solution As we said in Example 1, it often works to take u to be an expression that is being raised to a power. We usually also want to see the derivative of u as a factor in the integrand so that we can cancel terms involving x. In this case, $x^3 + 1$ is being raised to a power, so let's set $u = x^3 + 1$. Its derivative is $3x^2$; in the integrand is x^2, which is missing the factor 3, but missing or incorrect constant factors are not a problem.

$$u = x^3 + 1$$

Write u as a function of x.

$$\frac{du}{dx} = 3x^2$$

Take the derivative of u with respect to x.

$$dx = \frac{1}{3x^2}\, du$$

Solve for dx: $dx = \dfrac{1}{du/dx}\, du$.

$$\int x^2(x^3 + 1)^2\, dx = \int x^2\, u^2 \frac{1}{3x^2}\, du$$

Substitute for u and dx.

$$= \int \frac{1}{3} u^2\, du$$

Cancel the terms with x.

$$= \frac{1}{9} u^3 + C$$

Take the antiderivative.

$$= \frac{1}{9}(x^3 + 1)^3 + C$$

Substitute for u in the answer.

Example 3 • An Expression in the Exponent

Evaluate $\int 3xe^{x^2}\, dx$.

Solution When we have an exponential with an expression in the exponent, it often works to substitute u for that expression. In this case let's set $u = x^2$:

$$u = x^2$$

$$\frac{du}{dx} = 2x$$

$$dx = \frac{1}{2x}\, du$$

Substituting into the integral, we get

$$\int 3xe^{x^2}\, dx = \int 3xe^u \frac{1}{2x}\, du = \int \frac{3}{2} e^u\, du = \frac{3}{2} e^u + C = \frac{3}{2} e^{x^2} + C$$

Example 4 • A Special Power

Evaluate $\int \dfrac{1}{2x + 5}\, dx$.

Solution We begin by rewriting the integrand as a power:

$$\int \frac{1}{2x + 5}\, dx = \int (2x + 5)^{-1}\, dx$$

Now we take our earlier advice and set u equal to the expression that is being raised to a power:

$$u = 2x + 5$$
$$\frac{du}{dx} = 2$$
$$dx = \frac{1}{2}\, du$$

Substituting, we have

$$\int \frac{1}{2x + 5}\, dx = \int \frac{1}{2}\, u^{-1}\, du = \frac{1}{2}\, \ln |u| + C = \frac{1}{2}\, \ln |2x + 5| + C$$

Example 5 • Choosing u

Evaluate $\int (x + 3)\sqrt{x^2 + 6x}\, dx$.

Solution There are two parenthetical expressions. Notice, however, that the derivative of the expression $(x^2 + 6x)$ is $2x + 6$, which is twice the term $(x + 3)$ in front of the radical. Recall that we would like the derivative of u to appear as a factor. So, let's set $u = x^2 + 6x$:

$$u = x^2 + 6x$$
$$\frac{du}{dx} = 2x + 6 = 2(x + 3)$$
$$dx = \frac{1}{2(x + 3)}\, du$$

Substituting into the integral, we get

$$\int (x + 3)\sqrt{x^2 + 6x}\, dx = \int (x + 3)\sqrt{u} \left[\frac{1}{2(x + 3)} \right] du$$

$$= \int \frac{1}{2}\sqrt{u}\, du = \frac{1}{2} \int u^{1/2}\, du$$

$$= \frac{1}{2}\, \frac{2}{3}\, u^{3/2} + C = \frac{1}{3}\, (x^2 + 6x)^{3/2} + C$$

Some cases require a little more work.

Example 6 • When the x Terms Do Not Cancel

Evaluate $\displaystyle\int \frac{2x}{(x-5)^2}\, dx$.

Solution We first rewrite

$$\int \frac{2x}{(x-5)^2}\, dx = \int 2x(x-5)^{-2}\, dx$$

This suggests that we should set $u = x - 5$.

$$
\begin{array}{|c|}
\hline
\\
u = x - 5 \\
\\
\dfrac{du}{dx} = 1 \\
\\
dx = du \\
\\
\hline
\end{array}
$$

Substituting, we have

$$\int \frac{2x}{(x-5)^2}\, dx = \int 2xu^{-2}\, du$$

Now, there is nothing in the integrand to cancel the x that appears. We can do the following: If, after substituting, there is still an x in the integrand, we go back to the expression for u, solve for x, and substitute the expression we obtain for x in the integrand. So, we take $u = x - 5$ and solve for $x = u + 5$. Substituting, we get

$$\int 2xu^{-2}\, du = \int 2(u+5)u^{-2}\, du$$

$$= 2\int (u^{-1} + 5u^{-2})\, du$$

$$= 2\ln |u| - \frac{10}{u} + C$$

$$= 2\ln |x - 5| - \frac{10}{x-5} + C$$

Example 7 • Application: Online Book Sales

The rate of increase in the total number of books sold online in the United States in the period 1997–2000 can be approximated by the logistic function

$$R(t) = \frac{83e^{1.97t}}{22 + e^{1.97t}} \text{ million books/year} \qquad (0 \le t \le 4)$$

(t is time in years since January, 1997).[4]

a. Find an expression for the total number of books sold online since January 1997.

b. How many books were sold online from January 1997 through December 2000?

---

[4]Based on a logistic regression. SOURCE: Ipsos-NPD Book Trends/*New York Times*, April 16, 2001, p. C1.

Solution

a. If we write the total number of books sold online to time t as $N(t)$, then we are told that

$$N'(t) = \frac{83e^{1.97t}}{22 + e^{1.97t}}$$

Thus,

$$N(t) = \int \frac{83e^{1.97t}}{22 + e^{1.97t}} \, dt$$

is the function we are after. To integrate the expression, take u to be the denominator of the integrand:

$$u = 22 + e^{1.97t}$$

$$\frac{du}{dt} = 1.97e^{1.97t}$$

$$dt = \frac{1}{1.97e^{1.97t}} \, du$$

$$N(t) = \int \frac{83e^{1.97t}}{22 + e^{1.97t}} \, dt$$

$$= \int \frac{83e^{1.97t}}{u} \cdot \frac{1}{1.97e^{1.97t}} \, du$$

$$= \frac{83}{1.97} \int \frac{1}{u} \, du$$

$$\approx 42.1 \ln |u| + C$$

$$= 42.1 \ln(22 + e^{1.97t}) + C \qquad \text{Why could we drop the absolute value?}$$

Now what is C? Since $N(t)$ represents the total number of books sold online *since time $t = 0$*, we have $N(0) = 0$ (since that is when we started counting). Thus,

$$0 = 42.1 \ln(22 + e^{1.97(0)}) + C$$

$$0 = 42.1 \ln(23) + C$$

$$C = -42.1 \ln(23) \approx -132$$

Therefore, the total number of books sold online since January 1997 is

$$N(t) = 42.1 \ln(22 + e^{1.97t}) - 132 \text{ million books}$$

b. Since December 2000 is represented by $t = 4$, the total number of books sold online from January 1997 through December 2000 is

$$N(4) = 42.1 \ln(22 + e^{1.97(4)}) - 132 \approx 200 \text{ million books.}[5]$$

---

[5]The actual number of books sold online was 161 million. This discrepancy results from a less than perfect fit of the logistic model to the actual data.

✳ *Before we go on . . .* You might wonder why we are writing a logistic function in the form we used above rather than in one of the "standard" forms $N/(1 + Ab^{-t})$ or $N/(1 + Ae^{-kt})$. Our only reason for doing this is to make the substitution work. To convert from the second standard form to the form we used in the example, multiply top and bottom by e^{kx}. (See Exercises 89 and 90 in Section 6.5 for further discussion.)

Shortcuts

If a and b are constants with $a \neq 0$, then we have the following formulas. (We have already seen one of them in Example 4. All of them can be obtained using the substitution $u = ax + b$. They will appear in the exercises.)

Shortcuts: Integrals of Expressions Involving $(ax + b)$

Rule

$$\int (ax + b)^n \, dx = \frac{(ax + b)^{n+1}}{a(n + 1)} + C$$

$$(\text{if } n \neq -1)$$

$$\int (ax + b)^{-1} \, dx = \frac{1}{a} \ln |ax + b| + C$$

$$\int e^{ax+b} \, dx = \frac{1}{a} e^{ax+b} + C$$

$$\int c^{ax+b} \, dx = \frac{1}{a \ln c} c^{ax+b} + C$$

Quick Example

$$\int (3x - 1)^2 \, dx = \frac{(3x - 1)^3}{3(3)} + C$$

$$= \frac{(3x - 1)^3}{9} + C$$

$$\int (3 - 2x)^{-1} \, dx = \frac{1}{(-2)} \ln |3 - 2x| + C$$

$$= -\frac{1}{2} \ln |3 - 2x| + C$$

$$\int e^{-x+4} \, dx = \frac{1}{(-1)} e^{-x+4} + C$$

$$= -e^{-x+4} + C$$

$$\int 2^{-3x+4} \, dx = \frac{1}{-3 \ln 2} 2^{-3x+4} + C$$

$$= -\frac{1}{3 \ln 2} 2^{-3x+4} + C$$

Guideline: When to Use Substitution and What to Use for u

Question If I am asked to calculate an antiderivative, how do I know when to use a substitution and when *not* to use one?

Answer Do *not* use substitution when integrating sums, differences, and/or constant multiples of powers of x and exponential functions, such as

$$2x^3 - \frac{4}{x^2} + \frac{1}{2x} + 3^x + \frac{2^x}{3}$$

To recognize when you should try a substitution, pretend that you were *differentiating* the given expression instead of integrating it. If differentiating the expression would require use of the chain rule, then integrating that expression may well require a substitution, as in, say, $x(3x^2 - 4)^3$ or $(x + 1)e^{x^2+2x-1}$. (In the first we have a *quantity* cubed, and in the second we have e raised to a *quantity*.)

Question If an integral seems to call for a substitution, what should I use for u?

Answer There are no set rules for deciding what to use for u, but the preceding examples show some common patterns:

- If you see a linear expression raised to a power, try setting u to be that linear expression. For example, in $(3x - 2)^{-3}$, set $u = 3x - 2$. (Alternatively, try using the shortcuts above.)

- If you see a constant raised to a linear expression, try setting u to be that linear expression. For example, in 3^{2x+1}, set $u = 2x + 1$. (Alternatively, try a shortcut.)

- If you see an expression raised to a power multiplied by the derivative of that expression (or a constant multiple of the derivative), try setting u to be that expression. For example, in $x^2(3x^3 - 4)^{-1}$, set $u = 3x^3 - 4$.

- If you see a constant raised to an expression, multiplied by the derivative of that expression (or a constant multiple of its derivative), try setting u to be that expression. For example, in $5(x + 1)e^{x^2+2x-1}$, set $u = x^2 + 2x - 1$.

- If you see an expression in the denominator and its derivative (or a constant multiple of its derivative) in the numerator, try setting u to be that expression. For example, in $2^{3x}/(3 - 2^{3x})$, set $u = 3 - 2^{3x}$.

Persistence often pays off: If a certain substitution does not work, try another approach or a different substitution.

6.2 EXERCISES

In Exercises 1–40, evaluate the given integrals.

1. $\int (3x + 1)^5 \, dx$

2. $\int (-x - 1)^7 \, dx$

3. $\int (-2x + 2)^{-2} \, dx$

4. $\int (2x)^{-1} \, dx$

5. $\int e^{-x} \, dx$

6. $\int e^{x/2} \, dx$

7. $\int 7.2\sqrt{3x - 4} \, dx$

8. $\int 4.4e^{(-3x+4)} \, dx$

9. $\int 1.2e^{(0.6x+2)} \, dx$

10. $\int 8.1\sqrt{-3x + 4} \, dx$

11. $\int x(3x^2 + 3)^3 \, dx$

12. $\int x(-x^2 - 1)^3 \, dx$

13. $\int x(x^2 + 1)^{1.3} \, dx$

14. $\int \frac{x}{(3x^2 - 1)^{0.4}} \, dx$

15. $\int (1 + 9.3e^{3.1x-2}) \, dx$

16. $\int (3.2 - 4e^{1.2x-3}) \, dx$

17. $\int \frac{e^x + e^{-x}}{2} \, dx$

18. $\int (e^{x/2} + e^{-x/2}) \, dx$

19. $\int 2x\sqrt{3x^2 - 1} \, dx$

20. $\int 3x\sqrt{-x^2 + 1} \, dx$

21. $\int xe^{-x^2+1} \, dx$

22. $\int xe^{2x^2-1} \, dx$

23. $\int (x + 1)e^{-(x^2+2x)} \, dx$

24. $\int (2x - 1)e^{2x^2-2x} \, dx$

25. $\int \frac{-2x - 1}{(x^2 + x + 1)^3} \, dx$

26. $\int \frac{x^3 - x^2}{3x^4 - 4x^3} \, dx$

27. $\int \frac{x^2 + x^5}{\sqrt{2x^3 + x^6 - 5}} \, dx$

28. $\int \frac{2(x^3 - x^4)}{(5x^4 - 4x^5)^5} \, dx$

29. $\int x(x - 2)^5 \, dx$

30. $\int x(x - 2)^{1/3} \, dx$

31. $\int 2x\sqrt{x + 1} \, dx$

32. $\int \frac{x}{\sqrt{x + 1}} \, dx$

33. $\int \frac{3e^{-1/x}}{x^2} \, dx$

34. $\int \frac{2e^{2/x}}{x^2} \, dx$

35. $\int \frac{e^{-0.05x}}{1 - e^{-0.05x}} \, dx$

36. $\int \frac{3e^{1.2x}}{2 + e^{1.2x}} \, dx$

37. $\int \frac{e^x - e^{-x}}{e^x + e^{-x}} \, dx$

38. $\int \frac{e^{x/2} + e^{-x/2}}{e^{x/2} - e^{-x/2}} \, dx$

39. $\int [(2x - 1)e^{2x^2-2x} + xe^{x^2}] \, dx$

40. $\int (xe^{-x^2+1} + e^{2x}) \, dx$

In Exercises 41–44, derive each equation, where a and b are constants with $a \neq 0$.

41. $\int (ax + b)^n \, dx = \dfrac{(ax + b)^{n+1}}{a(n + 1)} + C$ (if $n \neq -1$)

42. $\int (ax + b)^{-1} \, dx = \dfrac{1}{a} \ln |ax + b| + C$

43. $\int \sqrt{ax + b} \, dx = \dfrac{2}{3a} (ax + b)^{3/2} + C$

44. $\int e^{ax+b} \, dx = \dfrac{1}{a} e^{ax+b} + C$

In Exercises 45–58, use the shortcut formulas in the text and Exercises 41–44 to mentally calculate the integrals.

45. $\int e^{-x} \, dx$

46. $\int e^{x-1} \, dx$

47. $\int e^{2x-1} \, dx$

48. $\int e^{-3x} \, dx$

49. $\int (2x + 4)^2 \, dx$

50. $\int (3x - 2)^4 \, dx$

51. $\int \dfrac{1}{5x - 1} \, dx$

52. $\int (x - 1)^{-1} \, dx$

53. $\int (1.5x)^3 \, dx$

54. $\int e^{2.1x} \, dx$

55. $\int 1.5^{3x} \, dx$

56. $\int 4^{-2x} \, dx$

57. $\int (2^{3x+4} + 2^{-3x+4}) \, dx$

58. $\int (1.1^{-x+4} + 1.1^{x+4}) \, dx$

59. Find $f(x)$ if $f(0) = 0$ and the tangent line at $(x, f(x))$ has slope $x(x^2 + 1)^3$.

60. Find $f(x)$ if $f(1) = 0$ and the tangent line at $(x, f(x))$ has slope $x/(x^2 + 1)$.

61. Find $f(x)$ if $f(1) = 1/2$ and the tangent line at $(x, f(x))$ has slope xe^{x^2-1}.

62. Find $f(x)$ if $f(2) = 1$ and the tangent line at x has slope $(x - 1)e^{x^2-2x}$.

APPLICATIONS

63. Cost The marginal cost of producing the xth roll of film is given by $5 + 1/(x + 1)^2$. The total cost to produce one roll is $1000. Find the total cost function $C(x)$.

64. Cost The marginal cost of producing the xth box of CDs is given by $10 - x/(x^2 + 1)^2$. The total cost to produce two boxes is $1000. Find the total cost function $C(x)$.

65. Bottled-Water Sales The rate of U.S. sales of domestic bottled water for the period 1990–2000 can be approximated by

$$D(t) = 0.0166(t - 1990)^2 - 0.050(t - 1990) +$$
$$0.27 \text{ billion gallons per year}$$
$$(1990 \le t \le 2000)$$

where t is the year. Use an indefinite integral to compute the total sales $S(t)$ of domestic bottled water since 1990.

The model is based on regression of approximate data. Source: Beverage Marketing Corporation of New York/*New York Times,* June 21, 2001, p. C1.

66. Bottled-Water Sales The rate of U.S. sales of imported bottled water for the period 1990–2000 can be approximated by

$$I(t) = 0.0004(t - 1990)^2 + 0.013(t - 1990) +$$
$$0.06 \text{ billion gallons per year}$$
$$(1990 \le t \le 2000)$$

where t is the year. Use an indefinite integral to compute the total sales $M(t)$ of imported bottled water since 1990.

The model is based on regression of approximate data. Source: Beverage Marketing Corporation of New York/*New York Times,* June 21, 2001, p. C1.

67. Motion in a Straight Line The velocity of a particle moving in a straight line is given by $v = t(t^2 + 1)^4 + t$.
   a. Find an expression for the position s after a time t.
   b. Given that $s = 1$ at time $t = 0$, find the constant of integration C and hence an expression for s in terms of t without any unknown constants.

68. Motion in a Straight Line The velocity of a particle moving in a straight line is given by $v = 3te^{t^2} + t$.
   a. Find an expression for the position s after a time t.
   b. Given that $s = 3$ at time $t = 0$, find the constant of integration C and hence an expression for s in terms of t without any unknown constants.

COMMUNICATION AND REASONING EXERCISES

69. Are there any circumstances in which you should use the substitution $u = x$? Illustrate your answer by giving an example that shows the effect of this substitution.

70. Give an example of an integral that can be calculated by using the substitution $u = x^2 + 1$ and then carry out the calculation.

71. Give an example of an integral that can be calculated by using the power rule for antiderivatives and also by using the substitution $u = x^2 + x$ and then carry out the calculations.

72. At what stage of a calculation using a u substitution should you substitute back for u in terms of x: before or after taking the antiderivative?

73. You are asked to calculate $\int u/(u^2 + 1) \, du$. What is wrong with the substitution $u = u^2 + 1$?

74. What is wrong with the following "calculation" of $\int 1/(x^2 - 1) \, dx$?

$$\int \frac{1}{x^2 - 1} = \int \frac{1}{u} \qquad \text{Using the substitution } u = x^2 - 1$$
$$= \ln |u| + C$$
$$= \ln |x^2 - 1| + C$$

75. Show that *none* of the following substitutions work for $\int e^{-x^2} \, dx$: $u = -x$, $u = x^2$, and $u = -x^2$. (The antiderivative of e^{-x^2} involves the *error function* erf(x); see Exercise 94 in Section 6.5.)

76. Show that *none* of the following substitutions work for $\int \sqrt{1 - x^2} \, dx$: $u = 1 - x^2$, $u = x^2$, and $u = -x^2$. (The antiderivative of $\sqrt{1 - x^2}$ involves inverse trigonometric functions, a discussion of which is beyond the scope of this book.)

6.3   *The Definite Integral as a Sum: A Numerical Approach*

In Sections 6.1 and 6.2, we discussed the indefinite integral. There is a related, historically older concept called the **definite integral.** Let's introduce this new idea with an example. (We'll drop hints now and then about how the two types of integral are related. In Section 6.5 we discuss the exact relationship, which is one of the most important results in calculus.)

In Section 6.1 we used antiderivatives to answer questions of the form "Given the marginal cost, compute the total cost" (see Example 5 in Section 6.1). In this section we approach such questions more directly, using a numerical approach.

Example 1 • Total Cost

Your cell phone company offers you an innovative pricing scheme. When you make a call, the *marginal* cost of the tth minute of the call is

$$c(t) = \frac{20}{t + 100} \text{ dollars per minute}$$

Use a numerical calculation to estimate the cost of a 60-minute phone call.

Solution   As we pointed out above, we *could* solve this problem using an antiderivative, similar to Example 5 in Section 6.1. However, let's forget about antiderivatives for now and do the computation numerically. As we did when first computing derivatives, we approach the answer by making better and better approximations.

Let's start with a very crude estimate. The marginal cost at the beginning of your call is $c(0) = \$0.20$ per minute. If this cost were to remain constant for the length of your call, the total cost of the call would be

cost of call = cost per minute $\times$ number of minutes $\approx 0.20 \times 60 = \$12$

But the marginal cost does not remain constant. It goes down over the course of the call. (Why?) We can obtain a much more accurate estimate of the cost by looking at the call minute by minute. We estimate the cost of each minute, using the marginal cost at the beginning of that minute. The first minute costs (approximately) $c(0) = \$0.20$, but the second minute costs (approximately) $c(1) \approx \$0.198$. Adding together the costs of all 60 minutes gives us the estimate

$$c(0) \times 1 + c(1) \times 1 + \cdots + c(59) \times 1 \approx \$9.44 \quad \text{Minute-by-minute calculation}$$

If we assume that the phone company is honest about $c(t)$ being the marginal cost and is actually calculating your cost continuously, we get an even better estimate of the cost by looking at the call second by second. For example, the marginal cost at the beginning of the first second is $\$0.20$ per minute, and the first second is $\frac{1}{60}$ of a minute, so the first second costs you approximately $\$0.20 \times \frac{1}{60}$. The marginal cost at the beginning of the next second is $c\left(\frac{1}{60}\right) = \$0.199\,967$, so the cost is approximately $\$0.199\,967 \times \frac{1}{60}$. Adding together the costs of all 3600 seconds in the hour gives us the estimate

$$c(0) \times \left(\frac{1}{60}\right) + c\left(\frac{1}{60}\right) \times \left(\frac{1}{60}\right) + c\left(\frac{2}{60}\right) \times \left(\frac{1}{60}\right) + \cdots + c\left(\frac{3599}{60}\right) \times \left(\frac{1}{60}\right) \approx \$9.40$$

Second-by-second calculation

✳ *Before we go on . . .* The minute-by-minute calculation above is tedious to do by hand, and no one in their right mind would even *attempt* to do the second-by-second calculation by hand. Below we discuss ways of doing these calculations with the aid of technology.

The type of calculation done in this example is useful in many applications. Let's look at the general case and give the result a name.

In general, we have a function f (such as the function c in the example), and we consider an interval $[a, b]$ of possible values of the independent variable x (the time interval $[0, 60]$ in the example). We break the interval $[a, b]$ into some number of segments of equal length. In Example 1 we considered $[0, 60]$ first as a single segment, then broke it into 60 segments, and then into 3600 segments. Write n for the number of segments.

Next, we label the endpoints of these segments x_0 for a, x_1 for the end of the first segment, x_2 for the end of the second segment, and so on, until we get to x_n, the end of the nth segment, so that $x_n = b$. Thus,

$$a = x_0 < x_1 < \cdots < x_n = b$$

The first segment is the interval $[x_0, x_1]$, the second segment is $[x_1, x_2]$, and so on, until we get to the last segment, which is $[x_{n-1}, x_n]$. In Example 1, when n was 60, the segments were $[0, 1]$, $[1, 2]$, and so on, with the last segment being $[59, 60]$. In the general case, we are dividing the interval $[a, b]$ into n segments of equal length, so each segment has length $(b - a)/n$. We write Δx for $(b - a)/n$ (Figure 3). The widths of the intervals in the example when $n = 60$ were $\Delta x = 1$.

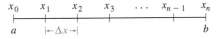

Figure 3

Having established this notation, we can write the calculation we want to do as follows: For each segment $[x_{k-1}, x_k]$, compute $f(x_{k-1})$, the value of the function f at the left endpoint. Multiply this value by the length of the interval, which is Δx. Then add together all n of these products to get the number

$$f(x_0)\Delta x + f(x_1)\Delta x + \cdots + f(x_{n-1})\Delta x$$

This sum is called a (left) **Riemann**[6] **sum** for f. (In Example 1 we computed three different Riemann sums.)

Since sums are often used in mathematics, mathematicians have developed a short-hand notation for them. We write

$$\sum_{k=0}^{n-1} f(x_k)\Delta x = f(x_0)\Delta x + f(x_1)\Delta x + \cdots + f(x_{n-1})\Delta x$$

The symbol $\sum$ is the Greek letter sigma and stands for **summation.** The letter k here is called the index of summation, and we can think of it as counting off the segments. We read the notation as "the sum from $k = 0$ to $n - 1$ of the quantities $f(x_k)\Delta x$." Think of it as a set of instructions:

---

[6]After Georg Friedrich Bernhard Riemann (1826–1866).

Set $k = 0$ and calculate $f(x_0)\Delta x$.

Set $k = 1$ and calculate $f(x_1)\Delta x$.

$\vdots$

Set $k = n - 1$ and calculate $f(x_{n-1})\Delta x$.

Now sum all the quantities so calculated.

Telephone Example ($n = 60$)

Cost of the 1st minute: $c(0) \times 1$

Cost of the 2nd minute: $c(1) \times 1$

Cost of the 60th minute: $c(59) \times 1$

Riemann Sum

If f is a continuous function, the **left Riemann sum** with n equal subdivisions for f over the interval $[a, b]$ is defined to be

$$\text{Riemann sum} = \sum_{k=0}^{n-1} f(x_k)\Delta x$$

$$= f(x_0)\Delta x + f(x_1)\Delta x + \cdots + f(x_{n-1})\Delta x$$

$$= [f(x_0) + f(x_1) + \cdots + f(x_{n-1})]\Delta x$$

where $a = x_0 < x_1 < \cdots < x_n = b$ are the subdivisions and $\Delta x = (b - a)/n$.

Quick Example

In Example 1 we computed three Riemann sums:

$n = 1$: Riemann sum $= c(0)\Delta t = 0.20 \times 60 = \$12$

$n = 60$: Riemann sum $= [c(t_0) + c(t_1) + \cdots + c(t_{n-1})]\Delta t$

$$= [c(0) + c(1) + \cdots + c(59)](1) \approx \$9.44$$

$n = 3600$: Riemann sum $= [c(t_0) + c(t_1) + \cdots + c(t_{n-1})]\Delta t$

$$= [c(0) + c(1/60) + \cdots + c(3599/60)](1/60) \approx \$9.40$$

As in Example 1 we were most interested in what happens to the Riemann sum when we let n get very large. When f is continuous,[7] its Riemann sums will always approach a limit as n goes to infinity. (This is not meant to be obvious. Proofs may be found in advanced calculus texts.) We give the limit a name.

The Definite Integral

If f is a continuous function, the **definite integral of f from a to b** is defined to be

$$\int_a^b f(x)\, dx = \lim_{n \to \infty} \sum_{k=0}^{n-1} f(x_k)\Delta x$$

The function f is called the **integrand,** the numbers a and b are the **limits of integration,** and the variable x is the **variable of integration.** A Riemann sum with a large number of subdivisions may be used to approximate the definite integral.

Quick Example

We approximated the definite integral for Example 1 using the Riemann sum with $n = 3600$:

$$\int_0^{60} c(t)\, dt = \int_0^{60} \frac{20}{t + 100}\, dt \approx 9.40$$

---

[7]And for some other functions as well.

If you run the LEFTSUM program repeatedly with larger and larger values of n, you will notice that the Riemann sum appears to get closer and closer to a fixed value, $9.400\,07\ldots$, so we can say with some certainty that the given integral is approximately $9.400\,07$.

The TI-83 also has a built-in function `fnInt`, which finds a very accurate approximation of a definite integral, but it uses a more sophisticated technique than we are discussing here.

Excel

We first need to calculate the numbers $c(0)$, $c(0.6)$, and so on. To enter the numbers $0, 0.6, 1.2, \ldots$ in the first column, proceed as follows: Enter 0 in cell A2 and 0.6 in cell A3, select A2:A3, grab the lower right (fill) handle, and drag down column A until you reach 59.4 (cell A101). Now enter the formula for $c(t)$ in cell B2, as

```
=20/(A2+100)
```

Copy this formula to cells B3 through B101:

	A	B	C
1	t	c(t)	
2	0	=20/(A2+100)	
3	0.6		
4	1.2		
5			
99			
100	58.8		
101	59.4		

Now to complete our calculation, enter in C2 the formula

```
=SUM(B2:B101)*0.6
```

for the Riemann sum, giving us the result:

	A	B	C
1	t	c(t)	Riemann Sum
2	0	0.2	9.42260915
3	0.6	0.198807157	
4	1.2	0.197628458	
5			
99			
100	58.8	0.125944584	
101	59.4	0.125470514	

Thus, we get the estimate of 9.4226 for the Riemann sum.

Web Site
The path

Web Site $\rightarrow$ Online Utilities $\rightarrow$ Numerical Integration Utility

will take you to a page that evaluates Riemann sums. Enter the function, the limits of integration, and n into the appropriate places and click on Left Sum to calculate the Riemann sum that we are using here. For a downloadable Excel spreadsheet that computes and also graphs Riemann sums, follow

Web Site $\rightarrow$ Online Utilities $\rightarrow$ Riemann Sum Grapher

---

APPLICATIONS: THE DEFINITE INTEGRAL AS TOTAL

In Example 1 we saw that if $c(t)$ is the *marginal* cost of a cell phone call, in dollars per minute, then the *total* cost of a 60-minute call is $\int_0^{60} c(t)\, dt$ dollars. If, instead of computing the total cost of the call over the interval $[0, 60]$ minutes, we computed the cost over the interval $[a, b]$ minutes, we would have obtained a total cost of $\int_a^b c(t)\, dt$ dollars. For example, the total cost of the call from $t = 5$ minutes to $t = 10$ minutes is $\int_5^{10} c(t)\, dt$ dollars.

The Definite Integral as a Total

If $r(x)$ is the rate of change of a quantity Q, then the total, or accumulated change, of the quantity as x changes from a to b is given by

$$\text{total change in quantity } Q = \int_a^b r(x)\, dx \qquad \text{Total change in } Q \text{ over } [a, b]$$

Units Note that we measure the rate of change r in units of Q per unit of x, and we measure the total change $\int_a^b r(x)\, dx$ in units of Q.

Quick Examples

1. If, at time t hours, you are selling wall posters at a rate of $s(t)$ posters per hour, then

$$\text{total number of posters sold from hour 3 to hour 5} = \int_3^5 s(t)\, dt$$

2. If, t days into the surf season, the water temperature at your favorite surf spot is changing at a rate of $0.05t^2 + 4$ degrees per day, then the total change in the water temperature from the start of the season ($t = 0$) to the start of day 10 is given by

$$\text{total change in temperature} = \int_0^{10} (0.05t^2 + 4)\, dt \text{ degrees}$$

3. If a skateboard is moving at $v(t)$ feet per second at time t seconds, then the total (net) distance the skateboard has moved during the interval $[2, 5]$ seconds is

$$\text{total distance moved} = \int_2^5 v(t)\, dt \text{ ft}$$

There is nothing particularly special about $1 - x^2$ in Example 2. In general, we can say the following.

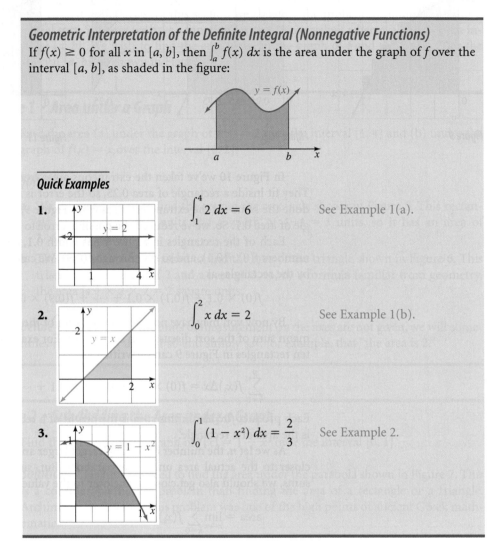

Geometric Interpretation of the Definite Integral (Nonnegative Functions)

If $f(x) \geq 0$ for all x in $[a, b]$, then $\int_a^b f(x)\, dx$ is the area under the graph of f over the interval $[a, b]$, as shaded in the figure:

$y = f(x)$

a b x

Quick Examples

1. $y = 2$

$$\int_1^4 2\, dx = 6$$ See Example 1(a).

2. $y = x$

$$\int_0^2 x\, dx = 2$$ See Example 1(b).

3. $y = 1 - x^2$

$$\int_0^1 (1 - x^2)\, dx = \frac{2}{3}$$ See Example 2.

In fact, to make the idea of area precise, we *define* the area under the graph to be the value of the integral.

Question Why did we consider only nonnegative functions?

Answer If $f(x_k) < 0$, then $f(x_k)\Delta x < 0$ and is the *negative* of the area of the rectangle shown in Figure 12.

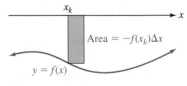

x_k

Area $= -f(x_k)\Delta x$

$y = f(x)$

Figure 12

The result is that, in regions where $f(x) < 0$, the area below the x axis and above the graph of f is *subtracted* from the value of the definite integral. This leads us to the following statement.

Geometric Interpretation of the Definite Integral (All Functions)

$\int_a^b f(x)\,dx$ is the area between $x = a$ and $x = b$ that is above the x axis and below the graph of f, minus the area that is below the x axis and above the graph of f:

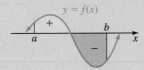

$$\int_a^b f(x)\,dx = \text{area above } x \text{ axis} - \text{area below } x \text{ axis}$$

Quick Example

$$\int_{-1}^{1} x\,dx = 0 \qquad \text{Since the areas above and below the } x \text{ axis are equal}$$

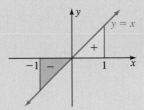

Another interpretation that you sometimes see is that the definite integral computes a *signed* area, where area above the x axis is positive and area below the x axis is negative.

Example 3 • Total Cost

Recall the following example from Section 6.3. Your cell phone company offers you an innovative pricing scheme. When you make a call, the *marginal* cost of the tth minute of the call is

$$c(t) = \frac{20}{t + 100} \text{ dollars per minute}$$

How is the total cost to make a 60-minute phone call related to the graph of c?

Solution We saw in Section 6.3 that the total cost to make a 60-minute phone call is given by the definite integral

$$\int_0^{60} c(t)\,dt$$

Now we have seen that this integral also represents the area under the graph of c (which is always positive) over the interval $[0, 60]$; that is, the total cost is the area of the region shown in Figure 13.

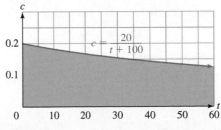

Figure 13

Question How can we tell from the graph that the total cost is represented by the shaded area?

Answer Take a look at how some Riemann sum approximations relate to the graph.

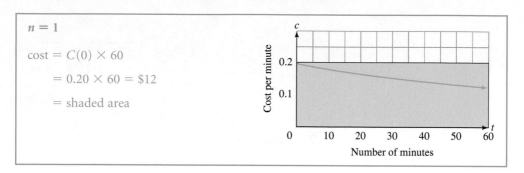

The diagrams illustrate how adding the costs for successive periods amounts to adding the areas of the rectangles representing the Riemann sum.

6.4 EXERCISES

In Exercises 1–10, use geometry (not Riemann sums) to compute the integrals.

1. $\int_0^1 1 \, dx$

2. $\int_0^2 5 \, dx$

3. $\int_0^1 x \, dx$

4. $\int_1^2 x \, dx$

5. $\int_0^1 \frac{1}{2} x \, dx$

6. $\int_{-2}^2 \frac{1}{2} x \, dx$

7. $\int_2^4 (x - 2) \, dx$

8. $\int_3^6 (x - 3) \, dx$

9. $\int_{-1}^1 x^3 \, dx$

10. $\int_1^2 \frac{1}{2} x \, dx$

T In Exercises 11–16, use technology to estimate the (signed) areas, using Riemann sums with $n = 10$ and $n = 100$. Round all answers to five significant digits.

11. Bounded by the line $y = x$, the x axis, and the lines $x = 0$ and $x = 1$

12. Bounded by the line $y = 2x$, the x axis, and the lines $x = 1$ and $x = 2$

13. Bounded by the curve $y = x^2 - 1$, the x axis, and the lines $x = 0$ and $x = 4$

14. Bounded by the curve $y = 1 - x^2$, the x axis, and the lines $x = -1$ and $x = 2$

15. Bounded by the x axis, the curve $y = e^{x^2}$, and the lines $x = 0$ and $x = 1$

16. Bounded by the x axis, the curve $y = e^{x^2-1}$, and the lines $x = 0$ and $x = 2$

APPLICATIONS

17. SUV Sales The following graph shows the number of sport utility vehicles (SUVs) sold in the United States each year from 1998 through 2003:

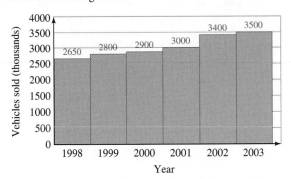

What is the area under this graph? What does this number represent?

2000–2003 figures are estimates. Sources: Oak Ridge National Laboratory; Light Vehicle MPG and Market Shares System; AutoPacific; *The U.S. Car and Light Truck Market,* 1999, pp. 24, 120, 121.

18. Net Sales The following graph shows Nokia's annual sales from 1995 through 2001:

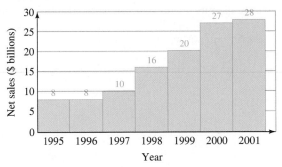

What is the area under this graph? What does this number represent?

Source: Nokia/*New York Times,* February 6, 2002, p. A3.

19. Total Profit Except for a single quarter in 2001, Amazon.com lost money each quarter from the first quarter of 1999 through the first quarter of 2002, as shown in the graph to the right. (Each bar represents the net income for one quarter.)

What is the net area between this graph and the *t* axis for $8 \le t \le 13$? (Each bar represents one quarter. Subtract the area under the axis as usual.) What does this number represent?

Source: Amazon.com/*New York Times,* April 24, 2002, p. C1.

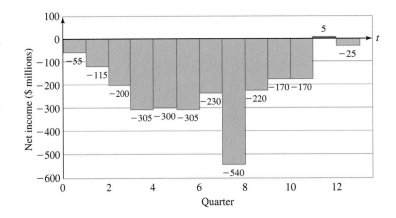

20. Toxic Cleanup Funding The graph to the right shows the annual change in the U.S. government Superfund cleanup trust fund balance from 1990 through 2003.

What is the net area between this graph and the time axis for $4 \le t \le 13$? (Each bar represents 1 year. Subtract the area under the axis as usual.) What does this number represent?

Source: Democratic staff of the House Committee on Transportation and Infrastructure, from EPA and White House data/*New York Times,* February 24, 2002, p. 24.

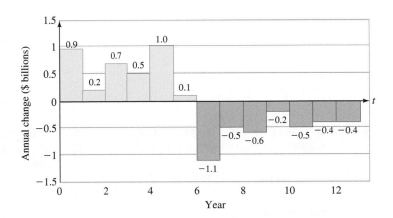

21. Surveying My uncle intends to build a kidney-shaped swimming pool in his small yard, and the town zoning board will approve the project only if the total area of the pool does not exceed 500 square feet. The following diagram shows the planned swimming pool, with measurements of its width at the indicated points. Will my uncle's plans be approved? (Use a Riemann sum to approximate the area.)

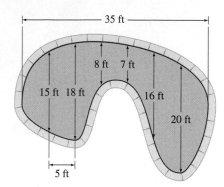

22. Pollution An aerial photograph of an ocean oil spill shows the pattern in the accompanying diagram. Assuming that the oil slick has a uniform depth of 0.01 meter, how many cubic meters of oil would you estimate to be in the spill? (Volume = area × thickness)

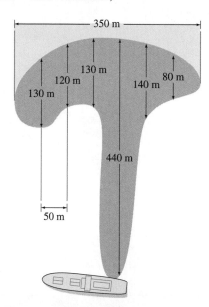

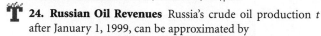 **23. Saudi Oil Revenues** The spot price of crude oil t days after January 1, 1999, can be approximated by

$$P(t) = 32 - 10\left(\frac{t}{365} - 1\right)^2 \text{ dollars per barrel}$$
$$(0 \le t \le 730)$$

Saudi Arabia's crude oil production over the same period can be approximated by

$$Q(t) = 78 - 70.8\left(\frac{t}{365} - 1\right)^2 \text{ million barrels per day}$$
$$(0 \le t \le 730)$$

a. Graph the revenue function $R(t) = P(t)Q(t)$, indicating the area that represents $\int_0^{365} R(t)\, dt$. What does this area mean?

b. Estimate the area in part (a), using a Riemann sum with $n = 200$. (Round the answer to three significant digits.) Interpret the answer.

SOURCE: Deutsche Bank/*New York Times*, November 21, 2001, p. A3.

24. Russian Oil Revenues Russia's crude oil production t after January 1, 1999, can be approximated by

$$Q(t) = 4 + 0.2\left(\frac{t}{365} - 1\right) \text{ million barrels per day}$$
$$(0 \le t \le 730)$$

a. Using the model for the spot price given in Exercise 23, graph the revenue function $R(t) = P(t)Q(t)$, indicating the area representing $\int_{365}^{730} R(t)\, dt$.

b. Estimate the area in part (a), using a Riemann sum with $n = 200$. (Round the answer to three significant digits.) Interpret the answer.

SOURCE: Deutsche Bank/*New York Times*, November 21, 2001, p. A3.

COMMUNICATION AND REASONING EXERCISES

25. The definite integral counts area under the x axis as negative. Give an example showing how this can be useful in applications.

26. Sketch the graph of a nonconstant function whose Riemann sum with $n = 1$ gives the exact value of the definite integral.

27. Sketch the graph of a nonconstant function whose Riemann sums with $n = 1, 5$, and 10 are all zero.

28. If $\int_a^b f(x)\, dx = 0$, what can you say about the graph of f?

29. Sketch the graphs of two distinct functions $f(x)$ and $g(x)$ such that $\int_a^b f(x)\, dx = \int_a^b g(x)\, dx$.

30. We have always used the *left* Riemann sum. The **right** Riemann sum is similar, except that we evaluate the function f on the right endpoint of each interval. Thus, the right Riemann sum is

$$\sum_{k=1}^{n} f(x_k)\, \Delta x$$

Let f be a decreasing function on the interval $[a, b]$. Demonstrate by means of a sketch that the left sum is always greater than or equal to the right sum. What can you say about *increasing* functions?

31. Let f be an increasing function. Draw a picture and demonstrate that the difference between the right and left sums for $\int_a^b f(x)\, dx$ is $[f(b) - f(a)]\Delta x$. Conclude that the difference goes to zero as $n \to +\infty$.

32. Another approximation of the integral is the **midpoint** approximation, in which we compute the sum

$$\sum_{k=1}^{n} f(\bar{x}_k)\Delta x$$

where $\bar{x}_k = (x_{k-1} + x_k)/2$ is the point midway between the left and right endpoints of the interval $[x_{k-1}, x_k]$. Why is it true that the midpoint approximation is exact if f is linear? (Draw a picture.)

6.5 The Definite Integral: An Algebraic Approach and the Fundamental Theorem of Calculus

So far we have calculated definite integrals by approximating them by Riemann sums. In this section we see an algebraic way of calculating many definite integrals, using antiderivatives.

The connection between antiderivatives and definite integrals is given by the Fundamental Theorem of Calculus. Here is the central idea behind the theorem: Suppose f is a continuous function defined on $[a, b]$. We may define a new function with domain $[a, b]$ by letting

$$F(x) = \int_a^x f(t)\, dt$$

Consider the following examples.

Example 1 • The Area Function

Use geometry to find explicit formulas for the following area functions:

a. $A(x) = \int_0^x 3\, dt \quad (x > 0)$ **b.** $A(x) = \int_0^x t\, dt \quad (x > 0)$

Solution

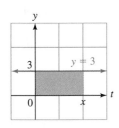

Figure 14

a. $A(x)$ is the area of the rectangle shown in Figure 14. This rectangle has width x and height 3, so its area is

$$A(x) = 3x$$

b. $A(x)$ is the area of the triangle shown in Figure 15.

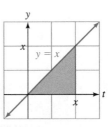

Figure 15

This triangle has base x and height x, so its area is $\frac{1}{2} x^2$. Therefore,

$$A(x) = \frac{1}{2} x^2$$

Before we go on . . . Notice that, in both (a) and (b), $A(x)$ is an *antiderivative* of the integrand: In part (a), $A'(x) = 3$, which is the function being integrated. In part (b), $A'(x) = x$, which again is the function being integrated (except that t has been replaced with x).

Example 2 • Graphing the Integral

Graph the function $F(x) = \int_0^x (1 - t^2)\, dt$ for $0 \le x \le 2$.

Solution   We can use technology to get a table of approximate values of F and draw the graph.

Graphing Calculator

The simplest way to graph a function and its integral on the TI-83 is to use the calculator's built-in numerical integration function `fnInt`. In the `Y=` window, enter

$$Y_1 = 1 - X^2$$
$$Y_2 = \text{fnInt}(Y_1(T), T, 0, X) \qquad \text{Format for } \int_0^x Y_1(t)\, dt$$

This defines Y_1 to be $f(x) = 1 - x^2$ and Y_2 to be F. Graphing these two functions in the window defined by Xmin = 0, Xmax = 2, Ymin = −3, and Ymax = 3 gives Figure 16. One of the graphs is the derivative of the other. Can you see which?

   Warning: It will take a while to graph F. You can speed things considerably, at the price of getting a slightly cruder graph, by increasing the value of the window variable Xres (the largest possible value is 8, and the smallest value is 1).

Figure 16

Excel

To see these graphs in Excel, first set up the spreadsheet to generate a table of 200 values of $f(t) = 1 - t^2$ as follows. (As in Section 6.1, you can use the `Fill`→`Series` dialog box from the `Edit` menu to enter a series in the first column, starting at 0, with a step value of 0.01, stopping at 2.)

	A	B	C
1	x	f(x)	
2	0	=1−A2^2	
3	0.01		
4	0.02		
5			
200			
201	1.99		
202	2		

Now use column C to compute approximate values of $F(x)$, as shown:

	A	B	C	D
1	x	f(x)	F(x)	
2	0	1	0	
3	0.01	0.9999	=0.01*SUM(B$2:B2)	
4	0.02	0.9996		
5				
200				
201	1.99	−2.9601		
202	2	−3		

	A	B	C
1	x	f(x)	F(x)
2	0	1	0
3	0.01	0.9999	0.01
4	0.02	0.9996	0.019999
5			
200			
201	1.99	−2.9601	−0.617099
202	2	−3	−0.6467

To graph both f and F, select cells A1 through C202 and ask Excel for an XY scatter graph. The result should look similar to Figure 16.

✳ ***Before we go on . . .*** Notice that in Figure 16 the area function F begins at zero when $x = 0$, increases as long as f is positive, and begins to decrease when f becomes negative. Remember that area below the x axis is subtracted from the value of the integral.

Notice also that, in the Excel worksheet above, once we have the values of F, we can find an approximation of F'. Since $\Delta x = 0.01$, we could approximate $F'(0.01)$, using the formula

```
=(C3-C2)/0.01
```

in cell D3. We could then copy down column D. However, the result would be a copy of column B, the values of f. Why does this happen?

Examples 1 and 2 suggest the first part of the following result. The second part will allow us to calculate integrals algebraically.

The Fundamental Theorem of Calculus (FTC)

Let f be a continuous function defined on the interval $[a, b]$.

1. If $A(x) = \int_a^x f(t)\, dt$, then $A'(x) = f(x)$; that is, A is an antiderivative of f.

2. If F is any continuous antiderivative of f and is defined on $[a, b]$, then

$$\int_a^b f(x)\, dx = F(b) - F(a)$$

Quick Examples

1. Example of 1: If $A(x) = \int_0^x e^{t^2}\, dt$, then $A'(x) = e^{x^2}$.

2. Example of 2: Since $F(x) = x^2$ is an antiderivative of $f(x) = 2x$,

$$\int_0^1 2x\, dx = F(1) - F(0) = 1^2 - 0^2 = 1$$

We'll sketch a proof of this result at the end of the section. Part 2 tells us that to calculate a definite integral we should first find an antiderivative.

Example 3 • Using the FTC to Calculate a Definite Integral

Calculate $\int_0^1 (1 - x^2)\, dx$.

Solution To use part 2 of the FTC, we need to find an antiderivative of $1 - x^2$. But we know that

$$\int (1 - x^2)\, dx = x - \frac{x^3}{3} + C$$

We need only one antiderivative, so let's take $F(x) = x - x^3/3$. The FTC tells us that

$$\int_0^1 (1 - x^2)\, dx = F(1) - F(0) = \left(1 - \frac{1}{3}\right) - (0) = \frac{2}{3}$$

which is the value we estimated in Section 6.4.

✱ **Before we go on . . .** A useful piece of notation is often used here. We write[9]

$$[F(x)]_a^b = F(b) - F(a)$$

Thus, we can rewrite the computation above as

$$\int_0^1 (1 - x^2)\, dx = \left[x - \frac{x^3}{3} \right]_0^1 = \left(1 - \frac{1}{3} \right) - (0) = \frac{2}{3}$$

Example 4 • More Use of the FTC

Compute the following definite integrals.

a. $\displaystyle\int_0^1 (2x^3 + 10x + 1)\, dx$ **b.** $\displaystyle\int_1^5 \left(\frac{1}{x^2} + \frac{1}{x} \right) dx$

Solution

a. $\displaystyle\int_0^1 (2x^3 + 10x + 1)\, dx = \left[\frac{1}{2}x^4 + 5x^2 + x \right]_0^1$

$$= \left(\frac{1}{2} + 5 + 1 \right) - 0 = \frac{13}{2}$$

b. $\displaystyle\int_1^5 \left(\frac{1}{x^2} + \frac{1}{x} \right) dx = \int_1^5 (x^{-2} + x^{-1})\, dx$

$$= [-x^{-1} + \ln |x|]_1^5$$

$$= \left(-\frac{1}{5} + \ln 5 \right) - (-1 + \ln 1) = \frac{4}{5} + \ln 5$$

When calculating a definite integral, we may have to use substitution to find the necessary antiderivative. We could substitute, evaluate the indefinite integral with respect to u, express the answer in terms of x, and then evaluate at the limits of integration. However, there is a shortcut, as we see in Example 5.

Example 5 • Using the FTC with Substitution

Evaluate $\displaystyle\int_1^2 (2x - 1)e^{2x^2 - 2x}\, dx$

Solution The shortcut we promised is to put *everything* in terms of u, including the limits of integration.

$$u = 2x^2 - 2x$$

$$\frac{du}{dx} = 4x - 2$$

$$dx = \frac{1}{4x - 2}\, du$$

When $x = 1$, $u = 0$.   Substitute $x = 1$ in the formula for u.
When $x = 2$, $u = 4$.   Substitute $x = 2$ in the formula for u.

---

[9]Several notations are in use, actually. Another common notation is $F(x)\big|_a^b$.

We get the value $u = 0$, for example, by substituting $x = 1$ in the equation $u = 2x^2 - 2x$. We can now rewrite the integral:

$$\int_1^2 (2x - 1)e^{2x^2 - 2x}\, dx = \int_0^4 (2x - 1)e^u \frac{1}{4x - 2}\, du$$

$$= \int_0^4 \frac{1}{2} e^u\, du$$

$$= \left[\frac{1}{2} e^u\right]_0^4 = \frac{1}{2} e^4 - \frac{1}{2}$$

✴ Before we go on . . . The alternative, longer calculation is first to calculate the indefinite integral:

$$\int (2x - 1)e^{2x^2 - 2x}\, dx = \int \frac{1}{2} e^u\, du = \frac{1}{2} e^u + C = \frac{1}{2} e^{2x^2 - 2x} + C$$

Then we can say that

$$\int_1^2 (2x - 1)e^{2x^2 - 2x}\, dx = \left[\frac{1}{2} e^{2x^2 - 2x}\right]_1^2 = \frac{1}{2} e^4 - \frac{1}{2}$$

APPLICATIONS

Example 6 • Total Cost

In several of the preceding sections, we considered the following example. Your cell phone company offers you an innovative pricing scheme. When you make a call, the *marginal* cost of the tth minute of the call is

$$c(t) = \frac{20}{t + 100} \text{ dollars per minute}$$

If you make a cellular call that lasts 60 minutes, how much will it cost you?

Solution We need to calculate

$$\int_0^{60} \frac{20}{t + 100}\, dt$$

We can evaluate this integral either by using the substitution $u = t + 100$, or by using the shortcut formula:

$$\int \frac{1}{ax + b}\, dx = \frac{1}{a} \ln |ax + b| + C$$

from Section 6.2. Using the shortcut formula gives

$$\int_0^{60} \frac{20}{t + 100}\, dt = 20 \int_0^{60} \frac{1}{t + 100}\, dt$$

$$= 20[\ln |t + 100|]_0^{60}$$

$$= 20(\ln 160 - \ln 100) \approx \$9.40$$

Compare this with Example 1 of Section 6.3, where we found the same answer by approximating with Riemann sums.

Example 7 • Computing Area

Find the total area of the region enclosed by the graph of $y = xe^{x^2}$, the x axis, and the vertical lines $x = -1$ and $x = 1$.

Solution The region whose area we want is shown in Figure 17. Notice the symmetry of the graph. Also, half the region in which we are interested is above the x axis, and the other half is below. If we calculated the integral $\int_{-1}^{1} xe^{x^2} \, dx$, the result would be

area above x axis $-$ area below x axis $= 0$

which does not give us the total area. To prevent the area below the x axis from being combined with the area above the axis, we do the calculation in two parts, as illustrated in Figure 18.

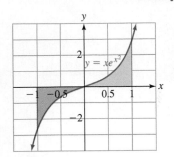

Figure 17

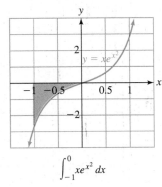

$$\int_{-1}^{0} xe^{x^2} \, dx$$

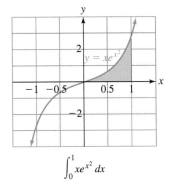

$$\int_{0}^{1} xe^{x^2} \, dx$$

Figure 18

(We broke the integral at $x = 0$ because that is where the graph crosses the x axis.) These integrals can be calculated, using the substitution $u = x^2$:

$$\int_{-1}^{0} xe^{x^2} \, dx = \frac{1}{2} [e^{x^2}]_{-1}^{0} = \frac{1}{2}(1 - e) \approx -0.859\,14 \qquad \text{Why is it negative?}$$

$$\int_{0}^{1} xe^{x^2} \, dx = \frac{1}{2} [e^{x^2}]_{0}^{1} = \frac{1}{2}(e - 1) \approx 0.859\,14$$

To obtain the total area, we should add the *absolute* values of these answers because we don't wish to count any area as negative. Thus,

total area $\approx 0.859\,14 + 0.859\,14 = 1.718\,28$

Proof of the Fundamental Theorem of Calculus

We end this section with a sketch of the proof of the Fundamental Theorem of Calculus. The FTC was first shown by the English mathematician Isaac Barrow (1630–1677), who was Newton's teacher. It was Newton, however, who recognized its real importance.

We begin with a continuous function f defined on an interval $[a, b]$, and we define the function

$$A(x) = \int_{a}^{x} f(t) \, dt$$

Remember that the definite integral is defined by a limit. Technically, one of the hardest parts of the proof is to show that, if f is continuous, then the limit always exists, so $A(x)$

is defined. This part of the proof is beyond the scope of this book, so we assume that it has been shown and proceed to calculate $A'(x)$ in order to verify part 1 of the FTC.

Recall that $A'(x)$ is also defined by a limit:

$$A'(x) = \lim_{h \to 0} \frac{A(x + h) - A(x)}{h}$$

We can think of $A(x + h)$ and $A(x)$ as the areas shown in Figure 19.[10] The figure shows that $A(x + h) - A(x)$ is the larger area minus the smaller area, which is the area of the "sliver" shown in Figure 20. (We take $h > 0$ for this discussion, but the case $h < 0$ is dealt with by a similar argument.)

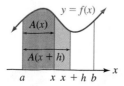

Figure 19

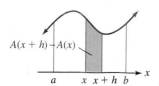

Figure 20

Thus, we can say that

$$A'(x) = \lim_{h \to 0} \frac{A(x + h) - A(x)}{h} = \lim_{h \to 0} \frac{\text{area of the sliver}}{h}$$

Here is an intuitive argument that the value of this limit is $f(x)$: The area of the sliver is approximately the same as the area of a rectangle of height $f(x)$ and width h, so the area of the sliver is approximately $f(x)h$. If we divide this area by h, we get

$$\frac{\text{area of the sliver}}{h} \approx f(x)$$

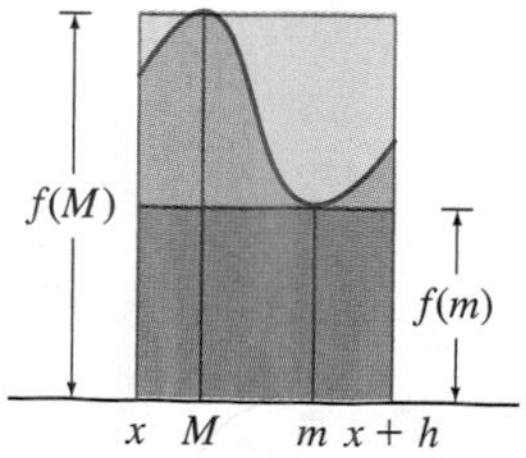

Here is a more careful argument. In Figure 21 we see an enlarged view of the sliver, showing also that there might be more "wobbles" in the function than in the picture we were looking at originally. In this figure we have also marked points M and m, where $f(M)$ is the maximum value of f on $[x, x + h]$ and $f(m)$ is the minimum value of f on $[x, x + h]$. We can see from the figure that the area of the sliver must be between the areas of two rectangles, the rectangles with width h and heights $f(m)$ and $f(M)$, respectively; that is,

$$f(m)h \le \text{area of the sliver} \le f(M)h$$

Dividing by h, we get

$$f(m) \le \frac{\text{area of the sliver}}{h} \le f(M)$$

Now consider what happens as h approaches 0. The two points m and M, caught between x and $x + h$, must both approach x. Since f is continuous, $f(m)$ and $f(M)$ must therefore both approach $f(x)$. The term in the middle, "area of the sliver/h," must therefore also approach $f(x)$. Thus,

$$A'(x) = \lim_{h \to 0} \frac{A(x + h) - A(x)}{h} = \lim_{h \to 0} \frac{\text{area of the sliver}}{h} = f(x)$$

This completes the proof of the first part of the FTC.

Figure 21

---

[10]To simplify the argument, we assume that $f(x) \ge 0$, as shown in the graph.

For the second part, let F be any antiderivative of f on $[a, b]$. Since F and A are both antiderivatives of f, they must differ by a constant (see Section 6.1); that is, $F(x) = A(x) + C$ for some constant C. Therefore,

$$F(b) - F(a) = [A(b) + C] - [A(a) + C]$$

$$= A(b) - A(a)$$

$$= \int_a^b f(x) \, dx - 0 = \int_a^b f(x) \, dx$$

6.5 EXERCISES

In Exercises 1–6, use geometry (not antiderivatives) to compute the given area function $A(x) = \int_a^x f(t) \, dt$.

1. $A(x) = \int_0^x 1 \, dt$; verify that $A'(x) = 1$.

2. $A(x) = \int_0^x t \, dt$; verify that $A'(x) = x$.

3. $A(x) = \int_2^x t \, dt$; verify that $A'(x) = x$.

4. $A(x) = \int_2^x 1 \, dt$; verify that $A'(x) = 1$.

5. $A(x) = \int_a^x 4 \, dt$; verify that $A'(x) = 4$.

6. $A(x) = \int_{-a}^x 4 \, dt$; verify that $A'(x) = 4$.

In Exercises 7–16, use antiderivatives to find the area function $A(x) = \int_a^x f(t) \, dt$ for the given $f(x)$ and a. Sketch both $f(x)$ and $A(x)$.

7. $f(x) = x$; $a = 0$ **8.** $f(x) = 2x$; $a = 1$

9. $f(x) = x^2$; $a = 0$ **10.** $f(x) = x^2 + 1$; $a = 0$

11. $f(x) = e^x$; $a = 0$ **12.** $f(x) = e^{-x}$; $a = 0$

13. $f(x) = \dfrac{1}{x}$; $a = 1$ **14.** $f(x) = \dfrac{1}{x + 1}$; $a = 0$

15. $f(x) = \begin{cases} x & \text{if } 0 \le x \le 1 \\ 2 & \text{if } x > 1 \end{cases}$; $a = 0$

16. $f(x) = \begin{cases} 2x & \text{if } 0 \le x \le 2 \\ 1 & \text{if } x > 2 \end{cases}$; $a = 0$

In Exercises 17–58, evaluate the integrals.

17. $\int_{-1}^1 (x^2 + 2) \, dx$ **18.** $\int_{-2}^1 (x - 2) \, dx$

19. $\int_0^1 (12x^5 + 5x^4 - 6x^2 + 4) \, dx$

20. $\int_0^1 (4x^3 - 3x^2 + 4x - 1) \, dx$

21. $\int_{-2}^2 (x^3 - 2x) \, dx$ **22.** $\int_{-1}^1 (2x^3 + x) \, dx$

23. $\int_1^3 \left(\dfrac{2}{x^2} + 3x \right) dx$ **24.** $\int_2^3 \left(x + \dfrac{1}{x} \right) dx$

25. $\int_0^1 (2.1x - 4.3x^{1.2}) \, dx$ **26.** $\int_{-1}^0 (4.3x^2 - 1) \, dx$

27. $\int_0^1 2e^x \, dx$ **28.** $\int_{-1}^0 3e^x \, dx$

29. $\int_0^1 \sqrt{x} \, dx$ **30.** $\int_{-1}^1 \sqrt[3]{x} \, dx$

31. $\int_0^1 2^x \, dx$ **32.** $\int_0^1 3^x \, dx$

33. $\int_0^1 18(3x + 1)^5 \, dx$ **34.** $\int_0^1 8(-x + 1)^7 \, dx$

35. $\int_{-1}^1 e^{2x-1} \, dx$ **36.** $\int_0^2 e^{-x+1} \, dx$

37. $\int_0^2 2^{-x+1} \, dx$ **38.** $\int_{-1}^1 3^{2x-1} \, dx$

39. $\int_0^{50} e^{-0.02x-1} \, dx$ **40.** $\int_{-20}^0 3e^{2.2x} \, dx$

41. $\int_{-1.1}^{1.1} e^{x+1} \, dx$ **42.** $\int_0^{\sqrt{2}} x\sqrt{2x^2 + 1} \, dx$

43. $\int_{-\sqrt{2}}^{\sqrt{2}} 3x\sqrt{2x^2 + 1} \, dx$ **44.** $\int_{-1.2}^{1.2} e^{-x-1} \, dx$

45. $\int_0^1 5xe^{x^2+2} \, dx$ **46.** $\int_0^2 \dfrac{3x}{x^2 + 2} \, dx$

47. $\int_2^3 \dfrac{x^2}{x^3 - 1} \, dx$ **48.** $\int_2^3 \dfrac{x}{2x^2 - 5} \, dx$

49. $\int_0^1 x(1.1)^{-x^2} \, dx$ **50.** $\int_0^1 x^2(2.1)^{x^3} \, dx$

51. $\int_1^2 \dfrac{e^{1/x}}{x^2} \, dx$ **52.** $\int_1^2 \dfrac{\sqrt{\ln x}}{x} \, dx$

53. $\int_0^2 \dfrac{x}{x + 1} \, dx$ **54.** $\int_{-1}^1 \dfrac{2x}{x + 2} \, dx$

55. $\int_1^2 x(x - 2)^5 \, dx$ **56.** $\int_1^2 x(x - 2)^{1/3} \, dx$

57. $\int_0^1 x\sqrt{2x + 1} \, dx$ **58.** $\int_{-1}^0 2x\sqrt{x + 1} \, dx$

In Exercises 59–66, calculate the total area of the described regions. Do not count the area beneath the x axis as negative.

59. Bounded by the line $y = x$, the x axis, and the lines $x = 0$ and $x = 1$

60. Bounded by the line $y = 2x$, the x axis, and the lines $x = 1$ and $x = 2$

61. Bounded by the curve $y = \sqrt{x}$, the x axis, and the lines $x = 0$ and $x = 4$

62. Bounded by the curve $y = 2\sqrt{x}$, the x axis, and the lines $x = 0$ and $x = 16$

63. Bounded by the curve $y = x^2 - 1$, the x axis, and the lines $x = 0$ and $x = 4$

64. Bounded by the curve $y = 1 - x^2$, the x axis, and the lines $x = -1$ and $x = 2$

65. Bounded by the x axis, the curve $y = xe^{x^2}$, and the lines $x = 0$ and $x = (\ln 2)^{1/2}$

66. Bounded by the x axis, the curve $y = xe^{x^2-1}$, and the lines $x = 0$ and $x = 1$

APPLICATIONS

67. Cost The marginal cost of producing the xth box of light bulbs is $5 + x^2/1000$ dollars. Determine how much is added to the total cost by a change in production from $x = 10$ to $x = 100$ boxes.

68. Revenue The marginal revenue of the xth box of Zip disks sold is $100e^{-0.001x}$ dollars. Find the revenue generated by selling items 101 through 1000.

69. Bottled-Water Sales The rate of U.S. sales of domestic bottled water for the period 1990–2000 can be approximated by

$$D(t) = 0.0166t^2 - 0.050t + 0.27 \text{ billion gallons per year}$$
$$(0 \le t \le 10)$$

where t is time in years since 1990. Use the FTC to estimate the total sales of domestic bottled water from 1995 to 2000. (Round your answer to the nearest 0.1 billion gallons.)

The model is based on regression of approximate data. Source: Beverage Marketing Corporation of New York/*New York Times*, June 21, 2001, p. C1.

70. Bottled-Water Sales The rate of U.S. sales of imported bottled water for the period 1990–2000 can be approximated by

$$I(t) = 0.0004t^2 + 0.013t + 0.06 \text{ billion gallons per year}$$
$$(0 \le t \le 10)$$

where t is time in years since 1990. Use the FTC to estimate the total sales of imported bottled water from 1995 to 2000. (Round your answer to the nearest 0.01 billion gallons.)

The model is based on regression of approximate data. Source: Beverage Marketing Corporation of New York/*New York Times*, June 21, 2001, p. C1.

71. Grants The rate of spending on grants by U.S. foundations was approximately

$$s(t) = 5.4e^{0.06t} \text{ billion dollars per year} \qquad (0 \le t \le 26)$$

where t is time in years since the start of 1976. Use this model to estimate, to the nearest \$1 billion, the total spending on grants from 1990 through 2001.

Based on a regression model. Data are adjusted for inflation. Source: The Foundation Center/*New York Times*, April 2, 2002, p. A21.

72. Global Warming The most abundant greenhouse gas is carbon dioxide (CO_2). According to a U.N. worst-case-scenario prediction, the rate of increase of the total amount of CO_2 in the atmosphere can be approximated by

$$C(t) \approx 0.98e^{0.0035t} \text{ parts per million per year}$$
$$(0 \le t \le 350)$$

where t is time in years since the start of 1750. Use this model to estimate, to the nearest 1 part per million, the total amount of CO_2 released into the atmosphere from 2000 through 2009.

Exponential regression is based on the 1750 figure and the 2100 U.N. prediction. Sources: Tom Boden/Oak Ridge National Laboratory, Scripps Institute of Oceanography/University of California, International Panel on Climate Change/*New York Times*, December 1, 1997, p. F1.

73. Displacement A car traveling down a road has a velocity of $v(t) = 60 - e^{-t/10}$ miles per hour at time t hours. Find the total distance it travels from time $t = 1$ hour to time $t = 6$. (Round your answer to the nearest mile.)

74. Displacement A ball thrown in the air has a velocity of $v(t) = 100 - 32t$ feet per second at time t seconds. Find the total displacement of the ball between times $t = 1$ second and $t = 7$ seconds and interpret your answer.

75. Military Spending Write $C(t)$ for the amount the U.S. military spent on personnel in year t, where t is measured in years since 2000. The rate of increase of $C(t)$ was about \$3.7 billion per year in 2000 and was projected to decrease to \$3 billion per year in 2007.
   a. Find a linear model for the rate of change $C'(t)$.
   b. Use your linear model to estimate the total change in annual military spending on personnel from 2000 to 2005.

While the cost per active-duty armed service member has been increasing, the number of armed service personnel has been decreasing. Annual costs are adjusted for inflation, and data are based on linear models for annual cost and personnel. Sources: U.S. Dept. of Defense; Stephen Daggett, military analyst, Congressional Research Service/*New York Times*, April 19, 2002, p. A21.

76. Military Spending in the 1990s Recall the function $C(t)$ from Exercise 75. The rate of change of $C(t)$ was about \$0.5 billion per year in 1990 and decreased to $-\$0.5$ billion per year in 1995.
   a. Find a linear model for the rate of change $C'(t)$, with t now measured in years since 1990.
   b. Use your linear model to estimate the total change in annual military spending on personnel from 1990 to 1995.

While the cost per active-duty armed service member has been increasing, the number of armed service personnel has been decreasing. Annual costs are adjusted for inflation, and data are based on linear models for annual cost and personnel. Sources: U.S. Dept. of Defense; Stephen Daggett, military analyst, Congressional Research Service/*New York Times*, April 19, 2002, p. A21.

77. Embryo Development The oxygen consumption of a bird embryo increases from the time the egg is laid through the time the chick hatches. In a typical galliform bird, the oxygen consumption can be approximated by

$$c(t) = -0.00271t^3 + 0.137t^2 - 0.892t + 0.149 \text{ milliliters per hour}$$
$$(8 \le t \le 30)$$

where t is the time (in days) since the egg was laid. (An egg will typically hatch at around $t = 28$.) Find the total amount of oxygen consumed during the ninth and tenth days ($t = 8$ to $t = 10$). Round your answer to the nearest milliliter.

The model approximates graphical data published in Roger S. Seymour, "The Brush Turkey," *Scientific American* (December 1991): 108–114.

78. Embryo Development The oxygen consumption of a turkey embryo increases from the time the egg is laid through the time the chick hatches. In a brush turkey, the oxygen consumption can be approximated by

$$c(t) = -0.00118t^3 + 0.119t^2 - 1.83t + 3.972 \text{ milliliters per hour}$$
$$(20 \le t \le 50)$$

where t is the time (in days) since the egg was laid. (An egg will typically hatch at around $t = 50$.) Find the total amount of oxygen consumed during the 21st and 22nd days ($t = 20$ to $t = 22$). Round your answer to the nearest milliliter.

The model approximates graphical data published in Roger S. Seymour, "The Brush Turkey," *Scientific American* (December 1991): 108–114.

79. Sales Weekly sales of your *Lord of the Rings* T-shirts have been falling by 5% per week. Assuming that you are now selling 50 T-shirts each week, how many shirts will you sell during the coming year? (Round your answer to the nearest shirt.)

80. Sales Annual sales of fountain pens in Littleville are presently 4000 per year and are increasing by 10% per year. How many fountain pens will be sold over the next 5 years?

81. Fuel Consumption The way Professor Waner drives, he burns gas at the rate of $1 - e^{-t}$ gallons each hour, t hours after a fill-up. Find the number of gallons of gas he burns in the first 10 hours after a fill-up.

82. Fuel Consumption The way Professor Costenoble drives, he burns gas at the rate of $1/(t + 1)$ gallons each hour, t hours after a fill-up. Find the number of gallons of gas he burns in the first 10 hours after a fill-up.

83. Big Brother The number of wiretaps authorized each year by U.S. courts from 1990 to 2000 can be approximated by

$$W(t) = \frac{1500e^{0.15t}}{0.77 + e^{0.15t}} \qquad (0 \le t \le 10)$$

where $t = 0$ represents June 1990.

a. Use a definite integral to estimate the total number of wiretaps from June 1990 to June 2000. (Round the answer to two significant digits.) (*Hint:* For help on evaluating the integral, see Example 7 in Section 6.2.)

b. The following graph shows the actual number of authorized wiretaps:

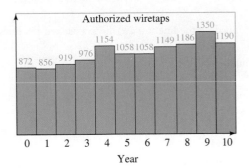

Does the integral in part (a) give an accurate estimate of the actual number to two significant digits? Explain.

Source: 2000 Wiretap Report, Administrative Office of the United States Courts: http://www.epic.org/privacy/wiretap/stats/2000_report/default.html

84. Big Brother The total number of wiretaps authorized each year by U.S. federal courts from 1990 to 2000 can be approximated by

$$W(t) = \frac{600e^{0.48t}}{1 + e^{0.48t}} \qquad (0 \le t \le 10)$$

where $t = 0$ represents June 1990.

a. Use a definite integral to estimate the total number of wiretaps from June 1990 to June 2000. (Round the answer to two significant digits.) (*Hint:* For help on evaluating the integral, see Example 7 in Section 6.2.)

b. The following graph shows the actual number of authorized wiretaps:

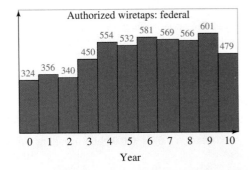

Does the integral in part (a) give an accurate estimate of the actual number to two significant digits? Explain.

Source: 2000 Wiretap Report, Administrative Office of the United States Courts: http://www.epic.org/privacy/wiretap/stats/2000_report/default.html

85. Health Care Spending The following chart shows approximate figures for annual spending on health care in the United States from 1981 to 1994, in billions of dollars:

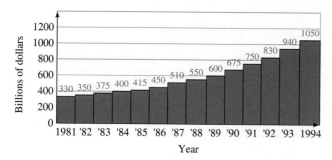

a. Let t be time in years since January 1981 and use the figures for 1981 ($t = 0$) and 1993 ($t = 12$) to construct an exponential model of the form $s(t) = Pe^{kt}$ for annual spending, where $s(t)$ is the annual spending (in billions of dollars) at time t.

b. Use your model to calculate the total amount spent on health care in the United States for the period shown and compare it with the result given by the data in the chart. (*Hint:* The period shown extends to January 1995.)

c. Use your model to predict total spending on health care in the United States from January 1995 to January 2000.

Sources: U.S. Dept. of Health and Human Services; Department of Commerce/*New York Times*, December 29, 1993, p. A12. The figures are estimates based on the graph.

86. Health Care Spending Repeat Exercise 85, but this time base your model on the figures for 1982 ($t = 1$) and 1994 ($t = 13$).

87. Total Cost Use the FTC to show that if $m(x)$ is the marginal cost at a production level of x items, then the cost function $C(x)$ is given by

$$C(x) = C(0) + \int_0^x m(t)\, dt$$

What term do we use for $C(0)$?

88. Total Sales The total cost of producing x items is given by

$$C(x) = 246.76 + \int_0^x 5t\, dt$$

Find the fixed cost and the marginal cost of producing the tenth item.

89. The Logistic Function and Sales
a. Show that the logistic function $f(x) = N/(1 + Ab^{-x})$ can be written in the form

$$f(x) = \frac{Nb^x}{A + b^x}$$

b. Use the result of part (a) and a suitable substitution to show that

$$\int \frac{N}{1 + Ab^{-x}}\, dx = \frac{N\ln(A + b^x)}{\ln b} + C$$

c. The number of books sold online each year in the United States in the period 1997–2000 can be modeled by

$$N(t) = \frac{82.8}{1 + 21.8(7.14)^{-t}} \text{ million books per year}$$

$$(0 \le t \le 4)$$

($t = 0$ represents January 1997). Use the result of part (b) to estimate the total online U.S. book sales for the years 1997 through 2000. (Round your answer to the nearest 1 million books.)

Based on a logistic regression model. Source: Ipsos-NPD Book Trends/*New York Times*, April 16, 2001, p. C1.

90. The Logistic Function and Education
a. Show that the logistic function $f(x) = N/(1 + Ae^{-kx})$ can be written in the form

$$f(x) = \frac{Ne^{kx}}{A + e^{kx}}$$

b. Use the result of part (a) and a suitable substitution to show that

$$\int \frac{N}{1 + Ae^{-kx}}\, dx = \frac{N\ln(A + e^{kx})}{k} + C$$

c. The number of high school graduates in the United States over the period 1993–1999 can be modeled by

$$N(t) = \frac{5}{1 + 1.081e^{-0.055t}} \text{ million graduates per year}$$

$$(0 \le t \le 7)$$

($t = 0$ represents January 1993). Use the result of part (b) to estimate the total number of high school graduates in the United States in the years 1993 through 1999. (Round your answer to the nearest 1 million graduates.)

Based on a logistic regression model. Source: Western Interstate Commission for Higher Education and the College Board/*New York Times*, February 17, 1999, p. B9.

91. Kinetic Energy The work done in accelerating an object from velocity v_0 to velocity v_1 is given by

$$W = \int_{v_0}^{v_1} v\, \frac{dp}{dv}\, dv$$

where p is its momentum, given by $p = mv$ ($m = $ mass). Assuming that m is a constant, show that

$$W = \frac{1}{2}mv_1^2 - \frac{1}{2}mv_0^2$$

The quantity $\frac{1}{2}mv^2$ is referred to as the **kinetic energy** of the object, so the work required to accelerate an object is given by its change in kinetic energy.

92. Einstein's Energy Equation According to the special theory of relativity, the apparent mass of an object depends on its velocity according to the formula

$$m = \frac{m_0}{(1 - v^2/c^2)^{1/2}}$$

where v is its velocity, m_0 is the *rest mass* of the object (that is, its mass when $v = 0$), and c is the velocity of light: approximately 3×10^8 meters per second.

a. Show that, if $p = mv$ is the momentum,

$$\frac{dp}{dv} = \frac{m_0}{(1 - v^2/c^2)^{3/2}}$$

b. Use the integral formula for W in Exercise 91, together with the result in part (a), to show that the work required to accelerate an object from a velocity of v_0 to v_1 is given by

$$W = \frac{m_0 c^2}{\sqrt{1 - v_1^2/c^2}} - \frac{m_0 c^2}{\sqrt{1 - v_0^2/c^2}}$$

We call the quantity $(m_0 c^2)/\sqrt{1 - (v^2/c^2)}$ the *total relativistic energy* of an object moving at velocity v. Thus, the work to accelerate an object from one velocity to another is given by the change in its total relativistic energy.

c. Deduce (as Albert Einstein did) that the total relativistic energy E of a body at rest with rest mass m is given by the famous equation

$$E = mc^2$$

93. The Natural Logarithm Returns Suppose you had never heard of the natural logarithmic function and were therefore stuck when you tried to find an antiderivative of $1/x$. Here is how you might proceed.

a. What does part 1 of the FTC give as an antiderivative of $1/x$?

b. Choosing $a = 1$, give this "new" function the name $M(x)$ and derive the following properties: $M(1) = 0$; $M'(x) = 1/x$.

c. Use the chain rule for derivatives to show that if a is any positive constant, then

$$\frac{d}{dx}[M(ax)] = \frac{1}{x}$$

d. Let

$$F(x) = M(ax) - M(x)$$

Deduce that $F'(x) = 0$ and hence that $F(x) = constant$.

e. By setting $x = 1$, find the value of this constant and hence deduce that

$$M(ax) = M(a) + M(x)$$

 94. The Error Function The **error function**, erf(x), is defined by

$$\text{erf}(x) = \frac{2}{\sqrt{\pi}} \int_0^x e^{-t^2}\, dt$$

This function is very important in statistics.

a. Find erf$'(x)$.

b. Use Riemann sums with $n = 100$ to approximate erf(1) and erf(2).

c. Find an antiderivative of e^{-x^2} in terms of erf(x).

d. Use the answers to parts (b) and (c) to approximate

$$\int_1^2 e^{-x^2}\, dx$$

COMMUNICATION AND REASONING EXERCISES

95. Explain how the indefinite integral and the definite integral are related.

96. Give an example of a nonzero function whose definite integral is zero.

97. Give a nontrivial example of a velocity function that will produce a displacement of 0 from time $t = 0$ to time $t = 10$.

98. Complete the following: The total sales from time a to time b are obtained from the marginal sales by taking its _____ _____ from _____ to _____.

99. Give an example of a decreasing function $f(x)$ with the property that $\int_a^b f(x)\, dx$ is positive for every choice of a and $b > a$.

100. Explain why, in computing the total change of a quantity from its rate of change, it is useful to have the definite integral subtract area below the x axis.

101. What does the FTC permit one to do?

102. Your friend has just told you that the function $f(x) = e^{-x^2}$ can't be integrated and hence has no antiderivative. Is your friend correct? Explain your answer.

103. Use the FTC to find an antiderivative F of

$$f(x) = \begin{cases} 0 & \text{if } x < 0 \\ 1 & \text{if } x \geq 0 \end{cases}$$

Is it true that $F'(x) = f(x)$ for *all* x?

104. According to the FTC as stated in this text, which functions are guaranteed to have antiderivatives? What does the answer to Exercise 103 tell you about the theorem as stated in this text?

CASE STUDY

Reproduced from the U.S. Department of Labor Web site

Wage Inflation

You are the assistant personnel manager at ABC Development Enterprises, a large corporation, and yesterday you received the following memo from the personnel manager.

To: SW
From: SC
Subject: Cost of labor

Yesterday, the CEO asked me to find some mathematical formulas to estimate (1) the trend in annual wage increases and (2) the average annual wage of assembly-line workers from the time they join the company to the time they retire. (She needs the information for next week's stockholder meeting.) So far, I have had very little luck: All I have been able to find is a table giving annual percent wage increases (attached). Also, I know that the average wage for an assembly-line worker in 1981 was $25,000 per year. Do you have any ideas?

Attachment[11]

Date	'81	'82	'83	'84	'85	'86	'87	'88
Annual Change (%)	9.3	8.0	5.3	5.0	4.2	4.0	3.2	3.4

Date	'89	'90	'91	'92	'93	'94	'95
Annual Change (%)	4.2	4.2	4.0	3.4	2.8	3.0	2.8

(So, for example, the wages increased 9.3% from 1981 to 1982.)

Getting to work, you decide that the first thing to do is fit these data to a mathematical curve (a *trendline*) that you can use to project future changes in wages. You graph the data to get a sense of what mathematical models might be appropriate (Figure 22).

Figure 22 • *Annual change in wages*

---

[11]The data show approximate year-to-year percent change in U.S. wages. Source: DataStream/*New York Times,* August 13, 1995, p. 26.

The graph suggests a decreasing trend, leveling off at about 3%. You recall that a variety of curves behave this way. One of the simplest is the curve

$$y = \frac{a}{t} + b \qquad (t \geq 1)$$

where a and b are constants (Figure 23).[12]

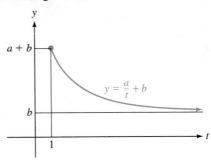

Figure 23

You let $t = 1$ correspond to the time of the first data point, 1981, and convert all percentages to decimals, giving the following table of data:

t	1	2	3	4	5	6	7	8
y	0.093	0.080	0.053	0.050	0.042	0.040	0.032	0.034

t	9	10	11	12	13	14	15
y	0.042	0.042	0.040	0.034	0.028	0.030	0.028

You then find the values of a and b that best fit the given data.[13] This gives you the following model for wage inflation (with figures rounded to five significant digits):

$$y = \frac{0.071\,813}{t} + 0.028\,647$$

(It is interesting that the model predicts wage inflation leveling off to about 2.86%.) Figure 24 shows the graph of y superimposed on the data.

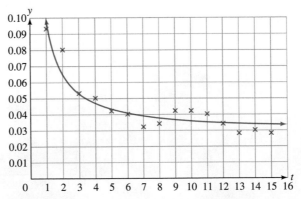

Figure 24

---

[12]There is a good mathematical reason for choosing a curve of this form: It is a first approximation (for $t \geq 1$) to a general rational function that approaches a constant as $t \to +\infty$.

[13]To do this you note that y is a linear function of $1/t$, namely, $y = a(1/t) + b$. Then you run a linear regression on the data $(1/t, y)$.

Now that you have a model for wage inflation, you must use it to find the annual wage. First, you realize that the model gives the *fractional rate of increase* of wages (since it is specified as a percentage, or fraction, of the total wage). In other words, if $w(t)$ represents a worker's annual wage at time t, Then

$$y = \frac{dw/dt}{w} = \frac{d}{dt}\,[\ln w] \qquad \text{By the chain rule for derivatives}$$

You find an equation for a worker's annual wage at time t by solving for w:

$$\ln w = \int y\, dt = \int \left(\frac{a}{t} + b\right) dt = a \ln t + bt + C$$

where a and b are as above and C is the constant of integration, so

$$w = e^{a \ln t + bt + C}$$

To compute C you substitute the initial data from the memo: $w(1) = 25{,}000$. Thus,

$$25{,}000 = e^{a \ln 1 + b + C} = e^{b+C} = e^{0.028647 + C}$$

Thus,

$$\ln(25{,}000) = 0.028\,647 + C$$

which gives

$$C = \ln(25{,}000) - 0.028\,647 \approx 10.098 \qquad \text{To five significant digits}$$

Now you can write down the following formula for the annual wage of an assembly-line worker as a function of t, the number of years since 1980:

$$w(t) = e^{a\ln t + bt + C} = e^{0.071813\ln t + 0.028647t + 10.098}$$

$$= e^{0.071813\ln t}e^{0.028647t}e^{10.098}$$

$$= t^{0.071813}e^{0.028647t}e^{10.098}$$

What remains is the calculation of the average annual wage. The average is the total wage earned over the worker's career divided by the number of years worked:

$$\overline{w} = \frac{1}{s-r}\int_r^s w(t)\, dt$$

where r is the time an employee begins working at the company and s is the time he or she retires. Substituting the formula for $w(t)$ gives

$$\overline{w} = \frac{1}{s-r}\int_r^s t^{0.071813}e^{0.028647t}e^{10.098}\, dt = \frac{e^{10.098}}{s-r}\int_r^s t^{0.071813}e^{0.028647t}\, dt$$

You cannot find an explicit antiderivative for the integrand, so you decide that the only way to compute it is numerically. You send the following memo to SC.

To: SC
From: SW
Subject: The formula you wanted

The average annual salary of an assembly-line worker here at ABC Development Enterprises is given by the formula

$$\overline{w} = \frac{e^{10.098}}{s - r} \int_r^s t^{0.071813} e^{0.028647t} \, dt$$

where r is the time in years after 1980 that a worker joins ABC and s is the time (in years after 1980) the worker retires. (The formula is valid only from 1981 on.) To calculate it easily (and impress the board members), I suggest you enter the following on your graphing calculator:

```
Y₁=(e^(10.098)/(S-R))fnInt(T^0.071813e^(0.028647T),T,R,S)
```

Then suppose, for example, a worker joined the company in 1983 ($r = 3$) and retired in 2001 ($s = 21$). All you do is enter

$3 \rightarrow R$

$21 \rightarrow S$

Y_1

and your calculator will give you the result: The average salary of the worker is $41,307.16.

Have a nice day.

SW

EXERCISES

1. Use the model developed above to compute the average annual income of a worker who joined the company in 1998 and left 3 years later.

2. What was the total amount paid by ABC Enterprises to the worker in Exercise 1?

3. What (if any) advantages are there to using a model for the annual wage inflation rate when the actual annual wage inflation rates are available?

4. The formula in the model was based on a 1981 salary of $25,000. Change the model to allow for an arbitrary 1981 salary of $$w_0$.

5. If we had used exponential regression to model the wage inflation data, we would have obtained

$$y = 0.071\,42e^{-0.067135t}$$

Graph this equation along with the actual wage data and the earlier model for y. Is this a better or worse model?

6. Use the actual data in the table to calculate the average salary of an assembly-line worker for the 6-year period from 1981 through the end of 1986 and compare it with the figure predicted by the model in the text.

CHAPTER 6 REVIEW TEST

1. Evaluate the following indefinite integrals.

a. $\int (x^2 - 10x + 2)\, dx$ **b.** $\int (e^x + \sqrt{x})\, dx$

c. $\int x(x^2 + 4)^{10}\, dx$ **d.** $\int \dfrac{x^2 + 1}{(x^3 + 3x + 2)^2}\, dx$

e. $\int 5e^{-2x}\, dx$ **f.** $\int xe^{-x^2/2}\, dx$

2. Evaluate the following definite integrals, using the Fundamental Theorem of Calculus.

a. $\int_0^1 (x - x^3)\, dx$ **b.** $\int_0^9 \dfrac{1}{x + 1}\, dx$

c. $\int_0^2 x^2 \sqrt{x^3 + 1}\, dx$ **d.** $\int_{-1}^1 3^{2x-2}\, dx$

3. Find the areas of the following regions. Do not count area below the x axis as negative.

a. The area bounded by $y = 4 - x^2$, the x axis, and the lines $x = -2$ and $x = 2$.

b. The area bounded by $y = 4 - x^2$, the x axis, and the lines $x = 0$ and $x = 5$.

c. The area bounded by $y = xe^{-x^2}$, the x axis, and the lines $x = 0$ and $x = 5$.

d. The area bounded by $y = |2x|$, the x axis, and the lines $x = -1$ and $x = 1$.

 4. Approximate the following definite integrals, using Riemann sums with $n = 10$, 100, and 1000.

a. $\int_0^1 e^{-x^2}\, dx$

b. $\int_1^3 x^{-x}\, dx$

OHaganBooks.com—ESTIMATING TOTALS

5. If OHaganBooks.com were to give away its latest best-seller, *A River Burns Through It*, the demand would be 100,000 books. The marginal demand for the book is $-20p$ at a price of p dollars.

a. What is the demand function for this book?

b. At what price does demand drop to zero?

6. When OHaganBooks.com was about to go online, it estimated that its weekly sales would begin at about 6400 books per week, with sales increasing at such a rate that weekly sales would double about every 2 weeks.

a. If these estimates had been correct, how many books would the company have sold in the first 5 weeks?

b. In fact, OHaganBooks.com modeled its weekly sales over a period of time after it went online with the function

$$s(t) = 6053 + \frac{4474e^{0.55t}}{e^{0.55t} + 14.01}$$

where t is the time in weeks after it went online. According to this model, how many books did it sell in the first 5 weeks?

7. An overworked employee at OHaganBooks.com goes to the top of the company's 100-foot-tall headquarters building and flings a book up into the air at a speed of 60 feet/second.

a. When will the book hit the ground 100 feet below? (Neglect air resistance.)

b. How fast will it be traveling when it hits the ground?

c. How high will the book go?

ADDITIONAL ONLINE REVIEW

If you follow the path

    Web Site → Everything for Calculus → Chapter 6

you will find the following additional resources to help you review:

A comprehensive chapter summary (including examples and interactive features)

Online section-by-section interactive tutorials

Section-by-section Excel tutorials

Additional review exercises (including interactive exercises and many with help)

A true/false chapter quiz

Web-based and Excel-based integration utilities

7

FURTHER INTEGRATION TECHNIQUES

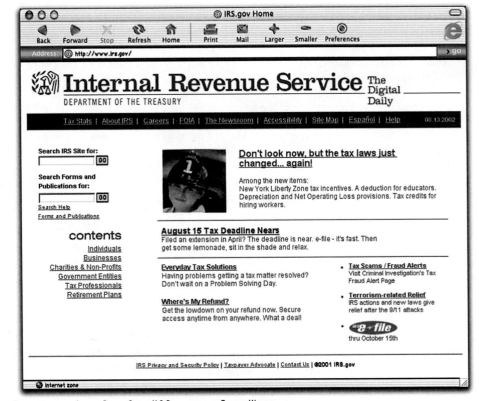

REPRODUCED FROM THE INTERNAL REVENUE SERVICE, U.S. DEPARTMENT OF THE TREASURY WEB SITE

INTERNET RESOURCES FOR THIS CHAPTER

At the Web site, follow the path
   Web Site → Everything for Calculus → Chapter 7
where you will find a detailed chapter summary you can print out, a true/false quiz, and a collection of sample test questions. You will also find an online numerical integration utility and other useful resources.

AND APPLICATIONS OF THE INTEGRAL

Introduction

In Chapter 6 we learned how to compute many integrals and saw some of the applications of the integral. In this chapter we look at some further techniques for computing integrals and then at more applications of the integral. We also see how to extend the definition of the definite integral to include integrals over infinite intervals and show how such integrals can be used for long-term forecasting. Finally, we introduce the beautiful theory of differential equations and some of its numerous applications.

7.1 Integration by Parts

Integration by parts is an integration technique that comes from the product rule for derivatives. The tabular method we present here has been around for some time and makes integration by parts quite simple, particularly in problems in which it has to be used several times. This particular version of the tabular method was developed and taught to us by Dan Rosen at Hofstra University.

We start with a little notation to simplify things while we introduce integration by parts (we use this notation only in the next few pages). If $f(x)$ is a function, denote its derivative by $D(f(x))$ and an antiderivative by $I(f(x))$. For example, if $f(x) = 2x^2$, then

$$D(f(x)) = 4x \quad \text{and} \quad I(f(x)) = \frac{2x^3}{3}$$

[If we wish, we can instead take $I(f(x)) = 2x^3/3 + 46$, but we opt to take the simplest antiderivative that we can.]

Integration by Parts

$$\int f(x)g(x)\,dx = f(x)I(g(x)) - \int D(f(x))I(g(x))\,dx$$

Quick Example
(Discussed more fully in Example 1)

$$\int x \cdot e^x\,dx = xI(e^x) - \int D(x)I(e^x)\,dx$$

$$= xe^x - \int 1 \cdot e^x\,dx \qquad\qquad I(e^x) = e^x; D(x) = 1$$

$$= xe^x - e^x + C \qquad\qquad \int e^x\,dx = e^x + C$$

As the Quick Example shows, although we could not immediately integrate $f(x)g(x) = x \cdot e^x$, we could easily integrate $D(f(x))I(g(x)) = 1 \cdot e^x = e^x$.

409

Question Where does the integration-by-parts formula come from?

Answer As we mentioned, it comes from the product rule for derivatives. We apply the product rule to the function $f(x)I(g(x))$:

$$D[f(x)I(g(x))] = D(f(x))I(g(x)) + f(x)D(I(g(x))) = D(f(x))I(g(x)) + f(x)g(x)$$

because $D(I(g(x)))$ is the derivative of the antiderivative of $g(x)$, which is $g(x)$. Integrating both sides gives

$$f(x)I(g(x)) = \int D(f(x))I(g(x)) \, dx + \int f(x)g(x) \, dx$$

A simple rearrangement of the terms gives us the integration-by-parts formula.

The integration-by-parts formula is easiest to use via the tabular method illustrated in Example 1, where we repeat the calculation we did above.

Example 1 • Integration by Parts

Calculate $\int xe^x \, dx$.

Solution First, the reason we *need* to use integration by parts to evaluate this integral is that none of the other techniques of integration that we've talked about up to now will help us. In particular, we cannot simply find antiderivatives of x and e^x and multiply them together. [You should check that $(x^2/2)e^x$ is *not* an antiderivative for xe^x.] However, as we saw above, this integral can be found by integration by parts. We want to find the integral of the *product* of x and e^x. We must make a decision: Which function will play the role of f and which will play the role of g in the integration-by-parts formula? Since the derivative of x is just 1, differentiating makes it simpler, so we try letting x be $f(x)$ and letting e^x be $g(x)$. We need to calculate $D(f(x))$ and $I(g(x))$, which we record in the following table:

	D	I
$+$	x	e^x
$-\int$	1 $\rightarrow$	e^x

$+x \cdot e^x - \int 1 \cdot e^x \, dx$

Below x in the D column, we put $D(x) = 1$; below e^x in the I column, we put $I(e^x) = e^x$. The arrow at an angle connecting x and $I(e^x)$ reminds us that the product $xI(e^x)$ appears in the answer; the plus sign on the left of the table reminds us that it is $+xI(e^x)$ that appears. The integral sign and the horizontal arrow connecting $D(x)$ and $I(e^x)$ remind us that the *integral* of the product $D(x)I(e^x)$ also appears in the answer; the minus sign on the left reminds us that we need to subtract this integral. Combining these two contributions, we get

$$\int xe^x \, dx = xe^x - \int e^x \, dx$$

The integral that appears on the right is much easier than the one we began with, so we can complete the problem:

$$\int xe^x \, dx = xe^x - \int e^x \, dx = xe^x - e^x + C$$

✳ ***Before we go on . . .*** What if we had made the opposite decision and put e^x in the D column and x in the I column? Then we would have had the following table:

	D	I
$+$	e^x	x
$-\int$	$e^x \quad \rightarrow$	$\dfrac{x^2}{2}$

This gives

$$\int xe^x \, dx = \frac{x^2}{2} e^x - \int \frac{x^2}{2} e^x \, dx$$

The integral on the right is harder than the one we started with, not easier! How do we know beforehand which way to go? We don't. We have to be willing to do a little trial and error. We try it one way, and if it doesn't make things simpler, we try it another way. *Remember, though, that the function we put in the I column must be one that we can integrate.*

Example 2 • Repeated Integration by Parts

Calculate $\int x^2 \, e^{-x} \, dx$.

Solution Again, we have a product—the integrand is the product of x^2 and e^{-x}. Since differentiating x^2 makes it simpler, we put it in the D column and get the following table:

	D	I
$+$	x^2	e^{-x}
$-\int$	$2x \quad \rightarrow$	$-e^{-x}$

This table gives us

$$\int x^2 \, e^{-x} \, dx = x^2(-e^{-x}) - \int 2x(-e^{-x}) \, dx$$

The last integral is simpler than the one we started with, but it still involves a product. It's a good candidate for another integration by parts. The table we would use would start with $2x$ in the D column and $-e^{-x}$ in the I column, which is exactly what we see in the last row of the table we've already made. Therefore, we *continue the process*, elongating the table above:

	D	I
$+$	x^2	e^{-x}
$-$	$2x$	$-e^{-x}$
$+\int$	$2 \quad \rightarrow$	e^{-x}

Notice how the signs on the left alternate. Here's why: To compute $\int 2x(-e^{-x})\,dx$, we would use the following table:

	D	**I**
$+$	$2x$	$-e^{-x}$
$-\int$	2	$\quad\searrow$ $\rightarrow\ e^{-x}$

However, we really need to compute its negative, $-\int 2x(-e^{-x})\,dx$, so we reverse all the signs.

Now, we still have to compute an integral (the integral of the product of the functions in the bottom row) to complete the computation. But why stop here? Let's continue the process one more step:

	D	**I**
$+$	x^2	e^{-x}
$-$	$2x$	$-e^{-x}$
$+$	2	e^{-x}
$-\int$	0	$\rightarrow\ -e^{-x}$

In the bottom line, we see that all that is left to integrate is $0(-e^{-x}) = 0$. Since the indefinite integral of 0 is C, we can read the answer from the table as

$$\int x^2 e^{-x}\,dx = x^2(-e^{-x}) - 2x(e^{-x}) + 2(-e^{-x}) + C$$

$$= -x^2 e^{-x} - 2xe^{-x} - 2e^{-x} + C = -e^{-x}(x^2 + 2x + 2) + C$$

Before we go on . . . Since that took several steps, let's check our work:

$$\frac{d}{dx}[-e^{-x}(x^2 + 2x + 2) + C] = e^{-x}(x^2 + 2x + 2) - e^{-x}(2x + 2) = x^2 e^{-x} \quad \checkmark$$

In Example 2 we saw a technique that we can summarize as follows.

Integrating a Polynomial Times a Function

If one of the factors in the integrand is a polynomial and the other factor is a function that can be integrated repeatedly, put the polynomial in the D column and keep differentiating until you get zero. Then complete the I column to the same depth and read off the answer.

For practice, redo Example 1, using this technique.

It is not always the case that the integrand is a polynomial times something easy to integrate, so we can't always expect to end up with a zero in the D column. In that case we hope that at some point we will be able to integrate the product of the functions in the last row. Here are some examples.

Example 3 • Polynomial Times a Logarithm

Calculate (a) $\int x \ln x \, dx$, (b) $\int (x^2 - x)\ln x \, dx$, and (c) $\int \ln x \, dx$.

Solution

a. This is a product and therefore a good candidate for integration by parts. Our first impulse is to differentiate x, but that would mean integrating $\ln x$, and we do not (yet) know how to do that. So we try it the other way around and hope for the best.

	D		*I*
+	$\ln x$		x
		$\searrow$	
$-\int$	$\dfrac{1}{x}$	$\rightarrow$	$\dfrac{x^2}{2}$

Why did we stop? If we continued the table, both columns would get more complicated. However, if we stop here, we get

$$\int x \ln x \, dx = (\ln x)\left(\frac{x^2}{2}\right) - \int\left(\frac{1}{x}\right)\left(\frac{x^2}{2}\right) dx = \frac{x^2}{2} \ln x - \frac{1}{2}\int x \, dx = \frac{x^2}{2} \ln x - \frac{x^2}{4} + C$$

b. We can use the same technique we used in part (a) to integrate any polynomial times the logarithm of x:

	D		*I*
+	$\ln x$		$x^2 - x$
		$\searrow$	
$-\int$	$\dfrac{1}{x}$	$\rightarrow$	$\dfrac{x^3}{3} - \dfrac{x^2}{2}$

$$\int (x^2 - x) \ln x \, dx = (\ln x)\left(\frac{x^3}{3} - \frac{x^2}{2}\right) - \int\left(\frac{1}{x}\right)\left(\frac{x^3}{3} - \frac{x^2}{2}\right) dx$$

$$= \left(\frac{x^3}{3} - \frac{x^2}{2}\right)\ln x - \int\left(\frac{x^2}{3} - \frac{x}{2}\right) dx = \left(\frac{x^3}{3} - \frac{x^2}{2}\right)\ln x - \frac{x^3}{9} + \frac{x^2}{4} + C$$

c. The integrand $\ln x$ does not look like a product. (It is the natural logarithm *of x, not* "the natural logarithm times *x*.") We can, however, *make* it into a product by thinking of it as $1 \cdot \ln x$. Since this is a polynomial times $\ln x$, we proceed as in parts (a) and (b):

	D		*I*
+	$\ln x$		1
		$\searrow$	
$-\int$	$\dfrac{1}{x}$	$\rightarrow$	x

The product of $1/x$ and x is just 1, and we certainly know how to integrate that. Thus, we have

$$\int \ln x \, dx = x \ln x - \int\left(\frac{1}{x}\right) x \, dx = x \ln x - \int 1 \, dx = x \ln x - x + C$$

Now we know what the integral of $\ln x$ is!

> ### Guideline: Whether to Use Integration by Parts and What Goes in the D and I Columns
>
> **Question** Will integration by parts always work to integrate a product?
>
> **Answer** No. Integration by parts often works for products in which one factor is a polynomial, but it will almost *never* work in the examples of products we saw when discussing substitution in Section 6.2. For example, although integration by parts can be used to compute $\int (x^2 - x)e^{2x-1}\, dx$ (put $x^2 - x$ in the D column and e^{2x-1} in the I column), it can*not* be used to compute $\int (2x - 1)e^{x^2-x}\, dx$ (put $u = x^2 - x$). Recognizing when to use integration by parts is best learned by experience.
>
> **Question** When using integration by parts, which expressions go in the D column and which in the I column?
>
> **Answer** Although there is no general rule, the following guidelines are useful:
>
> - To integrate a product in which one factor is a polynomial and the other can be integrated several times, put the polynomial in the D column and the other factor in the I column. Then differentiate the polynomial until you get zero.
>
> - If one of the factors is a polynomial but the other factor cannot be integrated easily, put the polynomial in the I column and the other factor in the D column. Stop when the product of the functions in the bottom row can be integrated.
>
> - If neither factor is a polynomial, put the factor that seems easiest to integrate in the I column and the other factor in the D column. Again, stop the table as soon as the product of the functions in the bottom row can be integrated.
>
> - If your method doesn't work, try switching the functions in the D and I columns or try breaking the integrand into a product in a different way. If none of this works, maybe integration by parts isn't the technique to use on this problem.

7.1 EXERCISES

In Exercises 1–38, evaluate the integrals.

1. $\int 2xe^x\, dx$

2. $\int 3xe^{-x}\, dx$

3. $\int (3x - 1)e^{-x}\, dx$

4. $\int (1 - x)e^x\, dx$

5. $\int (x^2 - 1)e^{2x}\, dx$

6. $\int (x^2 + 1)e^{-2x}\, dx$

7. $\int (x^2 + 1)e^{-2x+4}\, dx$

8. $\int (x^2 + 1)e^{3x+1}\, dx$

9. $\int (2 - x)2^x\, dx$

10. $\int (3x - 2)4^x\, dx$

11. $\int (x^2 - 1)3^{-x}\, dx$

12. $\int (1 - x^2)2^{-x}\, dx$

13. $\int \dfrac{x^2 - x}{e^x}\, dx$

14. $\int \dfrac{2x + 1}{e^{3x}}\, dx$

15. $\int x(x + 2)^6\, dx$

16. $\int x^2(x - 1)^6\, dx$

17. $\int \dfrac{x}{(x - 2)^3}\, dx$

18. $\int \dfrac{x}{(x - 1)^2}\, dx$

19. $\int x^3 \ln x\, dx$

20. $\int x^2 \ln x\, dx$

21. $\int (t^2 + 1)\ln(2t)\, dt$

22. $\int (t^2 - t)\ln(-t)\, dt$

23. $\int t^{1/3} \ln t\, dt$

24. $\int t^{-1/2} \ln t\, dt$

25. $\int \log_3 x\, dx$

26. $\int x \log_2 x\, dx$

27. $\int (xe^{2x} - 4e^{3x})\, dx$

28. $\int (x^2 e^{-x} + 2e^{-x+1})\, dx$

29. $\int (x^2 e^x - xe^{x2})\, dx$

30. $\int [(2x + 1)e^{x^2+x} - x^2 e^{2x+1}]\, dx$

31. $\displaystyle\int_0^1 (x + 1)e^x\, dx$

32. $\displaystyle\int_{-1}^1 (x^2 + x)e^{-x}\, dx$

33. $\displaystyle\int_0^1 x^2(x + 1)^{10}\, dx$

34. $\displaystyle\int_0^1 x^3(x + 1)^{10}\, dx$

35. $\displaystyle\int_1^2 x \ln(2x)\, dx$

36. $\displaystyle\int_1^2 x^2 \ln(3x)\, dx$

37. $\displaystyle\int_0^1 x \ln(x + 1)\, dx$

38. $\displaystyle\int_0^1 x^2 \ln(x + 1)\, dx$

39. Find the area bounded by the curve $y = xe^{-x}$, the x axis, and the lines $x = 0$ and $x = 10$.

40. Find the area bounded by the curve $y = x \ln x$, the x axis, and the lines $x = 1$ and $x = e$.

41. Find the area bounded by the curve $y = (x + 1)\ln x$, the x axis, and the lines $x = 1$ and $x = 2$.

42. Find the area bounded by the curve $y = (x - 1)e^x$, the x axis, and the lines $x = 0$ and $x = 2$.

APPLICATIONS

43. Displacement A rocket rising from the ground has a velocity of $2000te^{-t/120}$ feet/second, after t seconds. How far does it rise in the first 2 minutes?

44. Sales Weekly sales of graphing calculators can be modeled by the equation

$$s(t) = 10 - te^{-t/20}$$

where s is the number of calculators sold each week after t weeks. How many graphing calculators (to the nearest unit) will be sold in the first 20 weeks?

45. Total Cost The marginal cost of the xth box of light bulbs is $10 + [\ln(x + 1)]/(x + 1)^2$, and the fixed cost is $5000. Find the total cost to make x boxes of bulbs.

46. Total Revenue The marginal revenue for selling the xth box of light bulbs is $10 + 0.001x^2e^{-x/100}$. Find the total revenue generated by selling 200 boxes of bulbs.

47. Revenue You have been raising the price of your *Lord of the Rings* T-shirts by 50¢ per week, and sales have been falling continuously at a rate of 2% per week. Assuming you are now selling 50 T-shirts each week and charging $10 per T-shirt, how much revenue will you generate during the coming year? (Round your answer to the nearest dollar.) (*Hint:* Weekly revenue = weekly sales × price per T-shirt.)

48. Revenue Luckily, sales of your *Star Wars* T-shirts are now 50 T-shirts each week and increasing continuously at a rate of 5% per week. You are now charging $10 per T-shirt and are decreasing the price by 50¢ per week. How much revenue will you generate during the next 6 weeks?

49. Revenue About 70.5 million magazines were sold in the United States in 1982, and sales have been falling by about 2.5 million each year since then. Assume that the average price of magazines was $2.00 in 1982 and that this price has been increasing continuously at 5% per year. How much revenue was generated by sales of magazines in the period 1982–1992? (Give your answer to the nearest $1 million.)

These figures were for sales of 123 leading magazines at newsstands. Source: *New York Times,* December 6, 1993, p. D6.

50. Lost Revenue In 1982 about 35 million magazines placed on newsstands in the United States were not sold. This number has increased by about 0.6 million each year. Assume that the average price of magazines was $2 in 1982 and that this price has been increasing continuously at 5% per year. How much revenue was lost through unsold magazines in the period 1982–1992? (Give your answer to the nearest $1 million.) (*Hint:* Annual lost revenue = annual lost sales × price per magazine.)

These figures were for sales of 123 leading magazines at newsstands. Source: *New York Times,* December 6, 1993, p. D6.

COMMUNICATION AND REASONING EXERCISES

51. Your friend Janice claims that integration by parts allows one to integrate any product of two functions. Prove her wrong by giving an example of a product of two functions that cannot be integrated using integration by parts.

52. Complete the following sentence in words: The integral of $f(x)g(x)$ is the first times the integral of the second minus the integral of ____.

53. If $p(x)$ is a polynomial of degree n and $f(x)$ is some function of x, how many times do we generally have to integrate $f(x)$ to compute $\int p(x)f(x)\, dx$?

54. Use integration by parts to show that $\int (\ln x)^2\, dx = x(\ln x)^2 - 2x \ln x + 2x + C$.

55. Hermite's Identity If $f(x)$ is a polynomial of degree n, show that

$$\int_0^b f(x)e^{-x}\, dx = F(0) - F(b)e^{-b}$$

where $F(x) = f(x) + f'(x) + f''(x) + \cdots + f^{(n)}(x)$ (this is the sum of f and all of its derivatives).

56. Write a formula similar to Hermite's identity for $\int_0^b f(x)e^x\, dx$ when $f(x)$ is a polynomial of degree n.

7.2   *Area between Two Curves and Applications*

As we saw in Chapter 6, we can use the definite integral to calculate the area between the graph of a function and the x axis. With only a little more work, we can use it to calculate the area between two graphs. Figure 1 shows the graphs of two functions, $f(x)$ and $g(x)$, with $f(x) \geq g(x)$ for every x in the interval $[a, b]$.

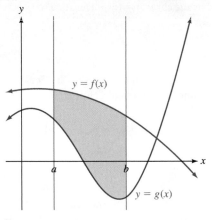

Figure 1

Question   How can we find the shaded area between the graphs of the two functions?

Answer   Pretty easily, actually, using the following formula.

Area between Two Graphs
If $f(x) \geq g(x)$ for all x in $[a, b]$, then the area of the region between the graphs of f and g and between $x = a$ and $x = b$ is given by

$$A = \int_a^b [f(x) - g(x)] \, dx$$

Let's look at an example and then discuss why the formula works.

Example 1 • The Area between Two Curves

Find the area of the region between $f(x) = -x^2 - 3x + 4$ and $g(x) = x^2 - 3x - 4$ and between $x = -1$ and $x = 1$.

Solution The area in question is shown in Figure 2.

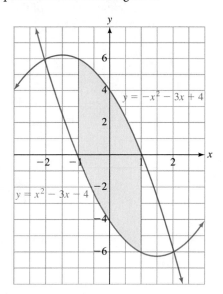

Figure 2

Since the graph of f lies above the graph of g in the interval $[-1, 1]$, we have $f(x) \geq g(x)$ for all x in $[-1, 1]$. Therefore, we can use the formula given above and calculate the area as follows:

$$A = \int_{-1}^{1} [f(x) - g(x)]\, dx = \int_{-1}^{1} (8 - 2x^2)\, dx = \left[8x - \frac{2}{3}x^3 \right]_{-1}^{1} = \frac{44}{3}$$

Question Why does the formula for the area between two curves work?

Answer Let's go back once again to the general case illustrated in Figure 1, where we were given two functions f and g with $f(x) \geq g(x)$ for every x in the interval $[a, b]$. We would like to avoid complicating the argument by the fact that the graph of g, or f, or both may dip below the x axis in the interval $[a, b]$ (as occurs in Figure 1 and also in Example 1). We can get around this issue as follows: Shift both graphs vertically upward by adding a big enough constant M to lift them both above the x axis in the interval $[a, b]$, as shown in Figure 3.

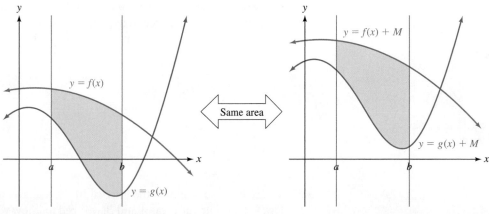

Figure 3

As the figure illustrates, the area of the region between the graphs is not affected, so we will calculate the area of the region shown on the right of Figure 3. That calculation is shown in Figure 4.

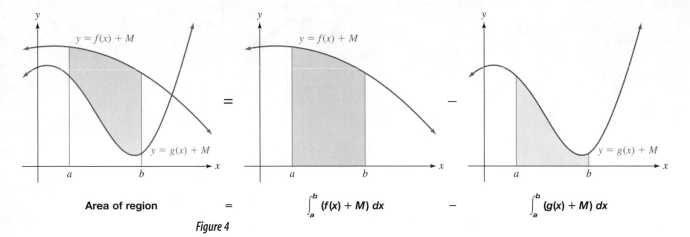

Area of region $\quad = \quad \displaystyle\int_a^b (f(x) + M)\, dx \quad - \quad \displaystyle\int_a^b (g(x) + M)\, dx$

Figure 4

From the figure, the area we want is

$$\int_a^b (f(x) + M)\, dx - \int_a^b (g(x) + M)\, dx = \int_a^b [(f(x) + M) - (g(x) + M)]\, dx$$
$$= \int_a^b [f(x) - g(x)]\, dx$$

which is the formula we gave originally.

So far, we've been assuming that $f(x) \geq g(x)$, so the graph of f never dips below the graph of g and so the graphs cannot cross (although they can touch). What if the two graphs *do* cross?

Example 2 • Regions Enclosed by Crossing Curves

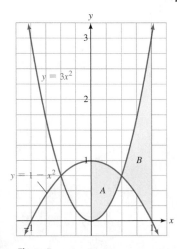

Figure 5

Find the area of the region between $y = 3x^2$ and $y = 1 - x^2$ and between $x = 0$ and $x = 1$.

Solution   The area we wish to calculate is shown in Figure 5. From the figure we can see that neither graph lies above the other over the whole interval. To get around this, we break the area into the two pieces on either side of the point at which the graphs cross and then compute each area separately. To do this we need to know exactly where that crossing point is. The crossing point is where $3x^2 = 1 - x^2$, so we solve for x:

$$3x^2 = 1 - x^2$$
$$4x^2 = 1$$
$$x^2 = \frac{1}{4}$$
$$x = \pm\frac{1}{2}$$

Since we are interested only in the interval $[0, 1]$, the crossing point is at $x = 1/2$.

Now, to compute the areas A and B, we need to know which graph is on top in each of these areas. We can see that from the figure, but what if the functions were more complicated and we could not easily draw the graphs? If we want to be sure, we can test the values of the two functions at some point in each region. But we really need not worry. If we make the wrong choice for the top function, the integral will yield the negative of the area (why?), so we can simply take the absolute value of the integral to get the area of the region in question. We have

$$A = \int_0^{1/2} [(1 - x^2) - 3x^2]\, dx = \int_0^{1/2} (1 - 4x^2)\, dx$$

$$= \left[x - \frac{4x^3}{3} \right]_0^{1/2} = \left(\frac{1}{2} - \frac{1}{6} \right) - (0 - 0) = \frac{1}{3}$$

and

$$B = \int_{1/2}^1 [3x^2 - (1 - x^2)]\, dx = \int_{1/2}^1 (4x^2 - 1)\, dx$$

$$= \left[\frac{4x^3}{3} - x \right]_{1/2}^1 = \left(\frac{4}{3} - 1 \right) - \left(\frac{1}{6} - \frac{1}{2} \right) = \frac{2}{3}$$

This gives a total area of $A + B = \dfrac{1}{3} + \dfrac{2}{3} = 1$.

✴ *Before we go on . . .* What would have happened if we had not broken the area into two pieces but just calculated the integral of the difference of the two functions? We would have calculated

$$\int_0^1 [(1 - x^2) - 3x^2]\, dx = \int_0^1 [1 - 4x^2]\, dx = \left[x - \frac{4x^3}{3} \right]_0^1 = -\frac{1}{3}$$

which is not even close to the right answer. What this integral calculated was actually $A - B$ rather than $A + B$. (Why?)

Example 3 • The Area Enclosed by Two Curves

Find the area enclosed by $y = x^2$ and $y = x^3$.

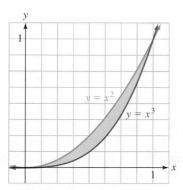

Figure 6

Solution This example has a new wrinkle: We are not told what interval to use for x. However, if we look at the graph in Figure 6, we see that the question can have only one meaning.

We are being asked to find the area of the shaded sliver, which is the only part of the picture that is actually *enclosed* by the two graphs. This sliver is bounded on either side by the two points where the graphs cross, so our first task is to find those points. They are the points where $x^2 = x^3$, so we solve for x:

$$x^2 = x^3$$

$$x^3 - x^2 = 0$$

$$x^2(x - 1) = 0$$

$$x = 0 \quad \text{or} \quad x = 1$$

Thus, we must integrate over the interval $[0, 1]$. Although we see from the diagram (or by substituting $x = \frac{1}{2}$) that the graph of $y = x^2$ is above that of $y = x^3$, let's pretend that we didn't notice that and calculate

$$\int_0^1 (x^3 - x^2)\, dx = \left[\frac{x^4}{4} - \frac{x^3}{3} \right]_0^1 = -\frac{1}{12}$$

This tells us that the required area is $\frac{1}{12}$ square unit and also that we had our integral reversed. Had we calculated $\int_0^1 (x^2 - x^3)\, dx$ instead, we would have found the correct answer, $\frac{1}{12}$.

We can summarize the procedure we used in Examples 2 and 3.

Finding the Area between the Graphs of f(x) and g(x)

1. Find all points of intersection by solving $f(x) = g(x)$ for x. This either determines the interval over which you will integrate or breaks up a given interval into regions between the intersection points.

2. Find the area of each region you found, by integrating the difference of the larger and the smaller function. (If you accidentally take the smaller minus the larger, the integral will give the negative of the area, so just take the absolute value.)

3. Add together the areas you found in step 2 to get the total area.

Consumers' Surplus

Consider a general demand curve presented, as is traditional in economics, as $p = D(q)$, where p is unit price and q is demand measured, say, in annual sales (Figure 7).

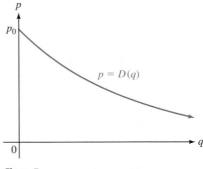

Figure 7

Thus, $D(q)$ is the price at which the demand will be q units per year. The price p_0 shown on the graph is the highest price that customers are willing to pay.

For a more concrete example, suppose the graph in Figure 7 is the demand curve for a particular new model of computer. When the computer first comes out and supplies are low (q is small), "early adopters" will be willing to pay a high price. This is the part of the graph on the left, near the p axis. As supplies increase and the price drops, more consumers will be willing to pay, and more computers will be sold. We can ask the following question.

Question How much are consumers willing to spend for the first $\bar{q}$ units?

Answer We can approximate consumers' willingness to spend as follows. We partition the interval $[0, \bar{q}]$ into n subintervals of equal length, as we did when discussing Riemann sums. Figure 8 shows a typical subinterval, $[q_{k-1}, q_k]$.

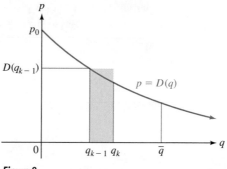

Figure 8

The price consumers are willing to pay for each of units q_{k-1} through q_k is approximately $D(q_{k-1})$, so the total that consumers are willing to spend for these units is approximately $D(q_{k-1})(q_k - q_{k-1}) = D(q_{k-1})\Delta q$, the area of the shaded region in Figure 8. Thus, the total amount that consumers are willing to spend for items 0 through $\bar{q}$ is

$$W \approx D(q_0)\Delta q + D(q_1)\Delta q + \cdots + D(q_{n-1})\Delta q = \sum_{k=0}^{n-1} D(q_k)\Delta q$$

which is a Riemann sum. The approximation becomes better the larger n becomes, and in the limit the Riemann sums converge to the integral

$$W = \int_0^{\bar{q}} D(q) \, dq$$

This is the total **consumers' willingness to spend** to buy the first $\bar{q}$ units. It is the area shaded in Figure 9.

Now suppose that the manufacturer simply sets the price at $\bar{p}$, with a corresponding demand of $\bar{q}$, so $D(\bar{q}) = \bar{p}$.

Question How much will consumers actually spend to buy these $\bar{q}$ items?

Answer That's easy: $\bar{p}\,\bar{q}$, the product of the unit price and the quantity sold. This is the **consumers' expenditure,** the area of the rectangle shown in Figure 10. Notice that we can write $\bar{p}\,\bar{q} = \int_0^{\bar{q}} \bar{p} \, dq$, as suggested by the figure.

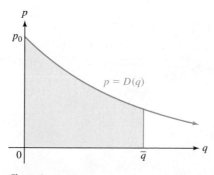

Figure 9

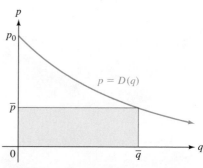

Figure 10

The difference between what consumers are willing to pay and what they actually pay is money in their pockets and is called the **consumers' surplus.**

Consumers' Surplus

If demand for an item is given by $p = D(q)$, the selling price is $\bar{p}$, and $\bar{q}$ is the corresponding demand [so that $D(\bar{q}) = \bar{p}$], then the **consumers' surplus** (CS) is the difference between their willingness to spend and their actual expenditure:

$$CS = \int_0^{\bar{q}} D(q) \, dq - \bar{p}\bar{q} = \int_0^{\bar{q}} (D(q) - \bar{p}) \, dq$$

Graphically, it is the area between the graphs of $p = D(q)$ and $p = \bar{p}$, as shown in the figure.

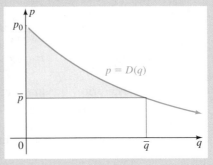

Example 4 • Consumers' Surplus

Your used-CD store has an exponential demand equation of the form

$$p = 15e^{-0.01q}$$

where q represents daily sales of used CDs and p is the price you charge per CD. Calculate the daily consumers' surplus if you sell your used CDs at $5 each.

Solution We are given $D(q) = 15e^{-0.01q}$ and $\bar{p} = 5$. We also need $\bar{q}$. By definition,

$$D(\bar{q}) = \bar{p} \quad \text{or} \quad 15e^{-0.01\bar{q}} = 5$$

which we must solve for $\bar{q}$:

$$e^{-0.01\bar{q}} = \frac{1}{3}$$

$$-0.01\bar{q} = \ln\left(\frac{1}{3}\right) = -\ln 3$$

$$\bar{q} = \frac{\ln 3}{0.01} \approx 109.8612$$

We now have

$$CS = \int_0^{\bar{q}} [D(q) - \bar{p}] \, dq$$

$$= \int_0^{109.8612} (15e^{-0.01q} - 5) \, dq$$

$$= \left[\frac{15}{-0.01} e^{-0.01q} - 5q \right]_0^{109.8612} \approx (-500 - 549.31) - (-1500 - 0)$$

$$= \$450.69 \text{ per day}$$

✳ ***Before we go on . . .*** We could have also calculated the consumers' willingness to spend and their actual expenditures. Their willingness to spend is

$$W = \int_0^{\bar{q}} D(q) \, dq$$

$$= \int_0^{109.8612} 15e^{-0.01q} \, dq$$

$$= \left[\frac{15}{-0.01} e^{-0.01q} \right]_0^{109.8612} \approx -500 - (-1500) \approx \$1000 \text{ per day}$$

Their actual expenditure is $\bar{p}\bar{q} = 5 \times 109.8612 = \$549.31$ per day. The difference is their surplus: $1000 - 549.31 = \$450.69$ per day.

Producers' Surplus

We can also calculate extra income earned by producers. Consider a supply equation of the form $p = S(q)$, where $S(q)$ is the price at which a supplier is willing to supply q items (per time period). Because a producer is generally willing to supply more units at a higher price per unit, a supply curve usually has a positive slope, as shown in Figure 11. The price p_0 is the lowest price that a producer is willing to charge.

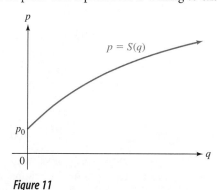

Figure 11

Question What is the minimum amount of money that producers are willing to receive in exchange for $\bar{q}$ items?

Answer Arguing as before, we say it is $\int_0^{\bar{q}} S(q) \, dq$.

Question If the producers charge $\bar{p}$ per item for $\bar{q}$ items, what is their total revenue?

Answer As before, the total revenue is $\bar{p}\bar{q} = \int_0^{\bar{q}} \bar{p} \, dq$.

The difference between the producers' total revenue and the minimum that they would have been willing to receive is the **producers' surplus.**

Producers' Surplus

The **producers' surplus** (PS) is the extra amount earned by producers who were willing to charge less than the selling price of $\bar{p}$ per unit and is given by

$$PS = \int_0^{\bar{q}} [\bar{p} - S(q)]\, dq$$

where $S(\bar{q}) = \bar{p}$. Graphically, it is the area of the region between the graphs of $p = \bar{p}$ and $p = S(q)$ for $0 \le q \le \bar{q}$, as in the figure.

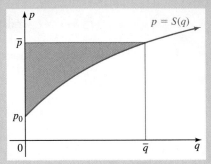

Example 5 • Producers' Surplus

My tie-dyed T-shirt enterprise has grown to the extent that I am now able to produce T-shirts in bulk, and several campus groups have begun placing orders. I have informed one group that I am prepared to supply $20\sqrt{p - 4}$ T-shirts at a price of p dollars per shirt. What is my total surplus if I sell T-shirts to the group at \$8 each?

Solution We need to calculate the producers' surplus when $\bar{p} = 8$. The supply equation is

$$q = 20\sqrt{p - 4}$$

but, to use the formula for producers' surplus, we need to express p as a function of q. First, we square both sides to remove the radical sign:

$$q^2 = 400(p - 4)$$

so

$$p - 4 = \frac{q^2}{400}$$

$$p = S(q) = \frac{q^2}{400} + 4$$

We now need the value of $\bar{q}$ corresponding to $\bar{p} = 8$. The easiest way to find it is to substitute $p = 8$ in the original equation, which gives

$$\bar{q} = 20\sqrt{8 - 4} = 20\sqrt{4} = 40$$

Thus,

$$PS = \int_0^{\bar{q}} [\bar{p} - S(q)]\, dq$$

$$= \int_0^{40} \left[8 - \left(\frac{q^2}{400} + 4 \right) \right] dq$$

$$= \int_0^{40} \left(4 - \frac{q^2}{400} \right) dq = \left[4q - \frac{q^3}{1200} \right]_0^{40} \approx \$106.67$$

✳ *Before we go on . . .* The amount I am willing to receive for these T-shirts is

$$\int_0^{40} S(q)\, dq = \int_0^{40} \left(\frac{q^2}{400} + 4 \right) dq = \left[\frac{q^3}{1200} + 4q \right]_0^{40} \approx \$213.33$$

By selling T-shirts for \$8 a piece, my total revenue will be $8 \times 40 = \$320$. The difference, $320 - 213.33 = \$106.67$, is my surplus.

Example 6 • Equilibrium

Continuing on with Example 5, a representative informs me that the campus group is prepared to order only $\sqrt{200(16 - p)}$ T-shirts at p dollars each. I would like to produce as many T-shirts for them as possible but avoid being left with unsold T-shirts. Given the supply curve from Example 5, what price should I charge per T-shirt, and what are the consumers' and producers' surpluses at that price?

Solution The price that guarantees neither a shortage nor a surplus of T-shirts is the **equilibrium price,** the price where supply equals demand. We have

Supply: $q = 20\sqrt{p - 4}$

Demand: $q = \sqrt{200(16 - p)}$

Equating these gives

$$20\sqrt{p - 4} = \sqrt{200(16 - p)}$$

$$400(p - 4) = 200(16 - p)$$

$$400p - 1600 = 3200 - 200p$$

$$600p = 4800$$

$$p = \$8 \text{ per T-shirt}$$

We therefore take $\bar{p} = 8$ (which happens to be the price we used in Example 5). We get the corresponding value for q by substituting $p = 8$ into either the demand or supply equation:

$$\bar{q} = 20\sqrt{8 - 4} = 40$$

Thus, $\bar{p} = 8$ and $\bar{q} = 40$.

We must now calculate the consumers' surplus and the producers' surplus. We calculated the producers' surplus for $\bar{p} = 8$ in Example 5:

$$PS = \$106.67$$

For the consumers' surplus, we must first express p as a function of q for the demand equation. Thus, we solve the demand equation for p as we did for the supply equation and we obtain

$$\text{Demand: } D(q) = 16 - \frac{q^2}{200}$$

Therefore,

$$CS = \int_0^{\bar{q}} [D(q) - \bar{p}] \, dq$$

$$= \int_0^{40} \left[\left(16 - \frac{q^2}{200} \right) - 8 \right] dq$$

$$= \int_0^{40} \left(8 - \frac{q^2}{200} \right) dq = \left[8q \frac{-q^3}{600} \right]_0^{40} \approx \$213.33$$

✳ ***Before we go on . . .*** Figure 12 shows both the consumers' surplus (top portion) and the producers' surplus (bottom portion).

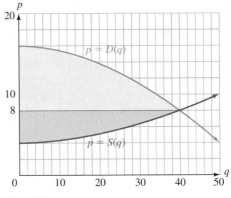

Figure 12

Since extra money in people's pockets is a Good Thing, the total of the consumers' and the producers' surpluses is called the **total social gain.** In this case it is

social gain = CS + PS = 213.33 + 106.67 = \$320.00

As you can see from Figure 12, the total social gain is also the area between two curves and equals

$$\int_0^{40} [D(q) - S(q)] \, dq$$

7.2 EXERCISES

In Exercises 1–24, find the areas of the indicated regions.

1. Between $y = x^2$ and $y = -1$ for x in $[-1, 1]$

2. Between $y = x^3$ and $y = -1$ for x in $[-1, 1]$

3. Between $y = -x$ and $y = x$ for x in $[0, 2]$

4. Between $y = -x$ and $y = x/2$ for x in $[0, 2]$

5. Between $y = x$ and $y = x^2$ for x in $[-1, 1]$

6. Between $y = x$ and $y = x^3$ for x in $[-1, 1]$

7. Between $y = e^x$ and $y = x$ for x in $[0, 1]$

8. Between $y = e^{-x}$ and $y = -x$ for x in $[0, 1]$

9. Between $y = (x - 1)^2$ and $y = -(x - 1)^2$ for x in $[0, 1]$

10. Between $y = x^2(x^3 + 1)^{10}$ and $y = -x(x^2 + 1)^{10}$ for x in $[0, 1]$

11. Enclosed by $y = x$ and $y = x^4$

12. Enclosed by $y = x$ and $y = -x^4$

13. Enclosed by $y = x^3$ and $y = x^4$

14. Enclosed by $y = x$ and $y = x^3$

15. Enclosed by $y = x^2$ and $y = x^4$

16. Enclosed by $y = x^4 - x^2$ and $y = x^2 - x^4$

17. Enclosed by $y = e^x$, $y = 2$, and the y axis

18. Enclosed by $y = e^{-x}$, $y = 3$, and the y axis

19. Enclosed by $y = \ln x$, $y = 2 - \ln x$, and $x = 4$

20. Enclosed by $y = \ln x$, $y = 1 - \ln x$, and $x = 4$

Ⓣ For Exercises 21–24, use technology to solve.

21. Enclosed by $y = e^x$ and $y = 2x + 1$

22. Enclosed by $y = e^x$ and $y = x + 2$

23. Enclosed by $y = \ln x$ and $y = \dfrac{x}{2} - \dfrac{1}{2}$

24. Enclosed by $y = \ln x$ and $y = x - 2$

In Exercises 25–36, calculate the consumers' surplus at the indicated unit price $\bar{p}$ for each demand equation.

25. $p = 10 - 2q; \bar{p} = 5$ **26.** $p = 100 - q; \bar{p} = 20$

27. $p = 100 - 3\sqrt{q}; \bar{p} = 76$ **28.** $p = 10 - 2q^{1/3}; \bar{p} = 6$

29. $p = 500e^{-2q}; \bar{p} = 100$ **30.** $p = 100 - e^{0.1q}; \bar{p} = 50$

31. $q = 100 - 2p; \bar{p} = 20$ **32.** $q = 50 - 3p; \bar{p} = 10$

33. $q = 100 - 0.25p^2; \bar{p} = 10$ **34.** $q = 20 - 0.05p^2; \bar{p} = 5$

35. $q = 500e^{-0.5p} - 50; \bar{p} = 1$ **36.** $q = 100 - e^{0.1p}; \bar{p} = 20$

In Exercises 37–48, calculate the producers' surplus at the indicated unit price $\bar{p}$ for each supply equation.

37. $p = 10 + 2q; \bar{p} = 20$ **38.** $p = 100 + q; \bar{p} = 200$

39. $p = 10 + 2q^{1/3}; \bar{p} = 12$ **40.** $p = 100 + 3\sqrt{q}; \bar{p} = 124$

41. $p = 500e^{0.5q}; \bar{p} = 1000$ **42.** $p = 100 + e^{0.01q}; \bar{p} = 120$

43. $q = 2p - 50; \bar{p} = 40$ **44.** $q = 4p - 1000; \bar{p} = 1000$

45. $q = 0.25p^2 - 10; \bar{p} = 10$ **46.** $q = 0.05p^2 - 20; \bar{p} = 50$

47. $q = 500e^{0.05p} - 50; \bar{p} = 10$ **48.** $q = 10(e^{0.1p} - 1); \bar{p} = 5$

APPLICATIONS

49. College Tuition A study of about 1800 U.S. colleges and universities resulted in the demand equation $q = 9859.39 - 2.17p$, where q is the enrollment at a college or university and p is the average annual tuition (plus fees) it charges. Officials at Enormous State University have developed a policy whereby the number of students it will accept each year at a tuition level of p dollars is given by $q = 100 + 0.5p$. Find the equilibrium tuition price $\bar{p}$ and the consumers' and producers' surpluses at this tuition level. What is the total social gain at the equilibrium price? (Round $\bar{p}$ to the nearest \$1 and all other answers to the nearest \$1000.)

Based on a study by A. L. Ostrosky, Jr., and J. V. Koch, as cited in their book, *Introduction to Mathematical Economics* (Prospect Heights, Ill.: Waveland Press, 1979): 133.

50. Fast Food A fast-food outlet finds that the demand equation for its new side dish, "Sweetdough Tidbit," is given by

$$p = \frac{128}{(q + 1)^2}$$

where p is the price (in cents) per serving and q is the number of servings that can be sold each hour at this price. At the same time, the franchise is prepared to sell $q = 0.5p - 1$ servings per hour at a price of p cents. Find the equilibrium price $\bar{p}$ and the consumers' and producers' surpluses at this price level. What is the total social gain at the equilibrium price?

51. Linear Demand Given a linear demand equation $q = -mp + b$ $(m > 0)$, find a formula for the consumers' surplus at a price level of $\bar{p}$ per unit.

52. Linear Supply Given a linear supply equation of the form $q = mp + b$ $(m > 0)$, find a formula for the producers' surplus at a price level of $\bar{p}$ per unit.

53. Foreign Trade Annual U.S. imports from China in the years 1988–1998 can be approximated by

$$I = 3.0 + 6.2t \text{ billion dollars per year} \qquad (0 \le t \le 10)$$

where t represents time in years since 1988. During the same period, annual U.S. exports to China could be approximated by

$$E = 4.1 + 0.94t \text{ billion dollars per year} \qquad (0 \le t \le 10)$$

a. What does the area between the graphs of I and E over the interval $[1, 10]$ represent?

b. Compute the area in part (a) and interpret your answer.

The models are based on linear regression. Source: U.S. Census Bureau/*New York Times*, April 8, 1999, p. C1.

54. Online Revenue America Online's total revenue (including revenue from subscriptions) was flowing in at a rate of approximately

$$R_t = 0.4t^2 + 0.1t + 2 \text{ billion dollars per year} \quad (0 \le t \le 2.5)$$

where t is time in years since January 1997. During the same period, revenue from subscriptions alone flowed in at a rate of approximately

$$R_s = 0.2t^2 + 0.2t + 1 \text{ billion dollars per year} \quad (0 \le t \le 2.5)$$

a. What does the area between the graphs of R_t and R_s over the interval $[0, 1]$ represent?

b. Compute the area in part (a), rounded to one significant digit, and interpret your answer.

The models are based on quadratic regression. Source: America Online/*New York Times*, July 4, 1999, p. BU6.

55. Swimming Pool Sales In-ground swimming pools were built in the United States at a rate of

$$g(t) = -0.45t^2 + 15t + 54 \text{ thousand pools per year}$$
$$(6 \le t \le 11)$$

where t is the time in years since 1990. At the same time, above-ground swimming pools were built at a rate of

$$b(t) = 0.36t^2 + 4.1t + 140 \text{ thousand pools per year}$$
$$(6 \le t \le 11)$$

a. What does the area between the graphs of g and b over the interval $[6, 11]$ represent?

b. Compute the area in part (a) and interpret your answer. (Round your answer to three significant digits.)

Based on a regression model. Source: PK Data/*New York Times*, July 5, 2001, p. C1.

56. **Oil Production** Saudi Arabia's crude oil production t days after January 1, 1999, can be approximated by

$$S(t) = 78 - 70.8\left(\frac{t}{365} - 1\right)^2 \text{ million barrels per day}$$
$$(0 \le t \le 730)$$

Russia's crude oil production for the same period can be approximated by

$$R(t) = 4 + 0.2\left(\frac{t}{365} - 1\right) \text{ million barrels per day}$$
$$(0 \le t \le 730)$$

a. What does the area between the graphs of S and R over the interval $[0, 365]$ represent?

b. Compute the area in part (a) and interpret your answer. (Round your answer to three significant digits.)

Source: Deutsche Bank/*New York Times*, November 21, 2001, p. A3.

57. **Health** The number of new cases of Alzheimer's disease in the United States in the period 2000–2050 is projected to be

$$N(t) = 0.38e^{0.017t} \text{ million per year} \qquad (0 \le t \le 50)$$

where t is the number of years since 2000. Over the same period, the number of new cases affecting people age 85 and older is projected to be

$$L(t) = 0.15e^{0.026t} \text{ million per year} \qquad (0 \le t \le 50)$$

a. Use these models to project the total number of cases of Alzheimer's disease among people under age 85 from 2010 to 2050. (Round your answer to two significant digits.)

b. How does the answer to part (a) relate to the area between two curves?

c. What does a comparison of the exponents in the formulas for N and L tell you about Alzheimer's disease?

Based on an exponential model. Source: Liesl E. Herbert, Laura A. Beckett, Paul A. Scherr, and Denis A. Evans, "Annual Incidence of Alzheimer's Disease in the United States Projected to the Years 2000 through 2050," *Alzheimer's Disease and Associated Disorders* 15(4), 2001 (new cases)/*New York Times*, March 31, 2002, p. 24.

58. **Health Care Spending** The rate of spending on health care (including hospital care) in the United States from 1960 to 2000 was approximately

$$S(t) = 0.13e^{0.059t} \text{ trillion dollars per year} \qquad (0 \le t \le 40)$$

where t is the number of years since 1960. At the same time, the rate of spending on hospital care was approximately

$$H(t) = 0.072e^{0.039t} \text{ trillion dollars per year} \qquad (0 \le t \le 40)$$

a. Use these models to estimate the total spent on health care other than hospital care from 1960 to 1980. (Round your answer to two significant digits.)

b. How does the answer to part (a) relate to the area between two curves?

c. What does a comparison of the exponents in the formulas for S and H tell you about health care spending?

Based on an exponential model. Source: U.S. Dept. of Health and Human Services/*New York Times*, January 8, 2002, p. A14.

59. **Subsidizing Emission Control** The marginal cost to the utilities industry of reducing sulfur emissions at several levels of reduction is shown in the following table:

Reduction (millions of tons)	8	10	12
Marginal Cost (dollars/ton)	270	360	780

These figures were produced in a computerized study of reducing sulfur emissions from the 1980 level by the given amounts. Source: Congress of the United States, Congressional Budget Office, *Curbing Acid Rain: Cost, Budget and Coal Market Effects* (Washington, D.C.: Government Printing Office, 1986): xx, xxii, 23, 80.

a. Show by substitution that these data can be modeled by

$$C'(q) = 1,000,000(41.25q^2 - 697.5q + 3210)$$

where q is the reduction in millions of tons of sulfur and $C'(q)$ is the marginal cost of reducing emissions (in dollars per million tons of reduction).[1]

b. If the government subsidizes sulfur emissions at a rate of \$400 per ton, find the total net cost to the utilities industry of removing 12 million tons of sulfur.

c. The answer to part (b) is represented by the area between two graphs. What are the equations of these graphs?

60. **Variable Emissions Subsidies** Referring to the marginal cost equation in Exercise 59, assume that the government subsidy for sulfur emissions is given by the formula

$$S'(q) = 500,000,000q$$

where q is the level of reduction in millions of tons and $S'(q)$ is the marginal subsidy in dollars per million tons.

a. Find the total net cost to the utilities industry of removing 12 million tons of sulfur. Interpret your answer.

b. The answer to part (a) is represented by the net area between two graphs. What are the equations of these graphs?

COMMUNICATION AND REASONING EXERCISES

61. The following graph shows Canada's annual exports and imports for the period 1997–2001:

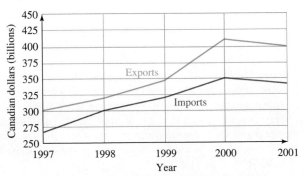

Source: http://strategis.ic.gc.ca, August 2002.

---

[1]You can obtain this equation directly from the data (as we did) by assuming that $C'(q) = aq^2 + bq + c$, substituting the values of $C'(q)$ for the given values of q and then solving for a, b, and c.

What does the area between the export and import curves represent?

62. The following graph shows a fictitious country's monthly exports and imports for the period 1997–2001:

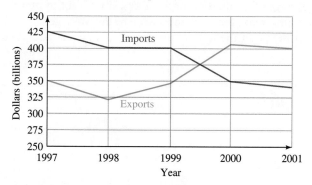

What does the total area enclosed by the export and import curves represent, and what does the definite integral of the difference, exports − imports, represent?

63. The following graph shows the daily revenue and cost in your Adopt-a-Chia operation t days from its inception:

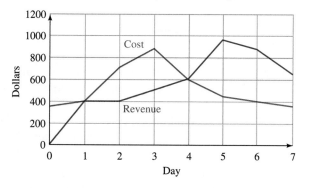

The area between the cost and revenue curves represents

(A) The accumulated loss through day 4 plus the accumulated profit for days 5 through 7.
(B) The accumulated profit for the week.
(C) The accumulated loss for the week.
(D) The accumulated cost through day 4 plus the accumulated revenue for days 5 through 7.

64. The following graph shows daily orders and inventory (stock on hand) for your *Mona Lisa* paint-by-number sets t days into last week:

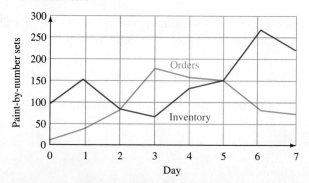

a. Which is greatest?
  (A) $\int_0^7$ (orders − inventory) dt
  (B) $\int_0^7$ (inventory − orders) dt
  (C) The area between the orders and inventory curves.
b. The answer to part (a) measures
  (A) The accumulated gap between orders and inventory.
  (B) The accumulated surplus through day 3 minus the accumulated shortage for days 3 through 5 plus the accumulated surplus through days 6 through 7.
  (C) The total net surplus.
  (D) The total net loss.

65. What is wrong with the following claim: "I purchased Consolidated Edison shares for \$40 at the beginning of January 2002. My total profit per share from this investment from January through July is represented by the area between the stock price curve and the purchase price curve, as shown on the following graph."
Source: http://money.excite.com, August 5, 2002.

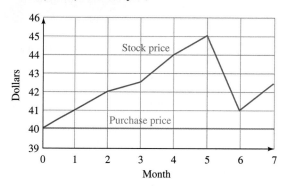

66. Your pharmaceutical company monitors the amount of medication in successive batches of 100-milligram tetracycline capsules and obtains the following graph:

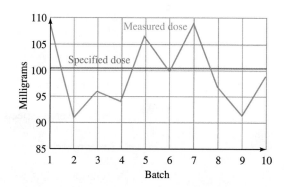

Your production manager claims that the batches of tetracycline conform to the exact dosage requirement because half of the area between the graphs is above the specified dose line and half is below it. Comment on this reasoning.

7.3 Averages and Moving Averages

Averages

We all know how to find the average of a collection of numbers. If we want to find the average of, say, 20 numbers, we simply add them up and divide by 20. More generally, if we want to find the **average,** or **mean,** of the n numbers $y_1, y_2, y_3, \ldots y_n$, we add them up and divide by n.

Average, or Mean, of a Collection of Values

$$\bar{y} = \frac{y_1 + y_2 + \cdots + y_n}{n}$$

Quick Example

The average of $\{0, 2, -1, 5\}$ is $\bar{y} = \dfrac{0 + 2 - 1 + 5}{4} = \dfrac{6}{4} = 1.5$.

But, we also use the word *average* in other senses. For example, we speak of the average speed of a car during a trip.

Example 1 • Average Speed

Over the course of 2 hours, my speed varied from 50 to 60 miles per hour, following the function $v(t) = 50 + 2.5t^2$ $(0 \le t \le 2)$. What was my average speed over those 2 hours?

Solution Recall that average speed is simply the total distance traveled divided by the time it took. Recall, also, that we can find the distance traveled by integrating the speed:

$$\text{distance traveled} = \int_0^2 v(t)\, dt$$

$$= \int_0^2 (50 + 2.5t^2)\, dt$$

$$= \left[50t + \frac{2.5}{3}t^3\right]_0^2 = 100 + \frac{20}{3} \approx 106.67 \text{ miles}$$

It took 2 hours to travel this distance, so the average speed was

$$\text{average speed} \approx \frac{106.67}{2} \approx 53.3 \text{ miles per hour}$$

In general, if we travel with velocity $v(t)$ from time $t = a$ to time $t = b$, we will travel a distance of $\int_a^b v(t)\, dt$ in time $b - a$, which gives an average velocity of

$$\text{average velocity} = \frac{1}{b - a} \int_a^b v(t)\, dt$$

Thinking of this calculation as finding the average value of the velocity function, we generalize and make the following definition.

Average Value of a Function

The **average,** or **mean,** of a function $f(x)$ on an interval $[a, b]$ is

$$\bar{f} = \frac{1}{b-a} \int_a^b f(x)\, dx$$

Quick Example

The average of $f(x) = x$ on $[1, 5]$ is

$$\bar{f} = \frac{1}{b-a} \int_a^b f(x)\, dx = \frac{1}{5-1} \int_1^5 x\, dx = \frac{1}{4} \left[\frac{x^2}{2} \right]_1^5 = \frac{1}{4} \left(\frac{25}{2} - \frac{1}{2} \right) = 3$$

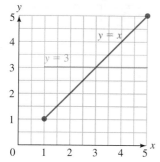

Figure 13

Note The average of a function has a geometric interpretation. Referring to the Quick Example, we can compare the graph of $y = f(x)$ with the graph of $y = 3$, both over the interval $[1, 5]$ (Figure 13). We can find the area under the graph of $f(x)$ by geometry or by calculus; it is 12. The area in the rectangle under $y = 3$ is also 12.

In general, the average $\bar{f}$ of a positive function over the interval $[a, b]$ gives the height of the rectangle over the interval $[a, b]$ that has the same area as the area under the graph of $f(x)$. The equality of these areas follows from the equation

$$(b - a)\bar{f} = \int_a^b f(x)\, dx$$

Example 2 • Average Balance

A savings account at the People's Credit Union pays 3% interest, compounded continuously, and at the end of the year you get a bonus of 1% of the average balance in the account during the year. If you deposit $10,000 at the beginning of the year, how much interest and how large a bonus will you get?

Solution We can use the continuous compound interest formula to calculate the amount of money you have in the account at time t:

$$A(t) = 10{,}000e^{0.03t}$$

where t is measured in years. At the end of 1 year, the account will have

$$A(1) = \$10{,}304.55$$

so you will have earned $304.55 interest. To compute the bonus, we need to find the average amount in the account, which is the average of $A(t)$ over the interval $[0, 1]$. Thus,

$$\bar{A} = \frac{1}{b-a} \int_a^b A(t)\, dt = \frac{1}{1-0} \int_0^1 10{,}000e^{0.03t}\, dt$$

$$= \frac{10{,}000}{0.03} \left[e^{0.03t} \right]_0^1 \approx \$10{,}151.51$$

The bonus is 1% of this, or $101.52.

✳ *Before we go on . . .* The 1% bonus was one-third of the total interest. Why did this happen? What fraction of the total interest would the bonus be if the interest rate was 4%, 5%, or 10%?

Moving Averages

Suppose you follow the performance of a company's stock by recording the daily closing prices. The graph of these prices may seem jagged or "jittery" due to almost random day-to-day fluctuations. To see any trends, you would like a way to "smooth out" these data. The **moving average** is one common way to do that.

Example 3 • Stock Prices

The following table shows Colossal Conglomerate's closing stock prices for 20 consecutive trading days:

Day	1	2	3	4	5	6	7	8	9	10
Price ($)	20	22	21	24	24	23	25	26	20	24

Day	11	12	13	14	15	16	17	18	19	20
Price ($)	26	26	25	27	28	27	29	27	25	24

Plot these prices and the 5-day moving average.

Solution The 5-day moving average is the average of each day's price together with the prices of the preceding 4 days. We can compute the 5-day moving averages starting on the fifth day. We get these numbers:

Day	1	2	3	4	5	6	7	8	9	10
Moving Average ($)					22.2	22.8	23.4	24.4	23.6	23.6

Day	11	12	13	14	15	16	17	18	19	20
Moving Average ($)	24.2	24.4	24.2	25.6	26.4	26.6	27.2	27.6	27.2	26.4

The closing stock prices and moving averages are plotted in Figure 14.

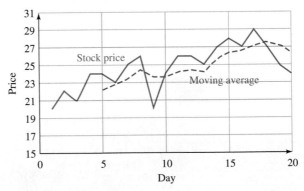

Figure 14

As you can see, the moving average is less volatile than the closing price. Since the moving average averages the stock's performance over 5 days, a single day's fluctuation is smoothed out. Look at day 9 in particular. The moving average also tends to lag behind the actual performance because it takes past history into account. Look at the downturns at days 6 and 18 in particular.

Graphing Calculator

To automate this calculation on a TI-83, first use

$$\texttt{seq(X,X,1,20)} \rightarrow \texttt{L}_1 \quad \boxed{\texttt{2nd}}\ \boxed{\texttt{STAT}} \rightarrow \texttt{OPS} \rightarrow 5$$

to enter the sequence of numbers 1 through 20 into the list $\texttt{L}_1$, representing the trading days. Then, using the list editor accessible through the $\texttt{STAT}$ menu, enter the daily stock prices in list $\texttt{L}_2$. You can now calculate the list of 5-day moving averages using the following command:

$$\texttt{seq((L}_2\texttt{(X)+L}_2\texttt{(X-1)+L}_2\texttt{(X-2)+L}_2\texttt{(X-3)+L}_2\texttt{(X-4))/5,X,5,20)} \rightarrow \texttt{L}_3$$

This has the effect of putting the moving averages into elements 1 through 15 of list $\texttt{L}_3$. If you wish to plot the moving average on the same graph as the daily prices, you will want the averages in $\texttt{L}_3$ to match up with the prices in $\texttt{L}_2$. One way to do this is to put four more entries at the beginning of $\texttt{L}_3$—say, copies of the first four entries of $\texttt{L}_2$. The following command accomplishes this:

$$\texttt{augment(seq(L}_2\texttt{(X),X,1,4),L}_3\texttt{)} \rightarrow \texttt{L}_3 \quad \boxed{\texttt{2nd}}\ \boxed{\texttt{STAT}} \rightarrow \texttt{OPS} \rightarrow 9$$

You can now graph the prices and moving averages by creating an xyLine scatter plot through the $\texttt{STAT PLOT}$ menu, with $\texttt{L}_1$ being the Xlist and $\texttt{L}_2$ being the Ylist for $\texttt{Plot1}$, and $\texttt{L}_1$ being the Xlist and $\texttt{L}_3$ the Ylist for $\texttt{Plot2}$.

Excel

Compute the moving averages in a column next to the daily prices, as shown here:

	A	B	C	D
1	Day	Price	Moving Average	
2	1	20		
3	2	22		
4	3	21		
5	4	24		
6	5	24	=AVERAGE(B2:B6)	
7	6	23		
8				
20				
21	20	24		

You can then create a facsimile of Figure 14 using a scatter plot.

 Before we go on . . . The period of 5 days is arbitrary. Using a longer period of time would smooth the data more but increase the lag. For data used as economic indicators, such as housing prices or retail sales, it is common to compute the 1-year moving average to smooth out seasonal variations.

It is also sometimes useful to compute moving averages of continuous functions. We may want to do this if we use a mathematical model of a large collection of data. Also, some physical systems have the effect of converting an input function (an electrical signal, for example) into its moving average. By an ***n-unit moving average*** of a function $f(x)$, we mean the function $\bar{f}$ for which $\bar{f}(x)$ is the average of the value of $f(x)$ on $[x - n, x]$. Using the formula for the average of a function, we get the following formula.

n-Unit Moving Average of a Function

The n-unit moving average of a function f is

$$\bar{f}(x) = \frac{1}{n} \int_{x-n}^{x} f(t)\, dt$$

Quick Example

The 2-unit moving average of $f(x) = x^2$ is

$$\bar{f}(x) = \frac{1}{2} \int_{x-2}^{x} t^2\, dt = \frac{1}{6}[t^3]_{x-2}^{x} = x^2 - 2x + \frac{4}{3}$$

The graphs of $f(x)$ and $\bar{f}(x)$ are shown in the figure.

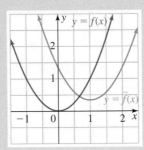

𝕋 Example 4 • Moving Average via Technology

Plot the 3-unit moving average of

$$f(x) = \frac{x}{1 + |x|} \qquad (-5 \le x \le 5)$$

Solution This function is a little tricky to integrate analytically because $|x|$ is defined differently for positive and negative values of x, so instead we use technology to approximate the integral. Here is how we could do it using a TI-83 graphing calculator. We enter the following:

```
Y1=X/(1+abs(X))
Y2=(1/3)fnInt(Y₁(T),T,X-3,X)
```

The Y_1 entry is $f(x)$, and the Y_2 entry is a numerical approximation of the 3-unit moving average of $f(x)$:

$$\bar{f}(x) = \frac{1}{3} \int_{x-3}^{x} \frac{t}{1 + |t|}\, dt$$

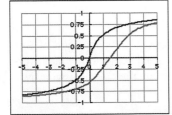

Figure 15

We set the viewing window ranges to $-5 \le x \le 5$ and $-1 \le y \le 1$ and plot these curves. (Be patient—the calculator has to do a numerical integration to obtain each point on the graph of the moving average.) The result is shown in Figure 15.

✴ **Before we go on . . .** Of course, once we've entered Y_1 and Y_2 as above, we can use the calculator to evaluate the moving average at any value of x. For instance, to calculate $\bar{f}(1.2)$, we enter $Y_2(1.2)$ on the home screen and find

$$\bar{f}(1.2) \approx -0.1196$$

7.3 EXERCISES

In Exercises 1–6, find the averages of the functions over the given intervals. Plot each function and its average on the same graph (as in Figure 13).

1. $f(x) = x^3$ over $[0, 2]$

2. $f(x) = x^3$ over $[-1, 1]$

3. $f(x) = x^3 - x$ over $[0, 2]$

4. $f(x) = x^3 - x$ over $[0, 1]$

5. $f(x) = e^{-x}$ over $[0, 2]$

6. $f(x) = e^x$ over $[-1, 1]$

In Exercises 7 and 8, complete the given table with the values of the 3-unit moving average of the given function.

7.

x	0	1	2	3	4	5	6	7
$r(x)$	3	5	10	3	2	5	6	7
$\bar{r}(x)$								

8.

x	0	1	2	3	4	5	6	7
$s(x)$	2	9	7	3	2	5	7	1
$\bar{s}(x)$								

In Exercises 9 and 10, some values of a function and its 3-unit moving average are given. Supply the missing information.

9.

x	0	1	2	3	4	5	6	7
$r(x)$	1	2			11		10	2
$\bar{r}(x)$			3	5		11		

10.

x	0	1	2	3	4	5	6	7
$s(x)$	1	5		1				
$\bar{s}(x)$			5		5	2	3	2

In Exercises 11–18, calculate the 5-unit moving average of each function. Plot each function and its moving average on the same graph, as in Example 4. (You may use graphing technology for these plots, but you should compute the moving averages analytically.)

11. $f(x) = x^3$

12. $f(x) = x^3 - x$

13. $f(x) = x^{2/3}$

14. $f(x) = x^{2/3} + x$

15. $f(x) = e^{0.5x}$

16. $f(x) = e^{-0.02x}$

17. $f(x) = \sqrt{x}$

18. $f(x) = x^{1/3}$

T In Exercises 19–22, use graphing technology to plot the given functions together with their 3-unit moving averages.

19. $f(x) = \dfrac{10x}{1 + 5|x|}$

20. $f(x) = \dfrac{1}{1 + e^x}$

21. $f(x) = \ln(1 + x^2)$

22. $f(x) = e^{1-x^2}$

APPLICATIONS

23. Employment The following table shows the approximate number of people employed in the United States during the period 1995–2001, on July 1 of each year ($t = 5$ represents 1995):

Year, t	5	6	7	8	9	10	11
Employment (millions)	117	120	123	125	130	132	132

The given values represent nonfarm employment and are approximate. SOURCE: Bureau of Labor Statistics/*New York Times*, December 17, 2001, p. C3.

What was the average number of people employed in the United States for the years 1996 through 2001?

24. Cell Phone Sales The following table shows the net sales (after-tax revenue) at Nokia, the Finnish cell phone company, for each year in the period 1995–2001 ($t = 5$ represents 1995):

Year, t	5	6	7	8	9	10	11
Net Sales ($ billions)	8	8	10	16	20	27	28

SOURCE: Nokia/*New York Times*, February 6, 2002, p. A3.

What were Nokia's average net sales for the years 1996 through 2000?

25. Television Advertising The cost, in millions of dollars, of a 30-second television ad during the Super Bowl in the years 1998 to 2001 can be approximated by

$$C(t) = 0.355t - 1.6 \text{ million dollars} \qquad (8 \le t \le 11)$$

($t = 8$ represents 1998). What was the average cost of a Super Bowl ad during the given period? SOURCE: *New York Times*, January 26, 2001, p. C1.

26. Television Advertising The cost, in millions of dollars, of a 30-second television ad during the Super Bowl in the years 1990 to 1998 can be approximated by

$$C(t) = 0.08t + 0.6 \text{ million dollars} \qquad (0 \le t \le 8)$$

($t = 0$ represents 1990). What was the average cost of a Super Bowl ad during the given period?

27. Investments If you invest $10,000 at 8% interest compounded continuously, what is the average amount in your account over 1 year?

28. Investments If you invest $10,000 at 12% interest compounded continuously, what is the average amount in your account over 1 year?

29. **Average Balance** Suppose you have an account (paying no interest) into which you deposit $3000 at the beginning of each month. You withdraw money continuously so that the amount in the account decreases linearly to zero by the end of the month. Find the average amount in the account over a period of several months. (Assume that the account starts at $0 at $t = 0$ months.)

30. **Average Balance** Suppose you have an account (paying no interest) into which you deposit $4000 at the beginning of each month. You withdraw $3000 during the course of each month, in such a way that the amount decreases linearly. Find the average amount in the account in the first 2 months. (Assume that the account starts at $0 at $t = 0$ months.)

31. **Employment** Refer to Exercise 23. Complete the following table by computing the 4-year moving average of employment in the United States.

Year, t	5	6	7	8	9	10	11
Employment (millions)	117	120	123	125	130	132	132
Moving Average (millions)							

How do the year-by-year changes in the moving average compare with those in the employment figures?

32. **Cell Phone Sales** Refer to Exercise 24. Complete the following table by computing the 4-year moving average of Nokia's net sales.

Year, t	5	6	7	8	9	10	11
Net Sales, $P(t)$ ($ billions)	8	8	10	16	20	27	28
Moving Average ($ billions)							

Is the average of the moving averages the same as the overall average? Explain.

33. **Vacation Spending** The following table shows approximate annual tourist expenditure, in millions of dollars, in Bermuda for the years 1976–1998:

1976	1977	1978	1979	1980	1981	1982	1983
150	180	180	200	270	300	280	310
1984	**1985**	**1986**	**1987**	**1988**	**1989**	**1990**	**1991**
340	350	350	430	460	450	470	490
1992	**1993**	**1994**	**1995**	**1996**	**1997**	**1998**	
450	450	520	530	470	470	470	

SOURCE: Bermuda Ministry of Finance/*New York Times*, April 28, 1999, p. C1.

a. Use technology to compute and plot the 5-year moving average of these data.

b. The graph of the moving average should appear almost linear over the range 1981–1991. Use the 1981 and 1991 moving average figures to give an estimate of the rate of change of tourist expenditure in Bermuda for the period 1981–1991. Interpret the result.

34. **Business Spending** The following table shows approximate annual international business expenditure, in millions of dollars, in Bermuda for the years 1976–1998:

1976	1977	1978	1979	1980	1981	1982	1983
60	90	60	70	100	120	150	160
1984	**1985**	**1986**	**1987**	**1988**	**1989**	**1990**	**1991**
160	200	240	260	260	310	320	330
1992	**1993**	**1994**	**1995**	**1996**	**1997**	**1998**	
340	360	400	440	500	600	760	

SOURCE: Bermuda Ministry of Finance/*New York Times*, April 28, 1999, p. C1.

a. Use technology to compute and plot the 5-year moving average of these data.

b. The graph of the moving average should appear almost linear over the range 1982–1990. Use the 1982 and 1990 moving average figures to give an estimate of the rate of change of business expenditure in Bermuda for the period 1982–1990. Interpret the result.

35. **SUVs** The average weight of a sport utility vehicle (SUV) can be approximated by

$$W = 3t^2 - 90t + 4200 \qquad (5 \le t \le 27)$$

where t is time in years ($t = 0$ represents 1970) and W is the average weight in pounds of an SUV produced in year t.

a. Compute the average weight of an SUV over the period 1975–1997, to the nearest 10 pounds.

b. Compute the 2-year moving average of W. (You need not simplify the answer.)

c. Without simplifying the answer in part (b), say what kind of function the moving average is.

SOURCE: The quadratic model is based on data published in the *New York Times*, November 30, 1997, p. 43.

36. **Sedans** The average weight of a sedan can be approximated by

$$W = 6t^2 - 240t + 4800 \qquad (5 \le t \le 27)$$

where t is time in years ($t = 0$ represents 1970) and W is the average weight of a sedan in pounds. Repeat Exercise 35 as applied to sedans.

SOURCE: The quadratic model is based on data published in the *New York Times*, November 30, 1997, p. 43.

37. **Medicare Spending** Annual federal spending on Medicare (in constant 2000 dollars) was projected to increase from $240 billion in 2000 to $600 billion in 2025.

a. Use this information to express s, the annual spending on Medicare (in billions of dollars), as a linear function of t, the number of years since 2000.

b. Find the 4-year moving average of your model.

c. What can you say about the slope of the moving average?

Data are rounded. Source: The Urban Institute's Analysis of the 1999 Trustee's Report: http://www.urban.org

38. Pasta Imports In 1990 the United States imported 290 million pounds of pasta. From 1990 to 2001, imports increased by an average of 40 million pounds per year.

a. Use these data to express q, the annual U.S. imports of pasta (in millions of pounds), as a linear function of t, the number of years since 1990.

b. Find the 4-year moving average of your model.

c. What can you say about the slope of the moving average?

Data are rounded. Sources: U.S. Dept. of Commerce/*New York Times*, September 5, 1995, p. D4; International Trade Administration (http://www.ita.doc.gov/), March 31, 2002.

39. Moving Average of a Linear Function Find a formula for the a-unit moving average of a general linear function $f(x) = mx + b$.

40. Moving Average of an Exponential Function Find a formula for the a-unit moving average of a general exponential function $f(x) = Ae^{kx}$.

41. Fair Weather The Cancun Royal Hotel's advertising brochure features the following chart, showing the year-round temperature:

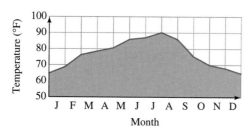

a. Estimate and plot the 2- and 3-month moving averages. (Use graphing technology, if available.)

b. What can you say about the 24-month moving average?

c. Comment on the limitations of a quadratic model for these data.

Source: Inspired by an exercise in the Harvard Consortium Calculus project (p. 76)

42. Foul Weather Repeat Exercise 41, using the following data from the Tough Traveler Lodge in Frigidville:

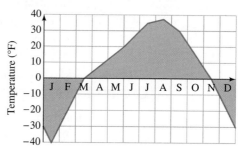

COMMUNICATION AND REASONING EXERCISES

43. What property does a (nonconstant) function have if its average value over an interval is zero? Sketch a graph of such a function.

44. Can the average value of a function f on an interval be greater than its value at every point in that interval? Explain.

45. Explain why it is sometimes more useful to consider the moving average of a stock price rather than the stock price itself.

46. Your monthly salary has been increasing steadily for the past year, and your average monthly salary over the past year was x dollars. Would you have earned more money if you were paid x dollars per month? Explain your answer.

47. Criticize the following claim: The average value of a function on an interval is midway between its highest and lowest value.

48. Your manager tells you that 12-month moving averages gives at least as much information as shorter-term moving averages and very often more. How would you argue that he is wrong?

49. Which of the following most closely approximates the original function? (a) Its 10-unit moving average, (b) its 1-unit moving average, or (c) its 0.8-unit moving average. Explain your answer.

50. Is an increasing function larger or smaller than its 1-unit moving average? Explain.

7.4 Continuous Income Streams

A company with a high volume of sales receives money almost continuously. For purposes of calculation, it is convenient to assume that the company literally does receive money continuously. In such a case we have a function $R(t)$ that represents the rate at which money is being received by the company at time t.

Example 1 • Continuous Income

An ice cream store's business peaks in late summer; the store's summer revenue is approximated by

$$R(t) = 300 + 4.5t - 0.05t^2 \text{ dollars per day} \qquad (0 \le t \le 92)$$

where t is measured in days after June 1. What is its total revenue for the months of June, July, and August?

Solution Let's approximate the total revenue by breaking up the interval $[0, 92]$ representing the 3 months into n subintervals $[t_{k-1}, t_k]$, each with length Δt. In the interval $[t_{k-1}, t_k]$, the store receives money at a rate of approximately $R(t_{k-1})$ dollars per day for Δt days, so it will receive a total of $R(t_{k-1})\Delta t$ dollars. Over the whole summer, then, the store will receive approximately

$$R(t_0)\Delta t + R(t_1)\Delta t + \cdots + R(t_{n-1})\Delta t \text{ dollars}$$

As we let n become large to better approximate the total revenue, this Riemann sum approaches the integral

$$\text{total revenue} = \int_0^{92} R(t) \, dt$$

Substituting the function we were given, we get

$$\text{total revenue} = \int_0^{92} (300 + 4.5t - 0.05t^2) \, dt$$

$$= \left[300t + 2.25t^2 - \frac{0.05}{3}t^3 \right]_0^{92} \approx \$33,666$$

Before we go on . . . We could approach this calculation another way: $R(t) = S'(t)$, where $S(t)$ is the total revenue earned up to day t. By the Fundamental Theorem of Calculus,

$$\text{total revenue} = S(92) - S(0) = \int_0^{92} R(t) \, dt$$

We did the calculation using Riemann sums mainly as practice for the next example.

Generalizing Example 1, we can say the following.

Total Value of a Continuous Income Stream

If the rate of receipt of income is $R(t)$ dollars per unit of time, then the total income received from time $t = a$ to $t = b$ is

$$\text{total value} = TV = \int_a^b R(t) \, dt$$

Example 2 • Future Value

Suppose the ice cream store (Example 1) deposits its receipts in an account paying 5% interest per year compounded continuously. How much money will it have in its account at the end of August?

Solution Now we have to take into account not only the revenue but also the interest it earns in the account. Again, we break the interval [0, 92] into n subintervals. During the interval $[t_{k-1}, t_k]$, approximately $R(t_{k-1})\Delta t$ dollars are deposited in the account. That money will earn interest until the end of August, a period of $92 - t_{k-1}$ days, or $(92 - t_{k-1})/365$ years. Remembering that 5% is the *annual* interest rate, the formula for continuous compounding tells us that by the end of August those $R(t_{k-1})\Delta t$ dollars will have turned into

$$R(t_{k-1})\Delta t\, e^{0.05(92-t_{k-1})/365} = R(t_{k-1})e^{0.05(92-t_{k-1})/365}\,\Delta t \text{ dollars}$$

Adding up the contributions from each subinterval, we see that the total in the account at the end of August will be approximately

$$R(t_0)e^{0.05(92-t_0)/365}\Delta t + R(t_1)e^{0.05(92-t_1)/365}\Delta t + \cdots + R(t_{n-1})e^{0.05(92-t_{n-1})/365}\Delta t$$

This is a Riemann sum; as n gets large, the sum approaches the integral

$$\text{future value} = FV = \int_0^{92} R(t)e^{0.05(92-t)/365}\, dt$$

Substituting $R(t) = 300 + 4.5t - 0.05t^2$, we obtain

$$FV = \int_0^{92} (300 + 4.5t - 0.05t^2)e^{0.05(92-t)/365}\, dt$$

$$\approx \$33,880 \qquad \text{Using technology or integration by parts}$$

Before we go on . . . The interest earned in the account was fairly small (compare this answer to Example 1). Not only was the money in the account for just 3 months, but also much of it was put in the account toward the end of that period, so had very little time to earn interest.

Generalizing again, we have the following.

Future Value of a Continuous Income Stream

If the rate of receipt of income from time $t = a$ to $t = b$ is $R(t)$ dollars per unit of time and the income is deposited as it is received in an account paying interest at rate r per unit of time, compounded continuously, then the amount of money in the account at time $t = b$ is

$$\text{future value} = FV = \int_a^b R(t)e^{r(b-t)}\, dt$$

Example 3 • Present Value

You are thinking of buying the ice cream store discussed in Examples 1 and 2. What is its income stream worth to you on June 1?

Solution Assuming that you have access to the same account paying 5% per year compounded continuously, the value of the income stream on June 1 is the amount of money that, if deposited June 1, would give you the same future value as the income stream will. If we let PV denote this present value, its value after 92 days will be

$$PVe^{0.05 \times 92/365}$$

If we equate this with the future value of the income stream, we get

$$PVe^{0.05 \times 92/365} = e^{0.05 \times 92/365} \int_0^{92} R(t)e^{-0.05t/365} \, dt$$

so

$$PV = \int_0^{92} R(t)e^{-0.05t/365} \, dt$$

Substituting the formula for $R(t)$ and integrating using technology or integration by parts, we get

$$PV \approx \$33,455$$

The general formula is the following.

Present Value of a Continuous Income Stream

If the rate of receipt of income from time $t = a$ to $t = b$ is $R(t)$ dollars per unit of time and the income is deposited as it is received in an account paying interest at rate r per unit of time, compounded continuously, then the value of the income stream at time $t = a$ is

$$\text{present value} = PV = \int_a^b R(t)e^{r(a-t)} \, dt$$

We can derive this formula from the relation

$$FV = PVe^{r(b-a)}$$

because the present value is the amount that would have to be deposited at time $t = a$ to give a future value of FV at time $t = b$.

These formulas are more general than we've said. They still work when $R(t) < 0$ if we interpret negative values as money flowing *out* rather than in. That is, we can use these formulas for income we receive or for payments that we make, or for situations where we sometimes receive money and sometimes pay it out. These formulas can also be used for flows of quantities other than money. For example, if we use an exponential model for population growth and we let $R(t)$ represent the rate of immigration [$R(t) > 0$] or emigration [$R(t) < 0$], then the future value formula gives the future population.

7.4 EXERCISES

In Exercises 1–6, find the total value of the given income stream and find its future value (at the end of the given interval), using the given interest rate.

1. $R(t) = 30,000$ $(0 \le t \le 10)$, at 7%

2. $R(t) = 40,000$ $(0 \le t \le 5)$, at 10%

3. $R(t) = 30,000 + 1000t$ $(0 \le t \le 10)$, at 7%

4. $R(t) = 40,000 + 2000t$ $(0 \le t \le 5)$, at 10%

5. $R(t) = 30,000e^{0.05t}$ $(0 \le t \le 10)$, at 7%

6. $R(t) = 40,000e^{0.04t}$ $(0 \le t \le 5)$, at 10%

In Exercises 7–12, find the total value of the given income stream and find its present value (at the beginning of the given interval), using the given interest rate.

7. $R(t) = 20,000$ $(0 \le t \le 5)$, at 8%

8. $R(t) = 50,000$ $(0 \le t \le 10)$, at 5%

9. $R(t) = 20,000 + 1000t$ $(0 \le t \le 5)$, at 8%

10. $R(t) = 50,000 + 2000t$ $(0 \le t \le 10)$, at 5%

11. $R(t) = 20,000e^{0.03t}$ $(0 \le t \le 5)$, at 8%

12. $R(t) = 50,000e^{0.06t}$ $(0 \le t \le 10)$, at 5%

APPLICATIONS

13. Revenue The rate of receipt of net sales (after-tax revenue) earned by Nokia, the Finnish cell phone company, from 1995 to 2001 can be approximated by

$$R(t) = 0.31t^2 - 1.1t + 4.4 \text{ billion dollars per year}$$
$$(5 \le t \le 11)$$

where t is time in years since January 1990. (Thus, $t = 5$ represents January 1995.) Estimate, to the nearest $1 billion, Nokia's total revenue from January 1995 to January 2001.

The model is based on a quadratic regression. Source: Nokia/*New York Times*, February 6, 2002, p. A3.

14. Revenue The rate of receipt of revenue by Nintendo from sales of Game Boy video games from 1996 to 2000 can be approximated by

$$R(t) = 0.19t^2 - 0.23t + 0.20 \text{ billion dollars per year}$$
$$(0 \le t \le 4)$$

where t is time in years since January 1996. Estimate, to the nearest $0.1 billion, Nintendo's total revenue from January 1996 to January 2000.

The model is based on a quadratic regression. Source: Gerard Klauer Mattison & Company/*New York Times*, March 25, 2000, p. C1.

15. Revenue The rate of receipt of revenue by Wal-Mart for 1995 to 2002 can be approximated by

$$R(t) = 83e^{0.15t} \text{ billion dollars per year} \qquad (0 \le t \le 7)$$

where t is time in years since January 1995. Estimate, to the nearest $1 billion, Wal-Mart's total revenue from January 1995 through January 2002.

The authors' model is based on data for Wal-Marts, Supercenters, Sam's Clubs, and Neighborhood Markets. Sources: Company reports/*New York Times*, February 24, 2002, p. 26.

16. Revenue The rate of receipt of revenue by Target for 1995 to 2001 can be approximated by

$$R(t) = 15e^{0.12t} \text{ billion dollars per year} \qquad (0 \le t \le 6)$$

where t is time in years since January 1995. Estimate, to the nearest $1 billion, Target's total revenue from January 1995 through January 2001.

The authors' model is based on data for Target stores only. Sources: Company reports/*New York Times*, February 24, 2002, p. 26.

17. Revenue Refer to Exercise 13. Suppose, from January 1995 on, Nokia invested its after-tax revenue in an investment yielding 2% compounded continuously. What, to the nearest $1 billion, would have been the total value of Nokia's after-tax revenues by January 2001?

18. Revenue Refer to Exercise 14. Suppose, from January 1996 on, Nintendo invested its revenue in an investment yielding 3% compounded continuously. What, to the nearest $0.1 billion, would have been the total value of Nintendo's revenue by January 2000?

19. Revenue Refer to Exercise 15. Suppose, from January 1995 on, Wal-Mart invested its revenue in an investment that depreciated continuously at a rate of 5% per year. What, to the nearest $1 billion, would have been the total value of Wal-Mart's revenues by January 2002?

20. Revenue Refer to Exercise 16. Suppose, from January 1995 on, Target invested its revenue in an investment that depreciated continuously at a rate of 3% per year. What, to the nearest $1 billion, would have been the total value of Target's revenue by January 2001?

21. Saving for Retirement You are saving for your retirement by investing $700 per month in an annuity with a guaranteed interest rate of 6% per year. With a continuous stream of investment and continuous compounding, how much will you have accumulated in the annuity by the time you retire in 45 years?

22. Saving for College When your first child is born, you begin to save for college by depositing $400 per month in an account paying 12% interest per year. With a continuous stream of investment and continuous compounding, how much will you have accumulated in the account by the time your child enters college 18 years later?

23. Saving for Retirement You begin saving for your retirement by investing $700 per month in an annuity with a guaranteed interest rate of 6% per year. You increase the amount you invest at the rate of 3% per year. With continuous investment and compounding, how much will you have accumulated in the annuity by the time you retire in 45 years?

24. Saving for College When your first child is born, you begin to save for college by depositing $400 per month in an account paying 12% interest per year. You increase the amount you save by 2% per year. With continuous investment and compounding, how much will have accumulated in the account by the time your child enters college 18 years later?

25. Bonds The U.S. Treasury issued a 30-year bond on October 15, 2001, paying 3.375% interest. So, if you bought $100,000 worth of these bonds, you would receive $3375 per year in interest for 30 years. An investor wishes to buy the rights to receive the interest on $100,000 worth of these bonds. The amount the investor is willing to pay is the present value of the interest payments, assuming a 4% rate of return. Assuming (incorrectly, but approximately) that the interest payments are made continuously, what will the investor pay?

Source: The Bureau of the Public Debt's Web site: http://www.publicdebt.treas.gov/.

26. Bonds Megabucks Corporation is issuing a 20-year bond paying 7% interest (see Exercise 25). An investor wishes to buy the rights to receive the interest on $50,000 worth of these bonds and seeks a 6% rate of return. Assuming that the interest payments are made continuously, what will the investor pay?

27. Valuing Future Income Inga was injured and can no longer work. As a result of a lawsuit, she is to be awarded the present value of the income she would have received over the next 20 years. Her income at the time she was injured was $100,000 per year, increasing by $5000 per year. What will be the amount of her award, assuming continuous income and a 5% interest rate?

28. Valuing Future Income Max was injured and can no longer work. As a result of a lawsuit, he is to be awarded the present value of the income he would have received over the next 30 years. His income at the time he was injured was $30,000 per year, increasing by $1500 per year. What will be the amount of his award, assuming continuous income and a 6% interest rate?

COMMUNICATION AND REASONING EXERCISES

29. Complete the following: The future value of a continuous income stream earning 0% interest is the same as the _____ value.

30. Complete the following: The present value of a continuous income stream earning 0% interest is the same as the _____ value.

31. Your study-group friend says that the future value of a continuous stream of income is always greater than the total value, assuming a positive rate of return. Is she correct? Why?

32. Your other study-group friend says that the present value of a continuous stream of income can sometimes be greater than the total value, depending on the (positive) interest rate. Is he correct? Explain

33. Arrange from smallest to largest: total value, future value, present value of a continuous stream of income (assuming a positive income and positive rate of return).

34. a. Arrange the following functions from smallest to largest, where $a \le t \le b$ and r and $R(t)$ are positive: $R(t)$, $R(t)e^{r(b-t)}$, $R(t)e^{r(a-t)}$.

 b. Use the result from part (a) to justify your answers in Exercises 31–33.

7.5 Improper Integrals and Applications

All the definite integrals we have seen so far have had the form $\int_a^b f(x)\,dx$, with a and b finite and $f(x)$ piecewise continuous and bounded on the closed interval $[a, b]$. We can relax these requirements somewhat. When we do so, we obtain what are called **improper integrals.** There are various types of improper integrals.

Integrals in Which a Limit of Integration Is Infinite

These integrals are written as

$$\int_a^{+\infty} f(x)\,dx \qquad \int_{-\infty}^b f(x)\,dx \qquad \text{or} \qquad \int_{-\infty}^{+\infty} f(x)\,dx$$

Let's concentrate for a moment on the first form, $\int_a^{+\infty} f(x)\,dx$. What does the $+\infty$ mean here? As it often does, it means that we are to take a limit as something gets large. Specifically, it means the limit as the upper bound of integration gets large.

Improper Integral with an Infinite Limit of Integration

We define

$$\int_a^{+\infty} f(x)\, dx = \lim_{M \to +\infty} \int_a^M f(x)\, dx$$

provided the limit exists. If the limit exists, we say that $\int_a^{+\infty} f(x)\, dx$ **converges.** Otherwise, we say that $\int_a^{+\infty} f(x)\, dx$ **diverges.** Similarly, we define

$$\int_{-\infty}^b f(x)\, dx = \lim_{M \to -\infty} \int_M^b f(x)\, dx$$

provided the limit exists. Finally, we define

$$\int_{-\infty}^{+\infty} f(x)\, dx = \int_{-\infty}^a f(x)\, dx + \int_a^{+\infty} f(x)\, dx$$

for some convenient a, provided *both* integrals on the right converge.

Quick Examples

1. $\displaystyle \int_1^{+\infty} \frac{dx}{x^2} = \lim_{M \to +\infty} \int_1^M \frac{dx}{x^2} = \lim_{M \to +\infty} \left[-\frac{1}{x} \right]_1^M = \lim_{M \to +\infty} \left(-\frac{1}{M} + 1 \right) = 1$ Converges

2. $\displaystyle \int_1^{+\infty} \frac{dx}{x} = \lim_{M \to +\infty} \int_1^M \frac{dx}{x} = \lim_{M \to +\infty} \left[\ln|x| \right]_1^M = \lim_{M \to +\infty} (\ln M - \ln 1) = +\infty$ Diverges

3. $\displaystyle \int_{-\infty}^{-1} \frac{dx}{x^2} = \lim_{M \to -\infty} \int_M^{-1} \frac{dx}{x^2} = \lim_{M \to -\infty} \left[-\frac{1}{x} \right]_M^{-1} = \lim_{M \to -\infty} \left(1 + \frac{1}{M} \right) = 1$ Converges

4. $\displaystyle \int_{-\infty}^{+\infty} e^{-x}\, dx = \int_{-\infty}^0 e^{-x}\, dx + \int_0^{+\infty} e^{-x}\, dx$

$$= \lim_{M \to -\infty} \int_M^0 e^{-x}\, dx + \lim_{M \to +\infty} \int_0^M e^{-x}\, dx$$

$$= \lim_{M \to -\infty} -\left[e^{-x} \right]_M^0 + \lim_{M \to +\infty} -\left[e^{-x} \right]_0^M$$

$$= \lim_{M \to -\infty} (e^{-M} - 1) + \lim_{M \to +\infty} (1 - e^{-M})$$

$$= +\infty + 1 \qquad\qquad\qquad\qquad \text{Diverges}$$

5. $\displaystyle \int_{-\infty}^{+\infty} xe^{-x^2}\, dx = \int_{-\infty}^0 xe^{-x^2}\, dx + \int_0^{+\infty} xe^{-x^2}\, dx$

$$= \lim_{M \to -\infty} \int_M^0 xe^{-x^2}\, dx + \lim_{M \to +\infty} \int_0^M xe^{-x^2}\, dx$$

$$= \lim_{M \to -\infty} \left[-\frac{1}{2} e^{-x^2} \right]_M^0 + \lim_{M \to +\infty} \left[-\frac{1}{2} e^{-x^2} \right]_0^M$$

$$= \lim_{M \to -\infty} \left(-\frac{1}{2} + \frac{1}{2} e^{-M^2} \right) + \lim_{M \to +\infty} \left(-\frac{1}{2} e^{-M^2} + \frac{1}{2} \right)$$

$$= -\frac{1}{2} + \frac{1}{2} = 0 \qquad\qquad\qquad\qquad \text{Converges}$$

Question We learned that the integral can be interpreted as the area under the curve. Is this still true for improper integrals?

Answer Yes. Figure 16 illustrates how we can represent an improper integral as the area of an infinite region.

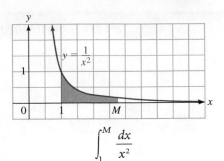
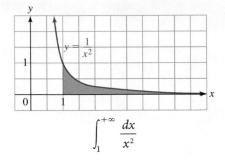

$$\int_1^M \frac{dx}{x^2}$$ $$\int_1^{+\infty} \frac{dx}{x^2}$$

Figure 16

On the left we see the area represented by $\int_1^M dx/x^2$. As M gets larger, the integral approaches $\int_1^{+\infty} dx/x^2$. In the picture, think of M being moved farther and farther along the x axis in the direction of increasing x, resulting in the region shown on the right.

Question   Wait! We calculated $\int_1^{+\infty} dx/x^2 = 1$. Does this mean that the infinitely long area in Figure 16 has an area of only 1 square unit?

Answer   That is exactly what it means. If you had enough paint to cover 1 square unit, you would never run out of paint while painting the region in Figure 16. This is one of the places where mathematics seems to contradict common sense. But common sense is notoriously unreliable when dealing with infinities.

Example 1 • Future Sales of VCRs

In 2001 sales of DVD players were starting to make inroads into the sales of VCRs, but VCRs were still selling well.[2] Approximately 15 million VCRs were expected to be sold in 2001. Suppose sales of VCRs decrease by 15% per year from 2001 on. How many VCRs, total, will be sold from 2001 on?

Solution   Recall that the total sales between two dates can be computed as the definite integral of annual sales. So, if we wanted the sales between the year 2001 and a year far in the future, we would compute $\int_0^M s(t)\, dt$ with a large M, where $s(t)$ is the annual sales t years after 2001. Since we want to know the *total* number of VCRs sold from 2001 on, we let $M \to +\infty$; that is, we compute $\int_0^{+\infty} s(t)\, dt$.

Since sales of VCRs are decreasing by 15% per year, we can model $s(t)$ by

$$s(t) = 15(0.85)^t \text{ million VCRs}$$

[2]SOURCE: "VCRs Outsell DVD Players over Holidays," *USA Today*, January 24, 2001.

where t is the number of years since 2001. Therefore,

$$\text{total sales from 2001 on} = \int_0^{+\infty} 15(0.85)^t \, dt$$

$$= \lim_{M \to +\infty} \int_0^M 15(0.85)^t \, dt$$

$$= \frac{15}{\ln 0.85} \lim_{M \to +\infty} [0.85^t]_0^M$$

$$= \frac{15}{\ln 0.85} \lim_{M \to +\infty} (0.85^M - 0.85^0)$$

$$= \frac{15}{\ln 0.85} (-1) \approx 92.3 \text{ million VCRs}$$

Integrals in Which the Integrand Becomes Infinite

We can sometimes compute integrals $\int_a^b f(x) \, dx$ in which $f(x)$ becomes infinite. The first case to consider is when $f(x)$ approaches $+\infty$ at either a or b.

Example 2 • Integrand Infinite at One Endpoint

Calculate $\int_0^1 \frac{1}{\sqrt{x}} \, dx$.

Solution Notice that the integrand approaches $+\infty$ as x approaches 0 from the right and is not defined at 0. This makes the integral an improper integral. Figure 17 shows the region whose area we are trying to calculate; it extends infinitely vertically rather than horizontally. Now, if $0 < r < 1$, the integral $\int_r^1 (1/\sqrt{x}) \, dx$ is a proper integral because we avoid the bad behavior at 0. This integral gives the area shown in Figure 18. If we let r approach 0 from the right, the area in Figure 18 will approach the area in Figure 17. So, we calculate

$$\int_0^1 \frac{1}{\sqrt{x}} \, dx = \lim_{r \to 0^+} \int_r^1 \frac{1}{\sqrt{x}} \, dx$$

$$= \lim_{r \to 0^+} [2\sqrt{x}]_r^1$$

$$= \lim_{r \to 0^+} (2 - 2\sqrt{r}) = 2$$

Thus, just as before, we have an infinitely long region with finite area.

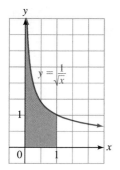

Figure 17

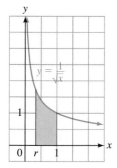

Figure 18

Generalizing, we make the following definition.

> ### Improper Integral in Which the Integrand Becomes Infinite
> If $f(x)$ is defined for all x with $a < x \le b$ but approaches $\pm\infty$ as x approaches a, we define
>
> $$\int_a^b f(x) \, dx = \lim_{r \to a^+} \int_r^b f(x) \, dx$$

provided the limit exists. Similarly, if $f(x)$ is defined for all x with $a \leq x < b$ but approaches $\pm\infty$ as x approaches b, we define

$$\int_a^b f(x) \, dx = \lim_{r \to b^-} \int_a^r f(x) \, dx$$

provided the limit exists. In either case, if the limit exists, we say that $\int_a^b f(x) \, dx$ **converges.** Otherwise, we say that $\int_a^b f(x) \, dx$ **diverges.**

Example 3 • Testing for Convergence

Does $\displaystyle\int_{-1}^3 \frac{x}{x^2 - 9} \, dx$ converge? If so, to what?

Solution We first check to see where, if anywhere, the integrand approaches $\pm\infty$. That will happen where the denominator becomes zero, so we solve $x^2 - 9 = 0$:

$$x^2 - 9 = 0$$
$$x^2 = 9$$
$$x = \pm 3$$

The solution $x = -3$ is outside the range of integration, so we ignore it. The solution $x = 3$ is, however, the right endpoint of the range of integration, so the integral is improper. We need to investigate the following limit:

$$\int_{-1}^3 \frac{x}{x^2 - 9} \, dx = \lim_{r \to 3^-} \int_{-1}^r \frac{x}{x^2 - 9} \, dx$$

Now, to calculate the integral, we use a substitution:

$$u = x^2 - 9$$

$$\frac{du}{dx} = 2x$$

$$dx = \frac{1}{2x} \, du$$

When $x = r, u = r^2 - 9$.

When $x = -1, u = (-1)^2 - 9 = -8$.

Thus,

$$\int_{-1}^r \frac{x}{x^2 - 9} \, dx = \int_{-8}^{r^2 - 9} \frac{1}{2u} \, du = \frac{1}{2} \left[\ln |u| \right]_{-8}^{r^2-9} = \frac{1}{2} \left(\ln |r^2 - 9| - \ln 8 \right)$$

Now we take the limit:

$$\int_{-1}^3 \frac{x}{x^2 - 9} \, dx = \lim_{r \to 3^-} \int_{-1}^r \frac{x}{x^2 - 9} \, dx = \lim_{r \to 3^-} \frac{1}{2} \left(\ln |r^2 - 9| - \ln 8 \right) = -\infty$$

because, as $r \to 3, r^2 - 9 \to 0$, and so $\ln |r^2 - 9| \to -\infty$. Thus, this integral diverges.

Example 4 • Integrand Infinite between the Endpoints

Does $\displaystyle\int_{-2}^{3} \frac{1}{x^2}\, dx$ converge? If so, to what?

Solution Again we check to see if there are any points at which the integrand approaches $\pm\infty$. There is such a point at $x = 0$. This is between the endpoints of the range of integration. To deal with this, we break the integral into two integrals:

$$\int_{-2}^{3} \frac{1}{x^2}\, dx = \int_{-2}^{0} \frac{1}{x^2}\, dx + \int_{0}^{3} \frac{1}{x^2}\, dx$$

Each integral on the right is an improper integral with the integrand approaching $\pm\infty$ at an endpoint. If both of the integrals on the right converge, we take the sum as the value of the integral on the left. So now we compute

$$\int_{-2}^{0} \frac{1}{x^2}\, dx = \lim_{r \to 0^-} \int_{-2}^{r} \frac{1}{x^2}\, dx = \lim_{r \to 0^-} \left[-\frac{1}{x} \right]_{-2}^{r}$$

$$= \lim_{r \to 0^-} \left(-\frac{1}{r} - \frac{1}{2} \right)$$

which diverges to $+\infty$. There is no need now to check $\int_{0}^{3} (1/x^2)\, dx$; since one of the two pieces of the integral diverges, we simply say that $\int_{-2}^{3} (1/x^2)\, dx$ diverges.

✴ **Before we go on . . .** What if we had been sloppy and not checked first whether the integrand approached $\pm\infty$ somewhere? Then we probably would have applied the Fundamental Theorem of Calculus (FTC) and done the following:

$$\int_{-2}^{3} \frac{1}{x^2}\, dx = \left[-\frac{1}{x} \right]_{-2}^{3} = \left(-\frac{1}{3} - \frac{1}{2} \right) = -\frac{5}{6} \qquad \text{X } \textit{WRONG}$$

Notice that the answer this "calculation" gives is patently ridiculous. Since $1/x^2 > 0$ for all x for which it is defined, any definite integral of $1/x^2$ *must* give a positive answer. *Moral:* Always check to see whether the integrand blows up anywhere in the range of integration. If it does, the FTC does not apply, and we must use the method of this example.

We end with an example of what to do if an integral is improper for more than one reason.

Example 5 • An Integral Improper in Two Ways

Does $\displaystyle\int_{0}^{+\infty} \frac{1}{\sqrt{x}}\, dx$ converge? If so, to what?

Solution This integral is improper for two reasons. First, the range of integration is infinite. Second, the integrand blows up at the endpoint 0. To separate these two problems, we break up the integral at some convenient point:

$$\int_{0}^{+\infty} \frac{1}{\sqrt{x}}\, dx = \int_{0}^{1} \frac{1}{\sqrt{x}}\, dx + \int_{1}^{+\infty} \frac{1}{\sqrt{x}}\, dx$$

We chose to break the integral at 1. Any positive number would have done, but 1 is generally easier to use in calculations.

We discussed the first piece, $\int_0^1 (1/\sqrt{x})\, dx$ earlier; it converges to 2. For the second piece, we have

$$\int_1^{+\infty} \frac{1}{\sqrt{x}}\, dx = \lim_{M\to+\infty} \int_1^M \frac{1}{\sqrt{x}}\, dx = \lim_{M\to+\infty} [2\sqrt{x}]_1^M = \lim_{M\to+\infty} (2\sqrt{M} - 2)$$

which diverges to $+\infty$. Since the second piece of the integral diverges, we conclude that $\int_0^{+\infty} (1/\sqrt{x})\, dx$ diverges.

7.5 EXERCISES

For some of the exercises in this section, you need to assume the fact that $\lim_{M\to+\infty} M^n e^{-M} = 0$ for all n.

In Exercises 1–26, decide whether each integral converges. If the integral converges, compute its value.

1. $\int_1^{+\infty} x\, dx$

2. $\int_0^{+\infty} e^{-x}\, dx$

3. $\int_{-2}^{+\infty} e^{-0.5x}\, dx$

4. $\int_1^{+\infty} \frac{1}{x^{1.5}}\, dx$

5. $\int_{-\infty}^2 e^x\, dx$

6. $\int_{-\infty}^{-1} \frac{1}{x^{1/3}}\, dx$

7. $\int_{-\infty}^{-2} \frac{1}{x^2}\, dx$

8. $\int_{-\infty}^0 e^{-x}\, dx$

9. $\int_0^{+\infty} x^2 e^{-6x}\, dx$

10. $\int_0^{+\infty} (2x - 4)e^{-x}\, dx$

11. $\int_0^5 \frac{2}{x^{1/3}}\, dx$

12. $\int_0^2 \frac{1}{x^2}\, dx$

13. $\int_{-1}^2 \frac{3}{(x + 1)^2}\, dx$

14. $\int_{-1}^2 \frac{3}{(x + 1)^{1/2}}\, dx$

15. $\int_{-1}^2 \frac{3x}{x^2 - 1}\, dx$

16. $\int_{-1}^2 \frac{3}{x^{1/3}}\, dx$

17. $\int_{-2}^2 \frac{1}{(x + 1)^{1/5}}\, dx$

18. $\int_{-2}^2 \frac{2x}{\sqrt{4 - x^2}}\, dx$

19. $\int_{-1}^1 \frac{2x}{x^2 - 1}\, dx$

20. $\int_{-1}^2 \frac{2x}{x^2 - 1}\, dx$

21. $\int_{-\infty}^{+\infty} xe^{-x^2}\, dx$

22. $\int_{-\infty}^{+\infty} xe^{1-x^2}\, dx$

23. $\int_0^{+\infty} \frac{1}{x \ln x}\, dx$

24. $\int_0^{+\infty} \ln x\, dx$

25. $\int_0^{+\infty} \frac{2x}{x^2 - 1}\, dx$

26. $\int_{-\infty}^0 \frac{2x}{x^2 - 1}\, dx$

APPLICATIONS

27. Advertising Revenue From June 2001 to June 2002, *GQ* magazine's advertising revenues can be approximated by

$$R(t) = 91.7(0.90)^t \text{ million dollars per year} \qquad (0 \le t \le 1)$$

where t is time in years since June 2001. By extrapolating this model into the indefinite future, project *GQ's* total advertising revenue from June 2001 on. (Round your answer to the nearest $1 million.)

Based on 6-month advertising revenue figures for June 2001 and June 2002. Source: *New York Times,* July, 29, 2002, p. C1.

28. Advertising Revenue From June 2001 to June 2002, *Esquire* magazine's advertising revenues can be approximated by

$$R(t) = 57.0(0.927)^t \text{ million dollars per year} \qquad (0 \le t \le 1)$$

where t is time in years since June 2001. By extrapolating this model into the indefinite future, project *Esquire's* total advertising revenue from June 2001 on. (Round your answer to the nearest $1 million.)

Based on 6-month advertising revenue figures for June 2001 and June 2002. Source: *New York Times,* July 29, 2002, p. C1.

29. Cigarette Sales According to U.S. tobacco industry reports to the Federal Trade Commission, the number of cigarettes sold domestically in 1999 decreased by 10.3% from the 1998 total of 485.5 billion cigarettes. Use an exponential model to forecast the total number of cigarettes sold from 1998 on. (Round your answer to the nearest 1 billion cigarettes.)

Source: Federal Trade Commission, 2001
http://www.ftc.gov/opa/2001/03/cigarette.htm

30. Sales Sales of the text *Calculus and You* have been declining continuously at a rate of 5% per year. Assuming that *Calculus and You* currently sells 5000 copies each year and that sales will continue this pattern of decline, calculate total future sales of the text.

31. Sales My financial adviser has predicted that annual sales of Frodo T-shirts will continue to decline by 10% each year. At the moment, I have 3200 of the shirts in stock and am selling them at a rate of 200 per year. Will I ever sell them all?

32. Revenue Alarmed about the sales prospects for my Frodo T-shirts (see Exercise 31), I will try to make up lost revenues by increasing the price by $1 each year. I now charge $10 per shirt. What is the total amount of revenue I can expect to earn from sales of my T-shirts, assuming the sales levels described in Exercise 31? (Round your answer to the nearest $1000.)

33. Education Let $N(t)$ be the number of high school students graduated in the United States in year t. This number is projected to change at a rate of about

$$N'(t) = 0.214t^{-0.91} \text{ million graduates per year} \qquad (0 \le t \le 21)$$

where t is time in years since 1990. In 1991 there were about 2.5 million high school students graduated. By extrapolating the model, what can you say about the number of high school students graduated in a year far in the future?

Based on a regression model. Source: U.S. Dept. of Education, 2002 (http://nces.ed.gov/).

34. Education, Martian Let $M(t)$ be the number of high school students graduated in the Republic of Mars in year t. This number is projected to change at a rate of about

$M'(t) = 0.321t^{-1.10}$ thousand graduates per year

$$(0 \le t \le 50)$$

where t is time in years since 2020. In 2021 there were about 1300 high school students graduated. By extrapolating the model, what can you say about the number of high school students graduated in a year far in the future?

35. Cell Phone Revenues The number of cell phone subscribers in China for the period 2000–2005 was projected to follow the equation

$$N(t) = 39t + 68 \text{ million subscribers}$$

in year t ($t = 0$ represents 2000). The average annual revenue per cell phone user was \$350 in 2000.

a. Assuming that, due to competition, the revenue per cell phone user decreases continuously at an annual rate of 10%, give a formula for the annual revenue in year t.

b. Using the model you obtained in part (a) as an estimate of the rate of change of total revenue, estimate the total revenue from 2000 into the indefinite future.

Based on a regression of projected figures (coefficients are rounded). Source: Intrinsic Technology/*New York Times*, November 24, 2000, p. C1.

36. Vid Phone Revenues The number of vid phone subscribers in the Republic of Mars for the period 2200–2300 was projected to follow the equation

$$N(t) = 18t - 10 \text{ thousand subscribers}$$

in year t ($t = 0$ represents 2200). The average annual revenue per vid phone user was $\overline{\overline{Z}}40$ in 2200.

$\overline{\overline{Z}}$ designates Zonars, the designated currency for the city-state of Utarek, Mars. Source: http://www.marsnext.com/comm/zonars.html

a. Assuming that, due to competition, the revenue per vid phone user decreases continuously at an annual rate of 20%, give a formula for the annual revenue in year t.

b. Using the model you obtained in part (a) as an estimate of the rate of change of total revenue, estimate the total revenue from 2200 into the indefinite future.

37. Foreign Investments According to data published by the World Bank, the annual flow of private investment to developing countries from more developed countries in the 1990s was approximately

$$q(t) = 10 + 1.56t^2$$

where $q(t)$ is the annual investment in billions of dollars and t is time in years since the start of 1986. Assuming a worldwide inflation rate of 5% per year, find the value of all private aid to developing countries from 1986 on in constant dollars. (The constant dollar value of $q(t)$ dollars t years from now is given by $q(t)e^{-rt}$, where r is the fractional rate of inflation.)

Source: The authors' approximation, based on data published by the World Bank/*New York Times*, December 17, 1993, p. D1.

38. Foreign Investments Repeat Exercise 37, using the following model for government loans and grants to developing countries:

$$q(t) = 38 + 0.6(t - 3)^2$$

Source: The authors' approximation, based on data published by the World Bank/*New York Times*, December 17, 1993, p. D1.

39. Online Book Sales The number of books each year sold online in the United States in the period 1997–2000 can be approximated by

$$N(t) = \frac{82.8(7.14)^t}{21.8 + (7.14)^t} \text{ million books per year}$$

($t = 0$ represents 1997). Investigate the integrals $\int_0^{+\infty} N(t)\, dt$ and $\int_{-\infty}^0 N(t)\, dt$ and interpret your answers.

The model is a logistic regression. Source: Ipsos-NPD Book Trends/*New York Times*, April 16, 2001, p. C1.

40. Mousse Sales The weekly demand for your company's Lo-Cal Mousse is modeled by the equation

$$q(t) = \frac{50e^{2t-1}}{1 + e^{2t-1}}$$

where t is time from now in weeks and $q(t)$ is the number of gallons sold each week. Investigate the integrals $\int_0^{+\infty} q(t)\, dt$ and $\int_{-\infty}^0 q(t)\, dt$ and interpret your answers.

The Normal Curve Exercises 41–44 require the use of a graphing calculator or computer programmed to do numerical integration. The *normal distribution curve*, which models the distributions of data in a wide range of applications, is given by the function

$$p(x) = \frac{1}{\sqrt{2\pi}\sigma} e^{-(x-\mu)^2/2\sigma^2}$$

where $\pi = 3.14159265\ldots$ and σ and μ are constants called the standard deviation and the mean, respectively. Its graph (for $\sigma = 1$ and $\mu = 2$) is shown in the figure:

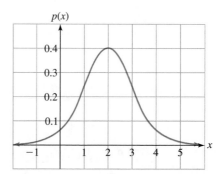

41. With $\sigma = 4$ and $\mu = 1$, approximate $\int_{-\infty}^{+\infty} p(x)\, dx$.

42. With $\sigma = 1$ and $\mu = 0$, approximate $\int_0^{+\infty} p(x)\, dx$.

43. With $\sigma = 1$ and $\mu = 0$, approximate $\int_1^{+\infty} p(x)\, dx$.

44. With $\sigma = 1$ and $\mu = 0$, approximate $\int_{-\infty}^1 p(x)\, dx$.

45. Variable Sales The value of your Chateau Petit Mont Blanc 1963 vintage burgundy is increasing continuously at an annual rate 40%, and you have a supply of 1000 bottles worth $85 each at today's prices. To ensure a steady income, you have decided to sell your wine at a diminishing rate—starting at 500 bottles per year and then decreasing this figure continuously at a fractional rate of 100% per year. How much income (to the nearest dollar) can you expect to generate by this scheme? (*Hint:* Use the formula for continuously compounded interest.)

46. Panic Sales Unfortunately, your large supply of Chateau Petit Mont Blanc is continuously turning to vinegar at a fractional rate of 60% per year! You have thus decided to sell off your Petit Mont Blanc at $50 per bottle, but the market is a little thin, and you can only sell 400 bottles each year. Since you have no way of knowing which bottles now contain vinegar until they are opened, you will have to give refunds for all the bottles of vinegar. What will be your net income before all the wine turns to vinegar?

47. Meteor Impacts The frequency of meteor impacts on Earth can be modeled by

$$n(k) = \frac{1}{5.6997k^{1.081}}$$

where $n(k) = N'(k)$ and $N(k)$ is the average number of meteors of energy less than or equal to k megatons that will hit Earth in 1 year. (A small nuclear bomb releases on the order of 1 megaton of energy.)

a. How many meteors of energy at least $k = 0.2$ hit Earth each year?

b. Investigate and interpret the integral $\int_0^1 n(k)\, dk$.

Source: The authors' model, based on data published by NASA International Near-Earth-Object Detection Workshop/*New York Times*, January 25, 1994, p. C1.

48. Meteor Impacts (Continuing Exercise 47)

a. Explain why the integral $\int_a^b kn(k)\, dk$ computes the total energy released each year by meteors with energies between a and b megatons.

b. Compute and interpret $\int_0^1 kn(k)\, dk$.

c. Compute and interpret $\int_1^{+\infty} kn(k)\, dk$.

49. The Gamma Function The gamma function is defined by the formula

$$\Gamma(x) = \int_0^{+\infty} t^{x-1} e^{-t}\, dt$$

a. Find $\Gamma(1)$ and $\Gamma(2)$.

b. Use integration by parts to show that for every positive integer n, $\Gamma(n + 1) = n\Gamma(n)$.

c. Deduce that $\Gamma(n) = (n - 1)!$ $[= (n - 1)(n - 2) \cdots 2 \cdot 1]$ for every positive integer n.

50. Laplace Transforms The Laplace Transform $F(x)$ of a function $f(t)$ is given by the formula

$$F(x) = \int_0^{+\infty} f(t)e^{-xt}\, dt \qquad (x > 0)$$

a. Find $F(x)$ for $f(t) = 1$ and for $f(t) = t$.

b. Find a formula for $F(x)$ if $f(t) = t^n$ $(n = 1, 2, 3, \ldots)$.

c. Find a formula for $F(x)$ if $f(t) = e^{at}$ (a constant).

COMMUNICATION AND REASONING EXERCISES

51. Why can't the Fundamental Theorem of Calculus (FTC) be used to evaluate $\int_{-1}^1 1/x\, dx$?

52. Why can't the FTC be used to evaluate $\int_1^{+\infty} 1/x^2\, dx$?

53. It sometimes happens that the FTC gives the correct answer for an improper integral. Does the FTC give the correct answer for improper integrals of the form

$$\int_{-a}^a \frac{1}{x^{1/r}}\, dx$$

if $r = 3, 5, 7, \ldots$?

54. Does the FTC give the correct answer for improper integrals of the form

$$\int_{-a}^a \frac{1}{x^r}\, dx$$

if $r = 3, 5, 7, \ldots$?

55. How could you use technology to approximate improper integrals? (Your discussion should refer to each type of improper integral.)

56. Use technology to approximate the integrals $\int_0^M e^{-(x-10)^2}\, dx$ for larger and larger values of M, using Riemann sums with 500 subdivisions. What do you find? Comment on the answer.

57. Make up an interesting application whose solution is $\int_{10}^{+\infty} 100te^{-0.2t}\, dt = \$1015.01$.

58. Make up an interesting application whose solution is $\int_{100}^{+\infty} 1/r^2\, dr = 0.01$.

7.6 Differential Equations and Applications

A **differential equation** is an equation that involves a derivative of an unknown function. A **first-order differential equation** involves only the first derivative of the unknown function. A **second-order differential equation** involves the second derivative of the unknown function (and possibly the first derivative). Higher-order differential equations are defined similarly. In this book we deal only with first-order differential equations.

To **solve** a differential equation means to find the unknown function. Many of the laws of science and other fields describe how things change. When expressed mathematically these laws take the form of equations involving derivatives—that is, differential equations. The field of differential equations is a large and very active area of study in mathematics, and we see only a small part of it in this section.

Example 1 • Motion

A drag racer accelerates from a stop so that its speed is $40t$ feet per second t seconds after starting. How far will the car go in 8 seconds?

Solution We can express this problem as a differential equation. We wish to find the car's position function $s(t)$. We are told about its speed, which is ds/dt. Precisely, we are told that

$$\frac{ds}{dt} = 40t$$

This is the differential equation we have to solve to find $s(t)$. But we know how to solve this kind of differential equation; we integrate:

$$s(t) = \int 40t \, dt = 20t^2 + C$$

We now have the **general solution** to the differential equation. By letting C take on different values, we get all the possible solutions. We can specify the one **particular solution** that gives the answer to our problem by imposing the **initial condition** that $s(0) = 0$. Substituting into $s(t) = 20t^2 + C$, we get

$$0 = s(0) = 20(0)^2 + C = C$$

so $C = 0$ and $s(t) = 20t^2$. To answer the question, the car travels $20(8)^2 = 1280$ feet in 8 seconds.

We did not have to work hard to solve the differential equation in Example 1. In fact, any differential equation of the form $dy/dx = f(x)$ can (in theory) be solved by integrating. (Whether we can actually carry out the integration is another matter!)

Simple Differential Equations
A **simple** differential equation has the form

$$\frac{dy}{dx} = f(x)$$

Its general solution is

$$y = \int f(x) \, dx$$

Quick Example
The differential equation

$$\frac{dy}{dx} = 2x^2 - 4x^3$$

is simple and has general solution

$$y = \int f(x) \, dx = \frac{2x^3}{3} - x^4 + C$$

Not all differential equations are simple, as Example 2 shows.

Example 2 • Separable Differential Equation

Consider the differential equation $\dfrac{dy}{dx} = \dfrac{x}{y^2}$.

a. Find the general solution.

b. Find the particular solution that satisfies the initial condition $y(0) = 2$.

Solution

a. This is not a simple differential equation because the right-hand side is a function of both x and y. We cannot solve this equation by just integrating because we would not know what to do with the y if we integrated the right-hand side with respect to x. The solution to this problem is to "separate" the variables.

Step 1 *Separate the variables algebraically.* We rewrite the equation as

$$y^2\, dy = x\, dx$$

Step 2 *Integrate both sides:*

$$\int y^2\, dy = \int x\, dx$$

giving

$$\frac{y^3}{3} = \frac{x^2}{2} + C$$

Step 3 *Solve for the dependent variable.* We solve for y:

$$y^3 = \frac{3}{2}x^2 + 3C = \frac{3}{2}x^2 + D$$

(rewriting $3C$ as D, an equally arbitrary constant), so

$$y = \left(\frac{3}{2}x^2 + D\right)^{1/3}$$

This is the general solution of the differential equation.

b. We now need to find the value for D that will give us the solution satisfying the condition $y(0) = 2$. Substituting 0 for x and 2 for y in the general solution, we get

$$2 = \left(\frac{3}{2}(0)^2 + D\right)^{1/3} = D^{1/3}$$

$$D = 2^3 = 8$$

Thus, the particular solution we are looking for is

$$y = \left(\frac{3}{2}x^2 + 8\right)^{1/3}$$

✳ *Before we go on . . .* We can check the general solution by calculating both sides of the differential equation and comparing.

$$\frac{dy}{dx} = \frac{d}{dx}\left(\frac{3}{2}x^2 + D\right)^{1/3} = x\left(\frac{3}{2}x^2 + D\right)^{-2/3}$$

$$\frac{x}{y^2} = \frac{x}{\left(\frac{3}{2}x^2 + 8\right)^{2/3}} = x\left(\frac{3}{2}x^2 + D\right)^{-2/3} \quad ✓$$

Question Above, we wrote $y^2\,dy$ and $x\,dx$. What do they mean?

Answer Although it is possible to give meaning to these symbols, for us they are just a notational convenience. We could have done the following instead:

$$y^2\,\frac{dy}{dx} = x$$

Now we integrate both sides with respect to x.

$$\int y^2\,\frac{dy}{dx}\,dx = \int x\,dx$$

On the left we can use substitution to write

$$\int y^2\,\frac{dy}{dx}\,dx = \int y^2\,dy$$

which brings us back to the equation

$$\int y^2\,dy = \int x\,dx$$

We were able to separate the variables in Example 2 because the right-hand side, x/y^2, was a *product* of a function of x and a function of y—namely,

$$\frac{x}{y^2} = x\left(\frac{1}{y^2}\right)$$

In general, we can say the following.

Separable Differential Equation

A **separable** differential equation has the form

$$\frac{dy}{dx} = f(x)g(y)$$

We solve a separable differential equation by separating the x's and the y's algebraically, writing

$$\frac{1}{g(y)}\,dy = f(x)\,dx$$

and then integrating:

$$\int \frac{1}{g(y)}\,dy = \int f(x)\,dx$$

Example 3 • Rising Medical Costs

Spending on Medicare from 2000 to 2025 was projected to rise continuously at an instantaneous rate of 3.7% per year.[3] Find a formula for Medicare spending y as a function of time t in years since 2000.

Solution We discussed this kind of growth when we first discussed exponential functions. By looking at this problem as one leading to a differential equation, we can see why the exponential function comes in.

When we say that Medicare spending y was going up continuously at an instantaneous rate of 3.7% per year, we mean that *the instantaneous rate of increase of y was 3.7% of its value,* or

$$\frac{dy}{dt} = 0.037y$$

This is a separable differential equation. Separating the variables gives

$$\frac{1}{y}\, dy = 0.037\, dt$$

Integrating both sides, we get

$$\int \frac{1}{y}\, dy = \int 0.037\, dt$$

so

$$\ln y = 0.037t + C$$

(We should write $\ln |y|$, but we know that the medical costs are positive.) We now solve for y:

$$y = e^{0.037t + C} = e^C e^{0.037t} = A e^{0.037t}$$

where A is a positive constant. This is the formula we used before for continuous percent growth.

✴ **Before we go on . . .** To determine A we need to know, for example, Medicare spending at time $t = 0$. The source cited estimates Medicare spending as $239.6 billion in 2000. Substituting $t = 0$ in the equation above gives

$$239.6 = A e^0 = A$$

Thus, projected Medicare spending is

$$y = 239.6 e^{0.037t} \text{ billion dollars}$$

t years after 2000.

---

[3]Spending is in constant 2000 dollars. Source: For projected data, the Urban Institute's Analysis of the 1999 Trustee's Report (http://www.urban.org).

Example 4 • Newton's Law of Cooling

Newton's law of cooling states that a hot object cools at a rate proportional to the difference between its temperature and the temperature of the surrounding environment. If a hot cup of coffee, at 170°F, is left to sit in a room at 70°F, how will the temperature of the coffee change over time?

Solution We let $H(t)$ denote the temperature of the coffee at time t. Newton's law of cooling tells us that $H(t)$ *decreases* at a rate proportional to the difference between $H(t)$ and 70°F, the temperature of the surrounding environment (also called the **ambient temperature**). In other words,

$$\frac{dH}{dt} = -k(H - 70)$$

where k is some positive constant.[4] Note that $H \geq 70$: The coffee will never cool to less than the ambient temperature. The variables here are H and t, which we can separate as follows:

$$\frac{dH}{H - 70} = -k\, dt$$

Integrating, we get

$$\int \frac{dH}{H - 70} = \int (-k)\, dt$$

so

$$\ln(H - 70) = -kt + C$$

(Note that $H - 70$ is positive, so we don't need absolute values.) We now solve for H:

$$H - 70 = e^{-kt+C}$$
$$= e^C e^{-kt}$$
$$= Ae^{-kt}$$
$$H(t) = 70 + Ae^{-kt}$$

where A is some positive constant. We can determine the constant A using the initial condition $H(0) = 170$:

$$170 = 70 + Ae^0 = 70 + A$$
$$A = 100$$

Therefore,

$$H(t) = 70 + 100e^{-kt}$$

---

[4]When we say that a quantity Q is *proportional* to a quantity R, we mean that $Q = kR$ for some constant k. The constant k is referred to as the **constant of proportionality.**

Question But what is k?

Answer The constant k determines the rate of cooling. Its value depends on the thing cooling, in this case the coffee and its container. Figure 19 shows two possible graphs, one with $k = 1$ and the other with $k = 0.1$.

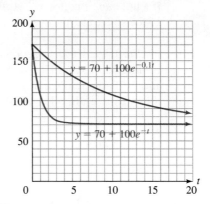

Figure 19

In any case we can see from the graph or the formula for $H(t)$ that the temperature of the coffee will approach the air temperature exponentially.

7.6 EXERCISES

In Exercises 1–10, find the general solution of each differential equation. Where possible, solve for y as a function of x.

1. $\dfrac{dy}{dx} = x^2 + \sqrt{x}$

2. $\dfrac{dy}{dx} = \dfrac{1}{x} + 3$

3. $\dfrac{dy}{dx} = \dfrac{x}{y}$

4. $\dfrac{dy}{dx} = \dfrac{y}{x}$

5. $\dfrac{dy}{dx} = xy$

6. $\dfrac{dy}{dx} = x^2 y$

7. $\dfrac{dy}{dx} = (x + 1)y^2$

8. $\dfrac{dy}{dx} = \dfrac{1}{(x + 1)y^2}$

9. $x\dfrac{dy}{dx} = \dfrac{1}{y}\ln x$

10. $\dfrac{1}{x}\dfrac{dy}{dx} = \dfrac{1}{y}\ln x$

In Exercises 11–20, for each differential equation, find the particular solution indicated.

11. $\dfrac{dy}{dx} = x^3 - 2x; y = 1$ when $x = 0$

12. $\dfrac{dy}{dx} = 2 - e^{-x}; y = 0$ when $x = 0$

13. $\dfrac{dy}{dx} = \dfrac{x^2}{y^2}; y = 2$ when $x = 0$

14. $\dfrac{dy}{dx} = \dfrac{y^2}{x^2}; y = \dfrac{1}{2}$ when $x = 1$

15. $x\dfrac{dy}{dx} = y; y(1) = 2$

16. $x^2 \dfrac{dy}{dx} = y; y(1) = 1$

17. $\dfrac{dy}{dx} = x(y + 1); y(0) = 0$

18. $\dfrac{dy}{dx} = \dfrac{y + 1}{x}; y(1) = 2$

19. $\dfrac{dy}{dx} = \dfrac{xy^2}{x^2 + 1}; y(0) = -1$

20. $\dfrac{dy}{dx} = \dfrac{xy}{(x^2 + 1)^2}; y(0) = 1$

APPLICATIONS

21. Sales Your monthly sales of Green Tea Ice Cream are falling at an instantaneous rate of 5%/month. If you currently sell 1000 quarts each month, find the differential equation describing your change in sales and then solve it to predict your monthly sales.

22. Profit Your monthly profit on sales of Avocado Ice Cream are rising at an instantaneous rate of 10%/month. If you currently make a profit of $15,000 each month, find the differential equation describing your change in profit and solve it to predict your monthly profits.

23. Cooling A bowl of clam chowder at 190°F is placed in a room whose air temperature is 75°F. After 10 minutes the soup has cooled to 150°F. Find the temperature of the chowder as a function of time. (Refer to Example 4 for Newton's law of cooling.)

24. Heating Newton's law of heating is just the same as his law of cooling: The rate of change of temperature is proportional to the difference between the temperature of an object and its surroundings, whether the object is hotter or colder than its surroundings. Suppose that a pie, at 20°F, is put in an oven at 350°F. After 15 minutes its temperature has risen to 80°F. Find the temperature of the pie as a function of time.

25. Market Saturation You have just introduced a new flat-screen monitor to the market. You predict that you will eventually sell 100,000 monitors and that your monthly rate of sales will be 10% of the difference between the saturation value and the total number you have sold up to that point. Find a differential equation for your total sales (as a function of the month) and solve. (What are your total sales at the moment when you first introduce the monitor?)

26. Market Saturation Repeat Exercise 25, assuming that monthly sales will be 5% of the difference between the saturation value (of 100,000 monitors) and the total sales to that point and assuming that you sell 5000 monitors to corporate customers before placing the monitor on the open market.

27. Approach to Equilibrium Extrasoft Toy Co. has just released its latest creation, a plush platypus named "Eggbert." The demand function for Eggbert dolls is $D(p) = 50,000 - 500p$ dolls each month when the price is p dollars. The supply function is $S(p) = 30,000 + 500p$ dolls each month when the price is p dollars. This makes the equilibrium price $20. The *Evans price adjustment model* assumes that if the price is set at a value other than the equilibrium price, it will change over time in such a way that its rate of change is proportional to the shortage $D(p) - S(p)$.
  a. Write the differential equation given by the Evans price adjustment model for the price p as a function of time.
  b. Find the general solution of the differential equation you wrote in part (a). (You will have two unknown constants, one being the constant of proportionality.)
  c. Find the particular solution in which Eggbert dolls are initially priced at $10 and the price rises to $12 after 1 month.

28. Approach to Equilibrium Spacely Sprockets has just released its latest model, the Dominator. The demand function is $D(p) = 10,000 - 1000p$ sprockets each year when the price is p dollars. The supply function is $S(p) = 8000 + 1000p$ sprockets each year when the price is p dollars.
  a. Using the Evans price adjustment model described in Exercise 27, write the differential equation for the price $p(t)$ as a function of time.
  b. Find the general solution of the differential equation you wrote in part (a).
  c. Find the particular solution in which Dominator sprockets are initially priced at $5 each but fall to $3 each after 1 year.

29. Determining Demand Nancy's Chocolates estimates that the elasticity of demand for its dark chocolate truffles is $E = 0.05p - 1.5$, where p is the price per pound. Nancy's sells 20 pounds of truffles each week when the price is $20 per

pound. Find the formula expressing the demand q as a function of p. Recall that the elasticity of demand is given by

$$E = -\frac{dq}{dp} \times \frac{p}{q}$$

30. Determining Demand Nancy's Chocolates estimates that the elasticity of demand for its chocolate strawberries is $E = 0.02p - 0.5$, where p is the price per pound. It sells 30 pounds of chocolate strawberries each week when the price is $30 per pound. Find the formula expressing the demand q as a function of p. Recall that the elasticity of demand is given by

$$E = -\frac{dq}{dp} \times \frac{p}{q}$$

31. Logistic Equation There are many examples of growth in which the rate of growth is slow at first, becomes faster, and then slows again as a limit is reached. This pattern can be described by the differential equation

$$\frac{dy}{dt} = ay(L - y)$$

where a is a constant and L is the limit of y. Show by substitution that

$$y = \frac{CL}{e^{-aLt} + C}$$

is a solution of this equation, where C is an arbitrary constant.

32. Logistic Equation Using separation of variables and integration with a table of integrals or a symbolic algebra program, solve the differential equation in Exercise 31 to derive the solution given there.

T Exercises 33–36 require the use of technology.

33. Market Saturation You have just introduced a new model of DVD player. You predict that the market will saturate at 2 million DVD players and that your total sales will be governed by the equation

$$\frac{dS}{dt} = \frac{1}{4}S(2 - S)$$

where S is the total sales in millions of DVD players and t is measured in months. If you give away 1000 DVD players when you first introduce them, what will S be? Sketch the graph of S as a function of t. About how long will it take to saturate the market? (See Exercise 31.)

34. Epidemics A certain epidemic of influenza is predicted to follow the function defined by

$$\frac{dA}{dt} = \frac{1}{10}A(20 - A)$$

where A is the number of people infected in millions and t is the number of months after the epidemic starts. If 20,000 cases are reported initially, find $A(t)$ and sketch its graph. When is A growing fastest? How many people will eventually be affected? (See Exercise 31.)

35. Growth of Tumors The growth of tumors in animals can be modeled by the Gompertz equation:

$$\frac{dy}{dt} = -ay \ln\left(\frac{y}{b}\right)$$

where y is the size of a tumor, t is time, and a and b are constants that depend on the type of tumor and the units of measurement.
a. Solve for y as a function of t.
b. If $a = 1$, $b = 10$, and $y(0) = 5$ cm³ (with t measured in days), find the specific solution and graph it.

36. Growth of Tumors Refer to Exercise 35. Suppose $a = 1$, $b = 10$, and $y(0) = 15$ cm³. Find the specific solution and graph it. Comparing its graph to the one obtained in Exercise 35, what can you say about tumor growth in these instances?

COMMUNICATION AND REASONING EXERCISES

37. What is the difference between a particular solution and the general solution of a differential equation? How do we get a particular solution from the general solution?

38. Why is there always an arbitrary constant in the general solution of a differential equation? Why are there not two or more arbitrary constants in a first-order differential equation?

39. Show by example that a **second-order** differential equation, one involving the second derivative y'', usually has two arbitrary constants in its general solution.

40. Find a differential equation that is not separable.

41. Find a differential equation whose general solution is $y = 4e^{-x} + 3x + C$.

42. Explain how, knowing the elasticity of demand as a function of either price or demand, you may find the demand equation (see Exercise 29).

CASE STUDY

Estimating Tax Revenues

You have just been hired by the incoming administration of your country as chief consultant for national tax policy, and you have been getting conflicting advice from the finance experts on your staff. Several of them have come up with plausible suggestions for new tax structures, and your job is to choose the plan that results in more revenue for the government.

Before you can evaluate their plans, you realize that it is essential to know your country's income distribution—that is, how many people earn how much money.[5] One might think that the most useful way of specifying income distribution would be to use a function that gives the exact number $f(x)$ of people who earn a given salary x. This would necessarily be a discrete function—it only makes sense if x happens to be a whole number of cents. There is, after all, no one earning a salary of exactly \$22,000.142 567! Furthermore, this function would behave rather erratically, since there are, for example, probably many more people making a salary of exactly \$30,000 than exactly \$30,000.01. Given these problems, it is far more convenient to start with the function defined by

$$N(x) = \text{total number of people earning between 0 and } x \text{ dollars}$$

Actually, you would want a "smoothed" version of this function. The graph of $N(x)$ might look like the one shown in Figure 20.

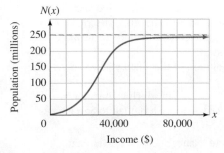

Figure 20

---

[5]To simplify our discussion, we are assuming that (1) all tax revenues are based on earned income and (2) everyone in the population we consider earns some income.

If we take the *derivative* of $N(x)$, we get an income distribution function. Its graph might look like the one shown in Figure 21.

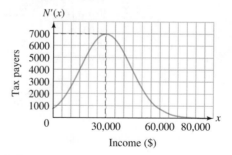

Figure 21

Since the derivative measures the rate of change, its value at x is the additional number of taxpayers per \$1 increase in salary. Thus, the fact that $N'(20,000) \approx 5500$ tells us that approximately 5500 people are earning a salary of between \$20,000 and \$20,001. In other words, N' shows the distribution of incomes among the population—hence the name *distribution function.*[6]

You thus send a memo to your experts requesting the income distribution function for the nation. After much collection of data, they tell you that the income distribution function is

$$N'(x) = 7000\, e^{-(x-30,000)^2/400,000,000}$$

This is in fact the function whose graph is shown in Figure 21 and is an example of a **normal distribution.** Notice that the curve is symmetric around the median income of \$30,000 and that about 7000 people are earning between \$30,000 and \$30,001 annually.[7]

Given this income distribution, your financial experts have come up with the two possible tax policies illustrated Figures 22 and 23.

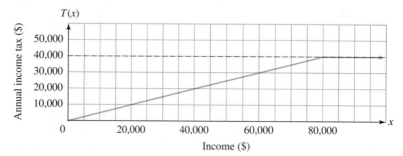

Figure 22 • *Alternative 1*

---

[6]A very similar idea is used in probability. See the optional chapter "Calculus Applied to Probability and Statistics" at the Web site.

[7]You might find it odd that you weren't given the original function N, but it will turn out that you don't need it. How would you compute it?

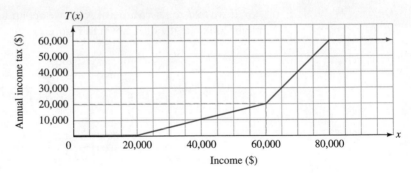

Figure 23 • *Alternative 2*

In the first alternative, all taxpayers pay half of their income in taxes, except that no one pays more than $40,000 in taxes. In the second alternative, there are four tax brackets, described by the following table:

Income ($)	Marginal Tax Rate (%)
0–20,000	0
20,000–60,000	50
60,000–80,000	200
Above 80,000	0

Now you must determine which alternative will generate more annual tax revenue.

Each of Figures 22 and 23 is the graph of a function, T. Rather than using the formulas for these particular functions, you begin by working with the general situation. You have an income distribution function N' and a tax function T, both functions of annual income. You need to find a formula for total tax revenues. First, you decide to use a cutoff so that you need to work only with incomes in some finite bracket $[0, M]$; you might use, for example, $M = \$10$ million (later you will let M approach $+\infty$). Next, you subdivide the interval $[0, M]$ into a large number of intervals of small width, Δx. If $[x_{k-1}, x_k]$ is a typical such interval, you wish to calculate the approximate tax revenue from people whose total incomes lie between x_{k-1} and x_k. You will then sum over k to get the total revenue.

You need to know how many people are making incomes between x_{k-1} and x_k. Because $N(x_k)$ people are making incomes *up to* x_k and $N(x_{k-1})$ people are making incomes up to x_{k-1}, the number of people making incomes between x_{k-1} and x_k is $N(x_k) - N(x_{k-1})$. Because x_k is very close to x_{k-1}, the incomes of these people are all approximately equal to x_{k-1} dollars, so each of these taxpayers is paying an annual tax of about $T(x_{k-1})$. This gives a tax revenue of

$$[N(x_k) - N(x_{k-1})]T(x_{k-1})$$

Now you do a clever thing. You write $x_k - x_{k-1} = \Delta x$ and replace $N(x_k) - N(x_{k-1})$ by

$$\frac{N(x_k) - N(x_{k-1})}{\Delta x}\Delta x$$

This gives you a tax revenue of about

$$\frac{N(x_k) - N(x_{k-1})}{\Delta x}T(x_{k-1})\Delta x$$

from wage earners in the bracket $[x_{k-1}, x_k]$. Summing over k gives an approximate total revenue of

$$\sum_{k=1}^{n} \frac{N(x_k) - N(x_{k-1})}{\Delta x} T(x_{k-1}) \Delta x$$

where n is the number of subintervals. The larger n is, the more accurate your estimate will be, so you take the limit of the sum as $n \to \infty$. When you do this, two things happen. First, the quantity

$$\frac{N(x_k) - N(x_{k-1})}{\Delta x}$$

approaches the derivative $N'(x_{k-1})$. Second, the sum, which you recognize as a Riemann sum, approaches the integral

$$\int_0^M N'(x) T(x) \, dx$$

You now take the limit as $M \to +\infty$ to get

$$\text{total tax revenue} = \int_0^{+\infty} N'(x) T(x) \, dx$$

This improper integral is fine in theory, but the actual calculation will have to be done numerically, so you stick with the upper limit of $10 million for now. You will have to check that it is reasonable at the end (notice that, by the graph of N', it appears that extremely few, if any, people earn that much). Now you already have a formula for $N'(x)$, but you still need to write formulas for the tax functions $T(x)$ for both alternatives.

Alternative 1 The graph in Figure 22 rises linearly from 0 to 40,000 as x ranges from 0 to 80,000 and then stays constant at 40,000. The slope of the first part is $40,000/80,000 = 1/2$. The taxation function is therefore

$$T(x) = \begin{cases} \dfrac{x}{2} & \text{if } 0 \leq x \leq 80,000 \\ 40,000 & \text{if } x \geq 80,000 \end{cases}$$

To perform the integration, you will therefore need to break the integral into two pieces, the first from 0 to 80,000 and the second from 80,000 to 10,000,000. In other words,

$$R_1 = \int_0^{80,000} \left(7000 e^{-(x-30,000)^2/400,000,000} \right) \frac{x}{2} \, dx + \int_{80,000}^{10,000,000} \left(7000 e^{-(x-30,000)^2/400,000,000} \right) 40,000 \, dx$$

You decide not to attempt this by hand![8] You use numerical integration software to obtain a grand total of $R_1 = \$3,732,760,000,000$, or $3.732 \, 76$ trillion (rounded to six significant digits).

---

[8]In fact, these integrals cannot be done in elementary terms at all.

Alternative 2 The graph in Figure 23 rises linearly from 0 to 20,000 as x ranges from 20,000 to 60,000, then rises from 20,000 to 60,000 as x ranges from 60,000 to 80,000, and then stays constant at 60,000. The slope of the first incline is 1/2, and the slope of the second incline is 2 (this is why the *marginal* tax rates are 50% and 200%, respectively). The taxation function is therefore

$$T(x) = \begin{cases} 0 & \text{if } 0 \le x \le 20{,}000 \\ \dfrac{x - 20{,}000}{2} & \text{if } 20{,}000 \le x \le 60{,}000 \\ 20{,}000 + 2(x - 60{,}000) & \text{if } 60{,}000 \le x \le 80{,}000 \\ 60{,}000 & \text{if } x \ge 80{,}000 \end{cases}$$

Values of x between 0 and 20,000 do not contribute to the integral, so

$$R_2 = \int_{20{,}000}^{60{,}000} (7000e^{-(x-30{,}000)^2/400{,}000{,}000})\left(\frac{x - 20{,}000}{2}\right) dx$$

$$+ \int_{60{,}000}^{80{,}000} (7000e^{-(x-30{,}000)^2/400{,}000{,}000})[20{,}000 + 2(x - 60{,}000)] \, dx$$

$$+ \int_{80{,}000}^{10{,}000{,}000} (7000e^{-(x-30{,}000)^2/400{,}000{,}000})60{,}000 \, dx$$

Numerical integration software gives $R_2 = \$1.520\ 16$ trillion—considerably less than Alternative 1. Thus, even though Alternative 2 taxes the wealthy more heavily, it yields less total revenue.

Now what about the cutoff at \$10 million annual income? If you try either integral again with an upper limit of \$100 million, you will see no change in either result to six significant digits. There simply are not enough taxpayers earning an income above \$10 million to make a difference. You conclude that your answers are sufficiently accurate and that the first alternative provides more tax revenue.

EXERCISES

 In Exercises 1–6, using technology, calculate the total tax revenue for a country with the given income distribution and tax policies (all currency in dollars).

1. $N'(x) = 3000e^{-(x-10{,}000)^2/10{,}000}$; 25% tax on all income

2. $N'(x) = 3000e^{-(x-10{,}000)^2/10{,}000}$; 45% tax on all income

3. $N'(x) = 5000e^{-(x-30{,}000)^2/100{,}000}$; no tax on an income below \$30,000, \$10,000 tax on any income of \$30,000 or above

4. $N'(x) = 5000e^{-(x-30{,}000)^2/100{,}000}$; no tax on an income below \$50,000, \$20,000 tax on any income of \$50,000 or above

5. $N'(x) = 7000\, e^{-(x-30{,}000)^2/400{,}000{,}000}$; $T(x)$ with the following graph:

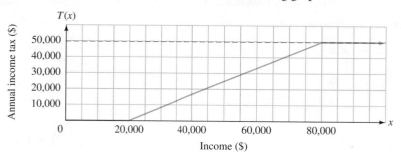

6. $N'(x) = 7000\, e^{-(x-30{,}000)^2/400{,}000{,}000}$; $T(x)$ with the following graph:

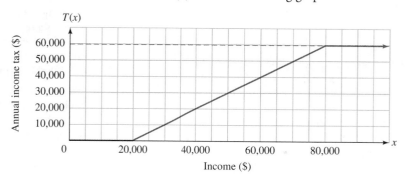

7. Let $P(x)$ be the number of people earning more than x dollars.

  a. What is $N(x) + P(x)$?

  b. Show that $P'(x) = -N'(x)$.

  c. Use integration by parts to show that if $T(0) = 0$, then the total tax revenue is

  $$\int_0^{+\infty} P(x)T'(x)\, dx$$

[*Note:* You may assume that $T'(x)$ is continuous, but the result is still true if we assume only that $T(x)$ is continuous and piecewise continuously differentiable.]

8. Income tax functions T are most often described, as in the text, by tax brackets and marginal tax rates.

  a. If one tax bracket is $a < x \le b$, show that $\int_a^b P(x)\, dx$ is the total income earned in the country that falls into that bracket (P as in Exercise 7).

  b. Use part (a) to explain directly why $\int_0^{+\infty} P(x)T'(x)\, dx$ gives the total tax revenue in the case where T is described by tax brackets and constant marginal tax rates in each bracket.

CHAPTER 7 REVIEW TEST

1. Evaluate the following integrals.

   a. $\int (x^2 + 2)e^x \, dx$

   b. $\int x^2 \ln(2x) \, dx$

   c. $\int_{-2}^{2} (x^3 + 1)e^{-x} \, dx$

   d. $\int_{1}^{e} x^2 \ln x \, dx$

   e. $\int_{1}^{\infty} \frac{1}{x^5} \, dx$

   f. $\int_{0}^{1} \frac{1}{\sqrt{1-x}} \, dx$

2. Find the areas of the following regions.
   a. Between $y = x^3$ and $y = 1 - x^3$ for x in $[0, 1]$
   b. Between $y = e^x$ and $y = e^{-x}$ for x in $[0, 2]$
   c. Enclosed by $y = 1 - x^2$ and $y = x^2$
   d. Between $y = x$ and $y = xe^{-x}$ for x in $[0, 2]$

3. Find the average values of the following functions over the indicated intervals.
   a. $f(x) = x^2 e^x$ over $[0, 1]$
   b. $f(x) = (x + 1)\ln x$ over $[1, 2e]$

4. Find the 2-unit moving averages of the following functions.
   a. $f(x) = x^{4/3}$ **b.** $f(x) = \ln x$

5. Solve the following differential equations.

   a. $\dfrac{dy}{dx} = x^2 y^2$ **b.** $\dfrac{dy}{dx} = xy + 2x$

   c. $xy \dfrac{dy}{dx} = 1; y(1) = 1$ **d.** $y(x^2 + 1)\dfrac{dy}{dx} = xy^2; y(0) = 2$

OHaganBooks.com—COPING WITH ACCUMULATION

6. Sales of the best-seller *A River Burns Through It* are dropping at OHaganBooks.com. To try to bolster sales, the company is dropping the price of the book, now $40, at a rate of $2 per week. As a result, this week OHaganBooks.com will sell 5000 copies, and it estimates that sales will fall continuously at a rate of 10% per week. How much revenue will it earn on sales of this book over the next 8 weeks?

7. OHaganBooks.com is about to start selling a new coffee table book, *Computer Designs of the Late Twentieth Century*. It estimates the demand curve to be $q = 1000\sqrt{200 - 2p}$, and its willingness to order books from the publisher is given by the supply curve $q = 1000\sqrt{10p - 400}$.
   a. Find the equilibrium price and demand.
   b. Find the consumers' and producers' surpluses at the equilibrium price.

8. OHaganBooks.com keeps its cash reserves in a bank account paying 6% compounded continuously. It starts a year with $1 million in reserves and does not withdraw or deposit any money.

   a. What is the average amount it will have in the account over the course of 2 years?

   b. Find the 1-month moving average of the amount it has in the account.

   c. Suppose instead that the account is initially empty, but OHaganBooks.com deposits money continuously into it starting at the rate of $100,000 per month and increasing continuously by $10,000 per month. How much money will the company have in the account at the end of 2 years?

   d. How much of the amount you found in part (c) was principal deposited, and how much was interest earned?

9. Megabucks Corporation is considering buying OHagan-Books.com. They estimate OHaganBooks.com's revenue stream at $50 million per year, growing continuously at a 10% rate. Assuming interest rates of 6%, how much is OHaganBooks.com's revenue for the next year worth now?

10. OHaganBooks.com is shopping around for a new bank. A junior executive at one bank offers them the following interesting deal: The bank will pay them interest continuously at a rate equal to 0.01% of the square of the amount of money they have in the account at any time. By considering what would happen if $10,000 was deposited in such an account, explain why the junior executive was fired shortly after this offer was made.

ADDITIONAL ONLINE REVIEW

If you follow the path

   Web Site → Everything for Calculus → Chapter 7

you will find the following additional resources to help you review.

A comprehensive chapter summary (including examples and interactive features)

Additional review exercises (including interactive exercises and many with help)

A true/false chapter quiz

Several useful graphing and numerical integration utilities

8

FUNCTIONS OF SEVERAL

CASE STUDY

Modeling Household Income

Millennium Real Estate Development Corporation is interested in developing housing projects for medium-sized families that have high household incomes. To decide which income bracket to target, the company has asked you, a paid consultant, for an analysis of the relationship of household size to household income and the effect of increasing household size on household income. How can you analyze the relevant data?

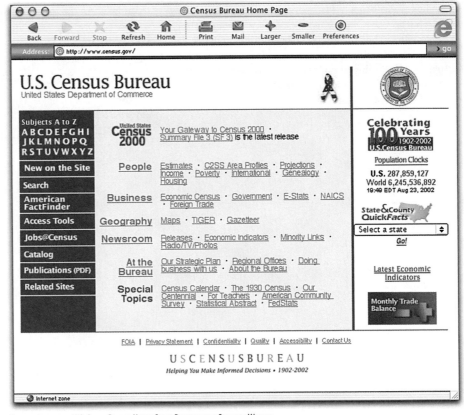

REPRODUCED FROM THE U.S. CENSUS BUREAU, UNITED STATES DEPARTMENT OF COMMERCE WEB SITE

INTERNET RESOURCES FOR THIS CHAPTER

At the Web site, follow the path

   Web Site → Everything for Calculus → Chapter 8

where you will find links to step-by-step tutorials for the main topics in this chapter, a detailed chapter summary, a true/false quiz, and a collection of sample test questions. You will also find an online surface grapher and other useful resources.

VARIABLES

Introduction

We have studied functions of a single variable extensively. But not every useful function is a function of only one variable. In fact, most are not. For example, if you operate an online bookstore, in competition with Amazon.com, BN.com, and Booksamillion.com, your sales may depend on those of your competitors. Your company's daily revenue might be modeled by a function such as

$$R(x, y, z) = 10{,}000 - 0.01x - 0.02y - 0.01z + 0.000\,01yz$$

where x, y, and z are the online daily revenues of Amazon.com, BN.com, and Booksamillion.com, respectively. Here, R is a function of three variables, since it *depends on x, y, and z*. As we will see, the techniques of calculus extend readily to such functions. Among the applications we look at is optimization: finding, where possible, the maximum or minimum of a function of two or more variables.

8.1 Functions of Several Variables from the Numerical and Algebraic Viewpoints

Recall that a function of one variable is a rule for manufacturing a new number $f(x)$ from a single independent variable x. A function of two or more variables is similar, but the new number now depends on more than one independent variable.

Function of Several Variables

A **real-valued function, f, of x, y, z, ...** is a rule for manufacturing a new number, written $f(x, y, z, \dots)$, from the values of a sequence of independent variables $(x, y, z, \dots)$. The function f is called a **real-valued function of two variables** if there are two independent variables, a **real-valued function of three variables** if there are three independent variables, and so on.

As with functions of one variable, functions of several variables can be represented numerically (using a table of values), algebraically (using a formula), and sometimes graphically* (using a graph).

___

Quick Examples

1. $f(x, y) = x - y$ Function of two variables
   $f(1, 2) = 1 - 2 = -1$ Substitute 1 for x and 2 for y.
   $f(2, -1) = 2 - (-1) = 3$ Substitute 2 for x and -1 for y.
   $f(y, x) = y - x$ Substitute y for x and x for y.

___

*See Section 8.2.

2. $g(x, y) = x^2 + y^2$ Function of two variables
     $g(-1, 3) = (-1)^2 + 3^2 = 10$ Substitute -1 for x and 3 for y.

3. $h(x, y, z) = x + y + xz$ Function of three variables
     $h(2, 2, -2) = 2 + 2 + 2(-2) = 0$ Substitute 2 for x, 2 for y, and -2 for z.

Figure 1 illustrates the concept of a function of two variables: In goes a pair of numbers and out comes a single number.

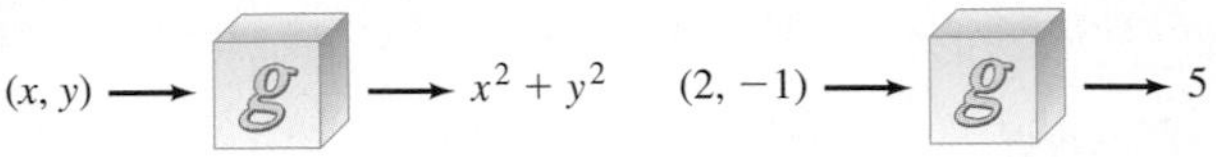

$(x, y) \longrightarrow$ [g] $\longrightarrow x^2 + y^2$ $(2, -1) \longrightarrow$ [g] $\longrightarrow 5$

Figure 1

Let's now look at a number of examples of interesting functions of several variables.

Example 1 • Cost Function

Suppose you own a company that makes two models of speakers: the Ultra Mini and the Big Stack. Your total monthly cost (in dollars) to make x Ultra Minis and y Big Stacks is given by

$$C(x, y) = 10,000 + 20x + 40y$$

What is the significance of each term in this formula?

Solution The terms have meanings similar to those we saw for linear cost functions of a single variable. Let's look at the terms one at a time.

- *Constant Term* Consider the monthly cost of making no speakers at all ($x = y = 0$). We find

$$C(0, 0) = 10,000 \qquad \text{Cost of making no speakers is \$10,000.}$$

Thus, the constant term 10,000 is the **fixed cost,** the amount you have to pay each month even if you make no speakers.

- *Coefficients of x and y* Suppose you make a certain number of Ultra Minis and Big Stacks one month and the next month you increase production by one Ultra Mini. The costs are

$$C(x, y) = 10,000 + 20x + 40y \qquad \text{First month}$$
$$C(x + 1, y) = 10,000 + 20(x + 1) + 40y \qquad \text{Second month}$$
$$= 10,000 + 20x + 20 + 40y$$
$$= C(x, y) + 20$$

Thus, each Ultra Mini adds $20 to the total cost. We say that $20 is the **marginal cost** of each Ultra Mini. Similarly, because of the term $40y$, each Big Stack adds $40 to the total cost. The marginal cost of each Big Stack is $40.

Before we go on . . . This is an example of a **linear function** of two variables. The coefficients of x and y play roles similar to that of the slope of a line. In particular, they give the rates of change of the function as each variable increases while the other stays constant (think about it). We say more about linear functions below.

Graphing Calculator

You can have a TI-83 compute $C(x, y)$ numerically as follows: In the Y= screen, enter

        Y₁=10000+20*X+40*Y

Then, to evaluate, say, $C(10, 30)$ (the cost to make 10 Ultra Minis and 30 Big Stacks), enter

        10→X

        30→Y

        Y₁

and the calculator will evaluate the function and give the answer, $C(10, 30) = 11,400$.

This procedure is too laborious if you want to calculate $f(x, y)$ for a number of different values of x and y.

Excel

Spreadsheets handle functions of several variables easily. The following setup shows how a table of values of C can be created, using values of x and y you enter:

	A	B	C	D
1	x	y	C(x,y)	
2	10	10	=10000+20*A2+40*B2	
3	20	30		
4	15	0		
5	0	30		
6	30	30		

A disadvantage of this layout is that entering values of x and y systematically in two columns is not easy. Can you find a way to remedy this? (See Example 3 for one method.)

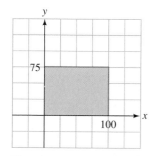

Figure 2

✴ **Before we go on . . .** Which values of x and y may we substitute into $C(x, y)$? Certainly, we must have $x \geq 0$ and $y \geq 0$ because it makes no sense to speak of manufacturing a negative number of speakers. Also, there is certainly some upper bound to the number of speakers that can be made in a month. The bound might take one of several forms. The number of each model may be bounded—say, $x \leq 100$ and $y \leq 75$. The inequalities $0 \leq x \leq 100$ and $0 \leq y \leq 75$ describe the region in the plane shaded in Figure 2.

Another possibility is that the *total* number of speakers is bounded—say, $x + y \leq 150$. This, together with $x \geq 0$ and $y \geq 0$, describes the region shaded in Figure 3. In either case the region shown represents the pairs (x, y) for which $C(x, y)$ is defined. Just as with a function of one variable, we call this region the **domain** of the function. As before, when the domain is not given explicitly, we agree to take the largest possible domain.

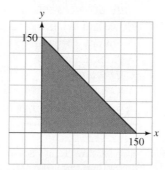

Figure 3

Example 2 • Faculty Salaries

David Katz came up with the following function for the salary of a professor with 10 years of teaching experience in a large university:

$$S(x, y, z) = 13{,}005 + 230x + 18y + 102z$$

Here, S is the salary in 1969–1970 in dollars per year, x is the number of books the professor has published, y is the number of articles published, and z is the number of "excellent" articles published.[1] What salary do you expect a professor with 10 years' experience earned in 1969–1970 if she published 2 books, 20 articles, and 3 "excellent" articles?

Solution All we need to do is calculate

$$S(2, 20, 3) = 13{,}005 + 230(2) + 18(20) + 102(3)$$

$$= \$14{,}131$$

✳ **Before we go on . . .** In Example 1, we gave a linear function of two variables. Here we have an example of a linear function of three variables. Katz came up with his model by surveying a large number of faculty members and then finding the linear function "best" fitting the data. Such models are called **multiple linear regression** models. In the case study at the end of this chapter, we will see a spreadsheet method of finding the coefficients of a multiple regression model from a set of observed data.

What does this model say about the value of a single book or a single article? If a book takes 15 times as long to write as an article, how would you recommend a professor spend her writing time?

Here are two simple kinds of functions of several variables.

Linear Function

A **linear function of the variables** $x_1, x_2, \ldots, x_n$ is a function of the form

$$f(x_1, x_2, \ldots, x_n) = a_0 + a_1x_1 + \cdots + a_nx_n \qquad (a_0, a_1, a_2, \ldots, a_n \text{ constants})$$

Quick Examples

1. $f(x, y) = 3x - 5y$ Linear function of x and y
2. $C(x, y) = 10{,}000 + 20x + 40y$ Example 1
3. $S(x, y, z) = 13{,}005 + 230x + 18y + 102z$ Example 2

Interaction Function

If we add to a linear function one or more terms of the form bx_ix_j (b a nonzero constant and $i \neq j$), we get a **second-order interaction function.**

Quick Examples

1. $C(x, y) = 10{,}000 + 20x + 40y + 0.1xy$
2. $R(x, y, z) = 10{,}000 - 0.01x - 0.02y - 0.01z + 0.000\,01yz$

---

[1] David A. Katz, "Faculty Salaries, Promotions and Productivity at a Large University," *American Economic Review* (June 1973): 469–477. Katz's equation actually included other variables, such as the number of dissertations supervised; our equation assumes that all of these are zero.

So far, we have been specifying functions of several variables **algebraically**—by using algebraic formulas. If you have ever studied statistics, you are probably familiar with statistical tables. These tables may also be viewed as representing functions **numerically,** as Example 3 shows.

Example 3 • Function Represented Numerically: Body Mass Index

The following table lists some values of the *body mass index*, which gives a measure of the massiveness of your body, taking height into account.[2] The variable *w* represents your weight in pounds, and *h* represents your height in inches. A body mass index of 25 or above is generally considered overweight.

h	$w \rightarrow$ 130	140	150	160	170	180	190	200	210
↓ 60	25.2	27.1	29.1	31.0	32.9	34.9	36.8	38.8	40.7
61	24.4	26.2	28.1	30.0	31.9	33.7	35.6	37.5	39.4
62	23.6	25.4	27.2	29.0	30.8	32.7	34.5	36.3	38.1
63	22.8	24.6	26.4	28.1	29.9	31.6	33.4	35.1	36.9
64	22.1	23.8	25.5	27.2	28.9	30.7	32.4	34.1	35.8
65	21.5	23.1	24.8	26.4	28.1	29.7	31.4	33.0	34.7
66	20.8	22.4	24.0	25.6	27.2	28.8	30.4	32.0	33.6
67	20.2	21.8	23.3	24.9	26.4	28.0	29.5	31.1	32.6
68	19.6	21.1	22.6	24.1	25.6	27.2	28.7	30.2	31.7
69	19.0	20.5	22.0	23.4	24.9	26.4	27.8	29.3	30.8
70	18.5	19.9	21.4	22.8	24.2	25.6	27.0	28.5	29.9
71	18.0	19.4	20.8	22.1	23.5	24.9	26.3	27.7	29.1
72	17.5	18.8	20.2	21.5	22.9	24.2	25.6	26.9	28.3
73	17.0	18.3	19.6	20.9	22.3	23.6	24.9	26.2	27.5
74	16.6	17.8	19.1	20.4	21.7	22.9	24.2	25.5	26.7
75	16.1	17.4	18.6	19.8	21.1	22.3	23.6	24.8	26.0
76	15.7	16.9	18.1	19.3	20.5	21.7	22.9	24.2	25.4

As the table shows, the value of the body mass index depends on two quantities: *w* and *h*. Let's write $M(w, h)$ for the body mass index function. What are $M(140, 62)$ and $M(210, 63)$?

Solution We can read the answers from the table:

$$M(140, 62) = 25.4 \qquad w = 140 \text{ lb}, h = 62 \text{ in.}$$

$$M(210, 63) = 36.9 \qquad w = 210 \text{ lb}, h = 63 \text{ in.}$$

---

[2]It is interesting that weight-lifting competitions are usually based on weight, rather than body mass index. As a consequence taller people are at a significant disadvantage in weight-lifting competitions because they must compete with shorter, stockier people of the same weight. (An extremely thin, very tall person can weigh as much as a muscular short person, although his body-mass index would be significantly lower.) SOURCES: *Shape Up America*/National Institute of Health/*New York Times,* May 2, 1999, p. WK3.

Excel

The function $M(w, h)$ is actually given by the formula

$$M(w, h) = \frac{0.45w}{(0.0254h)^2}$$

[The factor 0.45 converts the weight to kilograms, and 0.0254 converts the height to meters. If w is in kilograms and h is in meters, the formula is simpler: $M(w, h) = w/h^2$.] We can use this formula to re-create the table:

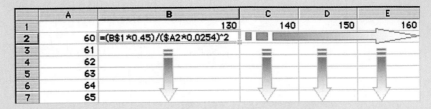

	A	B	C	D	E
1		130	140	150	160
2	60	=(B$1*0.45)/($A2*0.0254)^2			
3	61				
4	62				
5	63				
6	64				
7	65				

Here we have used B$1 instead of B1 for the w coordinate because we want all references to w use the same row (1). Similarly, we want all references to h to refer to the same column (A), so we used $A2 instead of A2.

Distance and Related Functions

Newton's Law of Gravity states that the gravitational force exerted by one particle on another depends on their masses and the distance between them. The distance between two particles in the xy plane can be expressed as a function of their coordinates.

Distance Formula

The distance between the points $P(x_1, y_1)$ and $Q(x_2, y_2)$ is

$$d = \sqrt{(x_2 - x_1)^2 + (y_2 - y_1)^2} = \sqrt{(\Delta x)^2 + (\Delta y)^2}$$

Derivation

The distance d is shown in the figure below:

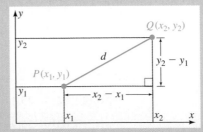

By the Pythagorean theorem applied to the right triangle shown, we get

$$d^2 = (x_2 - x_1)^2 + (y_2 - y_1)^2$$

Taking square roots (d is supposed to be a distance, so we take the positive square root), we get the distance formula. Notice that if we switch x_1 with x_2 or y_1 with y_2 we get the same result.

Quick Examples
1. The distance between the points $(3, -2)$ and $(-1, 1)$ is

$$d = \sqrt{(-1 - 3)^2 + (1 + 2)^2} = \sqrt{25} = 5 \qquad \text{Distance between points}$$

2. The distance from (x, y) to the origin $(0, 0)$ is

$$d = \sqrt{(x - 0)^2 + (y - 0)^2} = \sqrt{x^2 + y^2} \qquad \text{Distance to the origin}$$

The set of all points (x, y) whose distance from the origin $(0, 0)$ is a fixed quantity r is a circle centered at the origin with radius r. From Quick Example 2, we get the following equation for the circle centered at the origin with radius r:

$$\sqrt{x^2 + y^2} = r \qquad \text{Distance from the origin} = r$$

Squaring both sides gives the following equation, which we use in later sections.

Equation of the Circle of Radius r Centered at the Origin

$$x^2 + y^2 = r^2$$

Quick Examples
1. The circle of radius 1 centered at the origin has equation $x^2 + y^2 = 1$.
2. The circle of radius 2 centered at the origin has equation $x^2 + y^2 = 4$.
3. The circle of radius 3 centered at the origin has equation $x^2 + y^2 = 9$.

Now, let's return to Newton's Law of Gravity. According to Newton's law, the gravitational force exerted on a particle with mass m by another particle with mass M is given by the following function of distance:

$$F(r) = G\frac{Mm}{r^2}$$

Here, r is the distance between the two particles in meters, the masses M and m are given in kilograms, $G \approx 6.67 \times 10^{-11}$, and the resulting force is measured in newtons.[3]

Example 4 • Newton's Law of Gravity

Find the gravitational force exerted on a particle with mass m situated at the point (x, y) by another particle with mass M situated at the point (a, b). Express the answer as a function F of the coordinates of the particle with mass m.

Solution The formula above for gravitational force is expressed as a function of the distance, r, between the two particles. Since we are given the coordinates of the two particles, we can express r in terms of these coordinates using the formula for distance:

$$r = \sqrt{(x - a)^2 + (y - b)^2}$$

---

[3]A newton is the force that will cause a 1-kilogram mass to accelerate at 1 meter per second per second (m/sec²).

Substituting for r, we get

$$F(x, y) = G\frac{Mm}{(x - a)^2 + (y - b)^2}$$

✴ **Before we go on . . .** Notice that $F(a, b)$ is not defined because substituting $x = a$ and $y = b$ makes the denominator equal zero. Thus, the largest possible domain of F excludes the point (a, b). Since (a, b) is the only value of (x, y) for which F is not defined, we deduce that *the domain of F consists of all points (x, y) except for (a, b).* In other words, the domain of F is the whole xy plane with the single point (a, b) missing.[4]

Question Why have we expressed F as a function of x and y only and not also as a function of a and b?

Answer It's a matter of interpretation. When we write F as a function of x and y, we are thinking of a and b as *constants*. For example, (a, b) could be the coordinates of the Sun, which we often assume to be fixed in space, and (x, y) could be the coordinates of Earth, which is moving around the Sun. In that case it is most natural to think of x and y as variable and a and b as constant. In another context we may want to consider F as a function of four variables, x, y, a, and b.

8.1 EXERCISES

In Exercises 1–4, for each function evaluate **(a)** $f(0, 0)$, **(b)** $f(1, 0)$, **(c)** $f(0, -1)$, **(d)** $f(a, 2)$, **(e)** $f(y, x)$, and **(f)** $f(x + h, y + k)$.

1. $f(x, y) = x^2 + y^2 - x + 1$

2. $f(x, y) = x^2 - y - xy + 1$

3. $f(x, y) = 0.2x + 0.1y - 0.01xy$

4. $f(x, y) = 0.4x - 0.5y - 0.05xy$

In Exercises 5–8, for each function evaluate **(a)** $g(0, 0, 0)$, **(b)** $g(1, 0, 0)$, **(c)** $g(0, 1, 0)$, **(d)** $g(z, x, y)$, and **(e)** $g(x + h, y + k, z + l)$, provided such a value exists.

5. $g(x, y, z) = e^{x+y+z}$ **6.** $g(x, y, z) = \ln(x + y + z)$

7. $g(x, y, z) = \dfrac{xyz}{x^2 + y^2 + z^2}$ **8.** $g(x, y, z) = \dfrac{e^{xyz}}{x + y + z}$

9. Let $f(x, y, z) = 1.5 + 2.3x - 1.4y - 2.5z$. Complete the following sentences.

   a. f _____ by _____ units for every 1 unit of increase in x.

   b. f _____ by _____ units for every 1 unit of increase in y.

   c. _____ by 2.5 units for every _____.

10. Let $g(x, y, z) = 0.01x + 0.02y - 0.03z - 0.05$. Complete the following sentences.

   a. g _____ by _____ units for every 1 unit of increase in z.

   b. g _____ by _____ units for every 1 unit of increase in x.

   c. _____ by 0.02 unit for every _____.

In Exercises 11–18, classify each function as linear, interaction, or neither.

11. $L(x, y) = 3x - 2y + 6xy - 4y^2$

12. $L(x, y, z) = 3x - 2y + 6xz$

13. $P(x_1, x_2, x_3) = 0.4 + 2x_1 - x_3$

14. $Q(x_1, x_2) = 4x_2 - 0.5x_1 - x_1^2$

15. $f(x, y, z) = \dfrac{x + y - z}{3}$

16. $g(x, y, z) = \dfrac{xz - 3yz + z^2}{4z}$ $(z \neq 0)$

17. $g(x, y, z) = \dfrac{xz - 3yz + z^2y}{4z}$ $(z \neq 0)$

18. $f(x, y) = x + y + xy + x^2y$

In Exercises 19 and 20, use the given tabular representation of the function f to compute the quantities asked for.

19.

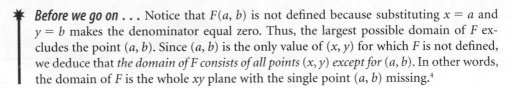

	$x \to$ 10	20	30	40
$y \downarrow$ 10	-1	107	162	-3
20	-6	194	294	-14
30	-11	281	426	-25
40	-16	368	558	-36

   a. $f(20, 10)$ **b.** $f(40, 20)$ **c.** $f(10, 20) - f(20, 10)$

---

[4]Mathematicians often refer to this as a "punctured plane."

20.

	x →			
	10	**20**	**30**	**40**
y ↓ **10**	162	107	−5	−7
20	294	194	−22	−30
30	426	281	−39	−53
40	558	368	−56	−76

a. $f(10, 30)$ **b.** $f(20, 10)$ **c.** $f(10, 40) + f(10, 20)$

In Exercises 21 and 22, use a spreadsheet or some other method to complete the given tables.

21. $P(x, y) = x - 0.3y + 0.45xy$

	x →			
	10	**20**	**30**	**40**
y ↓ **10**				
20				
30				
40				

22. $Q(x, y) = 0.4x + 0.1y - 0.06xy$

	x →			
	10	**20**	**30**	**40**
y ↓ **10**				
20				
30				
40				

23. The following statistical table lists some values of the Inverse F distribution ($\alpha = 0.5$):

n →										
d	**1**	**2**	**3**	**4**	**5**	**6**	**7**	**8**	**9**	**10**
↓ **1**	161.4	199.5	215.7	224.6	230.2	234.0	236.8	238.9	240.5	241.9
2	18.51	19.00	19.16	19.25	19.30	19.33	19.35	19.37	19.39	19.40
3	10.13	9.552	9.277	9.117	9.013	8.941	8.887	8.812	8.812	8.785
4	7.709	6.944	6.591	6.388	6.256	6.163	6.094	5.999	5.999	5.964
5	6.608	5.786	5.409	5.192	5.050	4.950	4.876	4.772	4.772	4.735
6	5.987	5.143	4.757	4.534	4.387	4.284	4.207	4.099	4.099	4.060
7	5.591	4.737	4.347	4.120	3.972	3.866	3.787	3.677	3.677	3.637
8	5.318	4.459	4.066	3.838	3.688	3.581	3.500	3.388	3.388	3.347
9	5.117	4.256	3.863	3.633	3.482	3.374	3.293	3.179	3.179	3.137
10	4.965	4.103	3.708	3.478	3.326	3.217	3.135	3.020	3.020	2.978

In Excel you can compute the value of this function at (n, d) by the formula

=FINV(0.05,n,d) The 0.05 is the value of alpha (α).

Use Excel to re-create this table.

24. The formula for the body mass index $M(w, h)$, if w is given in kilograms and h is given in meters, is

$$M(w, h) = \frac{w}{h^2}$$ See Example 3.

Use this formula to complete the following table in Excel:

	w →						
h	**100**	**110**	**120**	**130**	**140**	**150**	**160**
↓ **2**							
2.05							
2.1							
2.15							
2.2							
2.25							
2.3							
2.35							
2.4							
2.45							
2.5							

In Exercises 25–28, use either a graphing calculator or a spreadsheet to complete each table. Express all your answers as decimals rounded to four decimal places.

25.

x	y	$f(x, y) = x^2\sqrt{1 + xy}$
3	1	
1	15	
0.3	0.5	
56	4	

26.

x	y	$f(x, y) = x^2e^y$
0	2	
−1	5	
1.4	2.5	
11	9	

27.

x	y	$f(x, y) = x \ln(x^2 + y^2)$
3	1	
1.4	−1	
e	0	
0	e	

53. Volume The volume of an ellipsoid with cross-sectional radii a, b, and c is $V(a, b, c) = \frac{4}{3}\pi abc$.

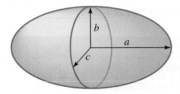

a. Find at least two sets of values for a, b, and c such that $V(a, b, c) = 1$.
b. Find the value of a such that $V(a, a, a) = 1$, and describe the resulting ellipsoid.

54. Volume The volume of a right elliptical cone with height h and radii a and b of its base is $V(a, b, h) = \frac{1}{3}\pi abh$.

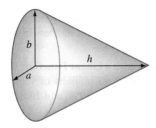

a. Find at least two sets of values for a, b, and h such that $V(a, b, h) = 1$.
b. Find the value of a such that $V(a, a, a) = 1$ and describe the resulting cone.

Exercises 55–58 involve Cobb–Douglas productivity functions. These functions have the form

$$P(x, y) = Kx^a y^{1-a}$$

where P stands for the number of items produced each year, x is the number of employees, and y is the annual operating budget. (The numbers K and a are constants that depend on the situation we are looking at, with $0 \leq a \leq 1$.)

55. Productivity How many items will be produced each year by a company with 100 employees and an annual operating budget of $500,000 if $K = 1000$ and $a = 0.5$? (Round your answer to one significant digit.)

56. Productivity How many items will be produced each year by a company with 50 employees and an annual operating budget of $1,000,000 if $K = 1000$ and $a = 0.5$? (Round your answer to one significant digit.)

57. Production (Cobb–Douglas Model) Two years ago my piano manufacturing plant employed 1000 workers, had an operating budget of $1 million, and turned out 100 pianos. Last year I slashed the operating budget to $10,000, and production dropped to 10 pianos.
a. Use the data for each of the 2 years and the Cobb–Douglas formula to obtain two equations in K and a.
b. Take logs of both sides in each equation and obtain two linear equations in a and log K.

c. Solve these equations to obtain values for a and K.
d. Use these values in the Cobb–Douglas formula to predict production if I increase the operating budget back to $1 million but lay off half the workforce.

58. Production (Cobb–Douglas Model) Repeat Exercise 57, using the following data: 2 years ago: 1000 employees, $1 million operating budget, 100 pianos; last year: 1000 employees, $100,000 operating budget, 10 pianos.

59. Pollution The burden of human-made aerosol sulfate in Earth's atmosphere, in grams per square meter (g/m²), is

$$B(x, n) = \frac{xn}{A}$$

where x is the total weight of aerosol sulfate emitted into the atmosphere each year and n is the number of years it remains in the atmosphere. A is Earth's surface area, approximately 5.1×10^{14} square meters.
a. Calculate the burden, given the 1995 estimated values of $x = 1.5 \times 10^{14}$ grams per year, and $n = 5$ days.
b. What does the function $W(x, n) = xn$ measure?

SOURCE: Robert J. Charlson and Tom M. L. Wigley, "Sulfate Aerosol and Climatic Change," *Scientific American* (February 1994): 48–57.

60. Pollution The amount of aerosol sulfate (in grams) was approximately 45×10^{12} grams in 1940 and has been increasing exponentially ever since, with a doubling time of approximately 20 years. Use the model from Exercise 59 to give a formula for the atmospheric burden of aerosol sulfate as a function of the time t in years since 1940 and the number of years n it remains in the atmosphere.

SOURCE: Robert J. Charlson and Tom M. L. Wigley, "Sulfate Aerosol and Climatic Change," *Scientific American* (February 1994): 48–57.

61. Alien Intelligence Frank Drake, an astronomer at the University of California at Santa Cruz, devised the following equation to estimate the number of planet-based civilizations in our Milky Way galaxy willing and able to communicate with Earth:

$$N(R, f_p, n_e, f_l, f_i, f_c, L) = R\, f_p n_e f_l f_i f_c\, L$$

where R = the number of new stars formed in our galaxy each year
f_p = the fraction of those stars that have planetary systems
n_e = the average number of planets in each such system that can support life
f_l = the fraction of such planets on which life actually evolves
f_i = the fraction of life-sustaining planets on which intelligent life evolves
f_c = the fraction of intelligent-life-bearing planets on which the intelligent beings develop the means and the will to communicate over interstellar distances
L = the average lifetime of such technological civilizations (in years)

a. What would be the effect on N if any one of the variables were doubled?

b. How would you modify the formula if you were interested only in the number of intelligent-life-bearing planets in our galaxy?

c. How could one convert this function into a linear function?

d. (For discussion) Try to come up with an estimate of N.

Source: *First Contact* (Plume Books/Penguin Group)/*New York Times,* October 6, 1992, p. C1.

62. More Alien Intelligence The formula given in Exercise 61 restricts attention to planet-based civilizations in our galaxy. Give a formula that includes intelligent planet-based aliens from the galaxy Andromeda. (Assume that all the variables used in the formula for the Milky Way have the same values for Andromeda.)

63. Level Curves The height of each point in a hilly region is given as a function of its coordinates by the formula

$$f(x, y) = y^2 - x^2$$

a. Use technology to plot the curves on which the height is 0, 1, and 2 on the same set of axes. These are called **level curves of f.**

b. Sketch the curve $f(x, y) = 3$ *without* using technology.

c. Sketch the curves $f(y, x) = 1$ and $f(y, x) = 2$ without using technology.

64. Isotherms The temperature (in degrees Fahrenheit) at each point in a region is given as a function of the coordinates by the formula

$$T(x, y) = 60.5(x - y^2)$$

a. Use technology to sketch the curves on which the temperature is 0°F, 30°F, and 90°F. These curves are called **isotherms.**

b. Sketch the isotherms corresponding to 20°F, 50°F, and 100°F *without* using technology.

c. What do the isotherms corresponding to negative temperatures look like?

COMMUNICATION AND REASONING EXERCISES

65. Let $f(x, y) = x/y$. How are $f(x, y)$ and $f(y, x)$ related?

66. Let $f(x, y) = x^2 y^3$. How are $f(x, y)$ and $f(-x, -y)$ related?

67. Give an example of a function of the two variables x and y with the property that interchanging x and y has no effect.

68. Give an example of a function f of the two variables x and y with the property that $f(x, y) = -f(y, x)$.

69. Give an example of a function f of the three variables x, y, and z with the property that $f(x, y, z) = f(y, x, z)$ and $f(-x, -y, -z) = -f(x, y, z)$.

70. Give an example of a function f of the three variables x, y, and z with the property that $f(x, y, z) = f(y, x, z)$ and $f(-x, -y, -z) = f(x, y, z)$.

71. Illustrate by means of an example how a real-valued function of the two variables x and y gives different real-valued functions of one variable when we restrict y to be different constants.

72. Illustrate by means of an example how a real-valued function of one variable x gives different real-valued functions of the two variables y and z when we substitute for x suitable functions of y and z.

73. If f is a linear function of x and y, show that if we restrict y to be a fixed constant then the resulting function of x is linear. Does the slope of this linear function depend on the choice of y?

74. If f is an interaction function of x and y, show that if we restrict y to be a fixed constant then the resulting function of x is linear. Does the slope of this linear function depend on the choice of y?

75. Suppose $C(x, y)$ represents the cost of x CDs and y cassettes. If $C(x, y + 1) < C(x + 1, y)$ for every $x \geq 0$ and $y \geq 0$, what does this tell you about the cost of CDs and cassettes?

76. Suppose $C(x, y)$ represents the cost of renting x DVDs and y video games. If $C(x + 2, y) < C(x, y + 1)$ for every $x \geq 0$ and $y \geq 0$, what does this tell you about the cost of renting DVDs and video games?

8.2 *Three-Dimensional Space and the Graph of a Function of Two Variables*

Just as functions of a single variable have graphs, so do functions of two or more variables. Recall that the graph of $f(x)$ consists of all points $(x, f(x))$ in the xy plane. By analogy, we would like to say that the graph of a function of *two* variables, $f(x, y)$, consists of all points of the form $(x, y, f(x, y))$. Thus, we need three axes: the x, y, and z axes. In other words, our graph will live in **three-dimensional space,** or **3-space.**[5]

---

[5]If we were dealing instead with a function of *three* variables, then we would need to go to *four-dimensional* space. Here we run into visualization problems (to say the least!) so we won't discuss the graphs of functions of three or more variables in this text.

Just as we had two mutually perpendicular axes in two-dimensional space (the *xy* plane), so we have three mutually perpendicular axes in three-dimensional space (Figure 4).

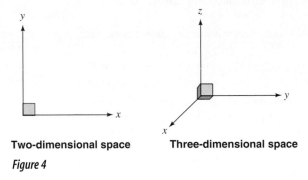

Two-dimensional space   **Three-dimensional space**

Figure 4

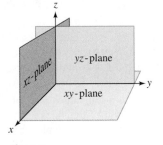

Figure 5

In both 2-space and 3-space, the axis labeled with the last letter goes up. Thus, the *z* direction is the "up" direction in 3-space, rather than the *y* direction.

Three important planes are associated with these axes: the *xy* plane, the *yz* plane, and the *xz* plane. These planes are shown in Figure 5. Any two of these planes intersect in one of the axes (for example, the *xy* and *xz* planes intersect in the *x* axis) and all three meet at the origin. Notice that the *xy* plane consists of all points with *z* coordinate = 0, the *xz* plane consists of all points with $y = 0$, and the *yz* plane consists of all points with $x = 0$.

In 3-space each point has *three* coordinates, as you might expect: the *x* coordinate, the *y* coordinate, and the *z* coordinate. To see how this works, look at the following examples.

Example 1 • *Plotting Points in Three Dimensions*

Locate the points $P(1, 2, 3)$, $Q(-1, 2, 3)$, $R(1, -1, 0)$, and $S(1, 2, -2)$ in 3-space.

Solution   To locate *P* the procedure is similar to the one we used in 2-space: Start at the origin, proceed 1 unit in the *x* direction, then proceed 2 units in the *y* direction, and, finally, proceed 3 units in the *z* direction. We wind up at the point *P* shown in Figure 6(a).

Here is another, extremely useful way of thinking about the location of *P*. First, look at the *x* and *y* coordinates, obtaining the point $(1, 2)$ in the *xy* plane. The point we want is then 3 units vertically above the point $(1, 2)$ because the *z* coordinate of a point is just its height. This strategy is shown in Figure 6(b).

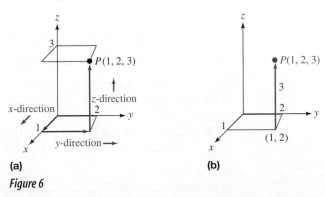

(a)   **(b)**

Figure 6

Plotting the points Q, R, and S is similar, using the convention that negative coordinates correspond to moves back, left, or down (see Figure 7).

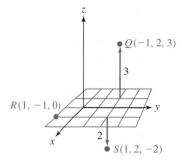

Figure 7

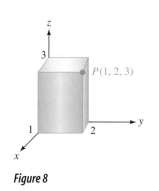

Figure 8

✴ **Before we go on . . .** Here is yet another approach to locating points in three dimensions. Consider again the point $P(1, 2, 3)$ and imagine a rectangular box situated with one of its corners at the origin, as shown in Figure 8. The corner opposite the origin is the point P.

Remember this:

The z coordinate of a point is its height above the xy plane.

Our next task is to describe the graph of a function $f(x, y)$ of two variables.

> ### Graph of a Function of Two Variables
> The **graph of the function f of two variables** is the set of all points $(x, y, f(x, y))$ in three-dimensional space, where we restrict the values of (x, y) to lie in the domain of f. In other words, the graph is the set of all the points (x, y, z) with $z = f(x, y)$.

For *every* point (x, y) in the domain of f, the z coordinate of the corresponding point on the graph is given by evaluating the function at (x, y). Thus, there will be a point on the graph above *every* point in the domain of f, so the graph is usually a *surface* of some sort.

Example 2 • Graph of a Function of Two Variables

Describe the graph of $f(x, y) = x^2 + y^2$.

Solution Your first thought might be to make a table of values. You could choose some values for x and y and then, for each such pair, calculate $z = x^2 + y^2$. For example, you might get the following table:

		$x \rightarrow$		
		−1	0	1
y	−1	2	1	2
↓	0	1	0	1
	1	2	1	2

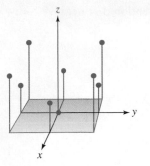

Figure 9

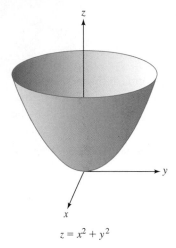

$z = x^2 + y^2$

Figure 10

This gives the following nine points on the graph of f: $(-1, -1, 2)$, $(-1, 0, 1)$, $(-1, 1, 2)$, $(0, -1, 1)$, $(0, 0, 0)$, $(0, 1, 1)$, $(1, -1, 2)$, $(1, 0, 1)$, and $(1, 1, 2)$. These points are shown in Figure 9. The points on the xy plane we chose for our table are the grid points in the xy plane, and the corresponding points on the graph are marked with solid dots. The problem is that this small number of points hardly tells us what the surface looks like, and even if we plotted more points, it is not clear that we would get anything more than a mass of dots on the page.

What can we do? There are several alternatives. One place to start is to use technology to draw the graph.[6] We then obtain something like Figure 10. This particular surface is called a **paraboloid.**

Question The name *paraboloid* suggests "parabola," and the surface is reminiscent of a parabola. Why is this?

Answer In answering this question, we will employ a useful technique for analyzing a graph. The technique is to slice through the surface with various planes. This will take a little while to explain.

If we slice vertically through this surface along the yz plane, we get the picture in Figure 11. The front edge, where we cut, looks like a parabola, and it is one. To see why, note that the yz plane is the set of points where $x = 0$. To get the intersection of $x = 0$ and $z = x^2 + y^2$, we substitute $x = 0$ in the second equation, getting $z = y^2$. This is the equation of a parabola in the yz plane.

Similarly, we can slice through the surface with the xz plane by setting $y = 0$. This gives the parabola $z = x^2$ in the xz plane (Figure 12).

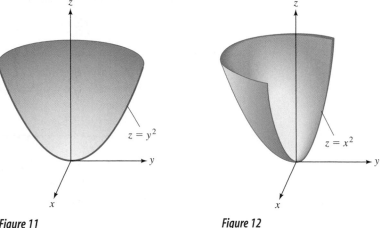

Figure 11   **Figure 12**

We can also look at horizontal slices through the surface—that is, slices by planes parallel to the xy plane. These are given by setting $z = c$ for various numbers c. For example, if we set $z = 1$, we will see only the points with height 1. Substituting in the equation $z = x^2 + y^2$ gives the equation

$$1 = x^2 + y^2$$

which is the equation of a circle of radius 1. If we set $z = 4$, we get the equation of a circle of radius 2:

$$4 = x^2 + y^2$$

---

[6]See Example 3 for a discussion of the use of a spreadsheet to draw a surface.

In general, if we slice through the surface at height $z = c$ we get a circle (of radius $\sqrt{c}$). Figure 13 shows several of these circles.

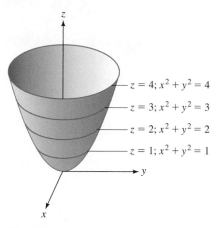

$z = 4; x^2 + y^2 = 4$

$z = 3; x^2 + y^2 = 3$

$z = 2; x^2 + y^2 = 2$

$z = 1; x^2 + y^2 = 1$

Figure 13

Looking at these circular slices, we see that this surface is the one we get by taking the parabola $z = x^2$ and spinning it around the z axis. This is an example of what is known as a **surface of revolution.**

✴ *Before we go on . . .* Notice that each horizontal slice through the surface was obtained by putting $z = constant$. This gave us an equation in x and y that described a curve. These curves are called the **level curves** of the surface $z = f(x, y)$. In this example the equations are of the form $x^2 + y^2 = constant$, and so the level curves are circles. Figure 14 shows the level curves for $c = 0, 1, 2, 3,$ and 4.

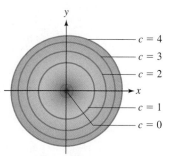

$c = 4$

$c = 3$

$c = 2$

$c = 1$

$c = 0$

Figure 14 • Level curves of the
paraboloid $z = x^2 + y^2$

The level curves give a contour map, or topographical map, of the surface. Each curve shows all the points on the surface at a particular height c. You can use this contour map to visualize the shape of the surface. In your mind's eye, move the contour at $c = 1$ to a height of 1 unit above the xy plane, the contour at $c = 2$ to a height of 2 units above the xy plane, and so on. You will end up with something like Figure 13.

The following summary includes the techniques we have just used, plus some additional ones.

Analyzing the Graph of a Function of Two Variables

If possible, use technology to render the graph of a given function $z = f(x, y)$. Given the function $z = f(x, y)$, you can analyze its graph as follows.

Step 1 Obtain the x, y, and z intercepts (the places where the surface crosses the coordinate axes).

x intercept(s): Set $y = 0$ and $z = 0$ and solve for x.

y intercept(s): Set $x = 0$ and $z = 0$ and solve for y.

z intercept: Set $x = 0$ and $y = 0$ and compute z.

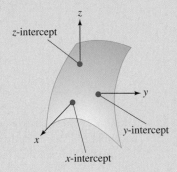

Step 2 Slice the surface along planes parallel to the xy, yz, and xz planes.

$z = constant$: Set $z = constant$ and analyze the resulting curves.
These are the curves resulting from horizontal slices, called **level curves.**

$x = constant$: Set $x = constant$ and analyze the resulting curves.
These are the curves resulting from slices parallel to the yz plane.

$y = constant$: Set $y = constant$ and analyze the resulting curves.
These are the curves resulting from slices parallel to the xz plane.

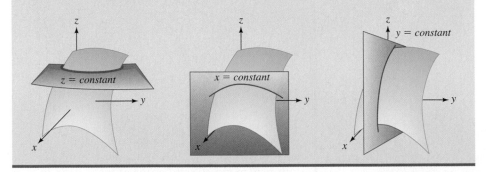

Spreadsheets often have built-in features to render surfaces such as the paraboloid in Example 2. In Example 3 we use Excel to graph another surface and then analyze it as above.

T *Example 3 • Analyzing a Surface*

Describe the graph of $f(x, y) = x^2 - y^2$.

Solution

Generating the Graph with Excel
First, set up a table showing a range of values of x and y and the corresponding values of the function (see Example 3 in Section 8.1):

	A	B	C	D	E	F	G	H
1		-3	-2	-1	0	1	2	3
2	-3	=B1^2-A2^2						
3	-2							
4	-1							
5	0							
6	1							
7	2							
8	3							

Next, we select the cells with the values (B2:H8) and insert a chart, with the `Surface` option selected and `Series in Columns` selected as the data option. The result is shown in Figure 15, which shows an example of a *saddle point* at the origin (we return to this idea in a later section).

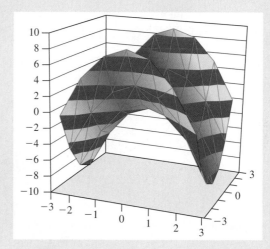

Figure 15

To analyze the graph for the features shown in the box above, replace $f(x, y)$ by z to obtain

$$z = x^2 - y^2$$

Step 1: Intercepts Setting any two of the variables x, y, and z equal to zero results in the third also being zero, so the x, y, and z intercepts are all zero. In other words, the surface touches all three axes in exactly one point, the origin.

Step 2: Slices Slices in various directions show more interesting features.

• *Slice by $x = c$* This gives $z = c^2 - y^2$, which is the equation of a parabola that opens downward. You can see two of these slices ($c = -3$, $c = 3$) as the front and back edges of the surface in Figure 15. [More are shown in Figure 16(a).]

- *Slice by* $y = c$ This gives $z = x^2 - c^2$, which is the equation of a parabola once again—this time, opening upward. You can see two of these slices ($c = -3, c = 3$) as the left and right edges of the surface in Figure 15. [More are shown in Figure 16(b).]

- *Slice by* $z = c$ This gives $x^2 - y^2 = c$, which is a hyperbola. The level curves for various values of c are visible in Figure 15 as the boundaries between the different shadings in the graph. [See Figure 16(c).] The case $c = 0$ is interesting: The equation $x^2 - y^2 = 0$ can be rewritten as $x = \pm y$ (why?), which represents two lines at right angles to each other.

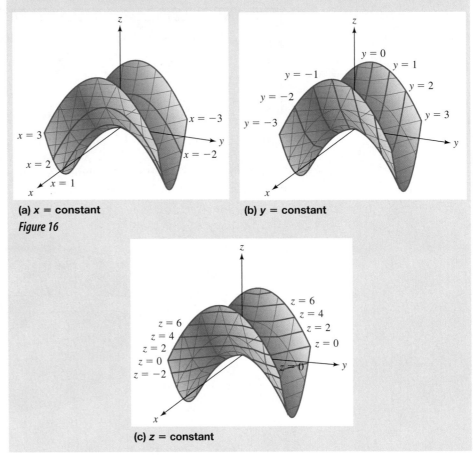

(a) x = constant

(b) y = constant

Figure 16

(c) z = constant

 Web Site

For a utility that does online graphing of three-dimensional surfaces, follow the path

Web Site → Everything for Calculus → Chapter 8 → Surface Graphing Utility

 Other Technology

To obtain really beautiful renderings of surfaces, you can use one of the commercial computer algebra software packages, such as Mathematica® or Maple®. These packages do much more than render surfaces and can be used, for example, to compute derivatives and antiderivatives, to solve equations algebraically, and to perform a variety of algebraic computations.

Example 4 • Graph of a Linear Function

Describe the graph of $g(x, y) = \dfrac{1}{2}x + \dfrac{1}{3}y - 1$.

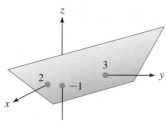

Figure 17

Solution Notice first that g is a linear function of x and y. Figure 17 shows a portion of the graph, which is a plane. We can get a good idea of what plane this is by looking at the x, y, and z intercepts.

- *x intercept* We set $y = 0$ and $z = 0$, which gives

$$0 = \frac{1}{2}x - 1$$

Hence, $x = 2$ is the place where the plane crosses the x axis.

- *y intercept* Setting $x = 0$ and $z = 0$ gives

$$0 = \frac{1}{3}y - 1$$

This equation gives $y = 3$ as the place where the plane crosses the y axis.

- *z intercept* We set $x = 0$ and $y = 0$ to get $z = -1$ as the place where the plane crosses the z axis.

Three points are enough to define a plane, so we can say that the plane is the one passing through the three points $(2, 0, 0)$, $(0, 3, 0)$, and $(0, 0, -1)$.

Before we go on . . . It can be shown that the graph of every linear function of two variables is a plane. What do the level curves look like?

8.2 EXERCISES

1. Sketch the cube with vertices $(0, 0, 0)$, $(1, 0, 0)$, $(0, 1, 0)$, $(0, 0, 1)$, $(1, 1, 0)$, $(1, 0, 1)$, $(0, 1, 1)$, and $(1, 1, 1)$.

2. Sketch the cube with vertices $(-1, -1, -1)$, $(1, -1, -1)$, $(-1, 1, -1)$, $(-1, -1, 1)$, $(1, 1, -1)$, $(1, -1, 1)$, $(-1, 1, 1)$, and $(1, 1, 1)$.

3. Sketch the pyramid with vertices $(1, 1, 0)$, $(1, -1, 0)$, $(-1, 1, 0)$, $(-1, -1, 0)$, and $(0, 0, 2)$.

4. Sketch the solid with vertices $(1, 1, 0)$, $(1, -1, 0)$, $(-1, 1, 0)$, $(-1, -1, 0)$, $(0, 0, -1)$, and $(0, 0, 1)$.

In Exercises 5–10, sketch the planes.

5. $z = -2$ **6.** $z = 4$

7. $y = 2$ **8.** $y = -3$

9. $x = -3$ **10.** $x = 2$

In Exercises 11–18, match each equation with one of the graphs **A–H**. (If necessary, use technology to render the surfaces.)

11. $f(x, y) = 1 - 3x + 2y$ **12.** $f(x, y) = 1 - \sqrt{x^2 + y^2}$

13. $f(x, y) = 1 - (x^2 + y^2)$ **14.** $f(x, y) = y^2 - x^2$

15. $f(x, y) = -\sqrt{1 - (x^2 + y^2)}$ **16.** $f(x, y) = 1 + (x^2 + y^2)$

17. $f(x, y) = \dfrac{1}{x^2 + y^2}$ **18.** $f(x, y) = 3x - 2y + 1$

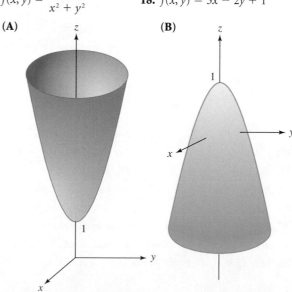

(A)   **(B)**

(C)

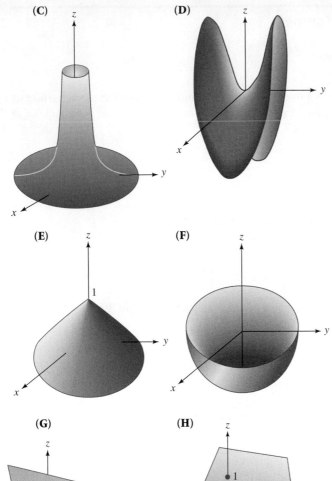

(D)

(E)

(F)

(G)

(H)

In Exercises 19–40, sketch the graphs of the functions.

19. $f(x, y) = 1 - x - y$ **20.** $f(x, y) = x + y - 2$

21. $g(x, y) = 2x + y - 2$ **22.** $g(x, y) = 3 - x + 2y$

23. $h(x, y) = x + 2$ **24.** $h(x, y) = 3 - y$

25. $r(x, y) = x + y$ **26.** $r(x, y) = x - y$

In Exercises 27–40, use of technology is suggested.

27. $s(x, y) = 2x^2 + 2y^2$. Show cross sections at $z = 1$ and $z = 2$.

28. $s(x, y) = -(x^2 + y^2)$. Show cross sections at $z = -1$ and $z = -2$.

29. $t(x, y) = x^2 + 2y^2$. Show cross sections at $x = 0$ and $z = 1$.

30. $t(x, y) = \frac{1}{2}x^2 + y^2$. Show cross sections at $x = 0$ and $z = 1$.

31. $f(x, y) = 2 + \sqrt{x^2 + y^2}$. Show cross sections at $z = 3$ and $y = 0$.

32. $f(x, y) = 2 - \sqrt{x^2 + y^2}$. Show cross sections at $z = 0$ and $y = 0$.

33. $f(x, y) = -2\sqrt{x^2 + y^2}$. Show cross sections at $z = -4$ and $y = 1$.

34. $f(x, y) = 2 + 2\sqrt{x^2 + y^2}$. Show cross sections at $z = 4$ and $y = 1$.

35. $f(x, y) = y^2$ **36.** $g(x, y) = x^2$

37. $h(x, y) = \frac{1}{y}$ **38.** $k(x, y) = e^y$

39. $f(x, y) = e^{-(x^2 + y^2)}$ **40.** $g(x, y) = \frac{1}{\sqrt{x^2 + y^2}}$

APPLICATIONS

41. Marginal Cost (Linear Model) Your weekly cost (in dollars) to manufacture x cars and y trucks is

$$C(x, y) = 240{,}000 + 6000x + 4000y$$

a. Describe the graph of the cost function C.
b. Describe the slice $x = 10$. What cost function does this slice describe?
c. Describe the level curve $z = 480{,}000$. What does this curve tell you about costs?

42. Marginal Cost (Linear Model) Your weekly cost (in dollars) to manufacture x bicycles and y tricycles is

$$C(x, y) = 24{,}000 + 60x + 20y$$

a. Describe the graph of the cost function C.
b. Describe the slice by $y = 100$. What cost function does this slice describe?
c. Describe the level curve $z = 72{,}000$. What does this curve tell you about costs?

43. Market Share (Cars and Light Trucks) Based on data from 1980–1998, the relationship between the domestic market shares of three major U.S. manufacturers of cars and light trucks is

$$x_3 = 0.66 - 2.2x_1 - 0.02x_2$$

where x_1, x_2, and x_3 are, respectively, the fractions of the market held by Chrysler, Ford, and General Motors. Thinking of General Motors' market share as a function of the shares of the other two manufacturers, describe the graph of the resulting function. How are the different slices by $x_1 = constant$ related to one another? What does this say about market share?

The model is based on a linear regression. Source: Ward's AutoInfoBank/*New York Times*, July 29, 1998, p. D6.

44. Market Share (Cereals) Based on data from 1993–1998, the relationship among the domestic market shares of three major manufacturers of breakfast cereal is

$$x_1 = -0.4 + 1.2x_2 + 2x_3$$

where x_1, x_2, and x_3 are, respectively, the fractions of the market held by Kellogg, General Mills, and General Foods. Thinking of Kellogg's market share as a function of shares of the other two manufacturers, describe the graph of the resulting function. How are the different slices by $x_2 = constant$ related to one another? What does this say about market share?

The model is based on a linear regression. SOURCE: Bloomberg Financial Markets/*New York Times*, November 28, 1998, p. C1.

45. Marginal Cost (Interaction Model) Your weekly cost (in dollars) to manufacture x cars and y trucks is

$$C(x, y) = 240,000 + 6000x + 4000y - 20xy$$

(Compare Exercise 41.)

a. Describe the slices $x = constant$ and $y = constant$.

b. Is the graph of the cost function a plane? How does your answer relate to part (a)?

c. What are the slopes of the slices $x = 10$ and $x = 20$? What does this say about cost?

46. Marginal Cost (Interaction Model) Repeat Exercise 45, using the weekly cost to manufacture x bicycles and y tricycles given by

$$C(x, y) = 24,000 + 60x + 20y + 0.3xy$$

(Compare Exercise 42.)

47. Housing Costs The cost C of building a house is related to the number k of carpenters used and the number e of electricians used by

$$C(k, e) = 15,000 + 50k^2 + 50e^2$$

Describe the level curves $C = 30,000$ and $C = 40,000$. What do these level curves represent?

SOURCE: Based on an exercise in A. L. Ostrosky, Jr., and J. V. Koch, *Introduction to Mathematical Economics* (Prospect Heights, Ill.: Waveland Press, 1979).

48. Housing Costs The cost C of building a house (in a different area from that in Exercise 47) is related to the number k of carpenters used and the number e of electricians used by

$$C(k, e) = 15,000 + 70k^2 + 40e^2$$

Describe the slices by the planes $k = 2$ and $e = 2$. What do these slices represent?

SOURCE: Based on an exercise in A. L. Ostrosky, Jr., and J. V. Koch, *Introduction to Mathematical Economics* (Prospect Heights, Ill.: Waveland Press, 1979).

49. Area The area of a rectangle of height h and width w is $A(h, w) = hw$. Sketch a few level curves of A. If the perimeter $h + w$ of the rectangle is constant, which h and w give the largest area? (We suggest you draw in the line $h + w = c$ for several values of c.)

50. Area The area of an ellipse with semiminor axis a and semimajor axis b is $A(a, b) = \pi ab$. Sketch the graph of A. If $a^2 + b^2$ is constant, what a and b give the largest area?

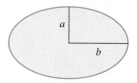

For Exercises 51–54, graphing technology is suggested.

51. Production (Cobb–Douglas Model) Graph the level curves at $z = 0$, 1, 2, and 3 of $P(x, y) = Kx^a y^{1-a}$ if $K = 1$ and $a = 0.5$. Here, x is the number of workers, y is the operating budget, and $P(x, y)$ is the productivity. Interpret the level curve at $z = 3$.

52. Production (Cobb–Douglas Model) Graph the level curves at $z = 0$, 1, 2, and 3 of $P(x, y) = Kx^a y^{1-a}$ if $K = 1$ and $a = 0.25$. Here, x is the number of workers, y is the operating budget, and $P(x, y)$ is the productivity. Interpret the level curve at $z = 0$.

53. Utility Suppose your newspaper is trying to decide between two competing desktop publishing software packages, Macro Publish and Turbo Publish. You estimate that if you purchase x copies of Macro Publish and y copies of Turbo Publish, your company's daily productivity will be

$$U(x, y) = 6x^{0.8}y^{0.2} + x$$

where $U(x, y)$ is measured in pages per day (U is called a *utility function*). Graph the level curves at $z = 0$, 10, 20, and 30. What does the level curve at $z = 0$ tell you?

54. Utility Suppose your small publishing company is trying to decide between two competing desktop publishing software packages, Macro Publish and Turbo Publish. You estimate that if you purchase x copies of Macro Publish and y copies of Turbo Publish, your company's daily productivity will be given by

$$U(x, y) = 5x^{0.2}y^{0.8} + x$$

where $U(x, y)$ is measured in pages per day. Graph the level curves at $z = 0$, 10, 20, and 30. Give a formula for the level curve at $z = 30$ specifying y as a function of x. What does this curve tell you?

COMMUNICATION AND REASONING EXERCISES

55. Complete the following: The graph of a linear function of two variables is a _____.

56. Complete the following: The level curves of a linear function of two variables are _____.

57. Why do we not sketch the graphs of functions of three or more variables?

58. The surface of a mountain can be thought of as the graph of what function?

59. Your study partner Slim just told you that the surface $z = f(x, y)$ you have been trying to graph must be a plane because you've already found that the slices $x = constant$ and $y = constant$ are all straight lines. Do you agree or disagree? Explain.

60. Your other study partner Shady claims that since the surface $z = f(x, y)$ you have been studying is a plane, it follows that all the slices $x = constant$ and $y = constant$ are straight lines. Do you agree or disagree? Explain.

61. Show that the distance between the points (x, y, z) and (a, b, c) is given by the following **three-dimensional distance formula:**

$$d = \sqrt{(x - a)^2 + (y - b)^2 + (z - c)^2}$$

$$\text{or } d = \sqrt{(\Delta x)^2 + (\Delta y)^2 + (\Delta z)^2}$$

The following diagram should be of assistance.

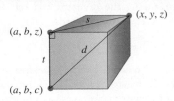

62. Use the result of Exercise 61 to show that the sphere of radius r centered at the origin has the equation

$$x^2 + y^2 + z^2 = r^2$$

63. Why is three-dimensional space used to represent the graph of a function of two variables?

64. Why is it that we can sketch the graphs of functions of two variables on the two-dimensional flat surfaces of these pages?

8.3 *Partial Derivatives*

Recall that if f is a function of x, then the derivative df/dx measures how fast f changes as x increases. If f is a function of two or more variables, we can ask how fast f changes as each variable increases while the others remain fixed. These rates of change are called the *partial derivatives of f*, and they measure how each variable contributes to the change in f. Here is a more precise definition.

Partial Derivatives

The **partial derivative of f with respect to x** is the derivative of f with respect to x, when all other variables are treated as constant. Similarly, the **partial derivative of f with respect to y** is the derivative of f with respect to y, with all other variables treated as constant, and so on for other variables. The partial derivatives are written as $\partial f/\partial x$, $\partial f/\partial y$, and so on. The symbol ∂ is used (instead of d) to remind us that there is more than one variable and that we are holding the other variables fixed.

Quick Examples

1. Let $f(x, y) = x^2 + y^2$.

$$\frac{\partial f}{\partial x} = 2x + 0 = 2x \qquad \text{Because } y^2 \text{ is treated as a constant}$$

$$\frac{\partial f}{\partial y} = 0 + 2y = 2y \qquad \text{Because } x^2 \text{ is treated as a constant}$$

2. Let $z = x^2 + xy$.

$$\frac{\partial z}{\partial x} = 2x + y \qquad\qquad \frac{\partial}{\partial x}[xy] = \frac{\partial}{\partial x}[x \cdot \text{constant}] = \text{constant} = y$$

$$\frac{\partial z}{\partial y} = 0 + x \qquad\qquad \frac{\partial}{\partial y}[xy] = \frac{\partial}{\partial x}[\text{constant} \cdot y] = \text{constant} = x$$

3. Let $f(x, y) = x^2y + y^2x - xy + y$.

$$\frac{\partial f}{\partial x} = 2xy + y^2 - y \qquad y \text{ is treated as a constant.}$$

$$\frac{\partial f}{\partial y} = x^2 + 2xy - x + 1 \quad x \text{ is treated as a constant.}$$

Interpretatation

$\partial f/\partial x$ is the rate at which f changes as x changes, for a fixed (constant) y.

$\partial f/\partial y$ is the rate at which f changes as y changes, for a fixed (constant) x.

Example 1 • Marginal Cost: Linear Model

We return to Example 1 from Section 8.1. Suppose you own a company that makes two models of speakers, the Ultra Mini and the Big Stack. Your total monthly cost (in dollars) to make x Ultra Minis and y Big Stacks is given by

$$C(x, y) = 10,000 + 20x + 40y$$

What is the significance $\partial C/\partial x$ and $\partial C/\partial y$?

Solution First, we compute these partial derivatives:

$$\frac{\partial C}{\partial x} = 20 \quad \text{and} \quad \frac{\partial C}{\partial y} = 40$$

We interpret the results as follows: $\partial C/\partial x = 20$ means that the cost is increasing at a rate of \$20 per additional Ultra Mini (if production of Big Stacks is held constant). $\partial C/\partial y = 40$ means that the cost is increasing at a rate of \$40 per additional Big Stack (if production of Ultra Minis is held constant). In other words, these are the **marginal costs** of each model of speaker: $\partial C/\partial x$ is the marginal cost of an Ultra Mini, and $\partial C/\partial y$ is the marginal cost of a Big Stack.

Before we go on . . . How much does the cost rise if you increase x by Δx and y by Δy? In this example the change in cost is given by

$$\Delta C = 20\Delta x + 40\Delta y = \frac{\partial C}{\partial x} \Delta x + \frac{\partial C}{\partial y} \Delta y$$

This suggests the **chain rule for several variables.** Part of this rule says that if x and y are both functions of t, then C is a function of t through them, and the rate of change of C with respect to t can be calculated as

$$\frac{dC}{dt} = \frac{\partial C}{\partial x} \cdot \frac{dx}{dt} + \frac{\partial C}{\partial y} \cdot \frac{dy}{dt}$$

We will not have a chance to use this interesting result in this book.

Example 2 • Marginal Cost: Interaction Model

Another possibility for the cost function in Example 1 is the interaction model

$$C(x, y) = 10,000 + 20x + 40y + 0.1xy$$

a. Now what are the marginal costs of the two models of speakers?

b. What is the marginal cost of manufacturing Big Stacks at a production level of 100 Ultra Minis and 50 Big Stacks per month?

Solution

a. We compute the partial derivatives:

$$\frac{\partial C}{\partial x} = 20 + 0.1y \quad \text{and} \quad \frac{\partial C}{\partial y} = 40 + 0.1x$$

Thus, the marginal cost of manufacturing Ultra Minis increases by $0.1, or 10¢, for each Big Stack that is manufactured. Similarly, the marginal cost of manufacturing Big Stacks increases by 10¢ for each Ultra Mini that is manufactured.

b. From part (a) the marginal cost of manufacturing Big Stacks is

$$\frac{\partial C}{\partial y} = 40 + 0.1x$$

At a production level of 100 Ultra Minis and 50 Big Stacks per month, $x = 100$ and $y = 50$. Thus,

$$\left. \frac{\partial C}{\partial y} \right|_{(100,50)} = 40 + 0.1(100) = \$50 \text{ per Big Stack}$$

Partial derivatives of functions of three variables are obtained in the same way as those for functions of two variables, as Example 3 shows.

Example 3 • Function of Three Variables

Calculate $\frac{\partial f}{\partial x}, \frac{\partial f}{\partial y}$, and $\frac{\partial f}{\partial z}$ if $f(x, y, z) = xy^2z^3 - xy$.

Solution Although we now have three variables, the calculation remains the same: $\partial f / \partial x$ is the derivative of f with respect to x, with *both* other variables, y and z, held constant:

$$\frac{\partial f}{\partial x} = y^2z^3 - y$$

Similarly, $\partial f / \partial y$ is the derivative of f with respect to y, with both x and z held constant:

$$\frac{\partial f}{\partial y} = 2xyz^3 - x$$

Finally, to find $\partial f / \partial z$, we hold both x and y constant and take the derivative with respect to z.

$$\frac{\partial f}{\partial z} = 3xy^2z^2$$

⭲ *Before we go on . . .* The procedure is the same for any number of variables: To get the partial derivative with respect to any one variable, we treat all the others as constants.

Geometric Interpretation of Partial Derivatives

Tangent line along slice through $y = b$

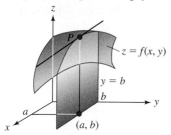

$\dfrac{\partial f}{\partial x}\Big|_{(a, b)}$ **is the slope of the tangent line at the point** $P(a, b, f(a, b))$ **along the slice through** $y = b$.

Figure 18

Question We know that if f is a function of one variable x, then the derivative df/dx gives the slopes of the tangent lines to its graph. What about the *partial* derivatives of a function of several variables?

Answer Suppose f is a function of x and y. By definition $\partial f/\partial x$ is the derivative of the function of x we get by holding y fixed. If we evaluate this derivative at the point (a, b), we are holding y fixed at the value b, taking the ordinary derivative of the resulting function of x, and evaluating this at $x = a$. Now, holding y fixed at b amounts to slicing through the graph of f along the plane $y = b$, resulting in a curve. Thus, the partial derivative is the slope of the tangent line to this curve at the point where $x = a$ and $y = b$, along the plane $y = b$ (Figure 18). This fits with our interpretation of $\partial f/\partial x$ as the rate of increase of f with increasing x when y is held fixed at b. The other partial derivative $\partial f/\partial y\big|_{(a, b)}$ is, similarly, the slope of the tangent line at the same point $P(a, b, f(a, b))$ but along the slice by the plane $x = a$. You should draw the corresponding picture for this on your own.

Example 4 • Geometry of the Interaction Model

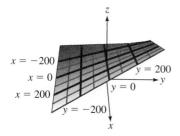

Figure 19

Let's return to the interaction model of cost in Example 2:

$$C(x, y) = 10{,}000 + 20x + 40y + 0.1xy$$

A portion of the surface $z = 10{,}000 + 20x + 40y + 0.1xy$ is shown in Figure 19, together with some of the slices through $x = constant$ and $y = constant$. Since these slices give straight lines, they are already tangent to the surface, and their slopes are given by the partial derivatives (which represent the marginal costs when x and y are nonnegative):

$$y = -200: \text{ slope} = \frac{\partial z}{\partial x}\Big|_{(x, -200)} = 20 + 0.1(-200) = 0$$

$$y = 0: \quad \text{slope} = \frac{\partial z}{\partial x}\Big|_{(x, 0)} = 20 + 0.1(0) = 20$$

$$y = 200: \quad \text{slope} = \frac{\partial z}{\partial x}\Big|_{(x, 200)} = 20 + 0.1(200) = 40$$

$$x = -200: \text{ slope} = \frac{\partial z}{\partial y}\Big|_{(-200, y)} = 40 + 0.1(-200) = 20$$

$$x = 0: \quad \text{slope} = \frac{\partial z}{\partial y}\Big|_{(0, y)} = 40 + 0.1(0) = 40$$

$$x = 200: \quad \text{slope} = \frac{\partial z}{\partial y}\Big|_{(200, y)} = 40 + 0.1(200) = 60$$

Second-Order Partial Derivatives

Just as for functions of a single variable, we can calculate second derivatives. Suppose, for example, that we have a function of x and y, say, $f(x, y) = x^2 - x^2y^2$. We know that

$$\frac{\partial f}{\partial x} = 2x - 2xy^2$$

If we want to take the partial derivative with respect to x once again, there is nothing to stop us:

$$\frac{\partial}{\partial x}\left[\frac{\partial f}{\partial x}\right] = 2 - 2y^2 \qquad \text{Take } \frac{\partial}{\partial x} \text{ of } \frac{\partial f}{\partial x}.$$

(The symbol $\partial/\partial x$ means "the partial derivative with respect to x," just as d/dx stands for "the derivative with respect to x.") This is called the **second partial derivative** and is written $\partial^2 f/\partial x^2$. We get the following derivatives similarly:

$$\frac{\partial f}{\partial y} = -2x^2y$$

$$\frac{\partial^2 f}{\partial y^2} = -2x^2 \qquad \text{Take } \frac{\partial}{\partial y} \text{ of } \frac{\partial f}{\partial y}.$$

Now what if we instead take the partial derivative with respect to y of $\partial f/\partial x$?

$$\frac{\partial^2 f}{\partial y \partial x} = \frac{\partial}{\partial y}\left[\frac{\partial f}{\partial x}\right] \qquad \text{Take } \frac{\partial}{\partial y} \text{ of } \frac{\partial f}{\partial x}.$$

$$= \frac{\partial}{\partial y}[2x - 2xy^2] = -4xy$$

Here, $\partial^2 f/\partial y \partial x$ means "first take the partial derivative with respect to x and then with respect to y" and is called a **mixed partial derivative.** If we differentiate in the opposite order, we get

$$\frac{\partial^2 f}{\partial x \partial y} = \frac{\partial}{\partial x}\left[\frac{\partial f}{\partial y}\right] = \frac{\partial}{\partial x}[-2x^2y] = -4xy$$

the same as $\partial^2 f/\partial y \partial x$. This is no coincidence: The mixed partial derivatives $\partial^2 f/\partial x \partial y$ and $\partial^2 f/\partial y \partial x$ are always the same as long as the first partial derivatives are both differentiable functions of x and y and the mixed partial derivatives are continuous. Since all the functions we will use are of this type, we can take the derivatives in any order we like when calculating mixed derivatives.

Before we go to the exercises, here is another notation for partial derivatives:

$$f_x \text{ means } \frac{\partial f}{\partial x}$$

$$f_y \text{ means } \frac{\partial f}{\partial y}$$

$$f_{xy} \text{ means } (f_x)_y = \frac{\partial^2 f}{\partial y \partial x} \qquad \text{Note the order in which the derivatives are taken.}$$

$$f_{yx} \text{ means } (f_y)_x = \frac{\partial^2 f}{\partial x \partial y}$$

We will sometimes use this more convenient notation, especially for second-order partial derivatives.

8.3 EXERCISES

In Exercises 1–18, calculate $\dfrac{\partial f}{\partial x}$, $\dfrac{\partial f}{\partial y}$, $\dfrac{\partial f}{\partial x}\Big|_{(1,\,-1)}$, and $\dfrac{\partial f}{\partial y}\Big|_{(1,\,-1)}$ when defined.

1. $f(x, y) = 10{,}000 - 40x + 20y$

2. $f(x, y) = 1000 + 5x - 4y$

3. $f(x, y) = 3x^2 - y^3 + x - 1$

4. $f(x, y) = x^{1/2} - 2y^4 + y + 6$

5. $f(x, y) = 10{,}000 - 40x + 20y + 10xy$

6. $f(x, y) = 1000 + 5x - 4y - 3xy$

7. $f(x, y) = 3x^2 y$ **8.** $f(x, y) = x^4 y^2 - x$

9. $f(x, y) = x^2 y^3 - x^3 y^2 - xy$ **10.** $f(x, y) = x^{-1}y^2 + xy^2 + xy$

11. $f(x, y) = (2xy + 1)^3$ **12.** $f(x, y) = \dfrac{1}{(xy + 1)^2}$

13. $f(x, y) = e^{x+y}$ **14.** $f(x, y) = e^{2x+y}$

15. $f(x, y) = 5x^{0.6}y^{0.4}$ **16.** $f(x, y) = -2x^{0.1}y^{0.9}$

17. $f(x, y) = e^{0.2xy}$ **18.** $f(x, y) = xe^{xy}$

In Exercises 19–28, find $\partial^2 f/\partial x^2$, $\partial^2 f/\partial y^2$, $\partial^2 f/\partial x \partial y$, and $\partial^2 f/\partial y \partial x$ and evaluate them at $(1, -1)$ if possible.

19. $f(x, y) = 10{,}000 - 40x + 20y$

20. $f(x, y) = 1000 + 5x - 4y$

21. $f(x, y) = 10{,}000 - 40x + 20y + 10xy$

22. $f(x, y) = 1000 + 5x - 4y - 3xy$

23. $f(x, y) = 3x^2 y$ **24.** $f(x, y) = x^4 y^2 - x$

25. $f(x, y) = e^{x+y}$ **26.** $f(x, y) = e^{2x+y}$

27. $f(x, y) = 5x^{0.6}y^{0.4}$ **28.** $f(x, y) = -2x^{0.1}y^{0.9}$

In Exercises 29–40, find $\partial f/\partial x$, $\partial f/\partial y$, $\partial f/\partial z$ and their values at $(0, -1, 1)$ if possible.

29. $f(x, y, z) = xyz$ **30.** $f(x, y, z) = xy + xz - yz$

31. $f(x, y, z) = -\dfrac{4}{x + y + z^2}$ **32.** $f(x, y, z) = \dfrac{6}{x^2 + y^2 + z^2}$

33. $f(x, y, z) = xe^{yz} + ye^{xz}$ **34.** $f(x, y, z) = xye^z + xe^{yz} + e^{xyz}$

35. $f(x, y, z) = x^{0.1}y^{0.4}z^{0.5}$ **36.** $f(x, y, z) = 2x^{0.2}y^{0.8} + z^2$

37. $f(x, y, z) = e^{xyz}$ **38.** $f(x, y, z) = \ln(x + y + z)$

39. $f(x, y, z) = \dfrac{2000z}{1 + y^{0.3}}$ **40.** $f(x, y, z) = \dfrac{e^{0.2x}}{1 + e^{-0.1y}}$

APPLICATIONS

41. Marginal Cost (Linear Model) Your weekly cost (in dollars) to manufacture x cars and y trucks is

$$C(x, y) = 240{,}000 + 6000x + 4000y$$

Calculate and interpret $\partial C/\partial x$ and $\partial C/\partial y$.

42. Marginal Cost (Linear Model) Your weekly cost (in dollars) to manufacture x bicycles and y tricycles is

$$C(x, y) = 24{,}000 + 60x + 20y$$

Calculate and interpret $\partial C/\partial x$ and $\partial C/\partial y$.

43. Market Share (Cars and Light Trucks) Based on data from 1980–1998, the relationship between the domestic market shares of three major U.S. manufacturers of cars and light trucks is

$$x_3 = 0.66 - 2.2x_1 - 0.02x_2$$

where x_1, x_2, and x_3 are, respectively, the fractions of the market held by Chrysler, Ford, and General Motors. Calculate $\partial x_3/\partial x_1$ and $\partial x_1/\partial x_3$. What do they signify, and how are they related to each other?

The model is based on a linear regression. Source: Ward's AutoInfoBank/*New York Times*, July 29, 1998, p. D6.

44. Market Share (Cereals) Based on data from 1993–1998, the relationship between the domestic market shares of three major manufacturers of breakfast cereal is

$$x_1 = -0.4 + 1.2x_2 + 2x_3$$

where x_1, x_2, and x_3 are, respectively, the fractions of the market held by Kellogg, General Mills, and General Foods. Compute and interpret $\partial x_2/\partial x_3$.

The models are based on a linear regression. Source: Bloomberg Financial Markets/*New York Times*, November 28, 1998, p. C1.

45. Marginal Cost (Interaction Model) Your weekly cost (in dollars) to manufacture x cars and y trucks is

$$C(x, y) = 240{,}000 + 6000x + 4000y - 20xy$$

(Compare Exercise 41.) Compute the marginal cost of manufacturing cars at a production level of 10 cars and 20 trucks.

46. Marginal Cost (Interaction Model) Your weekly cost (in dollars) to manufacture x bicycles and y tricycles is

$$C(x, y) = 24{,}000 + 60x + 20y + 0.3xy$$

(Compare Exercise 42.) Compute the marginal cost of manufacturing tricycles at a production level of 10 bicycles and 20 tricycles.

47. Brand Loyalty The fraction of Mazda car owners who chose another new Mazda can be modeled by the following function:

$$M(c, f, g, h, t) = 1.1 - 3.8c + 2.2f + 1.9g - 1.7h - 1.3t$$

Here, c is the fraction of Chrysler car owners who remained loyal to Chrysler, f is the fraction of Ford car owners remaining loyal to Ford, g the corresponding figure for General Motors, h the corresponding figure for Honda, and t for Toyota.

a. Calculate $\partial M/\partial c$ and $\partial M/\partial f$ and interpret the answers.

b. In 1995 it was observed that $c = 0.56$, $f = 0.56$, $g = 0.72$, $h = 0.50$, and $t = 0.43$. According to the model, what percentage of Mazda owners remained loyal to Mazda? (Round your answer to the nearest percentage point.)

The model is an approximation of a linear regression based on data from the period 1988–1995. Source: Chrysler, Maritz Market Research, Consumer Attitude Research, and Strategic Vision/*New York Times*, November 3, 1995, p. D2.

48. Brand Loyalty The fraction of Mazda car owners who chose another new Mazda can be modeled by the following function:

$$M(c, f) = 9.4 + 7.8c + 3.6c^2 - 38f - 22cf + 43f^2$$

where c is the fraction of Chrysler car owners who remained loyal to Chrysler and f is the fraction of Ford car owners remaining loyal to Ford.

a. Calculate $\partial M/\partial c$ and $\partial M/\partial f$ evaluated at the point $(0.7, 0.7)$, and interpret the answers.

b. In 1995 it was observed that $c = 0.56$, and $f = 0.56$. According to the model, what percentage of Mazda owners remained loyal to Mazda? (Round your answer to the nearest percentage point.)

The model is an approximation of a second-order regression based on data from the period 1988–1995. SOURCE: Chrysler, Maritz Market Research, Consumer Attitude Research, and Strategic Vision/*New York Times,* November 3, 1995, p. D2.

49. Family Income The following model is based on statistical data on the median family incomes of black and white families in the United States for the period 1950–2000:

$$z = 13{,}000 + 350t + 9900x + 220xt$$

where $\quad z =$ median family income

$\qquad t =$ year ($t = 0$ represents 1950)

$$x = \begin{cases} 0 & \text{if the income was for a black family} \\ 1 & \text{if the income was for a white family} \end{cases}$$

a. Use the model to estimate the median income of a black family in 1960.

b. Use the model to estimate the median income of a white family in 1960.

c. According to the model, how fast was the median income for a black family increasing in 1960?

d. According to the model, how fast was the median income for a white family increasing in 1960?

e. Do the answers to parts (c) and (d) suggest that the income gap between white and black families was widening or narrowing in the second half of the twentieth century?

Incomes are in constant 1997 dollars. The model is a multiple regression model and coefficients are rounded to two significant digits. SOURCE: *Statistical Abstract of the United States, 1999,* U.S. Census Bureau/*New York Times,* December 19, 1999, p. WK5.

50. Life Expectancy The following model is based on life expectancy for men and women in the United States for the period 1900–2000:

$$z = 50.9 + 0.325t - 1.95x - 0.055xt$$

where $\quad z =$ life expectancy of a person born in the United States

$\qquad t =$ year of birth ($t = 0$ represents 1900)

$$x = \begin{cases} 0 & \text{if the person was a female} \\ 1 & \text{if the person was a male} \end{cases}$$

a. Use the model to estimate, to the nearest year, the life expectancy of a female born in 1950.

b. Use the model to estimate, to the nearest year, the life expectancy of a male born in 1950.

c. According to the model, how fast was the life expectancy for a female increasing in 1950?

d. According to the model, how fast was the life expectancy for a male increasing in 1950?

e. Do the answers you have given suggest that the gap between male and female life expectancy was widening or narrowing in the twentieth century?

The model is a multiple regression model and coefficients are rounded to 3 significant digits. SOURCE: *Statistical Abstract of the United States, 1999,* U.S. Census Bureau/*New York Times,* December 19, 1999, p. WK5.

51. Marginal Cost Your weekly cost (in dollars) to manufacture x cars and y trucks is

$$C(x, y) = 200{,}000 + 6000x + 4000y - 100{,}000e^{-0.01(x+y)}$$

What is the marginal cost of a car? Of a truck? How do these marginal costs behave as total production increases?

52. Marginal Cost Your weekly cost (in dollars) to manufacture x bicycles and y tricycles is

$$C(x, y) = 20{,}000 + 60x + 20y + 50\sqrt{xy}$$

What is the marginal cost of a bicycle? Of a tricycle? How do these marginal costs behave as x and y increase?

53. Average Cost If you average your costs over your total production, you get the **average cost,** written $\overline{C}$:

$$\overline{C}(x, y) = \frac{C(x, y)}{x + y}$$

Find the average cost for the cost function in Exercise 51. Then find the marginal average cost of a car and the marginal average cost of a truck at a production level of 50 cars and 50 trucks. Interpret your answers.

54. Average Cost Find the average cost for the cost function in Exercise 52 (see Exercise 53). Then find the marginal average cost of a bicycle and the marginal average cost of a tricycle at a production level of five bicycles and five tricycles. Interpret your answers.

55. Marginal Revenue As manager of an auto dealership, you offer a car rental company the following deal: You will charge \$15,000 per car and \$10,000 per truck, but you will then give the company a discount of \$5000 times the square root of the total number of vehicles it buys from you. Looking at your marginal revenue, is this a good deal for the rental company?

56. Marginal Revenue As marketing director for a bicycle manufacturer, you come up with the following scheme: You will offer to sell a dealer x bicycles and y tricycles for

$$R(x, y) = 3500 - 3500e^{-0.02x - 0.01y} \text{ dollars}$$

Find your marginal revenue for bicycles and for tricycles. Are you likely to be fired for your suggestion?

57. Research Productivity Here we apply a variant of the Cobb–Douglas function to the modeling of research productivity. A mathematical model of research productivity at a particular physics laboratory is

$$P = 0.04x^{0.4}y^{0.2}z^{0.4}$$

where P is the annual number of groundbreaking research papers produced by the staff, x is the number of physicists on the research team, y is the laboratory's annual research budget, and z is the annual National Science Foundation subsidy to the laboratory. Find the rate of increase of research papers per government-subsidy dollar at a subsidy level of $1 million per year and a staff level of 10 physicists if the annual budget is $100,000.

58. Research Productivity A major drug company estimates that the annual number P of patents for new drugs developed by its research team is best modeled by the formula

$$P = 0.3x^{0.3}y^{0.4}z^{0.3}$$

where x is the number of research biochemists on the payroll, y is the annual research budget, and z is size of the bonus awarded to discoverers of new drugs. Assuming that the company has 12 biochemists on the staff, has an annual research budget of $500,000, and pays $40,000 bonuses to developers of new drugs, calculate the rate of growth in the annual number of patents per new research staff member.

59. Utility Your newspaper is trying to decide between two competing desktop publishing software packages, Macro Publish and Turbo Publish. You estimate that if you purchase x copies of Macro Publish and y copies of Turbo Publish, your company's daily productivity will be

$$U(x, y) = 6x^{0.8}y^{0.2} + x$$

where $U(x, y)$ is measured in pages per day.

a. Calculate $\left.\dfrac{\partial U}{\partial x}\right|_{(10, 5)}$ and $\left.\dfrac{\partial U}{\partial y}\right|_{(10, 5)}$ to two decimal places and interpret the results.

b. What does the ratio $\left.\dfrac{\partial U}{\partial x}\right|_{(10, 5)} / \left.\dfrac{\partial U}{\partial y}\right|_{(10, 5)}$ tell about the usefulness of these products?

60. Grades A production formula for a student's performance on a difficult English examination is given by

$$g(t, x) = 4tx - 0.2t^2 - x^2$$

where g is the grade the student can expect to get, t is the number of hours of study for the examination, and x is the student's grade-point average.

a. Calculate $\left.\dfrac{\partial g}{\partial t}\right|_{(10, 3)}$ and $\left.\dfrac{\partial g}{\partial x}\right|_{(10, 3)}$ and interpret the results.

b. What does the ratio $\left.\dfrac{\partial g}{\partial t}\right|_{(10, 3)} / \left.\dfrac{\partial g}{\partial x}\right|_{(10, 3)}$ tell about the relative merits of study and grade point average?

Based on an exercise in A. L. Ostrosky, Jr., and J. V. Koch, *Introduction to Mathematical Economics* (Prospect Heights, Ill.: Waveland Press, 1979).

61. Electrostatic Repulsion If positive electric charges of Q and q coulombs are situated at positions (a, b, c) and (x, y, z), respectively, then the force of repulsion they experience is given by

$$F = K\frac{Qq}{(x - a)^2 + (y - b)^2 + (z - c)^2}$$

where $K \approx 9 \times 10^9$, F is given in newtons (N), and all positions are measured in meters. Assume that a charge of 10 coulombs is situated at the origin and that a second charge of 5 coulombs is situated at $(2, 3, 3)$ and moving in the y direction at 1 meter/second. How fast is the electrostatic force it experiences decreasing? Round answer to one significant digit.

62. Electrostatic Repulsion Repeat Exercise 61, assuming that a charge of 10 coulombs is situated at the origin and that a second charge of 5 coulombs is situated at $(2, 3, 3)$ and moving in the negative z direction at 1 meter/second. Round answer to one significant digit.

63. Investments Recall that the compound interest formula for annual compounding is

$$A(P, r, t) = P(1 + r)^t$$

where A is the future value of an investment of P dollars after t years at an interest rate of r.

a. Calculate $\partial A/\partial P$, $\partial A/\partial r$, and $\partial A/\partial t$, all evaluated at $(100, 0.10, 10)$. (Round answers to two decimal places.) Interpret your answers.

b. What does the function $\left.\dfrac{\partial A}{\partial P}\right|_{(100, 0.10, t)}$ of t tell about your investment?

64. Investments Repeat Exercise 63, using the formula for continuous compounding:

$$A(P, r, t) = Pe^{rt}$$

65. Modeling with the Cobb–Douglas Production Formula Assume that you are given a production formula of the form

$$P(x, y) = Kx^a y^b \qquad (a + b = 1)$$

a. Obtain formulas for $\partial P/\partial x$ and $\partial P/\partial y$ and show that $\partial P/\partial x = \partial P/\partial y$ precisely when $x/y = a/b$.

b. Let x be the number of workers a firm employs and let y be its monthly operating budget in thousands of dollars. Assume that the firm currently employs 100 workers and has a monthly operating budget of $200,000. If each additional worker contributes as much to productivity as each additional $1000 per month, find values of a and b that model the firm's productivity.

66. Housing Costs The cost C of building a house is related to the number k of carpenters used and the number e of electricians used by

$$C(k, e) = 15,000 + 50k^2 + 60e^2$$

If three electricians are currently employed in building your new house and the marginal cost per additional electrician is the same as the marginal cost per additional carpenter, how many carpenters are being used? (Round your answer to the nearest carpenter.)

Based on an exercise in A. L. Ostrosky, Jr., and J. V. Koch, *Introduction to Mathematical Economics* (Prospect Heights, Ill.: Waveland Press, 1979).

67. Nutrient Diffusion Suppose 1 cubic centimeter of nutrient is placed at the center of a circular petri dish filled with water. We might wonder how the nutrient is distributed after a time of t seconds. According to the classical theory of diffusion, the concentration of nutrient (in parts of nutrient per part of water) after a time t is given by

$$u(r, t) = \frac{1}{4\pi Dt} e^{-r^2/(4Dt)}$$

Here D is the *diffusivity*, which we will take to be 1, and r is the distance from the center in centimeters. How fast is the concentration increasing at a distance of 1 centimeter from the center 3 seconds after the nutrient is introduced?

68. Nutrient Diffusion Refer to Exercise 67. How fast is the concentration increasing at a distance of 4 centimeters from the center 4 seconds after the nutrient is introduced?

COMMUNICATION AND REASONING EXERCISES

69. Given that $f(a, b) = r$, $f_x(a, b) = s$, and $f_y(a, b) = t$, complete the following: _____ is increasing at a rate of _____ units per unit of x, _____ is increasing at a rate of _____ units per unit of y, and the value of _____ is _____ when $x =$ _____ and $y =$ _____.

70. A firm's productivity depends on two variables, x and y. Currently, $x = a$ and $y = b$, and the firm's productivity is 4000 units. Productivity is increasing at a rate of 400 units per unit *decrease* in x and is decreasing at a rate of 300 units per unit *increase* in y. What does all of this information tell you about the firm's productivity function $g(x, y)$?

71. Complete the following: Let $f(x, y, z)$ be the cost to build a development of x cypods (one-bedroom units) in the city-state of Utarek, Mars, y argaats (two-bedroom units), and z orbici (singular: orbicus; three-bedroom units) in $\overline{Z}$ (zonars, the designated currency in Utarek). Then $\partial f/\partial z$ measures _____ and has units _____.
SOURCE: http://www.marsnext.com/comm/zonars.html

72. Complete the following: Let $f(t, x, y)$ be the projected number of citizens of the Principality State of Voodice, Luna, in year t since its founding, assuming the presence of x lunar vehicle factories and y domed settlements. Then $\partial f/\partial x$ measures _____ and has units _____.
SOURCE: http://www.voodice.info/

73. Give an example of a function $f(x, y)$ with $f(1, 1) = 10$, $f_x(1, 1) = -2$, and $f_y(1, 1) = 3$.

74. Give an example of a function $f(x, y, z)$ that has all of its partial derivatives nonzero constants.

75. The graph of $z = b + mx + ny$ (b, m, and n constants) is a plane.
   a. Explain the geometric significance of the numbers b, m, and n.
   b. Show that the equation of the plane passing through (h, k, l) with slope m in the x direction (in the sense of $\partial/\partial x$) and slope n in the y direction is

$$z = l + m(x - h) + n(y - k)$$

76. The **tangent plane** to the graph of $f(x, y)$ at $P(a, b, f(a, b))$ is the plane containing the lines tangent to the slice through the graph by $y = b$ (as in Figure 18) and the slice through the graph by $x = a$. Use the result of Exercise 75 to show that the equation of the tangent plane is

$$z = f(a, b) + f_x(a, b)(x - a) + f_y(a, b)(y - b)$$

8.4   Maxima and Minima

In Chapter 5 on applications of the derivative, we saw how to locate relative extrema of a function of a single variable. In this section we extend our methods to functions of two variables. Similar techniques work for functions of three or more variables.

Figure 20 shows a portion of the graph of the function $f(x, y) = 2(x^2 + y^2) - (x^4 + y^4) + 3$. The graph resembles a "flying carpet" and shows several interesting points, marked a, b, c, and d. The point a has coordinates $(0, 0, f(0, 0))$, is directly above the origin $(0, 0)$, and is the lowest point in its vicinity; water would puddle there. We say that f has a **relative minimum** at $(0, 0)$ because $f(0, 0)$ is smaller than $f(x, y)$ for any (x, y) near $(0, 0)$. Similarly, the point b is higher than any point in its vicinity. Thus, we say that f has a **relative maximum** at $(1, 1)$. The points c and d represent a new phenomenon and are called **saddle points.** They are neither relative maxima nor relative minima but seem to be a little of both.

To see more clearly what features a saddle point has, look at Figure 21, which shows a portion of the graph near the point d. It is easy to see where the term *saddle point* comes from. If we slice through the graph along $y = 1$, we get a curve on which d is the *lowest* point. Thus, d looks like a relative minimum along this slice. On the other hand, if we slice through the graph along $x = 0$, we get another curve, on which d is the *highest* point, so d looks like a relative maximum along this slice. This kind of behavior characterizes a saddle point: f has a **saddle point** at (r, s) if f has a relative minimum at (r, s)

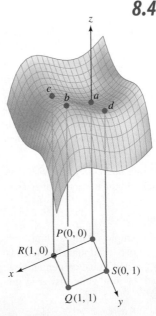

Figure 20

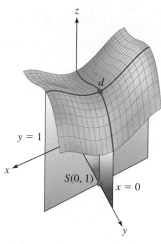

z

$y = 1$

x

$S(0, 1)$

d

$x = 0$

y

Figure 21

along some slice through that point and a relative maximum along another slice through that point. If you look at the other saddle point, c, in Figure 20, you see the same characteristics.

Although numerical information can help us locate the approximate position of relative extrema and saddle points, calculus permits us to locate these points accurately, just as we did for functions of a single variable. Look once again at Figure 20 and notice the following:

- The points P, Q, R, and S are all in the **interior** of the domain of f; that is, none of them lies on the boundary of the domain. Said another way, we can move some distance in any direction from any of these points without leaving the domain of f.

- The tangent lines along the slices through these points parallel to the x and y axes are *horizontal*. Thus, the partial derivatives $\partial f/\partial x$ and $\partial f/\partial y$ are zero when evaluated at any of the points P, Q, R, and S. This gives us a way of locating candidates for relative extrema and saddle points.

The following summary generalizes and also expands on some of what we have just said.

Relative and Absolute Maxima and Minima

The function $f(x, y, \ldots)$ has a **relative maximum** at $(r, s, \ldots)$ if $f(r, s, \ldots) \geq f(x, y, \ldots)$ for every point $(x, y, \ldots)$ near $(r, s, \ldots)$ in the domain of f. We say that $f(x, y, \ldots)$ has an **absolute maximum** at $(r, s, \ldots)$ if $f(r, s, \ldots) \geq f(x, y, \ldots)$ for every point $(x, y, \ldots)$ in the domain of f. The terms **relative minimum** and **absolute minimum** are defined in a similar way.

Locating Candidates for Relative Extrema and Saddle Points in the Interior of the Domain of f

1. Set $\partial f/\partial x = 0$, $\partial f/\partial y = 0$, $\ldots$, simultaneously, and solve for $x, y, \ldots$.

2. Check that the resulting points $(x, y, \ldots)$ are in the interior of the domain of f.

Points at which all the partial derivatives of f are zero are called **critical points.** Thus, the critical points are the only candidates for relative extrema and saddle points in the interior of the domain of f.*

Quick Examples

1. Let $f(x, y) = x^3 + (y - 1)^2$. Then $\partial f/\partial x = 3x^2$ and $\partial f/\partial y = 2(y - 1)$. Thus, we solve the system

$$3x^2 = 0 \quad \text{and} \quad 2(y - 1) = 0$$

The first equation gives $x = 0$, and the second gives $y = 1$, so the only critical point is $(0, 1)$. Since the domain of f is the whole Cartesian plane, the point $(0, 1)$ is interior and hence a candidate for a relative extremum or saddle point.†

---

*We'll be looking at extrema on the *boundary* of the domain of a function in Section 8.5. What we are calling critical points correspond to the *stationary* points of a function of one variable. We will not consider the analogs of the singular points.

†In fact, it is a saddle point. (Can you see why?)

2. Let $f(x, y) = e^{-(x^2+y^2)}$. Taking partial derivatives and setting them equal to zero gives

$$-2xe^{-(x^2+y^2)} = 0 \qquad \text{We set } \frac{\partial f}{\partial x} = 0.$$

$$-2ye^{-(x^2+y^2)} = 0 \qquad \text{We set } \frac{\partial f}{\partial y} = 0.$$

The first equation implies that $x = 0$,[‡] and the second implies that $y = 0$, so the only critical point is $(0, 0)$. This point is interior and hence a candidate for a relative extremum or saddle point.

[‡]Recall that if a product of two numbers is zero, then one or the other must be zero. In this case the number $e^{-(x^2+y^2)}$ can't be zero (since e^u is never zero), which gives the result claimed.

Example 1 • Locating Critical Points

Locate all critical points of $f(x, y) = x^2y - x^2 - 2y^2$. Graph the function to classify the critical points as relative maxima, minima, saddle points, or none of these.

Solution The partial derivatives are

$$f_x = 2xy - 2x = 2x(y - 1) \quad \text{and} \quad f_y = x^2 - 4y$$

Setting these equal to zero gives

$$x = 0 \text{ or } y = 1 \quad \text{and} \quad x^2 = 4y$$

We get a solution by choosing either $x = 0$ or $y = 1$ and substituting into $x^2 = 4y$.

Case 1: $x = 0$ Substituting into $x^2 = 4y$ gives $0 = 4y$ and hence $y = 0$. Thus, the critical point for this case is $(x, y) = (0, 0)$.

Case 2: $y = 1$ Substituting into $x^2 = 4y$ gives $x^2 = 4$ and hence $x = \pm 2$. Thus, we get two critical points for this case: $(2, 1)$ and $(-2, 1)$.

We now have three critical points altogether: $(0, 0)$, $(2, 1)$, and $(-2, 1)$. We get the corresponding points on the graph by substituting for x and y in the equation for f to get the z coordinates. The points are $(0, 0, 0)$, $(2, 1, -2)$, and $(-2, 1, -2)$.

Classifying the Critical Points Graphically

To classify the critical points graphically, we look at the graph of f shown in Figure 22. Examining the graph carefully, we see that the point $(0, 0, 0)$ is a relative maximum. As for the other two critical points, are they saddle points or relative maxima? They *seem* to be relative maxima along the y direction, but the slice in the x direction (through $x = 1$) seems to be horizontal. However, a diagonal slice (along $x = \pm y$) shows these two points as minima, and so they are saddle points. (If you don't believe this, we will get more evidence below and in a later example.)

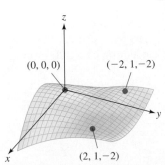

$(0, 0, 0)$ $(-2, 1, -2)$

$(2, 1, -2)$

Figure 22

Classifying the Critical Points Numerically

We can use a spreadsheet to classify the critical points numerically. Following is a tabular representation of the function obtained using Excel. (See Example 3 in Section 8.1 for information on using Excel to generate such a table.)

$x \rightarrow$

		−3	−2	−1	0	1	2	3
y	**−3**	−54	−34	−22	−18	−22	−34	−54
↓	**−2**	−35	−20	−11	−8	−11	−20	−35
	−1	−20	−10	−4	−2	−4	−10	−20
	0	−9	−4	−1	0	−1	−4	−9
	1	−2	−2	−2	−2	−2	−2	−2
	2	1	−4	−7	−8	−7	−4	1
	3	0	−10	−16	−18	−16	−10	0

The shaded and colored cells show rectangular neighborhoods of the three critical points $(0, 0)$, $(2, 1)$, and $(−2, 1)$. (Notice that they overlap.) The values of f at the points are at the centers of these rectangles. Looking at the gray neighborhood of $(x, y) = (0, 0)$, we see that $f(0, 0) = 0$ is the largest value of f in the shaded cells, suggesting that f has a maximum at $(0, 0)$. The pale orange shaded neighborhood of $(2, 1)$ on the right shows $f(2, 1) = −2$ as the maximum along some slices (for example, the vertical slice) and a minimum along the diagonal slice from top left to bottom right. This is what results in a saddle point on the graph. The point $(−2, 1)$ is similar, and thus f also has a saddle point at $(−2, 1)$.

Question Is there an algebraic way of deciding whether a given point is a relative maximum, relative minimum, or saddle point?

Answer There is a *second derivative test* for functions of two variables, stated as follows.

Second Derivative Test for Functions of Two Variables

Suppose (a, b) is a critical point in the interior of the domain of the function f of two variables. Let H be the quantity

$$H = f_{xx}(a, b)f_{yy}(a, b) - [f_{xy}(a, b)]^2 \qquad \text{\textit{H} is called the \textit{Hessian}.}$$

Then, if H is *positive*,

- f has a relative minimum at (a, b) if $f_{xx}(a, b) > 0$.
- f has a relative maximum at (a, b) if $f_{xx}(a, b) < 0$.

If H is *negative*,

- f has a saddle point at (a, b).

If $H = 0$, the test tells us nothing, so we need to look at the graph or a numerical table to see what is going on.

Quick Examples

1. Let $f(x, y) = x^2 - y^2$. Then

$$f_x = 2x \quad \text{and} \quad f_y = -2y$$

which gives $(0, 0)$ as the only critical point. Also,

$$f_{xx} = 2 \qquad f_{xy} = 0 \qquad f_{yy} = -2 \qquad \text{Note that these are constant.}$$

which gives $H = (2)(-2) - 0^2 = -2$. Since H is negative, we have a saddle point at $(0, 0)$.

2. Let $f(x, y) = x^2 + 2y^2 + 2xy + 4x$. Then

$$f_x = 2x + 2y + 4 \quad \text{and} \quad f_y = 2x + 4y$$

Setting these equal to zero gives a system of two linear equations in two unknowns:

$$x + y = -2 \quad \text{and} \quad x + 2y = 0$$

This system has solution $(-4, 2)$, so this is our only critical point. The second partial derivatives are $f_{xx} = 2, f_{xy} = 2$, and $f_{yy} = 4$, so $H = (2)(4) - 2^2 = 4$. Since $H > 0$ and $f_{xx} > 0$, we have a relative minimum at $(-4, 2)$.

Note

There is a second derivative test for functions of three or more variables, but it is considerably more complicated. We stick with functions of two variables for the most part in this book. The justification of the second derivative test is beyond the scope of this book.

Let's use the second derivative test to analyze the "flying carpet" function we saw at the beginning of this section.

Example 2 • Using the Second Derivative Test

Locate and classify all critical points of

$$f(x, y) = 2(x^2 + y^2) - (x^4 + y^4) + 3$$

Solution We first calculate the first-order partial derivatives:

$$f_x = 4x - 4x^3 = 4x(1 - x^2)$$

$$f_y = 4y - 4y^3 = 4y(1 - y^2)$$

Setting these equal to zero gives the following simultaneous equations:

$$4x(1 - x^2) = 0$$

$$4y(1 - y^2) = 0$$

The first equation has solutions $x = 0, 1,$ or -1, and the second has solutions $y = 0,$ 1, or -1. Since these are simultaneous equations, we need values of x and y that satisfy *both* equations. We can choose any one of the values for x that satisfies the first and any one of the values for y that satisfies the second. This gives us a total of *nine* critical points:

$$(0, 0) \quad (0, 1) \quad (0, -1) \quad (1, 0) \quad (1, 1) \quad (1, -1) \quad (-1, 0) \quad (-1, 1) \quad (-1, -1)$$

Four of these are the points $P, Q, R,$ and S on the graph in Figure 20. The other five are also points at which f has extreme or saddle points, but Figure 20 shows only a small portion of the actual graph, and the remaining five points are out of range. (More of the graph of f is shown in Figure 23.)

We now need to apply the second derivative test to each point in turn. First, we calculate all the second derivatives we need:

$$f_x = 4x - 4x^3, \quad \text{so} \quad f_{xx} = 4 - 12x^2 \quad \text{and} \quad f_{xy} = 0$$

and

$$f_y = 4y - 4y^3, \quad \text{so} \quad f_{yy} = 4 - 12y^2$$

We now look at each critical point in turn.

The point $(0, 0)$ $f_{xx}(0, 0) = 4, f_{yy}(0, 0) = 4$, and $f_{xy}(0, 0) = 0$. Thus, $H = 4(4) - 0^2 = 16$. Since $H > 0$ and $f_{xx}(0, 0) > 0$, we have a relative minimum at $(0, 0, 3)$.

The point $(0, 1)$ $f_{xx}(0, 1) = 4, f_{yy}(0, 1) = -8$, and $f_{xy}(0, 1) = 0$. Thus, $H = 4(-8) - 0^2 = -32$. Since $H < 0$, we have a saddle point at $(0, 1, 4)$.

The point $(0, -1)$ $f_{xx}(0, -1) = 4, f_{yy}(0, -1) = -8$, and $f_{xy}(0, -1) = 0$. These are the same values we got for the last point, so we again have a saddle point at $(0, -1, 4)$.

The point $(1, 0)$ $f_{xx}(1, 0) = -8, f_{yy}(1, 0) = 4$, and $f_{xy}(1, 0) = 0$. Thus, $H = (-8)4 - 0^2 = -32$. Since $H < 0$, we have a saddle point at $(1, 0, 4)$.

The point $(1, 1)$ $f_{xx}(1, 1) = -8, f_{yy}(1, 1) = -8$, and $f_{xy}(1, 1) = 0$. Thus, $H = (-8)^2 - 0^2 = 64$. Since $H > 0$ and $f_{xx}(1, 1) < 0$, we have a relative maximum at $(1, 1, 5)$.

The point $(1, -1)$ These are the same values we got for the last point, so we again have a relative maximum at $(1, -1, 5)$.

The point $(-1, 0)$ These are the same values we got for the point $(1, 0)$, so we have a saddle point at $(-1, 0, 4)$.

The point $(-1, 1)$ These are the same values we got for the point $(1, 1)$, so we have a relative maximum at $(-1, 1, 5)$.

The point $(-1, -1)$ These are again the same values we got for the point $(1, 1)$, so we have a relative maximum at $(-1, -1, 5)$.

Figure 23 shows a larger portion of the graph of $f(x, y)$ including all nine critical points. We leave it to you to spot their locations on the graph.

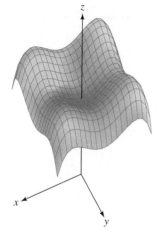

Figure 23

✳ **Before we go on . . .**
Question Which of the points we just found are *absolute* extrema?

Answer To tell whether a relative extremum is absolute, we need to know something about the entire graph of the function—not just the portion shown in Figure 23. If we graph more and more of the function, you will notice that the graph drops off without bound in all directions. Thus, there can be no absolute minima, whereas the relative maxima we found at the points $(\pm 1, \pm 1)$ are all absolute maxima.

Example 3 • The Second Derivative Test Again

Use the second derivative test to analyze the function $f(x, y) = x^2 y - x^2 - 2y^2$ discussed in Example 1 and confirm the results we got there.

Solution We saw in Example 1 that the first-order derivatives are

$$f_x = 2xy - 2x = 2x(y - 1) \quad \text{and} \quad f_y = x^2 - 4y$$

and the critical points are $(0, 0)$, $(2, 1)$, and $(-2, 1)$. We also need the second derivatives:

$$f_{xx} = 2y - 2 \qquad f_{xy} = 2x \qquad f_{yy} = -4$$

The point $(0, 0)$ $f_{xx}(0, 0) = -2, f_{xy}(0, 0) = 0$, and $f_{yy}(0, 0) = -4$, so $H = 8$. Since $H > 0$ and $f_{xx}(0, 0) < 0$, the second derivative test tells us that f has a relative maximum at $(0, 0)$.

The point $(2, 1)$ $f_{xx}(2, 1) = 0$, $f_{xy}(2, 1) = 4$, and $f_{yy}(2, 1) = -4$, so $H = -16$. Since $H < 0$, we know that f has a saddle point at $(2, 1)$.

The point $(-2, 1)$ $f_{xx}(-2, 1) = 0, f_{xy}(-2, 1) = -4$, and $f_{yy}(-2, 1) = -4$, so once again $H = -16$, and f has a saddle point at $(-2, 1)$.

Deriving the Regression Formulas

In Section 1.5 we presented the following set of formulas for the **regression,** or **best-fit, line** associated with a given set of data points (x_1, y_1), (x_2, y_2), . . . , (x_n, y_n).

Regression Line
The line that best fits the n data points (x_1, y_1), (x_2, y_2), . . . , (x_n, y_n) has the form

$$y = mx + b$$

where $\quad m = \dfrac{n(\sum xy) - (\sum x)(\sum y)}{n(\sum x^2) - (\sum x)^2}$

$$b = \frac{\sum y - m(\sum x)}{n}$$

n = number of data points

Question Where do these formulas come from?

Answer We first go back to the definition of the regression line: It is defined to be the line that minimizes the sum of the squares of the **residues,** measured by the vertical distances shown in Figure 24, which shows a regression line associated with $n = 5$ data points.

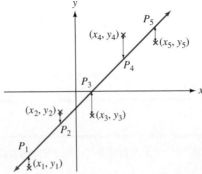

Figure 24

In the figure the points $P_1, \ldots, P_n$ on the regression line have coordinates $(x_1, mx_1 + b)$, $(x_2, mx_2 + b), \ldots, (x_n, mx_n + b)$. The residues are the quantities $y_{\text{observed}} - y_{\text{predicted}}$:

$$y_1 - (mx_1 + b), y_2 - (mx_2 + b), \ldots, y_n - (mx_n + b)$$

The sum of the squares of the residues is therefore

$$S(m, b) = [y_1 - (mx_1 + b)]^2 + [y_2 - (mx_2 + b)]^2 + \cdots + [y_n - (mx_n + b)]^2$$

and this is the quantity we must minimize by choosing m and b. Since we reason that there is a line that minimizes this quantity, there must be a relative minimum at that point. We will see in a moment that the function S has at most one critical point, which must therefore be the desired absolute minimum. To obtain the critical points of S, we set the partial derivatives equal to zero and solve:

$$S_m = 0: \quad -2x_1[y_1 - (mx_1 + b)] - \cdots - 2x_n[y_n - (mx_n + b)] = 0$$

$$S_b = 0: \quad -2[y_1 - (mx_1 + b)] - \cdots - 2[y_n - (mx_n + b)] = 0$$

Dividing by -2 and gathering terms allows us to rewrite the equations as

$$m(x_1^2 + \cdots + x_n^2) + b(x_1 + \cdots + x_n) = x_1y_1 + \cdots + x_ny_n$$

$$m(x_1 + \cdots + x_n) + nb = y_1 + \cdots + y_n$$

We can rewrite these equations more neatly using $\sum$ notation:

$$m(\textstyle\sum x^2) + b(\sum x) = \sum xy$$

$$m(\textstyle\sum x) + nb \quad = \sum y$$

This is a system of two linear equations in the two unknowns m and b. It may or may not have a unique solution. When there is a unique solution, we can conclude that the best-fit line is given by solving these two equations for m and b. Alternatively, there is a general formula for the solution of any system of two equations in two unknowns, and if we apply this formula to our two equations, we get the regression formulas above.

8.4 EXERCISES

In Exercises 1–4, classify each labeled point on the graph as one of the following:

(A) A relative maximum

(B) A relative minimum

(C) A saddle point

(D) A critical point but neither a relative extremum nor a saddle point

(E) None of the above

1.

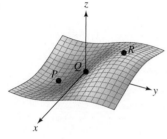

2.

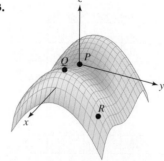

3.

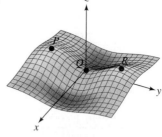

4.

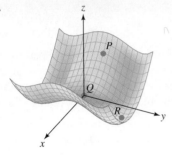

In Exercises 5–10, classify the shaded value in each table as one of the following:
(A) A relative maximum
(B) A relative minimum
(C) A saddle point
(D) Neither a relative extremum nor a saddle point

5. $x \rightarrow$

$y \downarrow$		-3	-2	-1	0	1	2
	-3	10	5	2	1	2	5
	-2	9	4	1	0	1	4
	-1	10	5	2	1	2	5
	0	13	8	5	4	5	8
	1	18	13	10	9	10	13
	2	25	20	17	16	17	20
	3	34	29	26	25	26	29

6. $x \rightarrow$

$y \downarrow$		-3	-2	-1	0	1	2
	-3	5	0	-3	-4	-3	0
	-2	8	3	0	-1	0	3
	-1	9	4	1	0	1	4
	0	8	3	0	-1	0	3
	1	5	0	-3	-4	-3	0
	2	0	-5	-8	-9	-8	-5
	3	-7	-12	-15	-16	-15	-12

7. $x \rightarrow$

$y \downarrow$		-3	-2	-1	0	1	2
	-3	5	0	-3	-4	-3	0
	-2	8	3	0	-1	0	3
	-1	9	4	1	0	1	4
	0	8	3	0	-1	0	3
	1	5	0	-3	-4	-3	0
	2	0	-5	-8	-9	-8	-5
	3	-7	-12	-15	-16	-15	-12

8. $x \rightarrow$

$y \downarrow$		-3	-2	-1	0	1	2
	-3	2	3	2	-1	-6	-13
	-2	3	4	3	0	-5	-12
	-1	2	3	2	-1	-6	-13
	0	-1	0	-1	-4	-9	-16
	1	-6	-5	-6	-9	-14	-21
	2	-13	-12	-13	-16	-21	-28
	3	-22	-21	-22	-25	-30	-37

9. $x \rightarrow$

$y \downarrow$		-3	-2	-1	0	1	2
	-3	4	5	4	1	-4	-11
	-2	3	4	3	0	-5	-12
	-1	4	5	4	1	-4	-11
	0	7	8	7	4	-1	-8
	1	12	13	12	9	4	-3
	2	19	20	19	16	11	4
	3	28	29	28	25	20	13

10. $x \rightarrow$

$y \downarrow$		-3	-2	-1	0	1	2
	-3	100	101	100	97	92	85
	-2	99	100	99	96	91	84
	-1	98	99	98	95	90	83
	0	91	92	91	88	83	76
	1	72	73	72	69	64	57
	2	35	36	35	32	27	20
	3	-26	-25	-26	-29	-34	-41

In Exercises 11–28, locate and classify all the critical points of the functions.

11. $f(x, y) = x^2 + y^2 + 1$

12. $f(x, y) = 4 - (x^2 + y^2)$

13. $g(x, y) = 1 - x^2 - x - y^2 + y$

14. $g(x, y) = x^2 + x + y^2 - y - 1$

15. $h(x, y) = x^2 y - 2x^2 - 4y^2$

16. $h(x, y) = x^2 + y^2 - y^2 x - 4$

17. $s(x, y) = e^{x^2 + y^2}$

18. $s(x, y) = e^{-(x^2 + y^2)}$

19. $t(x, y) = x^4 + 8xy^2 + 2y^4$

20. $t(x, y) = x^3 - 3xy + y^3$

21. $f(x, y) = x^2 + y - e^y$

22. $f(x, y) = xe^y$

23. $f(x, y) = e^{-(x^2 + y^2 + 2x)}$

24. $f(x, y) = e^{-(x^2+y^2-2x)}$

25. $f(x, y) = xy + \dfrac{2}{x} + \dfrac{2}{y}$

26. $f(x, y) = xy + \dfrac{4}{x} + \dfrac{2}{y}$

27. $g(x, y) = x^2 + y^2 + \dfrac{2}{xy}$

28. $g(x, y) = x^3 + y^3 + \dfrac{3}{xy}$

29. Refer to Exercise 11. Which (if any) of the critical points of $f(x, y) = x^2 + y^2 + 1$ are absolute extrema?

30. Refer to Exercise 12. Which (if any) of the critical points of $f(x, y) = 4 - (x^2 + y^2)$ are absolute extrema?

T 31. Refer to Exercise 15. Which (if any) of the critical points of $h(x, y) = x^2 y - 2x^2 - 4y^2$ are absolute extrema?

T 32. Refer to Exercise 16. Which (if any) of the critical points of $h(x, y) = x^2 + y^2 - y^2 x - 4$ are absolute extrema?

APPLICATIONS

33. Brand Loyalty Suppose the fraction of Mazda car owners who chose another new Mazda can be modeled by the function

$$M(c, f) = 11 + 8c + 4c^2 - 40f - 20cf + 40f^2$$

where c is the fraction of Chrysler car owners who remained loyal to Chrysler and f is the fraction of Ford car owners remaining loyal to Ford. Locate and classify all the critical points and interpret your answer.

This model is not accurate, although it was inspired by an approximation of a second-order regression based on data from the period 1988–1995. SOURCE: For original data, Chrysler, Maritz Market Research, Consumer Attitude Research, and Strategic Vision/*New York Times*, November 3, 1995, p. D2.

34. Brand Loyalty Repeat Exercise 33, using the function

$$M(c, f) = -10 - 8f - 4f^2 + 40c + 20fc - 40c^2$$

35. Pollution Control The cost of controlling emissions at a firm goes up rapidly as the amount of emissions reduced goes up. Here is a possible model:

$$C(x, y) = 4000 + 100x^2 + 50y^2$$

where x is the reduction in sulfur emissions, y is the reduction in lead emissions (in pounds of pollutant per day), and C is the daily cost to the firm (in dollars) of this reduction. Government clean-air subsidies amount to $500 per pound of sulfur and $100 per pound of lead removed. How many pounds of pollutant should the firm remove each day in order to minimize *net* cost (cost minus subsidy)?

36. Pollution Control Repeat Exercise 36, using the following information:

$$C(x, y) = 2000 + 200x^2 + 100y^2$$

with government subsidies amounting to $100 per pound of sulfur and $500 per pound of lead removed each day.

37. Revenue Your company manufactures two models of speakers, the Ultra Mini and the Big Stack. Demand for each depends partly on the price of the other. If one is expensive, then more people will buy the other. If p_1 is the price of the Ultra Mini, and p_2 is the price of the Big Stack, demand for the Ultra Mini is given by

$$q_1(p_1, p_2) = 100{,}000 - 100p_1 + 10p_2$$

where q_1 represents the number of Ultra Minis that will be sold in a year. The demand for the Big Stack is given by

$$q_2(p_1, p_2) = 150{,}000 + 10p_1 - 100p_2$$

Find the prices for the Ultra Mini and the Big Stack that will maximize your total revenue.

38. Revenue Repeat Exercise 37, using the following demand functions:

$$q_1(p_1, p_2) = 100{,}000 - 100p_1 + p_2$$

$$q_2(p_1, p_2) = 150{,}000 + p_1 - 100p_2$$

39. Luggage Dimensions American Airlines requires that the total outside dimensions (length + width + height) of a checked bag not exceed 62 inches. What are the dimensions of the largest volume bag that you can check on an American flight?

SOURCE: According to information on its Web site (http://www.aa.com/) as of August, 2002.

40. Luggage Dimensions American Airlines requires that the total outside dimensions (length + width + height) of a carry-on bag not exceed 45 inches. What are the dimensions of the largest volume bag that you can carry on an American flight?

SOURCE: According to information on its Web site (http://www.aa.com/) as of August, 2002.

41. Package Dimensions As of July 2002, for priority mail, the U.S. Postal Service (USPS) will accept only packages with a length plus girth no more than 108 inches. (See the figure.) What are the dimensions of the largest volume package that the USPS will accept? What is its volume?

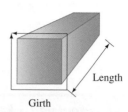

Length

Girth

42. Package Dimensions As of July 2002, United Parcel Service (UPS) will accept only packages with length no more than 108 inches and length plus girth no more than 130 inches. (See the figure for Exercise 41.) What are the dimensions of the largest volume package that UPS will accept? What is its volume?

COMMUNICATION AND REASONING EXERCISES

43. Sketch the graph of a function that has one extremum and no saddle points.

44. Sketch the graph of a function that has one saddle point and one extremum.

45. Sketch the graph of a function that has one relative extremum, no absolute extrema, and no saddle points.

46. Sketch the graph of a function that has infinitely many absolute maxima.

47. Let $H = f_{xx}(a, b)f_{yy}(a, b) - f_{xy}(a, b)^2$. What condition on H guarantees that f has a relative extremum at the point (a, b)?

48. Let H be as in Exercise 47. Give an example to show that it is possible to have $H = 0$ and a relative minimum at (a, b).

49. Suppose when the graph of $f(x, y)$ is sliced by a vertical plane through (a, b) parallel to either the xz plane or the yz plane, the resulting curve has a relative maximum at (a, b). Does this mean that f has a relative maximum at (a, b)? Explain your answer.

50. Suppose f has a relative maximum at (a, b). Does it follow that, if the graph of f is sliced by a vertical plane parallel to either the xz plane or the yz plane, the resulting curve has a relative maximum at (a, b)? Explain your answer.

51. **Average Cost** Let $C(x, y)$ be any cost function. Show that when the average cost is minimized, the marginal costs C_x and C_y both equal the average cost. Explain why this is reasonable.

52. **Average Profit** Let $P(x, y)$ be any profit function. Show that when the average profit is maximized, the marginal profits P_x and P_y both equal the average profit. Explain why this is reasonable.

53. The tangent plane to a graph was introduced in Exercise 76 in Section 8.3. Use the equation of the tangent plane given there to explain why the tangent plane is parallel to the xy plane at a relative maximum or minimum of $f(x, y)$.

54. Use the equation of the tangent plane given in Exercise 76 in Section 8.3 to explain why the tangent plane is parallel to the xy plane at a saddle point of $f(x, y)$.

8.5 *Constrained Maxima and Minima and Applications*

So far we have looked only at the relative extrema of f that lie in the interior of the domain of f. There may also be relative extrema on the boundary of the domain (just as the endpoints of the domain may be relative extrema for a function of one variable). This situation arises, for example, in optimization problems with constraints, similar to those we saw in Chapter 5 on applications of the derivative. Here is a typical example:

$$\text{Maximize } S = xy + 2xz + 2yz \text{ subject to } xyz = 4 \text{ and } x \geq 0, y \geq 0, z \geq 0.$$

There are two kinds of constraints in this example: equations and inequalities. The inequalities specify a restriction on the domain of S.

Our strategy for solving such problems is essentially the same as the strategy we used earlier. First, we use any equation constraints to eliminate variables. In the examples in this section, we are able to reduce to a function of only two variables. The inequality constraints then help define the domain of this function. Next, we locate any critical points in the interior of the domain. Finally, we look at the boundary of the domain. When there is no boundary to worry about, another method, called the *method of Lagrange multipliers*, comes in handy.

We first look at functions of two variables with restricted domains, so we can see how to handle the boundaries.

Example 1 • Restricted Domain

Find the maximum and minimum value of $f(x, y) = xy - x - 2y$ on the triangular region R with vertices $(0, 0)$, $(1, 0)$, and $(0, 2)$.

Solution   The domain of f is the region R shown shaded in Figure 25.

Step 1   *Locate critical points in the interior of the domain.* We take the partial derivatives as usual:

$$f_x = y - 1 \qquad f_y = x - 2 \qquad f_{xx} = 0 \qquad f_{xy} = 1 \qquad f_{yy} = 0$$

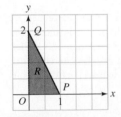

Figure 25

The only critical point is (2, 1). Since this lies outside the domain (the region R), we ignore it. Thus, there are no critical points in the interior of the domain of f.

Step 2 *Locate relative extrema on the boundary of the domain.* The boundary of the domain consists of three line segments, OP, OQ, and PQ. We deal with these one at a time.

- *Segment OP* This line segment has equation $y = 0$ with domain $0 \leq x \leq 1$. The behavior of the function f along this segment is given by substituting $y = 0$ into the expression for x:

$$f(x, 0) = -x \qquad \text{Substitute } y = 0 \text{ in the expression for } f.$$

We now find the relative extrema of this function of one variable by the methods we used in the chapter on applications of the derivative (Chapter 5). There are no critical points, and there are two endpoints, $x = 0$ and $x = 1$. Since $y = 0$ this gives us the following two candidates for relative extrema: $(0, 0, 0)$ and $(1, 0, -1)$.

- *Segment OQ* This line segment has equation $x = 0$ with $0 \leq y \leq 2$. Along this segment we see

$$f(0, y) = -2y \qquad \text{Substitute } x = 0 \text{ in the expression for } f.$$

We now locate the relative extrema of this function of one variable. Once again, there are only the endpoints $y = 0$ and $y = 2$. Since $x = 0$ this gives us the two candidates $(0, 0, 0)$ and $(0, 2, -4)$.

- *Segment PQ* This line segment has equation $y = -2x + 2$ with $0 \leq x \leq 1$. Along this segment we see

$$f(x, -2x + 2) = x(-2x + 2) - x - 2(-2x + 2) \qquad \text{Substitute } y = -2x + 2$$
$$= -2x^2 + 5x - 4 \qquad \text{in the expression for } f.$$

This function of x (whose graph is an upside-down parabola) has a stationary maximum when its derivative, $-4x + 5$, is zero, which occurs when $x = 5/4$. Since this is greater than 1, it lies outside the domain $0 \leq x \leq 1$, and we reject it. There are no other critical points, and the endpoints are $x = 0$ and $x = 1$. When $x = 0$, $y = -2(0) + 2 = 2$, giving the point $(0, 2, -4)$. When $x = 1$, $y = -2(1) + 2 = 0$, giving the point $(1, 0, -1)$.

Thus, the candidates for maxima and minima are $(0, 0, 0)$, $(1, 0, -1)$, and $(0, 2, -4)$ (which happen to lie over the corner points of the domain R). Since their z coordinates give the value of f, we see that f has an absolute maximum of 0 at the point $(0, 0)$ and an absolute minimum of -4 at the point $(0, 2)$.

Next we look at an example with a constraint equation.

Example 2 • Minimizing Area

Find the dimensions of an open-top rectangular box that has a volume of 4 cubic feet and the smallest possible surface area.

Solution Our first task is to rephrase this request as a mathematical optimization problem. Figure 26 shows a picture of the box with dimensions x, y, and z. We want to minimize the total surface area, which is given by

$$A = xy + 2xz + 2yz \qquad \text{base + sides + front and back}$$

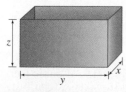

Figure 26

This is our **objective function.** We can't simply choose x, y, and z to all be zero, however, because the enclosed volume must be 4 cubic feet. So,

$$xyz = 4 \qquad \text{Constraint}$$

This is our **constraint** equation. Other unstated constraints are $x \geq 0$, $y \geq 0$, and $z \geq 0$. We now restate the problem as follows:

Minimize $A = xy + 2xz + 2yz$ subject to $xyz = 4$, $x \geq 0$, $y \geq 0$, $z \geq 0$.

As suggested in the discussion before Example 1, we proceed as follows.

Step 1 *Solve the constraint equation for one of the variables and then substitute in the objective function.* Solving the constraint equation for z gives

$$z = \frac{4}{xy}$$

Substituting this into the objective function gives a function of only two variables:

$$A = xy + \frac{8}{y} + \frac{8}{x}$$

Step 2 *Minimize the resulting function of two variables.* A word about the domain of A: Since we said that x and y must be nonnegative, we already know that (x, y) is restricted to the first quadrant. Moreover, since A is not defined if either x or y is zero, we exclude the x and y axes from the domain of A and are left with the interior of the first quadrant as our domain. Since this has no boundary, we look for critical points only in the interior:

$$A_x = y - \frac{8}{x^2} \qquad\qquad A_y = x - \frac{8}{y^2}$$

$$A_{xx} = \frac{16}{x^3} \qquad A_{xy} = 1 \qquad A_{yy} = \frac{16}{y^3}$$

We now equate the first partial derivatives to zero:

$$y = \frac{8}{x^2} \quad \text{and} \quad x = \frac{8}{y^2}$$

To solve for x and y, we substitute the first of these equations in the second, getting

$$x = \frac{x^4}{8}$$

$$x^4 - 8x = 0$$

$$x(x^3 - 8) = 0$$

The two solutions are $x = 0$, which we reject because x cannot be zero, and $x = 2$. Substituting $x = 2$ in $y = 8/x^2$ gives $y = 2$ also. Thus, the only critical point is $(2, 2)$. To apply the second derivative test, we compute

$$A_{xx}(2, 2) = 2 \quad A_{xy}(2, 2) = 1 \quad A_{yy}(2, 2) = 2$$

and find that $H = 3 > 0$, so we have a relative minimum at $(2, 2)$.

Question Is this relative minimum an absolute minimum?

Answer Yes. There must be a least surface area among all boxes that hold 4 cubic feet. (Why?) Since this would give a relative minimum of A and since the only possible relative minimum of A occurs at $(2, 2)$, this is the absolute minimum.

To finish the example, we get the value of z by substituting the values of x and y into the constraint equation $z = 4/(xy)$, which gives $z = 1$. Thus, the required dimensions of the box are

$$x = 2 \text{ ft} \qquad y = 2 \text{ ft} \qquad z = 1 \text{ ft}$$

In the next example, we return to the Ultra Mini and Big Stack speaker example introduced in Section 8.1 and look at an interaction model for profit.

Example 3 • Maximizing Profit

You own a company that makes two models of speakers, the Ultra Mini and the Big Stack. Each Ultra Mini requires 1 square foot of fabric and 3 feet of wire, and each Big Stack requires 5 square feet of fabric and 9 feet of wire. You have each week 100 square feet of fabric and 270 feet of wire to use. Your profit function is estimated to be

$$P(x, y) = 10x + 60y + 0.5xy$$

where x is the number of Ultra Minis, y is the number of Big Stacks, and P is your profit in dollars. Find the number of each model you should make each week to maximize your profit.

Solution The constraints in this problem come from the limited amount of fabric and wire available. The constraint on fabric is

$$x + 5y \le 100$$

The constraint on wire is

$$3x + 9y \le 270$$

There are also the constraints $x \ge 0$ and $y \ge 0$ that are often present in applications. Our problem is then the following:

$$\text{Maximize } P = 10x + 60y + 0.5xy \text{ subject to}$$
$$x + 5y \le 100$$
$$3x + 9y \le 270$$
$$x \ge 0, y \ge 0$$

If we graph the domain R, we get Figure 27. [The point of intersection $(75, 5)$ is found by solving the system of linear equations given by the equations of the two lines.] Since we have no equality constraints, we proceed as before.

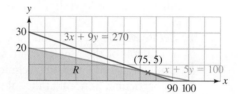

Figure 27

Step 1 *Locate critical points in the interior of the domain.*

$$P_x = 10 + 0.5y \quad \text{and} \quad P_y = 60 + 0.5x$$

Setting these equal to zero and solving for x and y gives the only critical point as $(-120, -20)$. However, this critical point is well outside the domain R, so we disregard it.

Step 2 *Locate relative extrema on the boundary of the domain.* The boundary of the domain consists of four line segments, which we consider one at a time.

- *Segment $y = 0, 0 \leq x \leq 90$* Here,

$$P(x, 0) = 10x + 60(0) + 0.5x(0) = 10x \quad \text{and} \quad P' = 10$$

and is therefore never zero. Thus, the only points we need to look at are the endpoints $x = 0$ and $x = 90$, which give $(0, 0, 0)$ and $(90, 0, 900)$.

- *Segment $x = 0, 0 \leq y \leq 20$* Here,

$$P(0, y) = 10(0) + 60y + 0.5(0)y = 60y$$

Again, there are no critical points, only the endpoints $y = 0$ and $y = 20$. This gives one more point, $(0, 20, 1200)$.

- *Segment $x = 100 - 5y, 5 \leq y \leq 20$* This gives

$$P(100 - 5y, y) = 10(100 - 5y) + 60y + 0.5(100 - 5y)y$$

$$= 1000 + 60y - 2.5y^2$$

$$P' = 60 - 5y$$

Setting the derivative equal to zero and solving, we get $y = 12$, which is within the domain $5 \leq y \leq 20$. The corresponding point on the graph is $(40, 12, 1360)$.

The two endpoints of this segment are $y = 5$ and $y = 20$. The point at $y = 20$ is $(0, 20, 1200)$, which we have already considered, and the other point is $(75, 5, 1237.5)$.

- *Segment $x = 90 - 3y, 0 \leq y \leq 5$* This gives

$$P(90 - 3y, y) = 10(90 - 3y) + 60y + 0.5(90 - 3y)y$$

$$= 900 + 75y + 1.5y^2$$

$$P' = 75 - 3y$$

Setting this equal to zero and solving for y, we get $y = 25$, which is outside the range $0 \leq y \leq 5$. Therefore, there are no critical points on this line segment. Its endpoints are $y = 0$ and $y = 5$, giving the points $(90, 0, 900)$ and $(75, 5, 1237.5)$, which we have already encountered.

Looking at the heights of all of the points we have found, the highest is $(40, 12, 1360)$, so this gives the largest profit. In other words, you should make 40 Ultra Minis and 12 Big Stacks each week, giving you the largest possible profit of $1360 per week.

✴ Before we go on . . . If you have studied linear programming, this example should remind you of the problems you solved by that technique. However, since the objective function is not linear, the techniques of linear programming fail to solve this problem. In particular, notice that the solution we found is *not* at a corner of the feasible region, as it must be in a linear programming problem, but is in the middle of one edge. The solution to another problem could be in the interior of the region.

The Method of Lagrange Multipliers

Suppose we have a constrained optimization problem in which it is difficult or impossible to solve a constraint equation for one of the variables. Then we can use the method of **Lagrange multipliers** to avoid this difficulty. We restrict attention to the case of a single constraint equation, although the method generalizes to any number of constraint equations.

Locating Relative Extrema Using the Method of Lagrange Multipliers

To locate the candidates for relative extrema of a function $f(x, y, \ldots)$ subject to the constraint $g(x, y, \ldots) = 0$, we solve the following system of equations for $x, y, \ldots,$ and λ:

$$f_x = \lambda g_x$$

$$f_y = \lambda g_y$$

$$\cdot$$
$$\cdot$$
$$\cdot$$

$$g = 0$$

The unknown λ is called a **Lagrange multiplier.** The points $(x, y, \ldots)$ that occur in solutions are then the candidates for the relative extrema of f subject to $g = 0$.

Example 4 • Using Lagrange Multipliers

Use the method of Lagrange multipliers to find the maximum value of $f(x, y) = 2xy$ subject to $x^2 + 4y^2 = 32$.

Solution We start by rewriting the problem in standard form:

Maximize $f(x, y) = 2xy$ subject to $x^2 + 4y^2 - 32 = 0$.

Here, $g(x, y) = x^2 + 4y^2 - 32$, and the system of equations we need to solve is thus

$$f_x = \lambda g_x \quad \text{or} \quad 2y = 2\lambda x$$

$$f_y = \lambda g_y \quad \text{or} \quad 2x = 8\lambda y$$

$$g = 0 \quad \text{or} \quad x^2 + 4y^2 - 32 = 0$$

A convenient way to solve such a system is to solve one of the equations for λ and then substitute in the remaining equations. We start by solving the first equation to obtain

$$\lambda = \frac{y}{x}$$

(A word of caution: Since we divided by x, we made the implicit assumption that $x \neq 0$, so before continuing we should check what happens if $x = 0$. But if $x = 0$, then the first equation, $2y = 2\lambda x$, tells us that $y = 0$ as well, and this contradicts the third equation: $x^2 + 4y^2 - 32 = 0$. Thus, we can rule out the possibility that $x = 0$.) Substituting in the remaining equations gives

$$x = 4\lambda y = \frac{4y^2}{x} \quad \text{or} \quad x^2 = 4y^2$$

$$x^2 + 4y^2 - 32 = 0$$

Notice how we have reduced the number of unknowns and also the number of equations by one. We can now substitute $x^2 = 4y^2$ in the last equation, obtaining

$$4y^2 + 4y^2 - 32 = 0$$
$$8y^2 = 32$$
$$y = \pm 2$$

We now substitute back to obtain

$$x^2 = 4y^2 = 16$$
$$x = \pm 4$$

We don't need the value of λ, so we won't solve for it. Thus, the candidates for relative extrema are given by $x = \pm 4$ and $y = \pm 2$, that is, the four points $(-4, -2)$, $(-4, 2)$, $(4, -2)$, and $(4, 2)$. Recall that we are seeking the values of x and y that give the maximum value for $f(x, y) = 2xy$. Since we now have only four points to choose from, we compare the values of f at these four points and conclude that the maximum value of f occurs when $(x, y) = (-4, -2)$ or $(4, 2)$.

Before we go on . . .

Question Something is suspicious here. We didn't check to see whether these candidates were relative extrema to begin with, let alone absolute extrema! How do we justify this omission?

Answer One of the difficulties with using the method of Lagrange multipliers is that it does not provide us with a test analogous to the second derivative test for functions of several variables. However, if you grant that the function in question does have an absolute maximum, then we require no test, since one of the candidates must give this maximum.

Question But how do we know that the given function has an absolute maximum?

Answer The best way to see this is by giving a geometric interpretation. The constraint $x^2 + 4y^2 = 32$ tells us that the point (x, y) must lie on the ellipse shown in Figure 28. The function $f(x, y) = 2xy$ gives the area of the rectangle shaded in Figure 28. Since there must be a largest such rectangle, the function f must have an absolute maximum for at least one pair of coordinates (x, y).

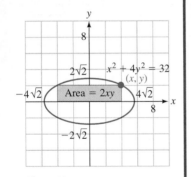

Figure 28

Question When can I use the method of Lagrange multipliers? When should I use it?

Answer We have discussed the method only when there is a single equality constraint. There is a generalization, which we do not discuss, that works when there are more equality constraints (we need to introduce one multiplier for each constraint). So, if you have a problem with more than one equality constraint or with any inequality constraints, you must use the method we discussed earlier in this section. On the other hand, if you have one equality constraint and it would be difficult to solve it for one of the variables, then you should use Lagrange multipliers. We have noted in the exercises where you might use Lagrange multipliers.

Question Why does the method of Lagrange multipliers work?

Answer An adequate answer is beyond the scope of this book.

Software packages like Excel have built-in algorithms that seek absolute extrema whether or not constraints are present. In Example 5 we use the Solver routine in Excel to redo Example 3.[7]

---

[7]If Solver does not appear in the Tools menu, you should first install it using your Excel installation software. (Solver is one of the Excel Add-Ins.)

18. At what point on the surface $z = (x^2 + x + y$
quantity $x^2 + y^2 + z^2$ a minimum? (The m
grange multipliers can be used here.)

APPLICATIONS

19. Cost Your bicycle factory makes two models, f
ten-speeds. Each week, your total cost (in doll
five-speeds and y ten-speeds is

$$C(x, y) = 10,000 + 50x + 70y - 0.$$

You want to make between 100 and 150 five-s
tween 80 and 120 ten-speeds. What combina
you the least? What combination will cost you

20. Cost Your bicycle factory makes two models, f
ten-speeds. Each week, your total cost (in doll
five-speeds and y ten-speeds is

$$C(x, y) = 10,000 + 50x + 70y - 0.4$$

You want to make between 100 and 150 five-s
tween 80 and 120 ten-speeds. What combina
you the least? What combination will cost you

21. Profit Your software company sells two opera
Walls and Doors. Your profit (in dollars) from s
of Walls and y copies of Doors is given by

$$P(x, y) = 20x + 40y - 0.1(x^2 + y$$

If you can sell a maximum of 200 copies of th
ing systems together, what combination will l
greatest profit?

22. Profit Your software company sells two opera
Walls and Doors. Your profit (in dollars) from s
of Walls and y copies of Doors is given by

$$P(x, y) = 20x + 40y - 0.1(x^2 + y$$

If you can sell a maximum of 400 copies of th
ing systems together, what combination will l
largest profit?

23. Temperature The temperature at the point
square with vertices $(0, 0)$, $(0, 1)$, $(1, 0)$, and $(1,$
$T(x, y) = x^2 + 2y^2$. Find the hottest and cold
the square.

24. Temperature The temperature at the point
square with vertices $(0, 0)$, $(0, 1)$, $(1, 0)$, and $(1,$
$T(x, y) = x^2 + 2y^2 - x$. Find the hottest and
on the square.

25. Temperature The temperature at the point $(x,$
$\{(x, y) | x^2 + y^2 \le 1\}$ is given by $T(x, y) = x^2 +$
the hottest and coldest points on the disc.

26. Temperature The temperature at the point $(x,$
$\{(x, y) | x^2 + y^2 \le 1\}$ is given by $T(x, y) = 2x^2$
hottest and coldest points on the disc.

Example 5 • Using Excel to Solve an Optimization Problem

Use Solver to solve the following problem:

Maximize	$P = 10x + 60y + 0.5xy$	Objective
subject to	$x + 5y \le 100$	Constraint 1
	$3x + 9y \le 270$	Constraint 2
	$x \ge 0, y \ge 0$	Constraint 3 and constraint 4

Solution

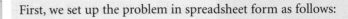

First, we set up the problem in spreadsheet form as follows:

	A	B	C	D	E	F
1	x	y	Objective	Constraints		
2	0	0	=10*A2+60*B2+0.5*A2*B2	=A2+5*B2	100	Constraint 1
3	Initial valus of x	Initial value of y		=3*A2+9*B	270	Constraint 2
4				=A2	0	Constraint 3
5				=B2	0	Constraint 4

A2 will be the cell that contains the value of x and B2 the cell that contains the value of y. (We have set them both to zero in anticipation that Solver will adjust them to satisfy the constraints and give an optimal solution.) Next, select `Solver` in the `Tools` menu to bring up the Solver dialog box. Here is the dialog box with all the necessary fields completed to solve the problem:

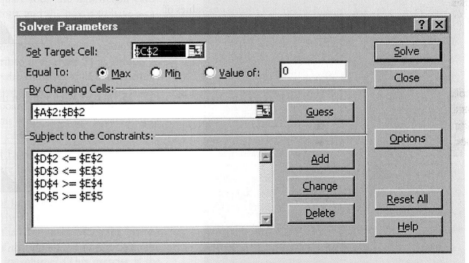

Notes

- The `Target Cell` refers to the cell that contains the objective function.
- `Max` is selected because we are maximizing the objective function.
- `Changing Cells` are obtained by selecting the cells that contain the current values of x and y.
- Constraints are added one at a time by clicking on the `Add` button and selecting the cells that contain the left- and right-hand sides of each inequality, as well as the type of inequality. (Equality constraints are also permitted.)

When we are done, we click on `Solve` and the (approximate) optimal solution appears in A2 and B2, with the maximum value of P appearing in cell C2. As we saw in Example 3, the optimal solution is $x = 40$, $y = 12$, $P = 1360$.

✳ ***Before we go or***
Question Can a
analytic meth

Answer Some
cally, and so s
tial in those
chance of rur

• The solutio

• The solutio

• Roundoff e

• Only one so

So, use Solver

8.5 EXERCISES

In all exercises for this section, a software pa
as Excel Solver can be used as a check on you
work. (Bear in mind, however, the cautions a
of Example 5.)

In Exercises 1–16, find the maximum and minimum v
each function and the points at which they occur.

1. $f(x, y) = x^2 + y^2; 0 \leq x \leq 2, 0 \leq y \leq 2$

2. $g(x, y) = \sqrt{x^2 + y^2}; 1 \leq x \leq 2, 1 \leq y \leq 2$

3. $h(x, y) = (x - 1)^2 + y^2; x^2 + y^2 \leq 4$

4. $k(x, y) = x^2 + (y - 1)^2; x^2 + y^2 \leq 9$

5. $f(x, y) = e^{x^2+y^2}; 4x^2 + y^2 \leq 4$

6. $g(x, y) = e^{-(x^2+y^2)}; x^2 + 4y^2 \leq 4$

7. $h(x, y) = e^{4x^2+y^2}; x^2 + y^2 \leq 1$

8. $k(x, y) = e^{-(x^2+4y^2)}; x^2 + y^2 \leq 4$

9. $f(x, y) = x + y + 1/(xy); x \geq \frac{1}{2}, y \geq \frac{1}{2}, x + y \leq 3$

10. $g(x, y) = x + y + 8/(xy); x \geq 1, y \geq 1, x + y \leq 6$

11. $h(x, y) = xy + 8/x + 8/y; x \geq 1, y \geq 1, xy \leq 9$

12. $k(x, y) = xy + 1/x + 4/y; x \geq 1, y \geq 1, xy \leq 10$

13. $f(x, y) = x^2 + 2x + y^2;$ on the region in the figure

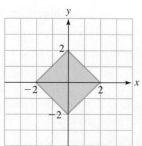

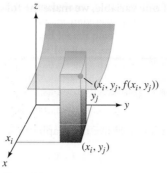

Figure 32

This gives us mn rectangles defined by $x_{i-1} \leq x \leq x_i$ and $y_{j-1} \leq y \leq y_j$. Over one of these rectangles, f is approximately equal to its value at one corner—say, $f(x_i, y_j)$. The volume under f over this small rectangle is then approximately the volume of the rectangular brick (size exaggerated) shown in Figure 32. This brick has height $f(x_i, y_j)$, and its base is Δx by Δy. Its volume is therefore $f(x_i, y_j) \Delta x \Delta y$. Adding together the volumes of all of the bricks over the small rectangles in R, we get

$$\iint_R f(x, y) \, dx \, dy \approx \sum_{j=1}^{n} \sum_{i=1}^{m} f(x_i, y_j) \, \Delta x \, \Delta y$$

This double sum is called a **double Riemann sum.** We define the double integral to be the limit of the Riemann sums as m and n go to infinity.

Algebraic Definition of the Double Integral

$$\iint_R f(x, y) \, dx \, dy = \lim_{n \to \infty} \lim_{m \to \infty} \sum_{j=1}^{n} \sum_{i=1}^{m} f(x_i, y_j) \, \Delta x \, \Delta y$$

Note

This definition is adequate (the limit exists) when f is continuous. More elaborate definitions are needed for badly behaved functions.

This definition also gives us a clue about how to compute a double integral. The innermost sum is $\sum_{i=1}^{m} f(x_i, y_j) \, \Delta x$, which is a Riemann sum for $\int_a^b f(x, y_j) \, dx$. The innermost limit is therefore

$$\lim_{m \to \infty} \sum_{i=1}^{m} f(x_i, y_j) \, \Delta x = \int_a^b f(x, y_j) \, dx$$

The outermost limit is then also a Riemann sum, and we get the following way of calculating double integrals.

Computing the Double Integral over a Rectangle

If R is the rectangle $a \leq x \leq b$ and $c \leq y \leq d$, then

$$\iint_R f(x, y) \, dx \, dy = \int_c^d \left[\int_a^b f(x, y) \, dx \right] dy = \int_a^b \left[\int_c^d f(x, y) \, dy \right] dx$$

The second formula comes from switching the order of summation in the double sum.

Quick Example

If R is the rectangle $1 \leq x \leq 2$ and $1 \leq y \leq 3$, then

$$\iint_R 1 \, dx \, dy = \int_1^3 \left[\int_1^2 1 \, dx \right] dy$$

$$= \int_1^3 [x]_{x=1}^2 \, dy \qquad \text{Evaluate the inner integral.}$$

$$= \int_1^3 1 \, dy \qquad [x]_{x=1}^2 = 2 - 1 = 1$$

$$= [y]_{y=1}^3 = 3 - 1 = 2$$

The Quick Example used a constant function for the integrand. Here is an example in which the integrand is not constant.

Example 1 • Double Integral over a Rectangle

Let R be the rectangle $0 \le x \le 1$ and $0 \le y \le 2$. Compute $\iint_R xy \, dx \, dy$.

Solution

$$\iint_R xy \, dx \, dy = \int_0^2 \int_0^1 xy \, dx \, dy$$

(We usually drop the parentheses like this.) As in the Quick Example, we compute this **iterated integral** from the inside out. First we compute

$$\int_0^1 xy \, dx$$

To do this computation, we do as we did when finding partial derivatives: We treat y as a constant. This gives

$$\int_0^1 xy \, dx = \left[\frac{x^2}{2} \cdot y \right]_{x=0}^1 = \frac{1}{2} y - 0 = \frac{y}{2}$$

We can now calculate the outer integral:

$$\int_0^2 \int_0^1 xy \, dx \, dy = \int_0^2 \frac{y}{2} \, dy = \left[\frac{y^2}{4} \right]_0^2 = 1$$

Before we go on . . . We could also reverse the order of integration:

$$\int_0^1 \int_0^2 xy \, dy \, dx = \int_0^1 \left(\left[x \cdot \frac{y^2}{2} \right]_{y=0}^2 \right) dx = \int_0^1 2x \, dx = [x^2]_0^1 = 1$$

Often we need to integrate over regions R that are not rectangular. There are two cases that come up. The first is a region like the one shown in Figure 33.

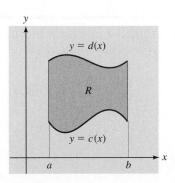

Figure 33

In this region the bottom and top sides are defined by functions $y = c(x)$ and $y = d(x)$, respectively, so that the whole region can be described by the inequalities $a \le x \le b$ and $c(x) \le y \le d(x)$. To evaluate a double integral over such a region, we have the following formula.

> ### Computing the Double Integral over a Nonrectangular Region
> If R is the region $a \leq x \leq b$ and $c(x) \leq y \leq d(x)$ (Figure 33), then we integrate over R according to the following equation:
>
> $$\iint_R f(x, y) \, dx \, dy = \int_a^b \left(\int_{c(x)}^{d(x)} f(x, y) \, dy \right) dx$$

Example 2 • Double Integral over a Nonrectangular Region

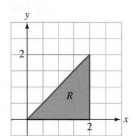

Figure 34

R is the triangle shown in Figure 34. Compute $\iint_R x \, dx \, dy$.

Solution R is the region described by $0 \leq x \leq 2, 0 \leq y \leq x$. We have

$$\iint_R x \, dx \, dy = \int_0^2 \int_0^x x \, dy \, dx$$

$$= \int_0^2 [xy]_{y=0}^x \, dx = \int_0^2 x^2 \, dx = \left[\frac{x^3}{3} \right]_0^2 = \frac{8}{3}$$

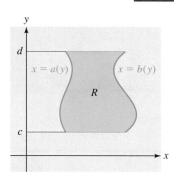

Figure 35

The second type of region is shown in Figure 35. This is the region described by $c \leq y \leq d$ and $a(y) \leq x \leq b(y)$. To evaluate a double integral over such a region, we have the following formula.

> ### Double Integral over a Nonrectangular Region (continued)
> If R is the region $c \leq y \leq d$ and $a(y) \leq x \leq b(y)$ (Figure 35), then we integrate over R according to the following equation:
>
> $$\iint_R f(x, y) \, dx \, dy = \int_c^d \left(\int_{a(y)}^{b(y)} f(x, y) \, dx \right) dy$$

Example 3 • Double Integral over a Nonrectangular Region

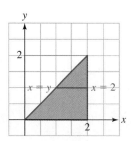

Figure 36

Redo Example 2, integrating in the opposite order.

Solution We can integrate in the opposite order if we can describe the region in Figure 34 in the way shown in Figure 35. In fact, it is the region $0 \leq y \leq 2$ and $y \leq x \leq 2$. To see this, we draw a horizontal line through the region, as in Figure 36. The line extends from $x = y$ on the left to $x = 2$ on the right, so $y \leq x \leq 2$. The possible heights for such a line are $0 \leq y \leq 2$. We can now compute the integral:

$$\iint_R x \, dx \, dy = \int_0^2 \int_y^2 x \, dx \, dy$$

$$= \int_0^2 \left[\frac{x^2}{2} \right]_{x=y}^2 \, dy = \int_0^2 \left(2 - \frac{y^2}{2} \right) dy = \left[2y - \frac{y^3}{6} \right]_0^2 = \frac{8}{3}$$

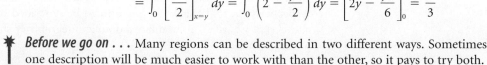

Before we go on . . . Many regions can be described in two different ways. Sometimes one description will be much easier to work with than the other, so it pays to try both.

APPLICATIONS

There are many applications of double integrals besides finding volumes. We can also use them to find *averages*. Remember that the average of $f(x)$ on $[a, b]$ is given by $\int_a^b f(x)\, dx$ divided by $(b - a)$, the length of the interval.

Average of a Function of Two Variables

The average of $f(x, y)$ on the region R is

$$\bar{f} = \frac{1}{A} \iint_R f(x, y)\, dx\, dy$$

Here, A is the area of R. We can compute the area A geometrically, by using the techniques from Chapter 7, or by computing

$$A = \iint_R 1\, dx\, dy$$

Quick Example

The average value of $f(x, y) = xy$ on the rectangle given by $0 \le x \le 1$ and $0 \le y \le 2$ is

$$\bar{f} = \frac{1}{2} \iint_R xy\, dx\, dy \qquad \text{The area of the rectangle is 2.}$$

$$= \frac{1}{2} \int_0^2 \int_0^1 xy\, dx\, dy$$

$$= \frac{1}{2} \cdot 1 = \frac{1}{2} \qquad \text{We calculated the integral in Example 1.}$$

Example 4 • Average Revenue

Your company is planning to price its new line of subcompact cars at between $10,000 and $15,000. The marketing department reports that if the company prices the cars at p dollars per car, the demand will be between $q = 20{,}000 - p$ and $q = 25{,}000 - p$ cars sold in the first year. What is the average of all the possible revenues your company could expect in the first year?

Solution Revenue is given by $R = pq$ as usual, and we are told that

$$10{,}000 \le p \le 15{,}000 \quad \text{and} \quad 20{,}000 - p \le q \le 25{,}000 - p$$

This domain D of prices and demands is shown in Figure 37. To average the revenue R over the domain D, we need to compute the area A of D. Using either calculus or geometry, we get $A = 25{,}000{,}000$. We then need to integrate R over D:

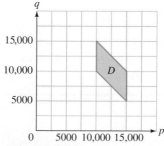

Figure 37

$$\iint_D pq\, dp\, dq = \int_{10{,}000}^{15{,}000} \int_{20{,}000 - p}^{25{,}000 - p} pq\, dq\, dp = \int_{10{,}000}^{15{,}000} \left[\frac{pq^2}{2} \right]_{q = 20{,}000 - p}^{25{,}000 - p} dp$$

$$= \frac{1}{2} \int_{10{,}000}^{15{,}000} [p(25{,}000 - p)^2 - p(20{,}000 - p)^2]\, dp$$

$$= \frac{1}{2} \int_{10{,}000}^{15{,}000} [225{,}000{,}000p - 10{,}000p^2]\, dp \approx 3{,}072{,}916{,}666{,}666{,}667$$

We get the following average:

$$\overline{R} = \frac{3,072,916,666,666,667}{25,000,000} \approx \$122,900,000$$

✸ **Before we go on . . .** To check that this is a reasonable answer, notice that the revenues at the corners of the domain are \$100,000,000 per year, \$150,000,000 per year (at two corners), and \$75,000,000 per year. Some of these are smaller than the average and some larger, as we would expect. You should also check that the maximum possible revenue is \$156,250,000 per year. (What is the minimum possible revenue?)

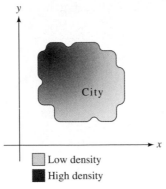

Low density
High density

Figure 38

Another useful application comes about when we consider density. For example, suppose $P(x, y)$ represents the population density (in people per square mile, say) in the city shown in Figure 38. If we break the city up into small rectangles (for example, city blocks), then the population in the small rectangle $x_{i-1} \leq x \leq x_i$ and $y_{j-1} \leq y \leq y_j$ is approximately $P(x_i, y_j)\,\Delta x\,\Delta y$. Adding up all of these population estimates we get

$$\text{total population} \approx \sum_{j=1}^{n} \sum_{i=1}^{m} P(x_i, y_j)\,\Delta x\,\Delta y$$

Since this is a double Riemann sum, when we take the limit as m and n go to infinity, we get the following calculation of the population of the city:

$$\text{total population} = \iint_{\text{city}} P(x, y)\,dx\,dy$$

Example 5 • Population

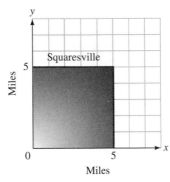

Figure 39

Squaresville is a city in the shape of a square 5 miles on a side. The population density at a distance of x miles east and y miles north of the southwest corner is $P(x, y) = x^2 + y^2$ thousand people per square mile. Find the total population of Squaresville.

Solution Squaresville is pictured in Figure 39, in which we put the origin in the southwest corner of the city. To compute the total population, we integrate the population density over the city S:

$$\text{population} = \iint_{\text{Squaresville}} P(x, y)\,dx\,dy = \int_0^5 \int_0^5 (x^2 + y^2)\,dx\,dy$$

$$= \int_0^5 \left[\frac{x^3}{3} + xy^2\right]_{x=0}^{5} dy = \int_0^5 \left[\frac{125}{3} + 5y^2\right] dy$$

$$= \frac{1250}{3} \approx 417 \text{ thousand people}$$

✸ **Before we go on . . .** Note that the average population density is the total population divided by the area of the city, which is about 17,000 people per square mile. Compare this calculation with the calculations of averages in Examples 3 and 4.

8.6 EXERCISES

In Exercises 1–16, compute the integrals.

1. $\int_0^1 \int_0^1 (x - 2y) \, dx \, dy$

2. $\int_{-1}^1 \int_0^2 (2x + 3y) \, dx \, dy$

3. $\int_0^1 \int_0^2 (ye^x - x - y) \, dx \, dy$

4. $\int_1^2 \int_2^3 \left(\frac{1}{x} + \frac{1}{y} \right) dx \, dy$

5. $\int_0^2 \int_0^3 e^{x+y} \, dx \, dy$

6. $\int_0^1 \int_0^1 e^{x-y} \, dx \, dy$

7. $\int_0^1 \int_0^{2-y} x \, dx \, dy$

8. $\int_0^1 \int_0^{2-y} y \, dx \, dy$

9. $\int_{-1}^1 \int_{y-1}^{y+1} e^{x+y} \, dx \, dy$

10. $\int_0^1 \int_y^{y+2} \frac{1}{\sqrt{x+y}} \, dx \, dy$

11. $\int_0^1 \int_{-x^2}^{x^2} x \, dy \, dx$

12. $\int_1^4 \int_{-\sqrt{x}}^{\sqrt{x}} \frac{1}{x} \, dy \, dx$

13. $\int_0^1 \int_0^x e^{x^2} \, dy \, dx$

14. $\int_0^1 \int_0^{x^2} e^{x^3+1} \, dy \, dx$

15. $\int_0^2 \int_{1-x}^{8-x} (x + y)^{1/3} \, dy \, dx$

16. $\int_1^2 \int_{1-2x}^{x^2} \frac{x+1}{(2x+y)^3} \, dy \, dx$

In Exercises 17–24, find $\iint_R f(x, y) \, dx \, dy$, where R is the indicated domain. (Remember that you often have a choice about the order of integration.)

17. $f(x, y) = 2$

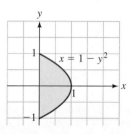

18. $f(x, y) = x$

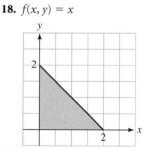

19. $f(x, y) = 1 + y$

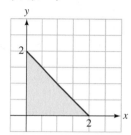

20. $f(x, y) = e^{x+y}$

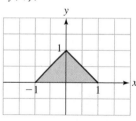

21. $f(x, y) = xy^2$

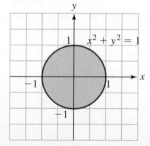

22. $f(x, y) = xy^2$

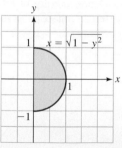

23. $f(x, y) = x^2 + y^2$

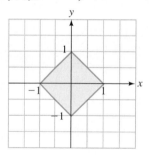

24. $f(x, y) = x^2$

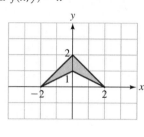

In Exercises 25–30, find the average value of the given function over the indicated domain.

25. $f(x, y) = y$

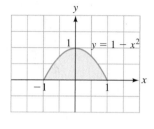

26. $f(x, y) = 2 + x$

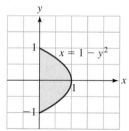

27. $f(x, y) = e^y$

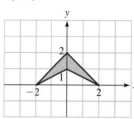

28. $f(x, y) = y$

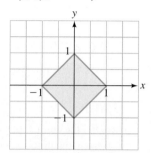

29. $f(x, y) = x^2$

30. $f(x, y) = x^2 + y^2$

In Exercises 31–36, sketch the region over which you are integrating and then write down the integral with the order of integration reversed (changing the limits of integration as necessary).

31. $\int_0^1 \int_0^{1-y} f(x, y) \, dx \, dy$

32. $\int_{-1}^1 \int_0^{1+y} f(x, y) \, dx \, dy$

33. $\int_{-1}^1 \int_0^{\sqrt{1+y}} f(x, y) \, dx \, dy$

34. $\int_{-1}^1 \int_0^{\sqrt{1-y}} f(x, y) \, dx \, dy$

35. $\int_1^2 \int_1^{4/x^2} f(x, y) \, dy \, dx$

36. $\int_1^{e^2} \int_0^{\ln x} f(x, y) \, dy \, dx$

37. Find the volume under the graph of $z = 1 - x^2$ over the region $0 \le x \le 1$ and $0 \le y \le 2$.

38. Find the volume under the graph of $z = 1 - x^2$ over the triangle $0 \le x \le 1$ and $0 \le y \le 1 - x$.

39. Find the volume of the tetrahedron shown in the figure. Its corners are $(0, 0, 0)$, $(1, 0, 0)$, $(0, 1, 0)$, and $(0, 0, 1)$.

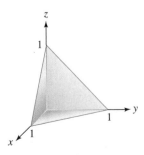

40. Find the volume of a tetrahedron with corners at $(0, 0, 0)$, $(a, 0, 0)$, $(0, b, 0)$, and $(0, 0, c)$.

APPLICATIONS

41. Productivity A productivity model at Handy Gadget Company is

$$P = 10,000x^{0.3}y^{0.7}$$

where P is the number of gadgets the company turns out each month, x is the number of employees at the company, and y is the monthly operating budget (in thousands of dollars). Because the company hires part-time workers, it uses anywhere between 45 and 55 workers each month, and its operating budget varies from $8000 to $12,000 per month. What is the average of the possible numbers of gadgets it can turn out each month? (Round the answer to the nearest 1000 gadgets.)

42. Productivity Repeat Exercise 41, using the productivity model

$$P = 10,000x^{0.7}y^{0.3}$$

43. Revenue Your latest CD-ROM of clip art is expected to sell between $q = 8000 - p^2$ and $q = 10,000 - p^2$ copies if priced at p dollars. You plan to set the price between $40 and $50. What are the maximum and minimum possible revenues you can make? What is the average of all the possible revenues you can make?

44. Revenue Your latest DVD drive is expected to sell between $q = 180,000 - p^2$ and $q = 200,000 - p^2$ units if priced at p dollars. You plan to set the price between $300 and $400. What are the maximum and minimum possible revenues you can make? What is the average of all the possible revenues you can make?

45. Revenue Your self-published novel has demand curves between $p = 15,000/q$ and $p = 20,000/q$. You expect to sell between 500 and 1000 copies. What are the maximum and minimum possible revenues you can make? What is the average of all the possible revenues you can make?

46. Revenue Your self-published book of poetry has demand curves between $p = 80,000/q^2$ and $p = 100,000/q^2$. You expect to sell between 50 and 100 copies. What are the maximum and minimum possible revenues you can make? What is the average of all the possible revenues you can make?

47. Population Density The town of West Podunk is shaped like a rectangle 20 miles from west to east and 30 miles from north to south (see the figure). It has a population density of $P(x, y) = e^{-0.1(x+y)}$ hundred people per square mile x miles east and y miles north of the southwest corner of town. What is the total population of the town?

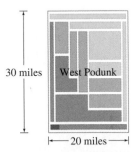

30 miles   West Podunk

← 20 miles →

48. Population Density The town of East Podunk is shaped like a triangle with an east–west base of 20 miles and a north–south height of 30 miles (see the figure). It has a population density of $P(x, y) = e^{-0.1(x+y)}$ hundred people per square mile x miles east and y miles north of the southwest corner of town. What is the total population of the town?

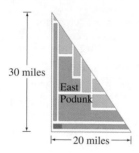

30 miles

East Podunk

|← 20 miles →|

49. Temperature The temperature at the point (x, y) on a square with vertices $(0, 0)$, $(0, 1)$, $(1, 0)$, and $(1, 1)$ is given by $T(x, y) = x^2 + 2y^2$. Find the average temperature on the square.

50. Temperature The temperature at the point (x, y) on a square with vertices $(0, 0)$, $(0, 1)$, $(1, 0)$, and $(1, 1)$ is given by $T(x, y) = x^2 + 2y^2 - x$. Find the average temperature on the square.

COMMUNICATION AND REASONING EXERCISES

51. Explain how double integrals can be used to compute the area between two curves in the xy plane.

52. Explain how double integrals can be used to compute the volume of solids in 3-space.

53. Complete the following: The first step in calculating an integral of the form $\int_a^b \int_{r(x)}^{s(x)} f(x, y) \, dy \, dx$ is to evaluate the integral _____, obtained by holding _____ constant and integrating with respect to _____.

54. If the units of $f(x, y)$ are zonars per square meter and x and y are given in meters, what are the units of $\int_a^b \int_{r(x)}^{s(x)} f(x, y) \, dy \, dz$?

55. If the units of $\int_a^b \int_{r(x)}^{s(x)} f(x, y) \, dy \, dx$ are paintings, the units of x are picassos, and the units of y are dalis, what are the units of $f(x, y)$?

56. Complete the following: If the region R is bounded on the left and right by vertical lines and on the top and bottom by the graphs of functions of x, then we integrate over R by first integrating with respect to _____ and then with respect to _____.

57. Show that if a, b, c, and d are constant, then

$$\int_a^b \int_c^d f(x)g(y) \, dx \, dy = \int_c^d f(x) \, dx \cdot \int_a^b g(y) \, dy$$

Test this result on the integral $\int_0^1 \int_1^2 ye^x \, dx \, dy$.

58. Refer to Exercise 57. If a, b, c, and d are constants, can $\int_a^b \int_c^d \frac{f(x)}{g(y)} \, dx \, dy$ be expressed as a product of two integrals? Explain.

CASE STUDY

Reproduced from the U.S. Census Bureau, United States Department of Commerce Web site

Modeling Household Income

Millennium Real Estate Development Corporation is interested in developing housing projects for medium-sized families that have high household incomes. To decide which income bracket to target, the company has asked you, a paid consultant, for an analysis of household income and household size in the United States. In particular, Millennium is interested in four issues:

1. The relationship between household size and household income and the effect of increasing household size on household income

2. The household size that corresponds to the highest household income

3. The change in the relationship between household size and household income over time

4. Some near-term projections of household income versus household size (to, say, 2005)

You decide that a good place to start would be with a visit to the Census Bureau's Web site at http://www.census.gov. After some time battling with search engines, you discover detailed information on household size versus household income for the period 1967–2000.[8] The following table summarizes the information on median household income.

---

[8]Household income is adjusted for inflation and given in 2000 dollars.

Household Size →

Year ↓		1	2	3	4	5	6	7
	1967	10,321	27,507	36,680	39,520	40,241	39,630	36,658
	1968	11,364	29,055	37,756	41,598	41,889	41,032	39,400
	1969	11,947	30,507	38,959	43,345	44,300	43,903	41,364
	1970	11,993	30,346	38,826	43,181	44,661	44,572	41,241
	1971	12,104	30,058	38,330	43,209	44,378	44,060	40,738
	1972	12,816	31,489	40,562	46,308	47,280	46,634	42,622
	1973	13,730	32,465	41,111	46,674	48,569	48,653	45,718
	1974	13,930	32,317	40,067	46,041	48,008	47,181	45,064
	1975	13,640	31,704	39,968	44,980	46,796	45,811	41,500
	1976	14,400	32,761	40,634	45,986	47,569	47,368	44,093
	1977	14,898	32,981	41,294	47,072	48,098	48,131	43,562
	1978	16,132	34,962	43,963	49,497	50,803	50,334	48,902
	1979	16,259	35,607	44,773	49,708	51,718	50,427	49,969
	1980	16,240	34,833	43,249	48,570	49,457	48,580	46,423
	1981	16,719	34,394	42,929	47,882	47,531	49,012	44,672
	1982	17,159	34,717	42,013	47,464	46,706	46,732	40,762
	1983	17,734	34,758	42,419	48,161	46,251	44,368	38,650
	1984	18,266	35,884	44,316	49,243	48,911	44,924	41,714
	1985	18,238	36,628	45,514	50,188	48,737	47,714	43,370
	1986	18,329	38,144	46,755	52,267	51,523	49,508	41,979
	1987	18,584	38,679	47,300	54,039	52,109	48,824	45,452
	1988	19,675	39,429	47,767	54,442	50,689	51,939	44,573
	1989	19,997	40,268	48,919	54,942	52,970	47,607	44,020
	1990	19,701	40,263	47,205	53,250	50,428	48,995	46,362
	1991	19,125	38,670	47,368	53,326	50,524	45,696	42,280
	1992	18,606	38,389	46,580	53,111	50,853	44,760	40,010
	1993	18,896	38,150	46,360	53,032	49,685	48,336	38,957
	1994	18,672	39,082	47,241	53,817	50,800	49,128	42,152
	1995	19,159	40,085	47,433	55,615	51,325	49,700	43,805
	1996	19,564	40,756	48,988	56,194	52,298	46,391	44,095
	1997	20,075	42,096	50,412	56,885	53,934	49,716	45,306
	1998	21,267	43,804	51,778	58,971	56,671	51,790	49,221
	1999	21,791	44,798	52,910	61,776	56,269	53,630	53,898
	2000	21,468	44,530	54,196	61,847	60,295	54,841	54,663

Source: U.S. Bureau of the Census (http://www.census.gov/hhes/income/histinc/h11.html) 2002.

You notice that the table is actually a numerical representation of the median household income I as a function of two variables, the household size n and the year t.

The numbers are a bit overwhelming, so you decide to use Excel to graph the data as a surface (Figure 40).

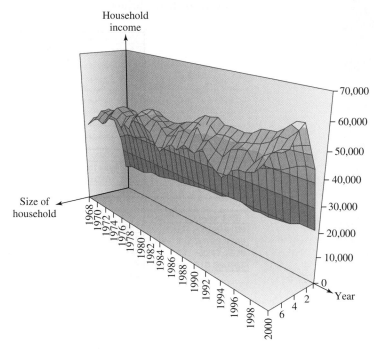

Figure 40

Now you definitely see two trends. First, household income peaks at around five or six people per household and then drops off at both ends. In fact, the slices through $t = constant$ look parabolic. Second, household income for all household sizes seems to increase more or less linearly with time (the slices through $n = constant$ are approximately linear).

At this point you realize that a mathematical model of these data would be useful; not only would it "smooth out the bumps," but it also would give you a way to complete the project for Millennium. Although technology can give you a regression model for data such as this, it is up to you to decide on the form of the model. It is in choosing an appropriate model that your analysis of the graph comes in handy. Since I appears to vary quadratically with the household size, you would like a general quadratic of the form

$$I = a + bn + cn^2$$

for each value of time t. Also, since I should vary linearly with time t, you would like

$$I = mt + k$$

for each value of n. Putting these together, you get the following candidate model:

$$I(n, t) = a_1 + a_2n + a_3n^2 + a_4t$$

where a_1, a_2, a_3, and a_4 are constants you need to determine.

You decide to use Excel to generate your model. The specific software tool you need is called the Analysis Toolpack, which comes with Excel. (It is found in the `Tools` menu as Data Analysis. If it is missing, select `Add-Ins` from the `Tools` menu and check `Analysis Toolpack`.)

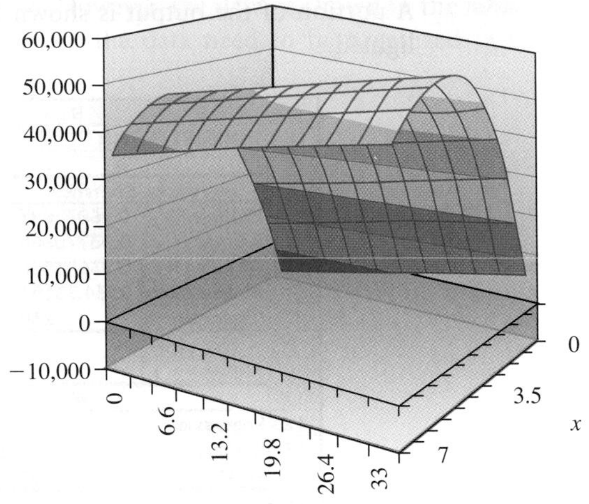

Figure 41

As comforting as these statistics are, nothing can be quite as persuasive as a graph. You turn to the graphing software of your choice and notice that the graph of the model appears to be a faithful representation of the data (Figure 41).

Now you get to work, using the model to address the questions posed by Millennium.

1. *The relationship between household size and household income and the effect of increasing household size on household income.* You already have a quantitative relationship in the form of the regression model. As for the second part of the question, the rate of change of median household income with respect to household size is given by the partial derivative:

$$\frac{\partial I}{\partial n} = 21,008 - 4278.2n \text{ dollars per additional family member}$$

Thus, for example, in a household of four people,

$$\frac{\partial I}{\partial n} = 21,008 - 4278.2(4) \approx \$3895 \text{ per additional family member}$$

On the other hand, when $n = 5$, one has

$$\frac{\partial I}{\partial n} = 21,008 - 4278.2(5) = -\$383 \text{ per additional family member}$$

Notice that the derivative is independent of time: The rate of change of average family income with respect to household size is independent of the date (according to your model).

2, 3. *The household size corresponding to the highest household income and the change in the relationship over time.* Although a glance at the graph shows you that there are no relative maxima, holding *t* constant (that is, on any given year) gives a relative maximum along the corresponding slice when

$$\frac{\partial I}{\partial n} = 0$$

or $21{,}008 - 4278.2n = 0$, which gives

$$n = \frac{21{,}008}{4278.2} \approx 4.91$$

In other words, households of five tend to have the highest household incomes. Again, $\partial I/\partial n$ does not depend on t, so this optimal household size seems independent of time t.

4. *Some near-term projections of household income versus household size.* As we have seen throughout the book, extrapolation can be a risky venture, however, *near-term* extrapolation from a good model can be reasonable. You enter the model in an Excel spreadsheet to obtain the following predicted median household incomes for the years 2001–2005:

Household Size →

Year ↓		1	2	3	4	5	6	7
	2001	25,061	39,652	49,964	55,998	57,755	55,232	48,432
	2002	25,413	40,004	50,316	56,351	58,107	55,585	48,784
	2003	25,765	40,356	50,668	56,703	58,459	55,937	49,136
	2004	26,117	40,708	51,020	57,055	58,811	56,289	49,488
	2005	26,469	41,060	51,372	57,407	59,163	56,641	49,840

EXERCISES

1. Use Excel to obtain an interaction model of the form

$$I(n, t) = a_1 + a_2 n + a_3 t + a_4 nt$$

Compare the fit of this model with that of the quadratic model above. Comment on the result.

2. How much is there to be gained by including a term of the form $a_5 t^2$ in the original model? (Perform the regression and analyze the result by referring to the P value for the resulting coefficient of t^2.)

3. The following table shows some data on U.S. population (in thousands) versus age and year. This table can be downloaded as an Excel file by following the path

Web Site → Everything for Calculus → Chapter 8 → Case Study Excel Data

Year(0 = 1990) →

		0	2	4	6	8	8.5
Age ↓	**2.5**	18,851	19,489	19,694	19,324	19,020	18,974
	7.5	18,058	18,285	18,742	19,425	19,912	19,931
	12.5	17,191	18,065	18,666	18,949	19,184	19,291
	17.5	17,763	17,170	17,707	18,644	19,460	19,554
	22.5	19,137	19,085	18,451	17,562	17,685	17,796
	27.5	21,233	20,152	19,142	18,993	18,621	18,513
	32.5	21,909	22,237	22,141	21,328	20,163	19,965
	37.5	19,980	21,092	21,973	22,550	22,600	22,589
	42.5	17,793	18,806	19,714	20,809	21,875	22,014
	47.5	13,823	15,362	16,685	18,438	18,850	19,007
	52.5	11,370	12,059	13,199	13,931	15,727	15,973
	57.5	10,474	10,487	10,937	11,362	12,408	12,631
	62.5	10,619	10,440	10,079	9,997	10,256	10,358
	67.5	10,076	9,973	9,963	9,895	9,575	9,515
	72.5	8,022	8,467	8,733	8,778	8,781	8,780
	77.5	6,146	6,392	6,575	6,873	7,195	7,238
	82.5	3,934	4,135	4,350	4,559	4,712	4,748
	87.5	2,050	2,170	2,287	2,395	2,533	2,560
	92.5	765	860	956	1,024	1,094	1,108
	97.5	206	231	249	287	317	324
	102.5	37	44	50	57	63	62

Use multiple regression to construct (a) a linear model and (b) an interaction model for the data (round all coefficients to four significant digits). Does the interaction model give a significantly better fit in terms of the multiple regression coefficient? Referring to the linear model, does the P value for the coefficient of y provide strong evidence that the population profile has been changing with time? (A P value of α indicates that we can be certain with a confidence level of $1 - \alpha$ that the associated coefficient is nonzero.)

4. Graph the data from Exercise 3 and decide whether the linear model gives a faithful representation of the actual data. If not, propose and construct an alternative model. How is the confidence level in the coefficient of y changed?

5. According to your model in Exercise 4, why does the age group with maximum population not change over time? Propose a model in which it does. Construct such a model and test the additional coefficient(s).

CHAPTER 8 REVIEW TEST

1.
  a. Let $g(x, y, z) = xy(x + y - z) + x^2$. Evaluate $g(0, 0, 0)$, $g(1, 0, 0)$, $g(0, 1, 0)$, $g(x, x, x)$, and $g(x, y + k, z)$.
  b. Let $f(x, y, z) = 2.72 - 0.32x - 3.21y + 12.5z$. Complete the following: f _____ by _____ units for every 1 unit of increase in x and _____ by _____ units for every unit of increase in z.
  c. Let $h(x, y) = 2x^2 + xy - x$. Complete the following table of values:

$y \downarrow$ \ $x \rightarrow$	-1	0	1
-1			
0			
1			

  d. Give a formula for a (single) function f with the property that $f(x, y) = -f(y, x)$ and $f(1, -1) = 3$.

2. For each of the given functions, compute the partial derivatives shown.
  a. $f(x, y) = x^2 + xy$; find f_x, f_y, and f_{yy}.
  b. $f(x, y) = e^{xy} + e^{3x^2 - y^2}$; find $\partial f / \partial x$ and $\partial^2 f / \partial x \partial y$.
  c. $f(x, y, z) = \dfrac{x}{x^2 + y^2 + z^2}$; find $\dfrac{\partial f}{\partial x}, \dfrac{\partial f}{\partial y}, \dfrac{\partial f}{\partial z}$, and $\dfrac{\partial f}{\partial x}\Big|_{(0,1,0)}$.
  d. $f(x, y, z) = x^2 + y^2 + z^2 + xyz$; find $f_{xx} + f_{yy} + f_{zz}$.

3. In each of the following, locate and classify all critical points.
  a. $f(x, y) = (x - 1)^2 + (2y - 3)^2$
  b. $g(x, y) = (x - 1)^2 - 3y^2 + 9$
  c. $h(x, y) = e^{xy}$
  d. $j(x, y) = xy + x^2$
  e. $f(x, y) = \ln(x^2 + y^2) - (x^2 + y^2)$

4. Solve the following constrained optimization problems.
  a. Find the point on the surface $z = \sqrt{x^2 + 2(y - 3)^2}$ closest to the origin.
  b. Find the location and values of the minimum and maximum of
  $$T(x, y) = (x - 1)^2 + y^2$$
  on the triangle with vertices $(0, 1)$, $(3, 0)$, and $(0, 0)$.
  c. Find the minimum and maximum values of
  $$F(x, y) = x^2 + 2y^2 - 2x - 4y + xy$$
  subject to $-1 \le x \le 1$ and $-2 \le y \le 2$.

5. Compute the given quantities.
  a. $\displaystyle\int_0^1 \int_0^2 2xy \, dx \, dy$
  b. $\displaystyle\int_1^2 \int_0^1 xye^{x+y} \, dx \, dy$
  c. $\displaystyle\int_0^2 \int_0^{2x} \dfrac{1}{x^2 + 1} \, dy \, dx$

  d. The average value of xye^{x+y} over the rectangle $0 \le x \le 1$, $1 \le y \le 2$.
  e. $\displaystyle\iint_R (x^2 - y^2) \, dx \, dy$, where R is the region shown in the figure:

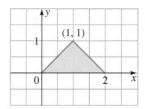

  f. The volume under the graph of $z = 1 - y$ over the region in the xy plane between the parabola $y = 1 - x^2$ and the x axis.

OHaganBooks.com—HANDLING DIFFERENT VARIABLES

6. OHaganBooks.com has two principal competitors: JungleBooks.com and FarmerBooks.com. Current Web-site traffic at OHaganBooks.com is estimated at 5000 hits per day. This number is predicted to decrease by 0.8 for every new customer of JungleBooks.com and by 0.6 for every new customer of FarmerBooks.com.
  a. Use this information to model the daily Web-site traffic at OHaganBooks.com as a linear function of the new customers of its two competitors.
  b. According to the model, if Junglebooks.com gets 100 new customers and OHaganBooks.com traffic drops to 4770 hits per day, how many new customers has FarmerBooks.com obtained?
  c. The model in part (a) did not take into account the growth of the total online consumer base. OHaganBooks.com expects to get approximately one additional hit per day for every 10,000 new Internet shoppers. Modify your model in part (a) so as to include this information.
  d. How many new Internet shoppers would it take to offset the effects on traffic at OHaganBooks.com of 100 new customers at each of its competitor sites?

7. To increase business at OHaganBooks.com, you have purchased banner ads at well-known Internet portals and advertised on television. The following interaction model shows the average number h of hits per day as a function of monthly expenditures x on banner ads and y on television advertising (x and y are in dollars):
  $$h(x, y) = 1800 + 0.05x + 0.08y + 0.000\,03xy$$
  a. Based on your model, how much traffic can you anticipate if you spend \$2000 each month for banner ads and \$3000 each month on television advertising?

b. Evaluate $\partial h/\partial y$, specify its units of measurement, and indicate whether it increases or decreases with increasing x.

c. How much should the company be spending on banner ads to obtain one hit per day for each $5 spent per month on television advertising?

d. One or more of the following statements is correct. Identify which one(s).

(A) If nothing is spent on television advertising, one more dollar spent each month in banner ads will buy approximately 0.05 hits per day at OHaganBooks.com.

(B) If nothing is spent on television advertising, one more hit each day at OHaganBooks.com will cost the company about 5¢ per month in banner ads.

(C) If nothing is spent on banner ads, one more hit each day at OHaganBooks.com will cost the company about 5¢ per month in banner ads.

(D) If nothing is spent on banner ads, one more dollar spent each month in banner ads will buy approximately 0.05 hits per day at OHaganBooks.com.

(E) Hits at OHaganBooks.com cost approximately 5¢ per month spent on banner ads, and this cost increases at a rate of 0.003¢ per month per hit.

8. The holiday season is now at its peak, and OHaganBooks.com has been understaffed and swamped with orders. The current backlog (orders unshipped for 2 or more days) has grown to a staggering 50,000, and new orders are coming in at a rate of 5000 per day. Research based on productivity data at OHaganBooks.com results in the following model:

$$P(x, y) = 1000x^{0.9}y^{0.1} \text{ additional orders filled per day}$$

where x is the number of additional personnel hired and y is the daily budget (excluding salaries) allocated to eliminating the backlog.

a. How many additional orders will be filled each day if the company hires ten additional employees and budgets an additional $1000 per day? (Round the answer to the nearest 100.)

b. In addition to the daily budget, extra staffing costs the company $150 per day for every new staff member hired. To fill at least 15,000 orders each day at a minimum total daily cost, how many new staff members should the company hire? (Use the method of Lagrange multipliers.)

9. To sell x paperbacks and y hardcover books in a week costs OHaganBooks.com

$$C(x, y) = 0.1x^2 + 0.2y^2 - 200x - 480y + 400,000 \text{ dollars}$$

If it sells between 900 and 1100 paperbacks and between 1000 and 1500 hardcover books each week, what combination will cost it the least, and what combination will cost it the most?

10. If OHaganBooks.com sells x paperback books and y hardcover books each week, it will make a weekly profit of
$$P(x, y) = 3x + 10y \text{ dollars}$$

If it sells between 1200 and 1500 paperback books and between 1800 and 2000 hardcover books each week, what is the average of all its possible weekly profits?

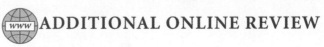

ADDITIONAL ONLINE REVIEW

If you follow the path

Web Site → Everything for Calculus → Chapter 8

you will find the following additional resources to help you review:

A comprehensive chapter summary (including examples and interactive features)

Additional review exercises (including interactive exercises and many with help)

A true/false chapter quiz

Several useful utilities, including three-dimensional graphers and regression utilities

9

TRIGONOMETRIC MODELS

CASE STUDY

Predicting Cocoa Inventories

As a consultant to Chocoholic Cocoa Company, you have been asked to model the worldwide production and consumption of cocoa, to determine how production cycles are affected by consumption, to estimate the trend in average world cocoa inventories, and to advise your cocoa producers whether to increase or decrease production. You have data showing the production and consumption of cocoa for the past 10 years. How will you analyze these data to prepare your report?

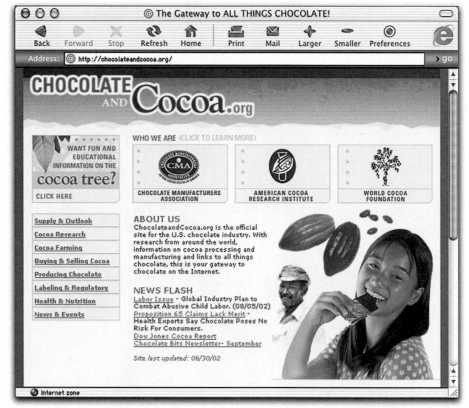

INTERNET RESOURCES FOR THIS CHAPTER

At the Web site, follow the path
    Web Site → Everything for Calculus → Chapter 9
where you will find links to step-by-step tutorials for the main topics in this chapter, a detailed chapter summary you can print out, a true/false quiz, and a collection of sample test questions. You will also find downloadable Excel tutorials for each section, an online numerical integration utility, and other resources.

Introduction

Cyclical behavior is common in the business world: There are seasonal fluctuations in the demand for surfing equipment, swimwear, snow shovels, and many other items. The nonlinear functions we have studied up to now cannot model this kind of behavior. To model cyclical behavior we need the **trigonometric functions.**

In the first section, we study the basic trigonometric functions—especially the **sine** and **cosine** functions from which all the trigonometric functions are built—and see how to model various kinds of periodic behavior using these functions. The rest of the chapter is devoted to the calculus of the trigonometric functions—their derivatives and integrals—and to its numerous applications.

9.1 Trigonometric Functions, Models, and Regression

The Sine Function

Figure 1 shows the approximate average daily high temperatures in New York's Central Park. If we draw the graph for several years, we get the repeating pattern shown in Figure 2, in which the x coordinate represents time in years, with $x = 0$ representing August 1, and the y coordinate represents the temperature in degrees Fahrenheit. This is an example of **cyclical,** or **periodic,** behavior.

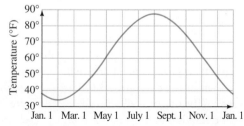

SOURCE: National Weather Service/*New York Times,* January 7, 1996, p. 36.

Figure 1

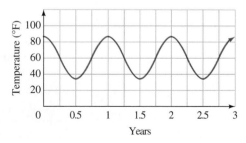

Figure 2

As we said in the introduction, cyclical behavior is also common in the business world. The graph in Figure 3 even suggests cyclical behavior in the European economy.

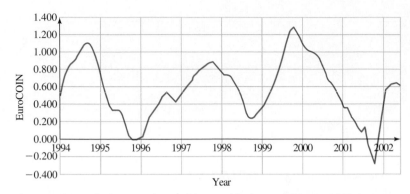

The EuroCOIN scale gives an estimate of the monthly change of Euro area GDP in percentage points per month. SOURCE: http://www.cepr.org/Data/eurocoin, August 2002.

Figure 3

From a mathematical point of view, the simplest models of cyclical behavior are the **sine** and **cosine** functions. An easy way to describe these functions is as follows. Imagine a bicycle wheel whose radius is 1 unit, with a marker attached to the rim of the rear wheel, as shown in Figure 4.

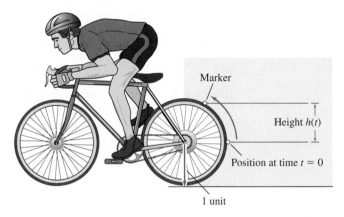

Figure 4

Now, we can measure the height $h(t)$ of the marker above the center of the wheel. As the wheel rotates, $h(t)$ fluctuates between -1 and $+1$. Suppose at time $t = 0$ the marker was at height zero as shown in the diagram, so $h(0) = 0$. Since the wheel has a radius of 1 unit, its circumference (the distance all around) is 2π, where $\pi = 3.14159265.\ldots$ If the cyclist happens to be moving at a speed of 1 unit per second, it will take the bicycle wheel 2π seconds to make one complete revolution. During the time interval $[0, 2\pi]$, the marker will first rise to a maximum height of $+1$, drop to a low point of -1, and then return to the starting position of 0 at $t = 2\pi$. This function $h(t)$ is called the **sine function,** denoted by $\sin(t)$. Figure 5 shows its graph.

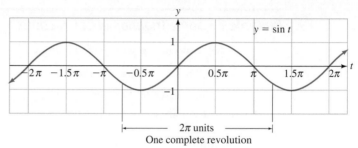

Technology formula: `sin(t)`

Figure 5

The Sine Function

"Bicycle Wheel" Definition

If a wheel of radius 1 unit rolls forward at a speed of 1 unit per second, then $\sin(t)$ is the height after t seconds of a marker on the rim of the wheel, starting in the position shown in Figure 4.

Geometric Definition

The **sine** of a real number t is the y coordinate (height) of the point P in the following diagram, where $|t|$ is the length of the arc shown.

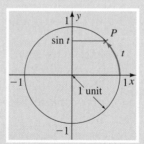

sin(t) = y coordinate of the point P

Quick Examples

From the graph, we see that

1. $\sin(\pi) = 0$

Graphing calculator: `sin(π)`
Excel: `sin(PI( ))`
(π is `PI( )` in Excel.)

2. $\sin\left(\dfrac{\pi}{2}\right) = 1$

Graphing calculator: `sin(π/2)`
Excel: `sin(PI( )/2)`

3. $\sin\left(\dfrac{3\pi}{2}\right) = -1$

Graphing calculator: `sin(3π/2)`
Excel: `sin(3*PI( )/2)`

T̃ *Example 1 • Some Trigonometric Functions*

Use technology to plot the following pairs of graphs on the same set of axes.

a. $f(x) = \sin(x)$; $g(x) = 2\sin(x)$

b. $f(x) = \sin(x)$; $g(x) = \sin(x + 1)$

c. $f(x) = \sin(x)$; $g(x) = \sin(2x)$

Solution

a. (*Important note:* If you are using a calculator, make sure it is set to *radians mode*, not degree mode.) We enter these functions as `sin(x)` and `2*sin(x)`, respectively. We use the range $-7 \le x \le 7$ for x suggested by the graph in Figure 5, but with larger range of y coordinates (why?): $-3 \le y \le 3$. The graphs are shown in Figure 6.

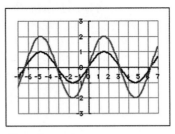

Figure 6

Here, $f(x) = \sin(x)$ is shown in blue, and $g(x) = 2\sin(x)$ in green. Notice that multiplication by 2 has doubled the **amplitude,** or the *distance it oscillates up and down.* Where the original sine curve oscillates between -1 and 1 (has amplitude 1), the new curve oscillates between -2 and 2. In general:

A sin(*x*) *has amplitude A*

b. We enter these functions as `sin(x)` and `sin(x+1)`, respectively, and we get Figure 7.

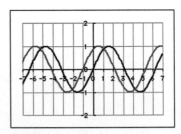

Figure 7

Once again, $f(x) = \sin(x)$ is shown in blue, and $g(x) = \sin(x + 1)$ is in green. The addition of 1 to the argument has shifted the graph to the left by 1 unit. In general:

Replacing x by x + c shifts the graph to the left c units

How would we shift the graph to the *right* 1 unit?

c. We enter these functions as `sin(x)` and `sin(2*x)`, respectively, and get the graph in Figure 8.

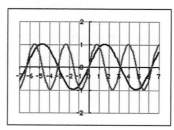

Figure 8

The graph of sin(2*x*) oscillates twice as fast as the graph of sin(*x*). In general:

Replacing x by bx multiplies the rate of oscillation by b

 Web Site

At the Web site, follow the path

    Web Site → Online Text → New Functions from Old: Scaled and Shifted Functions

for more discussion of the operations we just used to modify the graph of the sine function.

We can combine the operations in Example 1, and a vertical shift as well, to obtain the following.

The General Sine Function

The **general sine function** is

$$f(x) = A \sin[\omega(x - \alpha)] + C$$

Its graph is shown here:

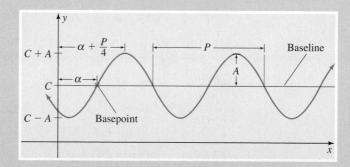

- *A* is the **amplitude** (the height of each peak above the baseline).
- *C* is the **vertical offset** (height of the baseline).
- *P* is the **period,** or **wavelength** (the length of each cycle), and is related to ω by

$$P = 2\pi/\omega \quad \text{or} \quad \omega = 2\pi/P$$

- ω is the **angular frequency** (the number of cycles in every interval of length 2π).
- α is the **phase shift.**

Example 2 • Electrical Current

The typical voltage V supplied by an electrical outlet in the United States is a sinusoidal function that oscillates between -165 volts and $+165$ volts with a frequency of 60 cycles per second. Find an equation for the voltage as a function of time t.

Solution What we are looking for is a function of the form

$$V(t) = A \sin[\omega(t - \alpha)] + C$$

Referring to the figure above, we can determine the constants.

- *Amplitude A and vertical offset C* Since the voltage oscillates between -165 volts and $+165$ volts, we see that $A = 165$ and $C = 0$.
- *Period P* Since the electric current oscillates 60 times in 1 second, the length of time it takes to oscillate once is $\frac{1}{60}$ second. Thus, the period is $P = \frac{1}{60}$.
- *Angular frequency ω* This is given by the formula

$$\omega = \frac{2\pi}{P} = 2\pi(60) = 120\pi$$

- *Phase shift α* The phase shift α tells us when the curve first crosses the t axis as it ascends. Since we are free to specify what time $t = 0$ represents, let's say that the curve crosses 0 when $t = 0$, so $\alpha = 0$. Thus,

$$V(t) = A \sin[\omega(t - \alpha)] + C = 165 \sin(120\pi t)$$

where t is time in seconds.

Example 3 • Cyclical Employment Patterns

An economist consulted by your employment agency indicates that the demand for temporary employment (measured in thousands of job applications per week) in your county can be roughly approximated by the function

$$d = 4.3 \sin(0.82t - 0.3) + 7.3$$

where t is time in years since January 2000. Calculate the amplitude, the vertical offset, the phase shift, the angular frequency, and the period, and interpret the results.

Solution To calculate these constants, we write

$$d = A \sin[\omega(t - \alpha)] + C = A \sin[\omega t - \omega\alpha] + C$$

$$= 4.3 \sin(0.82t - 0.3) + 7.3$$

and we see right away that $A = 4.3$ (the amplitude), $C = 7.3$ (vertical offset), and $\omega = 0.82$ (angular frequency). We also have

$$\omega\alpha = 0.3$$

$$\alpha = \frac{0.3}{\omega} = \frac{0.3}{0.82} \approx 0.37$$

(rounding to two significant digits; notice that all the constants are given to two digits). Finally, we get the period using the formula

$$P = \frac{2\pi}{\omega} = \frac{2\pi}{0.82} \approx 7.7$$

We can interpret these numbers as follows: The demand for temporary employment fluctuates in cycles of 7.7 years about a baseline of 7300 job applications per week. Every cycle, the demand peaks at 11,600 applications per week (4300 above the baseline) and dips to a low of 3000. In May 2000 ($t = 0.37$), the demand for employment was at the baseline level and rising.

The Cosine Function

Closely related to the sine function is the cosine function, defined as follows (refer to the definition of the sine function for comparison).

The Cosine Function

Geometric Definition

The **cosine** of a real number t is the x coordinate of the point P in the following diagram, in which $|t|$ is the length of the arc shown.

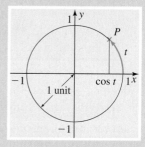

cos(t) = x coordinate of the point P

Graph of the Cosine Function

The graph of the cosine function is identical to the graph of the sine function, except that it is shifted $\pi/2$ units to the left.

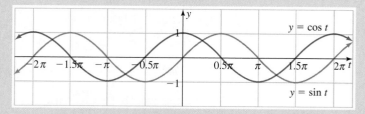

Technology formula: cos(t)

Notice that the coordinates of the point P in the diagram above are $(\cos t, \sin t)$ and that the distance from P to the origin is 1 unit. It follows from the Pythagorean theorem that the distance from a point (x, y) to the origin is $\sqrt{x^2 + y^2}$. Thus,

square of the distance from P to $(0, 0) = 1$

$$(\sin t)^2 + (\cos t)^2 = 1$$

We often write $(\sin t)^2$ as $\sin^2 t$ and similarly for the cosine, so we can rewrite the equation as

$$\sin^2 t + \cos^2 t = 1$$

This equation is one of the important relationships between the sine and cosine functions.

Fundamental Trigonometric Identities

Relationships between Sine and Cosine

The sine and cosine of a number t are related by

$$\sin^2 t + \cos^2 t = 1$$

We can obtain the cosine curve by shifting the sine curve to the left a distance of $\pi/2$. [See Example 1(b) for a shifted sine function.] Conversely, we can obtain the sine curve from the cosine curve by shifting it $\pi/2$ units to the right. These facts can be expressed as

$$\cos t = \sin(t + \pi/2)$$

$$\sin t = \cos(t - \pi/2)$$

Alternative Formulas

We can also obtain the cosine curve by first inverting the sine curve vertically (replace t by $-t$) and then shifting to the *right* a distance of $\pi/2$. This gives us two alternative formulas (which are easier to remember):

$$\cos t = \sin(\pi/2 - t) \qquad \underline{\text{Cos}}\text{ine is the sine of the } \underline{\text{com}}\text{plement.}$$

$$\sin t = \cos(\pi/2 - t)$$

Question Since we can rewrite the cosine function in terms of the sine function, who needs the cosine function anyway?

Answer Technically, nobody: We don't need the cosine function and can get by with only the sine function. On the other hand, it is convenient to have the cosine function because it starts at its highest point rather than at zero. These two functions and their relationship play important roles throughout mathematics.

The General Cosine Function

The general cosine function is

$$f(x) = A\cos[\omega(x - \alpha)] + C$$

Its graph is as follows:

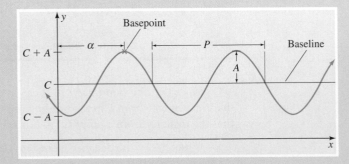

Note that the basepoint is at the highest point of the curve. All the constants have the same meaning as for the general sine curve:

- A is the **amplitude** (the height of each peak above the baseline).
- C is the **vertical offset** (height of the baseline).

- P is the **period,** or **wavelength** (the length of each cycle), and is related to ω by

    $P = 2\pi/\omega$ or $\omega = 2\pi/P$

- ω is the **angular frequency** (the number of cycles in every interval of length 2π).
- α is the **phase shift.**

Example 4 • Cash Flows into Stock Funds

The annual cash flow into stock funds (measured as a percentage of total assets) has fluctuated in cycles of approximately 40 years since 1955, when it was at a high point. The highs were roughly +15% of total assets, whereas the lows were roughly −10% of total assets.[1]

a. Model this cash flow with a cosine function of the time t in years, with $t = 0$ representing 1955.

b. Convert the answer in part (a) to a sine function model.

Solution

a. Cosine modeling is similar to sine modeling; we are seeking a function of the form

$$P(t) = A \cos[\omega(t - \alpha)] + C$$

- *Amplitude A and vertical offset C* The cash flow fluctuates between −10% and +15%. We can express this as a fluctuation of $A = 12.5$ about the average $C = 2.5$.
- *Period P* This is given as $P = 40$.
- *Angular frequency* ω We find ω from the formula

$$\omega = \frac{2\pi}{P} = \frac{2\pi}{40} = \frac{\pi}{20} \approx 0.157$$

- *Phase shift* α The basepoint is at the high point of the curve, and we are told that cash flow was at its high point at $t = 0$. Therefore, the basepoint occurs at $t = 0$, and so $\alpha = 0$.

    Putting the model together gives

    $$P(t) = A \cos[\omega(t - \alpha)] + C \approx 12.5 \cos(0.157t) + 2.5$$

where t is time in years since 1955.

b. To convert between a sine and cosine model, we can use one of the relationships given earlier. Let's use the formula

$$\cos x = \sin\left(x + \frac{\pi}{2}\right)$$

Therefore,

$$P(t) \approx 12.5 \cos(0.157t) + 2.5$$

$$= 12.5 \sin\left(0.157t + \frac{\pi}{2}\right) + 2.5$$

---

[1] SOURCE: Investment Company Institute/*New York Times,* February 2, 1997, p. F8.

Other Trigonometric Functions

The ratios and reciprocals of sine and cosine are given their own names.

Tangent, Cotangent, Secant, and Cosecant

Tangent $\quad \tan x = \dfrac{\sin x}{\cos x}$

Cotangent $\cot x = \mathrm{cotan}\, x = \dfrac{\cos x}{\sin x} = \dfrac{1}{\tan x}$

Secant $\quad \sec x = \dfrac{1}{\cos x}$

Cosecant $\quad \csc x = \mathrm{cosec}\, x = \dfrac{1}{\sin x}$

Trigonometric Regression

In the examples so far, we were given enough information to obtain a sine (or cosine) model directly. Often, however, we are given data that only *suggest* a sine curve. In such cases we can use regression to find the best-fit generalized sine (or cosine) curve.

🌴 Example 5 • Advertising

The cost, in millions of dollars, of a 30-second television ad during the Super Bowls from 1990 to 2001 can be modeled by the exponential function

$$C(t) = 0.522(1.14)^t$$

where $t = 0$ represents 1990.[2] This model indicates a 14% rate of inflation. If we adjust the actual costs for inflation at this rate, we obtain the following data (shown in constant 1990 dollars) and graph (Figure 9).[3]

t	0	1	2	3	4	5	6	7	8	9	10	11
Adjusted Cost	0.60	0.60	0.54	0.50	0.50	0.57	0.48	0.48	0.44	0.49	0.58	0.53

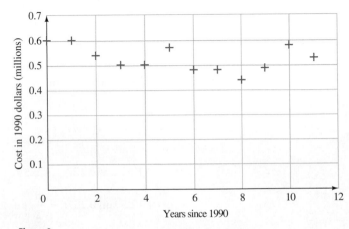

Figure 9

---

[2]The model was obtained using exponential regression. SOURCE: *New York Times*, January 26, 2001, p. C1.

[3]1991 and 1993 figures have been decreased by 0.06 and 0.04, respectively, to avoid regression errors on the TI-

a. Use technology to find the best-fit sine curve of the form $S(t) = A \sin[\omega(t - \alpha)] + C$.

b. Use your model to predict the (actual) cost of a Super Bowl ad in 2002.

Solution

a. We will solve this using the TI-83 and Excel.

Graphing Calculator

The TI-83 has a built-in sine regression utility. As with the other forms of regression discussed in Chapter 2, we start by entering the coordinates of the data points in the lists L_1 and L_2, then press $\boxed{\text{STAT}}$, select CALC, and choose option C: SinReg. Pressing $\boxed{\text{ENTER}}$ gives the sine regression equation in the home screen (we have rounded the coefficients to three decimal places):

$$S(t) \approx 0.052 \sin(1.338t + 0.977) + 0.518$$

Although this is not exactly in the form we want, we can rewrite it:

$$S(t) \approx 0.052 \sin\left[1.338\left(t + \frac{0.977}{1.338}\right)\right] + 0.518$$

$$\approx 0.052 \sin[1.338(t + 0.730)] + 0.518$$

The graph is shown in Figure 10a. This model gives a period of approximately

$$P = \frac{2\pi}{\omega} \approx \frac{2\pi}{1.338} \approx 4.70 \text{ years}$$

Excel

We set up our worksheet in the usual way, as shown below. For our initial guesses, let's roughly estimate the parameters from the graph. The amplitude is around $A = 0.1$, and the vertical offset is roughly $C = 0.55$. The period seems to be around $4-5$ years, so let's choose $P = 4.5$. This gives $\omega = 2\pi/P \approx 1.4$. Finally, let's take $\alpha = 0$ to start.

	B	C	D	E	F	G	H
1	C (Observed)	C (Predicted)	Residue^2	A	Omega	Alpha	C
2	0.6	=$E$2*SIN($F$2*(A2-$G$2))+$H$2	=(C2-B2)^2	0.1	1.4	0	0.55
3	0.6						
4	0.54						
5	0.5						
6	0.5						
7	0.57				SSE		
8	0.48				=SUM(D2:D13)		
9							

Solver gives us the following model (we have rounded the coefficients to three decimal places):

$$S(t) \approx 0.054 \sin[1.319(t + 0.607)] + 0.519$$

with a period of approximately

$$P = \frac{2\pi}{\omega} \approx \frac{2\pi}{1.319} \approx 4.76 \text{ years}$$

The graph is shown in Figure 10(b).

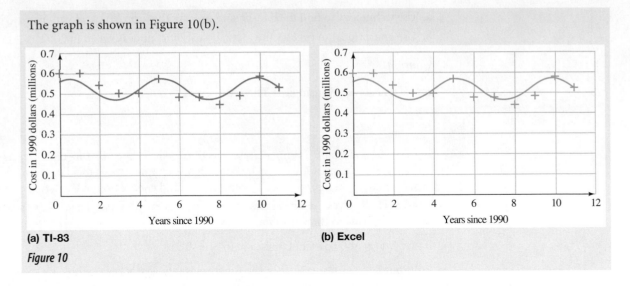

(a) TI-83

(b) Excel

Figure 10

Question Why do the results from TI-83 and Excel differ?

Answer Not all regression algorithms are identical, and it seems that the TI-83's algorithm is not quite as good as Excel's Solver at finding the best-fit sine curve. Indeed, the value for SSE for the TI-83 regression curve is about 0.015, whereas it is around 0.013 for the Excel curve, indicating a better fit.[4]

b. In 2002, $t = 12$. Substituting in the regression equation (unrounded coefficients) we get $S(12) \approx 0.468$ (TI-83) or $S(12) \approx 0.475$ (Excel). These figures are the *uninflated* cost in millions of 1990 dollars. To obtain the cost in 2002 dollars, we must subject this figure to 12 years' worth of inflation at 14%, obtaining (for the Excel figure)

$$\text{predicted cost} = 0.475(1.14)^{12} \approx \$2.29 \text{ million}[5]$$

Note that Solver estimated the period for us in Example 5. However, in many situations we know what to use for the period beforehand: For example, we expect sales of snow shovels to fluctuate according to an annual cycle. Thus, if we were using regression to fit a regression sine or cosine curve to snow shovel sales data, we would set $P = 1$ year and have Solver estimate only the remaining coefficients: A, C, and α.

Internet Topic: More on Trigonometric Functions
At the Web site, you will find further discussion of the graphs of the trigonometric functions, their relationship to right triangles, and some exercises. Follow the path
Web Site → Online Text → Trigonometric Functions and Calculus → The Six Trigonometric Functions

---

[4]Actually, the TI-83 algorithm seems problematic and tends to fail (giving an error message) on many sets of data. (See note 3.)

[5]The actual 2002 cost was $2.25 million.

9.1 EXERCISES

In Exercises 1–12, graph the given functions or pairs of functions on the same set of axes.

a. Sketch the curves without any technological help by consulting the discussion in Example 1.

b. Use technology to check your sketches.

1. $f(t) = \sin(t)$; $g(t) = 3\sin(t)$

2. $f(t) = \sin(t)$; $g(t) = 2.2\sin(t)$

3. $f(t) = \sin(t)$; $g(t) = \sin(t - \pi/4)$

4. $f(t) = \sin(t)$; $g(t) = \sin(t + \pi)$

5. $f(t) = \sin(t)$; $g(t) = \sin(2t)$

6. $f(t) = \sin(t)$; $g(t) = \sin(-t)$

7. $f(t) = 2\sin[3\pi(t - 0.5)] - 3$

8. $f(t) = 2\sin[3\pi(t + 1.5)] + 1.5$

9. $f(t) = \cos(t)$; $g(t) = 5\cos[3(t - 1.5\pi)]$

10. $f(t) = \cos(t)$; $g(t) = 3.1\cos(3t)$

11. $f(t) = \cos(t)$; $g(t) = -2.5\cos(t)$

12. $f(t) = \cos(t)$; $g(t) = 2\cos(t - \pi)$

In Exercises 13–18, model each curve with a sine function. (Note that not all are drawn with the same scale on the two axes.)

13.

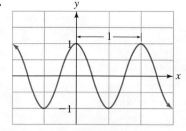

14.

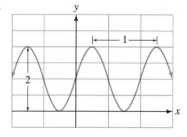

15.

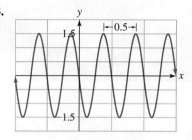

16.

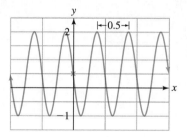

17.

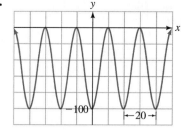

18.

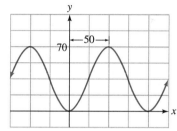

In Exercises 19–24, model each curve with a cosine function. (Note that not all are drawn with the same scale on the two axes.)

19.

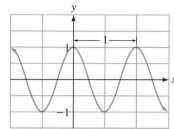

20.

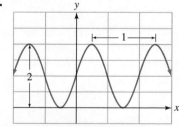

21.

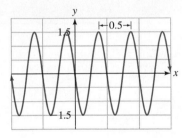

22.

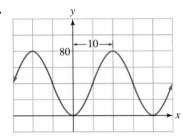

23.

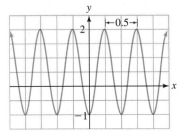

24.

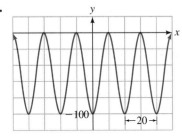

In Exercises 25–28, use the conversion formula $\cos x = \sin(\pi/2 - x)$ to replace each expression by a sine function.

25. $f(t) = 4.2 \cos(2\pi t) + 3$

26. $f(t) = 3 - \cos(t - 4)$

27. $g(x) = 4 - 1.3 \cos[2.3(x - 4)]$

28. $g(x) = 4.5 \cos[2\pi(3x - 1)] + 7$

Some Identities In Exercises 29 and 30, starting with the identity $\sin^2 x + \cos^2 x = 1$ and then dividing both sides of the equation by a suitable trigonometric function, derive the trigonometric identities.

29. $\sec^2 x = 1 + \tan^2 x$

30. $\csc^2 x = 1 + \cot^2 x$

Exercises 31–38 are based on the **addition formulas:**

$$\sin(x + y) = \sin x \cos y + \cos x \sin y$$

$$\sin(x - y) = \sin x \cos y - \cos x \sin y$$

$$\cos(x + y) = \cos x \cos y - \sin x \sin y$$

$$\cos(x - y) = \cos x \cos y + \sin x \sin y$$

31. Calculate $\sin(\pi/3)$, given that $\sin(\pi/6) = 1/2$ and $\cos(\pi/6) = \sqrt{3}/2$.

32. Calculate $\cos(\pi/3)$, given that $\sin(\pi/6) = 1/2$ and $\cos(\pi/6) = \sqrt{3}/2$.

33. Use the formula for $\sin(x + y)$ to obtain the identity $\sin(t + \pi/2) = \cos t$.

34. Use the formula for $\cos(x + y)$ to obtain the identity $\cos(t - \pi/2) = \sin t$.

35. Show that $\sin(\pi - x) = \sin x$.

36. Show that $\cos(\pi - x) = -\cos x$.

37. Use the addition formulas to express $\tan(x + \pi)$ in terms of $\tan(x)$.

38. Use the addition formulas to express $\cot(x + \pi)$ in terms of $\cot(x)$.

APPLICATIONS

39. Sunspot Activity The activity of the Sun (sunspots, solar flares, and coronal mass ejection) fluctuates in cycles of around 10–11 years. Sunspot activity can be modeled by the following function:

$$N(t) = 57.7 \sin[0.602(t - 1.43)] + 58.8$$

where t is the number of years since January 1, 1997, and $N(t)$ is the number of sunspots observed at time t.

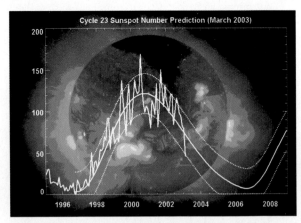

SOURCE: http://science.msfc.nasa.gov/ssl/pad/solar/images/ssn_predict_l.gif

a. What is the period of sunspot activity according to this model? (Round your answer to the nearest 0.1 year.)

b. What is the maximum number of sunspots observed? What is the minimum number? (Round your answers to the nearest sunspot.)

c. When, to the nearest year, is sunspot activity next expected to reach a high point?

The model is based on a regression obtained from predicted data for 1997–2006 and the mean historical period of sunspot activity from 1755 to 1995. Source: NASA Science Directorate; Marshall Space Flight Center (http://science.nasa.gov/ssl/pad/solar/predict.html), August 2002.

40. Solar Emissions The following model gives the flux of radio emission from the Sun:

$$F(t) = 49.6 \sin[0.602(t - 1.48)] + 111$$

where t is the number of years since January 1, 1997, and $F(t)$ is the flux of solar emissions of a specified wavelength at time t.

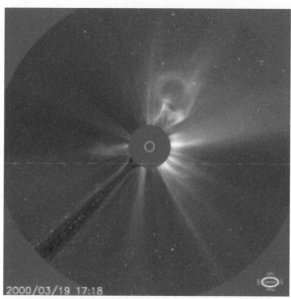

Source: http://science.nasa.gov/headlines/images/ssupdate/c2_19mar00_big.gif

a. What is the period of radio activity according to this model? (Round your answer to the nearest 0.1 year.)

b. What is the maximum flux of radio emissions? What is the minimum flux? (Round your answers to the nearest whole number.)

c. When to the nearest year is radio activity next expected to reach a low point?

Flux is measured at a wavelength of 10.7 cm. Source: NASA Science Directorate; Marshall Space Flight Center (http://science.nasa.gov/ssl/pad/solar/predict.html), August 2002.

 41. Computer Sales Sales of computers are subject to seasonal fluctuations. Computer City's sales of computers in 1995 and 1996 can be approximated by the function

$$s(t) = 0.106 \sin(1.39t + 1.61) + 0.455 \qquad (1 \le t \le 8)$$

where t is time in quarters ($t = 1$ represents the end of the first quarter of 1995) and $s(t)$ is computer sales (quarterly revenue) in billions of dollars.

a. Use technology to plot sales versus time from the end of the first quarter of 1995 through the end of the last quarter of 1996. Then use your graph to estimate the value of t and the quarter during which sales were lowest and highest.

b. Estimate Computer City's maximum and minimum quarterly revenue from computer sales.

c. Indicate how the answers to part (b) can be obtained directly from the equation for $s(t)$.

The model is based on a regression of data. Constants are rounded to three significant digits. Source: *New York Times,* January 8, 1997, p. D1.

 42. Computer Sales Repeat Exercise 41, using the following model for CompUSA's quarterly sales of computers:

$$s(t) = 0.0778 \sin(1.52t + 1.06) + 0.591$$

The model is based on a regression of data. Constants are rounded to three significant digits. Source: *New York Times,* January 8, 1997, p. D1.

43. Computer Sales (Based on Exercise 41, but no technology required) Computer City's sales of computers in 1995 and 1996 can be approximated by the function

$$s(t) = 0.106 \sin(1.39t + 1.61) + 0.455 \qquad (1 \le t \le 8)$$

where t is time in quarters ($t = 1$ represents the end of the first quarter of 1995) and $s(t)$ is computer sales (quarterly revenue) in billions of dollars. Calculate the amplitude, the vertical offset, the phase shift, the angular frequency, and the period and interpret the results.

The model is based on a regression of data. Constants are rounded to three significant digits. Source: *New York Times,* January 8, 1997, p. D1.

44. Computer Sales Repeat Exercise 43, using the following model for CompUSA's quarterly sales of computers:

$$s(t) = 0.0778 \sin(1.52t + 1.06) + 0.591$$

45. Biology Sigatoka leaf spot is a plant disease that affects bananas. In an infected plant, the percentage of leaf area affected varies from a low of around 5% at the start of each year to a high of around 20% at the middle of each year. Use the sine function to model the percentage of leaf area affected by Sigatoka leaf spot t weeks since the start of a year.

Based on graphical data. Source: American Phytopathological Society (http://www.apsnet.org/education/AdvancedPlantPath/Topics/Epidemiology/CyclicalNature.htm), July 2002.

46. Biology Apple powdery mildew is an epidemic that affects apple shoots. In a new infection, the percentage of apple shoots infected varies from a low of around 10% at the start of May to a high of around 60% 6 months later. Use the sine function to model the percentage of apple shoots affected by apple powdery mildew t months since the start of a year.

Based on graphical data. Source: American Phytopathological Society (http://www.apsnet.org/education/AdvancedPlantPath/Topics/Epidemiology/CyclicalNature.htm), July 2002.

47. Sales Fluctuations Sales of General Motors cars and light trucks in 1996 fluctuated from a high of $95 billion in October ($t = 0$) to a low of $80 billion in April ($t = 6$). Construct a sinusoidal model for the monthly sales $s(t)$ of General Motors.

These are rough figures based on the percentage of the market held by GM. Source: *New York Times,* January 9, 1997, p. D4.

48. **Sales Fluctuations** Sales of cypods (one-bedroom units) in the city-state of Utarek, Mars, fluctuate from a low of 5 units per week each February 1 ($t = 1$) to a high of 35 units per week each August 1 ($t = 7$). Use a sine function to model the weekly sales $s(t)$ of cypods, where t is time in months.
   SOURCE: See http://www.marsnext.com/comm/zonars.html

49. **Sales Fluctuations** Repeat Exercise 47, but this time use a cosine function for your model.

50. **Sales Fluctuations** Repeat Exercise 48, but this time use a cosine function for your model.

51. **Tides** The depth of water at my favorite surfing spot varies from 5 to 15 feet, depending on the time. Last Sunday high tide occurred at 5:00 A.M., and the next high tide occurred at 6:30 P.M. Use a sine function model to describe the depth of water as a function of time t in hours since midnight on Sunday morning.

52. **Tides** Repeat Exercise 51, using data from the depth of water at my second favorite surfing spot, where the tide last Sunday varied from a low of 6 feet at 4:00 A.M. to a high of 10 feet at noon.

53. **Inflation** The uninflated cost of Dugout brand snow shovels currently varies from a high of $10 on January 1 ($t = 0$) to a low of $5 on July 1 ($t = 0.5$).
   a. Assuming that this trend were to continue indefinitely, calculate the uninflated cost $u(t)$ of Dugout snow shovels as a function of time t in years. (Use a sine function.)
   b. Assuming a 4% annual rate of inflation in the cost of snow shovels, the cost of a snow shovel t years from now, adjusted for inflation, will be 1.04^t times the uninflated cost. Find the cost $c(t)$ of Dugout snow shovels as a function of time t.

54. **Deflation** Sales of my exclusive 1997 vintage Chateau Petit Mont Blanc vary from a high of ten bottles per day on April 1 ($t = 0.25$) to a low of four bottles per day on October 1.
   a. Assuming that this trend were to continue indefinitely, find the undeflated sales $u(t)$ of Chateau Petit Mont Blanc as a function of time t in years. (Use a sine function.)
   b. Regrettably, ever since that undercover exposé of my wine-making process, sales of Chateau Petit Mont Blanc have been declining at an annual rate of 12%. Using Exercise 53 as a guide, write down a model for the deflated sales $s(t)$ of Chateau Petit Mont Blanc t years from now.

   For Exercises 55–60, use technology to solve.

55. **Consumer Spending** The following table shows the annual percent change in consumer spending compared with the preceding year for even-numbered years ($t = 0$ represents 1990):

Year, t	0	2	4	6	8	10
Change in Spending (%)	7.3	8.0	7	4.5	5.5	7.2

Figures are approximate and are based on purchases of general merchandise. SOURCE: *New York Times*, February 6, 2001, p. C1.

   a. Plot the data and *roughly* estimate the period P and the parameters C, A, and α.
   b. Find the best-fit sine curve approximating the given data. [You may have to use your estimates from part (a) as initial guesses if you are using Solver.] Plot the given data together with the regression curve. (Round coefficients to three decimal places.)
   c. Complete the following: Based on the regression model, one might argue that the annual percent change in consumer spending shows a pattern that repeats itself every _____ years, from a low of __% to a high of __%. (Round answers to one decimal place.)

56. **Consumer Spending** Repeat Exercise 55, using the following data for the odd-numbered years (but not using the data for the even-numbered years):

Year, t	1	3	5	7	9	11
Change in Spending (%)	5.0	7.0	6.0	4.8	8.0	6.1

Some figures were from the second or third quarter of the stated year. SOURCE: *New York Times*, February 6, 2001, p. C1.

Music Musical sounds exhibit the same kind of periodic behavior as the trigonometric functions. High-pitched notes have short periods (less than 1/1000 second), whereas the lowest audible notes have periods of about 1/100 second. Some electronic synthesizers work by superimposing (adding) sinusoidal functions of different frequencies to create different textures. Exercises 57–60 show some examples of how superposition can be used to create interesting periodic functions.

57. **Sawtooth Wave**
   a. Graph the following functions in a window with $-7 \le x \le 7$ and $-1.5 \le y \le 1.5$.

   $$y_1 = \frac{2}{\pi} \cos x$$

   $$y_3 = \frac{2}{\pi} \cos x + \frac{2}{3\pi} \cos 3x$$

   $$y_5 = \frac{2}{\pi} \cos x + \frac{2}{3\pi} \cos 3x + \frac{2}{5\pi} \cos 5x$$

   b. Following the pattern established above, give a formula for y_{11} and graph it in the same window.
   c. How would you modify y_{11} to approximate a sawtooth wave with an amplitude of 3 and a period of 4π?

58. **Square Wave** Repeat Exercise 57, using sine functions in place of cosine functions in order to approximate a square wave.

59. **Harmony** If we add two sinusoidal functions with frequencies that are simple ratios of each other, the result is a pleasing sound. The following function models two notes an octave apart together with the intermediate fifth:

   $$y = \cos x + \cos 1.5x + \cos 2x$$

   Graph this function in the window $0 \le x \le 20$ and $-3 \le y \le 3$ and estimate the period of the resulting wave.

60. Discord If we add two sinusoidal functions with similar but unequal frequencies, the result is a function that "pulsates," or exhibits "beats." (Piano tuners and guitar players use this phenomenon to help them tune an instrument.) Graph the function

$$y = \cos(x) + \cos(0.9x)$$

in the window $-50 \leq x \leq 50$ and $-2 \leq y \leq 2$ and estimate the period of the resulting wave.

COMMUNICATION AND REASONING EXERCISES

61. What are the seasonal highs and lows for sales of a commodity modeled by a function of the form $s(t) = A \sin(2\pi t) + B$ (A, B constants)?

62. Your friend has come up with the following model for the choral society's Tupperware stock inventory: $r(t) = 4 \sin[2\pi(t - 2)/3] + 2.3$, where t is time in weeks and $r(t)$ is the number of items in stock. Comment on the model.

63. Your friend is telling everybody that all six trigonometric functions can be obtained from the single function $\sin x$. Is he correct? Explain your answer.

64. Another friend claims that all six trigonometric functions can be obtained from the single function $\cos x$. Is she correct? Explain your answer.

65. If weekly sales of sodas at a movie theater are given by $s(t) = A + B \cos(\omega t)$, what is the largest B can be? Explain your answer.

66. Complete the following: If the cost of an item is given by $c(t) = A + B \cos[\omega(t - \alpha)]$, then the cost fluctuates by _____ with a period of _____ about a base of _____, peaking at time $t = $ _____.

9.2 *Derivatives of Trigonometric Functions and Applications*

We start with the derivatives of the sine and cosine functions.

Derivatives of the Sine and Cosine Functions

$$\frac{d}{dx}[\sin x] = \cos x$$

$$\frac{d}{dx}[\cos x] = -\sin x \qquad \text{Notice the sign change.}$$

Quick Examples

1. $\dfrac{d}{dx}[x \cos x] = 1 \cdot \cos x + x \cdot (-\sin x)$ Product rule: $x \cos x$ is a product.*

$$= \cos x - x \sin x$$

2. $\dfrac{d}{dx}\left[\dfrac{x^2 + x}{\sin x}\right] = \dfrac{(2x + 1)(\sin x) - (x^2 + x)(\cos x)}{\sin^2 x}$ Quotient rule

---

*Apply the calculation thought experiment: If we were to compute $x \cos x$, the last operation we would perform is the multiplication of x and $\cos x$. Hence, $x \cos x$ is a product.

We justify these formulas at the end of the section, but we can see right away that they are plausible by examining Figure 11, which shows the graphs of the sine and cosine functions together with their derivatives.

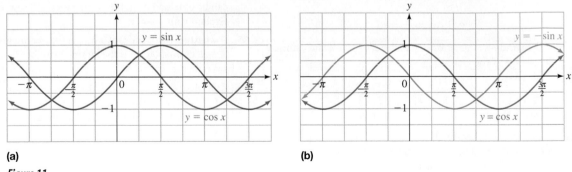

(a)

(b)

Figure 11

Notice, for instance, that in Figure 11(a) the graph of $\sin x$ is rising most rapidly when $x = 0$, corresponding to the maximum value of its derivative, $\cos x$. When $x = \pi/2$, the graph of $\sin x$ levels off, so that its derivative, $\cos x$, is zero. Another point to notice: Since periodic functions (such as sine and cosine) repeat their behavior, their derivatives must also be periodic.

Just as with logarithmic and exponential functions, the chain rule can be used to find more general derivatives.

Derivatives of Sines and Cosines of Functions

Original Rule	Generalized Rule	In Words
$\dfrac{d}{dx}[\sin x] = \cos x$	$\dfrac{d}{dx}[\sin u] = \cos u\,\dfrac{du}{dx}$	The derivative of the sine of a quantity is the cosine of that quantity, times the derivative of that quantity.
$\dfrac{d}{dx}[\cos x] = -\sin x$	$\dfrac{d}{dx}[\cos u] = -\sin u\,\dfrac{du}{dx}$	The derivative of the cosine of a quantity is negative sine of that quantity, times the derivative of that quantity.

Quick Examples

1. $\dfrac{d}{dx}[\sin(3x^2 - 1)] = \cos(3x^2 - 1)\,\dfrac{d}{dx}[3x^2 - 1]$ $u = 3x^2 - 1^*$

   $= 6x\cos(3x^2 - 1)$ We placed the $6x$ in front—see below.

2. $\dfrac{d}{dx}[\cos(x^3 + x)] = -\sin(x^3 + x)\,\dfrac{d}{dx}[x^3 + x]$ $u = x^3 + x$

   $= -(3x^2 + 1)\sin(x^3 + x)$

*If we were to evaluate $\sin(3x^2 - 1)$, the last operation we would perform is taking the sine of a quantity. Thus, the calculation thought experiment tells us that we are dealing with the *sine of a quantity*, and we use the generalized rule.

Note Avoid writing ambiguous expressions like $\cos(3x^2 - 1)(6x)$. Does this mean

$$\cos[(3x^2 - 1)(6x)]$$ The cosine of the quantity $(3x^2 - 1)(6x)$

or does it mean

$$[\cos(3x^2 - 1)](6x)$$ The product of $\cos(3x^2 - 1)$ and $6x$

To avoid the ambiguity, place the $6x$ in front of the cosine expression and write

$$6x\cos(3x^2 - 1)$$ The product of $6x$ and $\cos(3x^2 - 1)$

Example 1 • Derivatives of Trigonometric Functions

Find the derivatives of the following functions.

a. $f(x) = \sin^2 x$ **b.** $g(x) = \sin^2(x^2)$ **c.** $h(x) = e^{-x}\cos(2x)$

Solution

a. Recall that $\sin^2 x = (\sin x)^2$. The calculation thought experiment tells us that $f(x)$ is the square of a quantity.[6] Therefore, we use the chain rule (or generalized power rule) for differentiating the square of a quantity:

$$\frac{d}{dx}[u^2] = 2u\frac{du}{dx}$$

$$\frac{d}{dx}[(\sin x)^2] = 2(\sin x)\frac{d(\sin x)}{dx}$$ $u = \sin x$

$$= 2\sin x\cos x$$

Thus, $f'(x) = 2\sin x\cos x$.

b. We rewrite the function $g(x) = \sin^2(x^2)$ as $[\sin(x^2)]^2$. Since $g(x)$ is the square of a quantity, we have

$$\frac{d}{dx}[\sin^2(x^2)] = \frac{d}{dx}(s[\sin(x^2)]^2)$$ Rewrite $\sin^2(—)$ as $[\sin(—)]^2$.

$$= 2\sin(x^2)\frac{d[\sin(x^2)]}{dx}$$ $\frac{d}{dx}[u^2] = 2u\frac{du}{dx}$ with $u = \sin(x^2)$

$$= 2\sin(x^2)\cdot\cos(x^2)\cdot 2x$$ $\frac{d}{dx}\sin u = \cos u\frac{du}{dx}$ with $u = x^2$

Thus, $g'(x) = 4x\sin(x^2)\cos(x^2)$.

c. Since $h(x)$ is the product of e^{-x} and $\cos(2x)$, we use the product rule:

$$h'(x) = (-e^{-x})\cos(2x) + e^{-x}\frac{d}{dx}[\cos(2x)]$$

$$= (-e^{-x})\cos(2x) - e^{-x}\sin(2x)\frac{d}{dx}[2x]$$ $\frac{d}{dx}[\cos u] = -\sin u\frac{du}{dx}$

$$= -e^{-x}\cos(2x) - 2e^{-x}\sin(2x)$$

$$= -e^{-x}[\cos(2x) + 2\sin(2x)]$$

---

[6]Notice the difference between $\sin^2 x$ and $\sin(x^2)$. The first is the square of $\sin x$, whereas the second is the sine of the quantity x^2.

Question What about the derivatives of the other four trigonometric functions?

Answer The remaining trigonometric functions are tan, cot, sec, and csc. Each is a ratio of sines and cosines, so we can use the quotient rule to find their derivatives. For example, we can find the derivative of tan as follows:

$$\frac{d}{dx}[\tan x] = \frac{d}{dx}\left[\frac{\sin x}{\cos x}\right]$$

$$= \frac{(\cos x)(\cos x) - (\sin x)(-\sin x)}{\cos^2 x}$$

$$= \frac{\cos^2 x + \sin^2 x}{\cos^2 x} = \frac{1}{\cos^2 x} = \sec^2 x$$

We ask you to derive the other three derivatives in the exercises. Here is a list of the derivatives of all six trigonometric functions and their chain rule variants.

Derivatives of the Trigonometric Functions

Original Rule	**Generalized Rule**
$\dfrac{d}{dx}[\sin x] = \cos x$	$\dfrac{d}{dx}[\sin u] = \cos u\,\dfrac{du}{dx}$
$\dfrac{d}{dx}[\cos x] = -\sin x$	$\dfrac{d}{dx}[\cos u] = -\sin u\,\dfrac{du}{dx}$
$\dfrac{d}{dx}[\tan x] = \sec^2 x$	$\dfrac{d}{dx}[\tan u] = \sec^2 u\,\dfrac{du}{dx}$
$\dfrac{d}{dx}[\cot x] = -\csc^2 x$	$\dfrac{d}{dx}[\cot u] = -\csc^2 u\,\dfrac{du}{dx}$
$\dfrac{d}{dx}[\sec x] = \sec x \tan x$	$\dfrac{d}{dx}[\sec u] = \sec u \tan u\,\dfrac{du}{dx}$
$\dfrac{d}{dx}[\csc x] = -\csc x \cot x$	$\dfrac{d}{dx}[\csc u] = -\csc u \cot u\,\dfrac{du}{dx}$

Quick Examples

1. $\dfrac{d}{dx}[\tan(x^2 - 1)] = \sec^2(x^2 - 1)\dfrac{d(x^2 - 1)}{dx}$ $u = x^2 - 1$

$$= 2x\sec^2(x^2 - 1)$$

2. $\dfrac{d}{dx}[\csc(e^{3x})] = -\csc(e^{3x})\cot(e^{3x})\dfrac{d(e^{3x})}{dx}$ $u = e^{3x}$

$$= -3e^{3x}\csc(e^{3x})\cot(e^{3x})$$ The derivative of e^{3x} is $3e^{3x}$.

Example 2 • Gas Heating Demand

In Section 9.1 we saw that seasonal fluctuations in temperature suggested a sine function. For instance, we can use the function

$$T = 60 + 25\sin\left[\frac{\pi}{6}(x - 4)\right]$$ $T =$ temperature in °F; $x =$ months since January 1

to model a temperature that fluctuates between 35°F on February 1 ($x = 1$) and 85°F on August 1 ($x = 7$) (Figure 12). The demand for gas at a utility company can be expected to fluctuate in a similar way because demand grows with increased heating requirements. A reasonable model might therefore be

$$G = 400 - 100 \sin\left[\frac{\pi}{6}(x - 4)\right]$$

Why did we subtract the sine term?

where G is the demand for gas in cubic yards per day. Find and interpret $G'(10)$.

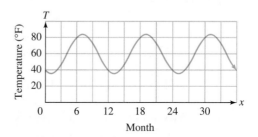

Figure 12

Solution First, we take the derivative of G:

$$G'(x) = -100 \cos\left[\frac{\pi}{6}(x - 4)\right] \cdot \frac{\pi}{6}$$

$$= -\frac{50\pi}{3} \cos\left[\frac{\pi}{6}(x - 4)\right] \text{ cubic yards per day per month}$$

Thus,

$$G'(10) = -\frac{50\pi}{3} \cos\left[\frac{\pi}{6}(10 - 4)\right]$$

$$= -\frac{50\pi}{3} \cos(\pi) = \frac{50\pi}{3}$$

Since $\cos \pi = -1$

Since the units of $G'(10)$ are cubic yards per day per month, we interpret the result as follows: On November 1 ($x = 10$), the daily demand for gas is increasing at a rate of $50\pi/3 \approx 52$ cubic yards per day per month. This is consistent with Figure 12, which shows the temperature decreasing on that date.

Now, where did these formulas for the derivatives of $\sin x$ and $\cos x$ come from? We first calculate the derivative of $\sin x$ from scratch, using the definition of the derivative:

$$\frac{d}{dx}[f(x)] = \lim_{h \to 0} \frac{f(x + h) - f(x)}{h}$$

$$\frac{d}{dx}[\sin x] = \lim_{h \to 0} \frac{\sin(x + h) - \sin x}{h}$$

We now use the addition formula in Exercise Set 9.1:

$$\sin(x + h) = \sin x \cos h + \cos x \sin h$$

Substituting this expression for $\sin(x + h)$ gives

$$\frac{d}{dx}[\sin x] = \lim_{h \to 0} \frac{\sin x \cos h + \cos x \sin h - \sin x}{h}$$

Grouping the first and third terms together and factoring out the term $\sin x$ gives

$$\frac{d}{dx}[\sin x] = \lim_{h \to 0} \frac{\sin x(\cos h - 1) + \cos x \sin h}{h}$$

$$= \lim_{h \to 0} \frac{\sin x(\cos h - 1)}{h} + \lim_{h \to 0} \frac{\cos x \sin h}{h} \qquad \text{Limit of a sum}$$

$$= \sin x \lim_{h \to 0} \frac{\cos h - 1}{h} + \cos x \lim_{h \to 0} \frac{\sin h}{h}$$

and we are left with two limits to evaluate. Calculating these limits analytically requires a little trigonometry.[7] Alternatively, we can get a good idea of what these two limits are by estimating them numerically or graphically. Figures 13 and 14 show the graphs of $(\cos h - 1)/h$ and $(\sin h)/h$, respectively.

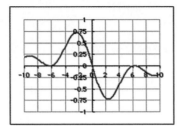

Figure 13   *Figure 14*

We find that

$$\lim_{h \to 0} \frac{\cos h - 1}{h} = 0 \quad \text{and} \quad \lim_{h \to 0} \frac{\sin h}{h} = 1$$

Therefore,

$$\frac{d}{dx}[\sin x] = (\sin x)(0) + (\cos x)(1) = \cos x$$

Question What about the derivative of the cosine function?

Answer Let's use the identity

$$\cos x = \sin\left(\frac{\pi}{2} - x\right)$$

from Section 9.1. If $y = \cos x = \sin(\pi/2 - x)$, then, using the chain rule, we have

$$\frac{dy}{dx} = \cos\left(\frac{\pi}{2} - x\right)\frac{d}{dx}\left[\frac{\pi}{2} - x\right]$$

$$= (-1)\cos\left(\frac{\pi}{2} - x\right)$$

$$= -\sin x \qquad\qquad \text{Using the identity } \cos\left(\frac{\pi}{2} - x\right) = \sin x$$

---

[7]You can find these calculations on the Web site by following the path

Web Site $\to$ Everything for Calculus $\to$ Chapter 9 $\to$ Proof of Some Trigonometric Limits

9.2 EXERCISES

In Exercises 1–28, find the derivatives of the given functions.

1. $f(x) = \sin x - \cos x$

2. $f(x) = \tan x - \sin x$

3. $g(x) = (\sin x)(\tan x)$

4. $g(x) = (\cos x)(\cot x)$

5. $h(x) = 2 \csc x - \sec x + 3x$

6. $h(x) = 2 \sec x + 3 \tan x + 3x$

7. $r(x) = x \cos x + x^2 + 1$

8. $r(x) = 2x \sin x - x^2$

9. $s(x) = (x^2 - x + 1)\tan x$

10. $s(x) = \dfrac{\tan x}{x^2 - 1}$

11. $t(x) = \dfrac{\cot x}{1 + \sec x}$

12. $t(x) = (1 + \sec x)(1 - \cos x)$

13. $k(x) = \cos^2 x$

14. $k(x) = \tan^2 x$

15. $j(x) = \sec^2 x$

16. $j(x) = \csc^2 x$

17. $p(x) = 2 + 5 \sin\left[\dfrac{\pi}{5}(x - 4)\right]$

18. $p(x) = 10 - 3 \cos\left[\dfrac{\pi}{6}(x + 3)\right]$

19. $u(x) = \cos(x^2 - x)$

20. $u(x) = \sin(3x^2 + x - 1)$

21. $v(x) = \sec(x^{2.2} + 1.2x - 1)$

22. $v(x) = \tan(x^{2.2} + 1.2x - 1)$

23. $w(x) = \sec x \tan(x^2 - 1)$

24. $w(x) = \cos x \sec(x^2 - 1)$

25. $y(x) = \cos(e^x) + e^x \cos x$

26. $y(x) = \sec(e^x)$

27. $z(x) = \ln|\sec x + \tan x|$

28. $z(x) = \ln|\csc x + \cot x|$

In Exercises 29–32, derive the given formulas from the derivatives of sine and cosine.

29. $\dfrac{d}{dx}[\sec x] = \sec x \tan x$

30. $\dfrac{d}{dx}[\cot x] = -\csc^2 x$

31. $\dfrac{d}{dx}[\csc x] = -\csc x \cot x$

32. $\dfrac{d}{dx}[\ln|\sec x|] = \tan x$

In Exercises 33–40, calculate the derivatives.

33. $\dfrac{d}{dx}[e^{-2x}\sin(3\pi x)]$

34. $\dfrac{d}{dx}[e^{5x}\sin(-4\pi x)]$

35. $\dfrac{d}{dx}[\sin(3x)]^{0.5}$

36. $\dfrac{d}{dx}\left[\cos\left(\dfrac{x^2}{x-1}\right)\right]$

37. $\dfrac{d}{dx}\left[\sec\left(\dfrac{x^3}{x^2-1}\right)\right]$

38. $\dfrac{d}{dx}\left[\left(\dfrac{\tan x}{2 + e^x}\right)^2\right]$

39. $\dfrac{d}{dx}\{[\ln|x|][\cot(2x - 1)]\}$

40. $\dfrac{d}{dx}[\ln|\sin x - 2xe^{-x}|]$

In Exercises 41 and 42, investigate the differentiability of the given functions at the given points. If $f'(a)$ exists, give its approximate value.

41. $f(x) = |\sin x|$
   a. $a = 0$ **b.** $a = 1$

42. $f(x) = |\sin(1 - x)|$
   a. $a = 0$ **b.** $a = 1$

In Exercises 43–48, estimate the limits (a) numerically and (b) using L'Hospital's rule.

43. $\lim\limits_{x \to 0} \dfrac{\sin^2 x}{x}$

44. $\lim\limits_{x \to 0} \dfrac{\sin x}{x^2}$

45. $\lim\limits_{x \to 0} \dfrac{\sin(2x)}{x}$

46. $\lim\limits_{x \to 0} \dfrac{\sin x}{\tan x}$

47. $\lim\limits_{x \to 0} \dfrac{\cos x - 1}{x^3}$

48. $\lim\limits_{x \to 0} \dfrac{\cos x - 1}{x^2}$

In Exercises 49–52, find the indicated derivative using implicit differentiation.

49. $x = \tan y$; find $\dfrac{dy}{dx}$.

50. $x = \cos y$; find $\dfrac{dy}{dx}$.

51. $x + y + \sin(xy) = 1$; find $\dfrac{dy}{dx}$.

52. $xy + x \cos y = x$; find $\dfrac{dy}{dx}$.

APPLICATIONS

53. Cost The cost in dollars of Dig-It brand snow shovels is given by

$$c(t) = 3.5 \sin[2\pi(t - 0.75)]$$

where t is time in years since January 1, 2002. How fast, in dollars per week, is the cost increasing each October 1?

54. Sales Daily sales of Doggy brand cookies can be modeled by

$$s(t) = 400 \cos[2\pi(t - 2)/7]$$

cartons, where t is time in days since Monday morning. How fast are sales changing on Thursday morning?

55. Sunspot Activity The activity of the Sun can be approximated by the following model of sunspot activity:

$$N(t) = 57.7 \sin[0.602(t - 1.43)] + 58.8$$

where t is the number of years since January 1, 1997, and $N(t)$ is the number of sunspots observed at time t. Compute and interpret $N'(6)$.

The model is based on a regression obtained from predicted data for 1997–2006 and the mean historical period of sunspot activity from 1755 to 1995. SOURCE: NASA Science Directorate; Marshall Space Flight Center (http://science.nasa.gov/ssl/pad/solar/predict.htm), August 2002

56. Solar Emissions The following model gives the flux of radio emission from the Sun:

$$F(t) = 49.6 \sin[0.602(t - 1.48)] + 111$$

where t is the number of years since January 1, 1997, and $F(t)$ is the average flux of solar emissions of a specified wavelength at time t. Compute and interpret $F'(5.5)$.

Flux is measured at a wavelength of 10.7 cm. SOURCE: NASA Science Directorate; Marshall Space Flight Center (http://science.nasa.gov/ssl/pad/solar/predict.htm), August 2002

57. Inflation Taking a 3.5% rate of inflation into account, the cost of DigIn brand snow shovels is given by

$$c(t) = 1.035^t[0.8 \sin(2\pi t) + 10.2]$$

where t is time in years since January 1, 2002. How fast, in dollars per week, is the cost of DigIn shovels increasing on January 1, 2003?

58. Deflation Sales, in bottles per day, of my exclusive mass-produced 2002 vintage Chateau Petit Mont Blanc follow the function

$$s(t) = 4.5e^{-0.2t} \sin(2\pi t)$$

where t is time in years since January 1, 2002. How fast were sales rising or falling on January 1, 2003?

59. Tides The depth of water at my favorite surfing spot varies from 5 to 15 feet, depending on the time. Last Sunday high tide occurred at 5:00 A.M., and the next high tide occurred at 6:30 P.M.

a. Obtain a cosine model describing the depth of water as a function of time t in hours since 5:00 A.M. on Sunday morning.

b. How fast was the tide rising (or falling) at noon on Sunday?

60. Tides Repeat Exercise 59, using data from the depth of water at my other favorite surfing spot, where the tide last Sunday varied from a low of 6 feet at 4:00 A.M. to a high of 10 feet at noon. (As in Exercise 59, take t as time in hours since 5:00 A.M.)

61. Tilt of Earth's Axis The tilt of Earth's axis from its plane of rotation about the Sun oscillates between approximately 22.5° and 24.5° with a period of approximately 40,000 years. We know that 500,000 years ago the tilt of Earth's axis was 24.5°.

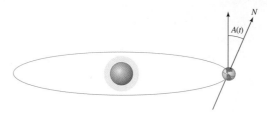

a. Which of the following functions best models the tilt of Earth's axis?

(A) $A(t) = 23.5 + 2 \sin\left(\dfrac{2\pi t + 500}{40}\right)$

(B) $A(t) = 23.5 + \cos\left(\dfrac{t + 500}{80\pi}\right)$

(C) $A(t) = 23.5 + \cos\left[\dfrac{2\pi(t + 500)}{40}\right]$

where $A(t)$ is the tilt in degrees and t is time in thousands of years, with $t = 0$ being the present time.

b. Use the model you selected in part (a) to estimate the rate at which the tilt was changing 150,000 years ago. (Round your answer to three decimal places and be sure to give the units of measurement.)

SOURCE: David Hodell, University of Florida/Juan Valesco/*New York Times*, February 16, 1999, p. F1.

62. Eccentricity of Earth's Orbit The eccentricity of Earth's orbit (that is, the deviation of Earth's orbit from a perfect circle) can be modeled by

$$E(t) = 0.025\left\{\cos\left[\dfrac{2\pi(t + 200)}{400}\right] + \cos\left[\dfrac{2\pi(t + 200)}{100}\right]\right\}$$

where $E(t)$ is the eccentricity and t is time in thousands of years, with $t = 0$ being the present time. What was the value of the eccentricity 200,000 years ago, and how fast was it changing?

This is a rough model based on the actual data. SOURCE: David Hodell, University of Florida/Juan Valesco/*New York Times*, February 16, 1999, p. F1.

COMMUNICATION AND REASONING EXERCISES

63. Complete the following: The rate of change of $f(x) = 3\sin(2x - 1) + 3$ oscillates between _____ and _____.

64. Complete the following: The rate of change of $g(x) = -3\cos(-x + 2) + 2x$ oscillates between _____ and _____.

65. Give two examples of a function $f(x)$ with the property that $f''(x) = -f(x)$.

66. Give two examples of a function $f(x)$ with the property that $f''(x) = -4f(x)$.

67. Give two examples of a function $f(x)$ with the property that $f'(x) = -f(x)$.

68. Give four examples of a function $f(x)$ with the property that $f^{(4)}(x) = f(x)$.

69. By referring to the graph of $f(x) = \cos x$, explain why $f'(x) = -\sin x$, rather than $\sin x$.

70. If A and B are constants, what is the relationship between $f(x) = A\cos x + B\sin x$ and its second derivative?

71. At what angle does the graph of $f(x) = \sin x$ depart from the origin?

72. At what angle does the graph of $f(x) = \cos x$ depart from the point $(0, 1)$?

9.3 Integrals of Trigonometric Functions and Applications

We saw in Section 6.1 that every calculation of a derivative gives us a calculation of an antiderivative. For instance, since we know that $\cos x$ is the derivative of $\sin x$, we can say that an antiderivative of $\cos x$ is $\sin x$:

$$\int \cos x \, dx = \sin x + C \qquad \text{An antiderivative of } \cos x \text{ is } \sin x.$$

The rules for the derivatives of sine, cosine, and tangent give us the following antiderivatives.

Indefinite Integrals of Some Trigonometric Functions

$$\int \cos x \, dx = \sin x + C \qquad \text{Because } \frac{d}{dx}[\sin x] = \cos x$$

$$\int \sin x \, dx = -\cos x + C \qquad \text{Because } \frac{d}{dx}[-\cos x] = \sin x$$

$$\int \sec^2 x \, dx = \tan x + C \qquad \text{Because } \frac{d}{dx}[\tan x] = \sec^2 x$$

Quick Examples

1. $\int (\sin x + \cos x) \, dx = -\cos x + \sin x + C$ Integral of sum = sum of integrals

2. $\int (4\sin x - \cos x) \, dx = -4\cos x - \sin x + C$ Integral of constant multiple

3. $\int (e^x - \sin x + \cos x) \, dx = e^x + \cos x + \sin x + C$

Example 1 • Substitution

Evaluate $\int (x + 3)\sin(x^2 + 6x)\, dx$.

Solution   There are two parenthetical expressions that we might replace with u. Notice, however, that the derivative of the expression $(x^2 + 6x)$ is $2x + 6$, which is twice the term $(x + 3)$ in front of the sine. Recall that we would like the derivative of u to appear as a factor. Thus, let's take $u = x^2 + 6x$.

$$
\begin{array}{|l|}
\hline
u = x^2 + 6x \\[2mm]
\dfrac{du}{dx} = 2x + 6 = 2(x + 3) \\[4mm]
dx = \dfrac{1}{2(x + 3)}\, du \\[2mm]
\hline
\end{array}
$$

Substituting into the integral, we get

$$
\int (x + 3)\sin(x^2 + 6x)\, dx = \int (x + 3)\sin u \left[\frac{1}{2(x + 3)} \right] du
$$

$$
= \int \frac{1}{2} \sin u\, du
$$

$$
= -\frac{1}{2} \cos u + C = -\frac{1}{2} \cos(x^2 + 6x) + C
$$

Example 2 • Definite Integrals

Compute the following.

a. $\displaystyle\int_0^\pi \sin x\, dx$ **b.** $\displaystyle\int_0^\pi x \sin(x^2)\, dx$

Solution

a. $\displaystyle\int_0^\pi \sin x\, dx = [-\cos x]_0^\pi = (-\cos \pi) - (-\cos 0)$

$$
= -(-1) - (-1) = 2
$$

Thus, the area under one "arch" of the sine curve is exactly 2 square units!

b. $\displaystyle\int_0^\pi x \sin(x^2)\, dx = \int_0^{\pi^2} \frac{1}{2} \sin u\, du$ After substituting $u = x^2$

$$
= \left[-\frac{1}{2} \cos u \right]_0^{\pi^2}
$$

$$
= \left[-\frac{1}{2} \cos(\pi^2) \right] - \left[-\frac{1}{2} \cos(0) \right]
$$

$$
= -\frac{1}{2} \cos(\pi^2) + \frac{1}{2} \qquad\qquad \cos(0) = 1
$$

We can approximate $\frac{1}{2} \cos(\pi^2)$ by a decimal or leave it in the above form, depending on what we want to do with the answer.

Antiderivatives of the Six Trigonometric Functions

The following table gives the indefinite integrals of the six trigonometric functions. (The first two we have already seen.)

Integrals of the Trigonometric Functions

$$\int \sin x \, dx = -\cos x + C$$

$$\int \cos x \, dx = \sin x + C$$

$$\int \tan x \, dx = -\ln |\cos x| + C \qquad\qquad\qquad \text{Shown below}$$

$$\int \cot x \, dx = \ln |\sin x| + C \qquad\qquad\qquad \text{See the exercise set.}$$

$$\int \sec x \, dx = \ln |\sec x + \tan x| + C \qquad\qquad \text{Shown below}$$

$$\int \csc x \, dx = -\ln |\csc x + \cot x| + C \qquad\qquad \text{See the exercise set.}$$

Question　Why does $\int \tan x \, dx = -\ln |\cos x| + C$?

Answer　We can write $\tan x$ as $(\sin x)/(\cos x)$ and put $u = \cos x$ in the integral:

$$\int \tan x \, dx = \int \frac{\sin x}{\cos x} \, dx$$

$$= -\int \frac{\sin x}{u} \, \frac{du}{\sin x}$$

$$= -\int \frac{du}{u}$$

$$= -\ln |u| + C$$

$$= -\ln |\cos x| + C$$

$$\boxed{\begin{aligned} u &= \cos x \\ \frac{du}{dx} &= -\sin x \\ dx &= -\frac{du}{\sin x} \end{aligned}}$$

Question　Why does $\int \sec x \, dx = \ln |\sec x + \tan x| + C$?

Answer　To get started we use a little "trick": Write $\sec x$ as

$$(\sec x)(\sec x + \tan x)/(\sec x + \tan x)$$

and put u equal to the denominator:

$$\int \sec x \, dx = \int \sec x \left(\frac{\sec x + \tan x}{\sec x + \tan x} \right) dx$$

$$= \int \sec x \, \frac{\sec x + \tan x}{u} \, \frac{du}{\sec x(\tan x + \sec x)}$$

$$= \int \frac{du}{u}$$

$$= \ln |u| + C$$

$$= \ln |\sec x + \tan x| + C$$

$$\boxed{\begin{aligned} u &= \sec x + \tan x \\ \frac{du}{dx} &= \sec x \tan x + \sec^2 x \\ &= \sec x(\tan x + \sec x) \\ dx &= \frac{du}{\sec x(\tan x + \sec x)} \end{aligned}}$$

Shortcuts

If a and b are constants with $a \neq 0$, then we have the following formulas. (All of them can be obtained using the substitution $u = ax + b$. They will appear in the exercises.)

Shortcuts: Integrals of Expressions Involving $(ax + b)$

Rule

$$\int \sin(ax + b) \, dx = -\frac{1}{a} \cos(ax + b) + C$$

$$\int \cos(ax + b) \, dx = \frac{1}{a} \sin(ax + b) + C$$

$$\int \tan(ax + b) \, dx = -\frac{1}{a} \ln |\cos(ax + b)| + C$$

$$\int \cot(ax + b) \, dx = \frac{1}{a} \ln |\sin(ax + b)| + C$$

$$\int \sec(ax + b) \, dx = \frac{1}{a} \ln |\sec(ax + b) + \tan(ax + b)| + C$$

$$\int \csc(ax + b) \, dx = -\frac{1}{a} \ln |\csc(ax + b) + \cot(ax + b)| + C$$

Quick Example

$$\int \sin(-4x) \, dx = \frac{1}{4} \cos(-4x) + C$$

$$\int \cos(x + 1) \, dx = \sin(x + 1) + C$$

$$\int \tan(-2x) \, dx = \frac{1}{2} \ln |\cos(-2x)| + C$$

$$\int \cot(3x - 1) \, dx = \frac{1}{3} \ln |\sin(3x - 1)| + C$$

$$\int \sec(9x) \, dx = \frac{1}{9} \ln |\sec(9x) + \tan(9x)| + C$$

$$\int \csc(x + 7) \, dx = -\ln |\csc(x + 7) + \cot(x + 7)| + C$$

Example 3 • Sales

The rate of sales of cypods (one-bedroom units) in the city-state of Utarek, Mars,[8] can be modeled by

$$s(t) = 7.5 \cos\left(\frac{\pi t}{6}\right) + 87.5 \text{ units per month}$$

where t is time in months since January 1. How many cypods are sold in a calendar year?

Solution Total sales over a calendar year are given by

$$\int_0^{12} s(t) \, dt = \int_0^{12} \left[7.5 \cos\left(\frac{\pi t}{6}\right) + 87.5\right] dt$$

$$= \left[7.5 \frac{6}{\pi} \sin\left(\frac{\pi t}{6}\right) + 87.5t\right]_0^{12} \qquad \text{We used a shortcut on the first term.}$$

$$= \left[7.5 \frac{6}{\pi} \sin(2\pi) + 87.5(12)\right] - \left[7.5 \frac{6}{\pi} \sin(0) + 87.5(0)\right]$$

$$= 87.5(12) \qquad\qquad\qquad \sin(2\pi) = \sin(0) = 0$$

$$= 1050 \text{ cypods}$$

---

[8]SOURCE: See http://www.marsnext.com/comm/zonars.html

✴ *Before we go on . . .* Would it have made any difference if we had computed total sales over the period [12, 24], [6, 18] or any interval of the form $[a, a + 12]$?

Using Integration by Parts with Trigonometric Functions

Example 4 • Integrating a Polynomial Times a Sine or Cosine

Calculate $\int (x^2 + 1)\sin(x + 1)\, dx$.

Solution We use the column method of integration by parts described in Section 7.1. Since differentiating $x^2 + 1$ makes it simpler, we put it in the D column and get the following table:

	D	I
$+$	$x^2 + 1$	$\sin(x + 1)$
	↘	
$-$	$2x$	$-\cos(x + 1)$
	↘	
$+$	2	$-\sin(x + 1)$
	↘	
$-\int$	0 →	$\cos(x + 1)$

[Notice that we used the shortcut formulas to repeatedly integrate $\sin(x + 1)$.] We can now read the answer from the table:

$$\int (x^2 + 1)\sin(x + 1)\, dx = (x^2 + 1)[-\cos(x + 1)] - 2x[-\sin(x + 1)] + 2[\cos(x + 1)] + C$$

$$= (-x^2 - 1 + 2)\cos(x + 1) + 2x\sin(x + 1) + C$$

$$= (-x^2 + 1)\cos(x + 1) + 2x\sin(x + 1) + C$$

Example 5 • Integrating an Exponential Times a Sine or Cosine

Calculate $\int e^x \sin x\, dx$.

Solution The integrand is the product of e^x and $\sin x$, so we put one in the D column and the other in the I column. For this example, it doesn't matter much which we put where.

	D	I
$+$	$\sin x$	e^x
	↘	
$-$	$\cos x$	e^x
	↘	
$+\int$	$-\sin x$ →	e^x

It looks like we're just spinning our wheels. Let's stop and see what we have:

$$\int e^x \sin x \, dx = e^x \sin x - e^x \cos x - \int e^x \sin x \, dx$$

At first glance, it appears that we are back where we started, still having to evaluate $\int e^x \sin x \, dx$. However, if we add this integral to both sides of the equation above, we can solve for it:

$$2 \int e^x \sin x \, dx = e^x \sin x - e^x \cos x + C$$

(Why $+C$?) So,

$$\int e^x \sin x \, dx = \frac{1}{2} e^x \sin x - \frac{1}{2} e^x \cos x + \frac{C}{2}$$

Since $C/2$ is just as arbitrary as C, we write C instead of $C/2$ and obtain

$$\int e^x \sin x \, dx = \frac{1}{2} e^x \sin x - \frac{1}{2} e^x \cos x + C$$

9.3 EXERCISES

In Exercises 1–28, evaluate the integrals.

1. $\int (\sin x - 2 \cos x) \, dx$

2. $\int (\cos x - \sin x) \, dx$

3. $\int (2 \cos x - 4.3 \sin x - 9.33) \, dx$

4. $\int \left(4.1 \sin x + \cos x - \frac{9.33}{x} \right) dx$

5. $\int \left(3.4 \sec^2 x + \frac{\cos x}{1.3} - 3.2e^x \right) dx$

6. $\int \left(\frac{3 \sec^2 x}{2} + 1.3 \sin x - \frac{e^x}{3.2} \right) dx$

7. $\int 7.6 \cos(3x - 4) \, dx$

8. $\int 4.4 \sin(-3x + 4) \, dx$

9. $\int x \sin(3x^2 - 4) \, dx$

10. $\int x \cos(-3x^2 + 4) \, dx$

11. $\int (4x + 2)\sin(x^2 + x) \, dx$

12. $\int (x + 1)[\cos(x^2 + 2x) + (x^2 + 2x)] \, dx$

13. $\int (x + x^2)\sec^2(3x^2 + 2x^3) \, dx$

14. $\int (4x + 2)\sec^2(x^2 + x) \, dx$

15. $\int (x^2)\tan(2x^3) \, dx$

16. $\int (4x)\tan(x^2) \, dx$

17. $\int 6 \sec(2x - 4) \, dx$

18. $\int 3 \csc(3x) \, dx$

19. $\int e^{2x} \cos(e^{2x} + 1) \, dx$

20. $\int e^{-x} \sin(e^{-x}) \, dx$

21. $\int_{-\pi}^{0} \sin x \, dx$

22. $\int_{\pi/2}^{\pi} \cos x \, dx$

23. $\int_{0}^{\pi/3} \tan x \, dx$

24. $\int_{\pi/6}^{\pi/2} \cot x \, dx$

25. $\int_{1}^{\sqrt{\pi+1}} x \cos(x^2 - 1) \, dx$

26. $\int_{0.5}^{(\pi+1)/2} \sin(2x - 1) \, dx$

27. $\int_{1/\pi}^{2/\pi} \frac{\sin\left(\frac{1}{x}\right)}{x^2} \, dx$

28. $\int_{0}^{\pi/3} \frac{\sin x}{\cos^2 x} \, dx$

In Exercises 29–32, derive each equation, where a and b are constants with $a \neq 0$.

29. $\int \cos(ax + b) \, dx = \frac{1}{a} \sin(ax + b) + C$

30. $\int \sin(ax + b) \, dx = -\frac{1}{a} \cos(ax + b) + C$

31. $\int \cot x \, dx = \ln |\sin x| + C$

32. $\int \csc x \, dx = -\ln |\csc x + \cot x| + C$

In Exercises 33–40, use the shortcut formulas given before Example 3 to mentally calculate the integrals.

33. $\int \sin(4x) \, dx$

34. $\int \cos(5x) \, dx$

35. $\int \cos(-x + 1) \, dx$

36. $\int \sin\left(\frac{1}{2}x\right) dx$

37. $\int \sin(-1.1x - 1) \, dx$

38. $\int \cos(4.2x - 1) \, dx$

39. $\int \cot(-4x) \, dx$

40. $\int \tan(6x) \, dx$

In Exercises 41–44, use geometry (not antiderivatives) to compute the integrals. (*Hint:* First draw the graph.)

41. $\int_{-\pi/2}^{\pi/2} \sin x \, dx$

42. $\int_{0}^{\pi} \cos x \, dx$

43. $\int_{0}^{2\pi} (1 + \sin x) \, dx$

44. $\int_{0}^{2\pi} (1 + \cos x) \, dx$

In Exercises 45–52, use integration by parts to evaluate the integrals.

45. $\int x \sin x \, dx$

46. $\int x^2 \cos x \, dx$

47. $\int x^2 \cos(2x) \, dx$

48. $\int (2x + 1)\sin(2x - 1) \, dx$

49. $\int e^{-x} \sin x \, dx$

50. $\int e^{2x} \cos x \, dx$

51. $\int_0^\pi x^2 \sin x \, dx$

52. $\int_0^{\pi/2} x \cos x \, dx$

Recall from Section 7.3 that the average of a function $f(x)$ on an interval $[a, b]$ is

$$\bar{f} = \frac{1}{b - a} \int_a^b f(x) \, dx$$

In Exercises 53 and 54, find the averages of the functions over the given intervals. Plot each function and its average on the same graph.

53. $f(x) = \sin x$ over $[0, \pi]$

54. $f(x) = \cos(2x)$ over $[0, \pi/4]$

In Exercises 55–58, decide whether each integral converges. (See Section 7.5.) If the integral converges, compute its value.

55. $\int_0^{+\infty} \sin x \, dx$

56. $\int_0^{+\infty} \cos x \, dx$

57. $\int_0^{+\infty} e^{-x} \cos x \, dx$

58. $\int_0^{+\infty} e^{-x} \sin x \, dx$

APPLICATIONS

59. Varying Cost The cost of producing a bottle of suntan lotion is changing at a rate of $0.04 - 0.1 \sin[(\pi/26)(t - 25)]$ dollars per week, t weeks after January 1. If it cost $1.50 to produce a bottle 12 weeks into the year, find the cost $C(t)$ at time t.

60. Varying Cost The cost of producing a box of holiday tree decorations is changing at a rate of $0.05 + 0.4 \cos[(\pi/6)(t - 11)]$ dollars per month, t months after January 1. If it cost $5 to produce a box on June 1, find the cost $C(t)$ at time t.

61. Pets My dog Miranda is running back and forth along a 12-foot stretch of garden in such a way that her velocity t seconds after she began is

$$v(t) = 3\pi \cos\left[\frac{\pi}{2}(t - 1)\right] \text{ feet/second}$$

How far is she from where she began 10 seconds after starting the run?

62. Pets My cat Prince Sadar is pacing back and forth along his favorite window ledge in such a way that his velocity t seconds after he began is

$$v(t) = -\frac{\pi}{2} \sin\left[\frac{\pi}{4}(t - 2)\right] \text{ feet/second}$$

How far is he from where he began 10 seconds after starting to pace?

For Exercises 63–68, recall from Section 7.3 that the average of a function $f(x)$ on an interval $[a, b]$ is

$$\bar{f} = \frac{1}{b - a} \int_a^b f(x) \, dx$$

63. Sunspot Activity The activity of the Sun (sunspots, solar flares, and coronal mass ejection) fluctuates in cycles of around 10–11 years. Sunspot activity can be modeled by the following function:

$$N(t) = 57.7 \sin[0.602(t - 1.43)] + 58.8$$

where t is the number of years since January 1, 1997, and $N(t)$ is the number of sunspots observed at time t. Estimate the average number of sunspots visible over the 2-year period 2002–2003. (Round your answer to the nearest whole number.)

The model is based on a regression obtained from predicted data for 1997–2006 and the mean historical period of sunspot activity from 1755 to 1995. Source: NASA Science Directorate; Marshall Space Flight Center (http://science.nasa.gov/ssl/pad/solar/predict.htm), August 2002.

64. Solar Emissions The following model gives the flux of radio emissions from the Sun:

$$F(t) = 49.6 \sin[0.602(t - 1.48)] + 111$$

where t is the number of years since January 1, 1997, and $F(t)$ is the flux of solar emissions of a specified wavelength at time t.

Estimate the average flux of radio emissions over the 5-year period 2001–2005. (Round your answer to the nearest whole number.)

Flux is measured at a wavelength of 10.7 cm. Source: NASA Science Directorate; Marshall Space Flight Center (http://science.nasa.gov/ssl/pad/solar/predict.htm), August 2002.

65. Biology Sigatoka leaf spot is a plant disease that affects bananas. In an infected plant, the percentage of leaf area affected varies from a low of around 5% at the start of each year to a high of around 20% at the middle of each year. Use a sine function model of the percentage of leaf area affected by Sigatoka leaf spot t weeks since the start of a year to estimate to the nearest 0.1% the average percentage of leaf area affected in the first quarter (13 weeks) of a year.

Based on graphical data. Source: American Phytopathological Society (http://www.apsnet.org/education/AdvancedPlantPath/Topics/Epidemiology/CyclicalNature.htm).

66. Biology Apple powdery mildew is an epidemic that affects apple shoots. In a new infection, the percentage of apple shoots infected varies from a low of around 10% at the start of May to a high of around 60% 6 months later. Use a sine function model of the percentage of apple shoots affected by apple powdery mildew t months since the start of a year to estimate to the nearest 0.1% the average percentage of apple shoots affected in the first 2 months of a year.

Based on graphical data. Source: American Phytopathological Society (http://www.apsnet.org/education/AdvancedPlantPath/Topics/Epidemiology/CyclicalNature.htm).

67. Electrical Current The typical voltage V supplied by an electrical outlet in the United States is given by

$$V(t) = 165 \cos(120\pi t)$$

where t is time in seconds.

a. Find the average voltage over the interval $[0, \frac{1}{6}]$. How many times does the voltage reach a maximum in 1 second? (This is referred to as the number of **cycles per second**.)

b. Plot the function $S(t) = [V(t)]^2$ over the interval $[0, \frac{1}{6}]$.

c. The **root mean square** voltage is given by the formula

$$V_{\text{rms}} = \sqrt{\bar{S}}$$

where $\bar{S}$ is the average value of $S(t)$ over one cycle. Estimate V_{rms}.

68. Tides The depth of water at my favorite surfing spot varies from 5 to 15 feet, depending on the time. Last Sunday high tide occurred at 5:00 A.M., and the next high tide occurred at 6:30 P.M. Use the cosine function to model the depth of water as a function of time t in hours since midnight on Sunday morning. What was the average depth of the water between 10:00 A.M. and 2:00 P.M.?

Income Streams Recall from Section 7.4 that the total income received from time $t = a$ to time $t = b$ from a continuous income stream of $R(t)$ dollars per year is

$$\text{total value} = TV = \int_a^b R(t)\, dt$$

In Exercises 69 and 70, find the total value of the given income stream over the given period.

69. $R(t) = 50{,}000 + 2000\pi \sin(2\pi t), 0 \le t \le 1$

70. $R(t) = 100{,}000 - 2000\pi \sin(\pi t), 0 \le t \le 1.5$

COMMUNICATION AND REASONING EXERCISES

71. What can you say about the definite integral of a sine or cosine function over a whole number of periods?

72. How are the derivative and antiderivative of $\sin x$ related?

73. What is the average value of $1 + 2 \cos x$ over a large interval?

74. What is the average value of $3 - \cos x$ over a large interval?

75. The acceleration of an object is given by $a = K \sin(\omega t - \alpha)$. What can you say about its displacement at time t?

76. Write down a function whose derivative is -2 times its antiderivative.

CASE STUDY

Predicting Cocoa Inventories

As a consultant to Chocoholic Cocoa Company, you have been asked to model the worldwide production and consumption of cocoa, to determine how production cycles are affected by consumption, to estimate the trend in average world cocoa inventories, and to advise your cocoa producers whether to increase or decrease production. You have data showing the production and consumption (in metric tons, or tonnes) of cocoa for the past 12 years (see Figure 15):

	1990	1991	1992	1993	1994	1995	1996	1997	1998	1999	2000	2001
Demand (consumption)	2200	2350	2300	2400	2450	2500	2600	2700	2800	2750	2950	2950
Supply (production)	2400	2500	2250	2390	2450	2600	2900	2700	2650	2750	3000	3150

Approximate figures are in thousands of tonnes. The 2001 supply figure is an estimate. SOURCE: EDF Man Cocoa Report, 2001/United Nations Conference on Trade and Development (http://www.unctad.org/infocomm/Diversification/bangkok/Zemek.PDF).

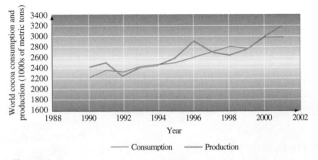

Figure 15

From the graph you deduce that the demand is roughly linear, but the supply tends to oscillate regularly above and below the demand graph fairly regularly. To see this oscillation more clearly, you reason that it would be best to subtract the demand from the supply, and so you enter the data in your worksheet and compute and graph the differences between the supply and demand (Figure 16).

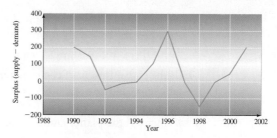

Figure 16

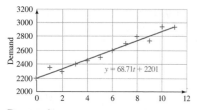

Demand

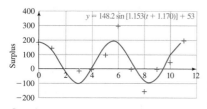

Surplus

Figure 17

The graph suggests a rough sine curve with amplitude around 150 and period around 6 years, so you use the techniques from Section 9.1 to construct a regression sine curve. Figure 17 shows the demand with its linear regression line and the surplus (supply − demand) with its sine regression curve ($t = 0$ represents 1990). Thus, your models for cocoa production and consumption (in thousands of metric tons) are

$$\text{consumption} = 68.71t + 2201$$
$$\text{production} = \text{demand} + \text{surplus} = 148.2 \sin[1.153(t + 1.170)] + 68.71t + 2254$$

These models allow you to make the following observations:

- Consumption of cocoa is increasing at an approximately constant rate of 68,710 tonnes per year.
- Production of cocoa fluctuates from about 95,000 tonnes below consumption to 201,000 tonnes above it every 5.4 years. (See the exercises.)

Interestingly—and this can also be seen in the raw data—production of cocoa exceeds consumption on average. This can be explained by the phenomenon that production anticipates a rising demand.

You now address the question: How does this pattern of production and consumption affect cocoa inventories? The accumulated cocoa inventory from 1990 to time t can be modeled by

$$A(t) = \int_0^t (148.2 \sin[1.153(x + 1.170)] + 53) \, dx$$
$$\approx [-128.5 \cos[1.153(x + 1.170)] + 53x]_0^t$$
$$\approx -128.5 \cos[1.153(t + 1.170)] + 53t + 28$$

The graph of $A(t)$ is shown in Figure 18. The graph clearly indicates a rising world inventory of cocoa and strongly suggests that production must be cut back substantially in order to reduce the surplus.

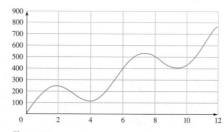

Figure 18

EXERCISES

1. Use the observed data to compute the actual accumulated cocoa inventory from the start of 1990 through 2001 and compare it with the value predicted by the model. (Use t to represent the time since the start of 1990 and round to the nearest 1000 tonnes.)
2. Justify the claim made above that production of cocoa fluctuates from 95,000 tonnes below consumption to 201,000 tonnes above it every 5.4 years.

3. What would be the maximum accumulated inventory if the production was given instead by the following formula?

$$P(t) = 148.2 \sin[1.153(t + 1.170)] + 68.71t + 2201$$

4. If the consumption of cocoa increased exponentially but the surplus was still as shown above, would you still advise cocoa producers to cut back on production? Explain.

5. Redo the entire analysis for the following (hypothetical) data on the sale of Fruit Punch on the planet Dune (in millions of tons).

	4000	4001	4002	4003	4004	4005	4006	4007	4008	4009	4010	4011
Demand (consumption)	2210	2350	2310	2410	2450	2510	2610	2700	2810	2760	2210	2350
Supply (production)	2340	2440	2190	2330	2390	2540	2840	2640	2590	2690	2340	2440

CHAPTER 9 REVIEW TEST

1. Model each of the following curves with a sine function. (The scales on the two axes may not be the same.)

a.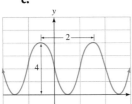

b.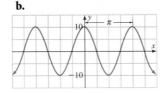

c.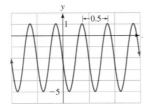

d.

2. Model the curves in Question 1 with cosine functions.

3. Find the derivatives of each of the following functions.
   a. $f(x) = \cos(x^2 - 1)$
   b. $f(x) = \sin(x^2 + 1)\cos(x^2 - 1)$
   c. $f(x) = \tan(2e^x - 1)$
   d. $f(x) = \sec\sqrt{x^2 - x}$
   e. $f(x) = \sin^2(x^2)$
   f. $f(x) = \cos^2[1 - \sin(2x)]$

4. Evaluate the following integrals.
   a. $\int 4 \cos(2x - 1)\, dx$
   b. $\int (x - 1)\sin(x^2 - 2x + 1)\, dx$
   c. $\int x \tan(x^2 + 1)\, dx$
   d. $\int_0^\pi \cos\left(x + \frac{\pi}{2}\right) dx$
   e. $\int_{\ln(\pi/2)}^{\ln(\pi)} e^x \sin(e^x)\, dx$
   f. $\int_\pi^{2\pi} \tan\left(\frac{x}{6}\right) dx$

5. Use integration by parts to evaluate the following integrals.
   a. $\int x^2 \sin x\, dx$
   b. $\int e^x \sin 2x\, dx$

OHaganBooks.com—UPS AND DOWNS

6. After several years in the business, OHaganBooks.com noticed that its sales showed seasonal fluctuations; weekly sales oscillated in a sine wave from a low of 9000 books per week to a high of 12,000 books per week, with the high point of the year being three quarters of the way through the year, in October. Model OHaganBooks.com's weekly sales as a function of t, the number of weeks into the year.

7. OHaganBooks.com recently hired a new sales manager, Mary Beth O'Connell, who predicts that the revenue from sales of the latest blockbuster *Elvish for Blockheads* will vary in accordance with annual releases of episodes of the movie series *Lord of the Rings Episodes 9–12*. She has come up with the following model (which includes the effect of diminishing sales):

$$R(t) = 20,000 + 15,000e^{-0.12t} \cos\left[\frac{\pi}{6}(t - 4)\right] \text{ dollars}$$
$$(0 \le t \le 72)$$

where t is time in months from now and $R(t)$ is the monthly revenue. How fast, to the nearest dollar, will the revenue be changing 10 months from now?

8. Refer to Question 7. Use technology or integration by parts to estimate, to the nearest $100, the total revenue from sales of *Elvish for Blockheads* over the next 10 months.

9. Having completed his doctorate in biophysics, Billy Sean O'Hagan will be accompanying the first manned mission to Mars. For reasons too complicated to explain (but having to do with the continuation of his doctoral research project and the timing of messages from his fiancée), during the voyage he will be consuming protein at a rate of

$$P(t) = 150 + 50 \sin\left[\frac{\pi}{2}(t - 1)\right] \text{ grams per day}$$

t days into the voyage. Find the total amount of protein he will consume as a function of time t.

🌐 ADDITIONAL ONLINE REVIEW

If you follow the path
   Web Site → Everything for Calculus → Chapter 9
you will find the following additional resources to help you review:

A comprehensive chapter summary (including examples and interactive features)

Online section-by-section interactive tutorials

Section-by-section Excel tutorials

Additional review exercises (including interactive exercises and many with help)

A true/false chapter quiz

Web-based and Excel-based integration utilities

Appendix A

ALGEBRA REVIEW

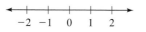

$$-2 \quad -1 \quad 0 \quad 1 \quad 2$$

Figure 1

Introduction

In this appendix we review some topics from algebra that you need to know to get the most out of this book. This appendix can be used either as a refresher course or as a reference.

There is one crucial fact you must always keep in mind: The letters used in algebraic expressions stand for numbers. All rules of algebra are just facts about the arithmetic of numbers. If you are not sure whether some algebraic manipulation you are about to do is legitimate, try it first with numbers. If it doesn't work with numbers, it doesn't work.

A.1 Real Numbers

The **real numbers** are the numbers that can be written in decimal notation, including those that require an infinite decimal expansion. The set of real numbers includes all integers, positive and negative; all fractions; and the irrational numbers, those with decimal expansions that don't simply repeat. Examples of irrational numbers are

$$\sqrt{2} = 1.414\,213\,562\,373\ldots \quad \text{and} \quad \pi = 3.141\,592\,653\,589\ldots$$

It is very useful to picture the real numbers as points on a line. As shown in Figure 1, larger numbers appear to the right, in the sense that if $a < b$ then the point corresponding to b is to the right of the point corresponding to a.

Intervals

Some subsets of the set of real numbers, called **intervals,** show up quite often, and so we have a compact notation for them.

INTERNET RESOURCES FOR THIS APPENDIX

At the Web site, follow the path
    Web Site → Everything for Calculus → Algebra Review
where you will find an interactive tutorial version of this algebra review.

Interval Notation

Here is a list of types of intervals, along with examples.

	Interval	Description	Picture	Example
Closed	$[a, b]$	Set of numbers x with $a \le x \le b$	(includes endpoints)	$[0, 10]$
Open	(a, b)	Set of numbers x with $a < x < b$	(excludes endpoints)	$(-1, 5)$
Half-Open	$(a, b]$	Set of numbers x with $a < x \le b$		$(-3, 1]$
	$[a, b)$	Set of numbers x with $a \le x < b$		$[0, 5)$
Infinite	$[a, +\infty)$	Set of numbers x with $a \le x$		$[10, +\infty)$
	$(a, +\infty)$	Set of numbers x with $a < x$		$(-3, +\infty)$
	$(-\infty, b]$	Set of numbers x with $x \le b$		$(-\infty, -3]$
	$(-\infty, b)$	Set of numbers x with $x < b$		$(-\infty, 10)$
	$(-\infty, +\infty)$	Set of all real numbers		$(-\infty, +\infty)$

Operations

There are five important operations on real numbers: addition, subtraction, multiplication, division, and exponentiation. *Exponentiation* means raising a real number to a power; for instance, $3^2 = 3 \cdot 3 = 9$; $2^3 = 2 \cdot 2 \cdot 2 = 8$.

A note on technology: Most graphing calculators and spreadsheets use an asterisk * for multiplication and a caret sign ^ for exponentiation. For instance, 3×5 is entered as 3*5, $3x$ as 3*x, and 3^2 as 3^2.

When we write an expression involving two or more operations, like

$$2 \cdot 3 + 4 \quad \text{or} \quad \frac{2 \cdot 3^2 - 5}{4 - (-1)}$$

we need to agree on the order in which to do the operations. Does $2 \cdot 3 + 4$ mean $(2 \cdot 3) + 4 = 10$ or $2 \cdot (3 + 4) = 14$? We all agree to use the following rules for the order in which we do the operations.

Standard Order of Operations
Parentheses and Fraction Bars
First, calculate the values of all expressions inside parentheses or brackets, working from the innermost parentheses out, before using them in other operations. In a fraction, calculate the numerator and denominator separately before doing the division.

Quick Examples
1. $6(2 + [3 - 5] - 4) = 6(2 + (-2) - 4) = 6(-4) = -24$

2. $\dfrac{4 - 2}{3(-2 + 1)} = \dfrac{2}{3(-1)} = \dfrac{2}{-3} = -\dfrac{2}{3}$

3. $3/(2 + 4) = \dfrac{3}{2 + 4} = \dfrac{3}{6} = \dfrac{1}{2}$

4. $(x + 4x)/(y + 3y) = 5x/(4y)$

Exponents
Next, perform exponentiation.

Quick Examples
1. $2 + 4^2 = 2 + 16 = 18$
2. $(2 + 4)^2 = 6^2 = 36$ } Note the difference.

3. $2\left(\dfrac{3}{4 - 5}\right)^2 = 2\left(\dfrac{3}{-1}\right)^2 = 2(-3)^2 = 2 \times 9 = 18$

4. $2(1 + 1/10)^2 = 2(1.1)^2 = 2 \times 1.21 = 2.42$

Multiplication and Division
Next, do all multiplications and divisions, from left to right.

Quick Examples
1. $2(3 - 5)/4 \cdot 2 = 2(-2)/4 \cdot 2$ Parentheses first
$\qquad\qquad\qquad = -4/4 \cdot 2$ Leftmost product
$\qquad\qquad\qquad = -1 \cdot 2 = -2$ Multiplications and divisions, left to right

2. $2(1 + 1/10)^2 \times 2/10 = 2(1.1)^2 \times 2/10$ Parentheses first
$\qquad\qquad\qquad\qquad = 2 \times 1.21 \times 2/10$ Exponent
$\qquad\qquad\qquad\qquad = 4.84/10 = 0.484$ Multiplications and divisions, left to right

3. $4\dfrac{2(4 - 2)}{3(-2 \cdot 5)} = 4\dfrac{2(2)}{3(-10)} = 4\dfrac{4}{-30} = \dfrac{16}{-30} = -\dfrac{8}{15}$

Addition and Subtraction
Last, do all additions and subtractions, from left to right.

Quick Examples
1. $2(3 - 5)^2 + 6 - 1 = 2(-2)^2 + 6 - 1 = 2(4) + 6 - 1 = 8 + 6 - 1 = 13$

2. $\left(\dfrac{1}{2}\right)^2 - (-1)^2 + 4 = \dfrac{1}{4} - 1 + 4 = -\dfrac{3}{4} + 4 = \dfrac{13}{4}$

3. $3/2 + 4 = 1.5 + 4 = 5.5$
4. $3/(2 + 4) = 3/6 = 1/2 = 0.5$ } Note the difference.

5. $4/2^2 + (4/2)^2 = 4/2^2 + 2^2 = 4/4 + 4 = 1 + 4 = 5$

Appendix A

Entering Formulas

Any good calculator or spreadsheet will respect the standard order of operations. However, we must be careful with division and exponentiation and use parentheses as necessary. The following table gives some examples of simple mathematical expressions and their equivalents in the functional format used in most graphing calculators, spreadsheets, and computer programs.

Mathematical Expression	Formula	Comments
$\dfrac{2}{3-x}$	2/(3−x)	Note the use of parentheses instead of the fraction bar. If we omit the parentheses, we get the expression shown next.
$\dfrac{2}{3}-x$	2/3−x	The calculator follows the usual order of operations.
$\dfrac{2}{3\times 5}$	2/(3*5)	Putting the denominator in parentheses ensures that the multiplication is carried out first. The asterisk * is usually used for multiplication in graphing calculators and computers.
$\dfrac{2}{x}\times 5$	(2/x)*5	Putting the fraction in parentheses ensures that it is calculated first. Some calculators will interpret 2/3*5 as $\dfrac{2}{3\times 5}$, but 2/3(5) as $\dfrac{2}{3}\times 5$.
$\dfrac{2-3}{4+5}$	(2−3)/(4+5)	Note once again the use of parentheses in place of the fraction bar.
2^3	2^3	The caret ^ is commonly used to denote exponentiation.
2^{3-x}	2^(3−x)	Be careful to use parentheses to tell the calculator where the exponent ends. Enclose the *entire exponent* in parentheses.
2^3-x	2^3−x	Without parentheses, the calculator will follow the usual order of operations: exponentiation and then subtraction.
3×2^{-4}	3*2^(−4)	On some calculators, the negation key is separate from the minus key.
$2^{-4\times 3}\times 5$	2^(−4*3)*5	Note once again how parentheses enclose the entire exponent.
$100\left(1+\dfrac{0.05}{12}\right)^{60}$	100*(1+0.05/12)^60	This is a typical calculation for compound interest.
$PV\left(1+\dfrac{r}{m}\right)^{mt}$	PV*(1+r/m)^(m*t)	This is the compound interest formula. *PV* is understood to be a single number (present value), *not* the product of P and V (or else we would have used P*V).
$\dfrac{2^{3-2}\times 5}{y-x}$	2^(3−2)*5/(y−x) or (2^(3−2)*5)/(y−x)	Notice again the use of parentheses to hold the denominator together. We could also have enclosed the numerator in parentheses, although this is optional. (Why?)
$\dfrac{2^y+1}{2-4^{3x}}$	(2^y+1)/(2−4^(3*x))	Here, it is necessary to enclose both the numerator and the denominator in parentheses.
$2^y+\dfrac{1}{2}-4^{3x}$	2^y+1/2−4^(3*x)	This is the effect of leaving out the parentheses around the numerator and denominator in the previous expression.

Accuracy and Rounding

When we use a calculator or computer, the results of our calculations are often given to far more decimal places than are useful. For example, suppose we are told that a square has an area of 2.0 square feet and we are asked how long its sides are. Each side is the square root of the area, which the calculator tells us is

$$\sqrt{2}\approx 1.414\ 213\ 562$$

However, the measurement of 2.0 square feet is likely accurate to only two digits, so our estimate of the lengths of the sides can be no more accurate than that. Therefore, we round the answer to two digits:

length of one side ≈ 1.4 feet

The digits that follow 1.4 are meaningless. The following guide makes these ideas more precise.

Significant Digits, Decimal Places, and Rounding

The number of **significant digits** in a decimal representation of a number is the number of digits that are not leading zeros after the decimal point (as in .0005) or trailing zeros before the decimal point (as in 5,400,000). We say that a value is **accurate to n significant digits** if only the first n significant digits are meaningful.

When to Round

After doing a computation in which all the quantities are accurate to no more than n significant digits, round the final result to n significant digits.

Quick Examples
1. 0.000 67 has two significant digits.   The 000 before 67 are leading zeros.
2. 0.000 670 has three significant digits.   The 0 after 67 is significant.
3. 5,400,000 has two or more significant digits.   We can't say how many of the zeros are trailing.*
4. 5,400,001 has seven significant digits.   The string of zeros is not trailing.
5. Rounding 63,918 to three significant digits gives 63,900.
6. Rounding 63,958 to three significant digits gives 64,000.
7. $\pi = 3.141\,592\,653\ldots$; $\dfrac{22}{7} = 3.142\,857\,142\ldots$. Therefore, $\dfrac{22}{7}$ is an approximation of π that is accurate to only three significant digits (3.14).
8. $4.02(1 + 0.02)^{1.4} \approx 4.13$ We rounded to three significant digits.

*If we obtained 5,400,000 by rounding 5,401,011, then it has three significant digits because the zero after the 4 is significant. On the other hand, if we obtained it by rounding 5,411,234, then it has only two significant digits. The use of scientific notation avoids this ambiguity: 5.40×10^6 (or 5.40 E6 on a calculator or computer) is accurate to three digits and 5.4×10^6 is accurate to two.

One more point, though: If, in a long calculation, you round the intermediate results, your final answer may be even less accurate than you think. As a general rule,

When calculating, don't round intermediate results. Rather, use the most accurate results obtainable or have your calculator or computer store them for you.

When you are done with the calculation, *then* round your answer to the appropriate number of digits of accuracy.

Appendix A

A.1 EXERCISES

In Exercises 1–24, calculate each expression, giving the answer as a whole number or a fraction in lowest terms.

1. $2[4 + (-1)](2 \cdot -4)$

2. $3 + [(4 - 2) \cdot 9]$

3. $20/(3*4)-1$

4. $2-(3*4)/10$

5. $\dfrac{3 + \{[3 + (-5)]\}}{3 - 2 \times 2}$

6. $\dfrac{12 - (1 - 4)}{2(5 - 1) \cdot 2 - 1}$

7. $(2-5*(-1))/1-2*(-1)$

8. $2-5*(-1)/(1-2*(-1))$

9. $2 \cdot (-1)^2/2$

10. $2 + 4 \cdot 3^2$

11. $2 \cdot 4^2 + 1$

12. $1 - 3 \cdot (-2)^2 \times 2$

13. $3\verb|^|2+2\verb|^|2+1$

14. $2\verb|^|(2\verb|^|2-2)$

15. $\dfrac{3 - 2(-3)^2}{-6(4 - 1)^2}$

16. $\dfrac{1 - 2(1 - 4)^2}{2(5 - 1)^2 \cdot 2}$

17. $10*(1+1/10)\verb|^|3$

18. $121/(1+1/10)\verb|^|2$

19. $3\left[\dfrac{-2 \cdot 3^2}{-(4 - 1)^2}\right]$

20. $-\left[\dfrac{8(1 - 4)^2}{-9(5 - 1)^2}\right]$

21. $3\left[1 - \left(-\dfrac{1}{2}\right)^2\right]^2 + 1$

22. $3\left[\dfrac{1}{9} - \left(\dfrac{2}{3}\right)^2\right]^2 + 1$

23. $(1/2)\verb|^|2-1/2\verb|^|2$

24. $2/(1\verb|^|2)-(2/1)\verb|^|2$

In Exercises 25–50, convert each expression into its technology formula equivalent as in the table in the text.

25. $3 \times (2 - 5)$

26. $4 + \dfrac{5}{9}$

27. $\dfrac{3}{2 - 5}$

28. $\dfrac{4 - 1}{3}$

29. $\dfrac{3 - 1}{8 + 6}$

30. $3 + \dfrac{3}{2 - 9}$

31. $3 - \dfrac{4 + 7}{8}$

32. $\dfrac{4 \times 2}{\left(\dfrac{2}{3}\right)}$

33. $\dfrac{2}{3 + x} - xy^2$

34. $3 + \dfrac{3 + x}{xy}$

35. $3.1x^3 - 4x^{-2} - \dfrac{60}{x^2 - 1}$

36. $2.1x^{-3} - x^{-1} + \dfrac{x^2 - 3}{2}$

37. $\dfrac{\left(\dfrac{2}{3}\right)}{5}$

38. $\dfrac{2}{\left(\dfrac{3}{5}\right)}$

39. $3^{4-5} \times 6$

40. $\dfrac{2}{3 + 5^{7-9}}$

41. $3\left(1 + \dfrac{4}{100}\right)^{-3}$

42. $3\left(\dfrac{1 + 4}{100}\right)^{-3}$

43. $3^{2x-1} + 4^x - 1$

44. $2^{x^2} - (2^{2x})^2$

45. 2^{2x^2-x+1}

46. $2^{2x^2-x} + 1$

47. $\dfrac{4e^{-2x}}{2 - 3e^{-2x}}$

48. $\dfrac{e^{2x} + e^{-2x}}{e^{2x} - e^{-2x}}$

49. $3\left[1 - \left(-\dfrac{1}{2}\right)^2\right]^2 + 1$

50. $3\left[\dfrac{1}{9} - \left(\dfrac{2}{3}\right)^2\right]^2 + 1$

A.2 Exponents and Radicals

In the last section we discussed exponentiation, or "raising to a power"—for example, $2^3 = 2 \cdot 2 \cdot 2$. In this section we discuss the algebra of exponentials more fully. First, we look at *integer* exponents: cases in which the powers are positive or negative whole numbers.

Integer Exponents

Positive Integer Exponents

If a is any real number and n is any positive integer, then by a^n we mean the quantity $a \cdot a \cdots \cdot a$ (n times); thus, $a^1 = a$, $a^2 = a \cdot a$, $a^5 = a \cdot a \cdot a \cdot a \cdot a$. In the expression a^n, the number n is called the **exponent,** and the number a is called the **base.**

Quick Examples

$$3^2 = 9 \qquad\qquad 2^3 = 8$$

$$0^{34} = 0 \qquad\qquad (-1)^5 = -1$$

$$10^3 = 1000 \qquad\qquad 10^5 = 100{,}000$$

Negative Integer Exponents

If a is any real number *other than zero* and n is any positive integer, then we define

$$a^{-n} = \frac{1}{a^n} = \frac{1}{a \cdot a \cdots \cdot a} \ (n \text{ times})$$

Quick Examples

$$2^{-3} = \frac{1}{2^3} = \frac{1}{8} \qquad\qquad 1^{-27} = \frac{1}{1^{27}} = 1$$

$$x^{-1} = \frac{1}{x^1} = \frac{1}{x} \qquad\qquad (-3)^{-2} = \frac{1}{(-3)^2} = \frac{1}{9}$$

$$y^7 y^{-2} = y^7 \, \frac{1}{y^2} = y^5 \qquad 0^{-2} \text{ is not defined.}$$

Zero Exponent

If a is any real number other than zero, then we define

$$a^0 = 1$$

Quick Examples

$$3^0 = 1 \qquad\qquad\qquad 1{,}000{,}000^0 = 1$$

$$0^0 \text{ is not defined.}$$

When combining exponential expressions, we use the following identities.

Exponent Identity	Quick Examples
1. $a^m a^n = a^{m+n}$	$2^3 2^2 = 2^{3+2} = 2^5 = 32$
	$x^3 x^{-4} = x^{3-4} = x^{-1} = \dfrac{1}{x}$
	$\dfrac{x^3}{x^2} = x^3 \dfrac{1}{x^2} = x^3 x^{-2} = x^1 = x$
2. $\dfrac{a^m}{a^n} = a^{m-n}$ (if $a \neq 0$)	$\dfrac{4^3}{4^2} = 4^{3-2} = 4^1 = 4$
	$\dfrac{x^3}{x^{-2}} = x^{3-(-2)} = x^5$
	$\dfrac{3^2}{3^4} = 3^{2-4} = 3^{-2} = \dfrac{1}{9}$
3. $(a^n)^m = a^{nm}$	$(3^2)^2 = 3^4 = 81$
	$(2^x)^2 = 2^{2x}$
4. $(ab)^n = a^n b^n$	$(4 \cdot 2)^2 = 4^2 2^2 = 64$
	$(-2y)^4 = (-2)^4 y^4 = 16y^4$
5. $\left(\dfrac{a}{b}\right)^n = \dfrac{a^n}{b^n}$ (if $b \neq 0$)	$\left(\dfrac{4}{3}\right)^2 = \dfrac{4^2}{3^2} = \dfrac{16}{9}$
	$\left(\dfrac{x}{-y}\right)^3 = \dfrac{x^3}{(-y)^3} = -\dfrac{x^3}{y^3}$

Caution

- In the first two identities, the bases of the expressions must be the same. For example, the first gives $3^2 3^4 = 3^6$, but does *not* apply to $3^2 4^2$.
- People sometimes invent their own identities, such as $a^m + a^n = a^{m+n}$, which is wrong! (Try it with $a = m = n = 1$.) If you wind up with something like $2^3 + 2^4$, you are stuck with it; there are no identities around to simplify it further. (You can factor out 2^3, but whether that is a simplification depends on what you are going to do with the expression next.)

Example 1 • Combining the Identities

a.
$$\frac{(x^2)^3}{x^3} = \frac{x^6}{x^3} \qquad \text{By identity 3}$$

$$= x^{6-3} \qquad \text{By identity 2}$$

$$= x^3$$

b.
$$\frac{(x^4 y)^3}{y} = \frac{(x^4)^3 y^3}{y} \qquad \text{By identity 4}$$

$$= \frac{x^{12} y^3}{y} \qquad \text{By identity 3}$$

$$= x^{12} y^{3-1} \qquad \text{By identity 2}$$

$$= x^{12} y^2$$

Example 2 • Eliminating Negative Exponents

Simplify the following and express the answer using no negative exponents.

a. $\dfrac{x^4 y^{-3}}{x^5 y^2}$ **b.** $\left(\dfrac{x^{-1}}{x^2 y}\right)^5$

Solution

a. $\dfrac{x^4 y^{-3}}{x^5 y^2} = x^{4-5} y^{-3-2} = x^{-1} y^{-5} = \dfrac{1}{xy^5}$

b. $\left(\dfrac{x^{-1}}{x^2 y}\right)^5 = \dfrac{(x^{-1})^5}{(x^2 y)^5} = \dfrac{x^{-5}}{x^{10} y^5} = \dfrac{1}{x^{15} y^5}$

Radicals

If a is any nonnegative real number, then its **square root** is the nonnegative number whose square is a. For example, the square root of 16 is 4, since $4^2 = 16$. We write the square root of n as $\sqrt{n}$. (Roots are also referred to as **radicals.**) It is important to remember that $\sqrt{n}$ is *never* negative. For instance, $\sqrt{9}$ is 3, and not -3, even though $(-3)^2 = 9$. If we want to speak of the "negative square root" of 9, we write it as $-\sqrt{9} = -3$. If we want to write both square roots at once, we write $\pm\sqrt{9} = \pm 3$.

The **cube root** of a real number a is the number whose cube is a. The cube root of a is written as $\sqrt[3]{a}$, so, for example, $\sqrt[3]{8} = 2$ (since $2^3 = 8$). Note that we can take the cube root of any number, positive, negative, or zero. For instance, the cube root of -8 is $\sqrt[3]{-8} = -2$ because $(-2)^3 = -8$. Unlike square roots, the cube root of a number may be negative. In fact, the cube root of a always has the same sign as a.

Higher roots are defined similarly. The **fourth root** of the *nonnegative* number a is defined as the nonnegative number whose fourth power is a and written as $\sqrt[4]{a}$. The **fifth root** of any number a is the number whose fifth power is a, and so on.

Note We cannot take an even-numbered root of a negative number, but we can take an odd-numbered root of any number. Even roots are always positive, whereas odd roots have the same sign as the number we start with.

Example 3 • Finding nth Roots

$\sqrt{4} = 2$	Since $2^2 = 4$
$\sqrt{16} = 4$	Since $4^2 = 16$
$\sqrt{1} = 1$	Since $1^2 = 1$
If $x \geq 0$, then $\sqrt{x^2} = x$	Since $x^2 = x^2$
$\sqrt{2} \approx 1.414\,213\,562$	$\sqrt{2}$ is not a whole number.
$\sqrt{1 + 1} = \sqrt{2} \approx 1.414\,213\,562$	First add, then take the square root.[1]
$\sqrt{9 + 16} = \sqrt{25} = 5$	Contrast with $\sqrt{9} + \sqrt{16} = 3 + 4 = 7$.

----

[1]In general, $\sqrt{a + b}$ means the square root of the *quantity* $(a + b)$. The radical sign acts like a pair of parentheses or a fraction bar, telling us to evaluate what is inside before taking the root. (See **Caution** in the following box.)

Appendix A

$$\sqrt[3]{27} = 3 \qquad\qquad \text{Since } 3^3 = 27$$

$$\sqrt[3]{-64} = -4 \qquad\quad \text{Since } (-4)^3 = -64$$

$$\sqrt[4]{16} = 2 \qquad\qquad \text{Since } 2^4 = 16$$

$\sqrt[4]{-16}$ is not defined.   Even-numbered root of a negative number

$\sqrt[5]{-1} = -1$, since $(-1)^5 = -1$.   Odd-numbered root of a negative number

$\sqrt[n]{-1} = -1$ if n is any odd number.

Question   In the example we saw that $\sqrt{x^2} = x$ if x is nonnegative. What happens if x is negative?

Answer   If x is negative, then x^2 is positive, and so $\sqrt{x^2}$ is still defined as the nonnegative number whose square is x^2. This number must be $|x|$, the **absolute value of x,** which is the nonnegative number with the same size as x. For instance, $|-3| = 3$, while $|3| = 3$, and $|0| = 0$. It follows that

$$\sqrt{x^2} = |x|$$

for every real number x, positive or negative. For instance,

$$\sqrt{(-3)^2} = \sqrt{9} = 3 = |-3| \quad \text{and} \quad \sqrt{3^2} = \sqrt{9} = 3 = |3|$$

In general, we find that

$$\sqrt[n]{x^n} = x \text{ if } n \text{ is odd} \quad \text{and} \quad \sqrt[n]{x^n} = |x| \text{ if } n \text{ is even.}$$

We use the following identities to evaluate radicals of products and quotients.

Radicals of Products and Quotients

If a and b are any real numbers (nonnegative in the case of even-numbered roots), then

$$\sqrt[n]{ab} = \sqrt[n]{a}\,\sqrt[n]{b} \qquad\qquad \text{radical of a product} = \text{product of radicals}$$

$$\sqrt[n]{\frac{a}{b}} = \frac{\sqrt[n]{a}}{\sqrt[n]{b}} \quad (\text{if } b \neq 0) \qquad \text{radical of a quotient} = \text{quotient of radicals}$$

Notes

- The first rule is similar to the rule $(ab)^2 = a^2b^2$ for the square of a product, and the second rule is similar to the rule $\left(\dfrac{a}{b}\right)^2 = \dfrac{a^2}{b^2}$ for the square of a quotient.

- **Caution**   There is no corresponding identity for addition:

$$\sqrt{a + b} \text{ is } not \text{ equal to } \sqrt{a} + \sqrt{b}$$

(consider $a = b = 1$, for example). Equating these expressions is a common error, so be careful!

Quick Examples

1. $\sqrt{9 \cdot 4} = \sqrt{9}\sqrt{4} = 3 \times 2 = 6$ Alternatively, $\sqrt{9 \cdot 4} = \sqrt{36} = 6$

2. $\sqrt{\dfrac{9}{4}} = \dfrac{\sqrt{9}}{\sqrt{4}} = \dfrac{3}{2}$

3. $\sqrt{4(3 + 13)} = \sqrt{4(16)} = \sqrt{4}\sqrt{16} = 2 \times 4 = 8$

4. $\sqrt[3]{-216} = \sqrt[3]{(-27)8} = \sqrt[3]{-27}\,\sqrt[3]{8} = (-3)2 = -6$

5. $\sqrt{x^3} = \sqrt{x^2 \cdot x} = \sqrt{x^2}\sqrt{x} = x\sqrt{x}$ (if $x \geq 0$)

6. $\sqrt{\dfrac{x^2 + y^2}{z^2}} = \dfrac{\sqrt{x^2 + y^2}}{\sqrt{z^2}} = \dfrac{\sqrt{x^2 + y^2}}{|z|}$ We can't simplify the numerator any further.

Rational Exponents

We already know what we mean by expressions such as x^4 and a^{-6}. The next step is to make sense of *rational* exponents: exponents of the form p/q with p and q integers as in $a^{1/2}$ and $3^{-2/3}$.

Question What should we mean by $a^{1/2}$?

Answer The overriding concern here is that all the exponent identities should remain true. In this case the identity to look at is the one that says that $(a^m)^n = a^{mn}$. This identity tells us that

$$(a^{1/2})^2 = a^1 = a$$

That is, $a^{1/2}$, when squared, gives us a. But that must mean that $a^{1/2}$ is the *square root* of a, or

$$a^{1/2} = \sqrt{a}$$

A similar argument tells us that, if q is any positive whole number, then

$$a^{1/q} = \sqrt[q]{a} \qquad \text{The } q\text{th root of } a$$

Notice that if a is negative, this makes sense only for q odd. To avoid this problem we usually stick to positive a.

Question If p and q are integers (q positive), what should we mean by $a^{p/q}$?

Answer By the exponent identities, $a^{p/q}$ should equal both $(a^p)^{1/q}$ and $(a^{1/q})^p$. The first is the qth root of a^p, and the second is the pth power of $a^{1/q}$, which gives us the following.

Conversion between Rational Exponents and Radicals

If a is any nonnegative number, then

$$a^{p/q} = \underset{\uparrow}{\sqrt[q]{a^p}} = \underset{\uparrow}{(\sqrt[q]{a})^p}$$

Exponential form   Radical form

In particular,

$$a^{1/q} = \sqrt[q]{a} \qquad \text{The } q\text{th root of } a$$

Notes

- If a is negative, all of this makes sense only if q is odd.
- All the exponent identities continue to work when we allow rational exponents p/q. In other words, we are free to use all the exponent identities even though the exponents are not integers.

Quick Examples

1. $4^{3/2} = (\sqrt{4})^3 = 2^3 = 8$

2. $8^{2/3} = (\sqrt[3]{8})^2 = 2^2 = 4$

3. $9^{-3/2} = \dfrac{1}{9^{3/2}} = \dfrac{1}{(\sqrt{9})^3} = \dfrac{1}{3^3} = \dfrac{1}{27}$

4. $\dfrac{\sqrt{3}}{\sqrt[3]{3}} = \dfrac{3^{1/2}}{3^{1/3}} = 3^{1/2-1/3} = 3^{1/6} = \sqrt[6]{3}$

5. $2^2 2^{7/2} = 2^2 2^{3+1/2} = 2^2 2^3 2^{1/2} = 2^5 2^{1/2} = 2^5 \sqrt{2} = 32\sqrt{2}$

Example 4 • Simplifying Algebraic Expressions

Simplify the following.

a. $\dfrac{(x^3)^{5/3}}{x^3}$ **b.** $\sqrt[4]{a^6}$ **c.** $\dfrac{(xy)^{-3}y^{3/2}}{x^{-2}\sqrt{y}}$

Solution

a. $\dfrac{(x^3)^{5/3}}{x^3} = \dfrac{x^5}{x^3} = x^2$

b. $\sqrt[4]{a^6} = a^{6/4} = a^{3/2} = a \cdot a^{1/2} = a\sqrt{a}$

c. $\dfrac{(xy)^{-3}y^{3/2}}{x^{-2}\sqrt{y}} = \dfrac{x^{-3}y^{-3}y^{-3/2}}{x^{-2}y^{1/2}} = \dfrac{1}{x^{-2+3}y^{1/2+3+3/2}} = \dfrac{1}{xy^5}$

Converting between Rational, Radical, and Exponential Form

In calculus we must often convert algebraic expressions involving powers of x, such as $3/(2x^2)$, into expressions where x does not appear in the denominator, such as $(3/2)x^{-2}$. Also, we must often convert expressions with radicals, such as $1/\sqrt{1 + x^2}$, into expressions with no radicals and all powers in the numerator, such as $(1 + x^2)^{-1/2}$. In these cases we are converting from **rational form** or **radical form** to **exponential form.**

Rational, Radical, and Exponential Form

An expression is in **rational form** if it is written with positive exponents only.

Quick Examples

1. $\dfrac{2}{3x^2}$ is in rational form.

2. $\dfrac{2x^{-1}}{3}$ is not in rational form because the exponent of x is negative.

3. $\dfrac{x}{6} + \dfrac{6}{x}$ is in rational form.

An expression is in **radical form** if it is written with integer powers and roots only.

Quick Examples

1. $\dfrac{2}{5\sqrt[3]{x}} + \dfrac{2}{x}$ is in radical form.

2. $\dfrac{2x^{-1/3}}{5} + 2x^{-1}$ is not in radical form because $x^{-1/3}$ appears.

3. $\dfrac{1}{\sqrt{1 + x^2}}$ is in radical form, but $(1 + x^2)^{-1/2}$ is not.

An expression is in **exponential form** if there are no radicals and all powers of unknowns occur in the numerator. We usually write such expressions as sums or differences of terms of the form

$$\text{constant} \times (\text{expression with } x)^p \qquad \text{As in } \frac{1}{3}x^{-3/2}$$

Quick Examples

1. $\dfrac{2}{3}x^4 - 3x^{-1/3}$ is in exponential form.

2. $\dfrac{x}{6} + \dfrac{6}{x}$ is not in exponential form because the second expression has x in the denominator.

3. $\sqrt[3]{x}$ is not in exponential form because it has a radical.

4. $(1 + x^2)^{-1/2}$ is in exponential form, but $1/\sqrt{1 + x^2}$ is not.

Example 5 • Converting from One Form to Another

Convert the following to rational form.

a. $\dfrac{1}{2}x^{-2} + \dfrac{4}{3}x^{-5}$ **b.** $\dfrac{2}{\sqrt{x}} - \dfrac{2}{x^{-4}}$

Convert the following to radical form.

c. $\dfrac{1}{2}x^{-1/2} + \dfrac{4}{3}x^{-5/4}$ **d.** $\dfrac{(3 + x)^{-1/3}}{5}$

Convert the following to exponential form.

e. $\dfrac{3}{4x^2} - \dfrac{x}{6} + \dfrac{6}{x}$ **f.** $\dfrac{2}{(x + 1)^2} - \dfrac{3}{4\sqrt[5]{2x - 1}}$

Solution
For parts (a) and (b), we eliminate negative exponents as we did in Example 2:

a. $\dfrac{1}{2}x^{-2} + \dfrac{4}{3}x^{-5} = \dfrac{1}{2} \cdot \dfrac{1}{x^2} + \dfrac{4}{3} \cdot \dfrac{1}{x^5} = \dfrac{1}{2x^2} + \dfrac{4}{3x^5}$

b. $\dfrac{2}{\sqrt{x}} - \dfrac{2}{x^{-4}} = \dfrac{2}{x^{1/2}} - 2x^4$

For parts (c) and (d), we rewrite all terms with fractional exponents as radicals:

c. $\dfrac{1}{2}x^{-1/2} + \dfrac{4}{3}x^{-5/4} = \dfrac{1}{2} \cdot \dfrac{1}{x^{1/2}} + \dfrac{4}{3} \cdot \dfrac{1}{x^{5/4}}$

$$= \dfrac{1}{2} \cdot \dfrac{1}{\sqrt{x}} + \dfrac{4}{3} \cdot \dfrac{1}{\sqrt[4]{x^5}} = \dfrac{1}{2\sqrt{x}} + \dfrac{4}{3\sqrt[4]{x^5}}$$

d. $\dfrac{(3+x)^{-1/3}}{5} = \dfrac{1}{5(3+x)^{1/3}} = \dfrac{1}{5\sqrt[3]{3+x}}$

For parts (e) and (f), we eliminate any radicals and move all expressions involving x to the numerator:

e. $\dfrac{3}{4x^2} - \dfrac{x}{6} + \dfrac{6}{x} = \dfrac{3}{4}x^{-2} - \dfrac{1}{6}x + 6x^{-1}$

d. $\dfrac{2}{(x+1)^2} - \dfrac{3}{4\sqrt[5]{2x-1}} = 2(x+1)^{-2} - \dfrac{3}{4(2x-1)^{1/5}}$

$$= 2(x+1)^{-2} - \dfrac{3}{4}(2x-1)^{-1/5}$$

Solving Equations with Exponents

Example 6 • Solving Equations

Solve the following equations.

a. $x^3 + 8 = 0$ **b.** $x^2 - \dfrac{1}{2} = 0$ **c.** $x^{3/2} - 64 = 0$

Solution

a. Subtracting 8 from both sides gives $x^3 = -8$. Taking the cube root of both sides gives $x = -2$.

b. Adding $\dfrac{1}{2}$ to both sides gives $x^2 = \dfrac{1}{2}$. Thus, $x = \pm\sqrt{\dfrac{1}{2}} = \pm\dfrac{1}{\sqrt{2}}$.

c. Adding 64 to both sides gives $x^{3/2} = 64$. Taking the reciprocal (2/3) power of both sides gives

$$(x^{3/2})^{2/3} = 64^{2/3}$$

$$x^1 = (\sqrt[3]{64})^2 = 4^2 = 16$$

$$x = 16$$

A.2 EXERCISES

In Exercises 1–16, evaluate the expressions.

1. 3^3

2. $(-2)^3$

3. $-(2 \cdot 3)^2$

4. $(4 \cdot 2)^2$

5. $\left(\dfrac{-2}{3}\right)^2$

6. $\left(\dfrac{3}{2}\right)^3$

7. $(-2)^{-3}$

8. -2^{-3}

9. $\left(\dfrac{1}{4}\right)^{-2}$

10. $\left(\dfrac{-2}{3}\right)^{-2}$

11. $2 \cdot 3^0$

12. $3 \cdot (-2)^0$

13. $2^3 2^2$

14. $3^2 3$

15. $2^2 2^{-1} 2^4 2^{-4}$

16. $5^2 5^{-3} 5^2 5^{-2}$

In Exercises 17–30, simplify each expression, expressing your answer in rational form.

17. $x^3 x^2$

18. $x^4 x^{-1}$

19. $-x^2 x^{-3} y$

20. $-xy^{-1}x^{-1}$

21. $\dfrac{x^3}{x^4}$

22. $\dfrac{y^5}{y^3}$

23. $\dfrac{x^2 y^2}{x^{-1} y}$

24. $\dfrac{x^{-1} y}{x^2 y^2}$

25. $\dfrac{(xy^{-1}z^3)^2}{x^2 y z^2}$

26. $\dfrac{x^2 y z^2}{(xyz^{-1})^{-1}}$

27. $\left(\dfrac{xy^{-2}z}{x^{-1}z}\right)^3$

28. $\left(\dfrac{x^2 y^{-1} z^0}{xyz}\right)^2$

29. $\left(\dfrac{x^{-1}y^{-2}z^2}{xy}\right)^{-2}$

30. $\left(\dfrac{xy^{-2}}{x^2 y^{-1}z}\right)^{-3}$

In Exercises 31–36, convert the expressions to rational form.

31. $3x^{-4}$

32. $\dfrac{1}{2}x^{-4}$

33. $\dfrac{3}{4}x^{-2/3}$

34. $\dfrac{4}{5}y^{-3/4}$

35. $1 - \dfrac{0.3}{x^{-2}} - \dfrac{6}{5}x^{-1}$

36. $\dfrac{1}{3x^{-4}} + \dfrac{0.1x^{-2}}{3}$

In Exercises 37–56, evaluate the expressions, rounding your answer to four significant digits where necessary.

37. $\sqrt{4}$

38. $\sqrt{5}$

39. $\sqrt{\dfrac{1}{4}}$

40. $\sqrt{\dfrac{1}{9}}$

41. $\sqrt{\dfrac{16}{9}}$

42. $\sqrt{\dfrac{9}{4}}$

43. $\dfrac{\sqrt{4}}{5}$

44. $\dfrac{6}{\sqrt{25}}$

45. $\sqrt{9} + \sqrt{16}$

46. $\sqrt{25} - \sqrt{16}$

47. $\sqrt{9 + 16}$

48. $\sqrt{25 - 16}$

49. $\sqrt[3]{8 - 27}$

50. $\sqrt[4]{81 - 16}$

51. $\sqrt[3]{27/8}$

52. $\sqrt[3]{8 \times 64}$

53. $\sqrt{(-2)^2}$

54. $\sqrt{(-1)^2}$

55. $\sqrt{\dfrac{1}{4}(1 + 15)}$

56. $\sqrt{\dfrac{1}{9}(3 + 33)}$

In Exercises 57–64, simplify the expressions, given that x, y, z, a, b, and c are positive real numbers.

57. $\sqrt{a^2 b^2}$

58. $\sqrt{\dfrac{a^2}{b^2}}$

59. $\sqrt{(x + 9)^2}$

60. $(\sqrt{x + 9})^2$

61. $\sqrt[3]{x^3(a^3 + b^3)}$

62. $\sqrt[4]{\dfrac{x^4}{a^4 b^4}}$

63. $\sqrt{\dfrac{4xy^3}{x^2 y}}$

64. $\sqrt{\dfrac{4(x^2 + y^2)}{c^2}}$

In Exercises 65–80, convert the expressions to exponential form.

65. $\sqrt{3}$

66. $\sqrt{8}$

67. $\sqrt{x^3}$

68. $\sqrt[3]{x^2}$

69. $\sqrt[3]{xy^2}$

70. $\sqrt{x^2 y}$

71. $\dfrac{x^2}{\sqrt{x}}$

72. $\dfrac{x}{\sqrt{x}}$

73. $\dfrac{3}{5x^2}$

74. $\dfrac{2}{5x^{-3}}$

75. $\dfrac{3x^{-1.2}}{2} - \dfrac{1}{3x^{2.1}}$

76. $\dfrac{2}{3x^{-1.2}} - \dfrac{x^{2.1}}{3}$

77. $\dfrac{2x}{3} - \dfrac{x^{0.1}}{2} + \dfrac{4}{3x^{1.1}}$

78. $\dfrac{4x^2}{3} + \dfrac{x^{3/2}}{6} - \dfrac{2}{3x^2}$

79. $\dfrac{1}{(x^2 + 1)^3} - \dfrac{3}{4\sqrt[3]{x^2 + 1}}$

80. $\dfrac{2}{3(x^2 + 1)^{-3}} - \dfrac{3\sqrt[3]{(x^2 + 1)^7}}{4}$

In Exercises 81–92, convert the expressions to radical form.

81. $2^{2/3}$

82. $3^{4/5}$

83. $x^{4/3}$

84. $y^{7/4}$

85. $(x^{1/2}y^{1/3})^{1/5}$

86. $x^{-1/3}y^{3/2}$

87. $-\dfrac{3}{2}x^{-1/4}$

88. $\dfrac{4}{5}x^{3/2}$

89. $0.2x^{-2/3} + \dfrac{3}{7x^{-1/2}}$

90. $\dfrac{3.1}{x^{-4/3}} - \dfrac{11}{7}x^{-1/7}$

91. $\dfrac{3}{4(1 - x)^{5/2}}$

92. $\dfrac{9}{4(1 - x)^{-7/3}}$

In Exercises 93–102, simplify the expressions.

93. $4^{-1/2}4^{7/2}$

94. $2^{1/a}/2^{2/a}$

95. $3^{2/3}3^{-1/6}$

96. $2^{1/3}2^{-1}2^{2/3}2^{-1/3}$

97. $\dfrac{x^{3/2}}{x^{5/2}}$

98. $\dfrac{y^{5/4}}{y^{3/4}}$

99. $\dfrac{x^{1/2}y^2}{x^{-1/2}y}$

100. $\dfrac{x^{-1/2}y}{x^2 y^{3/2}}$

101. $\left(\dfrac{x}{y}\right)^{1/3}\left(\dfrac{y}{x}\right)^{2/3}$

102. $\left(\dfrac{x}{y}\right)^{-1/3}\left(\dfrac{y}{x}\right)^{1/3}$

In Exercises 103–116, solve each equation for x, rounding your answer to four significant digits where necessary.

103. $x^2 - 16 = 0$

104. $x^2 - 1 = 0$

105. $x^2 - \dfrac{4}{9} = 0$

106. $x^2 - \dfrac{1}{10} = 0$

107. $x^2 - (1 + 2x)^2 = 0$

108. $x^2 - (2 - 3x)^2 = 0$

109. $x^5 + 32 = 0$

110. $x^4 - 81 = 0$

111. $x^{1/2} - 4 = 0$

112. $x^{1/3} - 2 = 0$

113. $1 - \dfrac{1}{x^2} = 0$

114. $\dfrac{2}{x^3} - \dfrac{6}{x^4} = 0$

115. $(x - 4)^{-1/3} = 2$

116. $(x - 4)^{2/3} + 1 = 5$

A.3 Multiplying and Factoring Algebraic Expressions

Multiplying Algebraic Expressions

Distributive Law

The **distributive law** for real numbers states that

$$a(b \pm c) = ab \pm ac$$

$$(a \pm b)c = ac \pm bc$$

for any real numbers a, b, and c.

___

Quick Examples

1. $2(x - 3)$ is *not* equal to $2x - 3$ but is equal to $2x - 2(3) = 2x - 6$.
2. $x(x + 1) = x^2 + x$
3. $2x(3x - 4) = 6x^2 - 8x$
4. $(x - 4)x^2 = x^3 - 4x^2$
5. $(x + 2)(x + 3) = (x + 2)x + (x + 2)3 = (x^2 + 2x) + (3x + 6) = x^2 + 5x + 6$
6. $(x + 2)(x - 3) = (x + 2)x - (x + 2)3 = (x^2 + 2x) - (3x + 6) = x^2 - x - 6$

There is a quicker way of expanding expressions like the last two, called the "FOIL" method (First, Outer, Inner, Last). Consider, for instance, the expression $(x + 1)(x - 2)$. The FOIL method says: Take the product of the first terms: $x \cdot x = x^2$, the product of the outer terms: $x \cdot (-2) = -2x$, the product of the inner terms: $1 \cdot x = x$, and the product of the last terms: $1 \cdot (-2) = -2$ and then add them all up, getting $x^2 - 2x + x - 2 = x^2 - x - 2$.

Example 1 • Using FOIL

a. $(x - 2)(2x + 5) = 2x^2 + 5x - 4x - 10 = 2x^2 + x - 10$

First Outer Inner Last

b. $(x^2 + 1)(x - 4) = x^3 - 4x^2 + x - 4$
c. $(a - b)(a + b) = a^2 + ab - ab - b^2 = a^2 - b^2$
d. $(a + b)^2 = (a + b)(a + b) = a^2 + ab + ab + b^2 = a^2 + 2ab + b^2$
e. $(a - b)^2 = (a - b)(a - b) = a^2 - ab - ab + b^2 = a^2 - 2ab + b^2$

The last three products in Example 1 are particularly important and worth memorizing.

Special Formulas

$(a - b)(a + b) = a^2 - b^2$	Difference of two squares
$(a + b)^2 = a^2 + 2ab + b^2$	Square of a sum
$(a - b)^2 = a^2 - 2ab + b^2$	Square of a difference

Quick Examples
1. $(2 - x)(2 + x) = 4 - x^2$
2. $(1 + a)(1 - a) = 1 - a^2$
3. $(x + 3)^2 = x^2 + 6x + 9$
4. $(4 - x)^2 = 16 - 8x + x^2$

Here are some longer examples that require the distributive law.

Example 2 • Multiplying Algebraic Expressions

a. $(x + 1)(x^2 + 3x - 4) = (x + 1)x^2 + (x + 1)3x - (x + 1)4$
$$= (x^3 + x^2) + (3x^2 + 3x) - (4x + 4)$$
$$= x^3 + 4x^2 - x - 4$$

b $(x^2 - \dfrac{1}{x} + 1)(2x + 5) = (x^2 - \dfrac{1}{x} + 1)2x + (x^2 - \dfrac{1}{x} + 1)5$
$$= (2x^3 - 2 + 2x) + (5x^2 - \frac{5}{x} + 5)$$
$$= 2x^3 + 5x^2 + 2x + 3 - \frac{5}{x}$$

c. $(x - y)(x - y)(x - y) = (x^2 - 2xy + y^2)(x - y)$
$$= (x^2 - 2xy + y^2)x - (x^2 - 2xy + y^2)y$$
$$= (x^3 - 2x^2y + xy^2) - (x^2y - 2xy^2 + y^3)$$
$$= x^3 - 3x^2y + 3xy^2 - y^3$$

Factoring Algebraic Expressions

We can think of factoring as applying the distributive law in reverse—for example,

$$2x^2 + x = x(2x + 1)$$

which can be checked by using the distributive law. Factoring is an art that you will learn with experience and the help of a few useful techniques.

Factoring Using a Common Factor

To use this technique, locate a **common factor**—a term that occurs as a factor in each of the expressions being added or subtracted (for example, x is a common factor in $2x^2 + x$, since it is a factor of both $2x^2$ and x). Once you have located a common factor, "factor it out" by applying the distributive law.

Quick Examples
1. $2x^3 - x^2 + x$ has x as a common factor, so

$$2x^3 - x^2 + x = x(2x^2 - x + 1)$$

2. $2x^2 + 4x$ has $2x$ as a common factor, so

$$2x^2 + 4x = 2x(x + 2)$$

3. $2x^2y + xy^2 - x^2y^2$ has xy as a common factor, so

$$2x^2y + xy^2 - x^2y^2 = xy(2x + y - xy)$$

4. $(x^2 + 1)(x + 2) - (x^2 + 1)(x + 3)$ has $x^2 + 1$ as a common factor, so

$$(x^2 + 1)(x + 2) - (x^2 + 1)(x + 3) = (x^2 + 1)[(x + 2) - (x + 3)]$$
$$= (x^2 + 1)(x + 2 - x - 3)$$
$$= (x^2 + 1)(-1) = -(x^2 + 1)$$

5. $12x(x^2 - 1)^5(x^3 + 1)^6 + 18x^2(x^2 - 1)^6(x^3 + 1)^5$ has $6x(x^2 - 1)^5(x^3 + 1)^5$ as a common factor, so

$$12x(x^2 - 1)^5(x^3 + 1)^6 + 18x^2(x^2 - 1)^6(x^3 + 1)^5$$
$$= 6x(x^2 - 1)^5(x^3 + 1)^5[2(x^3 + 1) + 3x(x^2 - 1)]$$
$$= 6x(x^2 - 1)^5(x^3 + 1)^5(2x^3 + 2 + 3x^3 - 3x)$$
$$= 6x(x^2 - 1)^5(x^3 + 1)^5(5x^3 - 3x + 2)$$

We would also like to be able to reverse calculations such as $(x + 2)(2x - 5) = 2x^2 - x - 10$. That is, starting with the expression $2x^2 - x - 10$, we would like to **factor** it to get the expression $(x + 2)(2x - 5)$. An expression of the form $ax^2 + bx + c$, where a, b, and c are real numbers, is called a **quadratic** expression in x. Given a quadratic expression $ax^2 + bx + c$, we would like to write it in the form $(dx + e)(fx + g)$ for some real numbers d, e, f, and g. There are some quadratics, such as $x^2 + x + 1$, that cannot be factored in this form at all. Here, we consider only quadratics that do factor, and in such a way that the numbers d, e, f, and g are integers (other cases are discussed in Section A.5). The usual technique of factoring such quadratics is a trial-and-error approach.

Factoring Quadratics by Trial and Error

To factor the quadratic $ax^2 + bx + c$, factor ax^2 as $(a_1x)(a_2x)$ (with a_1 positive) and c as c_1c_2 and then check whether $ax^2 + bx + c = (a_1x \pm c_1)(a_2x \pm c_2)$. If not, try other factorizations of a and c.

Quick Examples

1. To factor $x^2 - 6x + 5$, first factor x^2 as $(x)(x)$, and 5 as $(5)(1)$:

$(x + 5)(x + 1) = x^2 + 6x + 5$ No good

$(x - 5)(x - 1) = x^2 - 6x + 5$ Desired factorization

2. To factor $x^2 - 4x - 12$, first factor x^2 as $(x)(x)$, and -12 as $(1)(-12)$, $(2)(-6)$, or $(3)(-4)$. Trying them one by one gives

$(x + 1)(x - 12) = x^2 - 11x - 12$ No good

$(x - 1)(x + 12) = x^2 + 11x - 12$ No good

$(x + 2)(x - 6) = x^2 - 4x - 12$ Desired factorization

3. To factor $4x^2 - 25$, we can follow the above procedure, or recognize $4x^2 - 25$ as the difference of two squares:

$$4x^2 - 25 = (2x)^2 - 5^2 = (2x - 5)(2x + 5)$$

Note Not all quadratic expressions factor. In Section A.5 we look at a test that tells us whether or not a given quadratic factors.

Here are examples requiring either a little more work or a little more thought.

Example 3 • Factoring Quadratics

Factor the following.

a. $4x^2 - 5x - 6$ **b.** $x^4 - 5x^2 + 6$

Solution

a. Possible factorizations of $4x^2$ are $(2x)(2x)$ or $(x)(4x)$. Possible factorizations of -6 are $(1)(-6)$ or $(2)(-3)$. We now systematically try out all the possibilities until we come up with the correct one:

$(2x)(2x)$ and $(1)(-6)$:	$(2x + 1)(2x - 6) = 4x^2 - 10x - 6$	No good
$(2x)(2x)$ and $(2)(-3)$:	$(2x + 2)(2x - 3) = 4x^2 - 2x - 6$	No good
$(x)(4x)$ and $(1)(-6)$:	$(x + 1)(4x - 6) = 4x^2 - 2x - 6$	No good
$(x)(4x)$ and $(2)(-3)$:	$(x + 2)(4x - 3) = 4x^2 + 5x - 6$	Almost!
Change signs:	$(x - 2)(4x + 3) = 4x^2 - 5x - 6$	Correct

b. The expression $x^4 - 5x^2 + 6$ is not a quadratic, you say? Correct, it's a quartic (a fourth-degree expression). However, it looks rather like a quadratic. In fact, it is quadratic *in* x^2, meaning that it is

$$(x^2)^2 - 5(x^2) + 6 = y^2 - 5y + 6$$

where $y = x^2$. The quadratic $y^2 - 5y + 6$ factors as

$$y^2 - 5y + 6 = (y - 3)(y - 2)$$

so

$$x^4 - 5x^2 + 6 = (x^2 - 3)(x^2 - 2)$$

This is a sometimes useful technique.

Our last example is here to remind you why we should want to factor polynomials in the first place. We'll return to this in Section A.5.

Appendix A

Example 4 • Solving a Quadratic Equation by Factoring

Solve the equation $3x^2 + 4x - 4 = 0$.

Solution We first factor the left-hand side to get

$$(3x - 2)(x + 2) = 0$$

Thus, the product of the two quantities $(3x - 2)$ and $(x + 2)$ is zero. Now, if a product of two numbers is zero, one of the two must be zero. In other words, either $3x - 2 = 0$, giving $x = \dfrac{2}{3}$, or $x + 2 = 0$, giving $x = -2$; so, there are two solutions: $x = \dfrac{2}{3}$ and $x = -2$.

A.3 EXERCISES

In Exercises 1–22, expand each expression.

1. $x(4x + 6)$

2. $(4y - 2)y$

3. $(2x - y)y$

4. $x(3x + y)$

5. $(x + 1)(x - 3)$

6. $(y + 3)(y + 4)$

7. $(2y + 3)(y + 5)$

8. $(2x - 2)(3x - 4)$

9. $(2x - 3)^2$

10. $(3x + 1)^2$

11. $\left(x + \dfrac{1}{x}\right)^2$

12. $\left(y - \dfrac{1}{y}\right)^2$

13. $(2x - 3)(2x + 3)$

14. $(4 + 2x)(4 - 2x)$

15. $\left(y - \dfrac{1}{y}\right)\left(y + \dfrac{1}{y}\right)$

16. $(x - x^2)(x + x^2)$

17. $(x^2 + x - 1)(2x + 4)$

18. $(3x + 1)(2x^2 - x + 1)$

19. $(x^2 - 2x + 1)^2$

20. $(x + y - xy)^2$

21. $(y^3 + 2y^2 + y)(y^2 + 2y - 1)$

22. $(x^3 - 2x^2 + 4)(3x^2 - x + 2)$

In Exercises 23–30, factor each expression and simplify as much as possible.

23. $(x + 1)(x + 2) + (x + 1)(x + 3)$

24. $(x + 1)(x + 2)^2 + (x + 1)^2(x + 2)$

25. $(x^2 + 1)^5(x + 3)^4 + (x^2 + 1)^6(x + 3)^3$

26. $10x(x^2 + 1)^4(x^3 + 1)^5 + 15x^2(x^2 + 1)^5(x^3 + 1)^4$

27. $(x^3 + 1)\sqrt{x + 1} - (x^3 + 1)^2\sqrt{x + 1}$

28. $(x^2 + 1)\sqrt{x + 1} - \sqrt{(x + 1)^3}$

29. $\sqrt{(x + 1)^3} + \sqrt{(x + 1)^5}$

30. $(x^2 + 1)\sqrt[3]{(x + 1)^4} - \sqrt[3]{(x + 1)^7}$

In Exercises 31–48, (a) factor the given expression and (b) set the expression equal to zero and solve for the unknown (x in the odd-numbered exercises and y in the even-numbered exercises).

31. $2x + 3x^2$

32. $y^2 - 4y$

33. $6x^3 - 2x^2$

34. $3y^3 - 9y^2$

35. $x^2 - 8x + 7$

36. $y^2 + 6y + 8$

37. $x^2 + x - 12$

38. $y^2 + y - 6$

39. $2x^2 - 3x - 2$

40. $3y^2 - 8y - 3$

41. $6x^2 + 13x + 6$

42. $6y^2 + 17y + 12$

43. $12x^2 + x - 6$

44. $20y^2 + 7y - 3$

45. $x^2 + 4xy + 4y^2$

46. $4y^2 - 4xy + x^2$

47. $x^4 - 5x^2 + 4$

48. $y^4 + 2y^2 - 3$

A.4   Rational Expressions

Rational Expression

A **rational expression** is an algebraic expression of the form P/Q, where P and Q are simpler expressions (usually polynomials) and the denominator Q is not zero.

Quick Examples

1. $\dfrac{x^2 - 3x}{x}$ $\qquad\qquad P = x^2 - 3x, Q = x$

2. $\dfrac{x + \frac{1}{x} + 1}{2x^2y + 1}$ $P = x + \dfrac{1}{x} + 1, Q = 2x^2y + 1$

3. $3xy - x^2$ $P = 3xy - x^2, Q = 1$

Algebra of Rational Expressions

We manipulate rational expressions in the same way that we manipulate fractions, using the following rules.

Algebraic Rule		Quick Example
Product	$\dfrac{P}{Q} \cdot \dfrac{R}{S} = \dfrac{PR}{QS}$	$\dfrac{x+1}{x} \cdot \dfrac{x-1}{2x+1} = \dfrac{(x+1)(x-1)}{x(2x+1)} = \dfrac{x^2-1}{2x^2+x}$
Sum	$\dfrac{P}{Q} + \dfrac{R}{S} = \dfrac{PS + RQ}{QS}$	$\dfrac{2x-1}{3x+2} + \dfrac{1}{x} = \dfrac{(2x-1)x + 1(3x+2)}{(3x+2)x}$ $= \dfrac{2x^2 + 2x + 2}{3x^2 + 2x}$
Difference	$\dfrac{P}{Q} - \dfrac{R}{S} = \dfrac{PS - RQ}{QS}$	$\dfrac{x}{3x+2} - \dfrac{x-4}{x} = \dfrac{x^2 - (x-4)(3x+2)}{(3x+2)x}$ $= \dfrac{-2x^2 + 10x + 8}{3x^2 + 2x}$
Reciprocal	$\dfrac{1}{\left(\frac{P}{Q}\right)} = \dfrac{Q}{P}$	$\dfrac{1}{\left(\frac{2xy}{3x-1}\right)} = \dfrac{3x-1}{2xy}$
Quotient	$\dfrac{\left(\frac{P}{Q}\right)}{\left(\frac{R}{S}\right)} = \dfrac{P}{Q} \cdot \dfrac{S}{R} = \dfrac{PS}{QR}$	$\dfrac{\left(\frac{x}{x-1}\right)}{\left(\frac{y-1}{y}\right)} = \dfrac{xy}{(x-1)(y-1)} = \dfrac{xy}{xy-x-y+1}$
Cancellation	$\dfrac{PR}{QR} = \dfrac{P}{Q}$	$\dfrac{(x-1)(xy+4)}{(x^2y-8)(x-1)} = \dfrac{xy+4}{x^2y-8}$

Caution Cancellation of summands is *invalid*. For instance,

$$\frac{\cancel{x} + (2xy^2 - y)}{\cancel{x} + 4y} = \frac{2xy^2 - y}{4y} \quad \times \textit{WRONG} \qquad \text{Do \textit{not} cancel a summand.}$$

$$\frac{\cancel{x}(2xy^2 - y)}{4\cancel{x}y} = \frac{2xy^2 - y}{4y} \quad \checkmark \textit{CORRECT} \qquad \text{Do cancel a factor.}$$

Here are some examples that require several algebraic operations.

Example 1 • Simplifying Rational Expressions

a. $\dfrac{\left(\frac{1}{x+y} - \frac{1}{x}\right)}{y} = \dfrac{\left[\frac{x - (x+y)}{x(x+y)}\right]}{y} = \dfrac{\left[\frac{-y}{x(x+y)}\right]}{y} = \dfrac{-y}{xy(x+y)} = -\dfrac{1}{x(x+y)}$

b. $\dfrac{(x+1)(x+2)^2 - (x+1)^2(x+2)}{(x+2)^4} = \dfrac{(x+1)(x+2)[(x+2) - (x+1)]}{(x+2)^4}$

$= \dfrac{(x+1)(x+2)(x+2-x-1)}{(x+2)^4} = \dfrac{(x+1)(x+2)}{(x+2)^4} = \dfrac{x+1}{(x+2)^3}$

c. $\dfrac{2x\sqrt{x+1} - \dfrac{x^2}{\sqrt{x+1}}}{x+1} = \dfrac{\dfrac{2x(\sqrt{x+1})^2 - x^2}{\sqrt{x+1}}}{x+1} = \dfrac{2x(x+1) - x^2}{(x+1)\sqrt{x+1}}$

$$= \dfrac{2x^2 + 2x - x^2}{(x+1)\sqrt{x+1}} = \dfrac{x^2 + 2x}{\sqrt{(x+1)^3}} = \dfrac{x(x+2)}{\sqrt{(x+1)^3}}$$

A.4 EXERCISES

In Exercises 1–16, rewrite each expression as a single rational expression, simplified as much as possible.

1. $\dfrac{x-4}{x+1} \cdot \dfrac{2x+1}{x-1}$

2. $\dfrac{2x-3}{x-2} \cdot \dfrac{x+3}{x+1}$

3. $\dfrac{x-4}{x+1} + \dfrac{2x+1}{x-1}$

4. $\dfrac{2x-3}{x-2} + \dfrac{x+3}{x+1}$

5. $\dfrac{x^2}{x+1} - \dfrac{x-1}{x+1}$

6. $\dfrac{x^2-1}{x-2} - \dfrac{1}{x-1}$

7. $\dfrac{1}{\left(\dfrac{x}{x-1}\right)} + x - 1$

8. $\dfrac{2}{\left(\dfrac{x-2}{x^2}\right)} - \dfrac{1}{x-2}$

9. $\dfrac{1}{x}\left(\dfrac{x-3}{xy} + \dfrac{1}{y}\right)$

10. $\dfrac{y^2}{x}\left(\dfrac{2x-3}{y} + \dfrac{x}{y}\right)$

11. $\dfrac{(x+1)^2(x+2)^3 - (x+1)^3(x+2)^2}{(x+2)^6}$

12. $\dfrac{6x(x^2+1)^2(x^3+2)^3 - 9x^2(x^2+1)^3(x^3+2)^2}{(x^3+2)^6}$

13. $\dfrac{(x^2-1)\sqrt{x^2+1} - \dfrac{x^4}{\sqrt{x^2+1}}}{x^2+1}$

14. $\dfrac{x\sqrt{x^3-1} - \dfrac{3x^4}{\sqrt{x^3-1}}}{x^3-1}$

15. $\dfrac{\dfrac{1}{(x+y)^2} - \dfrac{1}{x^2}}{y}$

16. $\dfrac{\dfrac{1}{(x+y)^3} - \dfrac{1}{x^3}}{y}$

A.5   Solving Polynomial Equations

Polynomial Equation

A **polynomial equation** in one unknown is an equation that can be written in the form

$$ax^n + bx^{n-1} + \cdots + rx + s = 0$$

where $a, b, \ldots, r$, and s are constants.

   We call the largest exponent of x appearing in a nonzero term of a polynomial the **degree** of that polynomial.

Quick Examples

1. $3x + 1 = 0$ has degree 1 because the largest power of x that occurs is $x = x^1$. Degree 1 equations are called **linear equations.**

2. $x^2 - x - 1 = 0$ has degree 2 because the largest power of x that occurs is x^2. Degree 2 equations are also called **quadratic equations,** or just **quadratics.**

3. $x^3 = 2x^2 + 1$ is a degree 3 polynomial (or **cubic**) in disguise. It can be rewritten as $x^3 - 2x^2 - 1 = 0$, which is in the standard form for a degree 3 equation.

4. $x^4 - x = 0$ has degree 4. It is called a **quartic.**

Now comes the question: How do we solve these equations for x? This question was asked by mathematicians as early as 1600 B.C. Let's look at these equations one degree at a time.

Solution of Linear Equations

By definition a linear equation can be written in the form

$$ax + b = 0 \qquad\qquad a \text{ and } b \text{ are fixed numbers, and } a \neq 0.$$

Solving this is a nice mental exercise: Subtract b from both sides and then divide by a, getting $x = -b/a$. Don't bother memorizing this formula, just go ahead and solve linear equations as they arise. If you feel you need practice, see the exercises at the end of the section.

Solution of Quadratic Equations

By definition a quadratic equation has the form

$$ax^2 + bx + c = 0 \qquad\qquad a, b, \text{ and } c \text{ are fixed numbers, and } a \neq 0.^2$$

The solutions of this equation are also called the **roots** of $ax^2 + bx + c$. We're assuming that you saw quadratic equations somewhere in high school but may be a little hazy about the details of their solution. There are two ways of solving these equations—one works sometimes, and the other works every time.

Solving Quadratic Equations by Factoring (Works Sometimes)
If we can factor* a quadratic equation $ax^2 + bx + c = 0$, we can solve the equation by setting each factor equal to zero.

Quick Examples
1. $x^2 + 7x + 10 = 0$

$\qquad (x + 5)(x + 2) = 0$ Factor the left-hand side.

$\qquad x + 5 = 0 \quad\text{or}\quad x + 2 = 0$ If a product is zero, one or both factors is zero.

$\qquad$ Solutions: $x = -5$ and $x = -2$

2. $2x^2 - 5x - 12 = 0$

$\qquad (2x + 3)(x - 4) = 0$ Factor the left-hand side.

$\qquad 2x + 3 = 0 \quad\text{or}\quad x - 4 = 0$

$\qquad$ Solutions: $x = -3/2$ and $x = 4$

Test for Factoring
The quadratic $ax^2 + bx + c$, with $a, b,$ and c being integers, factors into an expression of the form $(rx + s)(tx + u)$ with $r, s, t,$ and u integers precisely when the quantity $b^2 - 4ac$ is a perfect square (that is, it is the square of an integer). If this happens, we say that the quadratic **factors over the integers.**

*See the section on factoring for a review of how to factor quadratics.

^{2}What happens if $a = 0$?

___
Quick Examples

1. $x^2 + x + 1$ has $a = 1$, $b = 1$, and $c = 1$, so $b^2 - 4ac = -3$, which is not a perfect square. Therefore, this quadratic does not factor over the integers.

2. $2x^2 - 5x - 12$ has $a = 2$, $b = -5$, and $c = -12$, so $b^2 - 4ac = 121$. Since $121 = 11^2$, this quadratic does factor over the integers (we factored it above).

___

Solving Quadratic Equations with the Quadratic Formula (Works Every Time)

The solutions of the general quadratic $ax^2 + bx + c = 0$ $(a \neq 0)$ are given by

$$x = \frac{-b \pm \sqrt{b^2 - 4ac}}{2a}$$

We call the quantity $\Delta = b^2 - 4ac$ the **discriminant** of the quadratic (Δ is the uppercase Greek letter delta), and we have the following general rules:

- If Δ is positive, there are two distinct real solutions.

- If Δ is zero, there is only one real solution: $x = -\dfrac{b}{2a}$. (Why?)

- If Δ is negative, there are no real solutions.

___
Quick Examples

1. $2x^2 - 5x - 12 = 0$ has $a = 2$, $b = -5$, and $c = -12$.

$$x = \frac{-b \pm \sqrt{b^2 - 4ac}}{2a} = \frac{5 \pm \sqrt{25 + 96}}{4} = \frac{5 \pm \sqrt{121}}{4} = \frac{5 \pm 11}{4}$$

$$= \frac{16}{4} \text{ or } -\frac{6}{4} = 4 \text{ or } -3/2 \qquad \Delta \text{ is positive in this example.}$$

2. $4x^2 = 12x - 9$ can be rewritten as $4x^2 - 12x + 9 = 0$, which has $a = 4$, $b = -12$, and $c = 9$.

$$x = \frac{-b \pm \sqrt{b^2 - 4ac}}{2a} = \frac{12 \pm \sqrt{144 - 144}}{8} = \frac{12 \pm 0}{8} = \frac{12}{8} = \frac{3}{2}$$

Δ is zero in this example.

3. $x^2 + 2x - 1 = 0$ has $a = 1$, $b = 2$, and $c = -1$.

$$x = \frac{-b \pm \sqrt{b^2 - 4ac}}{2a} = \frac{-2 \pm \sqrt{8}}{2} = \frac{-2 \pm 2\sqrt{2}}{2} = -1 \pm \sqrt{2}$$

The two solutions are $x = -1 + \sqrt{2} = 0.414\ldots$ and $x = -1 - \sqrt{2} = -2.414\ldots$. $\qquad \Delta$ is positive in this example.

4. $x^2 + x + 1 = 0$ has $a = 1$, $b = 1$, and $c = 1$. Since $\Delta = -3$ is negative, there are no real solutions.

Question This is all very useful, but where on earth does the quadratic formula come from?

Answer To see where it comes from, we will solve a general quadratic equation using "brute force." Start with the general quadratic equation

$$ax^2 + bx + c = 0$$

First, divide out the nonzero number a to get

$$x^2 + \frac{bx}{a} + \frac{c}{a} = 0$$

Now we **complete the square:** Add and subtract the quantity $\frac{b^2}{4a^2}$ to get

$$x^2 + \frac{bx}{a} + \frac{b^2}{4a^2} - \frac{b^2}{4a^2} + \frac{c}{a} = 0$$

We do this to get the first three terms to factor as a perfect square:

$$\left(x + \frac{b}{2a}\right)^2 - \frac{b^2}{4a^2} + \frac{c}{a} = 0$$

(Check this by multiplying out.) Adding $\frac{b^2}{4a^2} - \frac{c}{a}$ to both sides gives

$$\left(x + \frac{b}{2a}\right)^2 = \frac{b^2}{4a^2} - \frac{c}{a} = \frac{b^2 - 4ac}{4a^2}$$

Taking square roots gives

$$x + \frac{b}{2a} = \frac{\pm\sqrt{b^2 - 4ac}}{2a}$$

Finally, adding $-\frac{b}{2a}$ to both sides yields the result:

$$x = -\frac{b}{2a} + \frac{\pm\sqrt{b^2 - 4ac}}{2a}$$

$$= \frac{-b \pm \sqrt{b^2 - 4ac}}{2a}$$

Solution of Cubic Equations

By definition, a cubic equation can be written in the form

$$ax^3 + bx^2 + cx + d = 0 \quad a, b, c, \text{ and } d \text{ are fixed numbers, and } a \neq 0.$$

Now we get into something of a bind. Although there is a perfectly respectable formula for the solutions, it is very complicated and involves the use of complex numbers rather heavily.[3] So we discuss instead a much simpler method that *sometimes* works nicely. Here is the method in a nutshell.

---

[3] It was when this formula was discovered in the 16th century that complex numbers were first taken seriously. Although we would like to show you the formula, it is too large to fit in this footnote.

Solving Cubics by Finding One Factor

Start with a given cubic equation $ax^3 + bx^2 + cx + d = 0$.

Step 1 By trial and error, find one solution $x = s$. If a, b, c, and d are integers, the only possible *rational* solutions* are those of the form $s = \pm$(factor of d)/(factor of a).

Step 2 It will now be possible to factor the cubic as

$$ax^3 + bx^2 + cx + d = (x - s)(ax^2 + ex + f) = 0$$

To find $ax^2 + ex + f$, divide the cubic by $x - s$ using long division.†

Step 3 The factored equation says that either $x - s = 0$ or $ax^2 + ex + f = 0$. We already know that s is a solution, and now we see that the other solutions are the roots of the quadratic. Note that this quadratic may or may not have any real solutions, as usual.

Quick Example

To solve the cubic $x^3 - x^2 + x - 1 = 0$, we first find a single solution. Here, $a = 1$ and $d = -1$. Since the only factors of ± 1 are ± 1, the only possible rational solutions are $x = \pm 1$. By substitution we see that $x = 1$ is a solution. Thus, $(x - 1)$ is a factor. Dividing by $(x - 1)$ yields the quotient $(x^2 + 1)$. Thus,

$$x^3 - x^2 + x - 1 = (x - 1)(x^2 + 1) = 0$$

so either $x - 1 = 0$ or $x^2 + 1 = 0$.

Since the discriminant of the quadratic $x^2 + 1$ is negative, we don't get any real solutions from $x^2 + 1 = 0$, so the only real solution is $x = 1$.

Possible Outcomes When Solving a Cubic Equation

If you consider all the cases, there are three possible outcomes when solving a cubic equation:

1. One real solution (as in the Quick Example above)

2. Two real solutions (try, for example, $x^3 + x^2 - x - 1 = 0$)

3. Three real solutions (see the next example)

---

*There may be *irrational* solutions, however; for example, $x^3 - 2 = 0$ has the single solution $x = \sqrt[3]{2}$.
†Alternatively, use "synthetic division," a shortcut that would take us too far afield to describe.

Example 1 • Solving a Cubic

Solve the cubic $2x^3 - 3x^2 - 17x + 30 = 0$.

Solution First we look for a single solution. Here, $a = 2$ and $d = 30$. The factors of a are ± 1 and ± 2, and the factors of d are ± 1, ± 2, ± 3, ± 5, ± 6, ± 10, ± 15, and ± 30. This gives us a large number of possible ratios: ± 1, ± 2, ± 3, ± 5, ± 6, ± 10, ± 15, ± 30, $\pm 1/2$, $\pm 3/2$, $\pm 5/2$, and $\pm 15/2$. Undaunted, we first try $x = 1$ and $x = -1$, getting nowhere. So we move on to $x = 2$, and we hit the jackpot, since substituting $x = 2$ gives

$16 - 12 - 34 + 30 = 0$. Thus, $(x - 2)$ is a factor. Dividing yields the quotient $2x^2 + x - 15$. Here is the calculation:

$$
\begin{array}{r}
2x^2 + x - 15 \\
\hline
x - 2 \overline{\smash{)}\, 2x^3 - 3x^2 - 17x + 30} \\
2x^3 - 4x^2 \\
\hline
x^2 - 17x \\
x^2 - 2x \\
\hline
-15x + 30 \\
-15x + 30 \\
\hline
0
\end{array}
$$

Therefore,

$$2x^3 - 3x^2 - 17x + 30 = (x - 2)(2x^2 + x - 15) = 0$$

Setting the factors equal to zero gives either $x - 2 = 0$ or $2x^2 + x - 15 = 0$. We could solve the quadratic using the quadratic formula, but luckily, we notice that it factors as

$$2x^2 + x - 15 = (x + 3)(2x - 5)$$

Thus, the solutions are $x = 2$, $x = -3$, and $x = 5/2$.

Solution of Higher-Order Polynomial Equations

Logically speaking, our next step should be a discussion of quartics, then quintics (fifth-degree equations), and so on forever. Well, we've got to stop somewhere, and cubics may be as good a place as any. On the other hand, since we've gotten so far, we ought to at least tell you what is known about higher-order polynomials.

Quartics   Just as in the case of cubics, there is a formula to find the solutions of quartics.[4]

Quintics and Beyond   All good things must come to an end, we're afraid. It turns out that there is no "quintic formula." In other words, there is no single algebraic formula or collection of algebraic formulas that gives the solutions to all quintics. This question was settled by the Norwegian mathematician Niels Henrik Abel in 1824 after almost 300 years of controversy about this question. (In fact, several notable mathematicians had previously claimed to have devised formulas for solving the quintic, but these were all shot down by other mathematicians—this being one of the favorite pastimes of practitioners of our art.) The same negative answer applies to polynomial equations of degree 6 and higher. It's not that these equations don't have solutions, just that they can't be found using algebraic formulas.[5] However, there are certain special classes of polynomial equations that can be solved with algebraic methods. The way of identifying such equations was discovered around 1829 by the French mathematician Évariste Galois.[6]

---

[4]See, for example, L. E. Dickson, *First Course in the Theory of Equations* (New York: Wiley, 1922), or B. L. van der Waerden, *Modern Algebra* (New York: Frederick Ungar, 1953).

[5]What we mean by an "algebraic formula" is a formula in the coefficients using the operations of addition, subtraction, division, and the taking of radicals. Mathematicians call the use of such formulas in solving polynomial equations "solution by radicals." If you were a math major, you would eventually go on to study this under the heading of Galois theory.

[6]Both Abel and Galois died young. Abel died of tuberculosis at the age of 26, and Galois was killed in a duel at the age of 20.

A.5 EXERCISES

In Exercises 1–12, solve the equations for x (mentally, if possible).

1. $x + 1 = 0$

2. $x - 3 = 1$

3. $-x + 5 = 0$

4. $2x + 4 = 1$

5. $4x - 5 = 8$

6. $\frac{3}{4}x + 1 = 0$

7. $7x + 55 = 98$

8. $3x + 1 = x$

9. $x + 1 = 2x + 2$

10. $x + 1 = 3x + 1$

11. $ax + b = c \quad (a \neq 0)$

12. $x - 1 = cx + d \quad (c \neq 1)$

In Exercises 13–30, by any method, determine all real solutions of each equation. Check your answers by substitution.

13. $2x^2 + 7x - 4 = 0$

14. $x^2 + x + 1 = 0$

15. $x^2 - x + 1 = 0$

16. $2x^2 - 4x + 3 = 0$

17. $2x^2 - 5 = 0$

18. $3x^2 - 1 = 0$

19. $-x^2 - 2x - 1 = 0$

20. $2x^2 - x - 3 = 0$

21. $\frac{1}{2}x^2 - x - \frac{3}{2} = 0$

22. $-\frac{1}{2}x^2 - \frac{1}{2}x + 1 = 0$

23. $x^2 - x = 1$

24. $16x^2 = -24x - 9$

25. $x = 2 - \frac{1}{x}$

26. $x + 4 = \frac{1}{x - 2}$

27. $x^4 - 10x^2 + 9 = 0$

28. $x^4 - 2x^2 + 1 = 0$

29. $x^4 + x^2 - 1 = 0$

30. $x^3 + 2x^2 + x = 0$

In Exercises 31–44, find all real solutions of each equation.

31. $x^3 + 6x^2 + 11x + 6 = 0$

32. $x^3 - 6x^2 + 12x - 8 = 0$

33. $x^3 + 4x^2 + 4x + 3 = 0$

34. $y^3 + 64 = 0$

35. $x^3 - 1 = 0$

36. $x^3 - 27 = 0$

37. $y^3 + 3y^2 + 3y + 2 = 0$

38. $y^3 - 2y^2 - 2y - 3 = 0$

39. $x^3 - x^2 - 5x + 5 = 0$

40. $x^3 - x^2 - 3x + 3 = 0$

41. $2x^6 - x^4 - 2x^2 + 1 = 0$

42. $3x^6 - x^4 - 12x^2 + 4 = 0$

43. $(x^2 + 3x + 2)(x^2 - 5x + 6) = 0$

44. $(x^2 - 4x + 4)^2(x^2 + 6x + 5)^3 = 0$

A.6   Solving Miscellaneous Equations

Equations often arise in calculus that are not polynomial equations of low degree. Many of these complicated-looking equations can be solved easily if you remember the following fact, which we used in the previous section.

Solving an Equation of the Form $P \cdot Q = 0$

If a product is equal to 0, then at least one of the factors must be 0. That is, if $P \cdot Q = 0$, then either $P = 0$ or $Q = 0$.

Quick Examples

1. $x^5 - 4x^3 = 0$

$\quad x^3(x^2 - 4) = 0$ Factor the left-hand side.

$\quad$ Either $x^3 = 0$ or $x^2 - 4 = 0$. Either $P = 0$ or $Q = 0$.

$\quad x = 0, 2,$ or -2 Solve the individual equations.

2. $(x^2 - 1)(x + 2) + (x^2 - 1)(x + 4) = 0$

$\quad (x^2 - 1)[(x + 2) + (x + 4)] = 0$ Factor the left-hand side.

$\quad (x^2 - 1)(2x + 6) = 0$

$\quad$ Either $x^2 - 1 = 0$ or $2x + 6 = 0$. Either $P = 0$ or $Q = 0$.

$\quad x = -3, -1,$ or 1 Solve the individual equations.

Example 1 • Solving by Factoring

Solve $12x(x^2 - 4)^5(x^2 + 2)^6 + 12x(x^2 - 4)^6(x^2 + 2)^5 = 0$.

Solution Again, we start by factoring the left-hand side:

$$12x(x^2 - 4)^5(x^2 + 2)^6 + 12x(x^2 - 4)^6(x^2 + 2)^5$$

$$= 12x(x^2 - 4)^5(x^2 + 2)^5[(x^2 + 2) + (x^2 - 4)]$$

$$= 12x(x^2 - 4)^5(x^2 + 2)^5(2x^2 - 2)$$

$$= 24x(x^2 - 4)^5(x^2 + 2)^5(x^2 - 1)$$

Setting this equal to 0, we get

$$24x(x^2 - 4)^5(x^2 + 2)^5(x^2 - 1) = 0$$

which means that at least one of the factors of this product must be zero. Now it certainly cannot be the 24, but it could be the x: $x = 0$ is one solution. It could also be that

$$(x^2 - 4)^5 = 0 \quad \text{or} \quad x^2 - 4 = 0$$

which has solutions $x = \pm 2$. Could it be that $(x^2 + 2)^5 = 0$? If so, then $x^2 + 2 = 0$, but this is impossible because $x^2 + 2 \geq 2$ no matter what x is. Finally, it could be that $x^2 - 1 = 0$, which has solutions $x = \pm 1$. This gives us five solutions to the original equation:

$$x = -2, -1, 0, 1, \text{ or } 2$$

Example 2 • Solving by Factoring

Solve $(x^2 - 1)(x^2 - 4) = 10$.

Solution   Watch out! You may be tempted to say that $x^2 - 1 = 10$ or $x^2 - 4 = 10$, but this does not follow. If two numbers multiply to give you 10, what must they be? There are lots of possibilities: 2 and 5, 1 and 10, $-500{,}000$ and $-0.000\,02$ are just a few. The fact that the left-hand side is factored is nearly useless to us if we want to solve this equation. What we will have to do is multiply out, bring the 10 over to the left, and hope that we can factor what we get. Here goes:

$$x^4 - 5x^2 + 4 = 10$$

$$x^4 - 5x^2 - 6 = 0$$

$$(x^2 - 6)(x^2 + 1) = 0$$

(Here we used a sometimes useful trick that we mentioned in Section A.3: We treated x^2 like x and x^4 like x^2, so factoring $x^4 - 5x^2 - 6$ is essentially the same as factoring $x^2 - 5x - 6$.) *Now* we are allowed to say that one of the factors must be 0: $x^2 - 6 = 0$ has solutions $x = \pm\sqrt{6} = \pm 2.449\ldots$ and $x^2 + 1 = 0$ has no real solutions. Therefore, we get exactly two solutions, $x = \pm\sqrt{6} = \pm 2.449\ldots$.

To solve equations involving rational expressions, the following rule is very useful.

Solving an Equation of the Form P/Q = 0

If $\dfrac{P}{Q} = 0$, then $P = 0$.

How else could a fraction equal 0? If that is not convincing, multiply both sides by Q (which cannot be 0 if the quotient is defined).

Quick Example

Solve $\dfrac{(x + 1)(x + 2)^2 - (x + 1)^2(x + 2)}{(x + 2)^4} = 0$.

$(x + 1)(x + 2)^2 - (x + 1)^2(x + 2) = 0$ If $\dfrac{P}{Q} = 0$, then $P = 0$.

$(x + 1)(x + 2)[(x + 2) - (x + 1)] = 0$ Factor.

$(x + 1)(x + 2)(1) = 0$

Either $x + 1 = 0$ or $x + 2 = 0$

$x = -1$ or $x = -2$

$x = -1$ $x = -2$ does not make sense in the original equation: It makes the denominator 0. So it is not a solution, and $x = -1$ is the only solution.

Example 3 • Solving a Rational Equation

Solve $1 - \dfrac{1}{x^2} = 0$.

Solution   Write 1 as $\frac{1}{1}$, so that we now have a difference of two rational expressions:

$$\frac{1}{1} - \frac{1}{x^2} = 0$$

To combine these we can put both over a common denominator of x^2, which gives

$$\frac{x^2 - 1}{x^2} = 0$$

Now we can set the numerator, $x^2 - 1$, equal to zero. Thus,

$$x^2 - 1 = 0$$

$$(x - 1)(x + 1) = 0$$

giving $x = \pm 1$.

✸ *Before we go on . . .* This equation could also have been solved by writing

$$1 = \frac{1}{x^2}$$

and then multiplying both sides by x^2.

Example 4 • Solving Another Rational Equation

Solve $\dfrac{2x - 1}{x} + \dfrac{3}{x - 2} = 0$.

Solution   We *could* first perform the addition on the left and then set the top equal to 0, but here is another approach. Subtracting the second expression from both sides gives

$$\frac{2x - 1}{x} = \frac{-3}{x - 2}$$

Cross-multiplying [multiplying both sides by both denominators—that is, by $x(x - 2)$] now gives

$$(2x - 1)(x - 2) = -3x$$

$$2x^2 - 5x + 2 = -3x$$

Adding $3x$ to both sides gives the quadratic equation

$$2x^2 - 2x + 2 = 0$$

The discriminant is $(-2)^2 - 4 \cdot 2 \cdot 2 = -12 < 0$, so we conclude that there is no real solution.

✸ *Before we go on . . .* Notice that when we said that $(2x - 1)(x - 2) = -3x$, we were *not* allowed to conclude that $2x - 1 = -3x$ or $x - 2 = -3x$.

Appendix A

Example 5 • Solving a Rational Equation with Radicals

Solve $\dfrac{2x\sqrt{x+1} - \dfrac{x^2}{\sqrt{x+1}}}{x+1} = 0$.

Solution   Setting the top equal to 0 gives

$$2x\sqrt{x+1} - \frac{x^2}{\sqrt{x+1}} = 0$$

This still involves fractions. To get rid of the fractions, we could put everything over a common denominator $(\sqrt{x+1})$ and then set the top equal to 0, or we could multiply the whole equation by that common denominator in the first place to clear fractions. If we do the second, we get

$$2x(x+1) - x^2 = 0$$

$$2x^2 + 2x - x^2 = 0$$

$$x^2 + 2x = 0$$

Factoring,

$$x(x+2) = 0$$

so either $x = 0$ or $x + 2 = 0$, giving us $x = 0$ or $x = -2$. Again, one of these is not really a solution. The problem is that $x = -2$ cannot be substituted into $\sqrt{x+1}$ since we would then have to take the square root of -1, and we are not allowing ourselves to do that. Therefore, $x = 0$ is the only solution.

A.6 EXERCISES

Solve the following equations.

1. $x^4 - 3x^3 = 0$

2. $x^6 - 9x^4 = 0$

3. $x^4 - 4x^2 = -4$

4. $x^4 - x^2 = 6$

5. $(x+1)(x+2) + (x+1)(x+3) = 0$

6. $(x+1)(x+2)^2 + (x+1)^2(x+2) = 0$

7. $(x^2+1)^5(x+3)^4 + (x^2+1)^6(x+3)^3 = 0$

8. $10x(x^2+1)^4(x^3+1)^5 - 10x^2(x^2+1)^5(x^3+1)^4 = 0$

9. $(x^3+1)\sqrt{x+1} - (x^3+1)^2\sqrt{x+1} = 0$

10. $(x^2+1)\sqrt{x+1} - \sqrt{(x+1)^3} = 0$

11. $\sqrt{(x+1)^3} + \sqrt{(x+1)^5} = 0$

12. $(x^2+1)\sqrt[3]{(x+1)^4} - \sqrt[3]{(x+1)^7} = 0$

13. $(x+1)^2(2x+3) - (x+1)(2x+3)^2 = 0$

14. $(x^2-1)^2(x+2)^3 - (x^2-1)^3(x+2)^2 = 0$

15. $\dfrac{(x+1)^2(x+2)^3 - (x+1)^3(x+2)^2}{(x+2)^6} = 0$

16. $\dfrac{6x(x^2+1)^2(x^2+2)^4 - 8x(x^2+1)^3(x^2+2)^3}{(x^2+2)^8} = 0$

17. $\dfrac{2(x^2-1)\sqrt{x^2+1} - \dfrac{x^4}{\sqrt{x^2+1}}}{x^2+1} = 0$

18. $\dfrac{4x\sqrt{x^3-1} - \dfrac{3x^4}{\sqrt{x^3-1}}}{x^3-1} = 0$

19. $x - \dfrac{1}{x} = 0$

20. $1 - \dfrac{4}{x^2} = 0$

21. $\dfrac{1}{x} - \dfrac{9}{x^3} = 0$

22. $\dfrac{1}{x^2} - \dfrac{1}{x+1} = 0$

23. $\dfrac{x-4}{x+1} - \dfrac{x}{x-1} = 0$

24. $\dfrac{2x-3}{x-1} - \dfrac{2x+3}{x+1} = 0$

25. $\dfrac{x+4}{x+1} + \dfrac{x+4}{3x} = 0$

26. $\dfrac{2x-3}{x} - \dfrac{2x-3}{x+1} = 0$

ANSWERS TO SELECTED EXERCISES

Chapter 1

Exercises 1.1

1. a. 2 **b.** 0.5 **3. a.** −1.5 **b.** 8 **c.** −8 **5. a.** −7 **b.** −3 **c.** 1 **d.** $4y − 3$ **e.** $4(a + b) − 3$ **7. a.** 3 **b.** 6 **c.** 2 **d.** 6 **e.** $a^2 + 2a + 3$ **f.** $(x + h)^2 + 2(x + h) + 3$ **9. a.** 2 **b.** 0 **c.** 65/4 **d.** $x^2 + 1/x$ **e.** $(s + h)^2 + 1/(s + h)$ **f.** $(s + h)^2 + 1/(s + h) − (s^2 + 1/s)$ **11. a.** 1 **b.** 1 **c.** 0 **d.** 27 **13. a.** Yes; $f(4) = 63/16$ **b.** Not defined **c.** Not defined **15. a.** Not defined **b.** Not defined **c.** Yes; $f(−10) = 0$ **17. a.** $h(2x + h)$ **b.** $2x + h$ **19. a.** $−h(2x + h)$ **b.** $−(2x + h)$
21. $0.1*x^2-4*x+5$

x	0	1	2	3	4	5	6	7	8	9	10
f(x)	5	1.1	-2.6	-6.1	-9.4	-12.5	-15.4	-18.1	-20.6	-22.9	-25

23. $(x^2-1)/(x^2+1)$

x	0.5	1.5	2.5	3.5	4.5	5.5	6.5	7.5	8.5	9.5	10.5
h(x)	-0.6000	0.3846	0.7241	0.8491	0.9059	0.9360	0.9538	0.9651	0.9727	0.9781	0.9820

25. a. $P(5) = 117$, $P(10) = 132$, and $P(9.5) \approx 131$. Approximately 117 million people were employed in the United States on July 1, 1995; 132 million people on July 1, 2000; and 131 million people on January 1, 2000. **b.** [5, 11] **27. a.** $C(0) = 800$, $C(4) = 2800$, $C(5) = 4100$. Thus, there were 800 coffee shops in 1990, 2800 in 1994, and 4100 in 1995. **b.** 1996 **c.**

t	0	1	2	3	4	5	6	7	8	9	10
C(t)	800	1300	1800	2300	2800	4100	5400	6700	8000	9300	10,600

29. a. $(0.08*t+0.6)*(t<8)+(0.355*t-1.6)*(t>=8)$

t	0	1	2	3	4	5	6	7	8	9	10	11
C(t)	0.6	0.68	0.76	0.84	0.92	1	1.08	1.16	1.24	1.595	1.95	2.305

b. 0.355 **31.** $T(26,000) = \$600 + 0.15(26,000 − 6000) = \$3600.00$; $T(65,000) = \$3757.50 + 0.27(65,000 − 27,050) =$ $14,004 **33. a.** 358,600 **b.** 361,200 **c.** $6 **35. a.** $0 \le t \le 8$ **b.** $t \ge 0$ is not an appropriate domain because it would predict investments in South Africa into the indefinite future with no basis. (It would also lead to preposterous results for large values of t.) **37. a.** (2) **b.** $36.8 billion **39. a.** $12,000 **b.** $N(q) = 2000 + 100q^2 − 500q$; $N(20) = \$32,000$
41. a. $100*(1-12200/t^4.48)$ **b.**

t	9	10	11	12	13	14	15	16	17	18	19	20
p(t)	35.2	59.6	73.6	82.2	87.5	91.1	93.4	95.1	96.3	97.1	97.7	98.2

c. 82.2% **d.** 14 months **43.** t; m **45.** $y(x) = 4x^2 − 2$, or $f(x) = 4x^2 − 2$ **47.** $N = 200 + 10t$ (N = number of sound files, t = time in days) **49.** As the text reminds us: To evaluate f of a quantity (such as $x + h$) replace x everywhere by the *whole quantity* $x + h$, getting $f(x + h) = (x + h)^2 − 1$. **51.** False: Functions with infinitely many points in their domain [such as $f(x) = x^2$] cannot be specified numerically.

Exercises 1.2

1. a. 20 **b.** 30 **c.** 30 **d.** 20 **e.** 0 **3. a.** −1 **b.** 1.25 **c.** 0 **d.** 1 **e.** 0 **5. a.** (I) **b.** (IV) **c.** (V) **d.** (VI) **e.** (III) **f.** (II)

7.
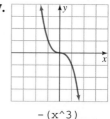
$-(x^3)$

9.

x^4

11.
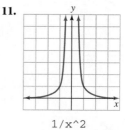
$1/x^2$

13. a. −1 **b.** 2 **c.** 2
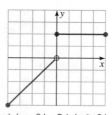
$x*(x<0)+2*(x\ge0)$

15. a. 1 **b.** 0 **c.** 1

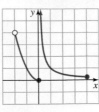

```
(x^2)*(x≤0)+
(1/x)*(0<x)
```

17. a. 0 **b.** 2 **c.** 3 **d.** 3

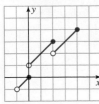

```
x*(x≤0)
+(x+1)*(0<x)*(x≤2)
+x*(2<x)
```

19. $f(6) \approx 2000$, $f(9) \approx 2800$, $f(7.5) \approx 2500$. In 1996, 2,000,000 SUVs were sold. In 1999, 2,800,000 were sold, and in the year beginning July 1997, 2,500,000 were sold. **21.** $f(6) - f(5)$; SUV sales increased more from 1995 to 1996 than from 1999 to 2000. **23. a.** $[-1.5, 1.5]$ **b.** $N(-0.5) \approx 131$, $N(0) \approx 132$, $N(1) \approx 132$. In July 1999 approximately 131 million people were employed. In January 2000 and January 2001, approximately 132 million people were employed. **c.** $[0.5, 1.5]$; employment was falling during the period July 2000–July 2001. **25. a.** (C) **b.** $20.80/shirt if the team buys 70 shirts **b.** Graph:

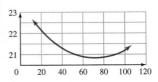

27. A quadratic model (B) is the best choice; the other models either predict perpetually increasing tourism or perpetually decreasing tourism. In fact, a parabola passes exactly through the three data points. **29. a.** `100*(1-12200/t^4.48)` **b.** Graph:

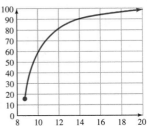

c. 82% **d.** 14 months **31.** 1999 Graph:

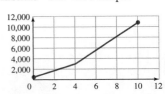

33. a. `(0.08*t+0.6)*(t<8)+(0.355*t-1.6)*(t>=8)` **b.** Graph:

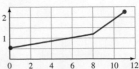

35. True. We can construct a table of values from any graph by reading off a set of values. **37.** False. In a numerically specified function, only certain values of the function are specified, giving only certain points on the graph. **39.** They are different portions of the graph of the associated equation $y = f(x)$. **41.** The graph of $g(x)$ is the same as the graph of $f(x)$ but shifted 5 units to the right.

Exercises 1.3

1.

x	−1	0	1
y	5	8	11

$m = 3$

3.

x	2	3	5
f(x)	−1	−2	−4

$m = -1$

5.

x	−2	0	2
f(x)	4	7	10

$m = 3/2$

7. $f(x) = -x/2 - 2$ **9.** $f(0) = -5$, $f(x) = -x - 5$ **11.** f is linear: $f(x) = 4x + 6$ **13.** g is linear: $g(x) = 2x - 1$ **15.** $-3/2$ **17.** $1/6$ **19.** Undefined **21.** 0 **23.** $-4/3$

25.

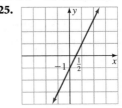

27.

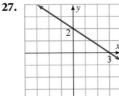

29.

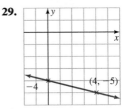

31.

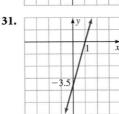

33.

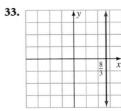

35.

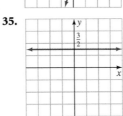

37.

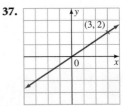

39. 2 **41.** 2 **43.** −2 **45.** Undefined **47** 1.5 **49.** −0.09 **51.** 1/2 **53.** $(d - b)/(c - a)$ **55. a.** 1 **b.** 1/2 **c.** 0 **d.** 3 **e.** −1/3 **f.** −1 **g.** Undefined **h.** −1/4 **i.** −2 **57.** $y = 3x$ **59.** $y = \frac{1}{4}x - 1$ **61.** $y = 10x - 203.5$ **63.** $y = -5x + 6$ **65.** $y = -3x + 2.25$ **67.** $y = -x + 12$ **69.** $y = 2x + 4$

71. Compute the corresponding successive changes Δx in x and Δy in y and compute the ratios $\Delta y / \Delta x$. If the answer is always the same number, then the values in the table come from a linear function.   **73.** $f(x) = -\dfrac{a}{b}x + \dfrac{c}{b}$. If $b = 0$, then $\dfrac{a}{b}$ is undefined, and y cannot be specified as a function of x. (The graph of the resulting equation would be a vertical line.)   **75.** Slope, 3   **77.** If m is positive, then y will increase as x increases; if m is negative, then y will decrease as x increases; if m is zero, then y will not change as x changes.   **79.** The slope increases because an increase in the y coordinate of the second point increases Δy while leaving Δx fixed.

Exercises 1.4

1. $C(x) = 1500x + 1200$ per day **a.** $5700 **b.** $1500 **c.** $1500
3. Fixed cost = $8000, marginal cost = $25 per bicycle
5. a. $C(x) = 0.4x + 70$, $R(x) = 0.5x$, $P(x) = 0.1x - 70$
b. $P(500) = -20$; a loss of $20 **c.** 700 copies   **7.** $q = -40p + 2000$ **9.** $q = -0.15p + 56.25$. The demand equation predicts that demand increases as the price goes down. However, the first-quarter 1998 price was lower than the third-quarter 1997 price, and yet the demand was also lower.   **11. a.** Demand: $q = -60p + 150$; supply: $q = 80p - 60$ **b.** $1.50 each
13. a. (1996, 125) and (1997, 135) or (1998, 140) and (1999, 150) **b.** The number of new in-ground pools increased most rapidly during the periods 1996–1997 and 1998–1999, when it rose by 10,000 new pools in a year.   **15.** $N = 400 + 50t$ million transactions. The slope gives the additional number of online shopping transactions per year and is measured in (millions of) transactions per year.   **17. a.** $s = 14.4t + 240$; Medicare spending is predicted to rise at a rate of $14.4 billion/year **b.** $816 billion   **19. a.** 2.5 feet/second **b.** 20 feet along the track **c.** After 6 seconds   **21. a.** 130 miles/hour **b.** $s = 130t - 1300$ **23.** $F = 1.8C + 32$; 86°F; 72°F; 14°F; 7°F   **25.** $I(N) = 0.05N + 50,000$; $N = $1,000,000$; marginal income is $m = $ 5¢/dollar of net profit.   **27. a.** $c = 0.05n + 1.7$ **b.** The slope represents the increase in daily circulation per additional newspaper that a company publishes. **c.** The model predicts that circulation would increase to 6.35 million.   **29.** $w = 2n - 58$; 42 billion pounds
31. $c = 0.075m - 1.5$; 0.75 pound   **33.** $T(r) = (1/4)r + 45$; $T(100) = 70°F$ **35.** $P(x) = 100x - 5132$, with domain $[0, 405]$. For profit, $x \geq 52$.   **37.** 5000 units   **39.** $FC/(SP - VC)$
41. $P(x) = 579.7x - 20,000$, with domain $x \geq 0$; $x = 34.50$ g/day for break even   **43.** Increasing by $355,000/year   **45. a.** $p = 10t + 15$ **b.** $p = 15t + 10$ **c.** $p = \begin{cases} 10t + 15 & \text{if } 0 \leq t < 1 \\ 15t + 10 & \text{if } 1 \leq t \leq 3 \end{cases}$

d. 40%   **47.** $C(t) = \begin{cases} -1400t + 30,000 & \text{if } 0 \leq t \leq 5 \\ 7400t - 14,000 & \text{if } 5 < t \leq 10 \end{cases}$; $C(3) = $ 25,800 students

49. $d(r) = \begin{cases} -40r + 74 & \text{if } 1.1 \leq r \leq 1.3 \\ \dfrac{130r}{3} - \dfrac{103}{3} & \text{if } 1.3 < r \leq 1.6 \end{cases}$; $d(1) = 34\%$

51. Bootlags per Martian yen; bootlags   **53.** (B)   **55.** It must increase by 10 units/day, including the third.   **57.** Increasing the number of items from the break even results in a profit: Since the

slope of the revenue graph is larger than the slope of the cost graph, it is higher than the cost graph to the right of the point of intersection, and hence corresponds to a profit.

Exercises 1.5

1. 6 **3.** 86 **5. a.** 0.5 (better fit) **b.** 0.75 **7. a.** 27.42 **b.** 27.16 (better fit)
9.

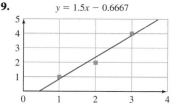
$y = 1.5x - 0.6667$

11.

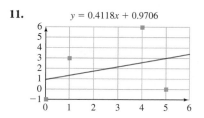
$y = 0.4118x + 0.9706$

13. a. $r = 0.9959$ (best, not perfect) **b.** $r = 0.9538$ **c.** $r = 0.3273$ (worst)   **15.** $y = -0.141t + 6.84$, $y(25) \approx 3.3 billion   **17.** $y = 1.46x + 6.97$, $y(10) \approx 22$

19. a.

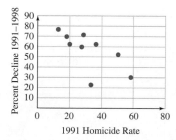
$y = -0.34t + 5.92$

b. Camera sales were declining at a rate of 0.34 million/year. **c.** Use of such a linear model to make predictions is not reasonable because the possible factors that may influence sales are not taken into account; 4.2 million cameras **d.** 5.8 million cameras. This is large compared with the other residues, confirming the point raised in part (c).   **21.** 7500 phone lines   **23.** Increasing at the rate of 0.27 year/year   **25. a.** The graph suggests that the percentage rate decreases as the size of the 1991 homicide rate increases.

b. Regression equation: $y \approx -0.9148x + 85.441$; $r \approx -0.689$. The slope is negative, supporting the observation that the percentage decreases with increasing 1991 homicide rate. The value of r suggests a reasonable fit. **c.** The percent decrease in the homicide rate declines by 0.91% for every 1-point increase in the

1991 homicide rate. **d.** 48.8%   **27.** The line that passes through (a, b) and (c, d) gives a sum-of-squares error SSE = 0, which is the smallest value possible.   **29.** 0   **31.** The regression line is the line passing through the given points.   **33.** No. The regression line through $(-1, 1)$, $(0, 0)$, and $(1, 1)$ passes through none of these points.

Chapter 1 Review Test

1. a.

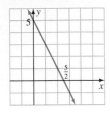

b.

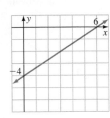

c.

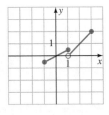

d.

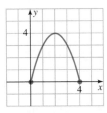

2. Absolute value, linear, linear, exponential, quadratic   **3. a.** $y = -3x + 11$ **b.** $y = -x + 1$ **c.** $y = (1/2)x + 3/2$ **d.** $y = (1/2)x - 1$ **4. a.** 500 hits: 10 books/day; 1000 hits: 20 books/day; 1500 hits: 32.5 books/day **b.** Book sales are increasing at a rate of 0.025 book/hit when the number of hits is between 1000 and 2000 each day **c.** 1400 hits/day   **5. a.** (A) **b.** (A) Leveling off, (B) Rising, (C) Rising (they begin to fall after 7 months), (D) Rising   **6. a.** $h = 0.05c + 1800$ **b.** 2100 hits/day **c.** $14,000/month   **7. a.** 2080 hits/day **b.** Probably not. This model predicts that Web-site traffic will start to decrease as advertising increases beyond $8500/month and then drop toward zero.   **8. a.** Cost: $C = 900 + 4x$; revenue: $R = 5.5x$; profit: $P = 1.5x - 900$ **b.** 600 novels/month **c.** 900 novels/month   **9. a.** $q = -60p + 950$ **b.** 50 novels/month **c.** $10, for a profit of $1200   **10. a.** $q = -53.3945p + 909.8165$ **b.** 483 novels/month

Chapter 2

Exercises 2.1

1. Vertex: $(-3/2, -1/4)$; y intercept: 2; x intercepts: $-2, -1$

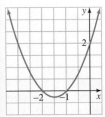

3. Vertex: $(2, 0)$; y intercept: -4; x intercept: 2

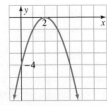

5. Vertex: $(-20, 900)$; y intercept: 500; x intercepts: $-50, 10$

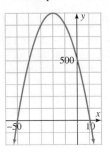

7. Vertex: $(-1/2, -5/4)$; y intercept: -1; x intercepts: $-1/2 \pm \sqrt{5}/2$

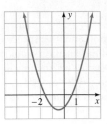

9. Vertex: $(0, 1)$; y intercept: 1; no x intercepts

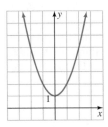

11. $R = -4p^2 + 100p$; maximum revenue when $p = 12.50

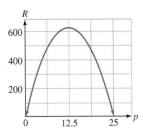

13. $R = -2p^2 + 400p$; maximum revenue when $p = 100

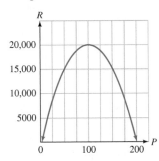

15. $y = -0.7955x^2 + 4.4591x - 1.6000$ **17.** $y = -1.1667x^2 - 6.1667x - 3$ **19. a.** Positive because the data suggest a curve that is concave up. **b.** (A) **c.** 1995; the parabola rises to the left of the vertex and thus predicts increasing sales as we go back in time, contradicting common sense.

21. 1985 $(t = 15)$; 3525 pounds

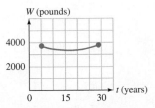

23. 5000 lb. The model is not trustworthy for vehicle weights larger than 5000 lb, because it predicts increasing fuel economy with increasing weight and 5000 is close to the upper limit of the domain of the function.   **25.** Maximum revenue when $p = $

$140, $R = 9800 **27.** Maximum revenue with 70 houses, $R =$ $9,800,000$ **29. a.** $q = -560x + 1400$; $R = -560x^2 + 1400x$ **b.** $P = -560x^2 + 1400x - 30$; $x = 1.25; $P = $845/mo$ **31.** $C = -200x + 620$; $P = -400x^2 + 1400x - 620$; $x =$ $1.75/hit$; $P = $605/mo$ **33. a.** $q = -10p + 400$ **b.** $R =$ $-10p^2 + 400p$ **c.** $C = -30p + 4200$ **d.** $P = -10p^2 + 430p -$ 4200; $p = 21.50 **35.** $R(t) = 95t^2 + 685t + 1350$ million dollars; $R(0) = 1350 million
37. $I(t) = 7.32t^2 - 34.5t + 401$ thousand dollars; $I(9) \approx$ $683,000$

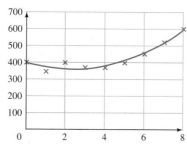

39. The x coordinate of the vertex represents the unit price that leads to the maximum revenue, the y coordinate of the vertex gives the maximum possible revenue, the x intercepts give the unit prices that result in zero revenue, and the y intercept gives the revenue resulting from zero unit price (which is obviously zero). **41.** Graph the data to see whether the points suggest a curve rather than a straight line. If the curve suggested by the graph is concave up or concave down, then a quadratic model would be a likely candidate.   **43.** If $q = mp + b$ (with $m < 0$), then the revenue is given by $R = pq = mp^2 + bp$. This is the equation of a parabola with $a = m < 0$ and so is concave down. Thus, the vertex is the highest point on the parabola, showing that there is a single highest value for R, namely, the y coordinate of the vertex.

45. Since $R = pq$, the demand must be given by $q = \dfrac{R}{p} =$ $\dfrac{-50p^2 + 60p}{p} = -50p + 60$.

Exercises 2.2

1. `4^x`

x	−3	−2	−1	0	1	2	3
f(x)	$\dfrac{1}{64}$	$\dfrac{1}{16}$	$\dfrac{1}{4}$	1	4	16	64

3. `3^(-x)`

x	−3	−2	−1	0	1	2	3
f(x)	27	9	3	1	$\dfrac{1}{3}$	$\dfrac{1}{9}$	$\dfrac{1}{27}$

5. `2*2^x or 2*(2^x)`

x	−3	−2	−1	0	1	2	3
g(x)	$\dfrac{1}{4}$	$\dfrac{1}{2}$	1	2	4	8	16

7. `-3*2^(-x)`

x	−3	−2	−1	0	1	2	3
g(x)	−24	−12	−6	−3	$-\dfrac{3}{2}$	$-\dfrac{3}{4}$	$-\dfrac{3}{8}$

9. `2^x-1`

x	−3	−2	−1	0	1	2	3
r(x)	$-\dfrac{7}{8}$	$-\dfrac{3}{4}$	$-\dfrac{1}{2}$	0	1	3	7

11. `2^(x-1)`

x	−3	−2	−1	0	1	2	3
s(x)	$\dfrac{1}{16}$	$\dfrac{1}{8}$	$\dfrac{1}{4}$	$\dfrac{1}{2}$	1	2	4

13.   **15.**   **17.**

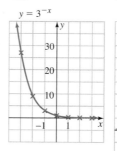
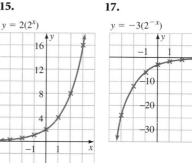

19. `e^(-2*x)` or `EXP(-2*x)`

x	−3	−2	−1	0	1	2	3
f(x)	403.4	54.60	7.389	1	0.1353	0.018 32	0.002 479

21. `1.01*2.02^(-4*x)`

x	−3	−2	−1	0	1	2	3
h(x)	4662	280.0	16.82	1.01	0.060 66	0.003 643	0.000 218 8

23. `50*(1+1/3.2)^(2*x)`

x	−3	−2	−1	0	1	2	3
r(x)	9.781	16.85	29.02	50	86.13	148.4	255.6

The following answers also show some common errors you should avoid.

25. `2^(x-1)`; *not* `2^x-1`
27. `2/(1-2^(-4*x))`; *not* `2/1-2^-4*x`; *not* `2/1-2^(-4*x)`
29. `(3+x)^(3*x)/(x+1)` or `((3+x)^(3*x))/(x+1)`; *not* `(3+x)^(3*x)/x+1`; *not* `(3+x^(3*x))/(x+1)`
31. `2*e^((1+x)/x)` or `2*EXP((1+x)/x)`; *not* `2*e^1+x/x`; *not* `2*e^(1+x)/x`; *not* `2*EXP(1+x)/x`

33. 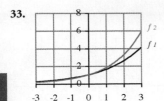 **35.**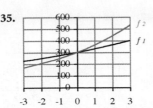

37. 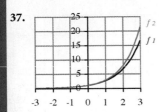 **39.**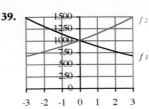

41. $f(x) = 500(0.5)^x$ **43.** $f(x) = 10(3)^x$ **45.** $f(x) = 500(0.45)^x$
47. $f(x) = -100(1.1)^x$ **49.** $y = 2.1213(1.4142^x)$ **51.** $y = 6.5975(1.3195^x)$ **53.** $y = 3.6742(0.9036^x)$ **55.** $y = 1.0442(1.7564)^x$ **57.** $y = 15.1735(1.4822)^x$ **59.** £6127
61. £204,006 **63.** 31.0 g, 9.25 g, 2.76 g **65.** 20,000 years
67. $y = 1000(2^{t/3})$; 65,536,000 bacteria after 2 days **69.** 53 mg
71. a. $P = 40t + 360$ **b.** $P = 360(1.1006)^t$; neither model is applicable. **73. a.** $y = 50,000(1.5^{t/2})$, $t = $ time in years since 2 years ago **b.** 91,856 tags **75. a.** $P = 180(1.011\ 20)^t$ million
b. 351 million **77.** $491.82
79. a.

Year	1950	2000	2050	2100
$C(t)$ parts per million	561	669	799	953

b. 2010 ($t = 260$)
81. a. $y \approx 5.907(0.939^t)$

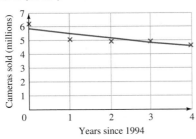

b. $y(5) \approx 4.3$ million cameras **c.** 5.7 million cameras, which is large compared with the other residuals and shows the danger of extrapolating based on past information alone. The possible factors that may influence sales, such as introduction of new products, are not taken into account in the model.
83. a. $y = 5.4433(1.0609)^t$

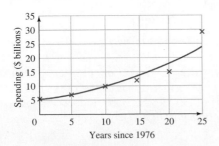

b. 6.09% **c.** $16 billion
85. (B) **87.** Exponential functions of the form $f(x) = A(b^x)$ $(b > 0)$ increase rapidly for large values of x. In real-life situations, such as population growth, this model is reliable only for relatively short periods of growth. Eventually, population growth tapers off due to pressures such as limited resources and overcrowding. **89.** Quadratic models are better with revenue and profit functions where demand depends linearly on the price. Exponential models are better with compound interest and population growth. **91.** Take the ratios y_2/y_1 and y_3/y_2. If they are the same, the points fit on an exponential curve. **93.** This reasoning is suspect—the bank need not use its computer resources to update all the accounts every minute but can instead use the continuous compounding formula to calculate the balance in any account at any time.

Exercises 2.3

1.

Logarithmic Form	$\log_{10} 10,000 = 4$	$\log_4 16 = 2$	$\log_3 27 = 3$	$\log_5 5 = 1$	$\log_7 1 = 0$	$\log_4 \frac{1}{16} = -2$

3.

Exponential Form	$(0.5)^2 = 0.25$	$5^0 = 1$	$10^{-1} = 0.1$	$4^3 = 64$	$2^8 = 256$	$2^{-2} = \frac{1}{4}$

5. 1.4650 **7.** -1.1460 **9.** -0.7324 **11.** 6.2657 **13.** $y = \log_3 x + 2$ **15.** $y = \log x + 2$ **17.** $y = \log_{1.0443} x - 28.1508$ **19.** 3.36 yr **21.** 11 yr **23.** 62,814 yr old **25.** 8 yr **27.** 151 mo **29.** 12 yr **31.** 13.08 yr **33. a.** $b = 3^{1/6} \approx 1.2009$ **b.** 3.8 mo **35.** 2360 million yr **37.** 3.89 days **39.** 3.2 h
41. a. $E(t) = 0.4214 \ln t + 1.962$ **b.** Yes; it predicts $3.24 billion. **c.** (A) **43.** $M(t) = 11.622 \ln t - 7.1358$. The model is unsuitable for large values of t because, for sufficiently large values of t, $M(t)$ will eventually become larger than 100%.
45. a. About 5.012×10^{16} joules of energy **b.** About 2.24%
d. 1000 **47. a.** 75 dB, 69 dB, 61 dB **b.** $D = 95 - 20 \log r$
c. 57,000 ft **49.** The logarithm of a negative number, were it defined, would be the power to which a base must be raised to give that negative number. But raising a base to a power never results in a negative number, so there can be no such number as the logarithm of a negative number. **51.** $\log_4 y$ **53.** Any logarithmic curve $y = \log_b t + C$ will eventually surpass 100% and hence not be suitable as a long-term predictor of market share. **55.** Time is increasing logarithmically with population; Solving $P = Ab^t$ for t gives $t = \log_b(P/A) = \log_b P - \log_b A$, which is of the form $t = \log_b P + C$. **57.** 8 **59.** x

Exercises 2.4

1. $N = 7, A = 6, b = 2; 7/(1+6*2^{-x})$

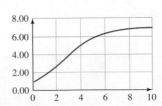

3. $N = 10, A = 4, b = 0.3; 10/(1+4*0.3^-x)$

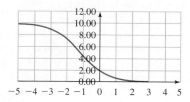

5. $N = 4, A = 7, b = 1.5; 4/(1+7*1.5^-x)$

7. $f(x) = \dfrac{200}{1 + 19(2^{-x})}$ **9.** $f(x) = \dfrac{6}{1 + 2^{-x}}$ **11.** (B)

13. (B)   **15.** (C)

17. $y = \dfrac{7.2}{1 + 2.4(1.05)^{-x}}$ **19.** $y = \dfrac{97}{1 + 2.2(0.942)^{-x}}$

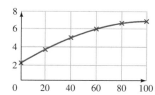   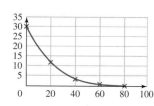

21. $N(t) = \dfrac{10,000}{1 + 9(1.25)^{-t}}; N(7) \approx 3463$ cases

23. $N(t) = \dfrac{3000}{1 + 29(2^{1/5})^{-t}}; t \approx 16$ days   **25. a.** (A) **b.** 16%/yr

27. a. 91% **b.** $P(x) \approx 14.33(1.05)^x$ **c.** \$38,000

29. a. $N(t) = \dfrac{82.8}{1 + 21.8(7.14)^{-t}}$. The model predicts that books

sales will level off at around 82.8 million books/yr. **b.** Not consistent; 15% of the market is represented by more than double the predicted value. This shows the difficulty in making long-term predictions from regression models obtained from a small

amount of data. **c.** 2001   **31. a.** $P(t) = \dfrac{5.7}{1 + 38.0(1.09)^{-t}}$; the

graph suggests a good fit.

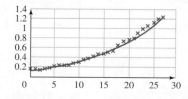

b. 9%/yr **c.** 5.7 million

33. $N(t) = \dfrac{5}{1 + 1.080(1.056)^{-t}}; t = 17$ or 2010.   **35.** Just as

diseases are communicated via the spread of a pathogen (such as a virus), new technology is communicated via the spread of information (such as advertising and publicity). Further, just as the spread of a disease is ultimately limited by the number of susceptible individuals, so the spread of a new technology is ultimately limited by the size of the potential market.   **37.** It can be used to predict where the sales of a new commodity might level off.

Chapter 2 Review Test

1. a.

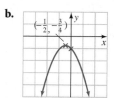

b.

2. a. $f(x) = \dfrac{2}{3}\, 3^x$ **b.** $f(x) = 10\sqrt{2}\left(\dfrac{1}{\sqrt{2}}\right)^x$ **3. a.** $-\dfrac{1}{2}\log_3 4$

b. ± 1 **c.** $\dfrac{1}{3}\log 1.05$ **d.** $\dfrac{\ln(A/P)}{m\ln(1 + i)}$ **4. a.** 10.2 yr **b.** 6 yr

c. 10.8 yr   **5. a.** $f(x) = \dfrac{900}{1 + 8(1.5)^{-x}}$

b. $f(x) = \dfrac{25}{1 + 4(1.1)^{-x}}$ **c.** $f(x) = \dfrac{20}{1 + 3(0.8)^{-x}}$

6. a. \$8500/mo; an average of approximately 2100 hits/day **b.** \$29,049/mo **c.** The fact that $-0.000\,005$, the coefficient of c^2, is negative   **7. a.** $R = -60p^2 + 950p$; $p = \$7.92$/novel, monthly revenue $= \$3760.42$ **b.** $P = -60p^2 + 1190p - 4700$; $p = \$9.92$/novel, monthly profit $= \$1200.42$ **8. a.** \$63,496.04 **b.** 7.0 h **c.** 32.8 h   **9.** (C)   **10. a.** 10,527 **b.** At 8.5 wk

Chapter 3

Exercises 3.1

1. -3 **3.** 0.3   **5.** $-\$25,000$/mo   **7.** -200 items/dollar **9.** $-\$1.25$/mo   **11.** 0.75 percentage point increase in unemployment per 1 percentage point increase in the deficit   **13.** 4 **15.** 2   **17.** 7/3

19.

h	Ave. Rate of Change
1	2
0.1	0.2
0.01	0.02
0.001	0.002
0.0001	0.0002

21.

h	Ave. Rate of Change
1	-0.1667
0.1	-0.2381
0.01	-0.2488
0.001	-0.2499
0.0001	$-0.249\,99$

23.

h	Ave. Rate of Change
1	9
0.1	8.1
0.01	8.01
0.001	8.001
0.0001	8.0001

25. a. 3 million people/yr. During the period 1995–2000, employment in the United States increased at an average rate of 3 million people/yr. **b.** 0 people/yr. During the period 2000–2001, the average rate of change of employment in the United States was 0 people/yr. **27. a.** 1998–2000; The number of companies that invested in venture capital each year was increasing most rapidly during the period 1998–2000, when it grew at an average rate of 650 companies/yr. **b.** 1999–2001; the number of companies that invested in venture capital each year was decreasing most rapidly during the period 1999–2001, when it decreased at an average rate of 50 companies/yr. **29. a.** 75 teams/yr **b.** Decreased **31. a.** [7, 8] and [10, 11]; −40,000 boats/yr; during the periods 1997–1998 and 2000–2001, the number of recreational boats decreased at an average rate of 40,000 boats/yr. **b.** −50/590 ≈ −0.0847; −25,000 boats/yr; over the period 1999–2001, the number of boats sold decreased at an average rate of 25,000 boats/yr, representing an 8.47% decrease over that period. **33. a.** 250 million transactions/yr, −150 million transactions/yr, 50 million transactions/yr. Over the period January 2000–January 2001, the (annual) number of online shopping transactions in the U.S. increased at an average rate of 250 million/yr. From January 2001 to January 2002, this number decreased at an average rate of 150 million/year. From January 2000 to January 2002, this number increased at an average rate of 50 million/yr. **b.** The average rate of change of $N(t)$ over [0, 2] is the average of the rates of change over [0, 1] and [1, 2]. **35. a.** (A) **b.** (C) **c.** (B) **d.** Approximately 0.04 million prisoners/yr; roughly equal
37. Answers will vary.

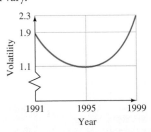

39. The index was increasing at an average rate of 300 points/day. **41.** 7.8 thousand pools/yr. The number of in-ground pools in the United States was growing at an average rate of about 7800 pools/yr over the period 1996–2000. **43.** $53.31 thousand/year. Between 1994 and 1998, the average after-tax income of the wealthiest 1% of United States taxpayers was increasing at an average rate of around $53,300/yr. **45. a.** 8.85 manatee deaths/100,000 boats; 23.05 manatee deaths/100,000 boats **b.** More boats result in more manatee deaths per additional boat. **47. a.** $305 million/yr; over the period 1997–1999, annual advertising revenues increased at an average rate of $305 million/yr. **b.** (A) **c.** $590 million/yr; the model projects annual advertising revenues to increase by $590 million/yr in 2000. **49.** The average rate of change of f over an interval [a, b] can be determined numerically, using a table of values, graphically, by measuring the slope of the corresponding line segment through two points on the graph, or algebraically, by using an algebraic formula for the function. Of these, the least precise is the graphical method, because it relies on reading coordinates of points on a graph. **51.** Answers will vary.

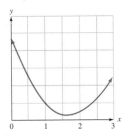

53. Six units of quantity A per unit of quantity C **55.** (A) **57.** Yes. Here is an example:

Year	2000	2001	2002	2003
Revenue ($ billion)	10	20	30	5

59. (A)

Exercises 3.2
1. 6 **3.** −5.5
5.

h	1	0.1	0.01
Ave. Rate	39	39.9	39.99

Instant. rate = 40 rupees/day

7.

h	1	0.1	0.01
Ave. Rate	140	66.2	60.602

Instant. rate = 60 rupees/day

9.

h	10	1
C_{ave}	4.799	4.7999

$C'(1000) = \$4.80$/item

11.

h	10	1
C_{ave}	99.91	99.90

$C'(100) = \$99.90$/item

13. a. R **b.** P **15. a.** P **b.** R **17. a.** Q **b.** P **19.** 1/2 **21.** 0 **23. a.** Q **b.** R **c.** P **25. a.** R **b.** Q **c.** P **27. a.** (1, 0) **b.** None **c.** (−2, 1) **29. a.** (−2, 0.3), (0, 0), (2, −0.3) **b.** None **c.** None **31.** $(a, f(a)); f'(a)$. **33. a.** (A) **b.** (C) **c.** (B) **d.** (B) **e.** (C) **35.** −2 **37.** −1.5 **39.** −5 **41.** 16 **43.** 0 **45.** −0.0025

47. a. 3 **b.** $y = 3x + 2$

49. a. $\dfrac{3}{4}$ **b.** $y = \dfrac{3}{4}x + 1$

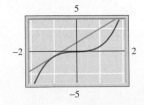

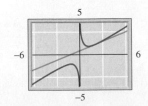

51. a. $\dfrac{1}{4}$ **b.** $y = \dfrac{1}{4}x + 1$

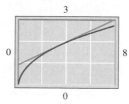

53. 1.000 **55.** 1.000 **57.** (B) **59.** (C) **61.** (A) **63.** (F)
65. a. None **b.** $x = -1.5, x = 0$

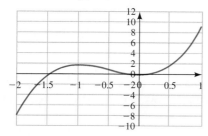

67. a. $x = 3$ **b.** None

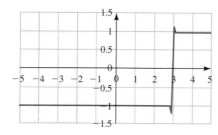

69.   **a.** $x = 1$ **b.** $x = 4.2$

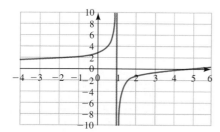

71. $q(100) = 50,000, q'(100) = -500$. A total of 50,000 pairs of sneakers can be sold at a price of $100, but the demand is decreasing at a rate of 500 pairs per $1 increase in the price.
73. a. Sales in 2000 were approximately 160,000 pools/year and increasing at a rate of 6000/yr. **b.** Decreasing because the slope is decreasing. **75. a.** (B) **b.** (B) **c.** (A) **d.** 1992 **e.** 0.05; in 1996 the total number of state prisoners was increasing at a rate of approximately 50,000 prisoners/yr. **77. a.** -96 ft/s **b.** -128 ft/s **79. a.** -100 million transactions/yr; over the period January 2001–January 2002, the (annual) number of online shopping transactions in the U.S. decreased at an average rate of 100 million/yr. **b.** 80 million transactions/yr; in January 2001 the (annual) number of online transactions was growing at a rate of about 80 million/yr. **c.** Online sales slowed and then decreased after January 2001. **81. a.** $305 million/yr

b. (A) **c.** $685 million/yr; in December 2000 AOL's advertising revenue was projected to be increasing at a rate of $685 million/yr. **83.** $A(0) = 4.5$ million; $A'(0) = 60,000$ **85. a.** 60% of children can speak at the age of 10 mo. At the age of 10 mo, this percentage is increasing by 18.2 percentage points/mo. **b.** As t increases, p approaches 100 percentage points (all children eventually learn to speak), and dp/dt approaches 0 because the percentage stops increasing. **87.** $S(5) \approx 109$, $\left.\dfrac{dS}{dt}\right|_{t=5} \approx 9.1$. After 5 wk, sales are 109 pairs of sneakers/wk, and sales are increasing at a rate of 9.1 pairs/wk each week. **89. a.** $P(50) \approx 62, P'(50) \approx 0.96$; 62% of U.S. households with an income of $50,000 have a computer. This percentage is increasing at a rate of 0.96 percentage points per $1000 increase in household income. **b.** P' decreases toward 0. **91.**

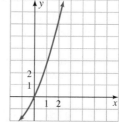

a. (D) **b.** 33 days after the egg was laid **c.** 50 days after the egg was laid **93.** $L(.95) = 31.2$ m and $L'(.95) = -304.2$ meters/warp. Thus, at a speed of warp 0.95, the spaceship has an observed length of 31.2 m, and its length is decreasing at a rate of 304.2 meters/unit warp, or 3.042 meters/increase in speed of 0.01 warp. **95.** Company B. Although the company is currently losing money, the derivative is positive, showing that the profit is increasing. Company A, on the other hand, has profits that are declining. **97.** (C) is the only graph in which the instantaneous rate of change on January 1 is greater than the 1-mo average rate of change. **99.** The tangent to the graph is horizontal at that point, and so the graph is almost horizontal near that point. **101.** Answers will vary.

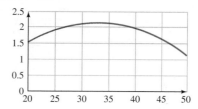

103. The difference quotient is not defined when $h = 0$ because there is no such number as 0/0. **105.** If $f(x) = mx + b$, then its average rate of change over any interval $[x, \ x + h]$ is $\dfrac{m(x + h) + b - (mx + b)}{h} = m$. Since this does not depend on h, the instantaneous rate is also equal to m. **107.** (A) because the average rate of change appears to be rising as we get closer to 5 from the left (see the bottom row).
109. (B)

111. Answers will vary.

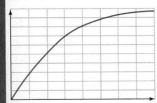

113. The derivative is positive and decreasing toward 0.
115. **117.**

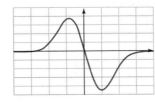

Exercises 3.3
1. 4; $2x$ **3.** 3; 3 **5.** 7; $6x + 1$ **7.** 4; $2 - 2x$ **9.** 14; $3x^2 + 2$
11. m; m **13.** 1; $1/x^2$ **15.** -1.2 **17.** 30.6 **19.** -7.1
21. 4.25 **23.** -0.6 **25.** $y = 4x - 7$ **27.** $y = -2x - 4$
29. $y = -3x - 1$ **31. a.** $f'(1) = 1/3$ **b.** Not differentiable at
0 **33. a.** Not differentiable at 1 **b.** Not differentiable at 0
35. a. Not differentiable at -1 **b.** $f'(1) \approx 1.1149$ **37. a.** $f'(1) =$
2 **b.** Not differentiable at 0 **39. a.** Not differentiable at -3
b. $f'(3) = 1$ **41. a.** Not differentiable at 1 **b.** Not differentiable at 0 **43.** 6000 pools/yr **45.** $s'(t) = -32t$; $s'(4) =$
-128 ft/s **47.** 62,000/yr each year **49.** $f'(8) = 26.6$ manatee
deaths per 100,000 boats. At a level of 800,000 boats, the number
of manatee deaths is increasing at a rate of 26.6 manatees per
100,000 additional boats. **51.** The algebraic method because it
gives the exact value of the derivative. The other two approaches
give only approximate values (except in some special cases).
53. Since the algebraic computation of $f'(a)$ is exact and not an
approximation, it makes no difference whether one uses the balanced difference quotient or the ordinary difference quotient in
the algebraic computation. **55.** The computation results in a
limit that cannot be evaluated.

Exercises 3.4
1. $5x^4$ **3.** $-4x^{-3}$ **5.** $-0.25x^{-0.75}$ **7.** $8x^3 + 9x^2$ **9.** $-1 - 1/x^2$

11. $\dfrac{dy}{dx} = 10(0) = 0$ (constant multiple and power rule)

13. $\dfrac{dy}{dx} = \dfrac{d}{dx}[x^2] + \dfrac{d}{dx}[x]$ (sum rule) $= 2x + 1$ (power rule)

15. $\dfrac{dy}{dx} = \dfrac{d}{dx}[4x^3] + \dfrac{d}{dx}[2x] - \dfrac{d}{dx}[1]$ (sum and difference) $=$

$4\dfrac{d}{dx}[x^3] + 2\dfrac{d}{dx}[x] - \dfrac{d}{dx}[1]$ (constant multiples) $= 12x^2 + 2$
(power rule)

17. $\dfrac{d}{dx}[x^{104} - 99x^2 + x] = \dfrac{d}{dx}[x^{104}] - \dfrac{d}{dx}[99x^2] + \dfrac{d}{dx}[x]$

(sums and differences) $= 104x^{103} - 99\dfrac{d}{dx}[x^2] + 1$ (constant

multiples and power rule) $= 104x^{103} - 198x + 1$ (power rule)
19. $f'(x) = 2x - 3$ **21.** $f'(x) = 1 + 0.5x^{-0.5}$ **23.** $g'(x) =$
$-2x^{-3} + 3x^{-2}$ **25.** $g'(x) = -\dfrac{1}{x^2} + \dfrac{2}{x^3}$ **27.** $h'(x) = -\dfrac{0.8}{x^{1.4}}$

29. $h'(x) = -\dfrac{2}{x^3} - \dfrac{6}{x^4}$ **31.** $r'(x) = -\dfrac{2}{3x^2} + \dfrac{0.1}{2x^{1.1}}$

33. $r'(x) = \dfrac{2}{3} - \dfrac{0.1}{2x^{0.9}} - \dfrac{4.4}{3x^{2.1}}$ **35.** $t'(x) = \dfrac{|x|}{x} - \dfrac{1}{x^2}$

37. $s'(x) = \dfrac{1}{2\sqrt{x}} - \dfrac{1}{2x\sqrt{x}}$ **39.** $s'(x) = 3x^2$

41. $t'(x) = 1 - 4x$ **43.** $2.6x^{0.3} + 1.2x^{-2.2}$ **45.** $1.2(1 - |x|/x)$

47. $3at^2 - 4a$ **49.** $5.15x^{9.3} - 99x^{-2}$ **51.** $-\dfrac{2.31}{t^{2.1}} - \dfrac{0.3}{t^{0.4}}$

53. $4\pi r^2$ **55.** 3 **57.** -2 **59.** -5

61. $y = 3x + 2$ **63.** $y = \dfrac{3}{4}x + 1$

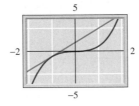 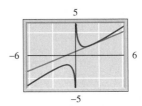

65. $y = \dfrac{1}{4}x + 1$

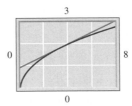

67. $x = -3/4$ **69.** No such values **71.** $x = 1, -1$
75. a. $s'(t) = 3.04t + 9.45$

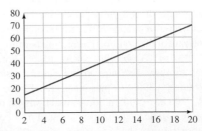

b. 1994–1995 ($t = 14$) **77.** $P'(t) = -5.2t + 13$; increasing
at a rate of 2.6 percentage points/yr **79. a.** $s'(t) = -32t$;
0, $-32, -64, -96, -128$ ft/s **b.** 5 s; downward at 160 ft/s
81. a. $n'(t) = -0.90t + 15$; increasing at a rate of 6000 pools/yr
b. (B) **83. a.** $f'(x) = 7.1x - 30.2$ manatees per 100,000 boats.

b. Increasing; the number of manatees killed per additional 100,000 boats increases as the number of boats increases. **c.** $f'(8) = 26.6$ manatees per additional 100,000 boats. At a level of 800,000 boats, the number of manatee deaths is increasing at a rate of 26.6 manatees per additional 100,000 boats. **85. a.** The excess of SUV sales in the United States over sales in Europe; the rate at which that difference was changing **b.** $404t - 410$; rising. The gap between SUV sales in the United States and Europe was increasing during the last 5 years of the 1990s. **c.** Not valid; the graphs use different y-axis scales; sales in Europe were increasing at an average rate of 62,000 vehicles/yr, while sales in the United States were increasing at an average rate of 460,000 vehicles/yr. However, SUV sales in Europe were rising at a greater *percentage* rate than in the United States **87.** After graphing the curve $y = 3x^2$, draw the line passing through $(-1, 3)$ with slope -6. **89.** The slope of the tangent line of g is twice the slope of the tangent line of f. **91.** $g'(x) = -f'(x)$ **93.** The left-hand side is not equal to the right-hand side. The *derivative* of the left-hand side is equal to the right-hand side, so your friend should have written $\dfrac{d}{dx}[3x^4 + 11x^5] = 12x^3 + 55x^4$. **95.** The derivative of a constant times a function is the constant times the derivative of the function so that $f'(x) = (2)(2x) = 4x$. Your enemy mistakenly computed the *derivative* of the constant times the derivative of the function. (The derivative of a product of two functions is *not* the product of the derivative of the two functions. The rule for taking the derivative of a product is discussed in Chapter 4.) **97.** Answers will vary.

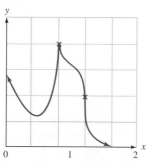

Exercises 3.5

1. $C'(1000) = \$4.80$/item **3.** $C'(100) = \$99.90$/item **5.** $C'(x) = 4$; $R'(x) = 8 - x/500$; $P'(x) = 4 - x/500$; $P'(x) = 0$ when $x = 2000$. Thus, at a production level of 2000, the profit is stationary (neither increasing nor decreasing) with respect to the production level. This may indicate a maximum profit at a production level of 2000. **7. a.** (B) **b.** (C) **c.** (C) **9. a.** $C'(x) = 1200 - 0.004x$; the cost is going up at a rate of \$1,199,984 per television commercial. The exact cost of airing the fifth television commercial is $C(5) - C(4) = \$1,199,982$. **b.** $\overline{C}(x) = 150/x + 1200 - 0.002x$; $\overline{C}(4) = \$1,237,492$ per television commercial. The average cost of airing the first four television commercials is \$1,237,492. **11.** The profit on the sale of 1000 DVDs is \$3000 and is decreasing at a rate of \$3 per additional DVD sold. **13.** $P \approx \$257.07$ and $dP/dx \approx 5.07$. Your current profit is \$257.07/mo, and this would increase at a rate of \$5.07 per addi-

tional magazine in sales. **15.** $P'(50) = \$350$. This means that, at an employment level of 50 workers, the firm's daily profit will increase at a rate of \$350 per additional worker it hires. **17. a.** (B) **b.** (B) **c.** (C) **19. a.** \$2.50/lb **b.** $R(q) = 20{,}000/q^{0.5}$ **c.** $R(400) = \$1000$; this is the monthly revenue that will result from setting the price at \$2.50/pound. $R'(400) = -\$1.25$/pound of tuna. Thus, at a demand level of 400 lb/mo, the revenue is decreasing at a rate of \$1.25/pound. **d.** The fishery should raise the price (to reduce the demand). **21. a.** $C(x) = $

$$500{,}000 + 685{,}000x - 10{,}000\sqrt{x};\ C'(x) = 685{,}000 - \frac{5000}{\sqrt{x}};$$

$$\overline{C}(x) = \frac{500{,}000}{x} + 685{,}000 - \frac{10{,}000}{\sqrt{x}}$$ **b.** $C'(3) = \$682{,}000/$

spot; $\overline{C}(3) = \$846{,}000$/spot. Since the marginal cost is less than the average cost, the cost of the fourth ad is lower than the average cost of the first three, so the average cost will decrease as x increases. **23. a.** $C'(q) = 200q$; $C'(10) = \$2000/1$-lb reduction in emissions **b.** $S'(q) = 500$; thus $S'(q) = C'(q)$ when $500 = 200q$, or $q = 2.5$ lb/day reduction. **c.** $N(q) = C(q) - S(q) = 100q^2 - 500q + 4000$; this is a parabola with lowest point (vertex) given by $q = 2.5$. The net cost at this production level is $N(2.5) = \$3375$/day. The value of q is the same as that for part (b). The net cost to the firm is minimized at the reduction level for which the cost of controlling emissions begins to increase faster than the subsidy. This is why we get the answer by setting these two rates of increase equal to each other. **25.** $M'(10) \approx 0.000\ 255\ 7$ mpg/mph; this means that, at a speed of 10 mph, the fuel economy is increasing at a rate of 0.000 255 7 mpg per 1-mph increase in speed. $M'(60) = 0$ mpg/mph; this means that at a speed of 60 mph the fuel economy is neither increasing nor decreasing with increasing speed. $M'(70) \approx -0.000\ 017\ 99$; this means that, at 70 mph, the fuel economy is decreasing at a rate of 0.000 017 99 mpg per 1-mph increase in speed. Thus, 60 mph is the most fuel-efficient speed for the car. **27.** (C) **29.** (D) **31.** (B) **33.** Cost is often measured as a function of the number of items x. Thus, $C(x)$ is the cost of producing (or purchasing, as the case may be) x items. **a.** The average cost function $\overline{C}(x)$ is given by $\overline{C}(x) = C(x)/x$. The marginal cost function is the derivative $C'(x)$ of the cost function. **b.** The average cost $\overline{C}(r)$ is the slope of the line through the origin and the point on the graph where $x = r$. The marginal cost of the rth unit is the slope of the tangent to the graph of the cost function at the point where $x = r$. **c.** The average cost function $\overline{C}(x)$ gives the average cost of producing the first x items. The marginal cost function $C'(x)$ is the rate at which cost is changing with respect to the number of items x, or the incremental cost per item, and approximates the cost of producing the $(x + 1)$st item. **35.** The marginal cost **37.** Not necessarily. For example, it may be the case that the marginal cost of the 101st item is larger than the average cost of the first 100 items (even though the marginal cost is decreasing). Thus, adding this additional item will *raise* the average cost. **39.** The circumstances described suggest that the average cost function is at a relatively low point at the current production level, and so it would be appropriate to advise the company to maintain current production levels; raising or lowering the production level will result in increasing average costs.

Exercises 3.6

1. 0 **3.** 4 **5.** Does not exist **7.** 1.5 **9.** 0.5 **11.** Diverges to $+\infty$ **13.** 0 **15.** 1 **17.** 0 **19. a.** -2 **b.** -1 **21. a.** 2 **b.** 1 **c.** 0 **d.** $+\infty$ **23. a.** 0 **b.** 2 **c.** -1 **d.** Does not exist. **e.** 2 **f.** $+\infty$ **25. a.** 1 **b.** 1 **c.** 2 **d.** Does not exist. **e.** 1 **f.** 2 **27. a.** 1 **b.** $+\infty$ **c.** $+\infty$ **d.** $+\infty$ **e.** Not defined **f.** -1 **29. a.** -1 **b.** $+\infty$ **c.** $-\infty$ **d.** Does not exist. **e.** 2 **f.** 1 **31.** $\lim\limits_{t\to 1^-} C(t) = 0.06$ and $\lim\limits_{t\to 1^+} C(t) = 0.08$, so $\lim\limits_{t\to 1} C(t)$ does not exist. **33.** $\lim\limits_{t\to+\infty} I(t) = +\infty$, $\lim\limits_{t\to+\infty}(I(t)/D(t)) \approx 0.024$. In the long term, sales of imported bottled water will rise without bound but will level off at around 2.4% of sales of domestic bottled water. In the real world, sales cannot rise without bound. Thus, the given models should not be extrapolated far into the future. **35. a.** $\lim\limits_{t\to+\infty} n(t) \approx 80$; online book sales can be expected to level off at 80 million/yr in the long term. **b.** $\lim\limits_{t\to+\infty} n'(t) \approx 0$; the annual number of books sold online can be expected to stop changing in the long term. **37.** 470; this suggests that students whose parents earn an exceptionally large income score an average of 470 on the SAT verbal test. **39.** To approximate $\lim\limits_{x\to a} f(x)$ numerically, choose values of x closer and closer to, and on either side of $x = a$, and evaluate $f(x)$ for each of them. The limit (if it exists) is then the number that these values of $f(x)$ approach. A disadvantage of this method is that it may never give the exact value of the limit but only an approximation. (However, we can make this as accurate as we like.) **41.** It is possible for $\lim\limits_{x\to a} f(x)$ to exist even though $f(a)$ is not defined. An example is $\lim\limits_{x\to 1} \dfrac{x^2 - 3x + 2}{x - 1}$. **43.** Any situation in which there is a sudden change can be modeled by a function in which $\lim\limits_{t\to a^+} f(t)$ is not the same as $\lim\limits_{t\to a^-} f(t)$. One example is the value of a stock market index before and after a crash: $\lim\limits_{t\to a^-} f(t)$ is the value immediately before the crash at time $t = a$, and $\lim\limits_{t\to a^+} f(t)$ is the value immediately after the crash. Another example might be the price of a commodity that is suddenly increased from one level to another. **45.** An example is $f(x) = (x - 1)(x - 2)$.

Exercises 3.7

1. Continuous on its domain **3.** Continuous on its domain **5.** Discontinuous at $x = 0$ **7.** Discontinuous at $x = -1$ **9.** Continuous on its domain **11.** Discontinuous at $x = -1$ and 0 **13.** (A), (B), (D), (E) **15.** 0 **17.** -1 **19.** No value possible **21.** -1 **23.** Continuous on its domain **25.** Continuous on its domain **27.** Discontinuity at $x = 0$ **29.** Discontinuity at $x = 0$ **31.** Continuous on its domain **33.** Not unless the domain of the function consists of all real numbers. (It is impossible for a function to be continuous at points not in its domain.) For example, $f(x) = 1/x$ is continuous on its domain—the set of nonzero real numbers—but not at $x = 0$. **35.** True. If the graph of a function has a break in its graph at any point a, then it cannot be continuous at the point a.

37. Answers will vary.

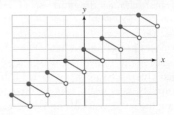

39. Answers will vary. The price of OHaganBooks.com stocks suddenly drops by $10 as news spreads of a government investigation. Let $f(x) = $ price of OHaganBooks.com stocks.

Exercises 3.8

1. 1 **3.** 2 **5.** 0 **7.** 6 **9.** 4 **11.** 2 **13.** 0 **15.** 0 **17.** 12 **19.** Diverges to $+\infty$ **21.** Does not exist; left and right (infinite) limits differ **23.** 3/2 **25.** 1/2 **27.** Diverges to $+\infty$ **29.** 0 **31.** 3/2 **33.** 1/2 **35.** Diverges to $-\infty$ **37.** 0 **39.** Discontinuity at $x = 0$ **41.** Continuous everywhere **43.** Discontinuity at $x = 0$ **45.** Discontinuity at $x = 0$ **47.** 1.59; if the trend continues indefinitely, the annual spending on police will be 1.59 times the annual spending on courts in the long run. **49.** $\lim\limits_{t\to+\infty} I(t) = +\infty$, $\lim\limits_{t\to+\infty}(I(t)/D(t)) \approx 0.024$. In the long term, sales of imported bottled water will rise without bound but will level off at around 2.4% of sales of domestic bottled water. In the real world, sales cannot rise without bound. Thus, the given models should not be extrapolated far into the future. **51.** $\lim\limits_{t\to+\infty} p(t) = 100$, $\lim\limits_{t\to+\infty} p'(t) = 0$. The percentage of children who learn to speak approaches 100% as their age increases, with the number of additional children learning to speak approaching zero. **53. a.** Yes; $\lim\limits_{t\to 8^-} C(t) = \lim\limits_{t\to 8^-} C(t) = 1.24$ **b.** No; $\lim\limits_{t\to 8^-} C'(t) = 0.08$ whereas $\lim\limits_{t\to 8^+} C'(t) = 0.355$. Until 1998 the cost of a Super Bowl ad was increasing at a rate of $80,000/yr. Immediately thereafter, it was increasing at a rate of $355,000/yr. **55.** To evaluate $\lim\limits_{x\to a} f(x)$ algebraically, first check whether $f(x)$ is a closed-form function. Then check whether $x = a$ is in its domain. If so, the limit is just $f(a)$; that is, it is obtained by substituting $x = a$. If not, then try to first simplify $f(x)$ in such a way as to transform it into a new function such that $x = a$ is in its domain and then substitute. A disadvantage of this method is that it is sometimes extremely difficult to evaluate limits algebraically, and rather sophisticated methods are often needed. **57.** She is wrong. Closed-form functions are continuous only at points in their domains, and $x = 2$ is not in the domain of the closed-form function $f(x) = 1/(x - 2)^2$. **59.** The statement may not be true, for instance, if $f(x) = \begin{cases} x + 2 & \text{if } x < 0 \\ 2x - 1 & \text{if } x \geq 0 \end{cases}$, then $f(0)$ is defined and equals -1, and yet $\lim\limits_{x\to 0} f(x)$ does not exist. The statement can be corrected by requiring that f be a closed-form function: "If f is a closed-form function and $f(a)$ is defined, then $\lim\limits_{x\to a} f(x)$ exists and equals $f(a)$." **61.** Answers will vary. For example,

$$f(x) = \begin{cases} 0 & \text{if } x \text{ is any number other than 1 or 2} \\ 1 & \text{if } x = 1 \text{ or 2} \end{cases}$$

Chapter 3 Review Test

1. a.

h	1	0.01	0.001
Ave. Rate of Change	-0.5	-0.9901	-0.9990

Slope ≈ -1

b.

h	1	0.01	0.001
Ave. Rate of Change	23	6.8404	6.7793

Slope ≈ 6.8

c.

h	1	0.01	0.001
Ave. Rate of Change	6.3891	2.0201	2.0020

Slope ≈ 2

d.

h	1	0.01	0.001
Ave. Rate of Change	0.6931	0.9950	0.9995

Slope ≈ 1

2. a. (i) P **(ii)** Q **(iii)** R **(iv)** S **b. (i)** None **(ii)** R **(iii)** Q **(iv)** P
c. (i) Q **(ii)** None **(iii)** None **(iv)** None **d. (i)** None **(ii)** R
(iii) P and S **(iv)** Q **3. a.** (B) **b.** (B) **c.** (B) **d.** (A) **e.** (C)
4. a. $2x + 1$ **b.** $-1/x^2$ **5. a.** $50x^4 + 2x^3 - 1$ **b.** $-50/x^6 -$
$2/x^5 + 1/x^2$ **c.** $9x^2 + x^{-2/3}$ **d.** $-4.2/x^{3.1} - 0.1/(2x^{0.9})$ **6. a.** $1 -$
$\dfrac{2}{x^3}$ **b.** $2 + \dfrac{1}{x^2}$ **c.** $-\dfrac{4}{3x^2} + \dfrac{0.2}{x^{1.1}} + \dfrac{1.1x^{0.1}}{3.2}$ **d.** $-4/x^2 +$
$1/4 - |x|/x$
7. a. `50*x^4+2*x^3-1` **b.** `-50/x^6-2/x^5+1/x^2`

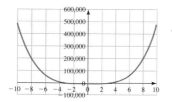 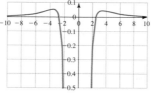

c. `9*x^2+1/(x^2)^(1/3)` **d.** `-4.2/x^3.1-0.1/(2*x^0.9)`

8. a. 500 books/wk **b.** [3, 4], [4, 5] **c.** [3, 5]; 650 books/wk
9. a. 274 books/wk
b. 636 books/wk
c. No; the function w begins
to decrease after $t = 14$.
10. a. 553 books/wk
b. 15 books/wk
c. Sales level off at around
10,527 books/wk, with a zero
rate of change.

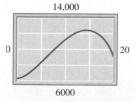 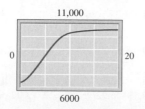

11. a. \$2.88/book **b.** \$3.715/book **c.** Approximately
$-\$0.000\,104$/book, per additional book sold **d.** At a sales level of
8000 books/wk, the cost is increasing at a rate of \$2.88/book (so
that the 8001st book costs approximately \$2.88 to sell), and it
costs an average of \$3.715/book to sell the first 8000 books. More-
over, the average cost is decreasing at a rate of \$0.000 104/book,
per additional book sold.

Chapter 4

Exercises 4.1

1. 3 **3.** $3x^2$ **5.** $2x + 3$ **7.** $210x^{1.1}$ **9.** $-2/x^2$ **11.** $2x/3$
13. $3(4x^2 - 1) + 3x(8x) = 36x^2 - 3$ **15.** $3x^2(1 - x^2) +$
$x^3(-2x) = 3x^2 - 5x^4$ **17.** $2(2x + 3) + (2x + 3)(2) = 8x + 12$

19. $\sqrt{x} + \dfrac{x}{2\sqrt{x}} = \sqrt{x} + \dfrac{\sqrt{x}}{2} = \dfrac{3\sqrt{x}}{2}$ **21.** $(x^2 - 1) +$
$2x(x + 1) = (x + 1)(3x - 1)$ **23.** $(x^{-0.5} + 4)(x - x^{-1}) +$
$(2x^{0.5} + 4x - 5)(1 + x^{-2})$ **25.** $8(2x^2 - 4x + 1)(x - 1)$
27. $\left(\dfrac{1}{3.2} - \dfrac{3.2}{x^2}\right)(x^2 + 1) + 2x\left(\dfrac{x}{3.2} + \dfrac{3.2}{x}\right)$ **29.** $2x(2x + 3)(7x +$
$2) + 2x^2(7x + 2) + 7x^2(2x + 3)$ **31.** $5.3(1 - x^{2.1})(x^{-2.3} - 3.4)$
$- 2.1x^{1.1}(5.3x - 1)(x^{-2.3} - 3.4) - 2.3x^{-3.3}(5.3x - 1)(1 - x^{2.1})$

33. $\dfrac{1}{2\sqrt{x}}\left(\sqrt{x} + \dfrac{1}{x^2}\right) + (\sqrt{x} + 1)\left(\dfrac{1}{2\sqrt{x}} - \dfrac{2}{x^3}\right)$

35. $\dfrac{2(3x - 1) - 3(2x + 4)}{(3x - 1)^2} = -\dfrac{14}{(3x - 1)^2}$

37. $\dfrac{(4x + 4)(3x - 1) - 3(2x^2 + 4x + 1)}{(3x - 1)^2} = \dfrac{6x^2 - 4x - 7}{(3x - 1)^2}$

39. $\dfrac{(2x - 4)(x^2 + x + 1) - (x^2 - 4x + 1)(2x + 1)}{(x^2 + x + 1)^2} = \dfrac{5x^2 - 5}{(x^2 + x + 1)^2}$

41. $\dfrac{(0.23x^{-0.77} - 5.7)(1 - x^{-2.9}) - 2.9x^{-3.9}(x^{0.23} - 5.7x)}{(1 - x^{-2.9})^2}$

43. $\dfrac{\frac{1}{2}x^{-1/2}(x^{1/2} - 1) - \frac{1}{2}x^{-1/2}(x^{1/2} + 1)}{(x^{1/2} - 1)^2} = \dfrac{-1}{\sqrt{x}(\sqrt{x} - 1)^2}$

45. $-3/x^4$

47. $\dfrac{[(x + 1) + (x + 3)](3x - 1) - 3(x + 3)(x + 1)}{(3x - 1)^2} = \dfrac{3x^2 - 2x - 13}{(3x - 1)^2}$

49. $\dfrac{[(x + 1)(x + 2) + (x + 3)(x + 2) + (x + 3)(x + 1)](3x - 1) - 3(x + 3)(x + 1)(x + 2)}{(3x - 1)^2}$

51. $4x^3 - 2x$ **53.** 64 **55.** 3 **57.** $y = 12x - 8$ **59.** $y =$
$x/4 + 1/2$ **61.** $y = -2$ **63.** $S'(5) = 10$ (sales are increasing at
a rate of 1000 units/mo); $p'(5) = -10$ (the price is dropping at a
rate of \$10/sound system/mo); $R'(5) = 900,000$ (revenue is in-
creasing at a rate of \$900,000/mo) **65.** Decreasing at a rate of
\$1/day **67.** Decreasing at a rate of approximately \$0.10/mo

69. $M'(x) = \dfrac{3000(3600x^{-2} - 1)}{(x + 3600x^{-1})^2}$; $M'(10) \approx 0.7670$ mpg/mph.

This means that at a speed of 10 mph the fuel economy is increasing at a rate of 0.7670 mpg/one mph increase in speed.
$M'(60) = 0$ mpg/mph. This means that at a speed of 60 mph the
fuel economy is neither increasing nor decreasing with increasing
speed. $M'(70) \approx -0.0540$. This means that at 70 mph the fuel

economy is decreasing at a rate of 0.0540 mpg/one mph increase in speed. 60 mph is the most fuel-efficient speed for the car (in Chapter 5 we discuss how to locate largest values in general).   **71.** $158.4 million, decreasing at a rate of $3259.20 million/yr   **73.** Cost: $10,000t/3 + 80,000$; Personnel: $-12,500t + 1,500,000$ (t since 1995). Increasing at a rate of $3420 million/yr.

75. $R'(p) = -\dfrac{5.625}{(1 + 0.125p)^2}$; $R'(4) = -2.5$ thousand organisms/h/1000 organisms. This means that the reproduction rate of organisms in a culture containing 4000 organisms is declining at a rate of 2500 organisms/h/1000 organisms.   **77.** Oxygen consumption is decreasing at a rate of 1634 mL/day. This is due to the fact that the number of eggs is decreasing, because $C'(25)$ is positive.   **79. a.** $W(t) - S(t)$ represents sales of non-sparkling water, and $S(t)/W(t)$ represents the fraction of bottled water that is sparkling. Its derivative is the rate of change of this fraction.   **b.** -0.023/yr. The percentage of total sales represented by sparkling water was declining by 2.3 points/yr in 1990.   **81.** The analysis is suspect because it seems to be asserting that the annual increase in revenue, which we can think of as dR/dt, is the product of the annual increases dp/dt in price and dq/dt in sales. However, since $R = pq$, the product rule implies that dR/dt is not the product of dp/dt and dq/dt but is instead

$$\frac{dR}{dt} = \frac{dp}{dt} \cdot q + p \cdot \frac{dq}{dt}$$

83. Answers will vary. $q = -p + 1000$ is one example.   **85.** Mine; it is increasing twice as fast as yours. The rate of change of revenue is given by $R'(t) = p'(t)q(t)$ since $q'(t) = 0$ Thus, $R'(t)$ does not depend on the selling price $p(t)$.   **87.** (A)

Exercises 4.2

1. $4(2x + 1)$ **3.** $-(x - 1)^{-2}$ **5.** $2(2 - x)^{-3}$ **7.** $(2x + 1)^{-0.5}$ **9.** $-4(4x - 1)^{-2}$ **11.** $-3/(3x - 1)^2$ **13.** $4(x^2 + 2x)^3(2x + 2)$ **15.** $-4x(2x^2 - 2)^{-2}$ **17.** $-5(2x - 3)(x^2 - 3x - 1)^{-6}$ **19.** $-6x/(x^2 + 1)^4$ **21.** $1.5(0.2x - 4.2)(0.1x^2 - 4.2x + 9.5)^{0.5}$

23. $4(2s - 0.5s^{-0.5})(s^2 - s^{0.5})^3$ **25.** $-\dfrac{x}{\sqrt{1 - x^2}}$

27. $-[(x + 1)(x^2 - 1)]^{-3/2}(3x - 1)(x + 1)$

29. $6.2(3.1x - 2) + 6.2/(3.1x - 2)^3$

31. $2[(6.4x - 1)^2 + (5.4x - 2)^3][12.8(6.4x - 1) + 16.2(5.4x - 2)^2]$

33. $-2(x^2 - 3x)^{-3}(2x - 3)(1 - x^2)^{0.5} - x(x^2 - 3x)^{-2}(1 - x^2)^{-0.5}$

35. $-56(x + 2)/(3x - 1)^3$ **37.** $3z^2(1 - z^2)/(1 + z^2)^4$

39. $3[(1 + 2x)^4 - (1 - x)^2]^2[8(1 + 2x)^3 + 2(1 - x)]$

41. $-0.43(x + 1)^{-1.1}[2 + (x + 1)^{-0.1}]^{3.3}$

43. $-\dfrac{\dfrac{1}{\sqrt{2x + 1}} - 2x}{(\sqrt{2x + 1} - x^2)^2}$

45. $54(1 + 2x)^2[1 + (1 + 2x)^3]^2\{1 + [1 + (1 + 2x)^3]^3\}^2$ **47.** $(100x^{99} - 99x^{-2})dx/dt$ **49.** $(-3r^{-4} + 0.5r^{-0.5})\,dr/dt$ **51.** $4\pi r^2\,dr/dt$ **53.** $-47/4$ **55.** $c = 100(0.42 + 0.02t)^2 - 160(0.42 + 0.02t) + 110$; $-$0.88/trade/mo

57. $\dfrac{dy}{dt} = \dfrac{dy}{dx}\dfrac{dx}{dt}$; $(-3) = 1.5\,\dfrac{dx}{dt}$. So, $\dfrac{dx}{dt} = \dfrac{-3}{1.5} = -2$ murders/100,000 residents/yr each year

59. a. $R(p) = -4p^2/3 + 80p$; $\left.\dfrac{dR}{dp}\right|_{q=60} = $40/$1$ increase in price   **b.** $-$0.75$/ruby   **c.** $-$30$/ruby. Thus, at a demand level of 60 rubies/wk, the weekly revenue is decreasing at a rate of $30/additional ruby demanded.   **61.** $\left.\dfrac{dP}{dn}\right|_{n=10} = 146,454.9$. At an employment level of ten engineers, Paramount will increase its profit at a rate of $146,454.90/additional engineer hired.   **63.** 0.000 158 manatees/boat, or 15.8 manatees/100,000 boats. Approximately 15.8 more manatees are killed each year for each additional 100,000 registered boats.   **65.** 12π mi^2/h   **67.** $200,000\pi$/wk $= $628,000$/wk   **69. a.** $q'(4) \approx 333$ units/mo   **b.** $dR/dq = 800$/unit   **c.** $dR/dt \approx $267,000$/mo   **71.** 3%/yr   **73.** 8%/yr   **75.** The glob squared, times the derivative of the glob   **77.** The derivative of a quantity cubed is three times the (original) quantity squared, times the derivative of the quantity. Thus, the correct answer is $3(3x^3 - x)^2(9x^2 - 1)$.   **79.** Following the calculation thought experiment, pretend that you were evaluating the function at a specific value of x. If the last operation you would perform is addition or subtraction, look at each summand separately. If the last operation is multiplication, use the product rule first; if it is division, use the quotient rule first; if it is any other operation (such as raising a quantity to a power or taking a radical of a quantity), use the chain rule first.   **81.** An example is $f(x) =$

$$\sqrt{x + \sqrt{x + \sqrt{x + \sqrt{x + \sqrt{x + 1}}}}}.$$

Exercises 4.3

1. $1/(x - 1)$ **3.** $1/(x \ln 2)$ **5.** $2x/(x^2 + 3)$ **7.** e^{x+3} **9.** $-e^{-x}$ **11.** $4^x \ln 4$ **13.** $2^{x^2 - 1}2x \ln 2$ **15.** $1 + \ln x$ **17.** $2x \ln x + (x^2 + 1)/x$ **19.** $10x(x^2 + 1)^4 \ln x + (x^2 + 1)^5/x$ **21.** $3/(3x - 1)$ **23.** $4x/(2x^2 + 1)$ **25.** $(2x - 0.63x^{-0.7})/(x^2 - 2.1x^{0.3})$ **27.** $-2/(-2x + 1) + 1/(x + 1)$ **29.** $3/(3x + 1) - 4/(4x - 2)$ **31.** $1/(x + 1) + 1/(x - 3) - 2/(2x + 9)$ **33.** $5.2/(4x - 2)$ **35.** $2/(x + 1) - 9/(3x - 4) - 1/(x - 9)$ **37.** $\dfrac{1}{(x + 1)\ln 2}$

39. $\dfrac{1 - \dfrac{1}{t^2}}{(t + \frac{1}{t})\ln 3}$ **41.** $\dfrac{2 \ln |x|}{x}$ **43.** $\dfrac{2}{x} - \dfrac{2 \ln(x - 1)}{x - 1}$

45. $e^x(1 + x)$ **47.** $1/(x + 1) + 3e^x(x^3 + 3x^2)$ **49.** $e^x(\ln |x| + 1/x)$ **51.** $2e^{2x+1}$ **53.** $(2x - 1)e^{x^2 - x + 1}$ **55.** $2xe^{2x-1}(1 + x)$ **57.** $4(e^{2x-1})^2$ **59.** $-4/(e^x - e^{-x})^2$

61. $5e^{5x-3}$ **63.** $-\dfrac{\ln x + 1}{(x \ln x)^2}$ **65.** $2(x - 1)$ **67.** $\dfrac{1}{x \ln x}$

69. $\dfrac{1}{2x \ln x}$ **71.** $y = (e/\ln 2)(x - 1) \approx 3.92(x - 1)$

73. $y = x$ **75.** $y = -[1/(2e)](x - 1) + e$ **77.** $451.00/yr   **79.** $446.02/yr   **81.** 277,000 people/yr   **83.** 0.000 283 g/yr

85. $P'(22) \approx -1.4$. This indicates that in ancient Rome the percentage of people surviving was decreasing at a rate of 1.4 percentage points/yr at age 22. That is, approximately 1.4% of the original birth cohort died that year.   **87. a.** (A) **b.** The verbal SAT increases by approximately 1 point.   **c.** $S'(x)$ decreases with increasing x so that as parental income increases the effect on SAT scores decreases.   **89.** $R(t) = 350e^{-0.1t}(39t + 68)$ million dollars; $R(2) \approx \$42$ billion $R'(2) \approx \$7$ billion/yr
91. 3,110,000 cases/wk; 11,200,000 cases/wk; 722,000 cases/wk

93. a. $W'(t) = -\dfrac{1500(0.77)(1.16)^{-t}(-1)\ln(1.16)}{[1 + 0.77(1.16)^{-t}]^2} \approx$

$\dfrac{171.425(1.16)^{-t}}{[1 + 0.77(1.16)^{-t}]^2}$; $W'(6) \approx 40.624 \approx 41$, to two significant

digits. The constants in the model are specified to two and three significant digits, so we cannot expect the answer to be accurate to more than two digits. In other words, all digits from the third on are probably meaningless. The answer tells one that in 1996 the number of authorized wiretaps was increasing at a rate of approximately 41 wiretaps/yr.   **b.** (A)

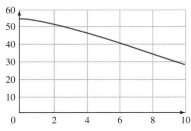

95. a.

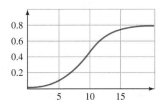

$p'(10) \approx 0.09$, so the percentage of firms using numeric control is increasing at a rate of 9 percentage points/yr after 10 yr. **b.** 0.80. Thus, in the long run, 80% of all firms will be using numeric control.   **c.** $p'(t) = 0.3816e^{4.46-0.477t}/(1 + e^{4.46-0.477t})^2$; $p'(10) = 0.0931$

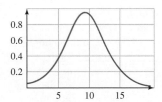

d. 0. Thus, in the long run the percentage of firms using numeric control will stop increasing.
97. e raised to the glob, times the derivative of the glob   **99.** 2 raised to the glob, times the derivative of the glob, times the natural logarithm of 2   **101.** The power rule does not apply when the exponent is not constant. The derivative of 3 raised to a quantity is 3 raised to the quantity, times the derivative of the quantity, times $\ln 3$. Thus, the correct answer is $3^{2x} 2 \ln 3$.   **103.** No. If

$N(t)$ is exponential, so is its derivative.   **105.** If $f(x) = e^{kx}$, then the fractional rate of change is $\dfrac{f'(x)}{f(x)} = \dfrac{ke^{kx}}{e^{kx}} = k$, the fractional growth rate.   **107.** If $A(t)$ is growing exponentially, then $A(t) = A_0 e^{kt}$ for constants A_0 and k. Its percentage rate of change is then
$\dfrac{A'(t)}{A(t)} = \dfrac{kA_0 e^{kt}}{A_0 e^{kt}} = k$, a constant.

Exercises 4.4

1. $-2/3$ **3.** x **5.** $(y - 2)/(3 - x)$ **7.** $-y$ **9.** $-\dfrac{y}{x(1 + \ln x)}$

11. $-x/y$ **13.** $-2xy/(x^2 - 2y)$ **15.** $-(6 + 9x^2 y)/(9x^3 - x^2)$
17. $3y/x$ **19.** $(p + 10p^2 q)/(2p - q - 10pq^2)$
21. $(ye^x - e^y)/(xe^y - e^x)$ **23.** $se^{st}/(2s - te^{st})$
25. $ye^x/(2e^x + y^3 e^y)$ **27.** $(y - y^2)/(-1 + 3y - y^2)$
29. $-y/(x + 2y - xye^y - y^2 e^y)$

31. $(x^3 + x)\sqrt{x^3 + 2}\left(\dfrac{3x^2 + 1}{x^3 + x} + \dfrac{1}{2}\dfrac{3x^2}{x^3 + 2}\right)$

33. $x^x(1 + \ln x)$ **35.** 1 **37.** -2 **39.** $-0.033\,14$ **41.** 31.73
43. -0.1898 **45.** 0 **47. a.** 500 T-shirts

b. $\left.\dfrac{dq}{dp}\right|_{p=5} = -125$ T-shirts/dollar. Thus, when the price is set at

\$5, the demand is dropping by 125 T-shirts/\$1 increase in price.

49. $\left.\dfrac{dk}{de}\right|_{e=15} = -0.307$ carpenter/electrician. This means that for

a \$200,000 house whose construction employs 15 electricians, adding one more electrician would cost as much as approximately 0.307 additional carpenter. In other words, one electrician is worth approximately 0.307 carpenter.   **51. a.** 22.93 hours. (The other root is rejected because it is larger than 30.)

b. $\dfrac{dt}{dx} = \dfrac{4t - 20x}{0.4t - 4x}$; $\left.\dfrac{dt}{dx}\right|_{x=3.0} \approx -11.2$ hours/grade point. This

means that for a 3.0 student who scores 80 on the examination, one grade point is worth approximately 11.2 hours.

53. $\left.\dfrac{dy}{dx}\right|_{x=100} = -\dfrac{3}{2}\dfrac{y}{x} \approx -\$848,528$/worker. The annual ex-

penditure to maintain production at 20,000 units/day is decreasing at a rate of \$848,528/additional worker employed. In other words, each additional worker is worth approximately \$848,528 to the company in annual expense.

55. $\dfrac{dr}{dy} = 2\dfrac{r}{y}$, so $\dfrac{dr}{dt} = 2\dfrac{r}{y}\dfrac{dy}{dt}$ by the chain rule.

57. Let $y = f(x)g(x)$. Then $\ln y = \ln f(x) + \ln g(x)$, and

$\dfrac{1}{y}\dfrac{dy}{dx} = \dfrac{f'(x)}{f(x)} + \dfrac{g'(x)}{g(x)}$, so $\dfrac{dy}{dx} = y\left[\dfrac{f'(x)}{f(x)} + \dfrac{g'(x)}{g(x)}\right] =$

$f(x)g(x)\left[\dfrac{f'(x)}{f(x)} + \dfrac{g'(x)}{g(x)}\right] = f'(x)g(x) + f(x)g'(x)$.   **59.** Writing

$y = f(x)$ specifies y as an explicit function of x. This can be regarded as an equation giving y as an *implicit* function of x. The procedure of finding dy/dx by implicit differentiation is then the same as finding the derivative of y as an explicit function of x: we take d/dx of both sides. **61.** True. The answer follows by differentiating both sides of the equation $y = f(x)$ with respect to y because we get $1 = f'(x) \cdot \dfrac{dx}{dy}$, giving $\dfrac{dx}{dy} = \dfrac{1}{f'(x)} = \dfrac{1}{dy/dx}$.

Chapter 4 Review Test

1. a. $e^x(x^2 + 2x - 1)$ **b.** $-4x/(x^2 - 1)^2$ **c.** $20x(x^2 - 1)^9$ **d.** $-20x/(x^2 - 1)^{11}$ **e.** $e^x(x^2 + 1)^9(x^2 + 20x + 1)$ **f.** $12(x - 1)^2/(3x + 1)^4$ **g.** $2xe^{x^2-1}$ **h.** $2x(x^2 + 2)\,e^{x^2-1}$ **i.** $2x/(x^2 - 1)$ **j.** $\dfrac{2x - 2x\ln(x^2 - 1)}{(x^2 - 1)^2}$ **2. a.** $\dfrac{2x - 1}{2y}$ **b.** $-\dfrac{2y}{2(x + y) - 1}$

c. $-y/x$ **d.** $\dfrac{y}{x(1 - y)}$ **3. a.** $x = (1 - \ln 2)/2$ **b.** $x = 0$

c. None **d.** $x = 1/3$ **4. a.** $R'(0) = p'(0)q(0) + p(0)q'(0) = (-1)(1000) + 20(200) = \$3000/\text{wk}$ (rising) **b.** 300 books/wk **c.** $R = pq$ gives $R' = p'q + pq'$. Thus, $R'/R = R'/(pq) = (p'q + pq')/pq = p'/p + q'/q$. **5. a.** Rising at a rate of 40 units/yr **b.** $\$110/\text{yr}$ **c.** $Q = P/E$ gives $Q' = (P'E - PE')/E^2$. Thus, $Q'/Q = Q'/(P/E) = (P'E - PE')/PE = P'/P - E'/E$.

6. a. $s'(t) = \dfrac{2{,}460.7e^{-0.55(t-4.8)}}{(1 + e^{-0.55(t-4.8)})^2}$ **b.** 553 books/wk **c.** 15 books/

wk **7.** 616.8 hits/day/wk **8. a.** -16.67 copies/\$1. The demand for the gift edition of *The Lord of the Rings* is dropping at a rate of about 16.67 copies/\$1 increase in the price. **b.** $dR/dp = q + p(dq/dp) \approx 1000 + 40(-16.67) \approx \$333$ per dollar is positive, so the price should be raised.

Chapter 5

Exercises 5.1

1. Absolute min.: $(-3, -1)$; relative max: $(-1, 1)$; relative min: $(1, 0)$; absolute max: $(3, 2)$ **3.** Absolute min: $(3, -1)$ and $(-3, -1)$; absolute max: $(1, 2)$ **5.** Absolute min: $(-3, 0)$ and $(1, 0)$; absolute max: $(-1, 2)$ and $(3, 2)$ **7.** Relative min: $(-1, 1)$ **9.** Absolute min: $(-3, -1)$; relative max: $(-2, 2)$; relative min: $(1, 0)$; absolute max: $(3, 3)$ **11.** Relative max: $(-3, 0)$; absolute min: $(-2, -1)$; stationary nonextreme point: $(1, 1)$ **13.** Stationary minimum at $x = -1$ **15.** Stationary minima at $x = -2$ and $x = 2$; stationary maximum at $x = 0$ **17.** Singular minimum at $x = 0$; stationary nonextreme point at $x = 1$ **19.** Stationary minimum at $x = -2$; singular nonextreme points at $x = -1$ and $x = 1$; stationary maximum at $x = 2$ **21.** Absolute max: $(0, 1)$; absolute min: $(2, -3)$; relative max: $(3, -2)$ **23.** Absolute min: $(-4, -16)$; absolute max: $(-2, 16)$; absolute min: $(2, -16)$; absolute max: $(4, 16)$ **25.** Absolute min: $(-2, -10)$; absolute max: $(2, 10)$ **27.** Absolute min: $(-2, -4)$; relative max: $(-1, 1)$; relative min: $(0, 0)$ **29.** Relative max: $(-1, 5)$; absolute min: $(3, -27)$ **31.** Absolute min: $(0, 0)$ **33.** Relative min: $(-2, 5/3)$; relative max: $(0, -1)$; relative min: $(2, 5/3)$ **35.** Relative max: $(0, 0)$; absolute min: $(1/3, -2\sqrt{3}/9)$ **37.** Relative max: $(0, 0)$; absolute min: $(1, -3)$ **39.** No relative extrema **41.** Absolute min: $(1, 1)$ **43.** Relative min: $(-1, 1 + 1/e)$; absolute min: $(0, 1)$; absolute max: $(1, e - 1)$ **45.** Relative max: $(-6, -24)$; relative min: $(-2, -8)$ **47.** Absolute max: $(1/\sqrt{2}, \sqrt{e/2})$; absolute min: $(-1/\sqrt{2}, -\sqrt{e/2})$ **49.** Relative min: $(0.15, -0.52)$ and

$(2.45, 8.22)$; relative max: $(1.40, 0.29)$ **51.** Absolute max: $(-5, 700)$: relative max: $(3.10, 28.19)$ and $(6, 40)$; absolute min: $(-2.10, -392.69)$; relative min: $(5, 0)$
53. Answers will vary. **55.** Answers will vary.

57. Not necessarily; it could be neither a relative maximum nor a relative minimum, as in the graph of $y = x^3$ at the origin.
59.

Exercises 5.2

1. $x = y = 5; P = 25$ **3.** $x = y = 3; S = 6$ **5.** $x = 2, y = 4$; $F = 20$ **7.** $x = 20, y = 10, z = 20; P = 4000$ **9.** 5×5
11. $5 \times 10 = 50\ \text{ft}^2$ **13.** $p = \$10$ **15.** $p = \$30$ **17. a.** \$1.41/lb **b.** 5000 lb **c.** \$7071.07/mo **19.** 34.5¢/lb, for an annual (per capita) revenue of \$5.95 **21.** \$42.50/ruby, for a weekly profit of \$408.33 **23. a.** 656 headsets, for a profit of \$28,120 **b.** \$143/headset **25.** 38,730 CD players, giving an average cost of \$28/CD player **27.** 2.5 lb **29.** $13\frac{1}{3}$ in. $\times 3\frac{1}{3}$ in. $\times 1\frac{1}{3}$ in. for a volume of $1600/27 \approx 59\ \text{in}^3$ **31.** $l = w = h \approx 20.67$ in., volume $\approx 8827\ \text{in.}^3$ **33.** $l = 30$ in, $w = 15$ in, $h = 30$ in. **35.** $l = 36$ in., $w = h = 18$ in., $V = 11{,}664\ \text{in.}^3$ **37. a.** 1.6 yr, or year 2001.6 **b.** $R_{\max} = \$28{,}241$ million **39.** $t = 2.5$, or midway through 1972.; $D(2.5)/S(2.5) \approx 4.09$. The number of new (approved) drugs per \$1 billion of spending on research and development reached a high of around four approved drugs per \$1 billion midway through 1972. **41.** 30 yr from now
43. 55 days **45.** 1600 copies. At this value of x, average profit equals marginal profit; beyond this the marginal profit is smaller than the average. **47.** Decreasing most rapidly in 1963; increasing most rapidly in 1989 **49.** Maximum when $t = 17$ days. This means that the embryo's oxygen consumption is increasing most rapidly 17 days after the egg is laid.
51. $h = r \approx 11.7$ cm **53.** 25 additional trees **55.** Third quarter of 1998 $(t \approx 1.6)$; 40.7 million books/yr
57.

1994 $(t \approx 14)$; $\dfrac{d}{dt}\left[\dfrac{S(t)}{W(t)}\right]_{t=14} \approx -0.072$. The fraction of

bottled-water sales due to sparkling water was decreasing most rapidly in 1994. At that time, it was decreasing at a rate of 7.2 percentage points/year.

59. You should sell them in 17 years' time, when they will be worth approximately \$3960. **61.** 71 employees **63.** (D) **65.** The problem is uninteresting because the company can accomplish the objective by cutting away the entire sheet of cardboard, resulting in a box with zero surface area. **67.** Not all absolute extrema occur at stationary points; some may occur at an endpoint or a singular point of the domain, as in Exercises 17, 18, 47, and 48. **69.** The minimum of dq/dp is the fastest that the demand is dropping in response to increasing price.

Exercises 5.3

1. 6 **3.** $4/x^3$ **5.** $-0.96x^{-1.6}$ **7.** $e^{-(x-1)}$ **9.** $2/x^3 + 1/x^2$
11. a. $a = -32$ ft/s^2 **b.** $a = -32$ ft/s^2 **13. a.** $a = 2/t^3 + 6/t^4$ ft/s^2 **b.** $a = 8$ ft/s^2 **15. a.** $a = -1/(4t^{3/2}) + 2$ ft/s^2 **b.** $a = 63/32$ ft/s^2 **17.** $(1, 0)$ **19.** $(1, 0)$ **21.** None **23.** $(-1, 0)$, $(1, 1)$ **25.** Points of inflection at $x = -1$ and $x = 1$ **27.** One point of inflection at $x = -2$ **29.** Points of inflection at $x = -2$, $x = 0$, $x = 2$ **31.** One point of inflection at $x = 0$
33. Points of inflection at $x = -2$ and $x = 2$
35. Absolute min: $(-1, 0)$; no points of inflection **37.** Relative max: $(-2, 21)$; relative min: $(1, -6)$; point of inflection: $(-1/2, 15/2)$

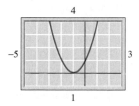

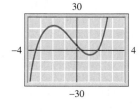

39. Absolute min: $(-4, -16)$ and $(2, -16)$; absolute max: $(-2, 16)$ and $(4, 16)$; point of inflection: $(0, 0)$ **41.** Absolute min: $(0, 0)$; points of inflection: $(1/3, 11/324)$ and $(1, 1/12)$

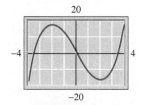

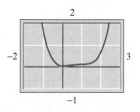

43. Relative min: $(-2, 5/3)$ and $(2, 5/3)$; relative max: at $(0, -1)$; vertical asymptotes: $x = \pm 1$ **45.** No extrema; points of inflection: $(0, 0)$, $(-3, -9/4)$, $(3, 9/4)$

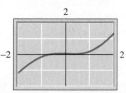

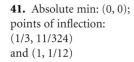

47. Relative min: $(1, 2)$; relative max: $(-1, -2)$; vertical asymptote: $y = 0$ **49.** Relative max: $(0,0)$; absolute min: $(1,-2)$

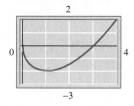

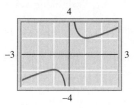

51. Absolute min: $(1, 1)$; vertical asymptote: $x = 0$ **53.** No relative extrema; points of inflection: at $(1,1)$ and $(-1, 1)$; vertical asymptote: at $x = 0$

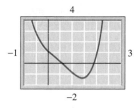

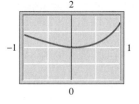

55. Absolute min: $(0, 1)$; absolute max: $(1, e - 1)$; relative max: $(-1, 1 + e^{-1})$ **57.** Absolute min: $(1.40, -1.49)$; points of inflection: $(0.21, 0.61)$ and $(0.79, -0.55)$

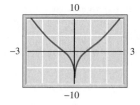

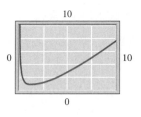

59. Relative min: $(-0.46, 0.73)$; relative max: $(0.91, 1.73)$; absolute min: $(3.73, -10.22)$; points of inflection: $(0.20, 1.22)$ and $(2.83, -5.74)$

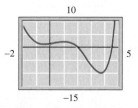

61. a. 2 yr into the epidemic **b.** 2 yr into the epidemic **63. a.** 2000 **b.** 2002 **c.** 1998 **65. a.** 465,000 prisoners **b.** Increasing at a rate of 41,500 prisoners/yr **c.** Accelerating at a rate of 3000 prisoners/yr/yr **67.** Point of inflection at $(9.333, 0.2873)$. Annual sales of bottled water were increasing at the slowest rate at $t = 9.333$ (that is, around October 1989) at which time approximately 287.3 million gals were being sold each year. **69. a.** There are no points of inflection in the graph of S. **b.** Since the graph is concave up, the derivative of S is increasing, and so the rate of *decrease* of SAT scores with increasing numbers of prisoners is diminishing. In other words, the apparent effect of more prisoners is diminishing. **71. a.** $\left.\dfrac{d^2n}{ds^2}\right|_{s=3} = -21.494$.

Thus, for a firm with annual sales of $3 million, the rate at which new patents are produced decreases with increasing firm size. This means that the returns (as measured in the number of new patents per increase of $1 million in sales) are diminishing as the firm size increases. **b.** $\dfrac{d^2n}{ds^2}\Big|_{s=7} = 13.474$. Thus, for a firm with annual sales of $7 million, the rate at which new patents are produced increases with increasing firm size by 13.474 new patents/$1 million increase in annual sales. **c.** There is a point of inflection when $s \approx 5.4587$ so that in a firm with sales of $5,458,700/year the number of new patents produced per additional $1 million in sales is a minimum. **73.** About $570/yr, after about 12 yr **75.** Increasing most rapidly in 17.64 yr; decreasing most rapidly now (at $t = 0$) **77. a.** Approximately 0.000 158 manatees/boat, or 15.8 manatees/100,000 boats

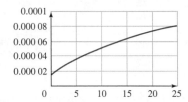

b. $M(t)/B(t)$ represents the number of manatees killed in year t per registered boat. $\lim_{t\to+\infty} M(t)/B(t)$ represents the number of manatees killed per registered boat in the long term. **c.** Concave down; (C) **79.** nonnegative

81.

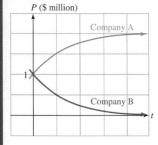

83. Sales were decreasing most rapidly in June 2002. **85.** At a point of inflection, the graph of a function changes either from concave up to concave down or vice versa. If it changes from concave up to concave down, then the derivative changes from increasing to decreasing and hence has a relative maximum. Similarly, if it changes from concave down to concave up, the derivative has a relative minimum.

Exercises 5.4

1. $P = 10,000$; $\dfrac{dP}{dt} = 1000$ **3.** Let R be the annual revenue of my company and let q be annual sales. $R = 7000$ and $dR/dt = -700$. Find dq/dt. **5.** Let p be the price of a pair of shoes and let q be the demand for shoes. $dp/dt = 5$. Find dq/dt. **7.** Let T be the average global temperature and let q be the number of Bermuda shorts sold each year. $T = 60$ and $dT/dt = 0.1$. Find dq/dt. **9. a.** $6/(100\pi) \approx 0.019$ km/s **b.** $6/(8\sqrt{\pi}) \approx 0.4231$ km/s **11.** 7.5 ft/s **13.** Decreasing at a rate of $1.66/player/wk. **15.** Monthly sales will drop at a rate of 26 T-shirts/mo.

17. Raise the price by 3¢/wk. **19.** The price is decreasing at a rate of approximately 31¢/lb/mo. **21.** $3/(4\pi) \approx 0.24$ ft/min **23.** $2300/\sqrt{4100} \approx 36$ mph **25.** The y coordinate is decreasing at a rate of 16 units/s. **27.** The daily operating budget is dropping at a rate of $2.40/yr. **29.** $1814/yr **31.** Their prior experience must increase at a rate of approximately 0.97 yr every year.
33. $\dfrac{2500}{9\pi}\left(\dfrac{3}{5000}\right)^{2/3} \approx 0.63$ m/s **35.** $\dfrac{\sqrt{1+128\pi}}{4\pi} \approx$
1.6 cm/s **37.** 0.5137 computer/household and increasing at a rate of 0.0230 computer/household/yr. **39.** The average SAT score was 904.71 and decreasing at a rate of 0.11/yr. **41.** Decreasing by 2 percentage points/yr **43.** The section is called "Related Rates" because the goal is to compute the rate of change of a quantity based on a knowledge of the rate of change of a related quantity. **45.** Answers will vary. One example: A rectangular solid has dimensions 2 cm × 5 cm × 10 cm, and each side is expanding at a rate of 3 cm/s. How fast is the volume increasing? **49.** Linear **51.** Let x = my grades and y = your grades. If $dx/dt = 2\, dy/dt$, then $dy/dt = (1/2)dx/dt$.

Exercises 5.5

1. $E = 1.5$; the demand is going down 1.5% per 1% increase in price at that price level; revenue is maximized when $p = 25$; weekly revenue at that price is $12,500. **3. a.** $E = 6/7$; the demand is going down 6% per 7% increase in price at that price level; thus, a price increase is in order. **b.** Revenue is maximized when $p = 100/3 \approx 33.33$. **c.** Demand would be $(100 - 100/3)^2 = (200/3)^2 \approx 4444$ cases/wk. **5. a.** $E = 1.71$. Thus, the demand is elastic at the given tuition level, showing that a decrease in tuition will result in an increase in revenue. **b.** They should charge an average of $2271.75/student, and this will result in an enrollment of about 4930 students, giving a revenue of about $11,199,000. **7. a.** $E = -\dfrac{mp}{mp+b}$ **b.** $p = -\dfrac{b}{2m}$
9. a. $E = r$ **b.** E is independent of p. **c.** If $r = 1$, then the revenue is not affected by the price. If $r > 1$, then the revenue is always elastic; if $r < 1$, the revenue is always inelastic. This is an unrealistic model because there should always be a price at which the revenue is a maximum. **11. a.** $E = 51$; the demand is going down 51% per 1% increase in price at that price level; thus, a large price decrease is advised. **b.** Revenue is maximized when $p = ¥0.50$. **c.** Demand would be $100e^{-3/4+1/2} \approx 78$ paint-by-number sets/mo. **13. a.** $q = -1500p + 6000$. **b.** $2/hamburger, giving a total weekly revenue of $6000. **15.** $E \approx 0.77$. At a family income level of $20,000, the fraction of children attending a live theatrical performance is increasing by 0.77% per 1% increase in household income. **17. a.** $E \approx 0.46$. The demand for computers is increasing by 0.46% per 1% increase in household income. **b.** E decreases as income increases. **c.** Unreliable; it predicts a likelihood greater than 1 at incomes of $123,000 and above. In a more appropriate model, one would expect the curve to level off at or below 1. **d.** $E \approx 0$ **19.** $\dfrac{Y}{Q} \cdot \dfrac{dQ}{dY} = \beta$. An increase in income of x% will result in an increase in demand of βx%. (Note that we should *not* take the negative here because we expect an increase in income to produce an *increase* in demand.)

21. a. $q = 1000e^{-0.3p}$ **b.** At $p = \$3$, $E = 0.9$; at $p = \$4$, $E = 1.2$; at $p = \$5$, $E = 1.5$ **c.** $p = \$3.33$ **d.** $p = \$5.36$. Selling at a lower price would increase demand, but you cannot sell more than 200 lb anyway. You should charge as much as you can and still be able to sell all 200 lb. **23.** the price is lowered **25.** Start with $R = pq$ and differentiate with respect to p to obtain $\dfrac{dR}{dp} = q + p\dfrac{dq}{dp}$. For a stationary point, $dR/dp = 0$, and so $q + p\dfrac{dq}{dp} = 0$.

Rearranging this result gives $p\dfrac{dq}{dp} = -q$ and hence $-\dfrac{dq}{dp} \cdot \dfrac{p}{q} = 1$, or $E = 1$, showing that stationary points of R correspond to points of unit elasticity. **27.** The distinction is best illustrated by an example. Suppose q is measured in weekly sales and p is the unit price in dollars. Then the quantity $-dq/dp$ measures the drop in weekly sales/\$1 increase in price. The elasticity of demand E, on the other hand, measures the *percentage* drop in sales per 1% increase in price. Thus, $-dq/dp$ measures absolute change, and E measures fractional, or percentage, change.

Chapter 5 Review Test

1. a. Relative max: $(-1, 5)$; absolute min: $(-2, -3)$ and $(1, -3)$ **b.** Relative min: $(2, 3)$; absolute min: $(-2, -1/9)$ **c.** Absolute min: $(1, 0)$ **d.** No extrema **e.** Absolute min: $(-2, -1/4)$ **f.** Absolute min: $(0, 2)$ **2. a.** Relative max: $x = 1$; point of inflection: $x = -1$ **b.** Relative min: $x = 2$; point of inflection: $x = -1$ **c.** Relative max: $x = -2$; relative min: $x = 1$; point of inflection: $x = -1$ **d.** Relative min: $x = -2.5$ and $x = 1.5$; relative max: $x = 3$; point of inflection: $x = 2$

3. a.

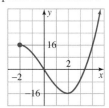

Relative max: $(-2, 16)$; absolute min: $(2, -16)$; inflection: $(0, 0)$; no horizontal or vertical asymptotes

b.

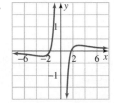

Relative min: $(-3, -2/9)$; relative max: $(3, 2/9)$; inflection: $(-3\sqrt{2}, -5\sqrt{2}/36)$ and $(3\sqrt{2}, 5\sqrt{2}/36)$; vertical asymptote: $x = 0$; horizontal asymptote: $y = 0$

c.

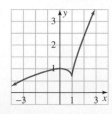

Relative max: $(0, 1)$; relative min: $(1, 2/3)$; no inflection points; no horizontal or vertical asymptotes

4. a. $E = \dfrac{2p^2 - 33p}{-p^2 + 33p + 9}$ **b.** $0.52, 2.03$; when the price is \$20, demand is dropping at a rate of 0.52% per 1% increase in the price; when the price is \$25, demand is dropping at a rate of 2.03% per 1% increase in the price. **c.** \$22.14/book **5. a.** Profit $= -p^3 + 42p^2 - 288p - 181$ **b.** \$24/copy **c.** \$3275 **d.** For maximum revenue, the company should charge \$22.14/copy. At this price, the cost is decreasing at a linear rate with increasing price, but the revenue is not decreasing (its derivative is zero). Thus, the profit is increasing with increasing price, suggesting that the maximum profit will occur at a higher price. **6. a.** $s'(t) = \dfrac{2460.7\, e^{-0.55(t-4.8)}}{[1 + e^{-0.55(t-4.8)}]^2}$; sales were growing fastest in week 5.

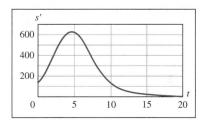

b. Point of inflection **c.** 10,527; if weekly sales continue as predicted by the model, they will level off at around 10,500 books/week in the long term. **d.** Zero; if weekly sales continue as predicted by the model, then the rate of change of sales approaches zero in the long term. **7.** $-1.5/(1 + 3\pi/2) \approx -0.263$ in./s

Chapter 6

Exercises 6.1

1. $x^6/6 + C$ **3.** $6x + C$ **5.** $x^2/2 + C$ **7.** $x^3/3 - x^2/2 + C$ **9.** $x + x^2/2 + C$ **11.** $-x^{-4}/4 + C$ **13.** $x^{3.3}/3.3 - x^{-0.3}/0.3 + C$ **15.** $u^3/3 - \ln|u| + C$ **17.** $\dfrac{2x^{3/2}}{3} + C$ **19.** $3x^5/5 + 2x^{-1} - x^{-4}/4 + 4x + C$ **21.** $\ln|x| - 2/x + 1/2x^2 + C$ **23.** $3x^{1.1}/1.1 - x^{5.3}/5.3 - 4.1x + C$ **25.** $\dfrac{x^{0.9}}{0.3} + \dfrac{40}{x^{0.1}} + C$ **27.** $2e^x + 5\ln|x| + x/4 + C$ **29.** $12.2x^{0.5} + x^{1.5}/9 - e^x + C$ **31.** $2^x/\ln 2 - 3^x/\ln 3 + C$ **33.** $100(1.1^x)/\ln(1.1) + C$ **35.** $-1/x - 1/x^2 + C$ **37.** $f(x) = x^2/2 + 1$ **39.** $f(x) = e^x - x - 1$ **41.** $C(x) = 5x - x^2/20,000 + 20,000$ **43.** $C(x) = 5x + x^2 + \ln\ x + 994$ **45.** $I(t) = -0.7t + 7.4; I(8) = 1.8$ **47.** $S(t) = 0.005\,53t^3 - 0.025t^2 + 0.27t$ billion gal **49. a.** $C'(t) = -0.1t + 3.7$ billion dollars/yr **b.** $C(t) = -0.05t^2 + 3.7t + 135$ billion dollars **51. a.** $s = t^3/3 + t + C$ **b.** $C = 1; s = t^3/3 + t + 1$ **53.** 320 ft/s downward **57.** $(1280)^{1/2} \approx 35.78$ ft/s **59. a.** 80 ft/s **b.** 60 ft/s **c.** 1.25 s **61.** $\sqrt{2} \approx 1.414$ times as fast **63.** They differ by a constant, $G(x) - F(x) = $ constant **65.** antiderivative, marginal **67.** $\int f(x)\, dx$ represents the total cost of manufacturing x items. The units of $\int f(x)\, dx$ are the product of the units of $f(x)$ and the units of x. **69.** $\int (f(x) + g(x))\, dx$ is, by definition, an antiderivative of $f(x) + g(x)$. Let $F(x)$ be an antiderivative of $f(x)$ and let $G(x)$ be an antiderivative of $g(x)$. Then, since the derivative of $F(x) + G(x)$ is $f(x) + g(x)$ (by the rule for sums of derivatives), this means that $F(x) + G(x)$ is an antiderivative of $f(x) + g(x)$. In symbols, $\int (f(x) + g(x)) \, dx = F(x) + G(x) + C = \int f(x) \, dx + \int g(x) \, dx$, the sum of the indefinite integrals. **71.** $\int x \cdot 1 \, dx = \int x \, dx = x^2/2 + C$, whereas $\int x \, dx \cdot \int 1 \, dx = (x^2/2 + D) \cdot (x + E)$, which is not the same as $x^2/2 + C$, no matter what values we choose for the constants C, D, and E. **73.** derivative, indefinite integral, indefinite integral, derivative

Exercises 6.2

1. $(3x + 1)^6/18 + C$ **3.** $(-2x + 2)^{-1}/2 + C$ **5.** $-e^{-x} + C$
7. $1.6(3x - 4)^{3/2} + C$ **9.** $2e^{(0.6x+2)} + C$ **11.** $(3x^2 + 3)^4/24 + C$ **13.** $(x^2 + 1)^{2.3}/4.6 + C$ **15.** $x + 3e^{3.1x-2} + C$
17. $(e^x - e^{-x})/2 + C$ **19.** $2(3x^2 - 1)^{3/2}/9 + C$
21. $-(1/2)e^{-x^2+1} + C$ **23.** $-(1/2)e^{-(x^2+2x)} + C$
25. $(x^2 + x + 1)^{-2}/2 + C$ **27.** $(2x^3 + x^6 - 5)^{1/2}/3 + C$
29. $(x - 2)^7/7 + (x - 2)^6/3 + C$
31. $4[(x + 1)^{5/2}/5 - (x + 1)^{3/2}/3] + C$
33. $3e^{-1/x} + C$ **35.** $20 \ln |1 - e^{-0.05x}| + C$
37. $\ln(e^x + e^{-x}) + C$ **39.** $(e^{2x^2-2x} + e^{x^2})/2 + C$
45. $-e^{-x} + C$ **47.** $(1/2)e^{2x-1} + C$ **49.** $(2x + 4)^3/6 + C$
51. $(1/5)\ln |5x - 1| + C$ **53.** $(1.5x)^4/6 + C$
55. $1.5^{3x}/[3 \ln(1.5)] + C$
57. $(2^{3x+4} - 2^{-3x+4})/(3 \ln 2) + C$ **59.** $f(x) = (x^2 + 1)^4/8 - 1/8$
61. $f(x) = (1/2)e^{x^2-1}$ **63.** $C(x) = 5x - 1/(x + 1) + 995.5$
65. $S(t) = 0.005\,53(t - 1990)^3 - 0.025(t - 1990)^2 + 0.27(t - 1990)$ billion gal **67. a.** $s = (t^2 + 1)^5/10 + t^2/2 + C$
b. $C = 9/10; s = (t^2 + 1)^5/10 + t^2/2 + 9/10$ **69.** None; the substitution $u = x$ simply replaces the letter x throughout by the letter u and thus does not change the integral at all. For instance, the integral $\int x(3x^2 + 1) \, dx$ becomes $\int u(3u^2 + 1) \, du$ if we substitute $u = x$. **71.** The integral $\int (2x + 1)(x^2 + x) \, dx$ can be calculated by expanding the integrand or by using the substitution $u = x^2 + x$, as follows. Expanding the integrand: $\int (2x + 1)(x^2 + x) \, dx = \int (2x^3 + 3x^2 + x) \, dx = x^4/2 + x^3 + x^2/2 + C$. Substituting $u = x^2 + x$: $\int (2x + 1)(x^2 + x) \, dx = \int u \, du = u^2/2 + C = (x^2 + x)^2/2 + C$. Since expanding the second answer gives the first, both methods result in the same solution. **73.** The purpose of substitution is to introduce a new variable that is defined in terms of the variable of integration. One cannot say $u = u^2 + 1$ because u is not a new variable. Instead, define $w = u^2 + 1$ (or any other letter different from u).

Exercises 6.3

1. 4 **3.** 6 **5.** 0.7456 **7.** 2.3129 **9.** 2.5048 **11. a.** 3.3045
b. 3.1604 **c.** 3.1436 **13. a.** 0.0275 **b.** 0.0258 **c.** 0.0256
15. $99.95 **17.** 3.8 billion gal **19.** 5801; this represents the total number of wiretaps authorized by U.S. courts from 1995 through 1999. **21.** 91.2 ft **23.** 198; 198 million books were sold online during the period 1997 through 2000. **25. a.** 99.4% **b.** 0 (to at least 15 decimal places) **27.** increases **29.** The total cost is $c(1) + c(2) + \cdots + c(60)$, which is represented by the Riemann sum approximation of $\int_1^{61} c(t) \, dt$ with $n = 60$. **31.** $[f(x_1) + f(x_2) + \cdots + f(x_n)]\Delta x = \sum_{k=1}^{n} f(x_k)\Delta x$

33. If increasing n does not change the value of the answer when rounded to three decimal places, then the answer is accurate to three decimal places.

Exercises 6.4

1. 1 **3.** 1/2 **5.** 1/4 **7.** 2 **9.** 0 **11.** 0.45, 0.495
13. 14.24, 17.0144 **15.** 1.3813, 1.4541 **17.** 18,250 million; this represents the total number of SUVs sold in the U.S. from 1998 through 2003. **19.** $-$580 million; this represents the total net income for the first quarter of 2001 through the first quarter of 2002. **21.** Yes. The Riemann sum gives an estimated area of 420 ft². **23. a.**

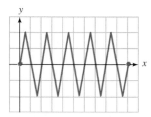

The area represents the total oil revenue earned by Saudi Arabia in 1999.
b. $\int_0^{365} R(t) \, dt \approx 590,000$. The total oil revenue earned by Saudi Arabia in 1999 was approximately $590 billion. **25.** Answers will vary. One example: Set $P(t)$ to be the rate of change of profit at time t. If $P(t)$ is negative, then the profit is decreasing, so the change in profit, represented by the definite integral of $P(t)$, is negative. **27.** Answers will vary, but an example is the following:

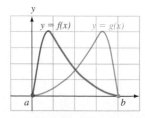

29. Answers will vary, but an example is the following:

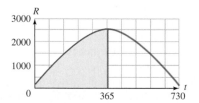

The following is a sketch of a typical example. The difference between the right and left sums for $\int_a^b f(x) \, dx$ is the area of the unshaded regions.

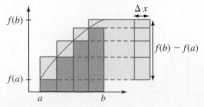

Exercises 6.5

1. $A(x) = x$ **3.** $A(x) = x^2/2 - 2$ **5.** $A(x) = 4(x - a)$
7. $A(x) = x^2/2; f(x):$ $A(x):$

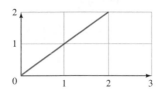   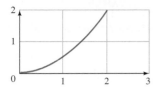

9. $A(x) = x^3/3; f(x):$ $A(x):$

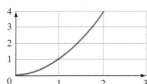   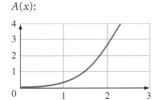

11. $A(x) = e^x - 1; f(x):$ $A(x):$

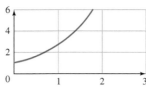   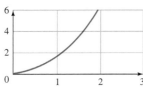

13. $A(x) = \ln x; f(x):$ $A(x):$

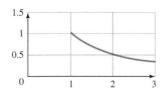   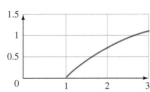

15. $A(x) = x^2/2$, if $0 \le x \le 1$; $1/2 + 2(x - 1)$, if $x > 1$;
$f(x):$

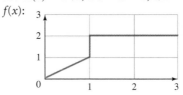

$A(x):$

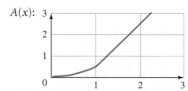

17. $14/3$ **19.** 5 **21.** 0 **23.** $40/3$ **25.** -0.9045
27. $2(e - 1)$ **29.** $2/3$ **31.** $1/\ln 2$ **33.** $4^6 - 1 = 4095$
35. $(e^1 - e^{-3})/2$ **37.** $3/(2 \ln 2)$ **39.** $50(e^{-1} - e^{-2})$ **41.** $e^{2.1} - e^{-0.1}$ **43.** 0 **45.** $(5/2)(e^3 - e^2)$ **47.** $(1/3)[\ln 26 - \ln 7]$
49. $0.1/[2.2 \ln(1.1)]$ **51.** $e - e^{1/2}$ **53.** $2 - \ln 3$ **55.** $-4/21$
57. $3^{5/2}/10 - 3^{3/2}/6 + 1/15$ **59.** $1/2$ **61.** $16/3$ **63.** $56/3$
65. $1/2$ **67.** $\$783$ **69.** 4.3 billion gal   **71.** $\int_{14}^{26} 5.4e^{0.06t}\, dt \approx$ $\$220$ billion   **73.** 296 mi   **75. a.** $C'(t) = -0.1t + 3.7$ billion dollars/yr **b.** $\$17.25$ billion   **77.** 61 mL   **79.** 907 T-shirts
81. 9 gal   **83. a.** $11,000$ **b.** To estimate the actual number of wiretaps over the given period from the chart, take half of the

1990 and 2000 numbers and add the numbers from 1991 through 1999. This gives 10,737, which agrees with the answer in part (a) when we round to two significant digits. Therefore, the integral in part (a) does give an accurate estimate.   **85. a.** $s(t) = 330e^{0.0872t}$
b. $\$9044$ billion (compared to $\$8225$ billion) **c.** $\$7011$ billion
89. c. 200 million books   **93. a.** $\int_a^x (1/t)\, dt$ **95.** They are related by the Fundamental Theorem of Calculus, which (briefly) states that the definite integral of a suitable function can be calculated by evaluating the indefinite integral at the two endpoints and subtracting.   **97.** An example is $v(t) = t - 5$.   **99.** An example is $f(x) = e^{-x}$.   **101.** Two things: (1) Calculate definite integrals by using antiderivatives; (2) Obtain formulas for an antiderivative of a function in terms of a definite integral, or area.

103. $F(x) = \begin{cases} 0 & \text{if } x < 0 \\ x & \text{if } x \ge 0 \end{cases}; F'(x) = f(x)$ for every $x \ne 0$ [since $F(x)$ is not differentiable at $x = 0$].

Chapter 6 Review Test

1. a. $x^3/3 - 5x^2 + 2x + C$ **b.** $e^x + (2/3)x^{3/2} + C$

c. $(1/22)(x^2 + 4)^{11} + C$ **d.** $-\dfrac{1}{3(x^3 + 3x + 2)} + C$

e. $-(5/2)e^{-2x} + C$ **f.** $-e^{-x^2/2} + C$ **2. a.** $1/4$ **b.** $\ln 10 \approx 2.303$
c. $52/9$ **d.** $40/(81 \ln 3)$ **3. a.** $32/3$ **b.** $97/3$ **c.** $(1 - e^{-25})/2$
d. 2 **4. a.** $0.777\ 816\ 82$, $0.749\ 978\ 60$ $0.747\ 140\ 13$ **b.** 0.786 $907\ 44$, $0.697\ 190\ 04$, $0.688\ 492\ 94$ **5. a.** $100,000 - 10p^2$
b. $\$100$ **6. a.** About 86,000 books **b.** About 35,800 books
7. a. 5 s **b.** 100 ft/s **c.** 156.25 ft above ground level

Chapter 7

Exercises 7.1

1. $2e^x(x - 1) + C$ **3.** $-e^{-x}(2 + 3x) + C$
5. $e^{2x}(2x^2 - 2x - 1)/4 + C$ **7.** $-e^{-2x+4}(2x^2 + 2x + 3)/4 + C$
9. $2^x[(2 - x)/\ln 2 + 1/(\ln 2)^2] + C$
11. $-3^{-x}[(x^2 - 1)/\ln 3 + 2x/(\ln 3)^2 + 2/(\ln 3)^3] + C$
13. $-e^{-x}(x^2 + x + 1) + C$
15. $(1/7)x(x + 2)^7 - (1/56)(x + 2)^8 + C$
17. $-x/[2(x - 2)^2] - 1/[2(x - 2)] + C$
19. $(x^4 \ln x)/4 - x^4/16 + C$ **21.** $(t^3/3 + t)\ln(2t) - t^3/9 - t + C$ **23.** $(3/4)t^{4/3}(\ln t - 3/4) + C$ **25.** $x \log_3 x - x/\ln 3 + C$
27. $e^{2x}(x/2 - 1/4) - 4e^{3x}/3 + C$ **29.** $e^x(x^2 - 2x + 2) - e^{x^2}/2$ $+ C$ **31.** e **33.** $38,229/286$ **35.** $(7/2)\ln 2 - 3/4$ **37.** $1/4$
39. $1 - 11e^{-10}$ **41.** $4 \ln 2 - 7/4$ **43.** $28,800,000(1 - 2e^{-1})$ ft
45. $5001 + 10x - 1/(x + 1) - [\ln(x + 1)]/(x + 1)$
47. $\$33,598$ **49.** $\$1478$ million   **51.** Answers will vary. Examples are xe^{x^2} and $e^{x^2} = 1 \cdot e^{x^2}$.   **53.** $n + 1$ times

Exercises 7.2

1. $8/3$ **3.** 4 **5.** 1 **7.** $e - 3/2$ **9.** $2/3$ **11.** $3/10$ **13.** $1/20$
15. $4/15$ **17.** $2 \ln 2 - 1$ **19.** $8 \ln 4 + 2e - 16$ **21.** 0.3222
23. 0.3222 **25.** $\$6.25$ **27.** $\$512$ **29.** $\$119.53$ **31.** $\$900$
33. $\$416.67$ **35.** $\$326.27$ **37.** $\$25$ **39.** $\$0.50$ **41.** $\$386.29$
43. $\$225$ **45.** $\$25.50$ **47.** $\$12,684.63$ **49.** $\bar{p} = \$3655$; $CS \approx$ $\$856,000$; $PS \approx \$3,715,000$. The total social gain is approximately $\$4,571,000$.   **51.** $CS = (1/2m)(b - m\bar{p})^2$ **53. a.** The area represents the accumulated U.S. trade deficit with China (total excess of imports over exports) for the 9-year period 1989–1998.

b. 250.47; the United States accumulated a \$250.47 billion trade deficit with China over the period 1989–1998. **55. a.** The area represents the difference between the total number of above-ground pools and in-ground pools built from 1996 to 2001. **b.** 268; from 1996 to 2001, 268,000 more above-ground pools than in-ground pools were built in the U.S. **57. a.** 12 billion cases **b.** This is the area of the region between the graphs of $N(t)$ and $L(t)$ for $10 \le t \le 50$. **c.** Since the exponent for $L(t)$ is larger, the fraction of new cases that occur among people age 85 and older is projected to rise. **59. a.** We verify, using the formula in the text, that $C'(8) = 270$ million, $C'(10) = 360$ million, and $C'(12) = 780$ million **b.** \$7,260 million **c.** The two equations are $y = 1,000,000(41.25q^2 - 697.5q + 3,210)$ and $y = 400,000,000$ for $0 \le q \le 12$. **61.** The area between the export and import curves represents Canada's accumulated trade surplus (that is, the total excess of exports over imports) from January 1997 through December 2001. **63.** (A) **65.** The claim is wrong because the area under a curve can represent income only if the curve is a graph of income *per unit time*. The value of a stock price is not income per unit time—the income can only be realized when the stock is sold, and it amounts to the current market price. The total net income (per share) from the given investment would be the stock price on the date of sale minus the purchase price of \$40.

Exercises 7.3

1. Average = 2

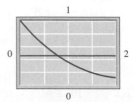

3. Average = 1

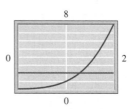

5. Average = $(1 - e^{-2})/2$

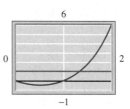

7.

x	0	1	2	3	4	5	6	7
$r(x)$	3	5	10	3	2	5	6	7
$\bar{r}(x)$			6	6	5	10/3	13/3	6

9.

x	0	1	2	3	4	5	6	7
$r(x)$	1	2	6	7	11	15	10	2
$\bar{r}(x)$			3	5	8	11	12	9

11. Moving average: $\bar{f}(x) = x^3 - (15/2)x^2 + 25x - 125/4$

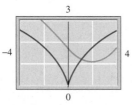

13. Moving average: $\bar{f}(x) = (3/25)[x^{5/3} - (x - 5)^{5/3}]$

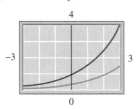

15. $\bar{f}(x) = (2/5)[e^{0.5x} - e^{0.5(x-5)}]$

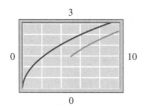

17. $\bar{f}(x) = (2/15)[x^{3/2} - (x - 5)^{3/2}]$

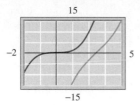

19.

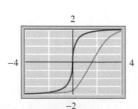

21.

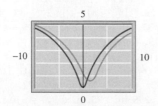

23. 127 million people   **25.** $1.7345 million   **27.** $10,410.88
29. $1500
31.

Year, t	5	6	7	8	9	10	11
Employment (millions)	117	120	123	125	130	132	132
Moving average (millions)				121.25	124.5	127.5	129.75

Some changes are larger, and others are smaller.
33. a.

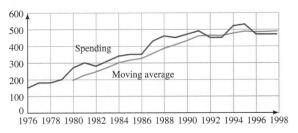

b. $23.4 million/yr; tourist spending in Bermuda was increasing at a rate of approximately $23.4 million/yr during the given period.   **35. a.** 3650 lb   **b.** $(1/2)[t^3 - (t - 2)^3 - 45[t^2 - (t - 2)^2] + 8400]$ **c.** Quadratic   **37.** **a.** $s = 14.4t + 240$ **b.** $\bar{s}(t) = 14.4t + 211.2$ **c.** The slope of the moving average is the same as the slope of the original function.   **39.** $\bar{f}(x) = mx + b - ma/2$
41. a.

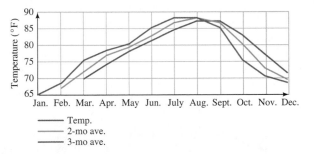

— Temp.
— 2-mo ave.
— 3-mo ave.

b. The 24-mo moving average is constant and equal to the year-long average of approximately 77°F.   **c.** A quadratic model could not be used to predict temperatures beyond the given 12-mo period because temperature patterns are periodic and parabolas are not.   **43.** The area above the x axis equals the area below the x axis. Example: $y = x$ on $[-1, 1]$.   **45.** The moving average "blurs" the effects of short-term oscillations in the price and shows the longer-term trend of the stock price.   **47.** This need not be the case; for instance, the function $f(x) = x^2$ on $[0, 1]$ has average value 1/3, whereas the value midway between the maximum and minimum is 1/2.   **49. c.** A shorter-term moving average most closely approximates the original function because it averages the function over a shorter period, and continuous functions change by only a small amount over a small period.

Exercises 7.4

1. $TV = \$300,000$; $FV = \$434,465.45$ **3.** $TV = \$350,000$; $FV = \$498,496.61$ **5.** $TV = \$389,232.76$; $FV = \$547,547.16$
7. $TV = \$100,000$; $PV = \$82,419.99$ **9.** $TV = \$112,500$; $PV =$

$92,037.48   **11.** $TV = \$107,889.50$; $PV = \$88,479.69$ **13.** $98 billion   **15.** $1028 billion   **17.** $103 billion   **19.** $893 billion   **21.** $1,943,162.44   **23.** $3,086,245.73   **25.** $58,961.74   **27.** $1,792,723.35   **29.** Total   **31.** She is correct, provided there is a positive rate of return, in which case the future value (which includes interest) is greater than the total value (which does not).   **33.** $PV < TV < FV$

Exercises 7.5

1. Diverges   **3.** Converges to $2e$ **5.** Converges to e^2 **7.** Converges to 1/2   **9.** Converges to 1/108   **11.** Converges to $3 \times 5^{2/3}$ **13.** Diverges   **15.** Diverges   **17.** Converges to $(5/4)(3^{4/5} - 1)$ **19.** Diverges   **21.** Converges to 0   **23.** Diverges   **25.** Diverges   **27.** $870 million   **29.** 4466 billion cigarettes   **31.** No; you will not sell more than 2000 of them.   **33.** The integral diverges, and so the number of graduates each year will rise without bound.   **35. a.** $R(t) = 350e^{-0.1t}(39t + 68)$ million dollars/yr   **b.** $1,603,000 million   **37.** $25,160 billion   **39.** $\int_0^{+\infty} N(t) \, dt$ diverges, indicating that there is no bound to the expected future total online sales of books. $\int_{-\infty}^0 N(t) \, dt$ converges to approximately 1.889, indicating that total online sales of books prior to 1997 amounted to approximately 1.889 million books.   **41.** 1   **43.** 0.1587   **45.** $70,833   **47. a.** 2.468 meteors on average   **b.** The integral diverges. We can interpret this as saying that the number of impacts by meteors smaller than 1 megaton is very large. (This makes sense because, for example, this number includes meteors no larger than a grain of dust.)   **49. a.** $\Gamma(1) = 1$; $\Gamma(2) = 1$ **51.** The integral does not converge, so the number given by the FTC is meaningless.   **53.** Yes; the integrals converge to zero, and the FTC also gives zero.   **55.** In all cases you need to rewrite the improper integral as a limit and use technology to evaluate the integral of which you are taking the limit. Evaluate for several values of the endpoint approaching the limit. In the case of an integral in which one of the limits of integration is infinite, you may have to instruct the calculator or computer to use more subdivisions as you approach $+\infty$.   **57.** Answers will vary.

Exercises 7.6

1. $y = x^3/3 + 2x^{3/2}/3 + C$ **3.** $y^2/2 = x^2/2 + C$ **5.** $y = Ae^{x^2/2}$ **7.** $y = -2/[(x + 1)^2 + C]$ **9.** $y = \pm\sqrt{(\ln x)^2 + C}$ **11.** $y = x^4/4 - x^2 + 1$ **13.** $y = (x^3 + 8)^{1/3}$ **15.** $y = 2x$ **17.** $y = e^{x^2/2} - 1$ **19.** $y = -2/[\ln(x^2 + 1) + 2]$ **21.** With $s(t) =$ monthly sales after t mo, $ds/dt = -0.05s$; $s = 1000$ when $t = 0$. Solution: $s = 1000e^{-0.05t}$ qt/m.   **23.** $H(t) = 75 + 115e^{-0.04274t}$ degrees Fahrenheit after t minutes   **25.** With $S(t) =$ total sales after t mo, $dS/dt = 0.1(100,000 - S)$; $S(0) = 0$. Solution: $S = 100,000(1 - e^{-0.1t})$ monitors after t mo.   **27. a.** $dp/dt = k[D(p) - S(p)] = k(20,000 - 1000p)$ **b.** $p = 20 - Ae^{-kt}$ **c.** $p = 20 - 10e^{-0.2231t}$ dollars after t mo   **29.** $q = 0.6078e^{-0.05p}p^{1.5}$ **33.** $S = (2/1999)/(e^{-0.5t} + 1/1999)$

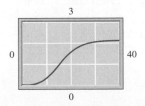

It will take about 27 mo to saturate the market.
35. a. $y = be^{Ae^{-at}}$; $A =$ constant **b.** $y = 10e^{-0.69315e^{-t}}$

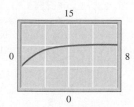

37. A general solution gives all possible solutions to the equation, using at least one arbitrary constant. A particular solution is one specific function that satisfies the equation. We obtain a particular solution by substituting specific values for any arbitrary constants in the general solution. **39.** Example: $y'' = x$ has general solution $y = (1/2)x^2 + Cx + D$ (integrate twice). **41.** $y' = -4e^{-x} + 3$

Chapter 7 Review Test

1. a. $(x^2 - 2x + 4)e^x + C$ **b.** $(1/3)x^3 \ln 2x - x^3/9 + C$ **c.** $-e^2 - 39/e^2$ **d.** $(2e^3 + 1)/9$ **e.** $1/4$ **f.** 2 **2. a.** $3/(2 \cdot 2^{1/3}) - 1/2$
b. $e^2 + 1/e^2 - 2$ **c.** $(2\sqrt{2})/3$ **d.** $1 + 3/e^2$ **3. a.** $e - 2$
b. $[2(e^2 + e)\ln 2 + e^2 + 5/4]/(2e - 1) \approx 5.105\,48$
4. a. $(3/14)[x^{7/3} - (x - 2)^{7/3}]$
b. $(1/2)[x \ln x - (x - 2)\ln(x - 2) - 2]$
5. a. $y = -3/(x^3 + C)$ **b.** $y = Ae^{x^2/2} - 2$ **c.** $y = \sqrt{2 \ln |x| + 1}$
d. $y = 2\sqrt{x^2 + 1}$ **6.** Approximately \$910,000
7. a. $\bar{p} = \$50$; $\bar{q} = 10,000$ **b.** $CS \approx \$333,000$; $PS \approx \$66,700$
8. a. \$1,062,500 **b.** $997,500e^{0.06t}$ **c.** \$5,549,000
d. Principal: \$5,280,000; interest: \$269,000 **9.** \$51 million
10. The amount in the account would be given by $y = 10,000/(1 - t)$, where t is time in years, so would approach infinity 1 yr after the deposit.

Chapter 8

Exercises 8.1

1. a. 1 **b.** 1 **c.** 2 **d.** $a^2 - a + 5$ **e.** $y^2 + x^2 - y + 1$
f. $(x + h)^2 + (y + k)^2 - (x + h) + 1$ **3. a.** 0 **b.** 0.2
c. -0.1 **d.** $0.18a + 0.2$ **e.** $0.1x + 0.2y - 0.01xy$
f. $0.2(x + h) + 0.1(y + k) - 0.01(x + h)(y + k)$ **5. a.** 1 **b.** e
c. e **d.** e^{x+y+z} **e.** $e^{x+h+y+k+z+l}$ **7. a.** Does not exist. **b.** 0 **c.** 0
d. $xyz/(x^2 + y^2 + z^2)$ **e.** $(x + h)(y + k)(z + l)/[(x + h)^2 + (y + k)^2 + (z + l)^2]$ **9. a.** increases, 2.3, **b.** decreases, 1.4,
c. f decreases, 1 unit increase in z **11.** Neither **13.** Linear
15. Linear **17.** Interaction **19. a.** 107 **b.** -14 **c.** -113
21.

$x \rightarrow$	10	20	30	40
y 10	52	107	162	217
$\downarrow$ 20	94	194	294	394
30	136	281	426	571
40	178	368	558	748

25. 18
4
0.0965
47,040
27. 6.9078
1.5193
5.4366
0
29. Let $z =$ annual sales of Z (in millions of dollars), $x =$ annual sales of X, and $y =$ annual sales of Y. The model is $z = -2.1x + 0.4y + 16.2$. **31.** $\sqrt{2}$ **33.** $\sqrt{a^2 + b^2}$ **35.** $1/2$
37. Circle with center $(2, -1)$ and radius 3 **39.** The marginal cost of cars is \$6000/car. The marginal cost of trucks is \$4000/truck. **41.** $C(x, y) = 10 + 0.03x + 0.04y$, where C is the cost in dollars, $x =$ no. of video clips sold per month, $y =$ no. of audio clips sold per month **43. a.** 29% **b.** 7%
c. 0.4-point drop **45. a.** CBS
b. ABC **c.** 7.9 **d.** $A(f, n, c) = \dfrac{f + 0.27c - 0.87n + 36.7}{3.1}$,
where f is Fox's prime-time rating **47. a.** \$9980 **b.** $R(z) = 9850 + 0.04z$ **49.** $s(c, t) = 1.25c - 2t$ **51.** $U(11, 10) - U(10, 10) \approx 5.75$. This means that if your company now has ten copies of Macro Publish and ten copies of Turbo Publish, then the purchase of one additional copy of Macro Publish will result in a productivity increase of approximately 5.75 pages/day.
53. a. Answers will vary. $(a, b, c) = (3, 1/4, 1/\pi)$; $(a, b, c) = (1/\pi, 3, 1/4)$ **b.** $a = \left(\dfrac{3}{4\pi}\right)^{1/3}$; the resulting ellipsoid is a sphere
with radius a. **55.** 7 million
57. a. $100 = K(1000)^a(1,000,000)^{1-a}$; $10 = K(1000)^a(10,000)^{1-a}$
b. $\log K - 3a = -4$; $\log K - a = -3$ **c.** $a = 0.5, K \approx 0.003\,162$
d. $P = 71$ pianos (to the nearest piano) **59. a.** 4×10^{-3} g/m^2
b. The total weight of sulfates in Earth's atmosphere
61. a. The value of N would be doubled.
b. $N(R, f_p, n_e, f_l, f_i, L) = Rf_p n_e f_l f_i L$, where here L is the average lifetime of an intelligent civilization **c.** Take the logarithm of both sides, since this would yield the linear function $\ln(N) = \ln(R) + \ln(f_p) + \ln(n_e) + \ln(f_l) + \ln(f_i) + \ln(f_c) + \ln(L)$.
63. a.   **b.**   **c.**

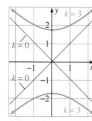 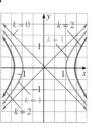

65. They are reciprocals of each other. **67.** For example, $f(x, y) = x^2 + y^2$. **69.** For example, $f(x, y, z) = xyz$. **71.** For example, let $f(x, y) = x + y$. Then setting $y = 3$ gives $f(x, 3) = x + 3$. This can be viewed as a function of the single variable x. Choosing other values for y gives other functions of x. **73.** If $f = ax + by + c$, then fixing $y = k$ gives $f = ax + (bk + c)$, a linear function with slope a and intercept $bk + c$. The slope is independent of the choice of $y = k$. **75.** CDs cost more than cassettes.

Exercises 8.2

1.

3.

5.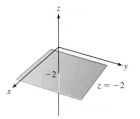

7.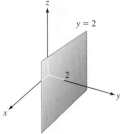

9.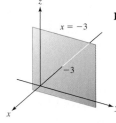

11. (H)   **13.** (B)   **15.** (F)   **17.** (C)

19.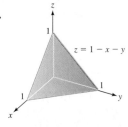

21.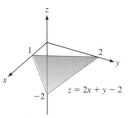

23.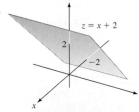

25.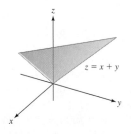

27.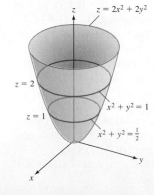

29. $z = x^2 + 2y^2$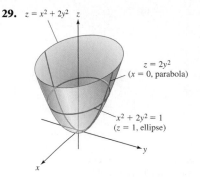

31.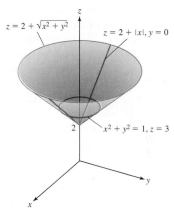

33.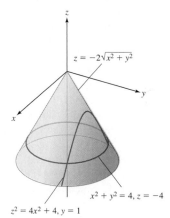

35.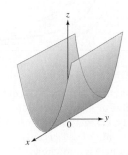

37. **39.**

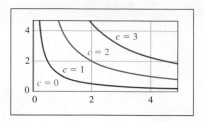

41. a. The graph is a plane with x intercept -40, y intercept -60, and z intercept 240,000. **b.** The slice $x = 10$ is the straight line with equation $z = 300,000 + 4000y$. It describes the cost function for the manufacture of trucks if car production is held fixed at ten cars per week. **c.** The level curve $z = 480,000$ is the straight line $6000x + 4000y = 240,000$. It describes the number of cars and trucks you can manufacture to maintain weekly costs at \$480,000. **43.** The graph is a plane with x_1 intercept 0.3, x_2 intercept 33, and x_3 intercept 0.66. The slices by x_1 = constant are straight lines that are parallel to each other. Thus, the rate of change of General Motors' share as a function of Ford's share does not depend on Chrysler's share. Specifically, GM's share decreases by 0.02 percentage point per 1 percentage-point increase in Ford's market share, regardless of Chrysler's share. **45. a.** The slices x = constant and y = constant are straight lines. **b.** No. Even though the slices x = constant and y = constant are straight lines, the level curves are not, and so the surface is not a plane. **c.** The slice $x = 10$ has a slope of 3800. The slice through $x = 20$ has a slope of 3600. Manufacturing more cars lowers the marginal cost of manufacturing trucks. **47.** Both level curves are quarter circles. (We see only the portion in the first quadrant because $e \geq 0$ and $k \geq 0$.) The level curve $C = 30,000$ represents the relationship between the number of electricians and the number of carpenters used in building a home that costs \$30,000; similarly for the level curve $C = 40,000$. **49.** The following figure shows several level curves together with several lines of the form $h + w = c$.

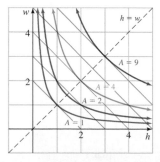

From the figure, thinking of the curves as contours on a map, we see that the largest value of A anywhere along any of the lines $h + w = c$ occurs midway along the line, when $h = w$. Thus, the largest area rectangle with a fixed perimeter occurs when $h = w$ (that is, when the rectangle is a square).

51.

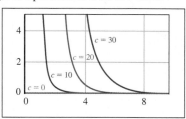

The level curve at $z = 3$ has the form $3 = x^{0.5}y^{0.5}$, or $y = 9/x$, and shows the relationship between the number of workers and the operating budget at a production level of 3 units.

53.

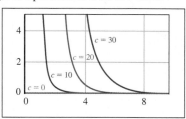

The level curve at $z = 0$ consists of the nonnegative y axis ($x = 0$) and tells us that zero utility corresponds to zero copies of Macro Publish, regardless of the number of copies of Turbo Publish. (Zero copies of Turbo Publish does not necessarily result in zero utility, according to the formula.)

55. Plane **57.** The graph of a function of three or more variables lives in four-dimensional (or higher) space, which makes it difficult to draw or visualize. **59.** Disagree: For example, the function $f(x, y) = xy$ has such slices, but its graph is not a plane. **63.** We need one dimension for each of the variables plus one dimension for the value of the function.

Exercises 8.3

1. $f_x(x, y) = -40$; $f_y(x, y) = 20$; $f_x(1, -1) = -40$; $f_y(1, -1) = 20$ **3.** $f_x(x, y) = 6x + 1$; $f_y(x, y) = -3y^2$; $f_x(1, -1) = 7$; $f_y(1, -1) = -3$ **5.** $f_x(x, y) = -40 + 10y$; $f_y(x, y) = 20 + 10x$; $f_x(1, -1) = -50$; $f_y(1, -1) = 30$ **7.** $f_x(x, y) = 6xy$; $f_y(x, y) = 3x^2$; $f_x(1, -1) = -6$; $f_y(1, -1) = 3$ **9.** $f_x(x, y) = 2xy^3 - 3x^2y^2 - y$; $f_y(x, y) = 3x^2y^2 - 2x^3y - x$; $f_x(1, -1) = -4$; $f_y(1, -1) = 4$ **11.** $f_x(x, y) = 6y(2xy + 1)^2$; $f_y(x, y) = 6x(2xy + 1)^2$; $f_x(1, -1) = -6$; $f_y(1, -1) = 6$ **13.** $f_x(x, y) = e^{x+y}$; $f_y(x, y) = e^{x+y}$; $f_x(1, -1) = 1$; $f_y(1, -1) = 1$ **15.** $f_x(x, y) = 3x^{-0.4}y^{0.4}$; $f_y(x, y) = 2x^{0.6}y^{-0.6}$; $f_x(1, -1)$ undefined; $f_y(1, -1)$ undefined **17.** $f_x(x, y) = 0.2ye^{0.2xy}$; $f_y(x, y) = 0.2xe^{0.2xy}$; $f_x(1, -1) = -0.2e^{-0.2}$; $f_y(1, -1) = 0.2e^{-0.2}$ **19.** $f_{xx}(x, y) = 0$; $f_{yy}(x, y) = 0$; $f_{xy}(x, y) = f_{yx}(x, y) = 0$; $f_{xx}(1, -1) = 0$; $f_{yy}(1, -1) = 0$; $f_{xy}(1, -1) = f_{yx}(1, -1) = 0$ **21.** $f_{xx}(x, y) = 0$; $f_{yy}(x, y) = 0$; $f_{xy}(x, y) = f_{yx}(x, y) = 10$; $f_{xx}(1, -1) = 0$; $f_{yy}(1, -1) = 0$; $f_{xy}(1, -1) = f_{yx}(1, -1) = 10$ **23.** $f_{xx}(x, y) = 6y$; $f_{yy}(x, y) = 0$; $f_{xy}(x, y) = f_{yx}(x, y) = 6x$; $f_{xx}(1, -1) = -6$; $f_{yy}(1, -1) = 0$; $f_{xy}(1, -1) = f_{yx}(1, -1) = 6$ **25.** $f_{xx}(x, y) = e^{x+y}$; $f_{yy}(x, y) = e^{x+y}$; $f_{xy}(x, y) = f_{yx}(x, y) = e^{x+y}$; $f_{xx}(1, -1) = 1$; $f_{yy}(1, -1) = 1$; $f_{xy}(1, -1) = f_{yx}(1, -1) = 1$ **27.** $f_{xx}(x, y) = -1.2x^{-1.4}y^{0.4}$; $f_{yy}(x, y) = -1.2x^{0.6}y^{-1.6}$; $f_{xy}(x, y) = f_{yx}(x, y) = 1.2x^{-0.4}y^{-0.6}$; $f_{xx}(1, -1)$ undefined; $f_{yy}(1, -1)$ undefined; $f_{xy}(1, -1)$ and $f_{yx}(1, -1)$ undefined **29.** $f_x(x, y, z) = yz$; $f_y(x, y, z) = xz$; $f_z(x, y, z) = xy$; $f_x(0, -1, 1) = -1$; $f_y(0, -1, 1) = 0$; $f_z(0, -1, 1) = 0$ **31.** $f_x(x, y, z) = 4/(x + y + z^2)^2$; $f_y(x, y, z) = 4/(x + y + z^2)^2$;

$f_z(x, y, z) = 8z/(x + y + z^2)^2$; $f_x(0, -1, 1)$ undefined; $f_y(0, -1, 1)$ undefined; $f_z(0, -1, 1)$ undefined **33.** $f_x(x, y, z) = e^{yz} + yze^{xz}$; $f_y(x, y, z) = xze^{yz} + e^{xz}$; $f_z(x, y, z) = xy(e^{yz} + e^{xz})$; $f_x(0, -1, 1) = e^{-1} - 1$; $f_y(0, -1, 1) = 1$; $f_z(0, -1, 1) = 0$ **35.** $f_x(x, y, z) = 0.1x^{-0.9}y^{0.4}z^{0.5}$; $f_y(x, y, z) = 0.4x^{0.1}y^{-0.6}z^{0.5}$; $f_z(x, y, z) = 0.5x^{0.1}y^{0.4}z^{-0.5}$; $f_x(0, -1, 1)$ undefined; $f_y(0, -1, 1)$ undefined; $f_z(0, -1, 1)$ undefined **37.** $f_x(x, y, z) = yze^{xyz}$; $f_y(x, y, z) = xze^{xyz}$; $f_z(x, y, z) = xye^{xyz}$; $f_x(0, -1, 1) = -1$; $f_y(0, -1, 1) = 0$ **39.** $f_x(x, y, z) = 0$; $f_y(x, y, z) = -\dfrac{600z}{y^{0.7}(1 + y^{0.3})^2}$; $f_z(x, y, z) = \dfrac{2000}{1 + y^{0.3}}$; $f_x(0, -1, 1)$ undefined; $f_y(0, -1, 1)$ undefined; $f_z(0, -1, 1)$ undefined **41.** $\partial C/\partial x = 6000$; the marginal cost to manufacture each car is \$6000. $\partial C/\partial y = 4000$; the marginal cost to manufacture each truck is \$4000. **43.** $\partial x_3/\partial x_1 = -2.2$; General Motors' market share decreases by 2.2 percentage points per 1 percentage-point increase in Chrysler's market share if Ford's share is unchanged. $\partial x_1/\partial x_3 = -1/2.2$; Chrysler's market share decreases by 1 percentage point per 2.2 percentage-point increase in General Motors' market share if Ford's share is unchanged. The two partial derivatives are reciprocals of each other. **45.** \$5600/car **47. a.** $\partial M/\partial c = -3.8$, $\partial M/\partial f = 2.2$. For every 1-point increase in the percentage of Chrysler owners who remain loyal, the percentage of Mazda owners who remain loyal decreases by 3.8 points. For every 1-point increase in the percentage of Ford owners who remain loyal, the percentage of Mazda owners who remain loyal increases by 2.2 points. **b.** 16% **49. a.** \$16,500 **b.** \$28,600 **c.** \$350/yr **d.** \$570/yr **e.** Widening **51.** The marginal cost of cars is $6000 + 1000e^{-0.01(x+y)}$ per car. The marginal cost of trucks is $4000 + 1000e^{-0.01(x+y)}$ per truck. Both marginal costs decrease as production rises. **53.** $\overline{C}(x, y) =$

$$\dfrac{200,000 + 6000x + 4000y - 100,000e^{-0.01(x+y)}}{x + y}; \overline{C}_x(50, 50) =$$

−\$2.64/car. This means that at a production level of 50 cars and 50 trucks per week, the average cost per vehicle is decreasing by \$2.64 for each additional car manufactured. $\overline{C}_y(50, 50) = $ −\$22.64/truck. This means that at a production level of 50 cars and 50 trucks per week, the average cost per vehicle is decreasing by \$22.64 for each additional truck manufactured. **55.** No; your marginal revenue from the sale of cars is $15,000 - (2500/\sqrt{x + y})$ per car and $10,000 - (2500/\sqrt{x + y})$ per truck from the sale of trucks. These increase with increasing x and y. In other words, you will earn more revenue per vehicle with increasing sales, and so the rental company will pay more for each additional vehicle it buys. **57.** $P_z(10, 100,000, 1,000,000) \approx$ 0.000 101 0 papers/dollar **59. a.** $U_x(10, 5) = 5.18$, $U_y(10, 5) = 2.09$. This means that, if ten copies of Macro Publish and five copies of Turbo Publish are purchased, the company's daily productivity is increasing at a rate of 5.18 pages/day for each additional copy of Macro purchased and by 2.09 pages/day for each additional copy of Turbo purchased. **b.** $\dfrac{U_x(10, 5)}{U_y(10, 5)} \approx 2.48$ is the ratio of the usefulness of one additional copy of Macro to one of Turbo. Thus, with ten copies of Macro and five copies of Turbo,

the company can expect approximately 2.48 times the productivity per additional copy of Macro compared to Turbo. **61.** 6×10^9 N/s **63. a.** $A_P(100, 0.1, 10) = 2.59$; $A_r(100, 0.1, 10) = 2357.95$; $A_t(100, 0.1, 10) = 24.72$. Thus, for a \$100 investment at 10% interest, after 10 years the accumulated amount is increasing at a rate of \$2.59 per \$1 of principal, at a rate of \$2357.95 per increase of 1 in r (note that this would correspond to an increase in the interest rate of 100%), and at a rate of \$24.72/yr. **b.** $A_P(100, 0.1, t)$ tells you the rate at which the accumulated amount in an account bearing 10% interest with a principal of \$100 is growing per \$1 increase in the principal, t years after the investment. **65. a.** $P_x = Ka(y/x)^b$ and $P_y = Kb(x/y)^a$. They are

equal precisely when $\dfrac{a}{b} = (x/y)^b(x/y)^a$. Substituting $b = 1 - a$

now gives $a/b = x/y$. **b.** The given information implies that $P_x(100, 200) = P_y(100, 200)$. By part (a) this occurs precisely when $a/b = x/y = 100/200 = 1/2$. But $b = 1 - a$, so $a/(1 - a) = 1/2$, giving $a = 1/3$ and $b = 2/3$. **67.** Decreasing at 0.0075 parts of nutrient per part of water per sec **69.** f is increasing at a rate of $\underline{s}$ units per unit of x, f is increasing at a rate of $\underline{t}$ units per unit of y, and the value of f is $\underline{r}$ when $x = \underline{a}$ and $y = \underline{b}$. **71.** the marginal cost of building an additional orbicus; zonars per unit **73.** One example is $f(x, y) = -2x + 3y + 9$. Another is $f(x, y) = xy - 3x + 2y + 10$. **75. a.** b is the z intercept of the plane; m is the slope of the intersection of the plane with the xz plane; n is the slope of the intersection of the plane with the yz plane. **b.** Write $z = b + rx + sy$. We are told that $\partial z/\partial x = m$, so $r = m$. Similarly, $s = n$. Thus, $z = b + mx + ny$. We are also told that the plane passes through (h, k, l). Substituting gives $l = b + mh + nk$. This gives b as $l - mh - nk$. Substituting in the equation for z therefore gives $z = l - mh - nk + mx + ny = l + m(x - h) + n(y - k)$, as required.

Exercises 8.4

1. P: relative minimum; Q: none of the above; R: relative maximum **3.** P: saddle point; Q: relative maximum; R: none of the above **5.** relative minimum **7.** Neither **9.** Saddle point **11.** Relative minimum at $(0, 0, 1)$ **13.** Relative maximum at $(-1/2, 1/2, 3/2)$ **15.** Relative maximum at $(0, 0, 0)$; saddle points at $(\pm 4, 2, -16)$ **17.** Relative minimum at $(0, 0, 1)$ **19.** Relative minimum at $(-2, \pm 2, -16)$; $(0, 0)$ a critical point that is not a relative extremum **21.** Saddle point at $(0, 0, -1)$ **23.** Relative maximum at $(-1, 0, e)$ **25.** Relative minimum at $(2^{1/3}, 2^{1/3}, 3(2^{2/3}))$ **27.** Relative minimum at $(1, 1, 4)$ and $(-1, -1, 4)$ **29.** Absolute minimum at $(0, 0, 1)$ **31.** None; the relative maximum at $(0, 0, 0)$ is not absolute. [Look at, say, $(10, 10)$.] **33.** Minimum of 1/3 at $(c, f) = (2/3, 2/3)$. Thus, at least 1/3 of all Mazda owners would choose another new Mazda, and this lowest loyalty occurs when 2/3 of Chrysler and Ford owners remain loyal to their brands. **35.** It should remove 2.5 lb of sulphur and 1 lb of lead each day. **37.** You should charge \$580.81 for the Ultra Mini and \$808.08 for the Big Stack. **39.** $l = w = h \approx 20.67$ in., volume ≈ 8827 in.³ **41.** 18 in. $\times$ 18 in. $\times$ 36 in., volume = 11,664 in.³

43.

45. Continues up indefinitely

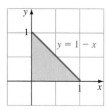

Continues down indefinitely
Function not defined on unit circle

47. H must be positive. **49.** No. In order for there to be a relative maximum at (a, b), *all* vertical planes through (a, b) should yield a curve with a relative maximum at (a, b). It could happen that a slice by another vertical plane through (a, b) (such as $x - a = y - b$) does not yield a curve with a relative maximum at (a, b). [An example is $f(x, y) = x^2 + y^2 - \sqrt{xy}$, at the point $(0, 0)$. Look at the slices through $x = 0$, $y = 0$, and $y = x$.]

51. $\overline{C}_x = \dfrac{\partial}{\partial x}\left[\dfrac{C}{x + y}\right] = \dfrac{(x + y)C_x - C}{(x + y)^2}$. If this is zero, then

$(x + y)C_x = C$, or $C_x = \dfrac{C}{x + y} = \overline{C}$.

Similarly, if $\overline{C}_y = 0$ then $C_y = \overline{C}$. This is reasonable because if the average cost is decreasing with increasing x, then the average cost is greater than the marginal cost C_x. Similarly, if the average cost is increasing with increasing x, then the average cost is less than the marginal cost C_x. Thus, if the average cost is stationary with increasing x, then the average cost equals the marginal cost C_x. (The situation is similar for the case of increasing y.) **53.** The equation of the tangent plane at the point (a, b) is $z = f(a, b) + f_x(a, b)(x - a) + f_y(a, b)(y - b)$. If f has a relative extremum at (a, b), then $f_x(a, b) = 0 = f_y(a, b)$. Substituting these into the equation of the tangent plane gives $z = f(a, b)$, a constant. But the graph of $z = constant$ is a plane parallel to the xy plane.

Exercises 8.5

1. Maximum value of 8 at $(2, 2)$, minimum value of 0 at $(0, 0)$
3. Maximum value of 9 at $(-2, 0)$, minimum value of 0 at $(1, 0)$
5. Maximum value of e^4 at $(0, \pm 2)$, minimum value of 1 at $(0, 0)$
7. Maximum value of e^4 at $(\pm 1, 0)$, minimum value of 1 at $(0, 0)$
9. Maximum value of 5 at $(1/2, 1/2)$, minimum value of 3 at $(1, 1)$ **11.** Maximum value of 161/9 at $(1, 9)$ and $(9, 1)$, minimum value of 12 at $(2, 2)$ **13.** Maximum value of 8 at $(2, 0)$, minimum value of -1 at $(-1, 0)$ **15.** Maximum value of 8 at $(2, 0)$, minimum value of 0 at $(0, 0)$ **17.** $(1/\sqrt{3}, 1/\sqrt{3}, 1/\sqrt{3})$, $(-1/\sqrt{3}, -1/\sqrt{3}, 1/\sqrt{3})$, $(1/\sqrt{3}, -1/\sqrt{3}, -1/\sqrt{3})$, $(-1/\sqrt{3}, 1/\sqrt{3}, -1/\sqrt{3})$ **19.** For minimum cost of \$16,600, make 100 five-speeds and 80 ten-speeds. For maximum cost of \$17,400, make 100 five-speeds and 120 ten-speeds. **21.** For a maximum profit of \$4500, sell 50 copies of Walls and 150 copies of Doors. **23.** Hottest point: $(1, 1)$; coldest point: $(0, 0)$ **25.** Hottest points: $(-1/2, \pm\sqrt{3}/2)$; coldest point: $(1/2, 0)$ **27.** $(0, 1/2, -1/2)$ **29.** $(-5/9, 5/9, 25/9)$ **31.** $l \times w \times h = 1 \times 1 \times 2$ **33.** $(2l/h)^{1/3} \times (2l/h)^{1/3} \times 2^{1/3}(h/l)^{2/3}$, where $l = $ cost of lightweight cardboard and $h = $ cost of heavy-duty cardboard per square foot **35.** 18 in. $\times$ 18 in. $\times$ 36 in.; volume $= 11,664$ in.^{3} **37.** $1 \times 1 \times 1/2$ **39.** 150 qt of vanilla and 100 qt of mocha, for

a profit of \$325 **41.** Offer 100 sections of Finite Math and no sections of Applied Calculus. **43.** Use 7.280 servings of Mixed Cereal and 1.374 servings of Tropical Fruit Dessert. **45.** Buy 100 of each. **47.** Yes. There may be relative extrema at points on the boundary of the domain of the function. The partial derivatives of the function need not be zero at such points. **49.** If the only constraint is an equality constraint and if it is impossible to eliminate one of the variables in the objective function by substitution (solving the constraint equation for a variable or some other method) **51.** In a linear programming problem, the objective function is linear, and so the partial derivatives can never be zero. (We are ignoring the simple case in which the objective function is constant.) It follows that the extrema cannot occur in the interior of the domain (since the partial derivatives must be zero at such points).

Exercises 8.6

1. $-1/2$ **3.** $e^2/2 - 7/2$ **5.** $(e^3 - 1)(e^2 - 1)$ **7.** 7/6
9. $[e^3 - e - e^{-1} + e^{-3}]/2$ **11.** 1/2 **13.** $(e - 1)/2$ **15.** 45/2
17. 8/3 **19.** 4/3 **21.** 0 **23.** 2/3 **25.** 2/3 **27.** $2(e - 2)$

29. 2/3

31. $\displaystyle\int_0^1 \int_0^{1-x} f(x, y)\, dy\, dx$

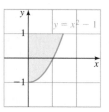

33. $\displaystyle\int_0^1 \int_{x^2-1}^1 f(x, y)\, dy\, dx$

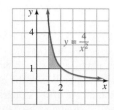

35. $\displaystyle\int_1^4 \int_1^{2/\sqrt{y}} f(x, y)\, dx\, dy$

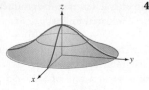

37. 4/3 **39.** 1/6 **41.** 162,000 gadgets **43.** Maximum revenue is \$375,500. Minimum revenue is \$256,000. Average revenue is \$312,750. **45.** Maximum revenue is \$20,000. Minimum revenue is \$15,000. Average revenue is \$17,500. **47.** 8216 **49.** 1° **51.** The area between the curves $y = r(x)$ and $y = s(x)$ and the vertical lines $x = a$ and $x = b$ is given by $\int_a^b \int_{r(x)}^{s(x)} dy\, dx$, assuming that $r(x) \le s(x)$ for $a \le x \le b$. **53.** The first step in calculating an integral of the form $\int_a^b \int_{r(x)}^{s(x)} f(x, y)\, dy\, dx$ is to evaluate the integral $\int_{r(x)}^{s(x)} f(x, y)\, dy$, obtained by holding x constant and integrating with respect to y. **55.** Paintings per picasso per dali

57. Left-hand side is $\int_a^b \int_c^d f(x)g(y)\, dx\, dy = \int_a^b \left[g(y) \int_c^d f(x)\, dx \right] dy$ [since $g(y)$ is treated as a constant in the inner integral] $= \left[\int_c^d f(x)\, dx \right] \int_a^b g(y)\, dy$ [since $\int_c^d f(x)\, dx$ is a constant and can therefore be taken outside the integral]. $\int_0^1 \int_1^2 ye^x\, dx\, dy = \frac{1}{2}(e^2 - e)$ no matter how we compute it.

Chapter 8 Review Test

1. a. 0; 1; 0; $x^3 + x^2$; $x(y + k)(x + y + k - z) + x^2$ **b.** Decreases by 0.32 unit; increases by 12.5 units. **c.** Reading left to right, starting at the top: 4, 0, 0, 3, 0, 1, 2, 0, 2 **d.** Answers will vary; two examples are $f(x, y) = 3(x - y)/2$ and $f(x, y) = 3(x - y)^3/8$. **2. a.** $f_x = 2x + y, f_y = x, f_{yy} = 0$ **b.** $\partial f/\partial x = ye^{xy} + 6xe^{3x^2 - y^2}; \partial^2 f/\partial x\partial y = (xy + 1)e^{xy} - 12xye^{3x^2 - y^2}$

c. $\dfrac{\partial f}{\partial x} = \dfrac{-x^2 + y^2 + z^2}{(x^2 + y^2 + z^2)^2}; \dfrac{\partial f}{\partial y} = -\dfrac{2xy}{(x^2 + y^2 + z^2)^2}; \dfrac{\partial f}{\partial z} = -\dfrac{2xz}{(x^2 + y^2 + z^2)^2}; \dfrac{\partial f}{\partial x}\bigg|_{(0, 1, 0)} = 1$ **d.** 6

3. a. Absolute minimum at $(1, 3/2)$ **b.** Saddle point at $(1, 0)$ **c.** Saddle point at $(0, 0)$ **d.** Saddle point at $(0, 0)$ **e.** Absolute maximum at each point on the circle $x^2 + y^2 = 1$
4. a. $(0, 2, \sqrt{2})$ **b.** Minimum of 0 at $(1, 0)$; maximum of 4 at $(3, 0)$ **c.** Minimum of $-16/7$ at $(4/7, 6/7)$; maximum of 21 at $(-1, -2)$ **5. a.** 2 **b.** e^2 **c.** $\ln 5$ **d.** e^2 **e.** 1 **f.** 4/5
6. a. $h(x, y) = 5000 - 0.8x - 0.6y$ hits/day ($x = $ number of new customers at JungleBooks.com; $y = $ number of new customers at FarmerBooks.com) **b.** 250 **c.** $h(x, y, z) = 5000 - 0.8x - 0.6y + 0.0001z$ ($z = $ number of new Internet shoppers) **d.** 1.4 million **7. a.** 2320 hits/day **b.** $0.08 + 0.000\,03x$ hits (daily) per dollar spent on television advertising each month; increases with increasing x **c.** $4000/mo **d.** (A) **8. a.** About 15,800 orders/day **b.** 11 **9.** Selling 1000 paperbacks and 1200 hardcover books costs the least. Selling 900 or 1100 paperbacks and 1500 hardcover books costs the most. **10.** $23,050

Chapter 9

Exercises 9.1

1.

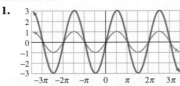

3.

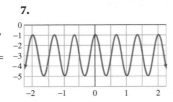

5.

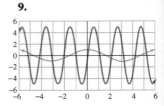

7.

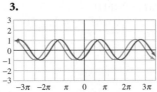

9.

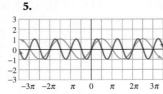

11.

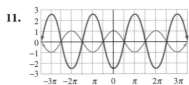

13. $f(x) = \sin(2\pi x) + 1$ **15.** $f(x) = 1.5\,\sin[4\pi(x - 0.25)]$
17. $f(x) = 50\,\sin[\pi(x - 5)/10] - 50$ **19.** $f(x) = \cos(2\pi x)$
21. $f(x) = 1.5\,\cos[4\pi(x - 0.375)]$
23. $f(x) = 40\,\cos[\pi(x - 10)/10] + 40$
25. $f(t) = 4.2\,\sin(\pi/2 - 2\pi t) + 3$
27. $g(x) = 4 - 1.3\,\sin[\pi/2 - 2.3(x - 4)]$ **31.** $\sqrt{3}/2$
37. $\tan(x + \pi) = \tan(x)$ **39. a.** $2\pi/0.602 \approx 10.4$ yr **b.** Maximum: 117; minimum: 1 **c.** 14.5 yr, or midway through 2011
41. a. Maximum sales occurred when $t \approx 4.5$ (during the first quarter of 1996). Minimum sales occurred when $t \approx 2.2$ (during the third quarter of 1995) and $t \approx 6.8$ (during the third quarter of 1996). **b.** Maximum quarterly revenues were $0.561 billion; minimum quarterly revenues were $0.349 billion. **c.** Maximum: $0.455 + 0.106 = 0.561$; minimum: $0.455 - 0.106 = 0.349$
43. Amplitude $= 0.106$; vertical offset $= 0.455$; phase shift $= -1.16$; angular frequency $= 1.39$; period $= 4.52$. In 1995 and 1996, quarterly revenue from the sale of computers at Computer City fluctuated in cycles of 4.52 quarters about a baseline of $0.455 billion. Every cycle, quarterly revenue peaked at $0.561 billion ($0.106 above the baseline) and dipped to a low of $0.349 billion. Revenue peaked early in the middle of the first quarter of 1996 [at $t = -1.16 + (5/4) \times 4.52 = 4.49$].
45. $P(t) = 7.5\,\sin[\pi(t - 13)/26] + 12.5$
47. $s(t) = 7.5\,\sin[\pi(t - 9)/6] + 87.5$
49. $s(t) = 7.5\,\cos(\pi t/6) + 87.5$
51. $d(t) = 5\,\sin[2\pi(t - 1.625)/13.5] + 10$
53. a. $u(t) = 2.5\,\sin[2\pi(t - 0.75)] + 7.5$
b. $c(t) = 1.04^t\{2.5\,\sin[2\pi(t - 0.75)] + 7.5\}$
55. a. $P \approx 8, C \approx 6, A \approx 2, \alpha \approx 8$ (answers will vary)

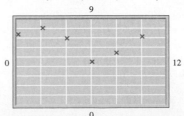

b. $C(t) = 6.437 + 1.755 \sin[0.636(t - 9.161)]$

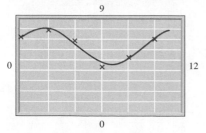

c. 9.9, 4.7%, 8.2%

57. a.

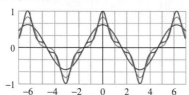

b. $y_{11} = \frac{2}{\pi} \cos x + \frac{2}{3\pi} \cos 3x + \frac{2}{5\pi} \cos 5x + \frac{2}{7\pi} \cos 7x + \frac{2}{9\pi} \cos 9x + \frac{2}{11\pi} \cos 11x$

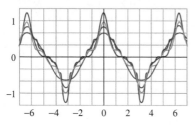

c. $y_{11} = \frac{6}{\pi} \cos \frac{x}{2} + \frac{6}{3\pi} \cos \frac{3x}{2} + \frac{6}{5\pi} \cos \frac{5x}{2} + \frac{6}{7\pi} \cos \frac{7x}{2} + \frac{6}{9\pi} \cos \frac{9x}{2} + \frac{6}{11\pi} \cos \frac{11x}{2}$ **59.** The period is approximately 12.6 units.

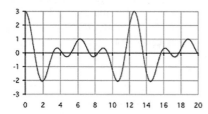

61. Lows: $B - A$; highs: $B + A$ **63.** He is correct. The other trigonometric functions can be obtained from the sine function by first using the formula $\cos x = \sin(x + \pi/2)$ to obtain cosine and then using the formulas $\tan x = \sin x/\cos x$, $\cot x = \cos x/\sin x$, $\sec x = 1/\cos x$, and $\csc x = 1/\sin x$ to obtain the rest. **65.** The largest B can be is A. Otherwise, if B is larger than A, the low figure for sales would have the negative value of $A - B$.

Exercises 9.2

1. $\cos x + \sin x$ **3.** $(\cos x)(\tan x) + (\sin x)(\sec^2 x)$
5. $-2 \csc x \cot x - \sec x \tan x + 3$ **7.** $\cos x - x \sin x + 2x$
9. $(2x - 1)\tan x + (x^2 - x + 1)\sec^2 x$
11. $-[\csc^2 x(1 + \sec x) + \cot x \sec x \tan x]/(1 + \sec x)^2$
13. $-2 \cos x \sin x$ **15.** $2 \sec^2 x \tan x$
17. $\pi \cos\left[\frac{\pi}{5}(x - 4)\right]$ **19.** $-(2x - 1)\sin(x^2 - x)$

21. $(2.2x^{1.2} + 1.2)\sec(x^{2.2} + 1.2x - 1) \tan(x^{2.2} + 1.2x - 1)$
23. $\sec x \tan x \tan(x^2 - 1) + 2x \sec x \sec^2(x^2 - 1)$
25. $e^x[-\sin(e^x) + \cos x - \sin x]$ **27.** $\sec x$
33. $e^{-2x}[-2 \sin(3\pi x) + 3\pi \cos(3\pi x)]$
35. $1.5[\sin(3x)]^{-0.5} \cos(3x)$
37. $\frac{x^4 - 3x^2}{(x^2 - 1)^2} \sec\left(\frac{x^3}{x^2 - 1}\right) \tan\left(\frac{x^3}{x^2 - 1}\right)$
39. $\frac{\cot(2x - 1)}{x} - 2 \ln|x| \csc^2(2x - 1)$ **41. a.** Not differentiable at 0 **b.** $f'(1) \approx 0.5403$ **43.** 0 **45.** 2 **47.** Does not exist **49.** $1/\sec^2 y$ **51.** $-[1 + y \cos(xy)]/[1 + x \cos(xy)]$
53. $c'(t) = 7\pi \cos[2\pi(t - 0.75)]; c'(0.75) \approx \$21.99/yr \approx \$0.42/wk$ **55.** $N'(6) \approx -32.12$. On January 1, 2003, the number of sunspots was decreasing at a rate of 32.12 sunspots per year. **57.** $c'(t) = 1.035^t[\ln(1.035)[(0.8 \sin(2\pi t) + 10.2] + 1.6\pi \cos(2\pi t)]; c'(1) = 1.035[10.2 \ln|1.035| + 1.6\pi] \approx \$5.57/yr$, or $\$0.11/wk$ **59. a.** $d(t) = 10 + 5 \cos(2\pi t/13.5)$
b. $d'(t) = -(10\pi/13.5)\sin(2\pi t/13.5); d'(7) \approx 0.270$. At noon, the tide was rising at a rate of 0.270 ft/h. **61. a.** (C)
b. Increasing at a rate of 0.157° per 1000 years **63.** $-6; 6$
65. Answers will vary. Example: $f(x) = \sin x; f(x) = \cos x$
67. Answers will vary. Example: $f(x) = e^{-x}; f(x) = -2e^{-x}$
69. The graph of $\cos x$ slopes down over the interval $(0, \pi)$, so that its derivative is negative over that interval. The function $-\sin x$, and not $\sin x$, has this property. **71.** The derivative of $\sin x$ is $\cos x$. When $x = 0$, this is $\cos(0) = 1$. Thus, the tangent to the graph of $\sin x$ at the point $(0, 0)$ has slope 1, which means it slopes upward at 45°.

Exercises 9.3

1. $-\cos x - 2 \sin x + C$ **3.** $2 \sin x + 4.3 \cos x - 9.33x + C$
5. $3.4 \tan x + (\sin x)/1.3 - 3.2e^x + C$ **7.** $(7.6/3)\sin(3x - 4) + C$ **9.** $-(1/6)\cos(3x^2 - 4) + C$ **11.** $-2\cos(x^2 + x) + C$
13. $(1/6)\tan(3x^2 + 2x^3) + C$ **15.** $-(1/6)\ln|\cos(2x^3)| + C$
17. $3 \ln|\sec(2x - 4) + \tan(2x - 4)| + C$ **19.** $(1/2)\sin(e^{2x} + 1) + C$ **21.** -2 **23.** $\ln(2)$ **25.** 0 **27.** 1 **33.** $-\frac{1}{4} \cos(4x) + C$
35. $-\sin(-x + 1) + C$ **37.** $[\cos(-1.1x - 1)]/1.1 + C$
39. $-\frac{1}{4} \ln|\sin(-4x)| + C$ **41.** 0 **43.** 2π **45.** $-x \cos x + \sin x + C$ **47.** $\left[\frac{x^2}{2} - \frac{1}{4}\right]\sin(2x) + \frac{x}{2} \cos(2x) + C$
49. $-\frac{1}{2} e^{-x}\cos x - \frac{1}{2} e^{-x} \sin x + C$ **51.** $\pi^2 - 4$
53. Average $= 2/\pi$

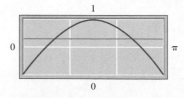

55. Diverges **57.** Converges to 1/2 **59.** $C(t) = 0.04t + \frac{2.6}{\pi} \cos\left[\frac{\pi}{26}(t - 25)\right] + 1.02$ **61.** 12 ft **63.** 79 sunspots

65. $P(t) = 7.5 \sin[(\pi/26(t - 13)] + 12.5;$ 7.7% **67. a.** Average voltage over $[0, 1/6]$ is zero; 60 cycles per second

b.

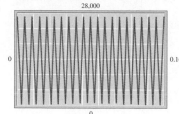

28,000

0

0.16

0

c. 116.673 V **69.** \$50,000 **71.** It is always zero. **73.** 1

75. $s = -\dfrac{K}{\omega^2} \sin(\omega t - \alpha) + Lt + M$ for constants L and M

Chapter 9 Review Test

1. a. $f(x) = 1 + 2 \sin x$ **b.** $f(x) = 10 \sin(2x + \pi/2)$ **c.** $f(x) = 2 + 2 \sin[\pi(x - 1)] = 2 + 2 \sin[\pi(x + 1)]$ **d.** $f(x) = -2 + 3 \sin[4\pi(x - 0.25)]$ **2. a.** $f(x) = 1 + 2 \cos(x - \pi/2)$
b. $f(x) = 10 \cos(2x)$ **c.** $f(x) = 2 + 2 \cos[\pi(x + 1/2)] = 2 + 2 \cos[\pi(x - 3/2)]$ **d.** $f(x) = -2 + 3 \cos[4\pi(x - 0.375)]$
3. a. $-2x \sin(x^2 - 1)$ **b.** $2x[\cos(x^2 + 1)\cos(x^2 - 1) - \sin(x^2 + 1)\sin(x^2 - 1)]$ **c.** $2e^x \sec^2(2e^x - 1)$

d. $\dfrac{(2x - 1)\sec\sqrt{x^2 - x}\,\tan\sqrt{x^2 - x}}{2\sqrt{x^2 - x}}$ **e.** $4x \sin(x^2)\cos(x^2)$

f. $4 \cos(2x)\cos[1 - \sin(2x)]\sin[1 - \sin(2x)]$

4. a. $2 \sin(2x - 1) + C$ **b.** $-\dfrac{1}{2}\cos(x^2 - 2x + 1) + C$

c. $-\dfrac{1}{2}\ln|(\cos(x^2 + 1)| + C$ **d.** -2 **e.** 1 **f.** $3 \ln 3$

5. a. $-x^2 \cos x + 2x \sin x + 2 \cos x + C$
b. $(1/5)e^x \sin 2x - (2/5)e^x \cos 2x + C$
6. $s(t) = 10{,}500 + 1500 \sin[(2\pi/52)t - \pi] = 10{,}500 + 1500 \sin(0.12083t - 3.14159)$ **7.** \$542/mo **8.** \$222,300

9. $150t - \dfrac{100}{\pi}\cos\left[\dfrac{\pi}{2}(t - 1)\right]$ g

Appendix A

A.1 Exercises

1. -48 **3.** 2/3 **5.** -1 **7.** 9 **9.** 1 **11.** 33 **13.** 14 **15.** 5/18
17. 1331/100 **19.** 6 **21.** 43/16 **23.** 0 **25.** $3*(2-5)$
27. $3/(2-5)$ **29.** $(3-1)/(8+6)$ **31.** $3-(4+7)/8$
33. $2/(3+x)-x*y^2$ **35.** $3.1*x^3-4*x^(-2)-60/(x^2-1)$ **37.** $(2/3)/5$ **39.** $3^(4-5)*6$
41. $3*(1+4/100)^(-3)$ **43.** $3^(2*x-1)+4^x-1$
45. $2^(2*x^2-x+1)$ **47.** $4*e^(-2*x)/(2-3*e^(-2*x))$ or $4*(e^(-2*x))/(2-3*e^(-2*x))$ **49.** $3*(1-(-1/2)^2)^2+1$

A.2 Exercises

1. 27 **3.** -36 **5.** 4/9 **7.** $-1/8$ **9.** 16 **11.** 2 **13.** 32

15. 2 **17.** x^5 **19.** $-\dfrac{y}{x}$ **21.** $\dfrac{1}{x}$ **23.** x^3y **25.** $\dfrac{z^4}{y^3}$

27. $\dfrac{x^6}{y^6}$ **29.** $\dfrac{x^4y^6}{z^4}$ **31.** $\dfrac{3}{x^4}$ **33.** $\dfrac{3}{4x^{2/3}}$ **35.** $1 - 0.3x^2 -$

$\dfrac{6}{5x}$ **37.** 2 **39.** 1/2 **41.** 4/3 **43.** 2/5 **45** 7 **47.** 5

49. -2.668 **51.** 3/2 **53.** 2 **55.** 2 **57.** ab **59.** $x + 9$

61. $x\sqrt[3]{a^3 + b^3}$ **63.** $\dfrac{2y}{\sqrt{x}}$ **65.** $3^{1/2}$ **67.** $x^{3/2}$ **69.** $(xy^2)^{1/3}$

71. $x^{3/2}$ **73.** $\dfrac{3}{5}x^{-2}$ **75.** $\dfrac{3}{2}x^{-1.2} - \dfrac{1}{3}x^{-2.1}$ **77.** $\dfrac{2}{3}x - \dfrac{1}{2}x^{0.1}$

$+ \dfrac{4}{3}x^{-1.1}$ **79.** $(x^2 + 1)^{-3} - \dfrac{3}{4}(x^2 + 1)^{-1/3}$ **81.** $\sqrt[3]{2^2}$

83. $\sqrt[3]{x^4}$ **85.** $\sqrt[5]{\sqrt{x}\sqrt[3]{y}}$ **87.** $\dfrac{-3}{2\sqrt[4]{x}}$ **89.** $\dfrac{0.2}{\sqrt[3]{x^2}} + \dfrac{3\sqrt{x}}{7}$

91. $\dfrac{3}{4\sqrt{(1 - x)^5}}$ **93.** 64 **95.** $\sqrt{3}$ **97.** $1/x$ **99.** xy

101. $\left(\dfrac{y}{x}\right)^{1/3}$ **103.** ± 4 **105.** $\pm 2/3$ **107.** $-1, -1/3$

109. -2 **111.** 16 **113.** ± 1 **115.** 33/8

A.3 Exercises

1. $4x^2 + 6x$ **3.** $2xy - y^2$ **5.** $x^2 - 2x - 3$ **7.** $2y^2 + 13y + 15$ **9.** $4x^2 - 12x + 9$ **11.** $x^2 + 2 + 1/x^2$ **13.** $4x^2 - 9$
15. $y^2 - 1/y^2$ **17.** $2x^3 + 6x^2 + 2x - 4$ **19.** $x^4 - 4x^3 + 6x^2 - 4x + 1$ **21.** $y^5 + 4y^4 + 4y^3 - y$ **23.** $(x + 1)(2x + 5)$
25. $(x^2 + 1)^5(x + 3)^3(x^2 + x + 4)$ **27.** $-x^3(x^3 + 1)\sqrt{x + 1}$
29. $(x + 2)\sqrt{(x + 1)^3}$ **31. a.** $x(2 + 3x)$ **b.** $x = 0, -2/3$
33. a. $2x^2(3x - 1)$ **b.** $x = 0, 1/3$ **35. a.** $(x - 1)(x - 7)$
b. $x = 1, 7$ **37. a.** $(x - 3)(x + 4)$ **b.** $x = 3, -4$
39. a. $(2x + 1)(x - 2)$ **b.** $x = -1/2, 2$ **41. a.** $(2x + 3)(3x + 2)$
b. $x = -3/2, -2/3$ **43. a.** $(3x - 2)(4x + 3)$ **b.** $x = 2/3, -3/4$
45. a. $(x + 2y)^2$ **b.** $x = -2y$ **47. a.** $(x^2 - 1)(x^2 - 4) = (x - 1)(x + 1)(x - 2)(x + 2)$ **b.** $x = \pm 1, \pm 2$

A.4 Exercises

1. $\dfrac{2x^2 - 7x - 4}{x^2 - 1}$ **3.** $\dfrac{3x^2 - 2x + 5}{x^2 - 1}$ **5.** $\dfrac{x^2 - x + 1}{x + 1}$

7. $\dfrac{x^2 - 1}{x}$ **9.** $\dfrac{2x - 3}{x^2y}$ **11.** $\dfrac{(x + 1)^2}{(x + 2)^4}$ **13.** $\dfrac{-1}{\sqrt{(x^2 + 1)^3}}$

15. $\dfrac{-(2x + y)}{x^2(x + y)^2}$

A.5 Exercises

1. -1 **3.** 5 **5.** 13/4 **7.** 43/7 **9.** -1 **11.** $(c - b)/a$

13. $-4, \dfrac{1}{2}$ **15.** No solutions **17.** $\pm\sqrt{\dfrac{5}{2}}$ **19.** -1

21. $-1, 3$ **23.** $\dfrac{1 \pm \sqrt{5}}{2}$ **25.** 1 **27.** $\pm 1, \pm 3$

29. $\pm\sqrt{\dfrac{-1 + \sqrt{5}}{2}}$ **31.** $-1, -2, -3$ **33.** -3 **35.** 1

37. -2 **39.** $1, \pm\sqrt{5}$ **41.** $\pm 1, \pm\dfrac{1}{\sqrt{2}}$ **43.** $-2, -1, 2, 3$

A.6 Exercises

1. 0, 3 **3.** $\pm\sqrt{2}$ **5.** $-1, -5/2$ **7.** -3 **9.** 0, -1 **11.** -1
(-2 is *not* a solution) **13.** $-2, -3/2, -1$ **15.** -1 **17.** $\pm\sqrt{2}$
19. ± 1 **21.** ± 3 **23.** 2/3 **25.** $-4, -1/4$

INDEX

INDEX OF COMPANIES, PRODUCTS, AND AGENCIES